EVS36—International Electric Vehicle Symposium and Exhibition (California, USA)

EVS36—International Electric Vehicle Symposium and Exhibition (California, USA)

Guest Editors

Joeri Van Mierlo
Genevieve Cullen

MDPI

Basel • Beijing • Wuhan • Barcelona • Belgrade • Novi Sad • Cluj • Manchester

Guest Editors
Joeri Van Mierlo
MOBI—Electromobility
Research Centre
Vrije Universiteit Brussel
Brussels
Belgium

Genevieve Cullen
Electric Drive Transportation
Association
Washington
USA

Editorial Office
MDPI AG
Grosspeteranlage 5
4052 Basel, Switzerland

This is a reprint of the Special Issue, published open access by the journal *World Electric Vehicle Journal* (ISSN 2032-6653), freely accessible at: https://www.mdpi.com/journal/wevj/special_issues/71SL43A62T.

For citation purposes, cite each article independently as indicated on the article page online and as indicated below:

Lastname, A.A.; Lastname, B.B. Article Title. *Journal Name* **Year**, *Volume Number*, Page Range.

ISBN 978-3-7258-3939-1 (Hbk)
ISBN 978-3-7258-3940-7 (PDF)
https://doi.org/10.3390/books978-3-7258-3940-7

Contents

Preface

This Special Issue gathers the best papers from EVS36, the 36th International Electric Vehicle Symposium and Exhibition, which was held from 11 to 14 June 2023 in Sacramento, California, USA.

The International Electric Vehicle Symposium (EVS) is the world's longest, largest, and highest-spec event in the field of electric vehicles, covering areas including pure electric vehicles, hybrid vehicles, fuel cell vehicles, and core components. With the support of the World Electric Vehicle Association, the symposium is hosted by three regional professional organizations under the World Electric Vehicle Association in North America (Electric Drive Transportation Association, EDTA), Europe (The European Association for Electromobility, E-Mobility Europe), and Asia (Electric Vehicle Association of Asia Pacific, EVAAP), in turn. EVS has already had a long history since its birth in Phoenix, Arizona, USA, in 1969.

The theme of EVS36 is "Driving the Transition to E-Mobility". The papers cover research, market, and government activities across all fields related to hybrid, battery, and fuel cell technologies, associated infrastructure, and services.

The authors of the best papers presented at EVS36 were invited to further extend their EVS36 paper, including their most recent research findings. After a second thorough round of peer review, these papers were published in this Special Issue of the *World Electric Vehicle Journal* (*WEVJ*), the official journal of the World Electric Vehicle Association (WEVA).

Joeri Van Mierlo and Genevieve Cullen
Guest Editors

World Electric Vehicle Journal

MDPI

Article

Demonstrating the Lessons Learned for Lightweighting EV Components through a Circular-Economy Approach

Floris Teunissen * and Esther van Bergen

Cenex Nederland, Overhoeksplein 2, 1031 KS Amsterdam, The Netherlands; esther.van.bergen@cenexgroup.nl
* Correspondence: floris.teunissen@cenexgroup.nl; Tel.: +31-6-1547276017897982

Abstract: LEVIS is an innovation project funded by the EU Horizon 2020 program. Its main objective is to develop lightweight multi-material solutions based on bio-based materials and carbon fiber thermoplastic composites for electric vehicle components and demonstrating the technical, operational, and economic feasibility of applying eco-design and circular-economy principles into the design process. The project demonstrates the application of these materials in four case studies: a suspension control arm, a battery box, a battery module housing, and a cross-car beam. All demonstrators achieved a 20%-to-40% reduction in component weight, but environmental assessment results varied significantly, with emissions changes ranging from an increase for suspension control arms to a 65.5% decrease for battery modules. Efficient use of materials, particularly in the battery box using hybrid solutions and bonding technologies, showed notable emissions reduction. In contrast, replacing steel with CFRPs in suspension control arms led to increased emissions, suggesting that CFRPs are more effective for replacing high-polluting materials like aluminum. Recycled carbon fibers proved more beneficial for low-polluting materials like steel. The environmental performance of technologies depends on the expected use of EVs and the electricity grid mix, with better outcomes in coal-reliant grids. Finally, no single recycling method is universally superior; the optimal method depends on the specific technologies and the energy required for recycled materials.

Keywords: lightweight materials; circular economy; electric vehicle; life-cycle assessment; life-cycle costing; bio-based materials; carbon fiber-reinforced plastics; recycling; pyrolysis; eco-design

Citation: Teunissen, F.; van Bergen, E. Demonstrating the Lessons Learned for Lightweighting EV Components through a Circular-Economy Approach. *World Electr. Veh. J.* **2024**, *15*, 415. https://doi.org/10.3390/wevj15090415

Academic Editors: Joeri Van Mierlo and Genevieve Cullen

Received: 29 July 2024
Revised: 27 August 2024
Accepted: 3 September 2024
Published: 11 September 2024

1. Introduction

The concept of a circular economy is gaining significant traction within the automotive industry, emerging as a crucial and strategic focus area. This shift is driven by concerns over resource scarcity, the need to control manufacturing and operational expenses, and the increasing emphasis on sustainability.

Given that a vehicle's weight directly impacts its energy efficiency and driving range, lightweight construction has become a key factor in accelerating the market expansion of electric vehicles (EVs). This, in turn, plays a vital role in achieving the greenhouse gas emission-reduction targets set for 2050 [1]. Additionally, the European Union has introduced regulations that establish environmental, social, and circular-economy standards pertinent to the automotive sector, with further regulations currently in the pipeline [2].

Fiber-reinforced polymers (FRPs) are lightweight structural materials extensively used in the automotive sector, primarily in sports and high-end luxury vehicles. Despite their benefits, challenges such as cost efficiency, production scalability, and end-of-life (EOL) management hinder their adoption in low-cost vehicle segments. Addressing these challenges could bridge the industrial gap, enabling mass production of FRPs for more affordable vehicles and thereby reducing EV production costs. Additionally, developing multi-material components that combine FRPs with metals offers a promising approach to lightweighting while maintaining necessary mechanical and functional performance.

These innovations are crucial in meeting the automotive industry's stringent CO_2 emission and recyclability requirements, especially under increasingly rigorous EU directives.

Building on previous Horizon projects, such as ALLIANCE [3], LoCoMaTech [4], ALIVE [5], ENLIGHT [6], GREENLIGHT [7], and ECOXY [8], LEVIS aims to further advance lightweighting and circularity in automotive manufacturing. For instance, ALLIANCE demonstrated significant weight reductions, which LEVIS aims to surpass by focusing on vehicle chassis components to achieve a 30% structural weight reduction. LoCoMaTech's advancements in low-cost aluminum body structures complement LEVIS's focus on multi-material parts combining thermoplastic composites with metals. ALIVE and ENLIGHT highlighted the potential of composite materials, and LEVIS extends it by integrating real-case demonstrators across the entire product value chain. GREENLIGHT and ECOXY's work on bio-based materials and recyclable composites inform LEVIS's cradle-to-cradle approach, targeting the recovery and reuse of materials to enhance sustainability.

Das (2011) [9] conducted a life-cycle assessment (LCA) of carbon fiber-reinforced polymer (CFRP) composites, demonstrating their potential to reduce the weight of automotive components while also considering the environmental impacts of their production and disposal. Despite the higher energy consumption in manufacturing compared to traditional materials like steel and aluminum, the overall life-cycle benefits—particularly in terms of fuel savings and reduced emissions during vehicle operation—could justify their use in lightweight applications, especially when higher weight savings could be achieved.

Agarwal et al. (2020) [10] and Gonçalves et al. (2022) [11] expanded on this by reviewing the progress and challenges in adopting lightweight materials for automotive applications. Agarwal et al. (2020) emphasized the potential of eco-friendly composite materials, particularly those combining lightweighting with environmental sustainability through the use of bio-based and recycled materials. Gonçalves et al. (2022) highlighted the need for comprehensive life-cycle evaluations of lightweight materials to fully understand their environmental implications, recommending future studies to improve inventory data, address greenhouse gas break-even points, and incorporate uncertainty and sensitivity analyses.

Prochatzki et al. (2023) [12] critically reviewed the state of the circular economy in the automotive sector, noting that both industry and research often equate circularity with recyclability, overlooking higher-quality strategies like reuse and repurposing. Their study emphasized the need for integrating circular-economy principles early in product development. This project addresses these gaps by applying eco-design principles during the design phase, aiming to enhance the circularity of automotive components.

The goal of this project is to showcase the viability of circularity within the automotive industry through the EU-funded initiative LEVIS. LEVIS has created multi-material solutions that utilize eco-friendly materials and carbon fiber thermoplastic composites, which are seamlessly integrated with metals. These materials are produced using scalable and cost-efficient manufacturing techniques. The project partners are demonstrating the effectiveness of these new technologies by applying them to four electric vehicle (EV) components: a suspension control arm (case study 1), a battery box (case study 2A), a battery module (case study 2B), and a cross-car beam (case study 3).

The primary focus of this study is to compare the environmental performance of the solutions developed during the LEVIS project. By conducting a thorough environmental assessment, this paper seeks to determine the environmental viability of these technologies. While cost-effectiveness assessments were also conducted during the project, they are mentioned only briefly here due to the confidentiality of detailed financial information.

In summary, this paper provides insights into the environmental benefits of the LEVIS project's multi-material solutions and their potential to contribute to a more sustainable and circular automotive industry.

2. Materials and Methods

2.1. Case Study 1: Suspension Control Arm

The suspension control arm connects movable parts like the wheel and knuckle to fixed parts such as the vehicle body and frame. It allows for vertical movement for suspension and rotation for steering. Its design and function vary based on vehicle handling and comfort targets. Each car has two front suspension control arms with bushings and a ball joint connecting them to the cradle and steering knuckle. Typically, these components are pressed in; however, the ball joint can also be screwed or riveted.

The conventional suspension control arm (Figure 1a) is predominantly constructed from steel, featuring aluminum inserts and bushings. The objective of the redesign was to preserve durability and safety while achieving a reduction in weight. The original suspension control arm, designed for a traditional internal combustion engine (ICE) vehicle, has been in production for 15 years. This component serves as a benchmark in a LCA within the context of an EV to compare the use-phase emissions of the new design.

In the new design, the metal body is replaced with CFRP, utilizing acrylic-based resins and a combination of bio-derived and PAN-based carbon fibers through the Resin Transfer Molding (RTM) process (Figure 2b). RTM involves the injection of a monomer and catalyst into a mold filled with reinforcement material, leading to in situ polymerization. The project focuses on using polyamide and acrylic-based resin, along with bio-based carbon fiber, and optimizing mechanical performance with specific sizing compositions. The optimization methodology involves several advanced simulation methodologies and workflows. This includes an initial CAD design, after which conceptual prototypes will be manufactured to assess the final geometry of the part. Once the geometry is defined based on developments in materials, fabrication processes, and structural integrity, integrated simulation workflows will be applied to verify design compliance with specifications. These simulations include process simulations to optimize manufacturing and reduce waste; initial structural simulations to analyze stiffness and stress under critical loads; and advanced structural simulations to predict failure, fatigue, and the impact of environmental factors like temperature and humidity.

The preforming stage is optimized to achieve productivity goals by utilizing Automated Tape Layering (ATL) technology and developing binders that are compatible with thermoplastic matrices, RTM processes, and ATL heating methods.

The reengineered suspension control arm incorporates metal inserts for bushings through co-molding, with bushings press-fitted afterward. This approach achieves up to a 30% reduction in weight while maintaining structural integrity.

Similar to the original design, the metal components in the new suspension control arm are recycled. However, to improve recyclability, the CFRP components are recycled using pyrolysis. This process involves heating the CFRP in an oxygen-free environment, which breaks down the resin matrix and recovers carbon fibers, along with gas and oil byproducts that can be used as chemical feedstock. Despite the production of byproducts, pyrolysis preserves the carbon fibers' high mechanical properties [13].

2.2. Case Study 2A: Battery Box

The battery box houses 10 battery modules, providing protection during operation and accidents and supporting these modules by integrating them into the vehicle frame for safety. Its internal structure facilitates wiring and heat dissipation. This project focuses specifically on the structural design of the battery box, aiming to optimize weight reduction while maintaining or enhancing the mechanical strength and durability of the components. The study does not include the development of an active cooling system or wiring. However, the battery box must meet the same requirements as the benchmark battery box, meaning the demonstrator must be capable of supporting the same wiring and cooling system as the benchmark vehicle.

(a) (b)

Figure 1. Demo 1—suspension control arm: (**a**) benchmark product and (**b**) new design. (Reprinted with permission from Ref. [14]. Copyright 2024 LEVIS project).

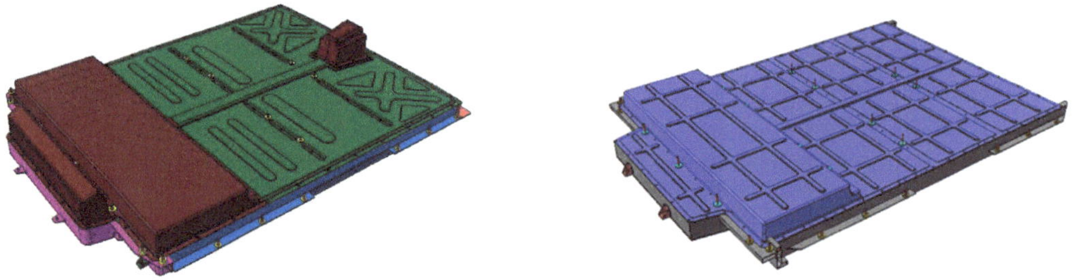

(a) (b)

Figure 2. Demo 2A—battery box: (**a**) benchmark product and (**b**) new design. (Reprinted with permission from Ref. [14]. Copyright 2024 LEVIS project).

The benchmark battery box (Figure 2a) is primarily constructed from aluminum, with some plastic and rubber components. It is used in a D-class EV from China, with a battery capacity ranging from 80 to 100 kWh.

To achieve weight reduction, hybrid solutions were implemented. CFRP patches and structures were bonded to aluminum beams and profiles to maintain structural integrity while reducing weight. Multi-material designs optimize performance and costs, with thermoplastic materials being a preferred solution. In situ consolidation enables composite and metal–composite joint development in a single step, enhancing stress distribution, fatigue strength, and production rates, without adding weight. Laser texturing on metallic parts is used for surface treatment, studied in EU projects [15–17], and currently at Technology Readiness Level 5 [18–20].

The original upper cover design combines aluminum and plastics, whereas the new version utilizes compression-molded CFRP with recycled carbon fibers. Compression molding of carbon fiber non-woven mats enables the production of high-strength, complex parts by molding chopped or continuous reinforced prepreg sheets under compression. Given that this component has lower mechanical requirements, short, chopped carbon fibers are sufficient, leading to a 30% weight reduction without compromising structural integrity.

The metal components of the new design (Figure 2b) are recycled similarly to the benchmark, but the CFRP components use different recycling methods: chemical recycling for the upper cover and pyrolysis for other CFRP parts. To facilitate recycling, debonding on-demand technologies are employed. Various joining technologies and surface treatments, such as adhesive tape layers and laser texturing, are used to enhance disassembly capabilities for maintenance, repair, and overhaul (MRO). A heating-activated on-demand disassembly technique optimizes recovery and recyclability for metal–composite joints.

2.3. Case Study 2B: Battery Modules

The primary role of the battery module is to store and distribute energy, with the module housing providing thermal, mechanical, and electrical protection for the cells, while also facilitating electrical distribution through the busbar system. The benchmark design (Figure 3a) uses plastics and epoxy glass, while the new design (Figure 3b) introduces a modular box made from Glass Fiber-Reinforced Plastics (GFRPs) with recycled polymers (PA6) and integrated sensors for health monitoring. The Structural Health Monitoring (SHM) system includes piezoelectric and temperature sensors to detect potential failures, thereby extending the module's service life.

(a) (b)

Figure 3. Demo 2B—battery module: (**a**) benchmark product and (**b**) new design. (Reprinted with permission from Ref. [14]. Copyright 2024 LEVIS project).

The busbars use novel lamination processes that reduce energy consumption during production. Aluminum tools optimize heat transfer between the press and busbar, saving 90% of the time taken compared to standard processes. GFRP modules are recycled mechanically by shredding the fibers and incorporating them into new composites. The new design achieves a 47% weight reduction.

2.4. Case Study 3: Cross-Car Beam

The cross-car beam, located in the dashboard area, focuses on the steering column carrier (Figure 4a, golden part). The benchmark is entirely steel, with welded brackets. The new design (Figure 4b) is a one-piece CFRP part using recycled carbon fibers through injection molding, achieving a 28% weight reduction. CFRP is mechanically recycled at the end of life.

(a) (b)

Figure 4. Demo 3—cross-car beam: (**a**) benchmark product and (**b**) new design. (Reprinted with permission from Ref. [14]. Copyright 2024 LEVIS project).

2.5. Environmental and Economic Assessment

2.5.1. Methodology

The environmental performance of the newly designed EV components is assessed against current industry standards using a life-cycle assessment (LCA). A life-cycle costing (LCC) methodology is employed to provide insights into the cost implications of material choices, production, manufacturing, and end-of-life processes, comparing them to current equivalents to determine financial feasibility.

A comparative LCA is conducted to evaluate the environmental performance of the newly designed EV components relative to existing industry applications. This LCA adheres to the framework set out in the ISO 14040–14044 standards [21]. The initial step in the analysis involves defining a benchmark product and conducting a sustainability assessment, which considers the environmental impact throughout the product's life cycle—from material extraction and component production to transportation, use, and end-of-life processes.

The new EV components are then analyzed using the same LCA methodology. Since the technologies are in early development stages, the initial LCA relied on laboratory-scale data provided by industry partners, which may not fully represent industrial-scale applications. Consequently, the initial analysis used laboratory data, which often yield lower production efficiency per unit of energy used, or estimates to simulate industrial-scale production. To enable a fair comparison, further research was conducted to predict the environmental impact of these new components at the industrial level, incorporating more comprehensive and representative data, as well as advanced modeling techniques. The results presented are based on industrial-scale data that were modeled and calculated rather than directly measured.

2.5.2. Functional Unit

The lifespan of the vehicle may vary for each benchmark vehicle; however, for the purposes of this study, it is kept consistent across all demonstrators. The functional unit is defined as follows:

> *The functional unit for this study is the installation and usage of a product that lasts the entire lifespan of a C-class electric vehicle driving a WLTP cycle, maintaining the vehicle's functionality and occupant safety. The average lifespan is considered to be 160,000 km.*

This definition means that, for the suspension control arm, the LCA assumes that two products are needed to perform the function of the EV, while for the battery module, ten products are required. All other components are considered single products.

2.5.3. Life-Cycle Inventory

Raw Materials and Manufacturing

The raw materials are directly linked to the Bills of Materials (BOMs) of the designs. These BOMs, provided by the project partners, detail the various parts of the demonstrators and benchmark products, along with the types of materials used. Additionally, the BOMs include the manufacturing processes employed to shape the parts from the raw materials. All materials are associated with datasets from the GaBi database. It is assumed that all materials used in the benchmark and demo are sourced from the EU-28. These datasets cover the extraction of raw materials. Material flows representing less than 1% of the total mass are excluded.

Data on energy consumption, resource utilization, waste generation, and emissions from manufacturing processes are obtained from project partners or the relevant literature. When primary data are unavailable, GaBi datasets are utilized to estimate manufacturing emissions. If neither primary nor GaBi data are available, the analysis follows a hierarchical approach: it first relies on the literature data, then calculations, and finally estimates, with processes potentially being omitted only if no other data source is sufficient.

Use Phase

The benchmark demonstrators do not have a "direct" use phase where they consume energy independently. However, they influence the energy consumption of a vehicle by their weight. The energy consumption associated with the benchmark demonstrator is calculated using the following formula:

$$EC = (ERV \times m \times mileage_{use})/1000$$

where ERV = energy reduction value (kWh/(100 km × 100 kg)), m = vehicle mass reduction (kg), EC = energy consumption through mass (kWh), and $mileage_{use}$ = lifetime vehicle (km).

The ERV values (see Table 1) are extracted from the literature based on Del Pero et al. (2020) [22] and are determined by the vehicle class and driving cycle. For this study, the World Light Test Procedure (WLTP) is used. Key assumptions made are the vehicle class for each specific demo case and the lifespan of the vehicle (160,000 km).

Table 1. ERV values for vehicle demonstrators.

Demonstrator	Vehicle Class	Milage	ERV (kWh/100 km × 100 kg)
Suspension control arm	B	160,000	0.56
Battery-holding set	D	160,000	0.66
Cross-car beam	C	160,000	0.58

End of Life (EOL)

Data for the EOL processes are collected similarly to the manufacturing processes. The allocation of recycling and reuse of materials is performed using the "value-corrected substitution" method. This method addresses the downcycling issue in LCA when dealing with products with high scrap ratios. To that end, key assumptions based on the price ratio are made regarding the recycling rate for each material mentioned below.

During production and EOL, large volumes of scrap are produced and recycled. However, the quality of recycled material is often lower than that of virgin material, meaning that scrap material cannot always replace virgin material on a one-to-one basis. The "value-corrected substitution" method uses the price ratio between different grades of scrap (based on their quality) and virgin material. The price ratios for the materials used in the model are as follows:

- ABS scrap ratio: 0.264 [23].
- Steel industrial scrap ratio: 0.21 [24].
- Pyrolysis ratio: 0.85.
- Mechanical ratio: 0.24 [25].
- Chemical ratio: 0.989 [25].
- Steel post-consumer ratio: 0.33 [24].
- Aluminum ratio: 0.42 [24].
- Copper ratio: 0.75 [24].

The energy requirements for the EOL processes for metals are sourced from GaBi datasets, which provide average values for energy use and emissions during the deconstruction of passenger vehicles. For the CFRPs and GFRPs used in the demonstrators, energy and emissions data for the pyrolysis process were supplied by project partners. For other EOL processes, such as chemical recycling, mechanical recycling, and incineration, the literature data [25] were utilized to determine the energy requirements, emissions, and scrap ratios.

2.6. Application of This Method to Other Vehicle Components

The project focuses on practically demonstrating technologies in specific case studies while also theoretically analyzing their potential impact on other vehicle components, in particular the Body-In-White (BIW), battery cells, and electrical motor. These replication studies systematically explore and validate innovative materials and manufacturing processes in BEV components through seven steps:

1. Establish criteria for component applications: Identify BEV components that can benefit from LEVIS materials for weight reduction without compromising structural integrity.
2. Identify alternative applications: Conduct workshops with LEVIS experts to brainstorm potential applications, classified by technical and economic feasibility.
3. Rank solutions with modified QFD: Use a Quality Function Deployment model to rank and score the top three material and manufacturing combinations for each component.
4. Material properties analysis: Compare functional requirements of selected components with LEVIS materials' capabilities to finalize suitable candidates.
5. Estimate weight reduction: Model the anticipated weight reduction for each component and the entire vehicle.
6. Emission reduction calculation: Calculate emission reductions based on the weight savings, considering the vehicle's life cycle.
7. Consolidate results: Summarize the selected solutions, estimated weight and emission reductions, and overall impact on vehicle performance and sustainability.

Steps 1 to 4 focus on identifying feasible LEVIS technologies for other vehicle components, leveraging the expertise of the LEVIS project partners. Once suitable candidates were identified, calculation methods were employed to estimate potential weight reduction and emission savings (steps 5 and 6).

To determine the weight of the replication components, an assumption was made where the average percentage reduction of the demonstrators was calculated. This varied according to different solutions (material + manufacturing process). Further weight reduction is possible, considering the impacts of the initial weight reduction of the Body-In-White due to lighter components. The initial weight reduction positively impacts the vehicle's range, allowing for a smaller battery pack, which in turn saves weight in battery cells. This also reduces motor torque and power requirements, enabling motor weight reduction. This also has been taken into account for the replication studies [26].

The emission reduction calculation in the replication studies focuses on quantifying the potential environmental benefits achieved through weight reduction. Baseline emissions are calculated based on the initial weight of the fully loaded vehicle (2230 kg) and its components, using conventional materials and manufacturing processes. This includes emissions from material extraction, manufacturing, vehicle operation, and disposal. Each component's weight reduction contributed to lowering overall emissions. The LCA results are used to calculate potential emission reductions. The expected emissions per kilogram of produced product encompass emissions associated with material use, manufacturing, and end-of-life processes and credits. The potential total emission reduction for the BIW was calculated using the weight reduction values and the Global-Warming Potential (GWP) per kilogram reduction. Similarly, the potential emission reduction from the weight reduction in the battery cells is calculated using a combination of EcoInvent data [27] and LCA results [28]. Emissions per kilogram for each component of the battery structure are provided by the LCA, and by applying the weight reduction values and the GWP per kilogram reduction, the potential total emission reduction for the battery structure is determined.

Use-phase emissions are calculated based on LEVIS motor efficiency studies [26], showing that the benchmark vehicle uses an average of 183.5 Wh per km. With a 2% increase in motor efficiency per 100 kg of vehicle weight saved, this leads to a total of 4.34% potential increase in motor efficiency. Emissions per Wh of electricity are derived from the GaBi database [24], indicating 0.33 kg CO_2 equivalent per Wh of electricity. The formula to calculate the CO_2 reduction is as follows:

$$\mathrm{GWP}_{\mathrm{usephase}} = 0.33 \times 183.5 \times \mathrm{Vehicle}_{\mathrm{lifespan}}$$

$$GWP_{reduction} = GWP_{usephasebenchmark} - (GWP_{usephasedemonstrastor} \times (1 - 0.0434))$$

The vehicle lifespan ($Vehicle_{lifespan}$) is assumed to be 160,000 km.

The total emissions saved ($GWP_{reduction}$) are then compared against the total emissions of an electric vehicle [27,29].

3. Results

This paper presents the environmental results from the project.

3.1. Life Cycle-Assessment Results

This study aims to achieve a 25% reduction in greenhouse gas (GHG) emissions at the component level by utilizing new materials, advanced manufacturing processes, weight reduction strategies, and end-of-life (EoL) approaches. The weight reduction targets for all components have been successfully met, and the study now focuses on evaluating the potential GHG emissions savings resulting from these changes. The outcomes of the climate change impact assessment, conducted using the ReCiPe method, are detailed in Table 2.

Table 2. Results LCA, life-cycle emission savings of components in percentage. New design compared to benchmark product [30].

	Modifications Compared to Benchmark	Changes in Weight	Changes in Climate Change Impact (kg CO_2-eq)	Changes in Climate Change (%)
Demo 1. Suspension control arm	Body material was replaced with CFRP.	−26%	+3.9	+12.1%
Demo 2A. Battery box		−31%	−400.16 (EU) −1196.17 (CN)	−28.7% (EU) −54.6% (CN)
Demo 2B. Battery module	Redesign and use of GFRP with recycled polymers.	−47%	−656.8	−65.5%
Demo 3. Cross-car beam	Replacing several different parts of the steering column carrier with a one-piece solid part made from CFRP through injection molding.	−26%	−1.32	−11.6%

Suspension control arm:

The GHG emissions of the suspension control arm increased by 12.1% compared to the benchmark product (see Table 2). Figure 5 compares the benchmark and demonstrator suspension control arms in terms of their impact on climate change across various life-cycle phases. As illustrated, the demonstrator has a slightly higher overall impact on climate change. The reduction in climate change impact during the use phase, due to weight reduction, is negligible when compared to the significant increase in impacts caused by material flows. The amount of aluminum and rubber has remained the same in the suspension control arm demonstrator. However, a significant portion of the steel in the benchmark model was replaced with CFRP, which contains 5% bio-carbon fiber. Although this change has resulted in approximately a 26% weight reduction (see Table 2), the climate-change impacts of carbon fibers are significantly higher than those of steel. This results in a 73.51 kg CO_2-eq impact on climate change for the demonstrator, compared to a 16.86 kg CO_2-eq impact for the benchmark (for the environmental impact of raw materials phase). The substantial impact of carbon fiber is primarily due to the significant amounts of electricity required for its production [30].

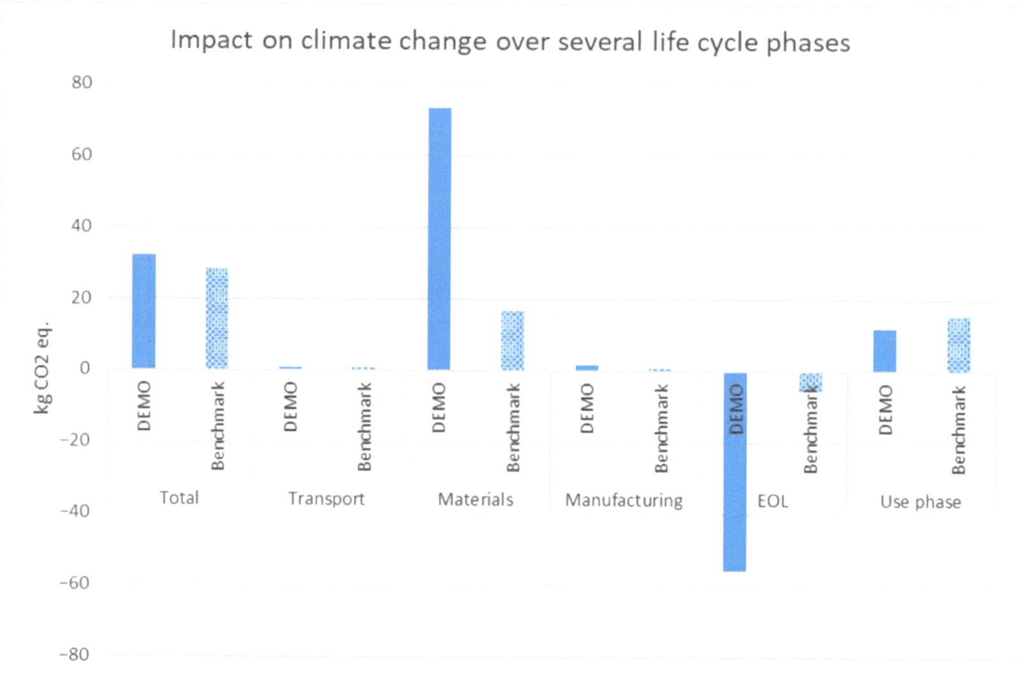

Figure 5. Impact on climate change over life-cycle phases of suspension control arm demo vs. benchmark. (Reprinted with permission from Ref. [14]. Copyright 2024 LEVIS project).

It is noteworthy that the suspension control-arm LCA was based on theoretical industry data. When laboratory data were used for the LCA on a pilot scale, the manufacturing phase, particularly the RTM manufacturing process, emerged as the most dominant source of climate-change emissions. A previous analysis based on lab data projected that the suspension control arm would emit more than twice as much GHG emissions as the benchmark. This highlights the significant positive environmental impact achieved through the optimization of materials and manufacturing processes.

Battery box:

The battery box of an automotive vehicle was analyzed to evaluate its environmental impact. For the battery box, two benchmark scenarios were considered: one where the production takes place in China (using only Chinese datasets from GaBi, Benchmark-CN) and one where the production takes place in Europe (where only European datasets from GaBi were used, benchmark EU). The sensitivity analysis contemplates the effect of relocating the manufacturing phase to Europe instead of China. The results indicate a significant decrease of 28% when using EU datasets and 54% when using Chinese datasets (Table 2) of GHG emissions over the entire life cycle. The emissions from the Chinese benchmark are a lot greater compared to the EU version, which highlights the importance of location and clean energy production for manufacturing and mining materials.

The reduction in emissions is primarily attributed to the manufacturing phase, with a decrease in emissions during the use phase also observed (Figure 6). The emissions during the production of the battery box demonstrator (including material mining, etc.) are significantly lower than those during the production of the benchmark product. Several factors contribute to this reduction. Firstly, the demonstrator uses less material, resulting in lighter aluminum, which in turn requires less aluminum for production, thereby reducing emissions. Secondly, the end-of-life procedures applied to the CFRP hybrid solutions and the upper cover yield relatively high credits during the end-of-life phase.

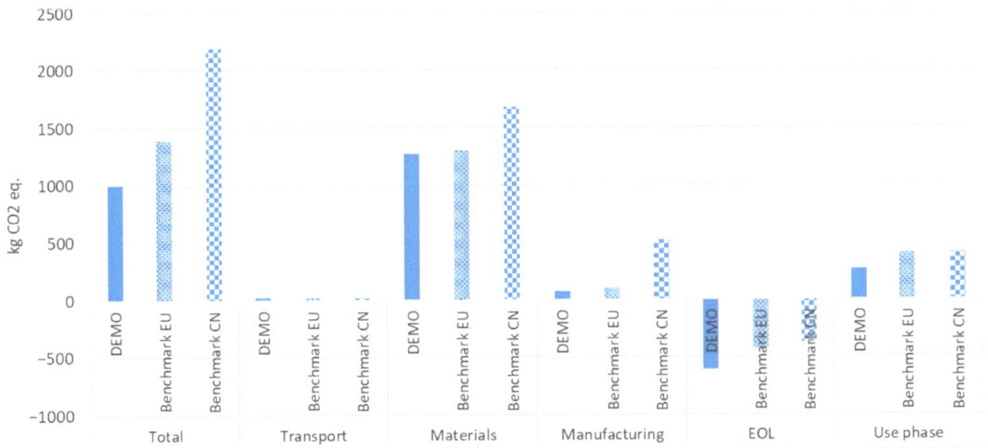

Figure 6. Impact on climate change over life-cycle phases of battery box demo vs. benchmarks. (Reprinted with permission from Ref. [14]. Copyright 2024 LEVIS project).

Battery Module:

The battery module demonstrates a substantial 47% reduction in GHG emissions over its entire life cycle, as shown in Figure 7. This reduction is primarily due to the improvements in the manufacturing and material phases. While some of the savings are attributed to the use phase, the battery module's design achieves the most significant weight reduction among all components, significantly contributing to its overall impact.

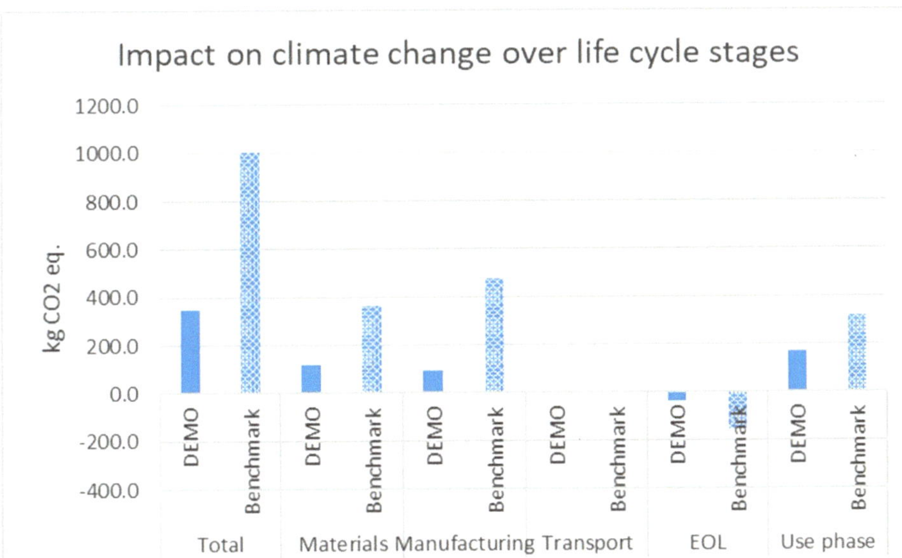

Figure 7. Impact on climate change (kg CO_2 eq.) over all the life-cycle phases of the battery module. (Reprinted with permission from Ref. [14]. Copyright 2024 LEVIS project).

The primary material contributing to the total emissions of the demonstrator battery module is the PA6 used for the GFRP. Glass fiber has a significantly lower carbon footprint compared to carbon fiber, thus explaining the relatively low emissions contribution compared to other demonstrators in this report. The battery module achieves substantial emission reductions primarily due to the improved design of the busbars, as it significantly reduces copper usage. In the benchmark LCA, the main contributor to emissions is the amount of copper used. Therefore, the significant reduction in copper for the demonstrator results in a substantial decrease in emissions.

For the manufacturing of the battery module, the lamination process is clearly the main contributor to emissions. This process requires a significant amount of electricity and is associated with the busbars. The demonstrator's busbars feature an efficient design that significantly reduces the amount of copper needed for both the busbars and the lamination process. Consequently, the demonstrator requires substantially less energy for this process compared to the benchmark.

Cross-car beam:

The cross-car beam shows a slight 10% reduction in GHG emissions (Figure 8) over its entire life cycle. While use-phase emissions have decreased, material emissions have increased due to the use of carbon fibers. A significant portion of steel in the benchmark was replaced by CFRP, resulting in a 28% weight reduction. However, the climate-change impacts of carbon fibers are much higher than those of steel, leading to a 10.80 kg CO_2-eq impact on climate change compared to the benchmark's 6.20 kg CO_2-eq (for the material phase). This high impact is primarily due to the significant amounts of electricity required for carbon fiber production, particularly during the oxidation and carbonization processes [22]. Although a large part of the material impact is offset during the EOL phase, the overall climate-change impact of the demonstrator remains higher than that of the benchmark.

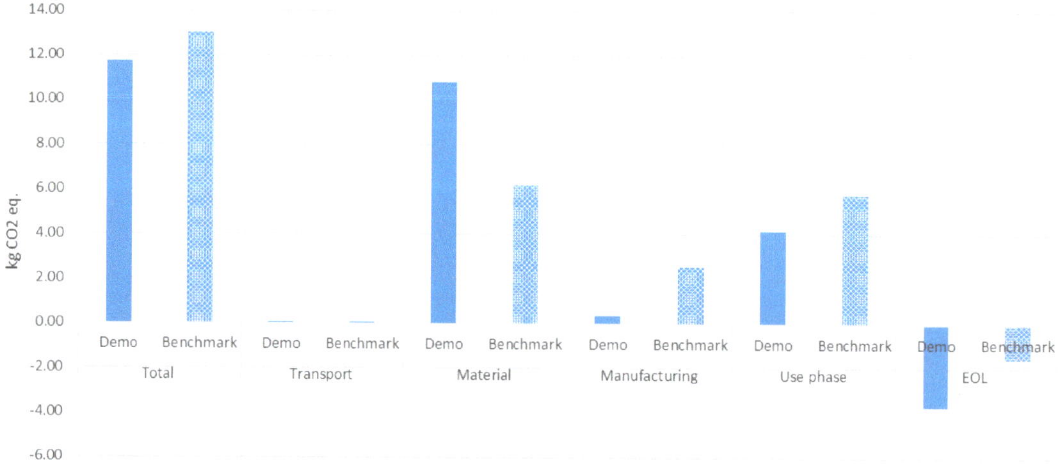

Figure 8. Impact on climate change over life-cycle phases of cross-car beam demo vs. benchmark. (Reprinted with permission from Ref. [14]. Copyright 2024 LEVIS project).

The injection molding process is the largest contributor to the demonstrator's emissions. Despite CFRP not being the largest part of the cross-car beam by mass, its processing contributes more emissions per kilogram compared to the stamping and bending of steel.

3.2. Sensitivity Analysis

3.2.1. Electricity Grid Mix

The summarized impacts in Table 2 are based on the electricity grid mix in the EU. All demonstrators and benchmarks were also evaluated using the electricity grid mixes of the

US and China. Among these, the Chinese electricity grid mix resulted in the highest total impact for both the benchmark and the demonstrators across all components. However, the benchmark was more significantly affected. This suggests that the improvements in the demonstrators compared to the benchmark would be greater than those shown in Table 2 when considering the Chinese electricity grid mix. Specifically, for the suspension control arm (Demo 1), the demonstrator even shows an improvement compared to the benchmark (Figure 9). For the US electricity grid mix, the improvements fell between those observed for the EU and China. Thus, it can be concluded that manufacturing and using electric cars in China and the US has even more positive climate-change impacts than in the EU.

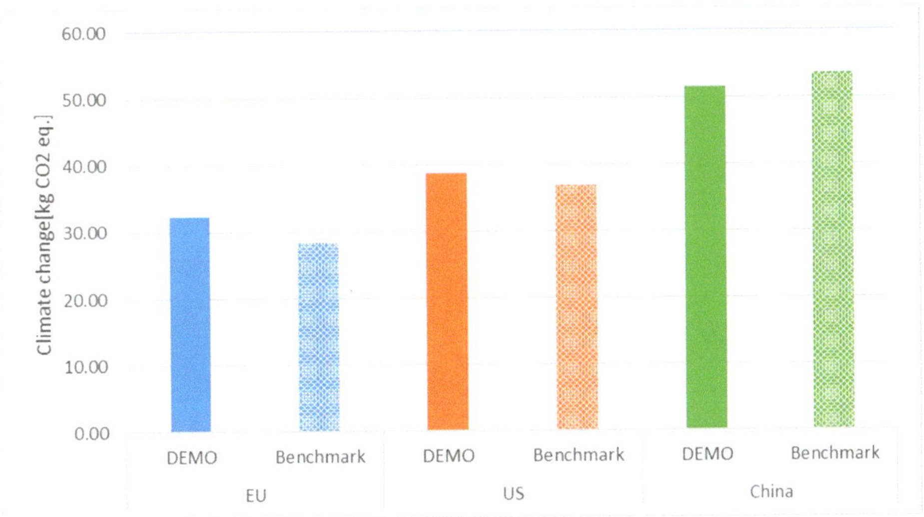

Figure 9. Effect of electricity grid mix on total climate-change impact of suspension control arm demo vs. benchmark. (Reprinted with permission from Ref. [14]. Copyright 2024 LEVIS project).

3.2.2. Lifespan

The baseline calculations were based on a 160,000 km lifespan. In the sensitivity analysis, increasing the lifespan resulted in a lower rate of increase in climate-change impact for all demonstrators compared to the benchmarks. This implies that, for lifespans exceeding 160,000 km, the net reduction in GWP will be greater than the values shown in Table 2, favoring the demonstrator. For Demo 1, specifically, a lifespan of more than 328,900 km will result in a lower climate-change impact than the benchmark (see Figure 10).

3.2.3. End-of-Life Methodologies

This study examined how different recycling methods could impact the overall climate-change effects for each demonstrator. Table 3 lists the recycling methods employed for each demonstrator. For the suspension control arm, the choice of recycling method notably affects the climate-change impacts. Incineration and mechanical recycling result in much higher environmental impacts compared to pyrolysis, which was used in the baseline calculations. Conversely, chemical recycling results in an end-of-life impact that is nearly equivalent to the benchmark. Specifically, incinerating CFRP increases the total global-warming impact of Demo 1 by 188% (see Figure 11).

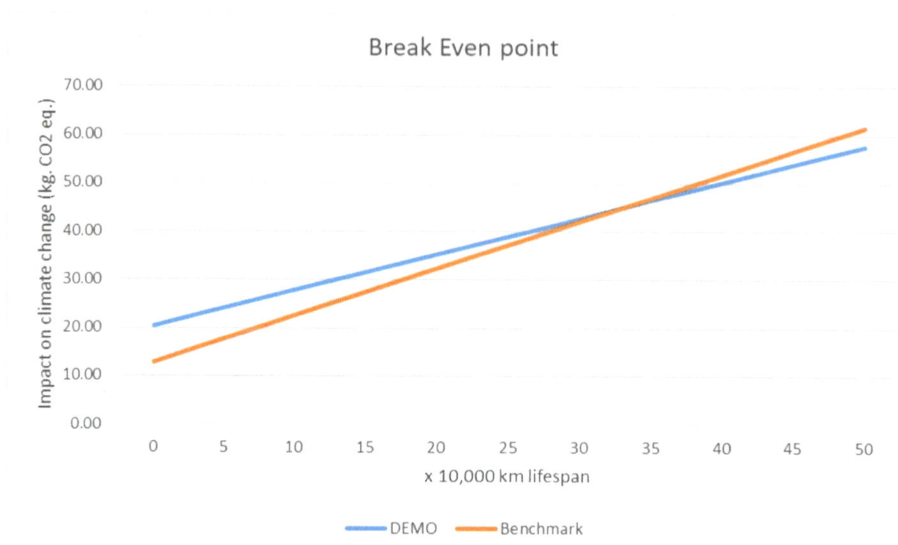

Figure 10. The effect of lifespan on total climate-change impact of suspension control arm demo vs. benchmark. (Reprinted with permission from Ref. [14]. Copyright 2024 LEVIS project).

Table 3. Recycling methods used per demonstrator.

Recycling Method	Pyrolysis	Chemical Recycling	Mechanical Recycling	Incineration
Demo 1. Suspension control arm	X			
Demo 2A. Battery box	Patches and internal beams	Upper cover		
Demo 2B. Battery module			X	
Demo 3. Cross-car beam			X	

In Demo 2A, although changing the recycling method increases the total climate change impact, it still remains significantly lower than the benchmark. For Demo 2B, mechanical recycling was used in the baseline calculations. In this scenario, both pyrolysis and chemical recycling slightly increase the total impact, while incineration reduces it (see Table 4). Pyrolysis and chemical recycling are both energy-intensive processes that yield high-quality recycled materials. However, when the emissions associated with producing the recycled material are relatively low, the energy required for these processes might not justify their use. In such cases, it might be more efficient to incinerate the material at the end of its life.

Table 4. Results of LCA and sensitivity analysis. CR = chemical recycling, L = LEVIS EOL process, and I = incineration. For the battery box, only the EU demonstrator was used for comparison.

Recycling Method	LCA	Grid Mix		Lifespan (Break-Even Point in 1000 km)	Recycling Method (Recycling Method with Best Outcome)
		US	CN		
Demo 1. Suspension control arm	+12.1%	+5.0%	+3.8%	328.8	+1.5% (CR)
Demo 2A. Battery box	−28.7% (EU)	−29.6%	−30.8%	N/A	−28.7% (L)
Demo 2B. Battery module	−65.5%	−62.7%	−59.0%	N/A	−67.1% (I)
Demo 3. Cross-car beam	−11.6%	−13.6%	−17.5%	29.7	−36.7% (CR)

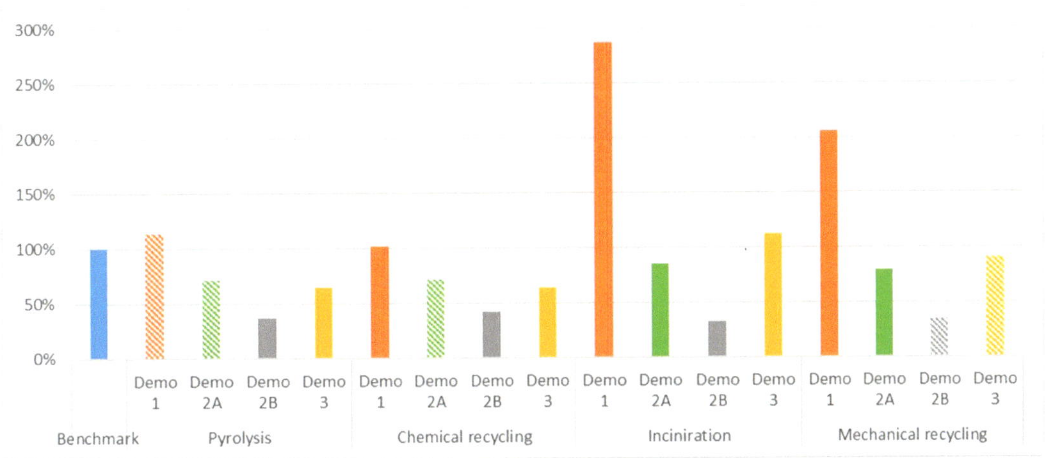

Figure 11. Results of the end-of-life sensitivity analysis of all the demonstrators compared in percentages to their benchmark. Patterned bars represent the actual end-of-life processes used in the LEVIS project. Filled bars represent theoretical recycling processes.

3.3. *Replication Studies Results*

The replication studies reveal significant potential for both weight reduction and environmental improvements. The final results indicate that the selected materials and processes can achieve a substantial weight reduction of 30.85% for the BIW. This weight reduction translates to a 4.25% decrease in energy consumption, alongside a reduction in battery and motor capacity by 16 kg and 9 kg, respectively (see Table 5).

Table 5. Results of the replication studies, showing the weight reduction and the emission reduction.

Component	Weight Reduction (kg)	Weight Reduction (%)	Emission Reduction (kg CO_2 eq.)	Emission Reduction (%)
BIW	107.80	30.85	415.66	22.04
Battery structure	37.76	31.00	186.15	1.44
Battery module	48.43	47.00	507.86	17.45
Battery cells	16.05	3.50	179.08	3.50
Electric motor	8.97	9.02	102.16	5.98
Use phase	-	-	419.68	4.34
Total EV	219	9.82	1810.57	5.98

Additionally, the application of these technologies significantly reduces the GWP of the vehicle structure. Although the GWP per kilogram is higher for most solutions compared to the benchmark, this is expected because these values do not account for the emission savings from weight reduction during the use phase. Due to the lighter weight of the new solutions, the overall emissions are generally lower compared to the benchmark. The current results indicate a notable 5.98% reduction in total emissions.

4. Discussion

Industry vs. Lab data

It is important to note that the current LCA results are based on modelled and calculated data. Future research should focus on implementing these technologies at a higher technology readiness level, with an LCA based on measured data to ensure more accurate and reliable results.

Life-Cycle Costing

An economic assessment was conducted as part of the project, but the details of this private report cannot be shared. One of the main goals of the project was to develop cost-efficient technologies that offer comparable economic results to currently available solutions. An LCC analysis was performed to study the economic performance of the demonstrators. The same model used for the LCA was applied here, with different inputs including raw material costs, manufacturing, use-phase savings, and end-of-life costs and credits. The results varied significantly: two demonstrators showed considerable cost savings, one showed cost increases, and one had similar economic performance to current solutions. Most cost increases were due to the relatively high prices of Elium resin and carbon fiber, as well as the expensive RTM process, which involves high labor and tooling costs. However, many cost savings were achieved through efficient design, resulting in less material use and therefore lower processing costs. Notable savings were also made from reduced electricity usage during the use phase. Shadow costs, which account for ecological effects, made up about 15% of the total LCC for some demonstrators and about 5% for others. These shadow costs were considered to evaluate the economic viability from an ecological perspective. Despite these findings, the results can vary greatly due to limitations in the assessment, particularly due to the lack of data. The demonstrators also have greater uncertainties in costs due to missing information on labor and tooling expenses. A more detailed economic analysis is necessary before commercializing these technologies or implementing them into other car components.

Eco-design

One of the key contributions of this project was the integration of eco-design principles into the design process of the demonstrators. To achieve this, a new toolkit, iEDGE (integrated Eco-Design Guideline and Evaluator) [31], was developed to assist designers in their assessments. iEDGE is designed to incorporate eco-design and circular-economy principles into the early stages of product design. Before starting the life cycle-assessment activities, the project partners set eco-design objectives for each component using the iEDGE toolkit. This toolkit was instrumental in defining requirements and key performance indicators (KPIs) across four critical areas: environmental, economic, technical, and social. Unlike traditional approaches that focus solely on improving environmental and social impacts, the iEDGE toolkit provides a comprehensive framework for identifying design improvements by considering all four main aspects of product design: environmental, economic, technical, and social. The toolkit is based on established methods and tools developed by universities and researchers specializing in eco-design, offering a balanced approach to achieving the project's goals. While it is challenging to measure the exact impact of the eco-design method on the final design and product circularity, comparing LCA results against the design objectives set by iEDGE provides some insights. The results indicate that the iEDGE toolkit effectively guided the design of the four demonstrators by highlighting key areas for environmental improvement. The areas identified by the toolkit were also the ones that showed the most improvement according to the LCA results. However, the project environment, while ideal for developing and testing the toolkit, may not have been perfect for its practical implementation. The toolkit was created in collaboration with the partners, incorporating their insights and needs. Yet, since the solutions were largely predetermined from the start of the project, it is unclear whether the toolkit had a significant effect on the outcome or if these outcomes would have been achieved regardless. To fully measure the toolkit's effectiveness, it should be used in a new design project from the very beginning of its development process. This would allow for a more accurate assessment of the toolkit's impact on the design and environmental performance of the products.

Data accessibility

Data accessibility posed a significant challenge in this study, particularly for the detailed modeling of EV components. To address this, a data-collection strategy (see Section 2.5.3) and sensitivity analyses were conducted to assess the robustness of our results under different scenarios. For example, the lifespan of the vehicle and the electricity grid mix for charging were addressed to evaluate how these changes might influence the overall environmental impact. While these strategies help to mitigate uncertainties, they also highlight the need for more comprehensive data collection in future research to further refine the conclusions drawn from this study.

5. Conclusions

All demonstrators achieved the goal of reducing component weight by 20% to 40%. However, the environmental assessment results showed significant variation, with emissions changes ranging from an increase for the suspension control arms to a 65.5% decrease for the battery module. Careful conclusions can be drawn from these results regarding the applied technologies:

Efficient use of materials

The hybrid solutions and bonding technologies used in the battery box demonstrated significant potential, achieving a total emission reduction of 28.7%. By strategically placing a small amount of carbon fiber in crucial parts of the battery box, the redesign reduced the need for a significant amount of high-polluting aluminum. In contrast, the suspension control arm showed an increase in emissions, suggesting that CFRPs are more effective when used to replace high-polluting materials like aluminum rather than relatively low-polluting materials like steel.

Recycled material use

Both the suspension control arm and the cross-car beam were initially made mostly of steel for the benchmark and were fully replaced with CFRPs for the demonstrator. The primary differences between them were the manufacturing technologies (RTM vs. injection molding) and the use of virgin versus recycled carbon fibers. The cross-car beam showed a better relative performance compared to the suspension control arm. This indicates that using recycled carbon fibers is more effective for replacing materials like steel, which have relatively low pollution levels.

Expected use of EV

According to this LCA model's assumptions, the solutions applied to the suspension control arm do not show a positive environmental performance compared to the benchmark. However, extending the lifespan of the demonstrator to over 328,800 km reduces its total climate change impact below that of the benchmark. Additionally, the type of electricity grid mix used for charging the EV significantly influences the climate-change impact. For instance, coal-reliant grids like China's show better a relative performance for the demonstrator. Therefore, the technologies implemented for the demonstrator are not necessarily worse than the benchmark; their performance depends on the specific circumstances and expected use of the EVs.

Recycling method

The sensitivity analysis revealed that no single recycling method is universally better than the others. The best method depends on the specific technologies used for the demonstrators. Notably, when the recycled material itself does not require much energy to produce, energy-intensive end-of-life processes with high-quality recycled materials (such as chemical recycling and pyrolysis) are not necessarily preferred over processes with low energy requirements and higher degradation, such as mechanical recycling or even incineration.

Author Contributions: Methodology, F.T.; software, F.T.; validation, F.T.; formal analysis, F.T.; writing—original draft preparation, F.T.; writing—review and editing, F.T.; visualization, F.T.; supervision, E.v.B.; project administration, E.v.B.; funding acquisition, E.v.B. All authors have read and agreed to the published version of the manuscript.

Funding: This research was part of the project LEVIS and was funded by European Union's Horizon 2020 Research and Innovation Programme under grant agreement No. 101006888.

Data Availability Statement: The original contributions presented in the study are included in the article, further inquiries can be directed to the corresponding author.

Acknowledgments: The authors would like to thank the project partners of LEVIS for their hard work, research, and development and for providing information, as it would be impossible to capture the lessons learned without their contributions.

Conflicts of Interest: Floris Teunissen and Esther van Bergen are employees of Cenex Nederland. The paper reflects the views of the scientists, and not the company.

References

1. European Commission. Available online: https://transport.ec.europa.eu/white-paper-2011_en (accessed on 26 October 2022).
2. Circularise. Available online: https://www.circularise.com/ (accessed on 21 October 2022).
3. ALLIANCE Project. Available online: https://alliance-heu-project.eu/project-description (accessed on 25 July 2024).
4. LoCoMaTech Project. Available online: http://www.locomatech.net/ProjectArea1/home (accessed on 25 July 2024).
5. ALIVE Project. Available online: https://cordis.europa.eu/project/id/314234/reporting (accessed on 25 July 2024).
6. ENLIGHT Project. Available online: https://cordis.europa.eu/project/id/314567/reporting (accessed on 25 July 2024).
7. GREENLIGHT Project. Available online: https://www.cbe.europa.eu/projects/greenlight (accessed on 25 July 2024).
8. ECOXY Project. Available online: https://ecoxy.eu/ (accessed on 25 July 2024).
9. Das, S. Life cycle assessment of carbon fiber-reinforced polymer composites. *Int. J. Life Cycle Assess.* **2011**, *16*, 268–282. [CrossRef]
10. Agarwal, J.; Sahoo, S.; Mohanty, S.; Nayak, S.K. Progress of novel techniques for lightweight automobile applications through innovative eco-friendly composite materials: A review. *J. Thermoplast. Compos. Mater.* **2020**, *33*, 978–1013. [CrossRef]
11. Gonçalves, M.; Monteiro, H.; Iten, M. Life Cycle Assessment studies on lightweight materials for automotive applications—An overview. *Energy Rep.* **2022**, *8*, 338–345. [CrossRef]
12. Prochatzki, G.; Mayer, R.; Haenel, J.; Schmidt, A.; Götze, U.; Ulber, M.; Fischer, A.; Arnold, M.G. A critical review of the current state of circular economy in the automotive sector. *J. Clean. Prod.* **2023**, *425*, 138787. [CrossRef]
13. Asmatulu, E.; Twomey, J.; Overcash, M. Recycling of fiberreinforced composites and direct structural composite recycling concept. *J. Compos. Mater.* **2014**, *48*, 593–608. [CrossRef]
14. Teunissen, F. LCA results of Levis demonstrators. *Zenodo* **2024**, *in press*.
15. Mera, L.; Blanco, L. The Role of Laser Texturing and Joint Strength as Multi- Material Manufacturing Enabler. In Proceedings of the Joining in Car Body Engineering (Automotive Circle) 2020, Bad Nauheim, Germany, 29 September–1 October 2020.
16. Blanco, L.; van der Straeten, K.; Peters, T. CFR-PA66-Steel Direct Joints with Enhanced Adhesion through Laser Texturing. In Proceedings of the 5th International Conference on Structural Adhesive Bonding, Porto, Portugal, 11–12 July 2019.
17. Rey, P. Manufacturing of Hybrid Metal/TP Composite Structural Car Parts Using Novel Technologies. In Proceedings of the AEE Expo 2019, Nuremberg, Germany, 5–6 June 2019.
18. ComMUnion Project. Available online: http://communionproject.eu/ (accessed on 12 April 2020).
19. Lay2form Project. Available online: http://lay2form-project.eu/ (accessed on 12 April 2020).
20. Flexhyjoin. Available online: https://www.flexhyjoin.eu/ (accessed on 12 April 2020).
21. *ISO/TC 207/SC 5*; Life Cycle Assessment. International Organization for Standardization: Geneva, Switzerland, 2022.
22. Del Pero, F.; Berzi, L.; Antonacci, A.; Delogu, M. Automotive lightweight design: Simulation modeling of mass-related consumption for electric vehicles. *Machines* **2020**, *8*, 51. [CrossRef]
23. Plasticker. Price of ABS. Retrieved from Plasticker: Plasticker.de. Available online: https://plasticker.de/ (accessed on 24 January 2024).
24. Gabi Sphera. Life Cycle Assessment LCA Sphera. Available online: https://gabi.sphera.com/topics/life-cycle-assessment-lca (accessed on 1 June 2022).
25. Stergiou, V.; Konstantopoulos, G.; Charitidis, C.A. Carbon fiber reinforced plastics in space: Life cycle assessment towards improved sustainability of space vehicles. *J. Compos. Sci.* **2022**, *6*, 144. [CrossRef]
26. Decelis, P.; Teunissen, F.; van Bergen, E. LEVIS Project D6.4 Replication study of LEVIS technologies. *Zenodo* **2024**, *in press*.
27. Eco-Invent. 2022. Available online: https://v391.ecoquery.ecoinvent.org/Details/LCI/17903086-ed44-4bc0-9c43-4251e40b5241/dd7f13f5-0658-489c-a19c-f2ff8a00bdd9 (accessed on 13 March 2024).
28. Greenvehicles LEVIS Project LCA Results. Available online: https://greenvehicles-levis.eu/app/uploads/2023/01/D6.1-Initial-LCA-Results-of-LEVIS-Demonstrators.pdf (accessed on 23 March 2023).
29. Sacchi, R.; Bauer, C.; Cox, B.; Mutel, C. When, where and how can the electrification of passenger cars reduce greenhouse gas emissions? *Renew. Sustain. Energy Rev.* **2022**, *162*, 112475. [CrossRef]
30. Groetsch, T.; Creighton, C.; Varley, R.; Kaluza, A.; Dér, A.; Cerdas, F.; Hermann, C. A modular LCA/LCC-modelling concept for evaluating material and process innovations in carbon fibre manufacturing. *Procedia CIRP* **2021**, *98*, 529–534. [CrossRef]
31. Greenvehicles LEVIS Project Eco Design Results. Available online: https://greenvehicles-levis.eu/app/uploads/2022/01/LEVIS_D1.3_Eco-design_Guidelines-and-Demonstrator-results_V1.1_cp.pdf (accessed on 23 March 2023).

World Electric Vehicle Journal

Article

Techno-Economic Suitability of Batteries for Different Mobile Applications—A Cell Selection Methodology Based on Cost Parity Pricing

Steffen Link *, Maximilian Stephan, Lukas Weymann and Tim Hettesheimer

Fraunhofer Institute for Systems and Innovation Research ISI, Breslauer Str. 48, 76139 Karlsruhe, Germany; maximilian.stephan@isi.fraunhofer.de (M.S.); lukas.weymann@isi.fraunhofer.de (L.W.); tim.hettesheimer@isi.fraunhofer.de (T.H.)

* Correspondence: steffen.link@isi.fraunhofer.de

Abstract: Rapid advancements in lithium-ion battery (LIB) technology have paved the way for the electrification of diverse applications, with continuous improvements in performance, substantial cost reductions, and the emergence of new manufacturers, formats, and cell chemistries. However, this diversity poses challenges in identifying the most suitable battery cells for specific applications. Here, we present a high-level techno-economic framework for cell selection, leveraging an extensive database of over 500 real-world cells, techno-economic analyses of emerging applications, and a Python-based modeling approach. We apply this method to three electrifiable mobile applications with distinct characteristics: battery electric cars, industrial forklifts, and regional passenger trains. Our results emphasize substantial variations in technical requirements, from power capability to energy density or longevity. We observe no particular differentiation according to cell formats, but tendencies for most suitable chemistries per application. No cell is suitable for all applications, particularly regarding the required maximum cell costs to ensure profitability, ranging from a few to several hundred Euros per kWh to achieve cost parity with a state-of-the-art reference technology. These findings highlight the importance of tailored cell selection strategies for decision makers to optimize performance and cost-effectiveness across different applications.

Keywords: battery cell selection; battery electric vehicle (BEV); techno-economic assessment; cost modeling; passenger trains; forklifts

Citation: Link, S.; Stephan, M.; Weymann, L.; Hettesheimer, T. Techno-Economic Suitability of Batteries for Different Mobile Applications—A Cell Selection Methodology Based on Cost Parity Pricing. *World Electr. Veh. J.* **2024**, *15*, 401. https://doi.org/10.3390/wevj15090401

Academic Editors: Joeri Van Mierlo and Genevieve Cullen

Received: 23 July 2024
Revised: 23 August 2024
Accepted: 30 August 2024
Published: 3 September 2024

1. Introduction

Batteries are central to reducing greenhouse gas (GHG) emissions in various sectors. Accordingly, the global demand for lithium-ion batteries (LIBs) has substantially increased and reached an estimated 1000 GWh market in 2023 [1], with annual growth rates of approximately 30–40% and projections for 2030 approaching over 4–6 TWh [1,2]. Herein, the electrification of passenger cars (i.e., battery electric vehicles (BEVs) and plug-in hybrid vehicles (PHEVs)) is widely recognized as the key driver and enabler of future developments worldwide, while stationary storage systems from home to industrial scales are gaining momentum. The resulting battery innovations, such as rapidly decreasing production costs, increasing energy densities, longer lifetimes, and improved fast-charging capability, already allow for the technically feasible and economically viable electrification of an increasing number of mobile and stationary applications [1,3,4].

Increasing electrification also entails increasing complexity, specialization, and heterogeneity, since each application has specific requirements and particularities that must be considered when selecting battery cells and designing battery systems. On the one hand, cell suppliers are aiming to find new applications for their battery portfolio, but can barely keep track of all the applications and their specifics. On the other hand, manufacturers and cell integrators for various applications are searching for the most suitable batteries among many choices; however, they can barely keep track of all the available LIB cells and their characteristics. In the future, a wide choice of alternative battery technologies could further increase the variety of cell chemistries and formats [5].

This paper builds on an approach presented at the EVS36 Symposium 2023 [6]. It proposes a techno-economic framework that enables high-level yet tailored cell selection for different applications. This approach relies on a Python-based model in which cell-specific cost-parity prices (CPP) can be determined for different applications. The cost-parity price of a battery cell is defined as the maximum cell price to achieve cost parity with an existing reference technology [7]. A battery cell with a cost-parity price that exceeds its actual price offers the potential for significant economic benefit through its implementation [6]. These results can be compared to those of an alternative, e.g., combustion-based applications [7]. The main benefit of this approach is that the same cell can be examined for its suitability in several applications very easily, instead of studying applications isolated and independently. Plus, it allows for more tailored and specific cell selection rather than universal selection methods, such as the well-known Ragone plot [8] or the ENPOLITE tool [9].

This paper is structured as follows. Section 2 introduces our three selected applications and identifies the current research gap. Section 3 starts with presenting the methodology in general and then examines the particularities per application and ecological considerations. Section 4 presents the results, while Section 5 complements the discussion. Section 6 closes with summarizing the results, and provides key implications and recommendations.

2. Background and Research Gap

In the literature, the costs of internal combustion engine vehicles (ICEV) and battery-electric vehicles (BEV) are typically compared and evaluated using lifetime analysis approaches, i.e., total cost of ownership (TCO) or Life Cycle Costing (LCC). Recent TCO calculation discussions for BEVs are shown in [10–12], for instance. As a standard, the result is a monetary price (TCO) or CO_2 footprint (LCA) per kilometer driven. The TCO results vary depending on the type of application being compared and the assumptions made, for example, regarding driving cycles [13] or cost estimations with high influence and sensitivity of battery costs in BEVs [14,15]. Generic price projections are often used for the battery costs on cell or pack level [10]. Moreover [7], for long-haul trucks, there is currently no study in place to determine which cells at what price can support a (cost) parity with a reference application in comparison with the battery-electric alternative (i.e., ICEV with BEV) while incorporating the technical suitability. With this paper, we intend to fill this gap by presenting a method to determine the cost parity for three selected mobile applications. The applications of battery-electric vehicles, trains, and forklifts were selected to represent the highly diverse requirements for the battery in mobile applications, while keeping the complexity within reasonable limits.

The electrification of passenger cars through battery-electric vehicles is very likely a no-regret policy to decarbonize individual transport [16], with a promising outlook due to continued advancements in technology, supportive policies, and increasing consumer acceptance. However, challenges remain for single-battery technologies, necessitating tradeoffs regarding battery performance, life, cost, and safety [17], as well as advanced system integration [17,18]. To reach competitiveness with ICEV, the EUCAR stats cell-level target costs of 70 EUR/kWh by 2030 [19], while the BATT4EU Strategic Research and Innovation Agenda sets 75 to 100 EUR/kWh as pack-level target costs for mobility applications by 2030 [20]. BNEF [21] found that volume-weighted average prices in 2023 were 89 USD/kWh (cell-level) and 128 USD/kWh (pack-level).

To decarbonize rail transport, diesel multiple units are being replaced with battery electric multiple units (BEMUs). Modern BEMUs use combined battery–catenary power, drawing energy from overhead catenary or lithium-ion batteries if no overhead catenary is available. They charge while under catenary, or at stops and during braking, making them efficient on routes with mixed electrification, particularly with non-electrified sections below 100 km [22]. Battery requirements are high, balancing energy density to maximize passenger capacity, power density for fast recharging, safety characteristics, and long lifetime, making lithium-iron-phosphate (LFP) an attractive cell chemistry as the cathode-active material. However, different manufacturers also rely on lithium-metal oxides like nickel–mangan–cobalt (NMC) as the cathode material or lithium–titanate (LTO) as a substitution of the commonly used anode material graphite [23,24].

For industrial applications and logistics, the economic comparison of forklift trucks over its lifecycle is key in the consideration for decision-makers [25]. In the past, the methodological focus was on comparing battery-powered and fuel-cell-powered forklift trucks from a technical point of view [26,27], in an economic or ecological utilization comparison [26,28–30], on impacts from the technology selection of the warehouse economics [31] and handling activities [32], or a combination of such different aspects [33]. Furthermore, different battery technologies are employed for the evaluation and comparison of forklift trucks. This entails a techno-economic comparison of the LIB with a conventional lead-acid battery (LAB) [31,34,35], life-cycle costing [36], as well as analyzing the utilization of LIBs in a second life use-case [37]. For several years, there has been a notable increase in the proportion of LIBs in the product portfolios of original equipment manufacturers (OEMs) of industrial applications and forklift trucks.

A review of the recent techno-economic analyses reveals a notable absence of consideration given to the large range of battery cells available. There is a paucity of analyses that take into account not only LIBs with defined cost and performance indicators, but also the broad array of (real-world) battery cell data based on their individual product specifications to make an economical best-fit technology selection for individual applications. At the same time, the total cost of ownership (TCO) metric is commonly used as an indicator for the cost-competitiveness of a technology [10,11,13–15,34,36,38], but cannot clearly answer the question of the price threshold at which a single cell becomes economically viable in battery-electric applications. It is our intention to address this gap by proposing a novel methodology for cell selection based on cost-parity pricing. In addition to the analysis in [6], this paper also presents a cross-application comparison of the results in three different mobile applications, and further supplements the CPP analysis with a calculated CO_2 footprint for each cell in the aforementioned applications.

3. Methodology

3.1. General

Our systematic approach involves three steps, as shown in Figure 1: energy simulation, battery sizing, and cost calculation.

First, we define the input data used in our energy simulation. This comprises the technical specifications of the respective product (e.g., passenger car, train, or forklift), synthetic and real-world load profiles (e.g., time-based standard speed profiles such as the WLTP driving cycle for cars or processed distance-based speed profiles from real driving conditions for trains), and other application-specific requirements that may affect energy consumption. Technical product specifications depend on the application and may involve information such as weight, rated power, or number of passengers. Given these input values, the energy simulation determines the respective energy consumption for an electrified product version. The input data are derived from public information based on desk research and assumptions based on expert consultations with various stakeholders in the car, train, or forklift truck industry, e.g., application manufacturers and operators.

Second, over 500 battery cells and their specific technical capabilities are fed into the battery sizing algorithm, which determines the required number of cells and the final battery system's capacity to fulfill all of the requirements and load profiles. The sizing algorithm from cell to pack level was implemented in accordance with [7]. Technical capabilities include, among others, battery chemistry, cell format, volumetric and gravimetric energy density, and C rates for charging and discharging. The cell database is generated based on publicly available data sheet information, and is available for download [39].

Third, the total cost of ownership (TCO) or levelized cost of energy (LCOE) are calculated based on the chosen battery capacity and simulated energy needs. These results are then compared to the TCO/LCOE of the next best alternative (e.g., gasoline cars). From a first-user perspective, cost calculations cover all relevant capital expenditures (CAPEX) and operational expenditures (OPEX). The cost-parity price for the battery system, including eventual replacements and battery scrappage, is then obtained using the TCO delta between the battery version and the respective alternative. The cost-parity price per cell is calculated based on the number of cells, including potential replacements. If the cost parity was negative, this battery cell would not be suitable for this application. A positive cost-parity price indicates that at this cost, given the specifications stated in the database and the assumed use-case conditions and load profiles, this cell would perform as well as the corresponding alternative. Thus, the cost-parity price indicates a techno-economic upper price limit. If the cell was available at a lower price, there would be an advantage for the user and thus an incentive to buy, making these battery cells more attractive.

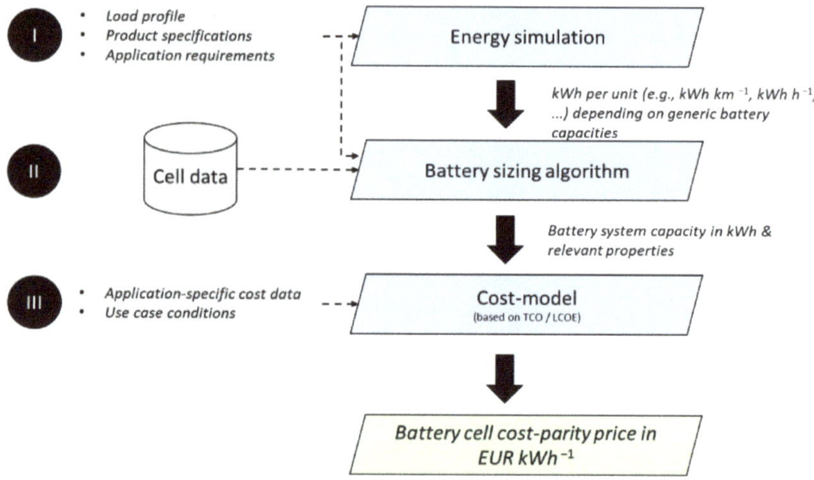

Figure 1. Schematic structure of the cost-parity model.

While the general methodology is described in Figure 1, the detailed procedure for each application is quite different and is described below. However, there are some similarities between the applications in the calculations. For instance, the battery size is mainly affected by weight and energy consumption for most mobile applications. Table 1 shows the main sources that are used for our calculations for the different steps in the cost-parity model. The calculations were carried out in accordance with [7] and with the relevant data for each application. For instance, the energy consumption during operation differs greatly, the battery capacity required for a driving profile taking into account the available installation space is highly application-individual, and costs like taxes for BEVs or worktime in logistics for forklift operation are added or even eliminated compared to the reference application. As generalized use-cases are required for this analysis, the limited data availability on typical driving scenarios and use-cases had to be filled by assumptions based on various studies and expert consultancy, as discussed in the following subchapters.

The methodology is applied to the three distinct mobility applications with different requirements to determine whether there are cells that are suitable for all the applications under consideration or whether different cells should be selected for these specific applications. This allows for an economic selection on the basis of the cost-parity price, taking into account the technical suitability of an LIB's cell chemistry and format.

Table 1. Main sources and model input data for own calculations in accordance with [7], adapted for each application use case.

		Main Sources for Each Application Type		
		Passenger Car	**Train**	**Forklift**
I	Energy simulation	[40]	[24,41,42]	[27,43]
II	Battery sizing algorithm	[39,40,44]	[39,44]	[34,39,44]
III	Cost model	[15,45,46]	[24]	[31,34]

3.2. Particularities for Passenger Cars

The passenger car model is built on the WLTP driving cycle as a global standard for measuring energy consumption, pollutant levels, CO_2 emissions, or the all-electric nature of fully electric cars. This driving cycle specifies a target speed over time, road gradients, and duration of stops along the route, covering urban, rural, and highway operations.

The vehicle simulation uses a quasi-static longitudinal dynamics model to determine the energy consumption and average speed for a range of vehicle masses based on the vehicle parameters and the WLTP driving cycle. The VW ID.3 is the reference vehicle, leading to 408 km as the target range. Given this target range and the mass-dependent energy consumption, the battery sizing algorithm determines the required battery size based on the cell-specific properties defined in the cell database. The cost model uses the VW Golf VIII to determine cost-parity prices.

3.3. Particularities for Passenger Trains

The passenger train model mimics a route-specific application and train configuration, since no standard case exists. Thus, we reference the commuting service from Nuremberg to Hof and vice versa, which is not yet electrified and also represents a typical route that would be suitable for battery operation (in terms of length, topography, and traffic frequency). This covers the full operation schedule such as four stops, idle times (1 min each), and station waiting times for turning around (20 min); route characteristics, such as section distances, speeds (max. 160 km/h), gradients, and maximum permissible weights (22.5 tonnes); and other characteristics, such as the number of daily runs (3), operating days per year (320), and the existence of overhead lines. The total distance is 167 km, which equals 95–105 min, while the middle section of approximately 90 km is not electrified.

Similar to cars, the vehicle simulation determines the energy consumption for a range of vehicle masses for this route using a quasi-static longitudinal dynamics model. A three-unit train with 115 tonnes of curb weight, an overall length of 70 m, a capacity for 410 people (220 seated and 190 standing), 1000 kW constant power, and 2600 kW peak serves as a reference. The combined installation volume (i.e., subfloor and rooftop) equals 7500 L as a reasonable installation space volume. Given the non-electrified middle section, the initial and final sections for charging via the overhead line, and the mass-dependent energy consumption, the battery sizing algorithm determines the required battery size based on the cell-specific properties defined in the cell database. Additionally, battery sizing covers the restriction that potential battery replacements must occur within the revision cycles of the train (every 8 years) to avoid unplanned downtime. Finally, the cost model uses an equivalent diesel train to determine cost-parity prices.

3.4. Particularities for Forklifts

The maximum load capacity of forklift trucks plays a decisive role in their utilization and energy consumption. In a preliminary step, we evaluated 30 publicly available VDI Guideline 2198 [43] type sheets from LAB counterbalanced forklift trucks and found an energy consumption of 4 to 10 kWh per hour for class 1 trucks [33] with 1 to 3 tonnes in maximum lift load. This value is not specifically tailored to real use-cases, as the guideline is primarily aimed at comparing the energy consumption of different vehicles, but it shows a linear increase in energy consumption when the lift load increases. We have therefore recreated a use-case from a study with original vehicle data [27].

Firstly, we defined a concrete use-case for class 1 forklifts trucks. As the driving cycles vary strongly from use-case to use-case and battery sizes are not tailored to specific driving cycles, a hypothetic cycle was defined. It involves realistic stand-by times, characteristics, energy consumption within a single shift, and a 15 shift week, 3 shifts per working day, respectively. Instead of mimicking an actual driving cycle, it was defined to cover various aspects of characteristic forklift driving cycles simultaneously. The basis of battery dimensioning is the characteristic energy consumption in terms of the timeframe of operation and possible downtime periods needed to charge the forklift battery. Considering the cell properties, the required battery size is determined, as well as the resulting battery volume and lifetime with respect to the forklift lifetime and battery installation space. Forklifts are one of the few rare cases where the gravimetric energy density of the battery is not only irrelevant, but rather is inverse to other mobile applications. As the battery is used as a counterweight, a large battery mass is desired. The cost of additional counterweight for low-weight battery systems were also implemented in the model. Compared to other mobile applications, the cost-parity comparison of an LIB electric forklift truck is compared to that of a status quo lead-acid (LAB) electric forklift truck with a maximum load capacity of 2 tonnes. As LABs require regular battery swapping in these use-cases, whereas LIBs can be easily charged in stand-by times, the model was extended by additionally including a measure for the economic benefit of the working time saved, in accordance with the finding of [31].

3.5. Ecological Considerations

The life cycle assessment (LCA) of batteries and mobile applications is an essential tool for the analysis and comparison of technologies in terms of greenhouse gas (GHG) emissions and the individual carbon footprint. One of the most discussed use cases for LCA approaches in recent times is that of the lithium-ion battery [47] in electric vehicles [48–51], for instance, with cradle-to-gate results ranging from 12 to 313 kg CO_{2eq} per kWh of battery capacity [51]. One limitation of the LCA approach is that, for the purpose of comparing battery-electric applications with other technologies, it is necessary to include differing components and to consider all parts of the application in a whole life cycle assessment LCA [49].

In favor of our systemic view on different applications, real-world cell datasheets, and cost-parity comparisons, we state the carbon footprint of the batteries for the considered applications. This encompasses, i.e., how often a cell has to be replaced or if it has reached its calendric or cycling end of life. We compare this impact to the cost-parity price or technical characteristics. Our environmental assessment includes only battery-related GHG emissions depending on (1) cell materials, (2) production, and (3) formation (cf. Figure 2). We disregard other GHG emissions, comparisons to the next best alternative, or potential credits from recycling and reuse. The carbon footprint (in kg CO_{2eq} per kWh) for each individual cell is calculated as a function of cell chemistry, cell format, specific energy (in Wh/kg) or cell weight (in kg), and capacity (in Ah).

Finally, the overall CO_2 footprint at the application level (in kg CO_{2eq}) is calculated as a function of battery size (in kWh) and possible battery replacements in accordance with the lifetime determined in each cell data sheet.

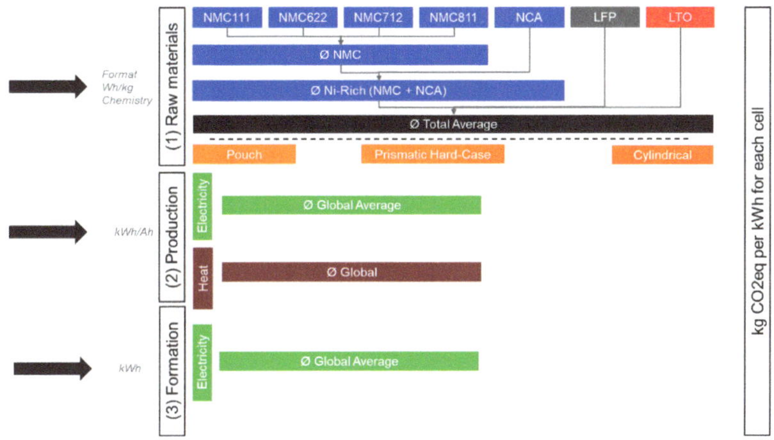

Figure 2. Overview of the CO_2 footprint methodology and characteristics.

4. Results

4.1. Energy Consumption and Battery Dimensioning

The energy consumption of a passenger car (apart from its use) depends particularly on its vehicle weight. A vehicle simulation was carried out for different vehicle weights to determine the influence of weight on consumption. The energy consumption ranges from 0.133 kWh/km with a vehicle weight of 1435 kg to 0.197 kWh/km with a weight of 3000 kg. As the vehicle weight increases, the energy consumption increases due to the increased rolling and acceleration resistance. A consumption of 0.149 kWh/km for the BEV was calculated with a battery size of 64.3 kWh, which fits well with measured real-world data for this kind of vehicle [40]. The consumption of the reference vehicle, Golf VIII, is approximately 5.6 L per 100 km [52].

The energy consumption of a passenger train at 25 °C is approximately 4.6–5.3 kWh/km, depending on the train mass and direction of travel. Recent studies and industry values indicate a range from 3 to 4 kWh/km in standard operations to 5–6.5 kWh/km in demanding operations, including all auxiliary consumers and heating in winter [24,41,42], indicating the good representativeness of our simulation model. In contrast, the simulated energy consumption of the diesel train is between 9.5 and 10.1 kWh/km, which equals approximately one liter per kilometer. The calculated gross battery capacity is typically approximately 800 kWh and 4500–11,700 kg. The median was 8200 kg, while the lower quartile was 6000 kg and the upper quartile was 9800 kg.

The energy consumption of a forklift truck depends on the maximum load capacity of a counterbalance forklift truck. As the maximum load capacity increases, the energy consumption increases partly due to the higher lifting load and the total weight of the forklift truck. A characteristic energy consumption of 3 kWh/h was derived from [27,53] for the battery dimensioning. The battery is dimensioned so that the LIB forklift truck can fulfill the same operating conditions as an LAB forklift truck. Considering a shift with 8 h of working time, the operative usage time of the forklift truck in a warehouse was estimated to be a maximum of 5 h [27]. Our model assumed that half of the remaining 3 h are available for charging, as usually charging stations have to be shared among the different forklifts of the fleet. While specific time losses (e.g., driving to a charging station, connecting the charging wire) were subtracted, only a time of 90 min per shift was considered to be used for charging. The required battery capacity is thus dependent on the charging rate of the cell and the available charging power. As the volumetric energy density of LIBs is much greater than that of LABs, the battery volume seems less important for battery dimensioning in electric forklift trucks. Nevertheless, a space requirement of 341 L was assumed, as derived from a commonly used 48 V 6 PzS 540 Ah lead-acid battery, i.e., [54], for 2-tonne forklift trucks. This serves as an installation space limit for the volumetric dimensioning of the LIB.

4.2. Cost Parity Analysis

For the cost comparison, a BEV similar to the VW ID.3 is compared with a corresponding combustion engine vehicle, the VW Golf VIII. The cost model considers all cost components of BEVs that differ from those of the VW Golf VIII: powertrain costs, taxes, maintenance, energy consumption, and battery costs. The first three cost components are independent of the cell selection. The specific energy of the cell influences the energy consumption. The costs for the battery depend on the required battery size, cycle stability, calendar life, and cell price.

The TCO breakdown in Figure 3 shows that the costs for the conventional vehicle amount to just over EUR 25,000. Assuming an average annual mileage of ~13,600 km in Germany, a large part of this cost is related to energy costs of approximately EUR 9700 as well as maintenance costs (EUR ~ 7850) and powertrain costs (EUR ~ 6900). Considering the tax advantage for BEVs, there is a remaining budget of almost EUR 15,000 for the battery layout to reach cost parity with the ICV. Above all, the maintenance and energy costs are significantly lower for BEVs.

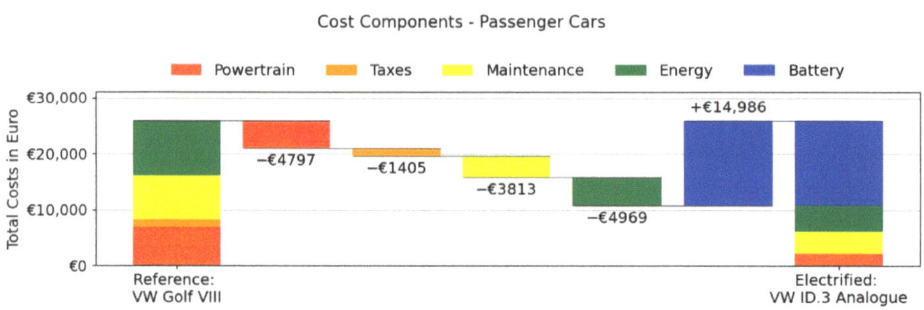

Figure 3. Cost comparison results for passenger cars over ten years.

For trains, the cost model compares the battery-electric version to its diesel equivalent over 30 years of service life (cf. Figure 4). Cost factors include powertrain (incl. chassis), maintenance, energy costs, and battery costs. Lower powertrain costs result from the cost advantages of electric versus diesel powertrains and are independent of cell selection. In contrast, energy consumption is influenced by the battery weight and thus depends on the cell selection. Battery costs depend on the calculated battery size, cycle stability, calendar life, and cell price. Finally, the cell-specific cost-parity price is calculated so that the total costs for the diesel equivalent version are matched by considering the total number of required cells.

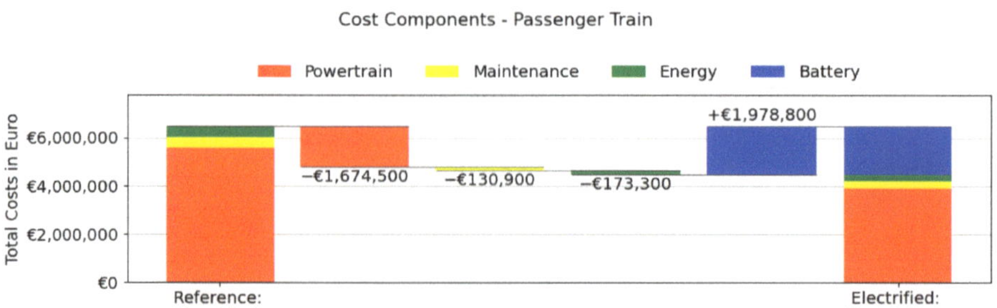

Figure 4. Cost comparison results for passenger trains over 30 years of service life.

For forklifts, the cost parity analysis takes into account the cost components of a battery-electric counterbalance forklift that change when comparing a lithium battery (LIB) to a lead-acid battery (LAB) forklift truck: battery maintenance, energy efficiency, labor time losses, (additional) counterweight, and the cost of the battery itself, as shown in Figure 5. The LAB's cost includes three battery replacements of the LAB and the initial purchase price. This corresponds to the assumed lifetime of the LAB of 6000 operating hours compared to the expected lifetime of 20,000 operating hours for an electric forklift [27]. The LIB's cost is determined by the required battery size (influenced by the charging power), cycle stability, calendar life, and cell price.

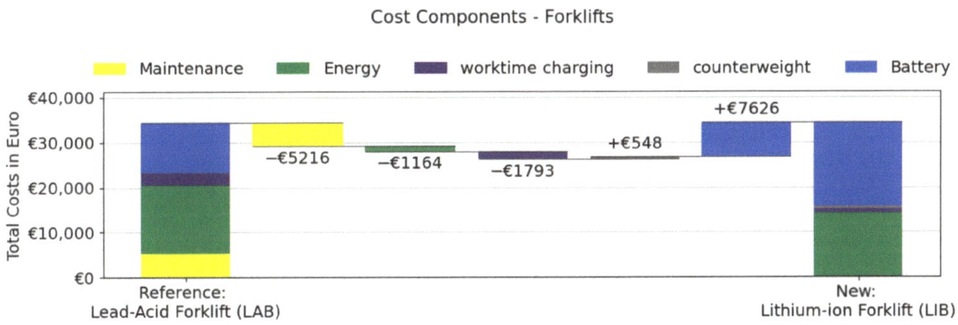

Figure 5. Cost comparison results for forklifts operating for more than 20,000 h.

Since no specific maintenance is required for LIBs compared to the maintenance of lead-acid batteries after every 1000 operating hours [27], cost benefits can be expected. Additional maintenance costs for the forklift itself (e.g., tire repair) were neglected, as this similarly affects both LIB and LAB forklifts [34]. The energy consumption during operation is considered equivalent for both technologies, but is influenced by the round-trip efficiency of each battery technology. The LIB technology eliminates the monetized loss of working time due to the necessary swapping of lead-acid batteries in three-shift operations. LIB forklifts need a counterweight for safe load handling because of the higher gravimetric energy density of LIBs, which is usually not needed for LABs. The costs of the counterweight are simplified, as the additional pure steel weight is multiplied by the weight difference that results from comparing the LAB and LIB.

4.3. Cost Parity and Technical Considerations

The following section shows exemplary cost-parity assessment results for passenger cars, trains, and forklifts. Although the weight of a battery system usually determines the additional energy demand, the available space for the battery is typically more limiting. Therefore, the plots show the cost-parity prices in EUR/kWh versus the required battery system volume in Figures 6–8. A dashed line indicates the available space from the reference case to facilitate comparison. Shapes and colors mark different cell formats and chemistries. For interpretation, the typical target direction to optimize is toward the top left quadrant, which signifies a high cost-parity price (indicating that the cell may be expensive due to its superior performance) while simultaneously requiring as little installation space as possible.

Figure 6 shows the results for the passenger car, considering all of the cells listed in the cell database. The battery volume of the VW ID.3 is plotted as a reference (dashed orange line). Different cell chemistries and cell formats are marked in color and with different markers. The differentiation of the cell chemistries could only be undertaken based on the nominal voltage of the cells, as further details are not included in the datasheet. In addition, the VW ID.3 cell is highlighted to indicate the status quo of the cells used in passenger cars.

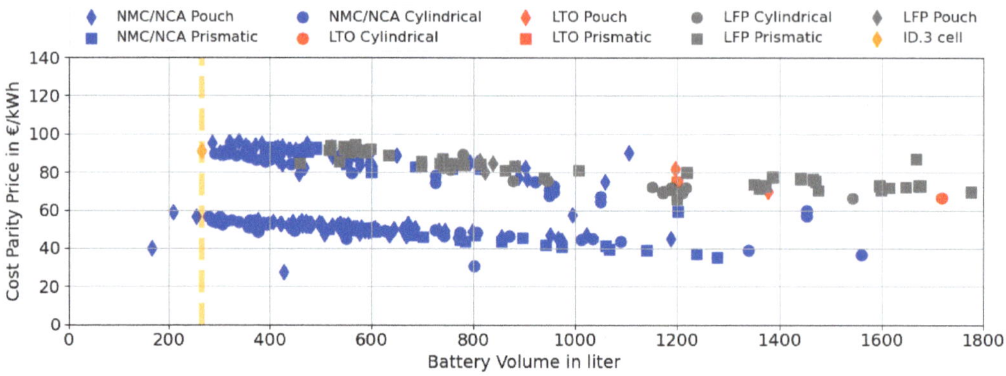

Figure 6. Results of the cost parity-based cell assessment for passenger cars. X-axis: required battery volume in liters. Y-axis: cost-parity price in EUR/kWh. Cell formats are represented by different shapes: pouch (diamond), cylindrical (circle), and prismatic hard-case (square). Cell chemistries are color-coded: Ni-rich cells (i.e., NMC and NCA) are in blue, LTO cells are in red, and LFP cells are in gray.

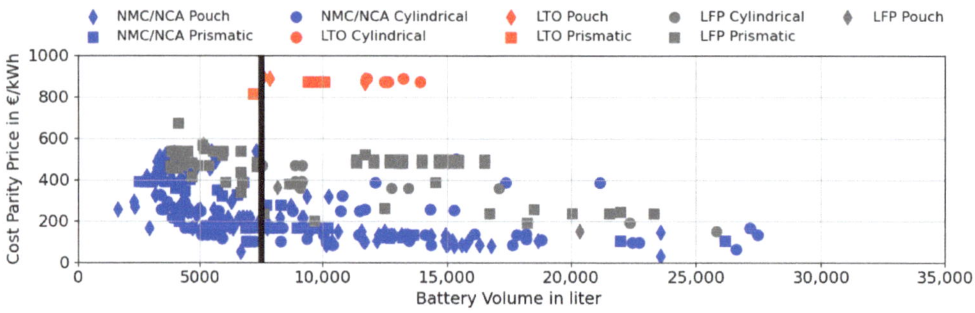

Figure 7. Results of the cost-parity-based cell assessment for passenger trains. X-axis: required battery volume in liters. Y-axis: cost-parity price in EUR/kWh. Cell formats are represented by different shapes: pouch (diamond), cylindrical (circle), and prismatic hard-case (square). Cell chemistries are color-coded: Ni-rich cells (i.e., NMC and NCA) are in blue, LTO cells are in red, and LFP cells are in gray.

The results show that LFP and LTO cells have the highest cost-parity due to their long battery life. However, these cells also require significantly more installation space. Very few cells require less installation space than the cell installed in the VW ID.3. Cost parity is reached for the VW ID.3 cell at approximately 60 EUR/kWh. This result is consistent with the higher prices for battery electric vehicles than for vehicles with combustion engines.

Figure 7 shows the final cost-parity-based cell assessment results for passenger trains. The black line indicates a reference volume of 7500 L. LTO cells have the highest cost parity due to their long lifetime and high C-rate, resulting in up to 900 EUR/kWh. LFP cells reach up to 680 EUR/kWh. In contrast, Ni-rich cells dominate the results, but only reach around 150–500 EUR/kWh. The available installation space is sufficient for many cells, indicating high practical feasibility. Assuming a cost parity price of 500 EUR/kWh for a battery size of approximately 800 kWh, the calculated acquisition costs would be approximately EUR 5.8 million, which is close to that reported in the other literature [24,41,42], which is approximately EUR 6.0–6.5 million.

Figure 8 shows the results of the cost-parity-based cell assessment for forklifts. The dashed purple line indicates the available installation volume of 341 L. The relative differences in the required installation space concerning the volumetric battery size are significantly greater. Only cells with medium to high energy densities can be accommodated in the installation space available for the battery. The suitable cells achieve cost-parity prices of less than 100 EUR/kWh for NMC cells and even 700 EUR/kWh for an LTO pouch cell. LFP cells are in the range of 200–600 EUR/kWh. This means that a broad range of LIB cells can achieve cost parity for forklift trucks in warehouse operation if these cells can be sourced at that price.

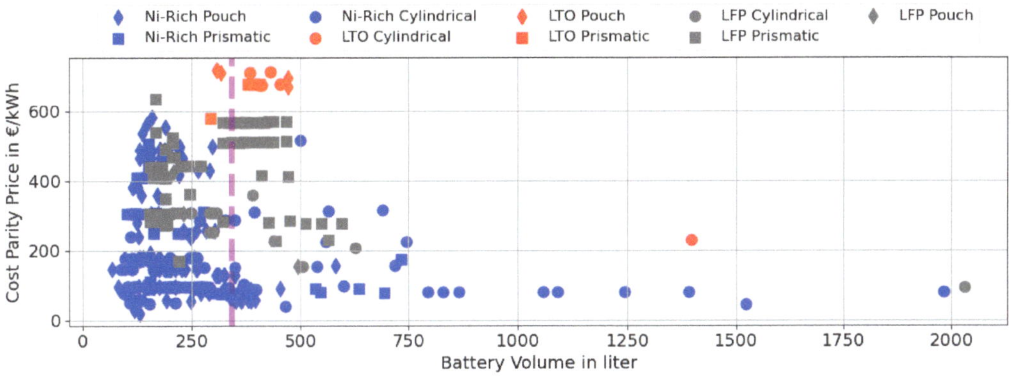

Figure 8. Results of the cost-parity-based cell assessment for forklifts. X-axis: required battery volume in liters. Y-axis: cost-parity price in EUR/kWh. Cell formats are represented by different shapes: pouch (diamond), cylindrical (circle), and prismatic hard-case (square). Cell chemistries are color-coded: Ni-rich cells (i.e., NMC and NCA) are in blue, LTO cells are in red, and LFP cells are in gray.

We found that LFP cells can be more expensive than NMC cells for achieving cost-parity. Some LFP and most LTO cells with lower energy densities are partly unsuitable for modeling, as a system-side fit is not always given concerning the available installation space. Thus, the battery cannot be sufficiently dimensioned with these cells to meet the needed capacity of 20 kWh. Although the methodology was only applied to a specific forklift application, it confirms the market's tendency to use LFPs instead of NMCs in industrial applications [55]. Since LFP cells and packs are already available on the market at a price of less than 150 EUR/kWh [21], using these cells for forklift applications may be more economically advantageous than using LABs. This result is also in accordance with recent studies on the economic competitiveness of LIB and LAB forklift trucks, which demonstrates that LIB generally offers economic benefits during the utilization phase [31,34].

4.4. Cost Parity and Ecological Considerations

Figure 9 shows the application-specific differences for cost-parity prices versus ecological footprint (CO_{2eq}). The actual battery size and the number of required batteries (i.e., battery replacements) strongly affect the results. The latter is affected by both calendar aging and the limited cycle life. We accumulate the total carbon footprint for each application to address the application-specific performance indicators, for instance, the size of the battery or accounting the entire application lifetime. For BEVs and forklifts, the impact is quite similar, from up to a few tonnes to approx. 50 tonnes CO_{2eq} absolute, whereas the operating lifetime and the required size of the battery in trains greatly multiplies its impact.

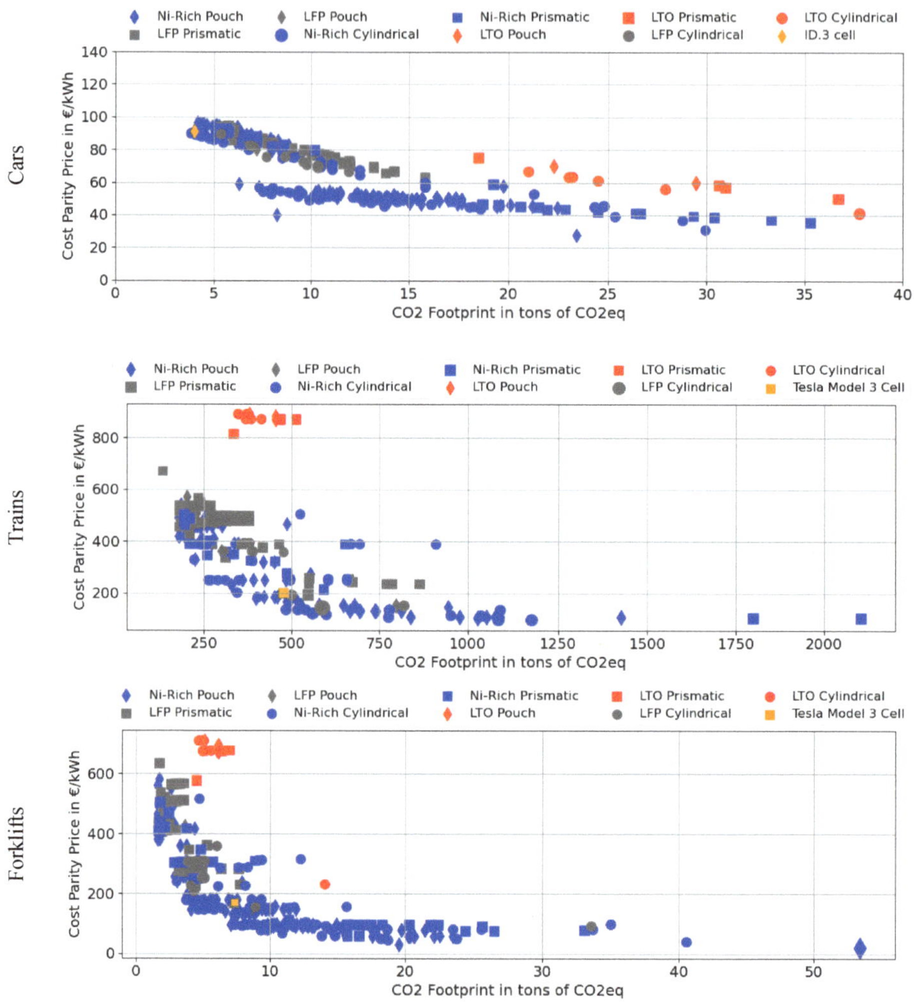

Figure 9. Comparison of cost-parity versus CO_2 impact on battery production. Upper: cars; middle: trains; lower: forklifts. X-Axis: CO_2 footprint in tonnes of CO_{2eq}. Y-Axis: cost-parity price in EUR/kWh. Cell formats are represented by different shapes: pouch (diamond), cylindrical (circle), and prismatic hard-case (square). Cell chemistries are color-coded: Ni-rich cells (i.e., NMC and NCA) are in blue, LTO cells are in red, and LFP cells are in gray.

We highlight the high spread in CO_{2eq} for Ni-rich cells (NMC and NCA) due to the large heterogeneity in their technical performance. However, Ni-rich cells can reach good trade-offs between high cost-parity and low CO2 footprint. Ni-rich cells and LFP represent the best economic-ecologic trade-off for cars, reaching approximately 80–100 EUR/kWh and 4.3–5.6 tCO_{2eq}, respectively. In contrast, LTO cells swing at approximately 40–80 EUR/kWh and over 18 tCO_{2eq}. LFP and LTO cells become favorable over longer time windows for trains and forklifts since fewer replacements are needed.

4.5. Cross-Application Comparison for Selected Cells

Figure 10 compares cell-specific cost-parity prices (CPP), highlighting that the same cell might require cost-parity prices ranging from a few to several hundred Euros for different applications. It should be noted that only those cells are plotted that do not exceed the volumetric limitation of the application, especially for trains and forklifts (for further details, please refer to Section 4.3). It becomes obvious that the CPP of trains and forklifts are consistently observed to be in close proximity to one another. The CPP of the LTO cells for trains is slightly higher than that for forklifts. In the case of NMC and LFP cells, there is minimal differentiation in the application comparison. However, when comparing cars and trains, the pronounced economic constraints on cell prices for battery-electric passenger cars is noteworthy. This aligns with the postulated significance of battery costs for EV market diffusion [19,20] and simultaneously demonstrates that trains and forklifts can achieve cost parity with current reference applications, even with elevated LIB cell costs [34].

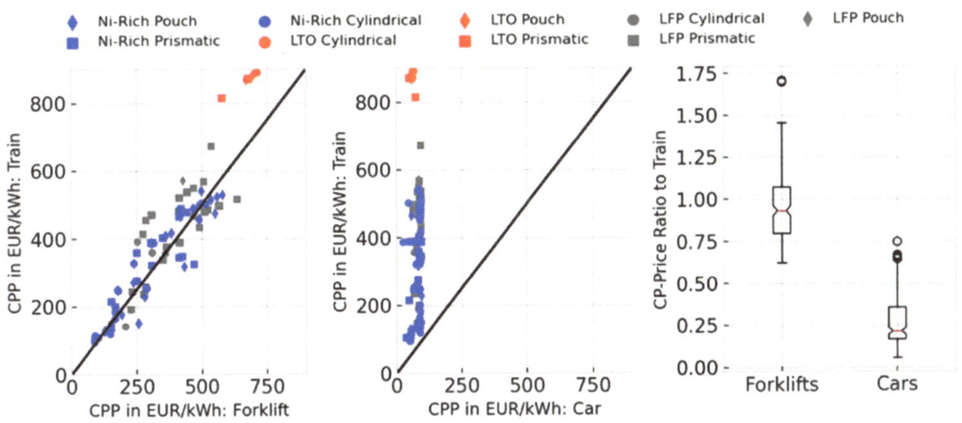

Figure 10. Left and middle: comparison of cell-specific cost-parity (CP) prices for forklifts and cars (x-axis) versus trains (y-axis). Cell formats are represented by different shapes: pouch (diamond), cylindrical (circle), and prismatic hard-case (square). Cell chemistries are color-coded: Ni-rich cells (i.e., NMC and NCA) are in blue, LTO cells are in red, and LFP cells are in gray. Right: cell-specific cost-parity price ratios for forklifts/cars versus trains.

A comparative analysis between forklifts and trains reveals similar required cost-parity prices (median: 93%). In contrast, when comparing cars to trains, cell-specific cost-parity prices must be substantially lower than for trains (median: 22%) to render them attractive for the automotive sector. However, we observe no particular differentiation according to cell chemistry or format.

5. Discussion

We extend the TCO to include the technical aspects of the application with the technical aspects of each individual LIB cell. Cell chemistry and formats are crucial due to their impact on performance, costs, and application suitability. As a result of the comparison, this not only provides a comparative value for an economic assessment, but also enables the economic selection of a technology—in this case, the battery cell—based on its technical suitability for an individual application and usage scenario in a techno-economic assessment approach. This supports cell integrators in their technology management, as well as providing insights for the further optimization of cell technologies for specific applications.

While our systematic approach involves the same three steps for each application, specific adaptations are necessary to tailor the procedure to certain particularities; see Section 3. First, load profiles for cars and forklifts are derived from existing standardized driving cycles, whereas a custom load profile imitating a specific route is devised for passenger trains. Second, train and car applications are compared against conventional vehicles with internal combustion engines as a reference, while a reference vehicle equipped with lead-acid batteries is employed for forklifts. Third, vehicle weight is the most influential factor on energy consumption for trains and cars, whereas this is the maximum lifting capacity for forklifts. Fourth, cost considerations vary, since CAPEX and OPEX items depend on the application.

Although we determine the cost-parity price specifically for each application, we note that there may be countless load profiles and utilization patterns behind each application instead of just one. We also refer to real-world uncertainties and usage patterns, leading to different energy requirements, as witnessed with the standard WLTP driving cycle [56]. Thus, future studies may include more differentiation within an application. However, our approach is effective for application-specific cell selection, aligning with common cost thresholds and industry trends, such as 100 EUR/kWh as the common threshold for BEV battery cells [1] or the extinction of LAB-powered forklifts. Our results indicate that the discussed BEV to ICEV parity from 2026 [15] or 2030 [19,20] onwards could already be met, especially with today's cell prices of below 100 EUR/kWh [21], which is significantly lower than the calculated prices for some suitable cells in the database.

Other limitations involve our battery cell database, assumptions for cost parity, and the variability of final retail prices. Firstly, our database relies on publicly available battery cell datasheets, encompassing only a fraction of all available cells. Some cells may already be outdated, while the latest cell generation is likely underrepresented due to unpublished data. Additionally, we highlight potential uncertainties when utilizing datasheet information. The values presented are mainly obtained from standardized test environments and conditions, which may not precisely depict real-world cell performance due to variable ambient conditions and specific charge–discharge load profiles inherent to applications and embedded use-cases. Secondly, our approach required us to scale cell-level costs to the system level and vice versa. However, no information was available on battery chemistry or format dependency, and we used the same scaling for all applications. The advanced system integration and engineering per application, potentially also cell-format- and chemistry-specific, is, however, a decisive aspect to enhance battery performance and lower costs [17,18]. Third, we emphasize that the calculated cost-parity price (i.e., the maximum allowed cell price) may substantially differ from cell retail prices that are affected by purchase quantities or supplier contracts.

6. Conclusions and Outlook

In this paper, a methodical approach for a cell assessment based on cost parity was presented and demonstrated using three different mobile applications: passenger cars, passenger trains, and forklifts with highly specific characteristics. The developed methodology allows for a high-level yet tailored matching of publicly available technical cell data, application-specific requirements, and use-case conditions to determine the cost parity price for each specific cell for a certain application. We draw three main conclusions from our analysis.

First, only a certain number of the considered battery cells are suitable for all applications. On the one hand, this is mainly related to low energy densities, meaning that the available installation space could be exceeded. On the other hand, low specific energies may lead to weight-based limitations. However, suitable cells have been identified for all of the considered applications.

Second, the calculated cost-parity prices differ greatly for different applications. We emphasize that costs are the primary criterion in selecting battery cells, but technical aspects are gaining importance. While prices well below 100 EUR/kWh are required for passenger cars, prices for trains and forklifts can be substantially higher, reaching up to 950 EUR/kWh or 750 EUR/kWh, respectively. There are fewer format-specific dependencies, but there are major differences between the chemistries. Herein, we showcase LTO chemistries with high lifetimes (cyclic and calendar) but low energy densities versus NMC chemistries with higher energy densities but usually lower lifetimes.

Third, ecological considerations during the battery cell selection depend on the application-specific lifetime (i.e., the number of required batteries) and battery size. The former is affected by both calendar aging and the limited cycle life, whereby different formats and chemistries may reach similar levels in long-term applications (i.e., forklifts and trains). In comparison, LFP and NMC batteries dominate LTO for cars by achieving substantially lower CO_2 footprints. The CO_2 footprint is employed solely for the purpose of relative comparison between the cells. With a view to global CO_2 labeling, a CO_2 price could be incorporated into the cost parity price at a subsequent stage. This becomes particularly pertinent when alternative battery technologies with high sustainability promises are included in the analysis and assessment alongside different LIB cells.

We highlight that the proposed cell selection methodology is a valuable decision-support tool for manufacturers/cell integrators and cell suppliers to solve the trade-off between technical restrictions and economic considerations for specific applications. For cell suppliers, this approach facilitates comparisons of their cells with others, enabling them to identify potential new applications or assess the impact of performance enhancements. For manufacturers and cell integrators, this approach facilitates the comparison of available cells, enabling them to identify the most suitable cells for their applications. Finally, we emphasize that our analysis is based on a single underlying usage pattern per application so that future studies may include more distinctions within an application, which is likely to cause even greater variation.

Author Contributions: Conceptualization, T.H. and S.L.; Methodology, S.L. and T.H.; Software, S.L., L.W. and M.S.; Formal analysis, T.H., S.L., L.W. and M.S.; Data curation, T.H., S.L., L.W. and M.S.; Writing—original draft preparation, T.H. and S.L.; Writing—review and editing, L.W. and M.S.; Project administration, T.H.; Funding acquisition, T.H. All authors have read and agreed to the published version of the manuscript.

Funding: This research was funded by the Federal Ministry of Education and Research Germany within the project "BetterBat" under grant number 03XP0362.

Data Availability Statement: The raw data supporting the conclusions of this article will be made available by the authors on request.

Acknowledgments: Special thanks to Olaf Teichert for modeling the passenger cars.

Conflicts of Interest: The authors declare no conflicts of interest.

References

1. Hettesheimer, T.; Neef, C.; Rosellón Inclán, I.; Link, S.; Schmaltz, T.; Schuckert, F.; Stephan, A.; Stephan, M.; Thielmann, A.; Weymann, L.; et al. Lithium-Ion Battery Roadmap-Industrialization Perspectives toward 2030. Fraunhofer ISI, Karlsruhe, 2023. Available online: https://www.isi.fraunhofer.de/en/competence-center/neue-technologien/geschaeftsfelder/industrielle-technologien/batterieforschung.html#Publications (accessed on 20 June 2024).
2. Degen, F.; Winter, M.; Bendig, D.; Tübke, J. Energy consumption of current and future production of lithium-ion and post lithium-ion battery cells. *Nat. Energy* **2023**, *8*, 1284–1295. [CrossRef]
3. Nykvist, B.; Olsson, O. The feasibility of heavy battery electric trucks. *Joule* **2021**, *5*, 901–913. [CrossRef]
4. Schmaltz, T.; Wicke, T.; Weymann, L.; Voß, P.; Neef, C.; Thielmann, A. Solid-State Battery Roadmap 2035+. Fraunhofer ISI, Karlsruhe, 2022. Available online: https://www.isi.fraunhofer.de/en/competence-center/neue-technologien/geschaeftsfelder/industrielle-technologien/batterieforschung.html#Publications (accessed on 20 June 2024).
5. Stephan, A.; Hettesheimer, T.; Neef, C.; Schmaltz, T.; Stephan, M.; Link, S.; Heizmann, J.L.; Thielmann, A. Alternative Battery Technologies Roadmap 2030+. Fraunhofer ISI, Karlsruhe, 2023. Available online: https://www.isi.fraunhofer.de/en/competence-center/neue-technologien/geschaeftsfelder/industrielle-technologien/batterieforschung.html#Publications (accessed on 12 April 2024).
6. Hettesheimer, T.; Link, S.; Weymann, L.; Stephan, M.; Teichert, O. Find the right cell for every applications: A methodical approach based on cost-parity calculation. In Proceedings of the EVS36 Symposium, Sacramento, CA, USA, 11–14 June 2023. Available online: https://evs36.com/wp-content/uploads/finalpapers/FinalPaper_Hettesheimer_Tim%20(2).pdf (accessed on 15 March 2024).
7. Teichert, O.; Link, S.; Schneider, J.; Wolff, S.; Lienkamp, M. Techno-economic cell selection for battery-electric long-haul trucks. *eTransportation* **2023**, *16*, 100225. [CrossRef]

8. Ragone, D.V. Review of Battery Systems for Electrically Powered Vehicles. In *SAE Technical Paper Series, Proceedings of the Mid-Year Meeting, Washington, DC, USA, 20 May 1968*; SAE International400 Commonwealth Drive: Warrendale, PA, USA, 1968.
9. Dechent, P.; Epp, A.; Jöst, D.; Preger, Y.; Attia, P.M.; Li, W.; Sauer, D.U. ENPOLITE: Comparing Lithium-Ion Cells across Energy, Power, Lifetime, and Temperature. *ACS Energy Lett.* **2021**, *6*, 2351–2355. [CrossRef]
10. Burke, A.; Zhao, J.; Fulton, L. Projections of the costs of light-duty battery-electric and fuel cell vehicles (2020–2040) and related economic issues. *Res. Transp. Econ.* **2024**, *105*, 101440. [CrossRef]
11. Figenbaum, E. Retrospective Total cost of ownership analysis of battery electric vehicles in Norway. *Transp. Res. Part D Transp. Environ.* **2022**, *105*, 103246. [CrossRef]
12. Hagman, J.; Ritzén, S.; Stier, J.J.; Susilo, Y. Total cost of ownership and its potential implications for battery electric vehicle diffusion. *Res. Transp. Bus. Manag.* **2016**, *18*, 11–17. [CrossRef]
13. García, A.; Monsalve, J.; Martinez-Boggio, S.; Tripathi, S. Techno-economic assessment of vehicle electrification in the six largest global automotive markets. *Energy Convers. Manag.* **2022**, *270*, 116273. [CrossRef]
14. Wang, Z.; Acha, S.; Bird, M.; Sunny, N.; Stettler, M.E.; Wu, B.; Shah, N. A total cost of ownership analysis of zero emission powertrain solutions for the heavy goods vehicle sector. *J. Clean. Prod.* **2024**, *434*, 139910. [CrossRef]
15. König, A.; Nicoletti, L.; Schröder, D.; Wolff, S.; Waclaw, A.; Lienkamp, M. An Overview of Parameter and Cost for Battery Electric Vehicles. *World Electr. Veh. J.* **2021**, *12*, 21. [CrossRef]
16. Jaramillo, P.; Kahn Ribeiro, S.; Newman, P.; Dhar, S.; Diemuodeke, O.E.; Kajino, T.; Lee, D.S.; Nugroho, S.; Ou, X.; Hammer, A.; et al. Transport. In *Climate Change 2022-Mitigation of Climate Change: Contribution of Working Group III to the Sixth Assessment Report of the Intergovernmental Panel on Climate Change*; IPCC Sixth Assessment Report; Mitigation of Climate Change, Ed.; Cambridge University Press: Cambridge, UK, 2023; pp. 1049–1160, ISBN 9781009157926.
17. Deng, J.; Bae, C.; Denlinger, A.; Miller, T. Electric Vehicles Batteries: Requirements and Challenges. *Joule* **2020**, *4*, 511–515. [CrossRef]
18. Fichtner, M. Recent Research and Progress in Batteries for Electric Vehicles. *Batter. Supercaps* **2022**, *5*, e202100224. [CrossRef]
19. EUCAR. Battery Requirements for Future Automotive Applications: EG BEV&FCEV. Available online: https://eucar.be/wp-content/uploads/2019/08/20190710-EG-BEV-FCEV-Battery-requirements-FINAL.pdf (accessed on 14 June 2024).
20. BATT4EU. Strategic Research & Innovation Agenda. February 2024. Available online: https://bepassociation.eu/our-work/sria/ (accessed on 14 June 2024).
21. BloombergNEF. Lithium-Ion Battery Pack Prices Hit Record Low of $139/kWh. Available online: https://about.bnef.com/blog/lithium-ion-battery-pack-prices-hit-record-low-of-139-kwh/ (accessed on 13 June 2024).
22. Pagenkopf, J.; Kaimer, S. Potentials of alternative propulsion systems for railway vehicles—A techno-economic evaluation. In Proceedings of the 2014 Ninth International Conference on Ecological Vehicles and Renewable Energies (EVER), Monte-Carlo, Monaco, 25–27 March 2014; IEEE: Piscataway, NJ, USA, 2014; pp. 1–8, ISBN 978-1-4799-3787-5.
23. McGordon, A.; Winnett, J.; Moeini, R.; Everson, J.; Meredith, J.; Dinh, T.Q.; Hughes, D.J. Battery Sizing for Hybrid and Electric Rail Vehicles. In Proceedings of the Fifth International Conference on Railway Technology: Research, Development and Maintenance, Montpellier, France, 22–25 August 2022; Pombo, J., Ed.; Civil-Comp Press: Edinburgh, UK, 2023; pp. 1–7.
24. Streuling, C.; Pagenkopf, J.; Schenker, M.; Lakeit, K. Techno-Economic Assessment of Battery Electric Trains and Recharging Infrastructure Alternatives Integrating Adjacent Renewable Energy Sources. *Sustainability* **2021**, *13*, 8234. [CrossRef]
25. Copetti, D.F.; Canha, L.N.; Botezeli, L.K.; Tolotti, D.; Schmidt, E.; Das Chagas Silva, J.V.; Jonas NeuhausSchneider, A.; Da Silva, G.K.; Eggers, D. Electric Forklifts: Technological Advancements and Impacts on Energy Transition and Sustainability. In Proceedings of the 2023 15th Seminar on Power Electronics and Control (SEPOC), Santa Maria, Brazil, 22–25 October 2023; IEEE: Piscataway, NJ, USA, 2023; pp. 1–6, ISBN 979-8-3503-1289-8.
26. Metzger, N.; Li, X. Technical and Economic Analysis of Fuel Cells for Forklift Applications. *ACS Omega* **2022**, *7*, 18267–18275. [CrossRef] [PubMed]
27. Günthner, W.A.; Micheli, R.; Micheli, R. *H2IntraDrive-Einsatz einer Wasserstoffbetriebenen Flurförderzeugflotte unter Produktionsbedingungen: Forschungsbericht zu dem Forschungsvorhaben der Forschungsstelle Lehrstuhl für Fördertechnik Materialfluss Logistik, Technische Universität München: Konsortialpartnern im Forschungsvorhaben sind BMW Group, Linde Material Handling: Im Rahmen des Nationalen Innovationsprogramms Wasserstoff-und Brennstoffzellentechnologie (NIP) = H2IntraDrive-Use of Hydrogen Powered Industrial Trucks under Production Conditions*; fml-Lehrstuhl für Fördertechnik Materialfluss Logistik Technische Universität München: Garching, Germany, 2015; ISBN 978-3-941702-58-5.
28. Renquist, J.V.; Dickman, B.; Bradley, T.H. Economic comparison of fuel cell powered forklifts to battery powered forklifts. *Int. J. Hydrogen Energy* **2012**, *37*, 12054–12059. [CrossRef]
29. Fuc, P.; Kurczewski, P.; Lewandowska, A.; Nowak, E.; Selech, J.; Ziolkowski, A. An environmental life cycle assessment of forklift operation: A well-to-wheel analysis. *Int. J. Life Cycle Assess.* **2016**, *21*, 1438–1451. [CrossRef]
30. Ramsden, T. An Evaluation of the Total Cost of Ownership of Fuel Cell Powered Material Handling Equipment NREL/TP-5600-56408. 2013. Available online: https://www.nrel.gov/docs/fy13osti/56408.pdf (accessed on 19 August 2024).
31. Alshaebi, A.; Dauod, H.; Weiss, J.; Yoon, S.W. Evaluation of Different Forklift Battery Systems Using Statistical Analysis and Discrete Event Simulation. In Proceedings of the IISE Annual Conference, Pittsburgh, PA, USA, 20–23 May 2017; pp. 1637–1642.
32. Modica, T.; Perotti, S.; Melacini, M. Green Warehousing: Exploration of Organisational Variables Fostering the Adoption of Energy-Efficient Material Handling Equipment. *Sustainability* **2021**, *13*, 13237. [CrossRef]

33. Gaines, L.L.; Elgowainy, A.; Wang, M.Q. Full Fuel-Cycle Comparison of Forklift Propulsion Systems. 2008. Available online: http://www.osti.gov/bridge (accessed on 19 August 2024).
34. Jiao, M.; Pan, F.; Huang, X.; Yuan, X. Evaluation on Total Cost of Ownership of Electric Forklifts with lithium-ion battery. In Proceedings of the 2021 IEEE 4th International Electrical and Energy Conference (CIEEC), Wuhan, China, 28–30 May 2021; IEEE: Piscataway, NJ, USA, 2021; pp. 2368–2372, ISBN 978-1-7281-7149-4.
35. Sauer, D.U. Batteriesysteme für Flurförderzeuge. In *Flurförderzeugtagung*; VDI: Düsseldorf, Germany, 2009; pp. 75–91.
36. Cicconi, P.; Postacchini, L.; Pallotta, E.; Monteriù, A.; Prist, M.; Bevilacqua, M.; Germani, M. A life cycle costing of compacted lithium titanium oxide batteries for industrial applications. *J. Power Sources* **2019**, *436*, 226837. [CrossRef]
37. Jiao, M.; Pan, F.; Huang, X.; Yuan, X. Application potential of second-life lithium-ion battery on forklift. In Proceedings of the 2021 IEEE 4th International Electrical and Energy Conference (CIEEC), Wuhan, China, 28–30 May 2021; IEEE: Piscataway, NJ, USA, 2021; pp. 2373–2377, ISBN 978-1-7281-7149-4.
38. Letmathe, P.; Suares, M. A consumer-oriented total cost of ownership model for different vehicle types in Germany. *Transp. Res. Part D Transp. Environ.* **2017**, *57*, 314–335. [CrossRef]
39. Link, S.; Teichert, O. Battery Cell Database. Available online: https://zenodo.org/records/10679242 (accessed on 20 April 2024).
40. Wassiliadis, N.; Steinsträter, M.; Schreiber, M.; Rosner, P.; Nicoletti, L.; Schmid, F.; Ank, M.; Teichert, O.; Wildfeuer, L.; Schneider, J.; et al. Quantifying the state of the art of electric powertrains in battery electric vehicles: Range, efficiency, and lifetime from component to system level of the Volkswagen ID.3. *eTransportation* **2022**, *12*, 100167. [CrossRef]
41. Klebsch, W.; Heininger, P.; Martin, J. *VDE Studie: Alternativen zu Dieseltriebzügen im SPNV—Einschätzung der Systemischen Potenziale*; Frankfurt am Main, VDE e.V.: Offenbach, Germany, 2019.
42. Pagenkopf, J.; Schirmer, T.; Böhm, M.; Streuling, C.; Herwartz, S. Marktanalyse Alternativer Antriebe im Deutschen Schienenpersonennahverkehr. Berlin. 2020. Available online: https://elib.dlr.de/134615/1/DLR_2020_Marktanalyse%20alternative%20Antriebe%20SPNV.pdf (accessed on 12 April 2024).
43. VDI. *Guideline 2198: Data Sheet for Industrial Trucks*; Beuth Verlag: Berlin, Germany, 2021.
44. Löbberding, H.; Wessel, S.; Offermanns, C.; Kehrer, M.; Rother, J.; Heimes, H.; Kampker, A. From Cell to Battery System in BEVs: Analysis of System Packing Efficiency and Cell Types. *World Electr. Veh. J.* **2020**, *11*, 77. [CrossRef]
45. Propfe, B.; Redelbach, M.; Santini, D.; Friedrich, H. Cost analysis of Plug-in Hybrid Electric Vehicles including Maintenance & Repair Costs and Resale Values. *World Electr. Veh. J.* **2012**, *5*, 886–895. [CrossRef]
46. Fries, M.; Kerler, M.; Rohr, S.; Schickram, S.; Sinning, M.; Lienkamp, M. An Overview of Costs for Vehicle Components, Fuels, Greenhouse Gas Emissions and Total Cost of Ownership: Update 2017. UC Davis. 2017. Available online: https://steps.ucdavis.edu/wp-content/uploads/2018/02/FRIES-MICHAEL-An-Overview-of-Costs-for-Vehicle-Components-Fuels-Greenhouse-Gas-Emissions-and-Total-Cost-of-Ownership-Update-2017-.pdf (accessed on 13 June 2024).
47. SAE. *Proceedings of the Mid-Year Meeting, Washington, DC, USA, 20 May 1968*; SAE Technical Paper Series; SAE International400 Commonwealth Drive: Warrendale, PA, USA, 1968.
48. Tolomeo, R.; de Feo, G.; Adami, R.; Sesti Osséo, L. Application of Life Cycle Assessment to Lithium Ion Batteries in the Automotive Sector. *Sustainability* **2020**, *12*, 4628. [CrossRef]
49. Syré, A.M.; Shyposha, P.; Freisem, L.; Pollak, A.; Göhlich, D. Comparative Life Cycle Assessment of Battery and Fuel Cell Electric Cars, Trucks, and Buses. *World Electr. Veh. J.* **2024**, *15*, 114. [CrossRef]
50. Lai, X.; Chen, Q.; Tang, X.; Zhou, Y.; Gao, F.; Guo, Y.; Bhagat, R.; Zheng, Y. Critical review of life cycle assessment of lithium-ion batteries for electric vehicles: A lifespan perspective. *eTransportation* **2022**, *12*, 100169. [CrossRef]
51. Arshad, F.; Lin, J.; Manurkar, N.; Fan, E.; Ahmad, A.; Tariq, M.-N.; Wu, F.; Chen, R.; Li, L. Life Cycle Assessment of Lithium-ion Batteries: A Critical Review. *Resour. Conserv. Recycl.* **2022**, *180*, 106164. [CrossRef]
52. Wieler, J.; Rudschied, W. VW Golf 8 im ADAC Test. Available online: https://www.adac.de/rund-ums-fahrzeug/autokatalog/marken-modelle/vw/vw-golf/ (accessed on 19 July 2024).
53. Expert Consultations with a Forklift Truck Manufacturer and End User. Personal communication. 2023.
54. AKKU SYS GmbH. 48V 6 PzS 540 Ah DIN B (1027 * 616 * 537mm L/B/H) Trog 57017028 Forklift Battery. Available online: https://staplerbatterie.center/a/48v-gabelstaplerbatterie-6-pzs-540-ah-din-b-1027-616-537mm-l-b-h-trog-57017028-inkl.-aquamatik-von-q-batteries/9885554/ (accessed on 19 July 2024).
55. Pillot, C. *The Rechargeable Battery Market 2021–2030*, 29th ed.; V1; Avicenne Energy: Paris, France, 2022.
56. European Environment Agency. Climate and Energy in the EU: Real-World CO_2 Emissions from New Cars and Vans. Available online: https://climate-energy.eea.europa.eu/topics/transport/real-world-emissions/data (accessed on 19 July 2024).

World Electric Vehicle Journal

MDPI

Article

Simulation and Testing of Self-Reconfigurable Battery Advanced Functions for Automotive Application

Rémy Thomas [1], Nicolas Léto [2], Jérome Lachaize [2], Sylvain Bacquet [3,*], Yan Lopez [1] and Leandro Cassarino [4]

1 CEA, Liten, Université Grenoble Alpes, F-38000 Grenoble, France; remy.thomas@cea.fr (R.T.); yan.lopez@cea.fr (Y.L.)
2 Vitesco Technologies, 40 Avenue du Général de Croutte, 31100 Toulouse, France; nicolas.leto@vitesco.com (N.L.); jerome.lachaize@vitesco.com (J.L.)
3 CEA, Leti, Université Grenoble Alpes, F-38000 Grenoble, France
4 CEA, CEA Tech Nouvelle Aquitaine, F-33600 Pessac, France; leandro.cassarino@cea.fr
* Correspondence: sylvain.bacquet@cea.fr

Abstract: This article presents the design and production work carried out jointly by Vitesco Technologies and the CEA in order to build a Self-Reconfigurable Battery (SRB) demonstrator representative of an electric vehicle traction battery pack. The literature demonstrates that the use of an SRB allows for individual bypassing or serialization of each cell in a battery pack, enabling control of the voltage output and dynamic balancing of the battery pack during all phases of vehicle use. The simulations and tests presented in this article confirm that the use of an SRB results in a 6% reduction in energy consumption compared to a Conventional Battery Pack (CBP) on a driving profile based on WLTP cycles. Additionally, an SRB enhances fast charging performance, with a charging time that is 22% faster than a CBP. Furthermore, it is shown that an SRB without a voltage inversion capability can still be connected directly to the AC grid for charging without the need for a dedicated converter, using only a single diode bridge rectifier for the whole system.

Keywords: battery management system; electric vehicle (EV); fast charge; self-reconfigurable battery

Citation: Thomas, R.; Léto, N.; Lachaize, J.; Bacquet, S.; Lopez, Y.; Cassarino, L. Simulation and Testing of Self-Reconfigurable Battery Advanced Functions for Automotive Application. *World Electr. Veh. J.* **2024**, *15*, 250. https://doi.org/10.3390/wevj15060250

Academic Editor: Peter Van den Bossche

Received: 15 March 2024
Revised: 21 May 2024
Accepted: 23 May 2024
Published: 8 June 2024

1. Introduction

As shown in many articles [1], battery-switching technologies of Self-Reconfigurable Batteries (SRBs) promise significant improvements in terms of autonomy and battery life, cell balancing [2–7], recharging capacity [8–11], improving the efficiency of vehicle drive trains [12–14] and even cell aging [15–17]. Their operating principle is as follows: switches are added to the power paths linking the cells to enable the number of active stages in series and parallel to be modulated dynamically [18], depending on the type of SRB. Some systems even incorporate H-bridges to enable the generation of alternating voltages, opening the way to motor control without an inverter [19–21] and direct recharging on the AC grid without a charger [10,22,23]. However, the increase in the number of switches raises the question of safety impact [24–27].

The aim of this study is to highlight the benefits of a simplified and cost-effective SRB architecture adapted from the design presented in [28]. The benefits include improved motor inverter efficiency, reduced charging time for DC fast charging and AC charging on the electricity grid.

In this study, fulfilling Vitesco Technologies requirements, the SRB architecture dynamically generates a strictly positive DC voltage to optimize the efficiency of the motor's inverter. It also allows the absorption of a rectified AC current to enable the battery to be charged from an AC voltage source without a charger, by means of a simple rectifier. Hence, only the dynamic modulation of the number of series stages is implemented. To maintain the capability of recharging on the AC grid, a simple rectifier diode bridge is added at the head of the architecture instead of H-bridges on cells, as it is not necessary to generate a sinusoidal waveform when discharging.

To assess the benefits, a Conventional Battery Pack (CBP) and the SRB system are compared by simulation in a complete simulation environment incorporating the various components of an electric vehicle power train. In addition, these simulations are validated on a real demonstrator. The comparison is based on different use cases. The first use case is a battery discharge following a driving profile based on cycles from the Worldwide harmonized Light vehicles Test Procedure (WLTP). Different batches of cells are used such as new and aged cells with 5% to 10% loss of State Of Health (SoH) in order to introduce dispersion conditions representative of an aged vehicle pack [29]. It is therefore possible to study the benefits of SRBs using an end-of-life battery pack or even a second-life battery pack. A comparison is then made using a fast charging use case to assess the benefits in terms of stored energy and charging time.

Finally, a direct charging experiment using a rectified AC current from the AC grid is presented to demonstrate the possibility of charging directly from the AC grid without an intermediate converter, despite the absence of the voltage inversion capability in the SRB architecture implemented.

The SRB demonstrator developed by Vitesco Technologies and the French Alternative Energies and Atomic Energy Commission (CEA) integrates a maximum of 120 cells in series with a maximum current of 125 A for charging and discharging. The system consists of a main controller that communicates with 20 modules. These modules are connected to each other in series in the power path. Each module independently controls the switching of six cells, each of which can be dynamically connected in series or bypassed. The system is flexible, making it easy to add or remove modules in series to adjust the maximum voltage. Initially, the system is composed of 14 Ah NMC cells. Then, in a second phase, 90 Ah cells are used. The total number of modules used is adjusted to match the number of cells available for each batch of cells. This setup makes it possible to validate the operation of cell switching in an electric vehicle application.

This article is an expansion of the work presented in EVS36 [30] with additional experimentations based on the introduction of a fourth batch of cells from a used BMW-i3 battery pack that has traveled 25,000 km. The different geometry of these cells meant that the demonstrator had to be modified. The driving profiles also had to be adapted to the new cell capacities. This expansion made it possible to present the results of an SRB system integrating cells directly representative of a real battery pack in use and resulting from a large-scale industrial process, which is unprecedented in the state of the art of reconfigurable battery packs.

2. Simulation

The main objectives are to demonstrate the benefits of the different capabilities listed in the introduction to this paper, by comparing the proposed SRB solution with equivalent CBP. With regard to optimizing the efficiency of the motor inverter, it is necessary to consider an SRB capable of supplying the necessary voltage up to the end-of-discharge conditions of the system. Thus, compared with a CBP for a given segment, the equivalent SRB to be considered must be made up of a larger number of cells in series. Lower-capacity cells are then required to maintain the relevance of the comparison from an energetic point of view. A process of adaptation is therefore necessary. This is why different battery configurations are simulated, in an environment representative of electric vehicles, to carry out the comparison.

The two batteries are dimensioned in order to provide the same energy at the wheel at the beginning of life, taking into account a capacity dispersion of 2% within the cells. In this simulation, the reference CBP is a 96S3P battery with 60 Ah cells, while the SRB has a 144S2P architecture with the same 60 Ah cells. To reduce the cost of this technology, a grouping of cells in series is also considered: instead of having one cell for each bypass switch/serial switch entity, groups of four cells in series for each bypass switch/serial switch entity are used. Consequently, the SRB in this simulation has 36 groups with each group consisting of four 1S2P cells and one bypass switch/serial switch entity.

2.1. Driving Simulation

The simulation environment representative of electric vehicles is described in detail in a previous study [31], where it was used to compare balancing solutions during driving cycles. For this study, the simulation model is improved by the use of representative energetic cells based on actual parameters as well as a representative dispersion of the cells in the pack.

Figure 1 highlights the benefits, in terms of driven distance, of the aged SRB compared with the aged CBP. The comparison is made based on a capacity reduced to 70% of the nominal capacity and a capacity dispersion of 4.8%. The cumulated losses in the battery pack are detailed for the SRB values, while the values in brackets correspond to the difference SRB minus CBP. Due to the control DC link voltage capability, one can see that a better global powertrain efficiency is achieved with a reduction of 51 Wh per 100 km. This increased efficiency and the SRB's ability to manage the dispersal of capacity result in a 6% (+24 km) increase in range, which is of the same order of magnitude as the gains seen in [13].

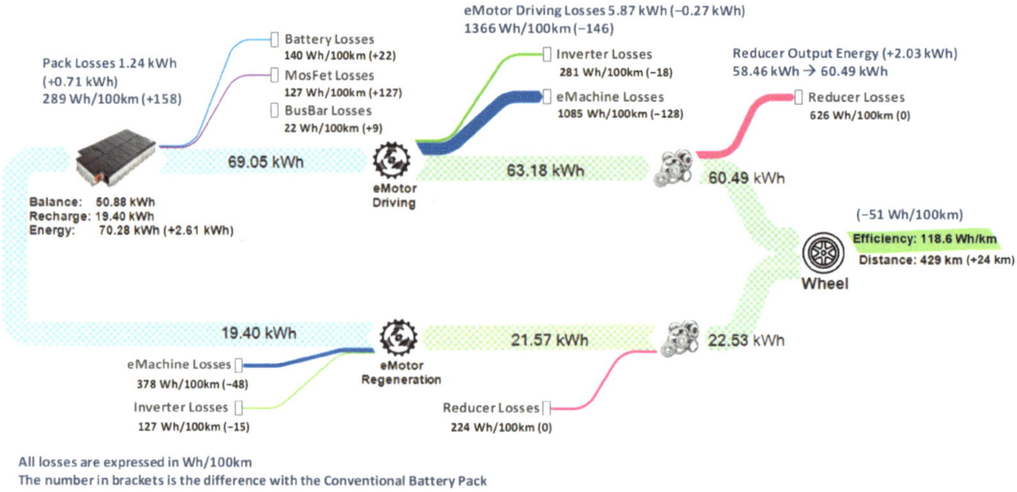

Figure 1. Sankey diagram for SRB/CBP comparison—driving use case.

2.2. Fast Charging Simulation

The CBP and SRB system configurations used for the previous driving profile are this time used to simulate 130 kW DC fast charging. This time, the nominal capacities are used to compare the solutions at the beginning of their lifetime. With regard to the SRB configuration defined above, the voltage set point of the DC bus output is fixed to 450 V. To take better account of the constraints of fast charging, the simulation model is updated with an improved thermal representativeness.

Figure 2 illustrates this comparison. By convention, the charging current and power are positive, as they are considered from the charger point of view. The power curves show that the maximum charge power can be maintained for a much longer time with the SRB system than with the CBP system. This is due to the faster rise in cell temperature in the CBP configuration, which leads to the limitation of the charging current.

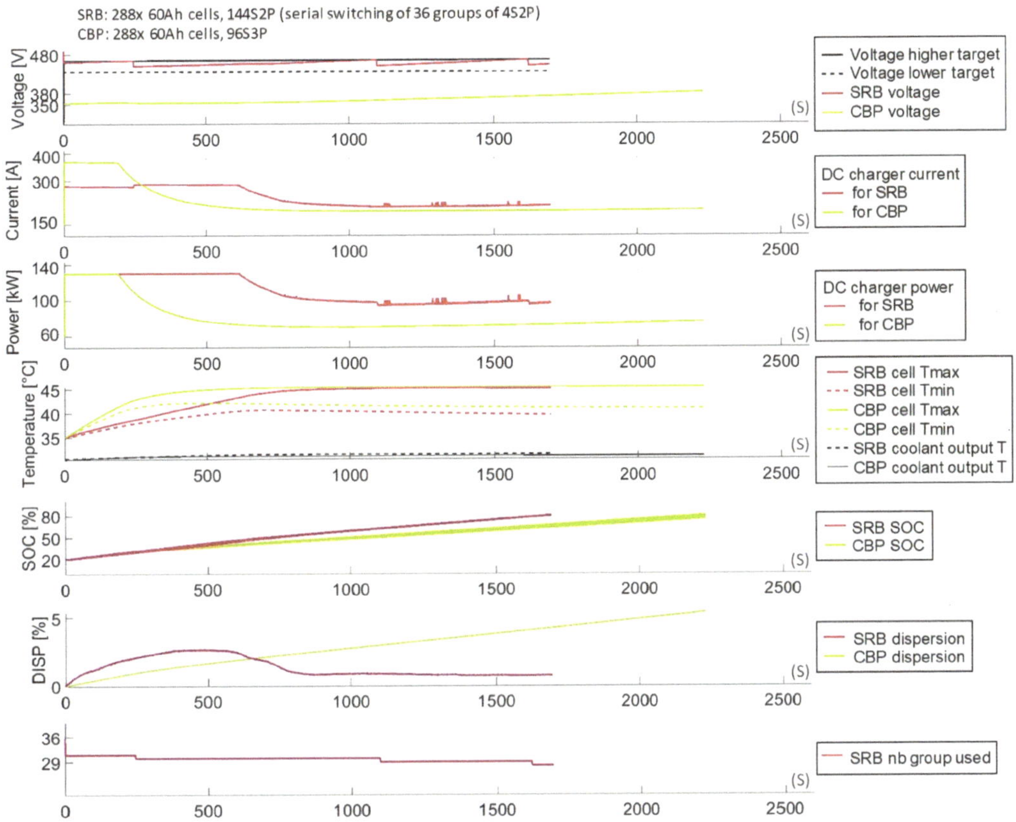

Figure 2. Simulation of fast DC charging for CBP and SRB.

At the beginning of the charge, almost all of the SRB cells are set in serial to reach the DC link voltage set point. Consequently, the cell-to-cell dispersion within the SRB is higher than the CBP's one when the charging power remains high. Nevertheless, when the charging power decreases, a smaller number of cells is required in serial. Then, power cell balancing is possible again, resulting in a very low cell-to-cell dispersion. This illustrates the great effectiveness of SRB power cell balancing.

Table 1 highlights the energies and losses involved in fast charging for both CBP and SRB. The losses are increased by 40% in the SRB compared to the CBP because of the additional electronic components and the higher recharging power allowed. Assuming identical charger efficiency, the DC charging efficiencies are, respectively, 97.0% and 98.2% for SRB and CBP. However, despite this increase in losses, the charging time from 20% to 80% state of charge is reduced from 37 min (2234 s) to 28 min (1687 s) due to the reduction in current required to charge the battery, which delays the rise in cell temperature. Charging time is therefore 24% faster with the SRB based on a battery capable of undergoing the same WLTP test than with the CPB configuration. This reduced fast charging time is a key element in the competitive BEV market [32]. In addition, the maximum current required from the charger is reduced.

Table 1. Energies and losses for fast DC charging.

	Battery Chemical Energy [kWh]	Battery Losses [kWh]	MosFet Losses [kWh]	BusBar Losses [kWh]	Battery Energy [kWh]
CBP	−50.20	0.80	0.00	0.12	−51.11
SRB	−50.19	0.83	0.60	0.10	−51.72

3. Experimental System Overview

The experimental system is an adaptation of the presented system in [28]. It consists of one master and twenty modules, each composed of 6 switchable cells. The modules provide voltage and temperature measurements through an isolated RS485 bus to the master. The master controller measures the overall voltage and current, processes the cell voltage and temperature measurements received from the modules and sends back the switching orders to be applied. In addition, this master controller includes high-level application management capabilities such as battery pack voltage regulation, dynamic cell balancing and safety features. Dynamic cell balancing is performed by alternating the cells used on the power path in order to manage the current drawn on each cell to provide the output power [33]. The state of charge of each series level is assessed by coulomb counting using a single current sensor located at the battery pack, combined with information on the bypass or series states of each level. For the test carried out in this study, the global amount of energy exchanged when operating WLTP cycles, or during the fast charging test, is assessed using the current sensor of the power lab equipment of the climatic chamber.

The architecture of the experimental SRB is illustrated in Figure 3. Unlike many reconfigurable batteries that rely on a phase-shifting carrier, this implementation uses a true real-time process to control the output voltage while using low-cost local controllers. This enables a faulty cell to be removed and replaced by the master controller in less than 100 µs from the time it is detected. It also allows the shape of a signal such as a disturbed electrical grid to be tracked as closely as possible, thereby reducing the size of the filtering components. In the case of DC discharge and DC fast charge, switch SW1 is closed and switch SW2 is open. For AC charging, switch SW1 is open and switch SW2 is closed to connect the SRB to the electrical grid via a rectifier diode bridge. The SRB then generates a rectified signal adjusted in real time to follow the waveform of the rectifier bridge with a slightly lower amplitude to create a charging current flowing through the cells. The internal global current sensor is used to control the current exchanged by adjusting the voltage differences between the SRB and the output of the rectifier diode bridge.

The CBP and SRB are compared using, for both cases, the self-reconfigurable electrical architecture, in order to use the same experimental setup. Hence, by serializing all the cells, the demonstrator behaves like a CBP, where all the cells of the batch are serialized without possible modification, while the SRB can control its output voltage by bypassing some cells. As the first cell in each module is used to power the switching electronics, the related serial levels are constantly bypassed from the power path to avoid interference in the comparison between SRB and CBP, in end-of-charge or end-of-discharge conditions and in the energy balance.

To charge and discharge the system, a secure test chamber with an 800 V 400 A power supply serving as both a source and sink is employed. Figure 4 shows the test bench setup in operation inside the test chamber. The experimental comparison enables the results obtained by simulation to be verified.

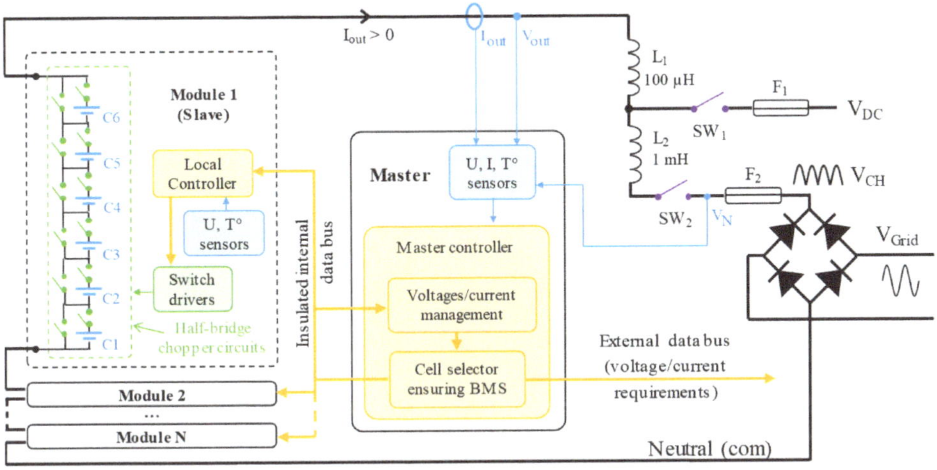

Figure 3. Experimental Self-Reconfigurable Battery architecture.

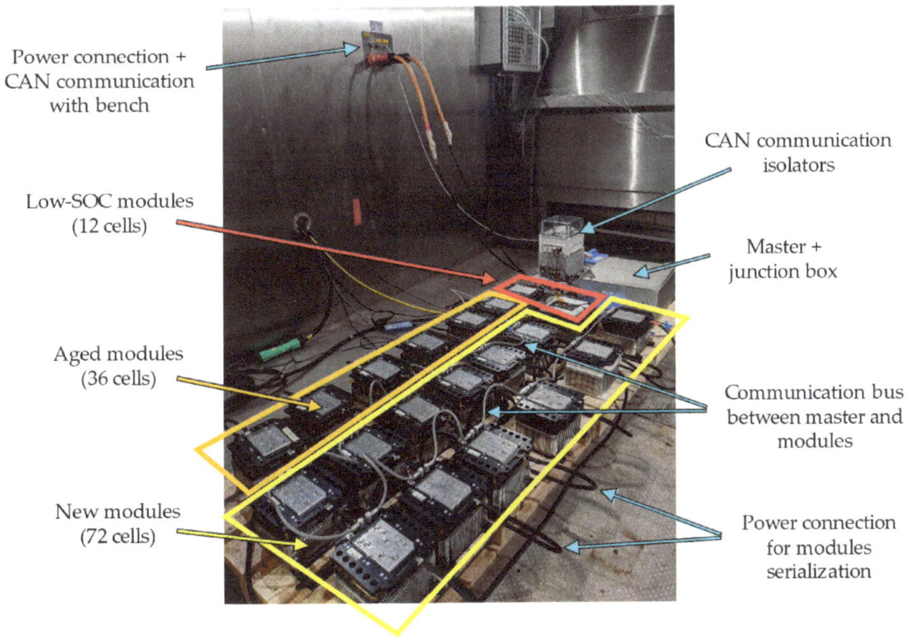

Figure 4. Self-Reconfigurable Battery test bench, 120 serial cells subdivided in 20 modules of 6 serial cells.

4. Experimental Comparisons with a Driving Profile Based on the Worldwide Harmonized Test Cycles for Light Vehicles (WLTC)

In the case of discharge comparisons over a driving cycle, one of the main objectives is to demonstrate the feasibility of optimizing inverter efficiency with a real SRB system. To this end, the inverter model is considered reliable and the simulation of its efficiency is used to generate the power profiles applied to the experimental configurations as well as the battery voltage set point profile for the SRB. The aim of the experiment is then to verify that the SRB is indeed capable of responding to the dynamic profiles of the voltage set point and that the gain in autonomy compared with a real CBP corresponds to those observed in the simulation. Additionally, another objective of the experiment is to compare the behavior of a CBP (in which cells cannot be bypassed) and that of an SRB with regard to cell-to-cell dispersion. Indeed, cell-to-cell dispersion has a significant impact on the performance of electric vehicles [34–36]. The comparison is carried out during a discharge imposed by a driving profile, which consists of several consecutive WLTCs.

Four batches of cells are created to address different use case scenarios. For each comparison, the same batch is used in both the CBP and the SRB. The first batch comprises 72 new NMC 14 Ah CALB cells. The second batch consists of 36 NMC 14 Ah CALB cells artificially aged by a laboratory cycling process. The aging process was carried out individually for each cell and therefore does not incorporate the aging divergence phenomenon that can be observed in a Conventional Battery Pack. The third batch is made of 36 cells from a mix of new NMC 14 Ah CALB cells and new NMC 14 Ah CALB cells slightly discharged to give a lower initial capacity. This batch is used to emulate a second life scenario using cells from different States Of Health (SOHs). The fourth batch is made of 42 NMC 94 Ah SDI cells from a real BMW i3 battery pack of around 25,000 km, in order to assess the benefits associated with the characteristics of the cells in a real commercial battery pack. A first batch of 60 new cells, a second and third batch of 30 cells and a fourth batch of 35 cells are finally used for comparison, considering that the first cell of each module is permanently bypassed from the power path. The characteristics of the different batches are summarized in Table 2.

Table 2. Batch characteristics.

Cell Batch ID	Use Case	Cell Details	Nb of Cells	Nb of Modules	Nb of Cells Used for Comparison
1	Beginning of life	New NMC 14 Ah (CALB)	72	12	60
2	Aged cells	Artificially aged NMC 14 Ah (CALB)	36	6	30
3	Second life	Mix of full and partially discharged NMC 14 Ah (CALB)	36	6	30
4	Cells with a lifetime of around 25,000 real-life km	BMW-i3 NMC 94 Ah (SDI)	42	7	35

To assess the dispersions involved in the experimental comparisons, cell capacities are estimated by coulomb counting with a specific charge and discharge cycle at a rate of 0.1 C between a minimum voltage of 3 V and a maximum voltage of 4.18 V. A dispersion of 0.95% for batch 1 of the new cells is observed, which remains realistic compared with what can be classically assessed for other cell references in the literature [37]. The dispersion of artificially aged cells in batch 2 is of the same order at 1.1%, which is quite low for cells representative of an aged pack [38]. This can be explained by the use of an individual aging process rather than a group process as in a Conventional Battery Pack where discrepancies can be observed with the increase in the number of cycles. The capacities of cells constituting batches 1 and 2 are, respectively, presented in Figure 5a,b, while the capacities of the cells in batches 3 and 4 are shown in Sections 4.3 and 4.4. The distinctive colors of each cell serve to enhance contrast during reading and have no other significance.

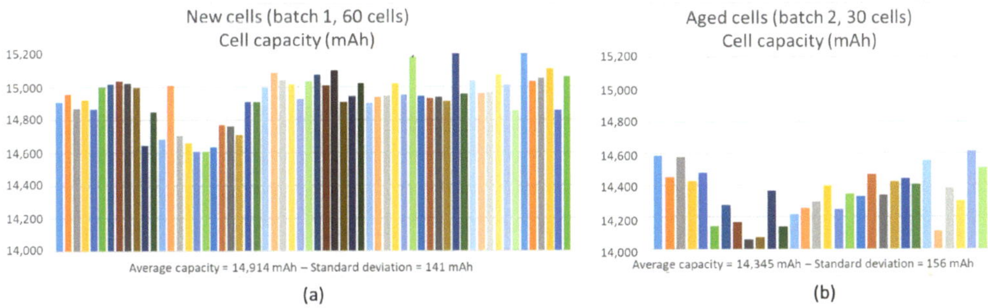

Figure 5. Capacity of cells constituting (**a**) batch 1 and (**b**) batch 2. Colors only to improve contrast.

The power profile generated from the simulation corresponds to the power consumed at the wheel. It takes into account the number of cells included in each batch to provide the corresponding power set point to be applied to the CBP and SRB packs. The optimum voltage set point for SRB tests is also generated for each batch according to the number of cells to be used.

Unlike the CBP, the SRB has the ability to balance the cells during charge and discharge. As the cells are always balanced, the SRB allows all the energy to be extracted from all the cells in the pack. In a CBP, this is not possible because the pack discharge must stop when a cell reaches its lower voltage limit, even if other cells still have energy. To ensure a fair comparison between CBP and SRB, all cells are fully charged and properly balanced before each test. All cells start at 4.18 V, and the test stops at the end of the driving profile or when the first of the cells under consideration reaches 3 V.

4.1. Driving Profile with New Cells (Batch 1)

Figure 6 shows the result of the driving profile with new cells (batch 1) in the SRB (top) and CBP (bottom) configurations. On the SRB profile, the battery output voltage set point (in orange) is perfectly stable at around 125 V, with the exception of two slots of around 200 s at times 1540 s and 3320 s, where the set point voltage changes dynamically to follow the parts of the WTLP cycle where power requirements are greater. The SRB voltage measurement (in blue) shows that the battery is perfectly in line with the set voltage, even during the most dynamic periods of the WLTP cycle, demonstrating the SRB's ability to generate the voltage profile required to optimize the inverter's efficiency in various driving phases. This optimization enables the SRB to complete the entire driving profile, unlike the CBP, which stops prematurely at 4364 s because a cell reaches the low voltage limit of 3 V.

Over a discharge period of 4364 s, which marks the end of the CBP's discharge, the SRB discharged 2886 Wh, while the CBP discharged 3060 Wh. The SRB therefore consumed 5.7% less energy than the CBP for the same distance traveled. The fact that the observed difference in consumption corresponds to the difference predicted by the simulation shows that the power profiles imposed are consistent with the simulation. At the end of the CBP driving profile, we observed that the cells were not well balanced, with a VCellMax–VCellMin difference of 180 mV, due to the dispersion of the cells' capacity in batch 1. In contrast, the SRB was able to balance the cells at all times, resulting in a VCellMax–VCellMin difference of less than 5 mV at the end of the driving profile. The figures of this test are reported in Table 3.

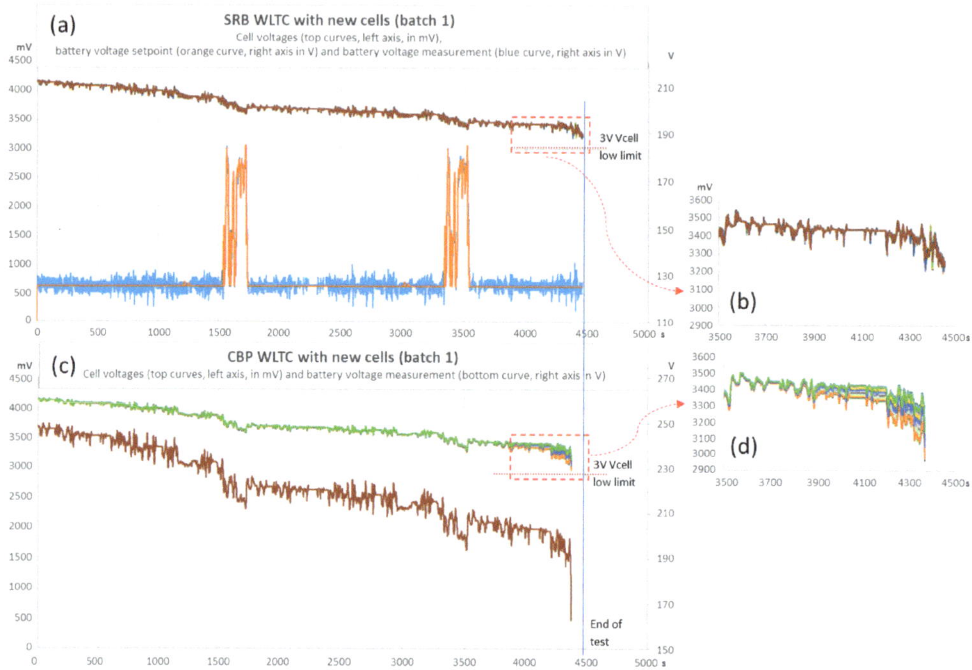

Figure 6. Driving profile with new cells (batch 1): (**a**) SRB configuration results; (**b**) zoom on SRB configuration results; (**c**) CBP configuration results; (**d**) zoom on CBP configuration results. Cell voltages colors only to improve contrast.

Table 3. Figures of WLTC test with new cells (batch 1).

	CBP	**SRB**
Run time	4364 s	4458 s
Remaining energy at end of cycle	0 Wh	104 Wh
Cell balance at end of cycle	Δ = 180 mV	Δ < 5 mV
Discharged–charged energies	@4364 s: 3060 Wh (batt low)	@4364 s: 2886 Wh @4455 s (end cycle): 2981 Wh (104 Wh left)

4.2. Driving Profile with Artificially Aged Cells (Batch 2)

Figure 7 shows the result of the driving profile with aged cells (batch 2) with the SRB (top) and CBP (bottom) configurations. The power profiles and the SRB set point profile are updated to take into account the reduction in the number of available cells compared to the first batch discharge experiment. The cells in batch 2 exhibited a lower average capacity, which resulted in neither the CBP nor the SRB configuration reaching the end of the test.

During a discharge period of 4027 s, the SRB discharged 1388 Wh while the CBP discharged 1468 Wh. The SRB consumed 5.45% less energy than the CBP for the same period, which once again shows that the power profiles applied are consistent with the simulation. The 1.1% capacity dispersion of artificially aged cells is not sufficient to introduce an additional gain in favor of the SRB, especially if the end of the CBP and SRB discharge occurs in a moderate power section of the WLTP cycle. Nevertheless, it can be observed that the emergence of cell voltage dispersion in the CBP occurs concurrently with the end of the second high-power phase of the WLTP cycle at 3500 s. The cell voltage imbalance of the CBP is also increased, as shown in Figure 7d. This suggests that a slightly greater dispersion in cell capacity could have had a greater impact on the results. The figures of this test are presented in Table 4.

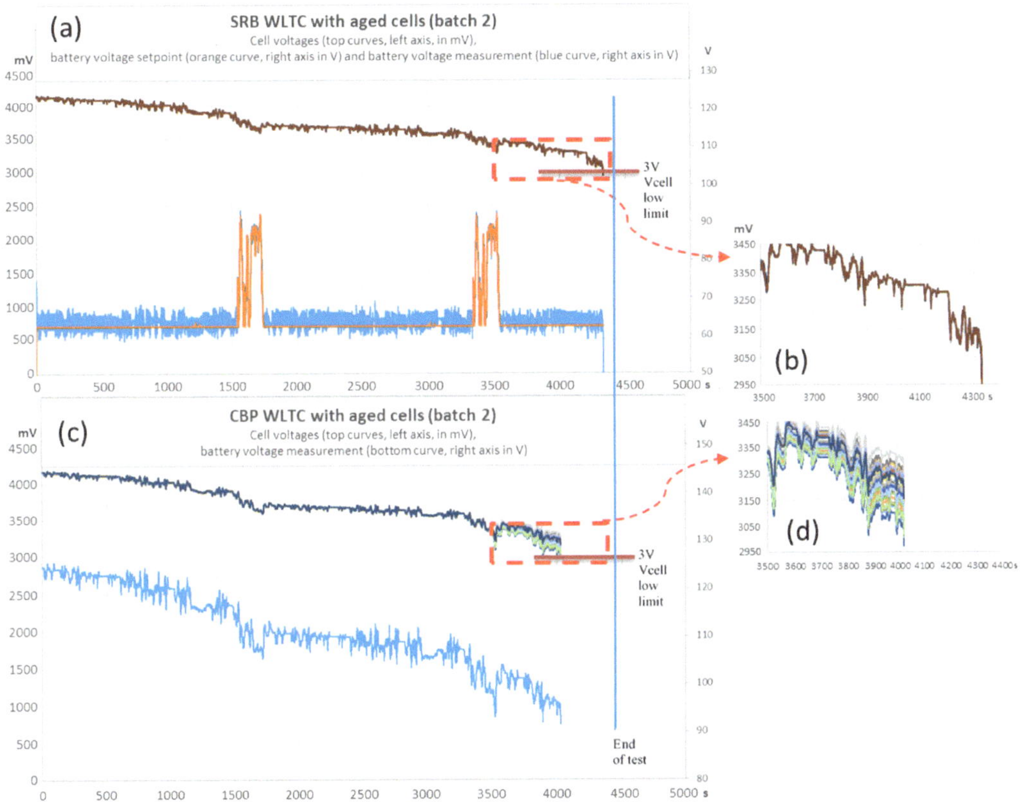

Figure 7. Driving profile with aged cells (batch 2): (**a**) SRB configuration results; (**b**) zoom on SRB configuration results; (**c**) CBP configuration results; (**d**) zoom on CBP configuration results. Cell voltages colors only to improve contrast.

Table 4. Figures of WLTC test with new cells (batch 2).

	CBP	SRB
Run time	4027 s	4329 s
Remaining energy at end of cycle	0 Wh	0 Wh
Cell balance at end of cycle	Δ = 220 mV	Δ < 5 mV
Discharged–charged energies	@4027 s: 1468 Wh	@4027 s: 1388 Wh @4329 s: 1436 Wh

4.3. Driving Profile with Second-Life Heterogeneous Cells (Batch 3)

The reparability of battery packs is becoming increasingly of interest to industry for the purpose of maintenance and optimizing system lifespan [39,40]. Heterogeneous cells in terms of capacity, and even different chemistries, can then constitute a reconditioned battery pack. To illustrate this, the operations of the CBP and the SRB are compared by conducting a driving profile with cells of heterogeneous capacities. Hence, a third batch of 30 cells is made using a mix of new cells at 100% state of charge and new cells with a downgraded capacity, as shown in Figure 8. Eight cells in batch 1 are slightly discharged to present a reduced capacity, introducing a dispersion of 3.15% for batch 3. For accuracy, the lowered capacities are set before each test using a voltage threshold and a discharging current below C/10.

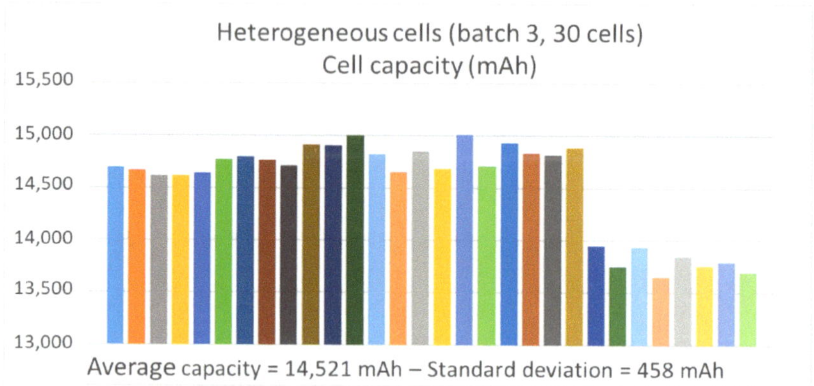

Figure 8. Heterogeneous capacity of cells constituting batch 3. Colors only to improve contrast.

The power profiles and the SRB set point profile are identical to those of the second batch discharge experiment, which had the same number of cells. The results of the driving profile applied to batch 3 are illustrated in Figure 9.

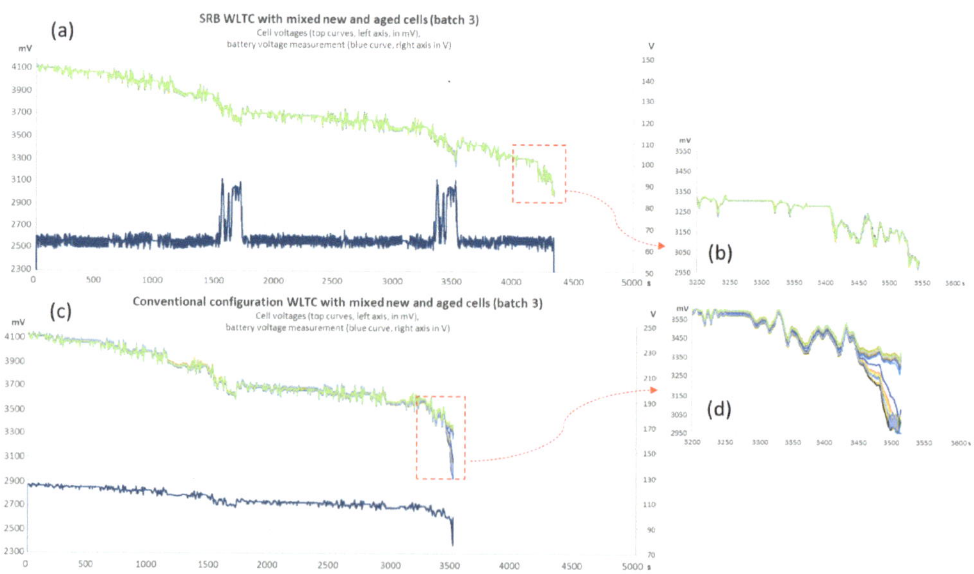

Figure 9. Driving profile with heterogeneous cells (batch 3): (**a**) SRB configuration results; (**b**) zoom on SRB configuration results; (**c**) CBP configuration results; (**d**) zoom on CBP configuration results. Cell voltages colors only to improve contrast.

In the CBP configuration, the weakest cell voltage drops before the others during the second high-power period of the WTLP cycle, causing the cycle to be halted at 3514 s, as shown in Figure 9c,d.

Figure 9a shows that the SRB is able to generate the output voltage corresponding to the voltage set point profile despite the use of a group of heterogeneous cells. The cells of the SRB are well balanced during the driving profile, while the battery voltage is controlled to maximize the inverter yield. The SRB allows weaker cells to be set aside while energy is extracted from all the other cells. Additionally, perfect balancing helps to avoid voltage drop due to high discharge current peaks, allowing the driving profile to continue. Hence, the system stops at 4347 s with perfectly balanced cells as shown in Figure 9b. The SRB consumes 5.3% less energy than the CBP for the same period of 3514 s. The figures of this test are presented in Table 5.

Table 5. Figures of WLTC test with cells from batch 3.

	CBP	SRB
Run time	3514 s	4347 s
Remaining energy at end of cycle	0 Wh	0 Wh
Cell balance at end of cycle	Δ = 428 mV	Δ = 21 mV
Discharged–charged energies	@3514 s: 1416 Wh	@3514 s: 1341.5 Wh @4347 s: 1437 Wh

4.4. Driving Profile with Cells from Real Battery Pack (Batch 4)

In industrial battery packs, cells with similar characteristics are assembled together to obtain the most homogenous batches possible during the manufacturing phase in order to maximize battery life [41]. In this test, the comparison is made using cells from a real 25,000 km industrial battery pack to show the benefits obtained from the slight dispersion of cells that can be found in real life. The driving profile is extended by repeating the same profile several times to match the increase in the battery capacity. The duration of the discharge process is calibrated to achieve a depth of discharge of approximately 10% of the battery's capacity for the two use cases where the discharged energies are compared. The profile is then repeated to fully discharge the battery in order to assess the energy remaining in each case.

The capacities of the fourth batch's cells used for this test are shown in Figure 10. The dispersion in capacity between cells in batch 4 is 0.37%. The capacity of the worst cell is approximately 91.67 A·h (98.57% of nominal capacity) with an average cell capacity of 92.29 A·h (99.23% of nominal capacity).

Test results are shown in Figure 11 with the SRB in the top part and the CBP in the bottom part. In both cases, the end of discharge is reached during a high-power phase. A state of charge of 10% is reached in 21,023 s in the CBP configuration, whereas the SRB configuration reaches 10% SOC in 21,869 s. In terms of energy, this leads to a reduced consumption of 4% when considering the energy consumed at 21,023 s for both configurations.

Regarding the total run time, the CBP configuration stops at 22,693 s, whereas the SRB configuration stops at 23,524 s, despite the fact that the SRB had to handle an additional power peak. In this context, the performance of the SRB dynamic balancing had no impact due to the very low imbalance of the cells from batch 4. Nevertheless, in terms of time duration, this leads to a gain of 3.66%, which is far from being insignificant considering the limited age of the battery pack. The figures of this test are presented in Table 6.

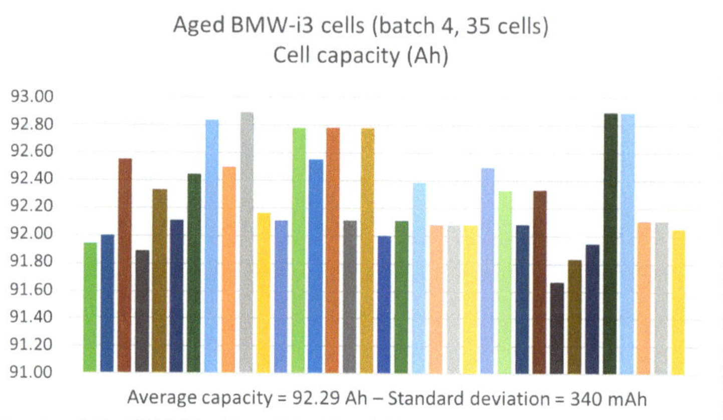

Figure 10. Capacity of cells from an aged BMW-i3 battery pack constituting batch 4. Colors only to improve contrast.

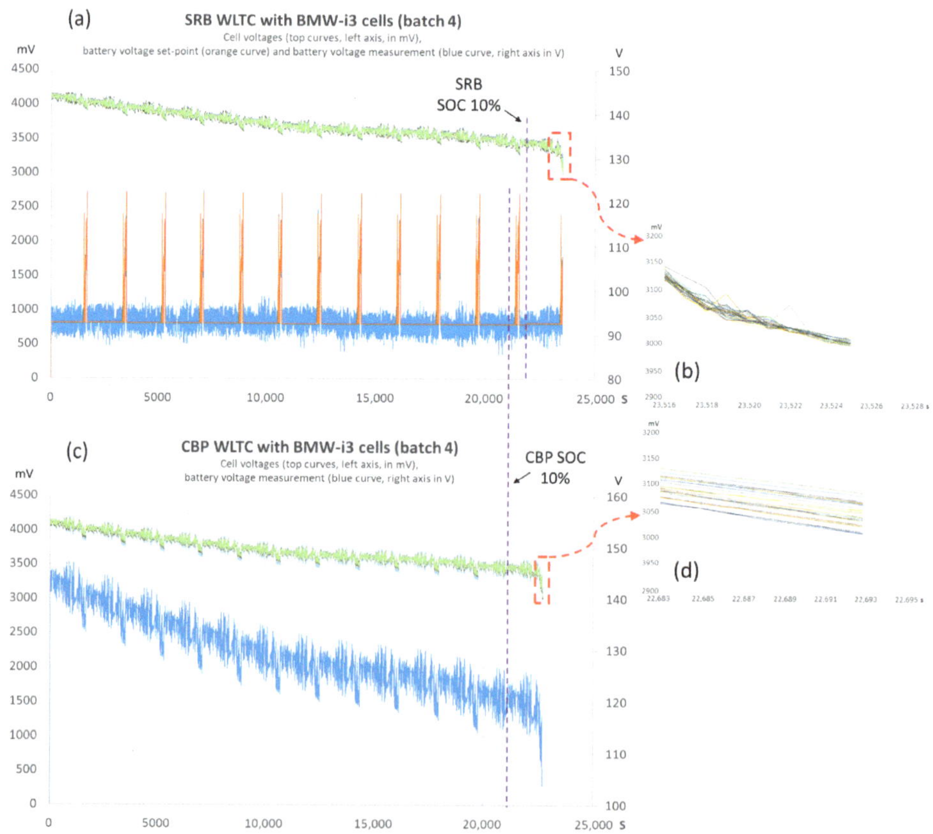

Figure 11. Driving profile with cells from a real BMW-i3 battery pack constituting batch 4: (**a**) SRB configuration results; (**b**) zoom on SRB configuration results; (**c**) CBP configuration results; (**d**) zoom on CBP configuration results. Cell voltages colors only to improve contrast.

Table 6. Figures of WLTC test with new cells (batch 4).

	CBP	SRB
Run time to 10%	21,023 s	21,869 s
Total run time	22,693 s	23,524 s
Remaining energy at end of cycle	989 Wh	976 Wh
Cell balance at end of cycle	Δ = 82 mV	Δ = 15 mV
Discharged–charged energies	@21,023 s: 10,192 Wh @22,693 s: 10,959 Wh	@21,023 s: 9783 W.h @21,869 s: 10,320 Wh @23,524 s: 11,076 Wh

4.5. Conclusion on Driving Profile

For each batch of cells tested, the conclusion is that the SRB manages to provide the dynamical output voltage corresponding to the voltage set point required to optimize the efficiency of the motor inverter. The fact that the observed difference in energy consumption corresponds to the difference predicted by the simulation shows that the imposed power profiles and the optimized voltage profile are consistent with the simulation presented in Section 2. In addition, good reproducibility is obtained with the different battery sizes.

Finally, the dynamic balancing capabilities of the SRB enable additional autonomy gains to be achieved by allowing the weakest cells to be set aside while energy is extracted from all the other cells. Perfectly balancing cells avoids voltage drop due to high discharge current peaks, enabling greater autonomy to be achieved with the driving profile.

5. Experimenting with Direct Charging of SRB on Electrical Grid

Removing the AC–DC inverters permits us to increase the charge yield [42]. To charge the Self-Reconfigurable Battery directly from the electrical grid without a charger, it is necessary to generate a perfectly synchronized voltage waveform. Furthermore, the voltage waveform of the electrical grid is never a perfect sinusoid; it has unpredictable distortions that must be taken into account when controlling the current exchanged with the battery. Due to the high control frequencies of the master controller, the SRB can produce an arbitrary voltage at its output. Hence, the output voltage of the SRB is directly adjusted in real time from the output of a charge current control loop that regulates the current exchanged with the electrical grid.

A Simulink algorithm is created to allow the SRB to be charged without a dedicated charger through a standard 16 A single-phase grid. The master controller contains a charge controller block based on this algorithm. This block receives as input the mean charge current set point and the instantaneous values of the grid voltage and the current exchanged. The output of this block is the number of cells to be connected in series, used to drive the output voltage in accordance with current regulation.

As the SRB developed for this study does not have the capacity to generate a negative voltage, a rectifier diode bridge is used to interface with the electrical grid. Filtering and safety elements are also added in series to the power circuit. The components used to connect the SRB to the power grid are illustrated in Figure 12, while the SRB architecture is illustrated in Figure 3.

At the beginning of the charging phase, the algorithm controls the SRB pack voltage until it matches the rectified grid voltage. When the signals are properly synchronized, a relay is closed to start charging. The start of SRB charging is shown in Figure 13 with the grid voltage (blue curve), the rectified grid voltage (green curve) and the charge current (pink curve). The test was carried out with a charging current of 3 A mean on a real electrical grid, so the sinusoidal curves are not perfect; nevertheless, the algorithm was able to perfectly match this voltage and its imperfections. Figure 14 shows a charge at 16 A. The shape of the current is always rectified and sinusoidal, regardless of the value of the current.

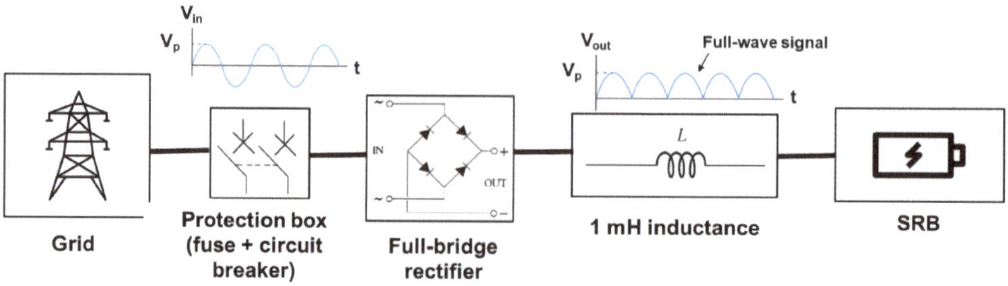

Figure 12. Elements between SRB and grid.

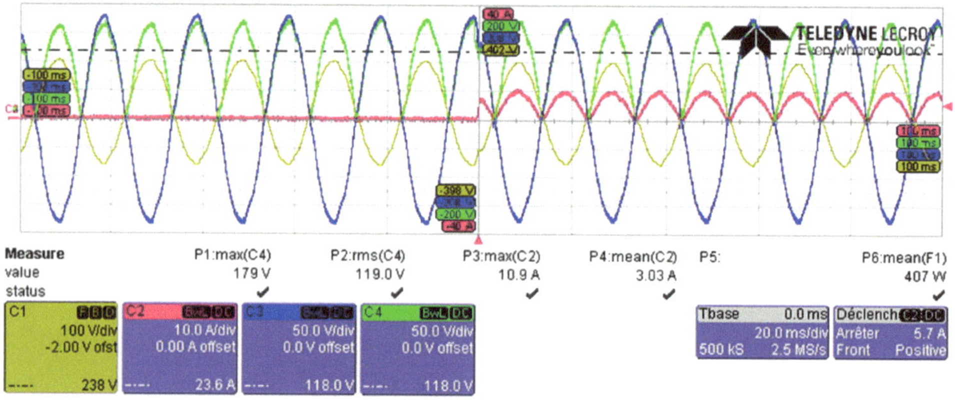

Figure 13. Start of SRB charging with electrical grid.

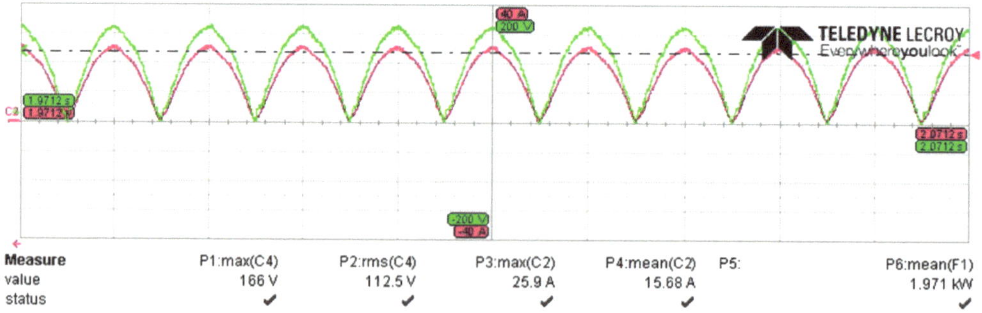

Figure 14. Rectified grid voltage (green) and charge current (pink) at 16 A mean.

6. Fast Charge Comparison

The objective of this test is to compare the behavior of the CBP and SRB during fast charging on a DC link. A first comparison is carried out with the same 12 modules using the batch 1 cells previously used for the WLTP tests. The first cell of each module is still permanently removed from the power path. A second comparison is carried out with eight modules from batch 4b, this time including the first stage of each module in the power path. Table 7 shows the batch characteristics used for each fast charge comparison.

Table 7. Batch characteristics used for fast charge comparison.

Cell Batch ID	Use Case	Cell Details	Nb of Cells	Nb of Modules	Nb of Cells Used for Comparison
1	Beginning of life	New NMC 14 Ah (CALB)	72	12	60
4b	Cells with a lifetime of around 25,000 real-life km	BMW-i3 NMC 94 Ah (SDI)	42	8	42

The charging current profile applied to the cells is adjusted according to the cell voltage, decreasing as the voltages approach end-of-charge conditions. To simplify control, the adjustment consists of three constant current amplitudes.

The Self-Reconfigurable Battery has the capability of individually bypassing cells at the end of their charge, whereas the Conventional Battery Pack must stop charging the whole system at the first cell in the end-of-charge state. Therefore, the criteria used to select the charging current level for each system must be different to take account of the differences in operation.

In the case of the CBP, the charging current level is affected by the voltage of the most charged cell, with different voltage thresholds to distinguish each level. However, a part of the cell voltage measured is related to the instantaneous current flowing through it due to its impedance. This voltage therefore decreases when the charging current is reduced. Hence, a hysteresis is added to the voltage thresholds to prevent a return to a higher charging current.

In the case of the SRB, only the highest voltage threshold 4.18 V is used to trigger a bypass of cells exceeding this value. The charging current level is then adjusted according to the number of cells remaining to be charged. The cells in bypass are put back into series when their voltage drops due to the relaxation effect. A hysteresis is applied to the voltage threshold used to reconnect the cells in series. This hysteresis is proportional to the amplitude of the charging current according to (1).

$$\text{Bypass to series Hysteresis [mV]} = \text{Current [A]}/1000 \quad (1)$$

The amplitude of the charge current for each level is adjusted between batch 1 and batch 4b to take account of the differences in terms of cell capacity. The cells in batch 1 are intended for power applications and accept a maximum continuous C-rate charge current of 10 C. Such a current is not recommended to preserve the health of the cells, but it is used in this study to illustrate the capabilities of the SRB electronics. It should be noted that in the case of batch 4b, the maximum current is limited by the current capacity of the test chamber. When the batch 4b cells are used, the experimental set-up has to change the test chamber, which reduces the capacity in terms of maximum current. As a result, the charge current rate is reduced for this batch, especially as the cells in batch 4b have a capacity 6.7 times greater than that of the cells in batch 1. This reduced charge rate enables the use of higher voltage thresholds for the CBP configuration. The fast charge conditions are summarized in Tables 8–10.

Table 8. Fast charge conditions for CBP using cells from batch 1.

CBP Condition	Batch 1 Charge Current (A)
VcellMax < 4.1 V	125 (8.9 C)
4 V < VcellMax < 4.15 V	24 (1.78 C)
4.14 V < VcellMax < 4.18 V	12 (0.86 C)

Table 9. Fast charge conditions for CBP using cells from batch 4b.

CBP Condition	Batch 4b Charge Current (A)
VcellMax < 4.15 V	90 (0.96 C)
4.05 V < VcellMax < 4.16 V	24 (0.25 C)
4.14 V < VcellMax < 4.18 V	12 (0.08 C)

Table 10. Fast charge conditions for SRB using cells from batch 1 and batch 4b.

SRB Condition	Batch 1 Charge Current (A)	Batch 4b Charge Current (A)
Series cell > 10 then 24 A	125 (8.9 C)	90 (0.96 C)
Series cell > 10 then 12 A	24 (1.78 C)	24 (0.25 C)
Series cell > 10 then stop	12 (0.86 C)	12 (0.08 C)

6.1. Comparison of Fast Charging with New Cells (Batch 1)

Figure 15 shows a comparison between the CPB and SRB during the fast charge phase with cells from batch 1.

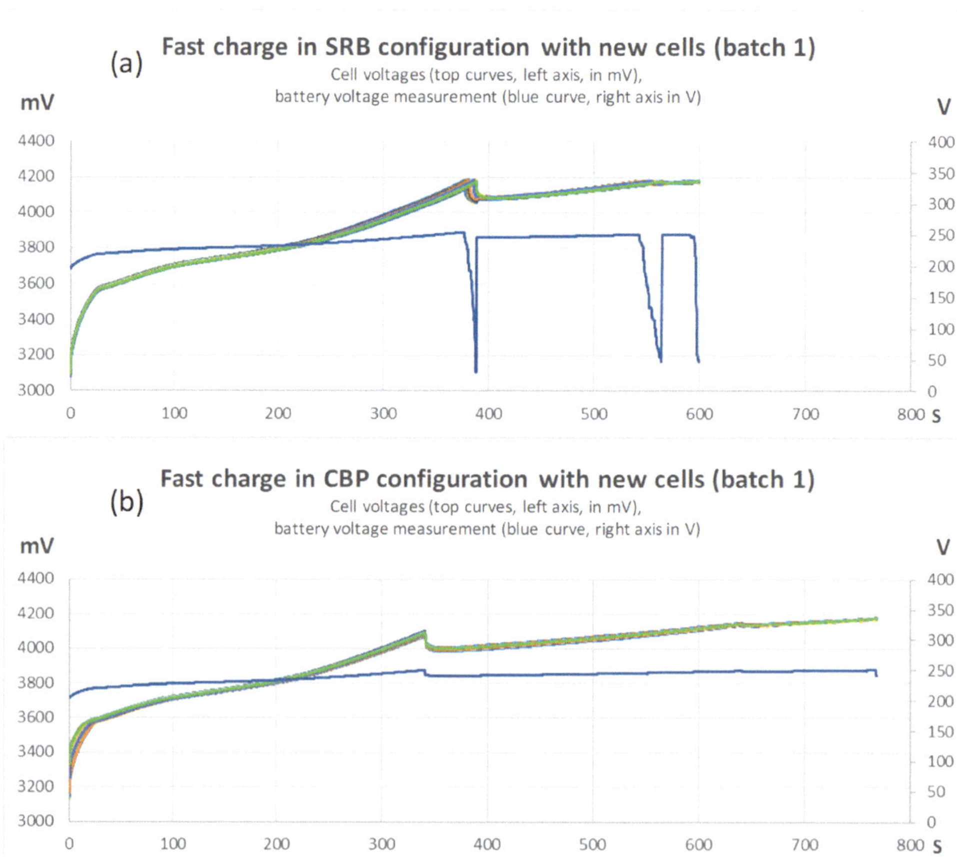

Figure 15. Comparison of charge for CBP and SRB configurations using cells from batch 1: (**a**) SRB configuration results; (**b**) CBP configuration results. Cell voltages colors only to improve contrast.

With the cells from batch 1, the state of charge 80% is reached before the first cell voltage threshold. Therefore, both the CBP and SRB reach this state of charge in around 332 s with a constant charging current of 125 A. The SRB is capable of bypassing a cell that reaches a voltage of 4.18 V and of reconnecting it in series when its voltage falls below a predefined threshold. In Figure 15a, the SRB voltage drops as cells are removed from the supply circuit when they reach this voltage threshold. When the number of cells in series falls below the 10-cell threshold, the charging current is reduced, which enables the SRB to return all cells to the power path, thus increasing the output voltage. One can note that the SRB takes advantage of the 80% to 100% state of charge interval to reach the full charge state in 22.8% less time than the CBP.

The total energy consumed by the charger is 3418 Wh for the SRB compared with 3342 Wh for the CBP, which represents a difference of 2.27%. This is due to the additional losses introduced by the series switches on the SRB. The figures of this test are presented in Table 11.

Table 11. Results of fast charging time for batch 1.

	CBP (s)	SRB (s)
SOC 20%	87	87
SOC 80%	332	334
Vcell max = 4.1 V	338	347
Vcell max = 4.15 V	502	366
SOC 100%	776	599

6.2. Comparison of Fast Charging with Cells from Real Battery Pack (Batch 4b)

Figure 16 shows a comparison between the CPB and SRB during the fast charge phase with cells from batch 4b. With the cells from batch 4b, the state of charge 80% is still reached before the first cell voltage threshold. Therefore, both the CBP and SRB reach this state of charge in 2908 s with a constant charging current of 90 A. It can be seen in Figure 16a that a longer time is needed for the SRB to fall below the 10-cell threshold, due to the reduced ratio between charging current and cell capacity. This longer time allows the discarded cells to be reconnected, resulting in a noisier SRB output voltage corresponding to alternating serial and bypass phases.

In this comparison, the SRB still takes advantage of the 80% to 100% state of charge interval to reach the full charge state 15.4% faster than the CBP. The total energy consumed by the charger is 17,645 Wh for the SRB compared with 17,007 Wh for the CBP. This represents a difference of 3.75%, this time including the first cells of each module from which energy is drawn to power the switching electronics. The figures of the fast charge comparison of batch 4b are presented in Table 12.

Table 12. Results of fast charging time for batch 4b.

	CBP (s)	SRB (s)
SOC 20%	727	727
SOC 80%	2908	2908
Vcell max = 4.15 V	2959	2978
Vcell max = 4.16 V	4069	3012
SOC 100%	6012	5087

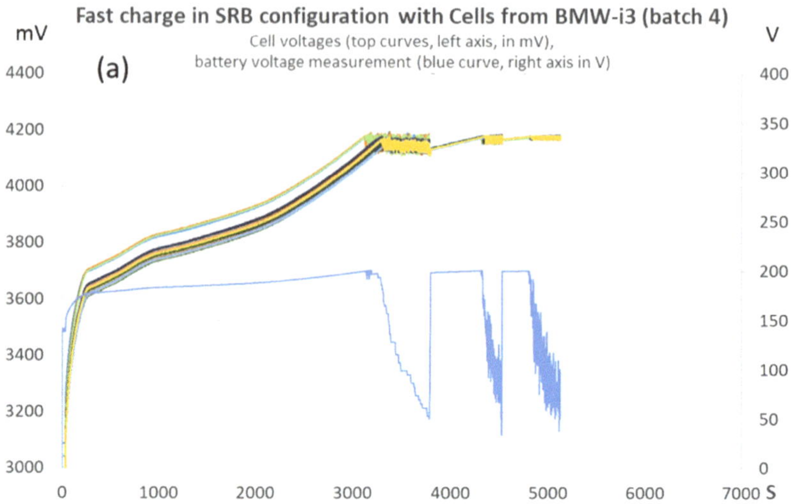

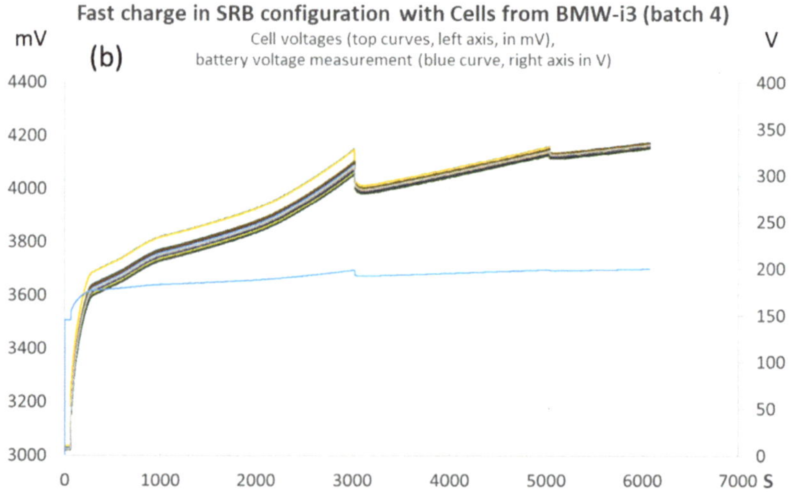

Figure 16. Comparison of charge for CBP and SRB configurations using cells from batch 4b: (**a**) SRB configuration results; (**b**) CBP configuration results. Cell voltages colors only to improve contrast.

7. Discussion

The experimental system is not far from being able to be integrated into a demonstrator vehicle, given the level of maturity above TRL 4 achieved by the developments initiated since 2012 at the CEA [43], as well as the size of the demonstrator in terms of on-board battery as well as power, current and voltage capacity.

The experimental validation of the results obtained by simulation demonstrates that it is possible to simulate the behavior of a real SRB device at vehicle scale with a good level of representativeness. This demonstration is especially important as it was carried out using the models and simulation tools developed by Vitesco for its industrial development processes.

With regard to the results obtained with the different batches of cells, it can be seen that the disparities between the cells do not affect the ability of the experimental SRB to generate the voltage profile required to optimize the inverter's efficiency.

The results also show that the impact of an SRB's balancing performance depends on the unbalanced characteristics of the cells and the power rating used at the end of the discharge. It can be seen that the higher the power demand at the end of discharge, the greater the impact on the SRB's balancing capacity. In this respect, it is important to note that perfectly balancing cell voltage before a power peak is not necessarily sufficient. Managing the voltage drops of the cells with the highest impedance during the power peak is also significant. In the case of this study, the proposed SRB system is able to replace in real time an active cell reaching its end-of-discharge voltage threshold. The swap with an inactive cell of any local controller is performed in a time interval of less than a few tens of microseconds, so that no disturbance of the output voltage occurs. In addition to the gains in autonomy presented in this study, we can therefore expect to see interesting gains in terms of power availability.

With regard to fast charging on a DC source, the main contribution of SRB systems is made from the time at which the conventional system reaches the first conditions requiring a drop in charging current. In this study, the conditions are voltage conditions, but they could also be over-temperature conditions in certain contexts as seen in the simulation or even conditions related to model-based parameters that the state of the art is beginning to use to optimize fast charging in conventional systems [44]. In this area, SRBs could gain an advantage from the prospect of improved cell parameter identification capabilities [45–47].

Despite all the advantages mentioned above, the adoption of reconfigurable batteries in the industrial sector is currently limited to start-up companies such as SwitchESS, with which the CEA has collaborated [48], or Bavertis [49]. The integration of reconfigurable batteries involves concentrating responsibility for a range of functions, such as the battery, the BMS, the charger and even the motor inverter. For Tier 1 manufacturers, this represents an important challenge, both economically and technically. Recently, the industry has taken a close interest in this technology, as shown by the interest expressed by Vitesco [50] and Stellantis [51].

8. Conclusions

This paper reports the results of simulation and experimentation of the advanced capabilities of a Self-Reconfigurable Battery, including battery output voltage control, active balancing during operation, AC grid charging without an inverter and fast DC charging. The performance of the SRB as a DC–DC power source is compared to that of a Conventional Battery Pack, revealing improved efficiency and faster charge rates. Specifically, the SRB increased the driving range by 6% and reduced charging time by 22%. In addition, the experimental results demonstrate the SRB's ability to operate heterogeneous cells correctly and extend battery life by reproducing the aging observed in real automotive batteries. It also demonstrates the capability of using unsorted cells or heterogeneous cells for second life.

To extend this study, given that the SRB introduces many more electronic components than the CBP, elements other than performance need to be taken into account to make a fair comparison between the two solutions, such as cost, reliability and safety.

As far as the cost aspect is concerned, estimating the economic benefit of reconfigurable batteries is complex because it has to integrate the benefits at the level of the overall system to be relevant. For example, improving the energy efficiency of the powertrain as a whole can reduce the cost of the battery for an equivalent range. Additionally, this improvement can also reduce the amount of CO_2 equivalent consumed, which is a very important cost criterion for manufacturers because of the regulations introduced in recent years. The cells are also less susceptible to aging, which extends the life of the pack and therefore saves the consumer a certain amount of money. Still on the economic aspect, the SRB's ability to integrate heterogeneous cells opens the way to the use of batteries that could be less expensive to produce because they would be less constrained in terms of homogeneity, or even could be from second life. In all the cases mentioned, dedicated detailed studies would be required to estimate the savings made.

From a reliability point of view, the large number of components used for switching raises questions, especially when SRBs are paradoxically highlighted for their ability to improve battery reliability by isolating faulty cells. However, SRBs have a certain potential because, although they use a large number of components, the mature manufacturing process of MOSFET and limited voltage range required make them more reliable than the increasingly sophisticated power components such as SIC and GAN switches used in power converters, even more so as the voltage levels of battery packs are becoming higher and higher to meet the constraints of rapid charging. The number of components used must therefore be considered in light of the Mean Time Between Failures (MTBF), by means of a rigorous analysis of the system's fault tree.

From the point of view of software constraints, the greater the complexity of the code, the more difficult and costly it is to ensure. The use of a distributed system such as the one presented in this study makes it possible to partition the software functions. In this way, the role of the software corresponding to the local controllers can be limited to ensuring the translation of the serial/bypass commands received from the central controller into local switch control with the appropriate transitions. In this way, it is possible to simplify the software used by the local controllers as much as possible in order to guarantee a satisfactory level of reliability and safety. This allows for mitigating the level of criticality applied to the central node software.

This project is a unique opportunity to work on these points, and the initial analysis shows that an SRB designed to automotive standards in a robust and safe manner could be a competitive solution if the powertrain is considered as a whole, including throughout its lifetime. Overall, this study highlights the impressive capabilities of an SRB and its potential for use in a variety of applications.

Author Contributions: Conceptualization, R.T., N.L., J.L., S.B., Y.L. and L.C.; methodology, R.T., N.L., J.L., S.B. and Y.L.; software, R.T., S.B. and L.C.; validation R.T., N.L., J.L., S.B., Y.L. and L.C.; writing—original draft preparation, R.T., N.L., J.L. and S.B.; writing—review and editing, R.T., N.L., J.L. and S.B.; supervision, N.L. and S.B.; project administration, N.L. and S.B. All authors have read and agreed to the published version of the manuscript.

Funding: This research has benefited from the support of the Automotive and Mobility Research Steering Committee (CORAM) funded by BPIFRANCE.

Data Availability Statement: The raw data supporting the conclusions of this article will be made available by the authors on request.

Acknowledgments: This work has benefited from technical support from A. Bouche-Pillon and M. Bey, for the use of the test chamber equipment, the generating of power cycles from the calculated power exchange files and the data logging.

Conflicts of Interest: Nicolas Léto, Jérome Lachaize are employees of Vitesco Technologies. The paper reflects the views of the scientists, and not the company.

Abbreviations

Symbols	Descriptions
BEV	Battery Electric Vehicle
CBP	Conventional Battery Pack
CEA	French Alternative Energies and Atomic Energy Commission
EVS36	36th Electrical Vehicle Symposium
SOC	state of charge
SOH	State Of Health
SRB	Self-Reconfigurable Battery
NMC	Nickel Manganese Cobalt oxides
WLTP	Worldwide harmonized Light vehicles Test Procedure
TRL	Technology Readiness Level
MTBF	Mean Time Between Failures

References

1. Komsiyska, L.; Buchberger, T.; Diehl, S.; Ehrensberger, M.; Hanzl, C.; Hartmann, C.; Hölzle, M.; Kleiner, J.; Lewerenz, M.; Liebhart, B.; et al. Critical Review of Intelligent Battery Systems: Challenges, Implementation, and Potential for Electric Vehicles. *Energies* **2021**, *14*, 5989. [CrossRef]
2. Thomas, R.; Lehmann, F.; Blatter, J.; Despesse, G.; Heiries, V. Performance Analysis of a Novel High Frequency Self-Reconfigurable Battery. *World Electr. Veh. J.* **2021**, *12*, 10. [CrossRef]
3. Baccari, S.; Tipaldi, M.; Mariani, V. Deep Reinforcement Learning for Cell Balancing in Electric Vehicles with Dynamic Reconfigurable Batteries. *IEEE Trans. Intell. Veh.* **2024**, 1–12. [CrossRef]
4. Wiedenmann, A.; Buberger, J.; Högerl, T.; Grupp, W.; Hohenegger, M.; Kuder, M.; Weyh, T.; Neve, A. Proactive Balancing for Reconfigurable Battery Systems with Automated Efficiency Analysis. In Proceedings of the 2023 International Conference on Smart Energy Systems and Technologies (SEST), Mugla, Turkiye, 4–6 September 2023. [CrossRef]
5. Lee, S.; Noh, G.; Ha, J.-I. Reconfigurable Power Circuits to Series or Parallel for Energy-Balanced Multicell Battery Pack. *IEEE Trans. Ind. Electron.* **2023**, *70*, 3641–3651. [CrossRef]
6. Li, Y.; Yin, P.; Chen, J. Active Equalization of Lithium-Ion Battery Based on Reconfigurable Topology. *Appl. Sci.* **2023**, *13*, 1154. [CrossRef]
7. Karnehm, D.; Bliemetsrieder, W.; Pohlmann, S.; Neve, A. Controlling Algorithm of Reconfigurable Battery for State of Charge Balancing Using Amortized Q-Learning. *Batteries* **2024**, *10*, 131. [CrossRef]
8. Lamprecht, A.; Narayanaswamy, S.; Steinhorst, S. Improving Fast Charging Efficiency of Reconfigurable Battery Packs. In Proceedings of the 2018 Design, Automation & Test in Europe Conference & Exhibition (DATE), Dresden, Germany, 19–23 March 2018. [CrossRef]
9. Balachandran, A.; Jonsson, T.; Eriksson, L. DC Charging Capabilities of Battery-Integrated Modular Multilevel Converters Based on Maximum Tractive Power. *Electricity* **2023**, *4*, 62–77. [CrossRef]
10. Buberger, J.; Hohenegger, M.; Estaller, J.; Wiedenmann, A.; Grupp, W.; Bliemetsrieder, W.; Kuder, M.; Lesnicar, A.; Weyh, T. Bidirectional Charging for BEVs with Reconfigurable Battery Systems via a Grid-Parallel Proportional-Resonant Controller. *Electricity* **2023**, *4*, 171–184. [CrossRef]
11. Prasad, R.; Namuduri, C.; Gopalakrishnan, S. On-Demand Battery Reconfiguration for 800V DC Fast Charging in Electric Vehicles. In Proceedings of the 2023 IEEE Energy Conversion Congress and Exposition (ECCE), Nashville, TN, USA, 29 October–2 November 2023. [CrossRef]
12. Sorokina, N.; Estaller, J.; Kersten, A.; Buberger, J.; Kuder, M.; Thiringer, T.; Eckerle, R.; Weyh, T. Inverter and Battery Drive Cycle Efficiency Comparisons of Multilevel and Two-Level Traction Inverters for Battery Electric Vehicles. In Proceedings of the 2021 IEEE International Conference on Environment and Electrical Engineering and 2021 IEEE Industrial and Commercial Power Systems Europe (EEEIC/I&CPS Europe), Bari, Italy, 7–10 September 2021. [CrossRef]
13. Xu, Y.; Kersten, A.; Ingelström, P.; Amirpour, S.; Klacar, S.; Sedarsky, D. Comparative Study of Efficiency Improvement with Adjustable DC-Link Voltage Powertrain Using DC-DC Converter and Quasi-Z-Source Inverter. In Proceedings of the 2023 IEEE Transportation Electrification Conference and Expo, Asia-Pacific (ITEC Asia-Pacific), Chiang Mai, Thailand, 28 November–1 December 2023. [CrossRef]
14. Li, Z.; Yang, A.; Chen, G.; Tashakor, N.; Zeng, Z.; Peterchev, A.V.; Goetz, S.M. A Rapidly Reconfigurable DC Battery for Increasing Flexibility and Efficiency of Electric Vehicle Drive Trains. *IEEE Trans. Transp. Electrif.* **2024**, 1. [CrossRef]
15. Blatter, J.; Heiries, V.; Thomas, R.; Despesse, G. Optimal Lifetime Management Strategy for Self-Reconfigurable Batteries. In Proceedings of the 2022 IEEE 95th Vehicular Technology Conference: (VTC2022-Spring), Helsinki, Finland, 19–22 June 2022. [CrossRef]
16. Kacetl, T.; Kacetl, J.; Tashakor, N.; Goetz, S. Ageing Mitigation and Loss Control in Reconfigurable Batteries in Series-Level Setups. In Proceedings of the EPE 2022 ECCE Europe, Hanover, Germany, 5 September 2022.
17. Škegro, A.; Zou, C.; Wik, T. Analysis of Potential Lifetime Extension through Dynamic Battery Reconfiguration. In Proceedings of the 2023 25th European Conference on Power Electronics and Applications (EPE'23 ECCE Europe), Aalborg, Denmark, 4–8 September 2023; pp. 1–11. [CrossRef]
18. Kuder, M.; Schneider, J.; Kersten, A.; Thiringer, T.; Eckerle, R.; Weyh, T. Battery Modular Multilevel Management (BM3) Converter Applied at Battery Cell Level for Electric Vehicles and Energy Storages. In Proceedings of the PCIM Europe Digital Days 2020; International Exhibition and Conference for Power Electronics, Intelligent Motion, Renewable Energy and Energy Management, Nuremburg, Germany, 7–8 July 2020; pp. 1–8.
19. D'Arco, S.; Quraan, M.; Tricoli, P.; Piegari, L. Low frequency operation of Modular Multilevel Converters with embedded battery cells for traction drives. In Proceedings of the 2016 International Symposium on Power Electronics, Electrical Drives, Automation and Motion, SPEEDAM, Anacapri, Italy, 22–24 June 2016; pp. 1375–1382. [CrossRef]
20. Davis, A.; Salameh, Z.M.; Eaves, S.S. Comparison of a synergetic battery pack drive system to a pulse width modulated AC induction motor drive for an electric vehicle. *IEEE Trans. Energy Convers.* **1999**, *14*, 245–250. [CrossRef]
21. Sorokina, N.; Högerl, T.; Wolfgang, B.; Hein, L.; Weyh, T.; Kuder, M. Investigation of Reconfigurable Battery Efficiency for an Application in an Electrical Sailplane. In Proceedings of the 2023 25th European Conference on Power Electronics and Applications (EPE'23 ECCE Europe), Aalborg, Denmark, 4–8 September 2023; pp. 1–8. [CrossRef]

22. Davis, A.; Salameh, Z.M.; Eaves, S.S. Evaluation of lithium-ion synergetic battery pack as battery charger. *IEEE Trans. Energy Convers.* **1999**, *14*, 830–835. [CrossRef]
23. Leite, R.S.; Afonso, J.L.; Monteiro, V. A Novel Multilevel Bidirectional Topology for On-Board EV Battery Chargers in Smart Grids. *Energies* **2018**, *11*, 3453. [CrossRef]
24. Vemuganti, H.P.; Sreenivasarao, D.; Ganjikunta, S.K.; Suryawanshi, H.M.; Abu-Rub, H. A Survey on Reduced Switch Count Multilevel Inverters. *IEEE Open J. Ind. Electron. Soc.* **2021**, *2*, 80–111. [CrossRef]
25. Schmid, M.; Gebauer, E.; Endisch, C. Structural Analysis in Reconfigurable Battery Systems for Active Fault Diagnosis. *IEEE Trans. Power Electron.* **2021**, *36*, 8672–8684. [CrossRef]
26. Xu, H.; Cheng, L.; Xu, S.; Liu, C.; Paizulamu, D. Operating Performance Evaluation and Improvement Method of Reconfigurable Battery Energy Storage System. In Proceedings of the 2022 12th International Conference on Power and Energy Systems (ICPES), Guangzhou, China, 23–25 December 2022. [CrossRef]
27. Bayati, M.; Tashakor, N.; Farahmandrad, M.; Abkenar, P.P.; Goetz, S. Fault-Tolerant Electric Vehicle Drivetrain with Reconfigurable Battery and Multiphase Machine. In Proceedings of the 2023 IEEE 2nd Industrial Electronics Society Annual On-Line Conference (ONCON), Online, 8–10 December 2023. [CrossRef]
28. Thomas, R.; Despesse, G.; Bacquet, S.; Fernandez, E.; Lopez, Y.; Ramahefa-Andry, P.; Cassarino, L. A High Frequency Self-Reconfigurable Battery for Arbitrary Waveform Generation. *World Electr. Veh. J.* **2021**, *12*, 8. [CrossRef]
29. Braco, E.; San Martín, I.; Sanchis, P.; Ursúa, A. Characterization and Capacity Dispersion of Lithium-Ion Second-Life Batteries from Electric Vehicles. In Proceedings of the 2019 IEEE International Conference on Environment and Electrical Engineering and 2019 IEEE Industrial and Commercial Power Systems Europe (EEEIC/I&CPS Europe), Genova, Italy, 11–14 June 2019. [CrossRef]
30. Bacquet, S.; Léto, N.; Lachaize, J.; Thomas, R.; Lopez, Y.; Cassarino, L. Simulation and testing of self reconfigurable battery advanced functions for automotive application. In Proceedings of the 36th International Electric Vehicle Symposium and Exhibition (EVS36), Sacramento, CA, USA, 11–14 June 2023.
31. Ayad, A.; Léto, N.; Schweizer Berberich, N.; Bornschlegel, S.; Lachaize, J.; Hevele, N.; Brockerhoff, P.; Lyubar, A. Active and Power Balancing Techniques: More Range and Longer Cell Lifetime in Electric Vehicles. In Proceedings of the 35th International Electric Vehicle Symposium and Exhibition (EVS35), Oslo, Norway, 11–15 June 2022.
32. Gnann, T.; Funke, S.; Jakobsson, N.; Plötz, P.; Sprei, F.; Bennehag, A. Fast charging infrastructure for electric vehicles: Today's situation and future needs. *Transp. Res. Part D Transp. Environ.* **2018**, *62*, 314–329. [CrossRef]
33. Ci, S.; Lin, N.; Wu, D. Reconfigurable Battery Techniques and Systems: A Survey. *IEEE Access* **2016**, *4*, 1175–1189. [CrossRef]
34. Kirkaldy, N.; Samieian, M.A.; Offer, G.J.; Marinescu, M.; Patel, Y. Lithium-ion battery degradation: Comprehensive cycle ageing data and analysis for commercial 21700 cells. *J. Power Sources* **2024**, *603*, 234185. [CrossRef]
35. Beck, D.; Dechent, P.; Junker, M.; Sauer, D.U.; Dubarry, M. Inhomogeneities and Cell-to-Cell Variations in Lithium-Ion Batteries, a Review. *Energies* **2021**, *14*, 3276. [CrossRef]
36. Naguib, M.; Kollmeyer, P.; Emadi, A. Lithium-Ion Battery Pack Robust State of Charge Estimation, Cell Inconsistency, and Balancing: Review. *IEEE Access* **2021**, *9*, 50570–50582. [CrossRef]
37. Schindler, M.; Sturm, J.; Ludwig, S.; Schmitt, J.; Jossen, A. Evolution of initial cell-to-cell variations during a three-year production cycle. *eTransportation* **2021**, *8*, 100102. [CrossRef]
38. Schuster, S.F.; Brand, M.J.; Berg, P.; Gleissenberger, M.; Jossen, A. Lithium-ion cell-to-cell variation during battery electric vehicle operation. *J. Power Sources* **2015**, *297*, 242–251. [CrossRef]
39. Martinez-Laserna, E.; Sarasketa-Zabala, E.; Stroe, D.I.; Swierczynski, M.; Warnecke, A.; Timmermans, J.M.; Goutam, S.; Rodriguez, P. Evaluation of lithium-ion battery second life performance and degradation. In Proceedings of the 2016 IEEE Energy Conversion Congress and Exposition (ECCE), Milwaukee, WI, USA, 18–22 September 2016; pp. 1–7. [CrossRef]
40. Hossain, E.; Murtaugh, D.; Mody, J.; Faruque, H.M.R.; Haque Sunny, M.S.; Mohammad, N. A Comprehensive Review on Second-Life Batteries: Current State, Manufacturing Considerations, Applications, Impacts, Barriers & Potential Solutions, Business Strategies, and Policies. *IEEE Access* **2019**, *7*, 73215–73252. [CrossRef]
41. Baumhöfer, T.; Brühl, M.; Rothgang, S.; Sauer, D.U. Production caused variation in capacity aging trend and correlation to initial cell performance. *J. Power Sources* **2014**, *247*, 332–338. [CrossRef]
42. Genovese, A.; Ortenzi, F.; Villante, C. On the Energy Efficiency of Quick DC Vehicle Battery Charging. *World Electr. Veh. J.* **2015**, *7*, 570–576. [CrossRef]
43. Despesse, G.; Sanjuan, S.L.; Gery, S. Battery Monitoring System Using Switching Battery Cells. In Proceedings of the RITF 2012—Research & Innovation for Transport Systems of the Future, Paris, France, 12–15 November 2012; pp. 12–15, hal-01681862.
44. Li, Y.; Vilathgamuwa, D.M.; Wikner, E.; Wei, Z.; Zhang, X.; Thiringer, T.; Wik, T.; Zou, C. Electrochemical Model-Based Fast Charging: Physical Constraint-Triggered PI Control. *IEEE Trans. Energy Convers.* **2021**, *36*, 3208–3220. [CrossRef]
45. Kersten, A.; Kuder, M.; Han, W.; Thiringer, T.; Lesnicar, A.; Weyh, T.; Eckerle, R. Online and On-Board Battery Impedance Estimation of Battery Cells, Modules or Packs in a Reconfigurable Battery System or Multilevel Inverter. In Proceedings of the IECON 2020 the 46th Annual Conference of the IEEE Industrial Electronics Society, Singapore, 18–21 October 2020. [CrossRef]
46. Schneider, D.; Liebhart, B.; Endisch, C. Active State and Parameter Estimation as Part of Intelligent Battery Systems. *J. Energy Storage* **2021**, *39*, 102638. [CrossRef]
47. Theiler, M.; Schneider, D.; Endisch, C. Experimental Investigation of State and Parameter Estimation within Reconfigurable Battery Systems. *Batteries* **2023**, *9*, 145. [CrossRef]

48. SwitchESS Cell-Wise Driving. Available online: https://switchess.com (accessed on 17 May 2024).
49. Bavertis. Available online: https://bavertis.com (accessed on 17 May 2024).
50. Concentrated Competence in Battery Management: Vitesco Technologies France Cooperates with CEA. Available online: https://www.leti-cea.com/cea-tech/leti/english/Pages/What's-On/Press%20release/Concentrated-competence-in-battery-management-Vitesco-Technologies-France-cooperates-with-CEA.aspx (accessed on 17 May 2024).
51. IBIS: Stellantis and Saft Reveal a Smarter, More Efficient Battery for Autos and Stationary Power. Available online: https://www.stellantis.com/en/news/press-releases/2023/july/ibis-stellantis-and-saft-reveal-a-smarter-more-efficient-battery-for-autos-and-stationary-power?adobe_mc_ref= (accessed on 17 May 2024).

World Electric Vehicle Journal

MDPI

Article

A Predictive Cabin Conditioning Strategy for Battery Electric Vehicles

Patrick Schutzeich [1,*], Stefan Pischinger [1], David Hemkemeyer [2], Kai Franke [1] and Paul Hamelbeck [1]

[1] Chair of Thermodynamics of Mobile Energy Conversion Systems (TME), RWTH Aachen University, Forckenbeckstraße 4, 52074 Aachen, Germany
[2] FEV, Neuenhofstraße 181, 52078 Aachen, Germany
* Correspondence: schutzeich@tme.rwth-aachen.de

Abstract: This paper is based on the work presented at EVS36 in Sacramento. The core of the work deals with the cabin climate control of battery electric vehicles (BEV) using model predictive control (MPC) approaches. These aim to reduce the energy demand for cabin air conditioning while maintaining comfort and air quality. The first step briefly overviews model predictive control approaches and the respective fundamentals. Afterward, the modeling for the system dynamics is explained. The challenge for the system model considering humid air is discussed, and the first implementation method is presented. With the added equations for the air quality and humidity, a logic to prevent window fogging was developed to improve safety. Ultimately, model-in-the-loop (MiL) investigations identified an energy-saving potential of up to 15.4% for cold and 39.7% for hot conditions compared to a rule-based strategy. In addition, the investigations carried out showed that it was also possible to improve indoor comfort by specifically influencing the air quality and humidity. Together with the safety criteria introduced to prevent window fogging, it was possible to present a strategy that can significantly improve thermal management for the cabin in modern BEVs.

Keywords: BEV; air conditioning; control system; energy efficiency; MPC; cabin comfort; air quality

Citation: Schutzeich, P.; Pischinger, S.; Hemkemeyer, D.; Franke, K.; Hamelbeck, P. A Predictive Cabin Conditioning Strategy for Battery Electric Vehicles. *World Electr. Veh. J.* **2024**, *15*, 224. https://doi.org/10.3390/wevj15060224

Academic Editors: Joeri Van Mierlo and Genevieve Cullen

Received: 21 March 2024
Revised: 29 April 2024
Accepted: 14 May 2024
Published: 22 May 2024

1. Introduction

At the UN Climate Change Conference in September 2015, 197 countries committed to limiting global warming to 1.5 °C compared to the pre-industrial age [1]. The steps required to achieve this were set out by the European Union (EU) in its Green Deal. This postulates the goal of reaching climate neutrality by 2050 [2]. Individual countries in Europe have set even stricter targets. The German government wants to achieve greenhouse gas neutrality by 2045. This is linked to the planned reduction of relevant emissions by 65% compared to 1990 levels until 2030 [3]. Many Europeans currently support the measures presented to achieve the targets [4].

Nevertheless, the planned projects will majorly impact mobility in Europe. In addition to revising CO_2 restrictions, investments in charging infrastructure are also intended to boost emission-free mobility in Europe [2,5]. In Germany, electric and fuel cell vehicles are also specifically promoted via an environmental bonus. In addition, the number of public charging points is to be increased to 1 million by 2030. The measures aim to bring up to 15 million fully electric cars onto German roads by 2030 [6].

Even if the development of the registrations for electrified vehicles within Europe (EU-27) looks very positive at first glance, they still fall short of expectations. Figure 1 shows these for the years 2010 to 2021. The number of battery electric vehicles (BEV) and plug-in hybrid electric vehicles has increased from around 7000 in 2011 to around 1.7 million in 2021 for the EU-27 states [7]. Despite the positive development, the share of electric vehicles will have to rise further to meet the EU's self-imposed targets. Customer acceptance and enthusiasm for electrified mobility must be further improved to achieve this. In particular, the range must be increased and charging times shortened to generate new purchasing impulses [8].

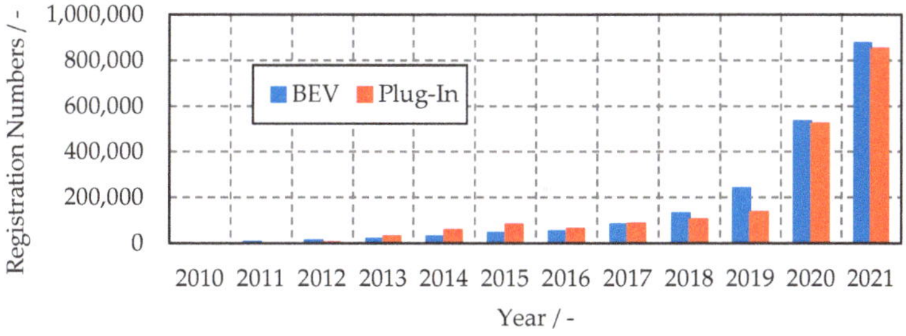

Figure 1. New registrations of electric cars, EU-27 (2011–2021) [7].

In particular, the cabin conditioning of BEVs strongly impacts the achievable driving range under hot and cold ambient conditions. In winter, heating the cabin can cause the achievable driving distance to deviate by up to 50% from the manufacturer's specifications [9–11]. The losses for short trips can increase to up to 70% [12,13]. This is mainly because the energy required must be taken directly from the battery to provide the necessary heating or cooling performance [14,15]. This further underlines the importance of an optimized thermal management strategy for the cabin conditioning of BEVs. The challenge is to minimize the energy required for cabin air conditioning while at the same time meeting the comfort requirements of the occupants.

One way of further developing the control system for heating and cooling the vehicle cabin is to use model predictive control approaches. These allow the control variables for a planned route to be calculated in advance and aim to achieve the desired interior temperature while minimizing energy consumption. A large amount of current research is mainly aimed at reducing energy requirements. Other important aspects, such as indoor air quality, passenger comfort, and humidity, should be considered. Nevertheless, using MPC has made it possible to demonstrate initial energy-saving potential [16–18]. This paper presents a model predictive control strategy for the energy-efficient air conditioning of BEVs. In this approach, the air quality is measured regarding CO_2 concentration, and comfort inside the cabin is ensured by evaluating the equivalent temperature. In addition to the already published investigations [19,20], a method is presented to consider air humidity and thus actively avoid windshield fogging. Comparable studies on model predictive control of cabin conditioning taking into account humidity were also presented in [21]. In contrast to the approaches presented there, using the acados framework [22–24] ensures that the MPC is suitable for vehicle operation. This was already demonstrated in the CEVOLVER project [25–28]. Furthermore, combining the consideration of humid air and the evaluation of the influence of the radiant surfaces through the equivalent temperature (EQT) ensures that many vehicle configurations can be investigated.

This publication focuses on expansion of the MPC by introducing a strategy to prevent window fogging. First, the basic principles of MPC controls are explained. Then, the necessary adaptations for implementing air humidity are presented. Finally, the first investigations on the control strategy are shown and evaluated. This conference paper is based on a contribution to the EVS 36 [29] and was significantly reworked for the publication within this journal. With the additions made, this publication addresses the challenges associated with humid air and shows ways in which these can be overcome efficiently and safely through an MPC strategy.

2. Fundamentals of the Model Predictive Control Approach

In engineering, model predictive control (MPC) is gaining popularity. It offers mathematical solutions for optimizing precisely formulated problems. Different decisions, consequences, and constraints can be considered to determine a suitable operating strategy for a defined system. The objective evaluation criterion is expressed by a cost function J and minimized [30]. This section briefly discusses the basic principles of MPC approaches. Afterward, the scope and specifications of the control strategy are explained.

This paper uses an MPC as an online optimization that determines the required control signals at a defined time step $\Delta\tau_s$. The optimization problem must be set up to satisfy the real-time constraints. This implies that the time between the initiation of the optimization process and the output of the control signals must be correspondingly short to the selected step size $\Delta\tau_s$ [30–32]. The standard formulation of this optimization problem is presented in the following equations:

$$\min_{c} J(c) = \sum_{k=0}^{N-1} l(x(k), u(k), k) \text{ Cost function} \tag{1}$$

under consideration of the following:

$$x(k+1) = f(x(k), u(k), k), k = 0, \ldots, N-1 \text{ System dynamic} \tag{2}$$

$$h(x(k), u(k), k) \leq 0, k = 0, \ldots, N-1 \text{ Inequality constraint} \tag{3}$$

$$G(x(k), u(k), k) = 0, k = 0, \ldots, N-1 \text{ Equality constraint} \tag{4}$$

$$x(0) = x_0 \text{ Initial conditions} \tag{5}$$

The complexity of the stated problem does not allow the optimization to be solved efficiently by analytical methods. For this reason, as is typical for many other technical applications, numerical solution methods are used [30]. A direct solution method was selected, characterized by the fact that the optimization problem is initially discretized on a defined time domain and can then be solved with static optimization methods. This solution method is widely used in MPC applications [30,33,34]. The multi-shooting method was chosen from the direct methods for its high accuracy while being well suited for real-time optimization [30,34]. This solution method includes a discrete dynamic optimization problem whose formulation is shown in Equations (1)–(5) [30]. This specific form of problem formulation is also known as non-linear programming (NLP) and is often used in this context. It considers the starting conditions of the system and the system dynamics, which a mathematical model and the constraints represent. These can be utilized, for example, to limit the state or control variables [35].

Cabin conditioning can be defined as a nonlinear multivariable system under constraints. For these kinds of systems, the nonlinear MPC approach is suitable. Considering the control objectives, the problem is solved on a fixed prediction horizon N_p. These include achieving a target temperature in the interior and minimizing the energy demand in the case of cabin air conditioning. For the entire prediction horizon, the vectors for the control variables are calculated for each time step. However, only the vector with the values for the current time is passed to the system to be controlled. After one time step $\Delta\tau_s$, the controller is reinitialized, and the optimization problem is solved again. It also shifts the prediction horizon at this time, which is why it is also called the sliding prediction horizon [35].

The acados [24] framework was used to implement the MPC in MATLAB Simulink. This approach has already been used for applications of nonlinear MPC in the field of powertrain conditioning, and its functionality has been proven [36].

3. Realization of the MPC Control Strategy in a Model-in-the-Loop (MiL) Environment

After a short introduction to the basics of MPC, this section describes its implementation in MATLAB Simulink. For this purpose, the structure of the model-in-the-loop environment will be discussed first. Subsequently, the system dynamics modeling is explained for the cabin conditioning, and a method to include air humidity is presented.

3.1. Structure of the MiL Environment

The MiL environment consists of three core elements—the plant model, the prediction function, and the MPC. The plant model represents the system behavior outside the control structure and provides important state variables to the MPC and the prediction function. The sensor interfaces replace the plant model during later implementation in the vehicle and are no longer needed. For this purpose, the interface must be adapted accordingly. The plant model contains the same mathematical descriptions as the MPC for the simulative considerations presented in this work. The model was validated using measurement data for a class A vehicle in different environmental conditions.

In the prediction function, the external conditions, such as the weather data, the route information, and the vehicle state, are determined and predicted for the planned trip. The predicted signals are then sent to the MPC along with the associated trajectories for the future. The latter is calculated from the current time step until the end of the prediction horizon. For the current executions, a perfect prediction is assumed. This means the actual system behavior does not deviate from the predicted behavior. This simplification is acceptable for the current evaluation of the potential of the functions. However, the effects cannot be neglected in principle [30–32,35]. This still needs to be investigated before the planned vehicle implementation.

The MPC processes the received information and sets up the optimization problem for the current time step using the acados framework [24]. After the optimization is successful, the control signals for the current time step are forwarded to the plant model.

3.2. Modelling of the System Dynamics—Cabin Conditioning

The MPC approach is mainly characterized by the fact that the system behavior is known and can thus be predicted. It follows that modeling the system dynamics is of the highest importance for the quality of the control strategy. On the one hand, the system's behavior must be reflected as accurately as possible; on the other hand, the computing time must be short; otherwise, the real-time capability cannot be guaranteed. In previous publications [19,20], the basic modeling of the cabin conditioning and the special features have already been discussed. For this reason, these shall be briefly summarized. Subsequently, the extension for the consideration of the air humidity is presented.

The overall objective of the model is to determine the heat-up and cool-down behavior of a vehicle cabin and the energy required. For this purpose, a single-zone model of the vehicle cabin was built. The selection of the modeling approach was based on an evaluation of different possibilities. The most complex but accurate way to describe the system characteristics of a vehicle cabin is through a CFD analysis [37,38]. The significant effort required for data processing and the high computing times make it impossible to use in an MPC control strategy. Multi-zone models are the second option for modeling the system behavior. These are mainly used to evaluate the vehicle cabin's energy consumption and calibrate the control approaches. This method also has computing times that cannot meet the real-time criteria for mathematical optimization [39,40]. A disadvantage of this simulation method is that the spatial resolution of the air flows and the temperature distribution cannot be investigated. This also means that more detailed analyses of occupant thermal comfort cannot be carried out. However, these disadvantages can be accepted, as the main purpose of the models is to determine the energy requirements for cabin air conditioning. The statements conclude that the one-zone models are the most suitable of the standard methods for modeling a vehicle cabin using an MPC. A vehicle cabin from an A-segment vehicle with an internal air volume of approximately 2 m^3 was used for this publication. The integrated interior corresponded to that of a basic model in the vehicle series.

The connection of the thermal masses in the interior is characteristic of the heating and cooling behavior of the air inside a vehicle cabin. Due to the chosen modeling approach, it is impossible to differentiate locally between the individual components (for example, steering wheel, dashboard, seats, etc.) in the simulation. For this reason, all thermal masses of the vehicle cabin are combined into one resulting mass. They are connected to the cabin air via convection with the heat flow $\dot{Q}_{Interior}$. In addition, the total losses via the vehicle shell $\dot{Q}_{Amb}$ and the solar radiation $\dot{Q}_{Solar}$ entering via the window surfaces must also be taken into account in the model. Finally, the incoming and outgoing air mass flow $\dot{m}_{Air,i}$ and the heat flow emitted by the passengers $\dot{Q}_{Pass}$ are accounted for. An overview of the considered boundaries is shown in Figure 2. The validation basis for this model was available measurement data for a class A vehicle.

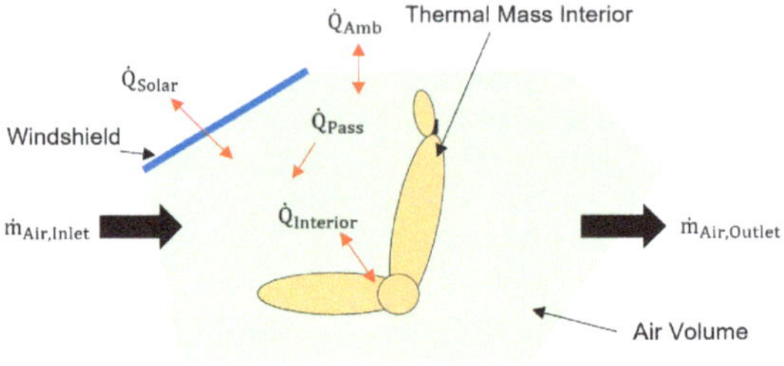

Figure 2. Schematic overview of the simulation model for the air volume of the vehicle cabin.

The representation of the HVAC extended the model. The airflow provided by the HVAC can be taken from the environment (fresh air mode) or the cabin interior (recirculation mode). The user can continuously adjust between the two modes. In addition to the described relationships, the impact of passengers on the cabin in terms of heating and CO_2 production is also considered. The latter influences the air quality. For the investigations in this work, a limit value of 1200 ppm in the interior was selected according to the recommendations in [41]. In the work published, the assumption was made that the air is always dry. For real systems, however, it significantly influences energy demand and comfort [42]. Preventing window fogging is also crucial due to its relevance to safety [21,42]. For this reason, the system modeling was extended by considering air humidity, which is explained in the following section.

In the first step, the state variables for the cabin were extended by modeling the air humidity and the amount of water inside. The water quantities of the incoming and outgoing airflow account for this. In addition, it must be considered that the occupants also emit water into the interior through breathing and sweating. The necessary parameters and formulas were taken directly from [42,43]. The biggest challenge in implementing the humidity was adapting the HVAC model. The active heating and cooling of the airflow directly affects the humidity. For example, cooling the air can cause water to precipitate and thus actively reduce the humidity of the air stream after the subsequent heating [42,43]. The extended functionality of the HVAC model is described in more detail below and visualized in Figure 3.

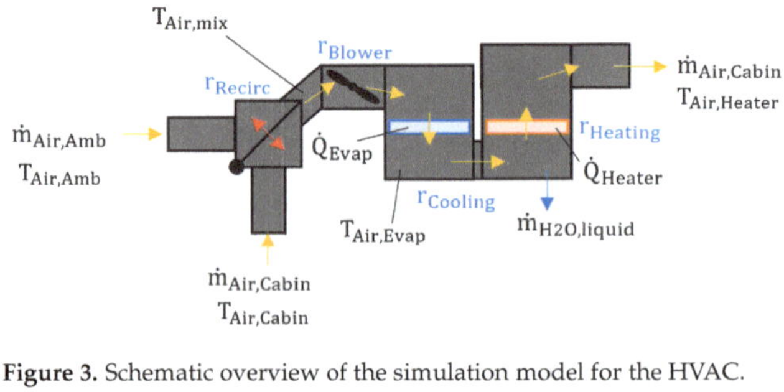

Figure 3. Schematic overview of the simulation model for the HVAC.

The HVAC system determines the conditions of the air entering the cabin. These properties can be adjusted via the control variables, which correspond to each component within the HVAC system. The summary of all control signals for the HVAC is presented in (6). The airflow itself is provided by a blower fan (~300 W), which is actuated by r_{Blower}. The control values for $r_{Cooling}$ and $r_{Heating}$ determine to which extend the air is heated up or cooled down by the system. The modeled vehicle was equipped with a direct air heater with a maximum output of 5 kW. The compressor within the refrigerant circuit for cooling the cabin has a maximum performance of 8 kW. The last variable $r_{Redirect}$ expresses which proportion of the air flow is redirected from the cabin to the HVAC. If the value is $r_{Redirect} = 1$, the cabin conditioning is running in recirculation mode. In contrast, a value of $r_{Recirc} = 0$ indicates a fresh mode operation of the HVAC.

$$u_{HVAC} = \begin{bmatrix} r_{Blower} \; r_{Recirc} \; r_{Cooling} \; r_{Heating} \end{bmatrix} \tag{6}$$

For the simplification of the calculations, the relation between the current air mass flow and the control variable is linearized to the following equation:

$$\dot{m} = \dot{m}_{max} \times r_{Blower} \tag{7}$$

The value $\dot{m}_{max}$ is defined as the maximum mass flow the blower can provide. The energy consumption of the blower P_{Blower} is defined by the following equation:

$$P_{Blower} = P_{Blower,Max} \times r_{Blower}{}^{2} \tag{8}$$

The relationship between the increased mass flow and energy consumption is quadratic due to the kinetic energy [21].

The airflow provided by the blower consists of a mixture of ambient and cabin air, which is determined by the control variable r_{Recirc}. The calculations used to determine the changes in the temperature of the mixed air flows $T_{Air,mix}$ as well as the resulting CO_2 concentration x_{mix,CO_2} and water concentration x_{mix,H_2O} are shown below [21].

$$T_{Air,mix} = \frac{r_{Recirc} \times c_{p,1+x,Cbn} \times T_{Air,Cabin} + (1 - r_{Recirc}) \times c_{p,1+x,Amb} \times T_{Air,Amb}}{r_{Recirc} \times c_{p,1+x,Cabin} + (1 - r_{Recirc}) \times c_{p,1+x,Amb}} \tag{9}$$

$$x_{mix,H_2O} = r_{Recirc} \times x_{Air,Cabin,H_2O} + (1 - r_{Recirc}) \times x_{Amb,H_2O} \tag{10}$$

$$x_{mix,CO_2} = r_{Recirc} \times x_{Air,Cabin,CO_2} + (1 - r_{Recirc}) \times x_{Amb,CO_2} \tag{11}$$

The specific heat capacity $c_{p,1+x,i}$ the heat capacity of humid and the other parameters are determined for either the ambient or the cabin proportion of the air flow [21,43]. All fluid properties for the model were taken from [44] or [45]

In addition, to calculate the values at the evaporator, a distinction was needed between the case with condensation and the case without condensation. Therefore, no direct calculation of the resulting temperature was possible. Instead, in the first step, the enthalpy of the exiting airflow was calculated [43].

$$h_{cool,out} = h_{cool,in} - \frac{\dot{H}_{cool,Max} \times r_{Cooling}}{\dot{m}} \tag{12}$$

The enthalpy of the entering air $h_{cool,in}$ is calculated using the following general formula [43].

$$h_i = c_{p,1+x,i} \times (T_i - T_0) + x_{H_2O,i} \times H_0 \tag{13}$$

To determine if condensation occurs within the evaporator, the water load of the incoming stream $x_{H_2O,in}$ is compared to the maximal water load at the evaporator outlet $x_{H_2O,dew}$ [43].

$$x_{H_2O,out} = \begin{cases} x_{H_2O,in} & \text{if } x_{H_2O,in} < x_{H_2O,dew} \\ x_{H_2O,dew} & \text{else} \end{cases} \tag{14}$$

Using this relationship, the amount of condensing water can also be calculated. It is assumed that the liquid water $\dot{m}_{H_2O,liquid}$ instantly leaves the system and can therefore not be reabsorbed later [43]. The variable $x_{H_2O,dew}$ is dependent on the enthalpy at the evaporator outlet. Using lookup tables, a function can be fitted to approximate the relationship between $x_{H_2O,dew}$ and $h_{cool,out}$ [21].

$$x_{H_2O,dew} = a_1 \times \exp(-a_2 \times h_{cool,out}) + a_3 + a_4 \times h_{cool,out} \tag{15}$$

The temperature of the airflow leaving the evaporator can now be calculated with Equation (18) [43].

$$T_{Air,Evap} = T_0 + \frac{h_{cool,out} - x_{H_2O,out} \times H_0}{c_{p,1+x,i}} \tag{16}$$

After the evaporator, the airflow enters the heater. The temperature changes within the heater are calculated using Equation (17) [43].

$$T_{Air,Heater} = \frac{\dot{h}_{Heating,Max}}{\dot{m}_{max} \times c_{p,1+x,Heating}} \times r_{Heating} + T_{Air,Evap} \tag{17}$$

The evaporator determines the temperature, the concentrations of CO_2, and the amount of water within the airflow, although they typically do not change during the heating process. The airflow properties entering the cabin can be calculated using these calculations. A typical function of the HVAC, which was not implemented in the model version presented, is an additional mixing chamber after the air heater. Together with an air bypass, this allows unconditioned air to be mixed with conditioned air and then enter the cabin. It represents an additional option for achieving the desired humidity and temperature. However, the elimination of this option does not mean a reduction in possible applications and was not featured in the HVAC system of the selected target vehicle. For future applications, this option can be added at any time.

As mentioned before, window fogging shall be prevented at any time. An obstructed view due to condensation on the windshield is a safety-related issue. Therefore, cabin humidity must be kept below the critical humidity for condensation on the windshield. This limit is calculated based on the windshield's temperature and external boundaries. The windshield temperature must be calculated using an approximation based on the heat flow between cabin air and the ambient [21].

For such an evaluation, the windshield heat transfer must be analyzed. As illustrated in Figure 4, the air flows inside and outside the glass area must be considered. The conductive heat transfer through the glass itself must also be calculated. For the condensation on the inside of the windshield, only the inner surface temperature $T_{Win,Internal}$ is relevant. This temperature can be determined with the following formula [21]:

$$T_{Win,Internal} = \frac{(\dot{\alpha}_{Win,External} + \lambda_{Win})\alpha_{Win,Internal} T_{Air,Cbn} + \alpha_{Externall}\lambda_{Win} T_{Air,Amb}}{(\alpha_{Win,External} + \lambda_{Win})\alpha_{Win,Internal} + \alpha_{External}\lambda_{Win}} \tag{18}$$

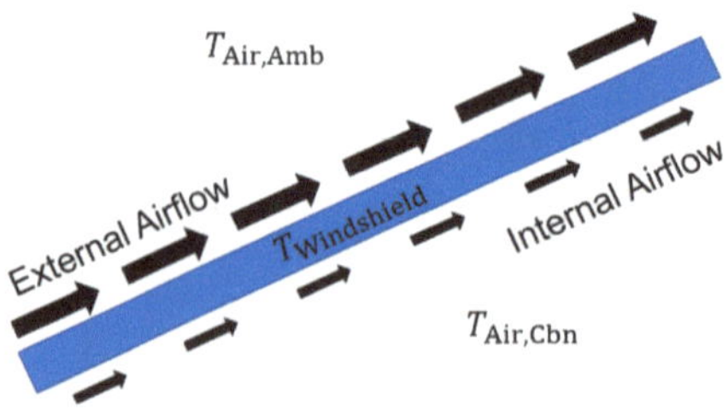

Figure 4. Schematic illustration of the modeling of the windshield temperature.

It uses the conductive heat transfer coefficient for glass λ_{Win} and the convective heat transfer at the outside ($\alpha_{\mathrm{Win,External}}$) and the inside ($\alpha_{\mathrm{Win,Internal}}$) of the windshield. To include the new variable in the optimization, a state variable for the windshield temperature has to be added to the system dynamics model. After that, the proper boundaries of the connected airflows must be represented. Therefore, ambient and cabin airflows define the air velocities inside and outside the windshield. Balancing the heat transfers between the external and internal air is critical to modeling the temperature of the cabin windshield. Comparing the internal surface temperature to the dew point limits the humidity to a level where window fogging is prevented [21].

A more detailed study of the mechanisms behind windshield fogging by Leriche et al. [46] showed that solar radiation is a significant impact factor. However, the influence of solar radiation is neglected in this first approximation. This is possible without risking unexpected condensation since solar radiation can only result in an increased window temperature compared to no solar radiation. Therefore, solar radiation should be addressed in this part of the model. The correlations of the windshield temperatures lead to a critical humidity value, which is then used as a threshold for the MPC strategy [21].

The methods presented here show some adjustments made to account for humidity. In addition, adaptations to the cost functions and the calibration of the models were also necessary. In the first step, limitations of the cabin air humidity were introduced to prevent window fogging. Therefore, the cost function was extended by a term calculating the difference between the dew point temperature at the current cabin conditions and the introduced internal windshield temperature. The difference is involved in a cost function term and calibrated with a scaling factor. As window fogging is relevant to safety, the scaling factor was initially set to a value that prioritizes humidity reduction quite early. In addition, an offset to the temperature differences between the dew point and the glass surface temperature was implemented to have an extra safety margin.

Furthermore, the model was extended by radiant panel heating. It was already demonstrated in [10] that this can save energy during heating. Therefore, it is necessary to evaluate the interior comfort separately in addition to the air temperature. This was fulfilled by the implementation of the equivalent temperature (EQT). The equivalent temperature can be understood as a theoretical evaluation parameter for thermal comfort in a defined environment. It describes the perception of a room climate, considering the air temperature, the effect of radiation, and the air velocity. This is essential to evaluate the influence of radiant heating on indoor comfort. For this purpose, a mean radiant temperature t_{mr} must first be determined (19) [47]:

$$t_{mr} = \sqrt[4]{(t_{sen} + 273.15)^4 + \frac{h_{Conv}}{\varepsilon \sigma_{SB}}(t_{sen} - t_a) + \frac{m_{sen} c_{p,sen}}{\varepsilon \sigma_{SB} A} \frac{dt}{d\tau}} \tag{19}$$

When determining this value, it is important to use a specific sensor and to have precise knowledge of its properties about radiant (ε) and convective (h_{Conv}) heat transfer. Together with the Stefan–Boltzmann (σ_{SB}) constant, the thermal capacity ($m_{sen} c_{p,sen}$), and the surface area (A), such a value can be calculated [47]. The evaluations in [10] show an example of how such a sensor is constructed and what properties it must have. These findings were used with existing measurement data to transfer this relationship in the simulation model.

To determine the equivalent temperature (20), the air velocity (v_a) inside the cabin and a clothing factor I_{cl} for the virtual occupants are also required. The latter describes how well they are insulated against heat transfer. The numerical values for the clothing factor can be taken from the tables in [48], for example.

$$t_{eq} = 0.55 \times t_a + 0.45 \times t_{mr} \frac{0.24 - 0.75\sqrt{v_a}}{1 + I_{cl}} \times (36.5 - t_a) \tag{20}$$

All necessary correlations can be found in [47–50]. Implementing the equivalent temperature to the system dynamics model can also be utilized for the MPC control strategy; by doing that, the cabin can be conditioned to a targeted perception of the environment instead of a fixed air temperature. For a more detailed description of the modeling, please refer to [19,20].

4. Investigations to Determine the Effectiveness of the MPC Approach

This investigation aims to demonstrate the operation of the MPC with humid air. Both cold and hot ambient conditions were considered. The focus should be on compliance with the fogging limits and the impact of humidity on energy consumption. For this purpose, the MPC was compared with a rule-based strategy. The exact boundary conditions are described below before the results are presented.

4.1. Description of the Rule-Based Control Strategy (RB) Used as the Basis for the Assessment

In [21], an operating strategy was presented based on the current state of the art. It can deal with the CO_2 concentration inside the cabin and adjust the recirculation rate accordingly. This was used as a basis and implemented for the comparison carried out. The principles of operation are briefly summarized in the next section.

In the first step, the airflow is controlled based on the deviation of the current cabin temperature from the desired set point temperature. The smaller the deviation, the lower the actuation of the cabin blower. However, a minimum airflow is always ensured. Reducing the air mass flow when the desired temperature is reached ensures stable control. The baseline control strategy aims to operate the system in recirculation mode as long as possible. It is possible to vary the recirculation rate between 0.1 and 0.9 continuously. The maximum value is maintained until the CO_2 limit of 1200 ppm is reached. From this point on, 90% of the air is drawn from the ambient. This means that the system is in fresh air mode. In this phase, the fan speed is also increased again to improve air quality. After reaching a lower threshold of 600 ppm, the system switches back to recirculation mode and resumes the original fan strategy. A PID control ensures the cooling and heating of the airflow.

4.2. Simulation Results at Cold Ambient Conditions

For cold ambient conditions, a heat-up of the cabin at -10 °C was simulated. For the duration, a period of 3600 s was selected. This equals a drive of two consecutive WLTC cycles. The desired temperature in the cabin was set to 20 °C, and as a target for the CO_2 concentration, 1200 ppm was chosen. A deviation from both target values was penalized by the cost function. The temperature set point was a target value for the equivalent temperature. The radiant heating panels were not activated for this first investigations. As additional ambient conditions, a relative humidity of 0.9 and a solar radiation of 0 W/m^2 were assumed.

Due to the different nature of the MPC and the RB strategy, it is not possible to ensure that the same temperature prevails in the interior at all times during the investigation. Both reached an indoor temperature of 18 °C within a short and comparable time. However, it can be seen that the MPC took longer to overcome the remaining 2 °C to reach the control target. Another aspect that stands out when comparing the temperature signals is that the MPC always remained slightly below the targeted 20 °C (control deviation less than 0.3 °C), while the RB approach sometimes exceeded this. This definitely has an influence on the energy and heating demands for cabin air conditioning. However, due to the small differences over time and the relatively low remaining control deviation of the MPC, this is accepted for this study. In future investigations, however, the cost function could be adjusted to achieve the target temperature more precisely, and additional rule-based comparison strategies could be used for the evaluation.

Figure 5 shows an overview of the control signals for the cabin blower (black), the recirculation rate (red), and the heating (blue). The individual signals are normalized to their maximum value and given as a percentage. The values of the MPC strategy (MPC) are shown with solid lines and those of the rule-based strategy (RB) with dashed lines. When examining the blower behavior, it is apparent that the RB variant reduces the fan output and therefore the air mass flow earlier, keeping it at a lower level on average. The recirculation rate for RB is almost always at the maximum value of 0.9, except for the phases in which the system switches to fresh air mode to improve air quality. These are also the only phases in which the fan output is increased slightly. In comparison, the MPC keeps the recirculation rate constant at a very high level. The somewhat higher fan speed ensures the required air quality. It can also be seen that the MPC generally requires less heating power. The heating requirement for the RB strategy is significantly higher, particularly during the fresh air operation phases. Similar observations were also made in [51].

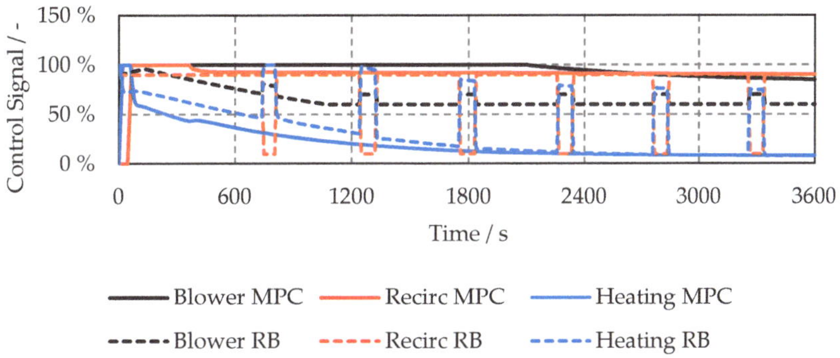

Figure 5. Comparison of the control signals for a cabin heat-up at −10 °C between the MPC and the RB approach.

The control signals' observations are confirmed when analyzing the energy demand for cabin conditioning in Figure 6. It is clear that energy can be saved by using the MPC. Overall, the savings amount to approximately 15.4%. Although more energy is invested in the fan's operation and thus in an increased air mass flow, significant benefits can be achieved by reducing the cooling requirement.

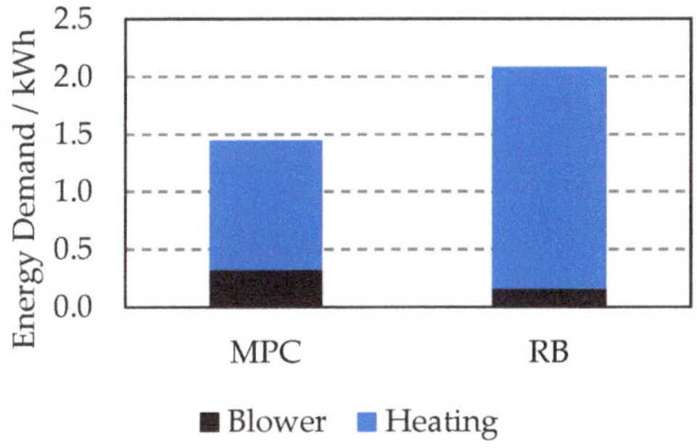

Figure 6. Comparison of the energy demand for a cabin heat-up at −10 °C between the MPC and the RB approach.

Figure 7 shows the CO_2 concentration curves inside the cabin for both strategies. The limit (red) can be adhered to by the MPC (blue) and the RB variant (black) for the majority of the scenario. When the recirculation rate (Figure 5) and the CO_2 concentration (Figure 7) for RB and the MPC are compared, the difference between the two approaches becomes clear. The maximum for the recirculation rate of the MPC is not limited to 0.9 and goes up to 1. The CO_2 concentration at the start rose faster due to the higher recirculation than the rule-based approach. After the limit of 1200 ppm was exceeded, the recirculation rate remained high for a certain period. After that, the MPC slowly reduced it to fall below the limit again. The MPC valued the energy costs at the beginning of the heat-up process higher than maintaining the CO_2 limits. This could be adapted by changing the cost function to meet higher requirements for air quality. Nevertheless, this is not considered efficient, as a slide overrun of the threshold should be classified as not critical for the passengers [41,52]. The CO_2 level for the rule-based approach increased much slower due to the maximum recirculation rate of 0.9. The fresh air mode was activated as soon as the limit was reached. The process was repeated multiple times during the simulation.

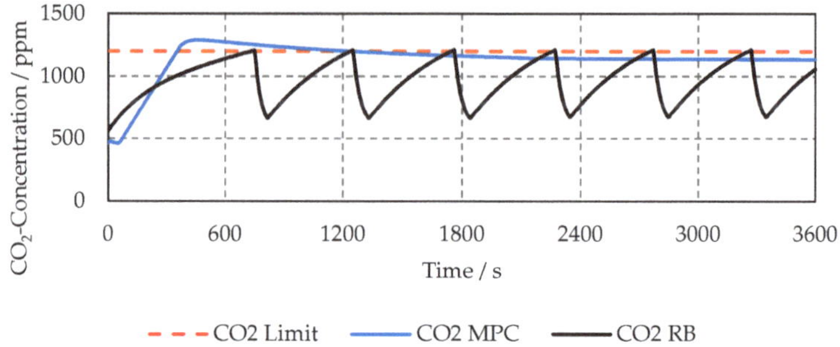

Figure 7. Comparison of the CO_2 concentration for a cabin heat-up at −10 °C between the MPC and the RB approach.

Due to the low response time of the electric air heaters, the air entering the cabin was already quite dry at the start of the simulation. Therefore, the humidity remained below the threshold value, indicating the risk of windshield fogging. In both cases, the humidity in the cabin was reduced quickly. Therefore, no other actions have to be considered. In other use cases, it could be observed that the MPC uses active dehumidification to stay below the humidity threshold. This method is very energy-intensive and only used when other measures cannot prevent window fogging. The results can also be seen in Figure 8. It can be noted that the humidity for the MPC application was slightly lower at the beginning of the simulation. This can be explained by the rule-based approach, obtaining 10% of the airflow from the environment at the start. The ambient humidity was very high. Therefore, the humidity inside the cabin also remained higher at this time. This effect could be neglected if the upper limit for the rule-based approach could also be set to 1. For more critical boundary conditions, this could lead to an increased risk of windshield fogging.

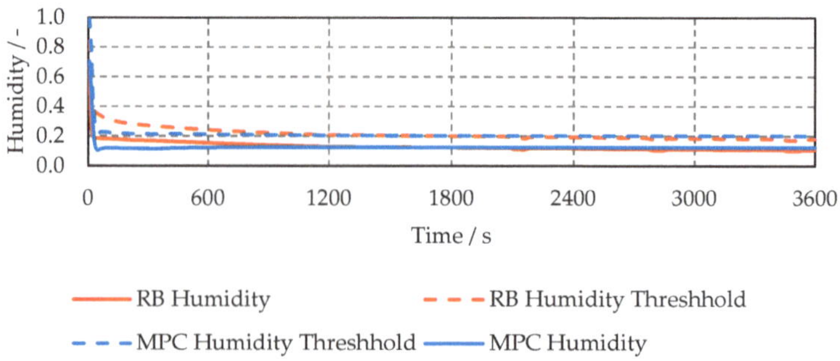

Figure 8. Humidity inside the cabin and upper humidity limit for a cabin heat-up at −10 °C.

4.3. Simulation Results at Hot Ambient Conditions

For hot conditions, an ambient temperature of 35 °C, relative humidity of 0.4, and solar radiation of 800 W/m^2 were chosen. All additional settings and boundaries were kept unchanged compared to the cold conditions. Once again, both control approaches could condition the cabin to the desired target temperature comparably quickly. The average control deviation required was below 0.3 °C for both methods. The differences between the indoor temperatures during the conditioning phase can also be identified for the hot scenario. However, these are even smaller than in the heating operation and are therefore also accepted for this publication.

Figure 9 shows the control signals for the control-based approach (RB) and the MPC for the cabin cooling process at 35 °C outside temperature. Similar to the cabin's heating, it is clear that the MPC approach used a higher fan speed. However, this was accompanied by a constantly higher recirculation rate, resulting in a lower power requirement for cabin cooling. Once again, it is noticeable that the cooling demand for RB rose sharply during the fresh air operation phase. As these tended to become longer in the course of the simulation, they had a significant effect on the system behavior.

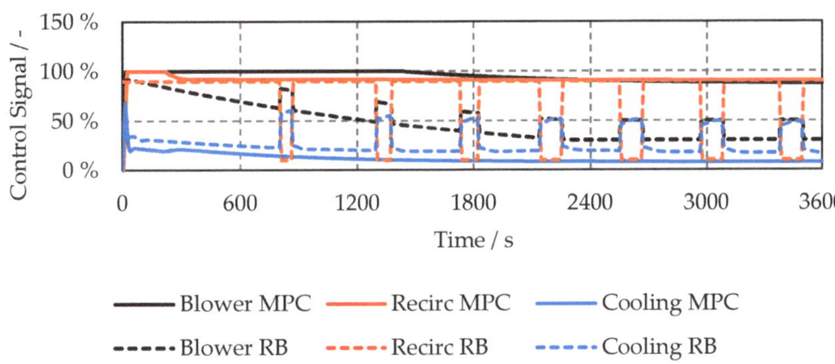

Figure 9. Comparison of the control signals for a cabin cool-down at 35 °C between the MPC and the RB approach.

This is also reflected in the energy demands for cabin air conditioning, illustrated in Figure 10. The MPC also used an increased air mass flow and a higher average recirculation rate for hot ambient temperatures to reduce the energy requirement. The simulation identified a savings potential of 37.9%.

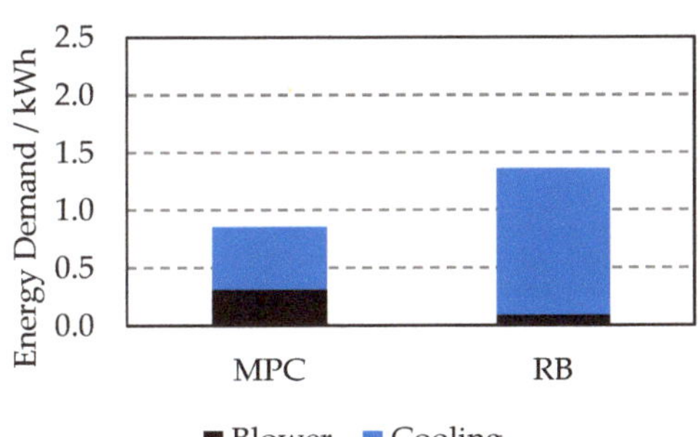

Figure 10. Comparison of the energy demand for a cabin cool-down at 35 °C between the MPC and the RB approach.

Figure 11 shows the CO_2 concentration for the MPC (blue) and the rule-based approach (black) in comparison to the limitation (red). Observing the CO_2 concentration and the respective recirculation rates presented in Figure 9, it becomes evident that the rule-based approach is forced to switch to fresh air mode more often than for cold conditions. This is an additional reason for the additional energy demand for the cabin's heating. Nevertheless, both strategies are capable of maintaining the limitation for the air quality at any time.

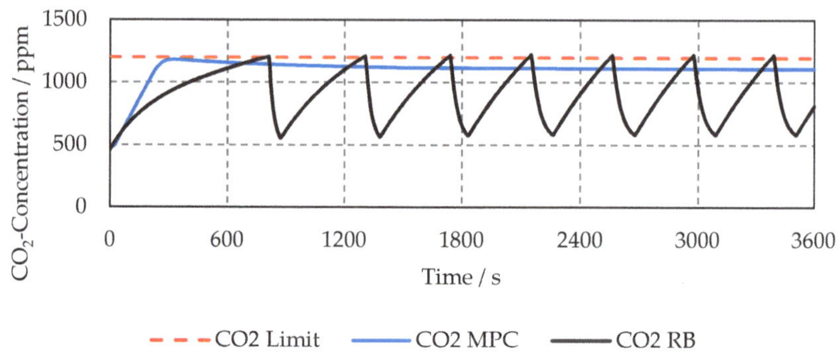

Figure 11. Comparison of the CO_2 concentration for a cabin cool-down at 35 °C between the MPC and the RB approach.

When looking at the humidity (Figure 12), the threshold values for preventing windshield fogging cannot be seen, as they are higher than 1. This seems unconvincing at first sight, as the relative humidity is limited to 1. The values must be interpreted as a theoretical representation of the limit value. It symbolizes that the humidity must theoretically exceed one at the given windshield temperature to cause window fogging. That means it is practically impossible to have window fogging at this status because of the high windshield temperature. In addition, it can be observed that the humidity inside the cabin increases every time the fresh air mode is engaged. Although this does not impact the windshield fogging for the investigated use case, it could be an issue for scenarios with more passengers or critical ambient conditions.

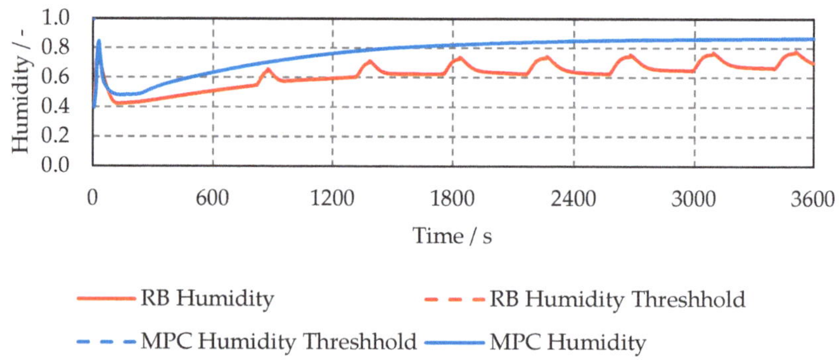

Figure 12. Humidity inside the cabin and upper humidity limit for a cabin cool-down at 35 °C.

5. Summary and Outlook

This paper presented a model predictive approach for the cabin conditioning of BEVs. The focus was on considering the air humidity in the control strategy and preventing windshield fogging. The necessary steps to change the existing system dynamics model were explained. The changes to the calculations within the air path and the implementation of the windshield area as an indication of the risk for window fogging were discussed in detail. It was shown that it is possible to include the air humidity appropriately and thus add a safety level for optimizing the control strategy for cabin conditioning. Also, the impact of air humidity on the energy demand was examined. In the presented MiL investigations, energy-saving potentials of up to 15.4% for cold and 37.9% for hot ambient conditions were achieved.

The presented findings show that the use of an MPC control strategy has a positive effect on the energy consumption of the cabin air conditioning and the thermal comfort in the passenger compartment. By using external information sources for the control strategies, future BEVs can benefit from the growing connectivity. Particularly in the important area of thermal management, MPC strategies can generate an increase in driving range, which can contribute to improving the acceptance of electric mobility.

Future studies should therefore be extended to other areas of thermal management, such as battery conditioning or the operation of complex heat pumps, in order to identify further energy-saving potential. The growing possibilities for creating powerful and fast-calculating models through the advance of machine learning and other methods could be further levers for potentiating the effectiveness of MPC in the future.

Author Contributions: Conceptualization, P.S. and P.H.; methodology, P.S. and P.H.; software, P.S. and P.H.; validation, P.S., K.F. and P.H.; formal analysis, P.S., S.P., D.H., K.F. and P.H.; investigation, P.S., K.F. and P.H.; resources, P.S., K.F. and P.H.; data curation, P.S.; writing—original draft preparation, P.S. and P.H.; writing—review and editing, P.S., S.P., D.H., K.F. and P.H.; visualization, P.S. and K.F.; supervision, S.P. and D.H.; project administration, P.S. and D.H.; funding acquisition, P.S., S.P. and D.H. All authors have read and agreed to the published version of the manuscript.

Funding: This project has received funding from the European Union's Horizon 2020 research and innovation programme under grant agreement 628 No 824295.

Data Availability Statement: The original contributions presented in the study are included in the article, further inquiries can be directed to the corresponding author.

Conflicts of Interest: David Hemkemeyer is employee of FEV Europe GmbH. The paper reflects the views of the scientists, and not the company.

References

1. Bundesministerium für Wirtschaft und Klimaschutz: Abkommen von Paris. Available online: https://www.bmwk.de/Redaktion/DE/Artikel/Industrie/klimaschutz-abkommen-von-paris.html (accessed on 29 February 2024).
2. Europäische Kommission. Umsetzung des europäischen Grünen Deals: Auf dem Weg zu einem Klimaneutralen Europa bis 2050. Available online: https://commission.europa.eu/strategy-and-policy/priorities-2019-2024/european-green-deal/delivering-european-green-deal_de (accessed on 29 February 2024).
3. Bundesregierung. Klimaschutzgesetz: Generationenvertrag für das Klima. 7 November 2022. Available online: https://www.bundesregierung.de/breg-de/schwerpunkte/klimaschutz/klimaschutzgesetz-2021-1913672 (accessed on 29 February 2024).
4. Europäische Kommission. Special Eurobarometer 513—Climate Change: Report. Available online: https://ec.europa.eu/clima/system/files/2021-07/report_2021_en.pdf (accessed on 29 February 2024).
5. Europäische Kommission. Mitteilung der Kommission an das Europäische Parlament, den Rat, den Europäischen Wirtschafts- und Sozialausschuss und den Ausschuss der Regionen: „Fit für 55": Auf dem Weg zur Klimaneutralität—Umsetzung des EU-Klimaziels für 2030. Available online: https://eur-lex.europa.eu/legal-content/DE/TXT/PDF/?uri=CELEX:52021DC0550&from=DE (accessed on 29 February 2024).
6. Bundesregierung. Nachhaltige Mobilität: Nicht Weniger Fortbewegen, Sondern Anders. 23 December 2022. Available online: https://www.bundesregierung.de/breg-de/schwerpunkte/klimaschutz/nachhaltige-mobilitaet-2044132 (accessed on 29 February 2024).
7. European Environment Agency. New Registrations of Electric Vehicles in Europe. Available online: https://www.eea.europa.eu/ims/new-registrations-of-electric-vehicles (accessed on 29 February 2024).
8. Götz, K.; Sunderer, G.; Birzle-Harder, B.; Deffner, J. *Attraktivität und Akzeptanz von Elektroautos: Ergebnisse aus dem Projekt OPTUM. Optimierung der Umweltentlastungspotenziale von Elektrofahrzeugen*; ISOE: Frankfurt am Main, Germany, 2012.
9. Tschöke, H.; Gutzmer, P.; Pfund, T. *Elektrifizierung des Antriebsstrangs: Grundlagen—Vom Mikro-Hybrid zum vollelektrischen Antrieb*; ATZ/MTZ-Fachbuch; Springer: Berlin/Heidelberg, Germany, 2019.
10. Hemkemeyer, D. Thermomanagement im Elektrischen Personenkraftwagen unter Nutzung der Abwärme des Antriebs. Ph.D. Thesis, RWTH Aachen University, Aachen, Germany, 9 October 2017.
11. Beetz, K.; Kohle, U.; Eberspach, G. Beheizungskonzepte für Fahrzeuge mit Alternativen Antrieben. *ATZ Automob. Z.* **2010**, *112*, 246–249. [CrossRef]
12. Allgemeiner Deutscher Automobil-Club E.V. Stromverbrauch von Sitzheizung und Co: Wie Hoch ist er Tatsächlich? Available online: https://www.adac.de/rund-ums-fahrzeug/ausstattung-technik-zubehoer/ausstattung/sitzheizung-verbrauch/ (accessed on 29 February 2024).
13. Rudschies, W. Elektroauto im Winter: So wirkt sich Kälte auf Verbrauch und Reichweite aus. Available online: https://www.adac.de/rund-ums-fahrzeug/elektromobilitaet/info/elektroauto-reichweite-winter/ (accessed on 29 February 2024).

14. Auer, M. *Ein Beitrag zur Erhöhung der Reichweite eines Batterieelektrischen Fahrzeugs durch Prädiktives Thermomanagement*; Springer: Wiesbaden, Germany, 2016.
15. VDI und VDE. Brennstoffzellen- und Batteriefahrzeuge: Bedeutung für die Elektromobilität. Last Updated: 7 June 2019. Available online: https://www.vdi.de/ueber-uns/presse/publikationen/details/brennstoffzellen-und-batteriefahrzeuge (accessed on 27 April 2022).
16. Chen, Y.; Kwak, K.H.; Kim, J.; Kim, Y.; Jung, D. Energy-efficient cabin climate control of electric vehicles using linear time-varying model predictive control. *Optim. Control Appl. Methods* **2023**, *44*, 773–797. [CrossRef]
17. Wang, H.; Meng, Y.; Zhang, Q.; Amini, M.R.; Kolmanovsky, I.; Sun, J.; Jennings, M. MPC-based Precision Cooling Strategy (PCS) for Efficient Thermal Management of Automotive Air Conditioning System. In Proceedings of the 2019 IEEE Conference on Control Technology and Applications (CCTA), Hong Kong, China, 19–21 August 2019; pp. 573–578.
18. Liu, Y.; Zhang, J. Electric Vehicle Battery Thermal and Cabin Climate Management Based on Model Predictive Control. *J. Mech. Des.* **2021**, *143*, 031705. [CrossRef]
19. Manns, P.; Hemkemeyer, D.; Linse, D. Predictive Cabin Climatization for Electric Vehicles. *ATZ Worldw.* **2022**, *124*, 36–39. [CrossRef]
20. Schutzeich, P.; Pischinger, S.; Hemkemeyer, D.; Wahl, A.; Franke, K. A Model Predictive Control Strategy for Advanced Passenger Compartment Air Conditioning in Vehicles with Electrified Powertrains. In Proceedings of the 2022 SAE Thermal Management Systems Symposium, Plymouth, MI, USA, 4–5 October 2022.
21. Schaut, S.; Sawodny, O. Thermal Management for the Cabin of a Battery Electric Vehicle Considering Passengers' Comfort. *IEEE Trans. Control Syst. Technol.* **2020**, *28*, 1476–1492. [CrossRef]
22. Verschueren, R.; Frison, G.; Kouzoupis, D.; Frey, J.; Van Duijkeren, N.; Zanelli, A.; Novoselnik, B.; Albin, T.; Quirynen, R.; Diehl, M. acados—A modular open-source framework for fast embedded optimal control. *Math. Program. Comput.* **2022**, *14*, 147–183. [CrossRef]
23. Systems Control And Optimization Laboratory. acados—Documentation. Last Updated: 2 January 2024. Available online: https://docs.acados.org/ (accessed on 29 February 2024).
24. Systems Control And Optimization Laboratory. acados—Embedded Workflow. Available online: https://docs.acados.org/embedded_workflow/index.html (accessed on 29 February 2024).
25. Cevolver Project Consortium. CEVOLVER: Connected Electric Vehicle Optimized for Life, Value, Efficiency and Range. Available online: https://cevolver.eu/ (accessed on 29 February 2024).
26. Cevolver Project Consortium. Integrated Energy & Thermal Management Validator: Poster. Available online: https://cevolver.eu/wp-content/uploads/2022/10/CEVOLVER_Poster_Validator1_FINAL.pdf (accessed on 29 February 2024).
27. Schutzeich, P.; Wahl, A. CEVOLVER—Thermal Management Improvements: Poster. Available online: https://cevolver.eu/wp-content/uploads/2022/10/CEVOLVER_Final_Event_Poster_Thermal_Management_FINAL.pdf (accessed on 29 February 2024).
28. Wahl, A.; Schutzeich, P. CEVOLVER—Advanced Thermal Management: An Enabler of Long Distance Capabilities? Available online: https://cevolver.eu/wp-content/uploads/2022/10/Thermal_Management_in_CEVOLVER_Final_Event_V8_compressed.pdf (accessed on 29 February 2024).
29. Schutzeich, P.; Hemkemeyer, D.; Franke, K.; Hamelbeck, P. A Predictive Cabin Conditioning Strategy for Battery Electric Vehicles. 36th International Electric Vehicle Symposium and Exhibition (EVS36). Available online: https://evs36.com/wp-content/uploads/finalpapers/FinalPaper_Schutzeich_Patrick.pdf (accessed on 13 December 2023).
30. Papageorgiou, M.; Leibold, M.; Buss, M. *Optimierung: Statische, Dynamische, Stochastische Verfahren für die Anwendung*; Springer: Berlin/Heidelberg, Germany, 2015; pp. 343–427.
31. Adamy, J. *Nichtlineare Systeme und Regelungen*; Springer: Berlin/Heidelberg, Germany, 2018; pp. 227–280.
32. Schwarzkopf, S. Echtzeitfähige Optimierungsbasierte Regelung von Stofftrennprozessen. Magdeburg, Otto-von-Guericke-Universität Magdeburg, Fakultät fur Elektrotechnik und Informationstechnik. Dissertation. 16 October 2012. Available online: https://d-nb.info/1054135266/34 (accessed on 24 January 2023).
33. Völz, A. *Modellprädiktive Regelung Nichtlinearer Systeme mit Unsicherheiten*; Springer: Wiesbaden, Germany, 2016.
34. Diehl, M.; Bock, H.G.; Schlöder, J.P. A Real-Time Iteration Scheme for Nonlinear Optimization in Optimal Feedback Control. *SIAM J. Control Optim.* **2005**, *43*, 1714–1736. [CrossRef]
35. Graichen, K. Methoden der Optimierung und optimalen Steuerung: Skriptum (Wintersemester 2012/2013). Last Updated: 5 August 2019. Available online: https://vdocuments.site/skriptum-methoden-der-optimierung-und-optimalen-steuerung-c-prof-dr-ing.html?page=1 (accessed on 25 January 2023).
36. Wahl, A.; Wellmann, C.; Krautwig, B.; Manns, P.; Chen, B.; Schernus, C.; Andert, J. Efficiency Increase through Model Predictive Thermal Control of Electric Vehicle Powertrains. *Energies* **2022**, *15*, 1476. [CrossRef]
37. Danca, P.; Bode, F.; Nastase, I.; Meslem, A. CFD simulation of a cabin thermal environment with and without human body—Thermal comfort evaluation. *E3S Web Conf.* **2018**, *32*, 1018. [CrossRef]
38. Chang, T.-B.; Sheu, J.-J.; Huang, J.-W.; Lin, Y.-S.; Chang, C.-C. Development of a CFD model for simulating vehicle cabin indoor air quality. *Transp. Res. Part D Transp. Environ.* **2018**, *62*, 433–440. [CrossRef]
39. Pathuri, R.; Patil, Y.; Nagarhalli, P.V. Deployment of 1D Simulation with Multi Air Zone Cabin Model for Air Conditioning System Development for Passenger Car. SAE Technical Paper Series. In Proceedings of the Symposium on International Automotive Technology 2015, Warrendale, PA, USA, 21–24 January 2015.

40. Poovendran, K.; Abel, D.; Reuscher, T.; Govender, V. Vehicle Cabin Thermal Multi-Zone Modelling for Control. In Proceedings of the 2020 2nd International Conference on Control Systems, Mathematical Modeling, Automation and Energy Efficiency (SUMMA), Lipetsk, Russia, 11–13 November 2020; pp. 489–495.
41. Angelova, R.A.; Markov, D.G.; Simova, I.; Velichkova, R.; Stankov, P. Accumulation of metabolic carbon dioxide (CO_2) in a vehicle cabin. *IOP Conf. Ser. Mater. Sci. Eng.* **2019**, *664*, 12010. [CrossRef]
42. Großmann, H.; Böttcher, C. *Pkw-Klimatisierung: Physikalische Grundlagen und Technische Umsetzung*; Springer: Berlin/Heidelberg, Germany, 2020.
43. Lucas, K. *Thermodynamik*; Springer: Berlin/Heidelberg, Germany, 2008.
44. Kretzschmar, H.-J.; Wagner, W. D2.1 Thermophysikalische Stoffwerte von Wasser. In *VDI-Wärmeatlas*; Stephan, P., Kabelac, S., Kind, M., Mewes, D., Schaber, K., Wetzel, T., Eds.; Springer: Berlin/Heidelberg, Germany, 2019; pp. 201–218.
45. Ingenieure, V.D. *VDI-Wärmeatlas*; Springer: Berlin/Heidelberg, Germany, 2013.
46. Leriche, M.; Roessner, W.; Reister, H.; Weigand, B. Numerical Investigation of Droplets Condensation on a Windshield: Prediction of Fogging Behavior. SAE Technical Paper Series. In Proceedings of the SAE 2015 World Congress & Exhibition, Detroit, MI, USA, 21–23 April 2015.
47. *DIN EN ISO 14505-2:2007-04*; Ergonomie der Thermischen Umgebung_-Beurteilung der Thermischen Umgebung in Fahrzeugen_-Teil_2: Bestimmung der Äquivalenttemperatur (ISO_14505-2:2006); German Version EN_ISO_14505-2:2006. ISO: Geneva, Switzerland, 2007.
48. Ye, G.; Yang, C.; Chen, Y.; Li, Y. A new approach for measuring predicted mean vote (PMV) and standard effective temperature (SET∗). *Build. Environ.* **2003**, *38*, 33–44. [CrossRef]
49. Shaw, E.W. Thermal Comfort: Analysis and applications in environmental engineering, by P. O. Fanger. 244 pp. DANISH TECHNICAL PRESS. Copenhagen, Denmark, 1970. Danish Kr. 76, 50. *R. Soc. Health J.* **1972**, *92*, 164. [CrossRef]
50. *ANSI/ASHRAE Standard 55-2023*; Thermal Environmental Conditions for Human Occupancy. American Society of Heating, Refrigerating and Air-Conditioning Engineers, Inc., American National Standards Institute: Peachtree Corners, GA, USA, 2023.
51. Arndt, M.; Sauer, M.; Wolz, M. Verbrauchssenkung durch verbesserte Klimaanlagen-Regelung. *ATZ Automob. Z.* **2017**, *109*, 404–410. [CrossRef]
52. Zhang, X.; Wargocki, P.; Lian, Z.; Thyregod, C. Effects of exposure to carbon dioxide and bioeffluents on perceived air quality, self-assessed acute health symptoms, and cognitive performance. *Indoor Air* **2017**, *27*, 47–64. [CrossRef] [PubMed]

World Electric Vehicle Journal

MDPI

Article

On the Aggregation and Monetization of Flexible Loads in the Context of EV Fleets

Kelaja Schert [1,*,†,‡], Florian Biedenbach [2], Thomas Müller [3], Michael Kluge [4] and Zoltán Nochta [5,‡]

1 SAP, Hasso-Plattner-Ring 7, 69190 Walldorf, Germany
2 FFE, Am Blütenanger 71, 80995 Munich, Germany; fbiedenbach@ffe.de
3 Coneva, Paul-Heyse-Straße 2-4, 80336 München, Germany; thomas@spine.energy
4 nextmove, Bouchéstraße 12, 12435 Berlin, Germany; m.kluge@nextmove.de
5 Department of Computer Science, Karlsruhe University of Applied Sciences, Moltkestraße 30, 76133 Karlsruhe, Germany; zoltan.nochta@h-ka.de
* Correspondence: k.schert@sap.com
† Current address: SAP, Dietmar-Hopp-Allee 16, 69190 Walldorf, Germany
‡ These authors contributed equally to this work.

Abstract: In this paper, we present an approach to the price-optimized charging of electric vehicles (EVs) based on energy flexibility. Fleet operators determine the minimum and the maximum power demand to charge EVs at a specific time and share this information as so-called power corridors (PCs) with an energy aggregator. The energy aggregator collects the predicted PCs from the fleet operators located in the same market area and aggregates the PCs. The energy provider periodically sends energy prices from the market to the energy aggregator, which purchases energy when its price is opportune. The energy aggregator calculates and delivers charge plans for each fleet operator involved and thus can pass along the purchase prices. The incentive design must ensure that fleet operators are better off by disclosing their flexibility data to the aggregator. This study can contribute to a new data-driven energy market communication system by providing insights on how to leverage the energy flexibility that EVs can offer to the energy system.

Keywords: EV; energy; optimization; smart charging; aggregator; flexibility

Citation: Schert, K.; Biedenbach, F.; Müller, T.; Kluge, M.; Nochta, Z. On the Aggregation and Monetization of Flexible Loads in the Context of EV Fleets. *World Electr. Veh. J.* **2024**, *15*, 216. https://doi.org/10.3390/wevj15050216

Academic Editors: Joeri Van Mierlo and Genevieve Cullen

Received: 4 December 2023
Revised: 18 March 2024
Accepted: 25 April 2024
Published: 14 May 2024

1. Introduction

In 2020, the road transport sector was responsible for 11.9% of greenhouse gas emissions worldwide [1]. To combat human-made climate change, a reduction in these emissions is urgently necessary. One possible strategy to reduce these emissions is the electrification of this sector, resulting in a yearly electrical energy demand of several hundred GWh in Europe [2]. Due to the generally high idle times of passenger cars, this total demand can be flexibly shifted. The charging processes can be scheduled when energy from volatile, renewable energy sources is available or when electricity prices are low. However, the question remains of how this can be implemented in practice. The major challenges are the determination of energy flexibility that fleet operators can offer and the optimization of the EV charging process according to the objective. Current research shows that existing policies of many countries prevent innovative approaches for flexibility trading [3]. Smart charging, i.e., advancing charging processes to times when electricity prices are low or renewable energy is available, is a common approach to running managed charging infrastructures. There are publications that examine smart charging on a theoretical [4–6] and practical basis [7]. The authors of [5] predict potential cost savings of 200 EUR/EV/year if smart charging based on variable prices is applied. Approaches to avoid over-coordination and herding effects have been discussed in the literature on price-based EV charging coordination [8]. One such approach, proposed by [9], involves spatial price differentiation to effectively incorporate distribution grid limitations into charging schedules. Another study by [10] emphasizes potential cost savings achieved by smart EV charging and the ability to feed energy back into the power grid (vehicle to grid, V2G). Various research projects have been working on the aggregation of vehicle fleets' energy consumption to charge them in a price-optimized way. The projects BDL and LamA in Germany can be mentioned [11] as examples. V2X Suisse is another example project in Switzerland [12]. Apart from research institutions, various companies are working on the development of commercial solutions for smart charging. Octopus Energy, for example, has implemented smart charging based on variable electricity tariffs for its customers in the UK using its platform Kraken [13]. The company enel X developed a platform-based solution for smart charging [14]. However, the aggregation process used by these companies is not transparent, and the solutions are proprietary. Open systems are not in the focus of related work.

This paper shows how power demand aggregation can be achieved and how it can be implemented independently of proprietary systems. Improving electrical fleet performance requires a clear objective and measurable variables. The concept of flexibility in general is considered domain-specific and thus difficult to define. In the case when systems should adapt to an external environment, like in our case, adapting the EV fleet to the price of energy, they can adapt better if the variables include flexibility in one or more dimensions [15].

Energy flexibility in our paper is considered as the possibility to adapt the power demand over time. Other definitions for energy flexibility are characterized by static approaches, considering the composition of parameters at a given time instant [16]. Approaches toward a dynamic flexibility function to control demand with penalty signals [15] are a common way to influence consumption behavior and propagate the paradigm shift toward a demand control energy system. The critics argue that penalty-based flexibility indexes depend on the interpretation of the energy providers. These improve their objectives with regard to CO_2 emissions or real-time prices without considering the actual amount of energy demanded by the consumers. Our approach presented in this paper is based on a bidirectional communication and data exchange between fleet operators, energy aggregators, and energy providers. Based on the information that the energy provider receives from energy suppliers and the grid operators, like market energy prices and grid peak times, the aggregated energy orders are being optimized. The goal is to better manage the overall energy and power demand of fleet operators by actively reacting to day-ahead and intraday market prices. This is realized by increasing and decreasing the fleet consumption over the day by controlling the individual charging sessions attuned. The availability of data is the key enabler for our approach to improve power-corridor predictions and the basis for a level playing field for exchanging flexible services between EV fleet operators and energy providers. Our research focuses on the utilization of information to improve the charging processes and costs of commercial EV fleet operators. For this purpose, we address the following research questions:

- What is the optimized usage of EVs in different scenarios like company fleets or rental fleets?
- How can our definition of the power corridor help optimize the energy consumption of EV fleets?
- What are the processes and algorithms required to aggregate and monetize flexible loads of EV fleets?
- What data need to be made available and by whom to feed the algorithms?
- What is required so that our results have an impact on the existing energy landscape?

2. Materials and Methods

2.1. Project Setup

A major goal of the project "*TRADE EVs II*" was to define a framework for addressing the above-mentioned questions. The project, with a duration of three years, was initiated by Elektrizitätswerke Schönau (EWS), Forschungsstelle für Energiewirtschaft e.V. (FFE), nextmove, and SAP in 2021. It involved three fleets with more than 400 EVs driven by employees of the project partners. The project built on the experience and results gained in the predecessor project, *"TRADE EVs I"* (TRADE EVs I, funding code: 01MX16002C), in which a charge schedule heuristic was deployed to optimize energy consumption [16] and a charging system prototype based on Open E-Mobility [17] was set up. TRADE EVs II extends the setup with an energy–flexibility aggregation system to establish the demand-side management for EV charging. In the project, we assessed two approaches for capturing EV data: the hardware-based approach used onboard units, and the software-based solution utilized telemetry services. Based on the accessible EV data, the charging system calculates the energy demand within the respective charging period. The data points considered are, for example, the state-of-charge (*SoC*), the battery model, and the charging priority of EVs.

The project was divided into two main work-streams called Concept and Application, as shown in Figure 1. The conceptual work started with the definition of use cases for controlled charging. The focus was thereby set on the use case of spot-market-optimized charging, in which charging processes are influenced by the current electricity spot-market prices. Subsequently, the concept was extended by integrating it with day-ahead markets, which resulted in the design of an aggregation algorithm and the interfaces required to establish a market communication process.

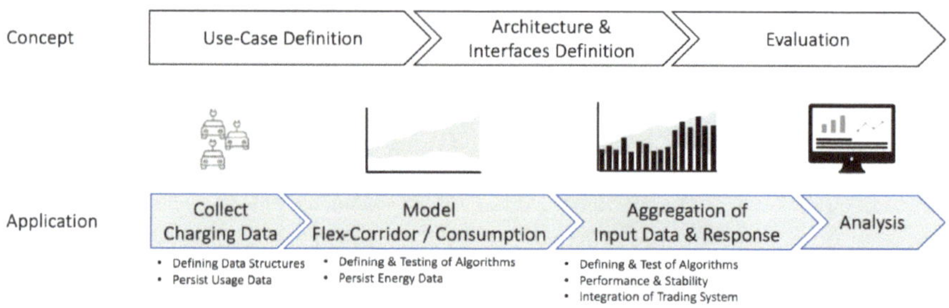

Figure 1. Sequence of project steps.

The application workstream started with the collection of charging data from the participating EVs. We developed a method to determine the flexibility of energy and power consumption of the EV fleets, which we termed "power corridor". In addition, we developed an algorithm to aggregate data about energy demand from different fleets and EV charging sites and to exchange price-related information.

2.2. Definitions and Basics

We assume that only the unidirectional charging of EVs is possible in the system. Hence, the power demand $P \geq 0$ holds at any point in time and energy consumption $E \geq 0$ for any time interval. For the mathematical modeling, we introduce the specific terms "power corridor" *PC*, "energy segment" *ES*, and "energy demand" *ED*. The charging system C can serve n ($n \epsilon \mathbb{N}$) electrical vehicles at the maximum (e.g., limited by the number of installed connectors). Accordingly, at any point of time t, k_t ($0 \leq k_t \leq n \mid k_t \epsilon \mathbb{Z}$) vehicles are supposedly connected. For example, in practice, it could be of interest to know or predict the number of charging EVs at C every 15 min. The connected (i.e., charging) vehicles are denoted as v_i ($i \leq k \mid i \epsilon \mathbb{N}$).

${P_{\min}}^t$ is the minimum power required by C to charge all connected EVs at time t (Equation (1)). Pausing/stopping all charging sessions at time t is equal to $P^t_{\min} = 0$ kW. Note that in practice, unused charging stations, e.g., while in stand-by-mode, could still draw power and consume energy:

$$ {P_{\min}}^t = \sum_{i=1}^{k} {P_{v_i}}^t \mid \min\left(P^t_{v_i}\right). \quad (1) $$

${P_{max}}^t \geq {P_{min}}^t$ is the maximum power that can be consumed by C while charging all connected vehicles at time t (Equation (2)). Note that ${P_{max}}^t$ can basically be limited by the connected EVs' aggregated maximum power demand to charge batteries but also by infrastructure restrictions at C, such as transformer capacity, fuse hierarchies, etc.

$$ {P_{max}}^t = \sum_{i=1}^{k} {P_{v_i}}^t \mid \max\left(P_{v_i}^t\right). \quad (2) $$

The power corridor PC^t is defined as a set of tuples that contain the maximal consumption power ${P_{max}}^t$ and the minimum required power ${P_{min}}^t$ of C at specific points in time (Equation (3)):

$$ PC^t = \left({P_{min}}^t, {P_{max}}^t\right). \quad (3) $$

The energy segment ES is defined as the maximum amount of energy, given the maximum and minimum power over time, P_{max}^t and P_{min}^t, that can be consumed within the time interval t_s (start) and t_e (end):

$$ ES = \int_{t_s}^{t_e} {P_{max}}^t - {P_{min}}^t dt \, , ES \in \mathbb{R}_0^+. \quad (4) $$

The energy demand ED_{v_i} foreseen for vehicle v_i is defined as the difference between the required SoC at departure SoC_{req} and the initial SoC upon arrival SoC_{start} within the time interval from connecting t_{si} and disconnecting t_{ei} the vehicle v_i. Note that the SoC is measured in kWh:

$$ ED_{v_i} = \left(SoC_{req}^{t_{ei}} - SoC_{start}^{t_{si}}\right). \quad (5) $$

The total energy demand ED of the charging infrastructure C within the time interval $[t_s, t_e]$ is calculated as the accumulated demands ED_{v_i} of the vehicles v_i that are connected to C:

$$ ED = \sum_{i=1}^{k} ED_{v_i} \mid t_{si}, t_{ei} \in [t_s, t_e]. \quad (6) $$

Figure 2 shows an example power corridor for charging a single EV. The EV is expected to be connected to the charging system between start time t_s and end time t_e. Within this time range, the required amount of energy for charging can be consumed, depicted as "Energy Demand" (in green). The illustrated power corridor defines boundaries of power that can be drawn by the EV during its stay. As the exemplary corridor has static P_{min} and P_{max} values at each point of time, the energy segment (in blue) has the shape of a regular rectangle. This would allow the fleet operator to delay (shift) the start of actual charging, as shown in Figure 2, depending on, e.g., the actual price of energy.

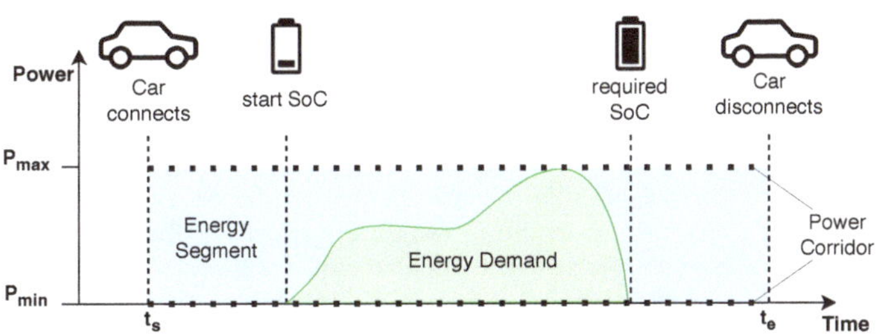

Figure 2. Schematic illustration of a power corridor. The power demand for charging the EV's battery starting by SoC_{start} to the required SoC_{req} level can be set between P_{min} and P_{max} within the time interval t_s and disconnection t_e of the EV.

As shown in Equation (7), the amount of demanded energy ED of C must be between 0 and ES for any time interval of interest. Otherwise, the charging demand cannot be fulfilled.

$$0 \leq ED \leq ES \; for \; [t_s, t_e]. \tag{7}$$

Based on (predicted or otherwise known) ES and ED values in a given scenario and situation, we consider the operator's Flexibility F of shifting demand as

$$F = \begin{cases} \frac{ES-ED}{ES} & \text{if } ES \neq 0 \\ 0 & \text{if } ES = 0 \vee ED = 0. \end{cases} \tag{8}$$

Accordingly, $F = 0$ if $ED = ES$ holds. The flexibility is increasing if

$$0 < ED \wedge ED < ES \longrightarrow F > 0 \; for \; [t_s, t_e]. \tag{9}$$

Table 1 shows two example calculations of two energy demands, $ED1$ and $ED2$, distributed over a seven-hour energy segment ES, accumulated from ES^t for each hour. The power of the charging sessions can be adapted dynamically. In both cases, the energy segment is $ES = 104$ kWh. In case $F = 0$, the power limit 30 kW of the infrastructure is the restricting factor at $t = 3$ and $t = 4$, so $F = 0$ because $ES < ED$. In the case $F = 0.33$, the energy demand ED_{vi} of the vehicles v_i is lower, so $ED < ES$ applies.

Table 1. Example illustration of two cases of how flexibility is calculated based on the given PC [P_{min}, P_{max}] in kW, energy segment ES in kWh, two different energy demands $ED1$ and $ED2$ in kWh, and flexibility F. The infrastructure has a power limit of 30 kW.

	Time							***ES***	***ED1***	***ED2***
$PC^t{}_{v_i}$	**1**	**2**	**3**	**4**	**5**	**6**	**7**		$F = 0$	$F = 0.33$
$PC^t{}_{v_1}$	[0, 0]	[0, 0]	[11, 11]	[11, 11]	[11, 11]	[0, 11]	[0, 11]		44	34
$PC^t{}_{v_2}$	[0, 0]	[0, 11]	[0, 11]	[0, 11]	[0, 11]	[0, 0]	[0, 0]		30	20
$PC^t{}_{v_3}$	[0, 11]	[0, 11]	[0, 11]	[0, 11]	[0, 0]	[0, 0]	[0, 0]		30	16
PC^t	[0, 11]	[0, 22]	[11, 30]	[11, 30]	[11, 22]	[0, 11]	[0, 11]			
ES^t	11	22	19	19	11	11	11	104	104	70

Energy segments ES forecasted with long timeframes hence hold a larger flexibility potential than ES with short timeframes and might be of substantial value for energy providers to realize demand-side management. The interface for exchanging this flexibility information is the precondition to create insights into how charging can be improved to save costs by grid-friendly operation.

Equations (7)–(9) are valid under the conditions that $P, ED, ES \geq 0$. By including renewable energy sources and bidirectional charging into the mathematical model, there is also the negative flexibility case imaginable if the energy demand is $ED < 0$:

$$0 > ED \wedge ED > -ES \longrightarrow F > 0 \; for \; [t_s, t_e], \tag{10}$$

$$PC = -ED \vee ED = -ES \longrightarrow F = 0 \; for \; [t_s, t_e]. \tag{11}$$

Other definitions of energy flexibility focus on the responsiveness of consumer behavior to signals like CO_2 intensity or the energy price. For example, they define a dynamic flexibility function to evaluate consumer behavior and how they react to the real-time energy situation. The calculated flexibility index can be used to apply penalties to influence the behavior of the consumers [15]. Our approach, in contrast, focuses on the transparent communication of energy demands and the power consumption the fleet operators are able to adjust for time. This enables the energy provider to allocate and plan the consumption and allows the aggregated fleets to receive the demanded power and energy by adapting consumption plans within their self-defined possibilities.

2.3. Challenges

Besides difficulties in predicting a fleet's energy consumption, the forecasting of local energy supply—especially for renewable energies—comes with challenges as well. This is partly due to analog measuring technology (missing digital data) and weather influences on energy generation. On the other hand, it is also due to static electricity tariffs, which cannot reflect the share of renewables and conceal information about the consumed energy. Providing dynamic tariffs can motivate fleet operators to shift demands and improve the sustainable charging behavior of self-interested charge point operators. In our setup, fleet operators need to specify the extent to which their power demand is defined with the PC, and the energy demand ED for the EVs. This holds another challenge because rational participants cannot be expected to prioritize the performance of the system over their own interests. Therefore, it is crucial to establish incentives that encourage the revelation and provision of flexibility among the participants. The incentive design must ensure that all are better off by disclosing their flexibility data, which means that they should receive benefits for revealing their information compared to withholding it. This allows the participants to adapt their behavior more flexibly while maximizing their utility. Ultimately, to ensure everyone's participation in the mechanism, it is essential to guarantee individual rationality, as well as the appropriate incentive and coordination mechanisms [15]. Data availability is the basis for improving the forecasting quality of the PC, as seen in the manufacturing industry, wherever even minor process adjustments can generate substantial value [18]. Slight variations in the power system's flexibility can also have a significant impact on economic results. To make the most of this flexibility, it is essential to have a clear understanding of the available flexibility resources.

2.4. Implementation Approach

Addressing the challenges according to flexible energy demand, we evaluate three different controlling scenarios, one for each of the three fleet types, small company fleet, rental car fleet, and large company fleet. All scenarios interact with the central aggregation system. The aggregator system transfers information between consumption facilities, generation facilities, and authorized market partners to generate value via the deliberate placement of energy purchase orders influenced by the different interests of the actors. Figure 3 shows the flow of actions that are conducted on a daily basis. The value is generated by the allocation of the forecasted energy demand within the flexible time range of the three consumers. With the incentive to charge when energy prices are low, the overall energy costs should be lowered.

In the first scenario, a smaller fleet with 15 EVs of the German energy provider EWS is involved. The EVs can use 10 AC charging points located at a company parking space. Each charge point (CP) is managed solely by its charging controller, which only communicates with the charging EV. In this scenario, the total load is set by the consumption of the EVs connected to the charging stations onsite. The forecast of the demanded charging energy at the site is trained based on the consumption data from the EV charging sessions on a daily basis. The prediction functions were continuously applied to increase the overall accuracy of the charging forecasts, for example, if new charging points and EVs are connected. EV drivers are aware that the charging session can be shifted to different timeslots during the parking period to avoid charging during price peaks.

The second scenario is the load-management scenario at nextmove, which has implemented peak shaving to operate more charge points in sequence than would be possible in parallel. The limitation of the connected load and local energy shortages have also been considered. The nextmove dataset has been provided from a rental fleet that contains 320 EVs of different usage types, such as business, private, and test drives. Currently, the fleet consists of 245 midsize battery EVs (35 kWh up to 64 kWh) and 75 large battery EVs (up to 120 kWh). The journeys were planable, and especially the business customers used the cars for frequent traveling. Most drivers use the rental to test an EV before buying it, which includes pushing it to its limits. For example, we observed that at the beginning of the rental period the *SoC* is much lower when the first charging session starts compared to the other charging sessions for the rest of the rental period. Within this scenario, we conducted experiments with push notifications and suggested charging when energy prices were low. In return, the EV drivers received a discount per kWh for their charging session. Wherever possible, in-car data have been used for the charging power prediction of individual cars. In the next step, these data were combined for several locations equipped with nextmove charging sites to calculate the energy demand for day-ahead activities. The rental station charging sites were already operated with a load management system to reflect the local grid's limitations and to adapt to the charging schedule received from the aggregator.

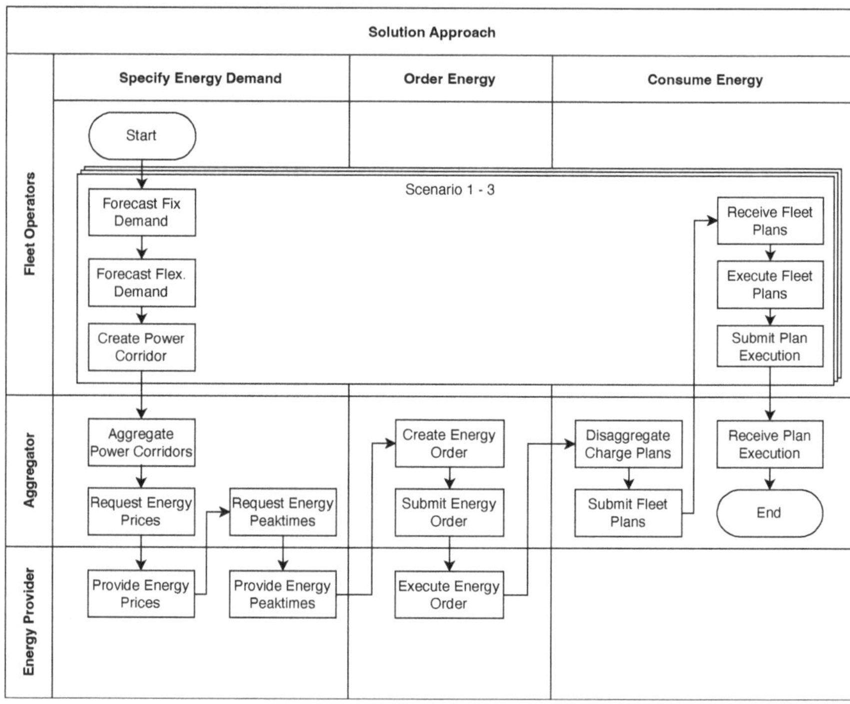

Figure 3. Flow chart of the solution approach per phase and actor of the demonstrator.

The third scenario at SAP is a smart-grid scenario, which integrates information from the local grid to actively steer the total consumption of a charging system with 81 installed charge points [17] serving 400 long-range employee EVs. This scenario integrates information from the local energy management system, which controls onsite photovoltaic (PV) and battery storage. Every 15 min, an optimization of the local consumption is triggered by a heuristic-based optimization model to minimize peak demand, load imbalance, and electricity costs [16]. The functionality to minimize the cost of electricity considers the availability of onsite photovoltaic energy generation as a complementary energy source but does not integrate external energy prices yet. This function requires additional data about fine-grained energy prices from the aggregator, which is planned as a prospective feature. The entire site can offer, by a simple estimation, a flexible energy potential from +20% to −20% of the planned fleet consumption (limited by the maximum allowed load of the site, 680 kW). The total charging capacity of all charge points is 1020 kW. Therefore, the infrastructure is always operated according to the site's maximal load. Additional local PV generation of 80 kWp and a 150 kWh stationary battery offer additional flexibility. Figure 4 shows a single charging plan for an EV, which is created by the optimizer to reduce the peak load in the grid at the SAP site.

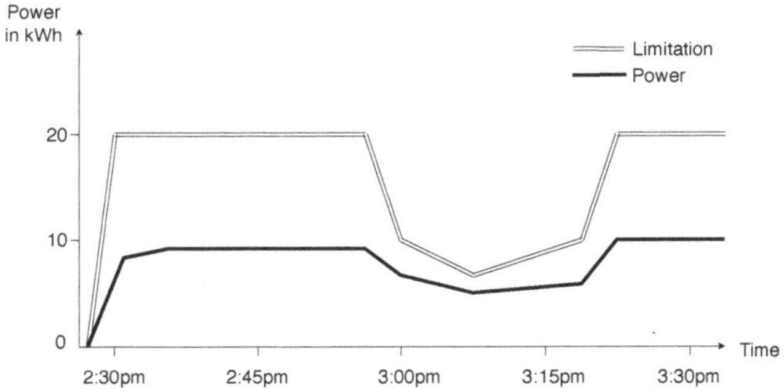

Figure 4. Definition of a CP charging plan based on the charging optimizer of a changing system. The CP charging plan provides the power limitation per charge point for every minute of the charging session. The actual power drawn from the EV for charging the battery is below the limitation.

For the implementation of the charging systems, we use open-source software [17]. All software systems are deployed as containerized applications on web services. The user interfaces are realized as desktop web applications, and there is also a mobile app for EV drivers. Each system runs independently of the other systems with separate persistence and application layers, therefore we are following decentralized architecture principles, which allows more specific conversions into marketable solutions.

2.5. Data Access for Optimization Data

Three different interfaces have been used by the fleet operators during the project to access real-time information from the charging sessions. Figure 5 shows the interfaces implemented for the charging system.

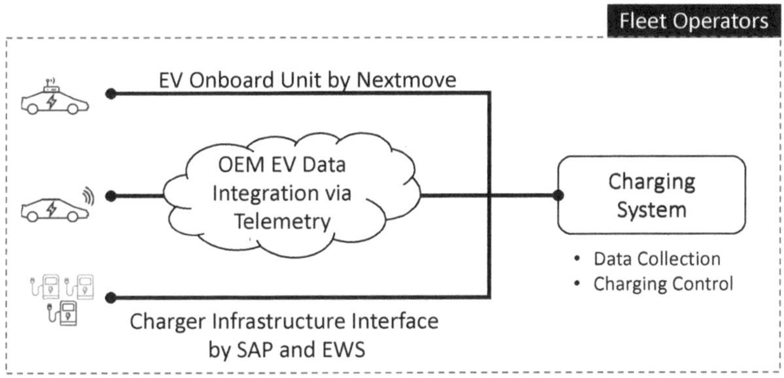

Figure 5. Different interfaces that are evaluated to access real-time charging session information.

2.5.1. Operations Based on Charge Point Data

All three scenarios use the open charge point protocol (OCPP) version 1.6. to exchange charging parameters for authentication and real-time charging session information to deploy charge plans. With data augmentation from an EV database and a user database, heuristical optimization problems like prioritization and the load management of charging sessions are implemented in the charging system [7,16]. The charge point data source is the basic data source for the charging systems in all three scenarios.

2.5.2. Hardware-Based Onboard Units for Real-Time Data

The onboard unit used for the project consists of a transmitter module using onboard diagnostics (OBD) as a data interface. During the project the onboard units support 51 different EV models from nextmove for real-time monitoring. The transmitter was implemented to be capable of obtaining over-the-air updates from the monitoring backend via its mobile connection to access the EV data interface. The price estimate for the developed onboard unit is approximately EUR 450 plus an additional data plan for connectivity. Due to firmware updates in the EV regarding in-car energy management, it was necessary to update during the project 300 units over-the-air. The availability of in-car real-time data depends on the car's state to prevent potential vampire losses during parking periods.

2.5.3. Software-Based Telematic Services for Real-Time Data

The enabling technology for software-based EV data access was realized with a telemetry service providing integration into the cloud services of the EV manufacturer for processing *SoC* information in real time. EV drivers from the SAP site in Mougins/France provided their consent for using the charging data for research purposes. For a yearly fee of EUR 60 per car, the service can be used without any hardware dependencies. Figure 6 shows a charging session with real-time optimization considering the *SoC* is provided by a telemetry service.

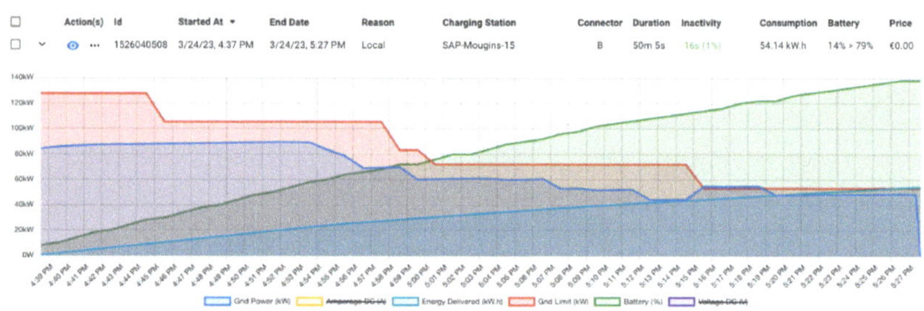

Figure 6. Example of a charging session in the demonstration charging system of SAP [17]. An increasing state of charge lowers the power consumption, and at 80%, the charging session ends.

3. Results

In this chapter, we detail the results of our experimental system setup. First, we outline the system architecture. Afterward, we present the evaluation process and describe the usage of EVs within the project.

3.1. System Architecture

The system mainly serves the needs of three types of entities called "fleet operator", "aggregator", and "energy provider". Each of these has its responsibilities and tasks. The architecture of the demonstrator in Figure 7 shows the entities' connected systems in a cascading pattern. Each fleet operator runs a charging system to control the energy consumption based on the charge plan for the own EV fleet. The aggregator operates an aggregation system that accumulates the demands from the connected fleet operators and communicates the aggregated flexible loads to the energy provider. In the trading system of the energy provider, the respective purchase orders are created and placed in the energy market.

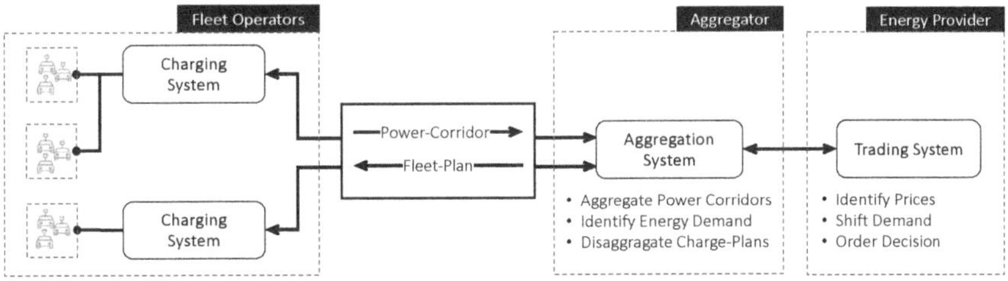

Figure 7. High-level architecture and main information flows between the involved roles within the demonstrator system.

Fleet operators have the task of charging the EVs of the users in an acceptable time while minimizing the cost of charging by considering CO_2 emissions, energy prices, and the local infrastructure situation. For the experimental setup, the fleet operators are obliged to share their flexible energy demand and corresponding power corridors in a given timeframe with the energy aggregator. In exchange, the fleet operator receives an EV charging plan from the energy provider, which is cost-optimized. This incentives the fleet operators to adapt the charging sessions of their fleets. The energy provider has the task of aggregating the power corridors and identifying the energy demand of the affected segments. On the energy provider level, the estimated power corridors received from the connected fleet operators are aggregated. Here, the aggregation includes the summation of power maxima and minima, as well as energy demands over the respective periods of time. Furthermore, the aggregation system generates a consistent view of flexibility originating from fleet operators, including slicing of energy demand segments appropriately (which may potentially overlap in different source fleets) and feasibility checking. A technical interface offers aggregated flexibility potentials to the trading system for corresponding procurement on electricity spot markets. According to the flexible energy demand, the trading system finally identifies current price levels and shifts the demand within the flexible range to make the best procurement decision. The best ordering decision is determined by input parameters, such as the current energy price, the grid capacity, and the situation of the charging systems, which are encoded in the aggregated representation of the received power corridors. The result of a procurement decision is a set of orders to be placed on the market and, in response, a set of transactions (trades) executed. All transactions on the market referring to the energy demand are ultimately composed into a schedule, which includes all the charge plans for the fleet operators. For each time slot (typically 15 min), the charge plans contain the total power to be delivered to the fleet operators. After obtaining the pool schedule from the trading system, the aggregation system disaggregates the pool charge plans according to the individual fleet operators' power corridors and energy demands. Herein, the result is a separate charge plan for each fleet operator, which will be propagated to the charging systems. In the next step, the energy provider will also be able to receive real-time consumption data from the charging systems to react to unforeseen changes in consumption, either by shifting loads between fleet operators or placing short-term order decisions on the intraday energy spot market. This mechanism helps minimize the imbalance (i.e., the mismatch between actual energy consumption and the charge plan backed by trades on the market), which would otherwise result in higher overall energy costs. Figure 8 shows an overview of the aggregation, trading, and disaggregation processes. The diagrams show the fleet charging power on the y-axis and the time on the x-axis. Summing up the flexible fleet demands results in the total energy demand of the aggregator (in green). P_{min} and P_{max} display the limits of power consumption that the fleet operators communicate to the aggregator. Based on the price signal and grid power peak information provided by the energy provider, the aggregator creates the price-optimized energy purchase orders according to the communicated power corridors (light blue). From the accumulated ordered energy, the charge plans for the fleet operators (black lines) are being disaggregated and sent to the fleet operators.

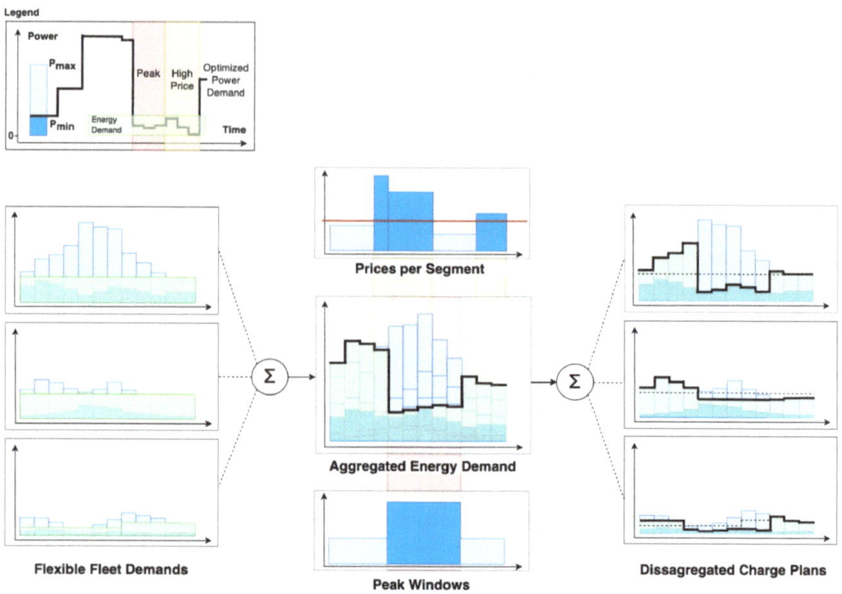

Figure 8. Energy aggregation process. The energy demand is aggregated to place purchasing orders, preferably at times with low prices and no peak loads. The disaggregation considers the minimum and maximum power values communicated by the fleet operators' charging systems.

3.2. Evaluation

The assessment of the implemented system is organized in three steps. The initial step focuses on testing the charging optimization for EVs to align with the local circumstances of the charging systems. The second step involves the collection of data from the charging systems, which will facilitate the forecast and the creation of a power corridor that is realistic to the EV fleet consumption toward the placement of an aggregated energy order in the energy market. In the third step, the breakdown of the centrally ordered energy quantity with real-time allocation processes for flexible demands is outlined. The first evaluation is the optimized usage of EVs in different scenarios depending on the usage of the EVs. In a large-company scenario (SAP), the EVs are regularly available, which leads to similar daily load profiles. For the rental-fleet scenario (next move), the fluctuation of the created monetary value by smart charging depends on the rental behavior and the battery size of the EVs, which are connected to the CPs onsite. For example, groups of transporter EVs are sometimes booked by customers for several weeks and are therefore not available for optimization of the fleet's charge plans. When the EV transporters are returned to the site again, this increases the flexibility of the load profile of the charging system significantly compared to proportionally more passenger EVs charging. Second, our definition of the power corridor allows the purchase of energy for fleet operators in the day-ahead market. Due to the day-ahead charging plan for the fleets, more market transparency can be provided and the aggregator has the possibility to place additional orders on the intraday market. The data created throughout the aggregation processes being evaluated and first simulations show the value of this approach [19]. Third, the algorithms and data required for aggregation and monetization of flexible loads are field-tested. The aggregation algorithm aggregates the data of the fleet consumption forecasts. The algorithm optimizes energy purchasing according to low-cost energy segments and peak windows in the power grid, and the disaggregation algorithm [19] that creates the charge plans for the fleets by calculating the amount of required power to serve the planned fleet demand. Finally, we proposed an overall approach that is already under more specific evaluations by other means and projects from [11–14].

3.3. Discussion

Optimizing the energy consumption of EV charging systems is not a trivial task. The difference between the grid limit and the grid power in Figure 6 shows that EVs do not simply charge up to the power of the assigned charge profile. Instead, each EV has its power plateaus on which it charges. These power plateaus, which are vehicle-model-dependent, are considered in the optimizer of the large-company scenario [7]. However, the power plateaus were not implemented in the rental-fleet and small-company scenarios. The differences in power plateaus allow the classification of EVs into three categories: small (with less than 35 kWh battery capacity), standard (35 kWh up to 64 kWh), and long-range (up to 120 kWh). These categories allow the further analysis of different consumption patterns. Further data analysis shows interdependencies with charge point models, car types, and real-time data to improve the optimization capabilities of the system. To identify the reasons for these different patterns, a survey has been conducted. Based on the test scenarios to forecast the flexible energy demand, customers have been surveyed on how their behavior affects the charging processes. The clustering of the data showed that most EV drivers picked the car to fit their driving scheme. The interview questions were as follows:

- Where is your main location to charge your EV?
- To what extend is your charging behavior affected by energy prices?

The analysis of the results shows that smaller EVs charge up to 80% at home, while standard EVs charge only up to 60% and long-range EV only up to 40% at home. According to these results, long-range EVs are the most relevant EVs for aggregation purposes at charging sites. However, most long-range EV users are not interested in electrical cost optimization at all because they do not need to charge offsite from home. These drivers are often business users and are triggered only by their individual charge demands, which the company pays. They usually use high-performance chargers during travel. The drivers of smaller EVs, on the other hand, are permanently looking for the next charging opportunity. This user group is really interested in the incentives a charging shift would offer them on a daily basis. But the greatest potential is among the standard EV users, which can delay a charging session to the next day. They have a larger battery but still connect often to the grid. Their battery size allows them to dynamically change their charging behavior, if there is a sufficient incentive available. This promises a potential field for development to provide end-user services and products offering optimized energy flexibility.

4. Conclusions

Our approach provides a framework that holds clearly defined areas of optimization for each in our research participating role: "fleet-operator", "aggregator", and "energy provider". Data availability has been identified as the limiting factor during the project to create substantial value from the data. The evaluation is performed based on the data transmitted from three charging systems which cover the presented scenarios: "small company fleet", "rental fleet", and "large company fleet". Data collection was implemented via OCPP, which provided 40,000 charging sessions over the last three years. We could record 8200 charging sessions that were optimized with *SoC* information that was gathered from OBD devices or telemetry services. Even when applying the load profile from the day ahead as an estimation of the power corridor, the purchase decisions of energy could already be improved by the aggregator by considering peak windows and prices, as described in Section 2.4.

The next step is to identify the predictors for charging behavior to improve the prediction accuracy for the power corridors and the flexible energy demand. Potential data sources could be booking systems with travel data, human resource systems with location and business car data, or facility management systems with data about the site infrastructure. Another open problem is to compare the data from the charging system forecasts with the actual energy consumption and the trading data, which can provide insights into how much value can be created with flexible energy consumption and how effective incentive systems can be designed. Viewing it from the business perspective, the consumption of cheaper energy is a promising result because the power corridor as a means for exchanging information between the roles of the fleet operator, aggregator, and energy provider creates transparency that shows improvement potentials of operational processes.

Author Contributions: Conceptualization, K.S.; methodology, K.S.; software, T.M.; validation, M.K.; formal analysis, K.S. and Z.N.; investigation, K.S. and M.K.; resources, F.B. and K.S.; data curation, K.S. and M.K.; writing—original draft preparation, K.S.; writing—review and editing, K.S. and Z.N.; visualization, K.S.; supervision, Z.N.; project administration, K.S.; funding acquisition, K.S. All authors have read and agreed to the published version of the manuscript.

Funding: This research was funded by the project TRADE EVs II of the Federal Ministry for Economic Affairs and Climate Action Germany grant number 01MV20006A.

Data Availability Statement: The data presented in this study are available on request from the corresponding author due to privacy.

Acknowledgments: Trading of Renewable Aggregated Distributed Energy by Electrical Vehicles, in short, TRADE EVs II, is a publicly funded project by the German Federal Ministry for Economic Affairs and Climate Action, Funding Code: 01MV20006A.

Conflicts of Interest: Kelaja Schert is an employee of SAP. Florian Biedenbach is an employee of FFE. Thomas Müller is an employee of Coneva. Michael Kluge is an employee of nextmove. Zoltan Nochta is an employee of SAP and HKA. The paper reflects the views of the scientists and not the companies.

Abbreviations

The following abbreviations are used in this manuscript:

AC	Alternating Current
EV	Electric Vehicle
CP	Charge Point
DC	Direct Current
OBD	On-board diagnostics
OCPP	Open Charge Point Protocol
PV	Photovoltaic
SoC	State of Charge
V2G	Vehicle to Grid

References

1. Ritchie, H.; Roser, M. Emissions by Sector. 2020. Available online: https://ourworldindata.org/emissions-by-sector (accessed on 16 November 2022).
2. IEA. Electricity Demand from the Electric Vehicle Fleet by Country and Region, 2030. 2022. Available online: https://www.iea.org/data-and-statistics/charts/electricity-demand-from-the-electric-vehicle-fleet-by-country-and-region-2030 (accessed on 16 November 2022).
3. Mlecnik, E.; Parker, J.; Ma, Z.; Corchero, C.; Knotzer, A.; Pernetti, R. Policy challenges for the development of energy flexibility services. *Energy Policy* **2020**, *137*, 111147. [CrossRef]
4. Nour, M.; Said, S.M.; Ali, A.; Farkas, C. Smart Charging of Electric Vehicles According to Electricity Price. In Proceedings of the 2019 International Conference on Innovative Trends in Computer Engineering (ITCE), Aswan, Egypt, 2–4 February 2019; pp. 432–437. [CrossRef]
5. Biedenbach, F.; Ziemsky, V. Opportunity or Risk? Model-Based Optimization of Electric Vehicle Charging Costs for Different Types of Variable Tariffs and Regulatory Scenarios from a Consumer Perspective. In Proceedings of the CIRED Porto Workshop 2022 E-Mobility and Power Distribution Systems, Porto, Portugal, 2–3 June 2022.
6. Spencer, S.I.; Fu, Z.; Apostolaki-Iosifidou, E.; Lipman, T.E. Evaluating smart charging strategies using real-world data from optimized plugin electric vehicles. *Transp. Res. Part Transp. Environ.* **2021**, *100*, 103023. [CrossRef]
7. Fleck, T.; Gohlke, S.; Nochta, Z. A System for the Efficient Charging of EV Fleets. *World Electr. Veh. J.* **2023**, *14*, 335. [CrossRef]

8. Flath, C.M.; Ilg, J.P.; Gottwalt, S.; Schmeck, H.; Weinhardt, C. Improving Electric Vehicle Charging Coordination through Area Pricing. *Transp. Sci.* **2014**, *48*, 619–634. [CrossRef]
9. Schuller, A.; Dietz, B.; Flath, C.M.; Weinhardt, C. Charging Strategies for Battery Electric Vehicles: Economic Benchmark and V2G Potential. *IEEE Trans. Power Syst.* **2014**, *29*, 2014–2022. [CrossRef]
10. Johnsen, D.; Strommenger, D. Gesteuertes Laden von Elektrofahrzeugen über Preisanreize—Anwendungsbeispiele und Handlungsbedarf. TÜV Rheinland Consulting GmbH Institut für Innovation und Technik (iit) in der VDI/VDE Innovation + Technik GmbH, December 2022. Available online: https://vdivde-it.de/sites/default/files/document/gesteuertes-laden-von-elektrofahrzeugen.pdf (accessed on 3 December 2023).
11. Pioneering Trial Involving Bidirectional Electric Vehicle Charging. *novatlantis*, 19 February 2022. Available online: https://novatlantis.ch/wp-content/uploads/2022/01/novatlantis_V2X_Press_Release_EN.pdf (accessed on 13 March 2022).
12. Case Study: Intelligent Octopus. Available online: https://www.octopusintelligence.com/competitive-intelligence-case-studies/ (accessed on 15 March 2023).
13. Enel X VPP FCAS Market Leadership. Enel X. Available online: https://www.enelx.com/au/en/resources/enel-x-vpp-fcas-leadership (accessed on 15 March 2023).
14. Ströhle, P. Integrating Consumer Flexibility in Smart Grid and Mobility Systems—An Online Optimization and Online Mechanism Design Approach. 2014. Available online: https://publikationen.bibliothek.kit.edu/1000045609 (accessed on 4 August 2022).
15. Junker, R.G.; Azar, A.G.; Lopes, R.A.; Lindberg, K.B.; Reynders, G.; Relan, R.; Madsen, H. Characterizing the energy flexibility of buildings and districts. *Appl. Energy* **2018**, *225*, 175–182. [CrossRef]
16. Frendo, O. Improving Smart Charging for Electric Vehicle Fleets by Integrating Battery and Prediction Models. 2021. Available online: https://madoc.bib.uni-mannheim.de/58770 (accessed on 19 December 2022).
17. SAP Labs France. Open e-Mobility. 2022. Available online: https://github.com/sap-labs-france/ev-server (accessed on 5 February 2023).
18. Jordan, W.C.; Graves, S.C. Principles on the Benefits of Manufacturing Process Flexibility. *Manag. Sci.* **1995**, *41*, 577–594. [CrossRef]
19. Schert, K.; Nochta, Z. Integration of electric fleet virtual power plants in energy markets. In Proceedings of the in 7th E-Mobility Power System Integration Symposium (EMOB 2023), Copenhagen, Denmark, 25 September 2023; Institution of Engineering and Technology: London, UK, 2023; pp. 83–90.

World Electric Vehicle Journal

Review

OCPP Interoperability: A Unified Future of Charging

Silke R. Kirchner

Ampure, 1333 Mayflower Ave., Monrovia, CA 91016, USA; silke.kirchner@ampure.com; Tel.: +1-(626)-862-9524

Abstract: Electric vehicle (EV) adoption grows steadily on a global scale, yet there is no consistent experience for EV drivers to charge their vehicles, which hinders the important EV mass market adoption. The Open Charge Point Protocol (OCPP) is the solution to this challenge, as it provides standardization and open communication between EV infrastructure components. The interplay of the OCPP with open cross-functional communication standards boosters driver experience on the one hand, while the charging station itself is integrated into a renewable energy ecosystem. This paper presents a deep dive into the combination of the OCPP with the OpenADR protocol, the Open Smart Charging Protocol (OSCP), the ISO 15118, and eRoaming protocols to explore possibilities and limitations. Furthermore, we suggest LoRa communication as an alternative to IP-based communication for deep-in building applications. Hence, this paper reveals the next important steps towards a successful EV mass market transition powered by user-friendliness and green energy.

Keywords: EV infrastructure; standardization; interoperability; communication protocols; NEVI

Citation: Kirchner, S.R. OCPP Interoperability: A Unified Future of Charging. *World Electr. Veh. J.* **2024**, *15*, 191. https://doi.org/10.3390/wevj15050191

Academic Editors: Joeri Van Mierlo and Genevieve Cullen

Received: 27 March 2024
Revised: 20 April 2024
Accepted: 22 April 2024
Published: 29 April 2024

1. Introduction

User-friendliness and large scale zero-emission vehicle infrastructure deployment is critical to achieving the White House net-zero emissions target by 2050 [1]. However, a recent survey of EV users reported substantial frustration with chargers being too slow, too crowded, or not operable [1,2]. In combination with range anxiety during long distance travel [3], a skeptical attitude toward EVs has evolved, which hinders a commitment to private or commercial EV ownership and prevents a profitable EV mass market adoption.

Interoperability within the EV infrastructure provides the solution to that challenge by nurturing a consistent and familiar EV driver experience powered by a reliable and green 'distributed energy resources' (DERs) energy ecosystem [4].

Interoperability in this case is two dimensional:

(1) Consumer facing;
(2) Technical or systems facing.

Consumer-facing interoperability includes the physical accessibility of a charger, universal payment methods at every charging station (CS), or 'one-matches-all' coupler hardware to be consistent with current re-fueling experiences for internal combustion engine (ICE) vehicles.

Technical interoperability encompasses standardized technical protocols and testing procedures with the aim of achieving consumer-facing interoperability and excellent user experience on a large scale. Standardized and generic data communications between different chargers and their respective central management systems (CMSs), together with a uniform data exchange between CMSs and third-party backends, such as e-mobility service providers (eMSPs) and capacity providers (CP) (counting utilities, distributed systems operators (DSOs), and cloud-based energy management systems (EMSs)), are the heart of interoperability.

It is with this consideration that Ampure (formerly Webasto Charging Systems) chargers strongly support and utilize the major de facto open-charger-to-cloud communication protocol in the US, the Open Charge Point Protocol (OCPP). Ampure chargers provide an interoperable, flexible, and expandable infrastructure platform that can integrate with a broad range of eMSPs, charging station operators (CSOs), automakers (OEMs), and CPs, fostering critical consumer interoperability and friendliness (Figure 1).

Figure 1. Left: Our charger is connected via OCPP1.6J to a cloud-based EMS, enabling capacity-based smart charging. **Right**: Deep-in building application of our industrial chargers facilitating LoRa. Reprinted from Ref. [5]. Most application protocols for data communication, such as the OCPP, require IP-based communication. For deep-in building charging technologies that demand low bandwidth and long-range (for example the transfer of battery status data in warehouses), a LoRa connectivity shall provide a great alternative to Wi-Fi, BLE, and cellular. LoRa facilitates long-range communication up to 10 miles and consumes ultra-low power. Reprinted from Ref. [6]. Via a LoRa gateway and encryption techniques, battery data can securely be transferred to the backend system.

2. Overview of Today's EVSE Protocol Landscape

This paper provides a unique and important overview of the Open Charge Point Protocol (OCPP) [7] by the Open Charge Alliance (OCA), and the interaction of the OCPP with cross-functional open standards, such as the OpenADR protocol [8], the Open Smart Charging Protocol (OSCP) [9], the ISO 15118 standard series [10], and eRoaming protocols (Figure 2) [11–14].

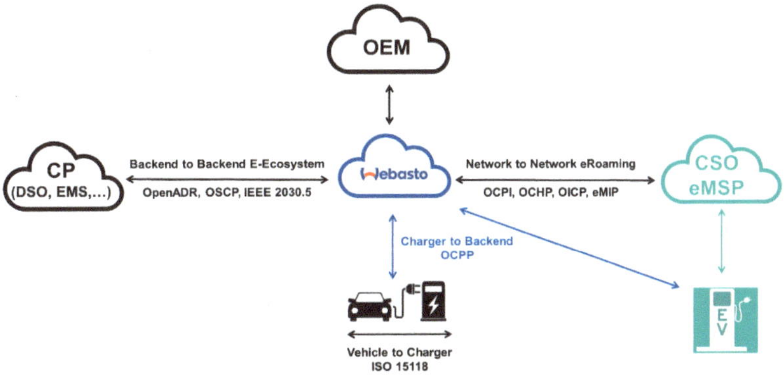

Figure 2. Communication protocols at a glance. The OCPP is the major open de facto communication protocol for charger-to-backend communication. Cross-functional backend communication is multi-faceted and facilitates different protocols for different needs. For the communication between capacity providers (CPs) and central management systems (CMSs), protocols such as the OpenADR protocol or the Open Smart Charging Protocol (OSCP) are in place. For eRoaming, which requires CMS to CSO/eMSPs backend communication, the Open Charge Point Interface (OCPI) protocol, the Open Clearing House Protocol (OCHP), the eMobility Interoperation Protocol (eMIP), and the Open InterCharge Protocol (OICP) can be used to serve hub-based or bilateral eRoaming structures. The ISO 15118 is an international standard series that contains specifications for secure, local, and bidirectional communication between EVs and chargers.

The intention is to facilitate an understanding of how the landscape of open and standardized application protocols boosts driver experience, while the growing EV infrastructure is integrated into a green and stable energy ecosystem. This review outlines the relevant characteristics of a protocol for its mass adoption success, possibilities, and limitations of open standards.

3. An Excerpt of Today's Leading Communication Protocols in the EV Industry

3.1. The OCPP, the OpenADR Protocol, and the OSCP Are Fundamental Contributors to a Scalable and Clean Electric Transportation Ecosystem

The major de facto and open protocol for charger-to-backend communication in the US (and globally) is the Open Charge Point Protocol (OCPP), which was initiated and has been maintained by the Open Charge Alliance (OCA).

The large-scale electrification of vehicles, fleets, and marine ports presents a threat to the grid, and as such it is crucial to bring chargers into the equation of energy demand and response systems. The combination of the OCPP with the Open Automated Demand Response (OpenADR) protocol or the Open Smart Charging Protocol (OSCP) turns a charger into a flexibility provider that can react to changes in demand response (DR) within a distributed energy resource (DER) energy ecosystem. Accordingly, an uninformed charging process can be converted into a smart technique, which is able to throttle or postpone a charging process based on currently and locally available grid capacity.

3.1.1. The Open Charge Point Protocol (OCPP)

The IP-based Open Charge Point Protocol (OCPP) is the major de facto and open communication protocol between a charging station (CS) and its respective central management system (CMS, Figure 3). The kick-off of a global and open protocol to standardize charger-to-backend communication in the EV industry was initiated by the E-Laad Foundation (now ElaadNL) in the year 2009, and it has been maintained and continuously developed by the members of the Open Charge Alliance. Due to the active support and contribution of major stakeholders and experts in the industry over decades, the open protocol has grown into a globally acknowledged communication protocol.

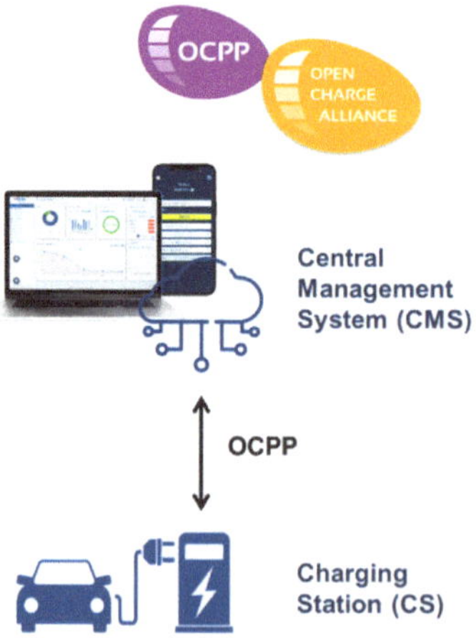

Figure 3. The OCPP is the major de facto open communication protocol between a charger and its CMS in the US and globally. The OCPP enables any CMS to connect with any charger, regardless of the vendor or manufacturer, if the CMS and the charger are compliant with the same OCPP version 1.6J or 2.0.1.

The OCPP enables any central management system to connect with any charger, regardless of the vendor or manufacturer, if both are compliant with the same OCPP version 1.6J or 2.0.1 [7]. This key attribute allows for a "mix and match" of chargers and a vendor agnostic infrastructure deployment, which is required for a rapid and sustainable growth of the charging industry. In addition, the real-life application of the OCPP with versatile charger networks gives insight into potential interoperability gaps, which can be flushed out and improved by the ongoing development of the protocol. The advantage of the Open Charge Alliance and the OCPP is the constant development and integration of new features and improvements based on real-world desires, needs, and lessons learned within the EV community. At the same time, this advantage also brings challenges and limitations. In comparison with proprietary and closed communication protocols, the OCPP does leave some room for technical interpretation between participants. Test tools, test labs, and global interoperability testing events, such as the "Plugfest" organized by the Open Charge Alliance and CharIn, mitigate the interoperability risk and are on the rise.

The dominant protocol version in the field is 1.6J; however, the industry has moved on to version 2.0.1 to benefit from the extended feature set such as demand response, load balancing, and tariff management, which are crucial functionalities of a modern and stable EV infrastructure. The OCPP 2.0.1 was released in March 2020 and serves Level 2 and DCFC techniques (GB/T, CHAdeMO, and CCS). It enables extended functionalities in the availability of chargers, payment, and reservation methods, smart charging options, and certificate management [7]. In addition, version 2.0.1 is required for a successful connection to the important ISO 15118 standard series [10], which enables Plug and Charge, and vehicle-to-grid applications. While the OCPP 1.6J and 2.0.1 are not backward compatible, all new versions, such as OCPP 2.1, which is in the release pipeline and includes for example generic interfaces for payment terminal integration, will be backward compatible moving forward.

OCPP interoperability unifies the charger network and, as such, substantially enhances the driver experience, with less stranded assets within a charging radius. Any stranded charger can be picked up by any operator using the same OCPP backend configuration.

Furthermore, an operator has the flexibility to purchase equipment from multiple vendors, which allows the operator to be manufacturer agnostic. Such interoperability fuels a fair market competition in the EVSE space, granting access to newcomers and not being solely dominated by a few established majority holders in the market.

3.1.2. Combining the OCPP with the OpenADR Protocol to Convert a Charger into a Smart Load Flexibility Provider

The production of renewable energy has become more and more decentralized, with individual businesses or households contributing to energy production through solar, wind turbines, and electric energy storage (EES) systems. An energy consumer has become an energy "prosumer", who produces and consumes renewable energy. In general, with so many active and different energy contributors, there is a huge desire for all of them to communicate and work together effectively to ensure grid safety and reliability [8,15].

To that aim, capacity providers—including utilities or distribution systems operators (DSOs)—use the OpenADR standard, which is maintained by the OpenADR Alliance, to enable a bidirectional IP-based communication between their top node(s) and aggregators or end devices.

The OpenADR protocol allows the coordination of end device responses to changes in currently and locally available energy supply/demand [16]. The protocol encompasses event messages, reports, and registration services, as well as availability schedules for dynamic price- and capacity-based programs [17]. The combination of the OCPP and the OpenADR protocol equips EV chargers with the capability to react to locally and currently available DER grid capacity, and makes a charging process flexible and smart (Figure 4).

While the OpenADR protocol standardizes the messaging and DR information exchange between a capacity provider's backend and the charger's central management system, the OCPP contains all required action commands to trigger the desired charger reaction. Such a charger reaction can be postponing a charging process, the consideration of priority charging, and the optimization of charging schedules [17].

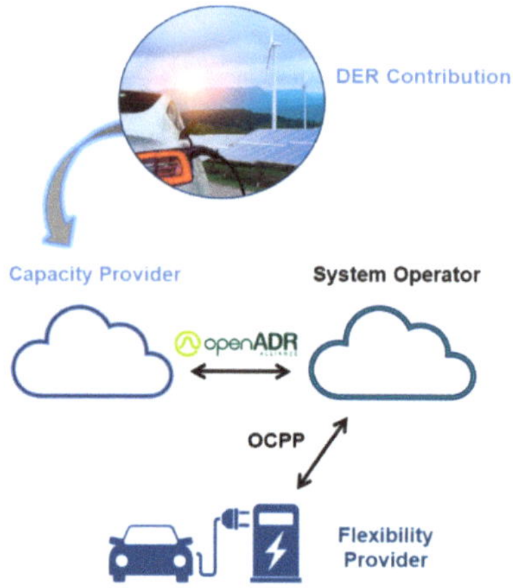

Figure 4. The line-up of the OCPP and the OpenADR standard turns a charger into an efficient load flexibility provider, which is integrated into a green DER energy ecosystem. While the OpenADR protocol standardizes the messaging and DR information exchange between a capacity provider's backend and the charger's central management system, the OCPP action commands initiate the desired reaction of the charger.

On the consumer side, the interaction of the OCPP and the OpenADR protocol saves cost per consumed energy unit (kWh) while maximizing the amount of renewable energy used for EV charging.

3.1.3. Combining the OCPP with the OSCP to Convert a Charger into a Smart Load Flexibility Provider

Similar to the OpenADR protocol, the Open Smart Charging Protocol or OSCP [9] takes the integration of EVs into a larger, dynamic, and flexible energy ecosystem (including photovoltaics, stationary batteries, heat pumps, etc.) into consideration. The OSCP standardizes the communication between the capacity provider, which can be a cloud-based EMS for example, and the charger's central management system, while also taking a 24 h prediction of the local available grid capacity into consideration [9]. Such communication capabilities of a charging station with the grid turns an operator into a flexibility provider, capable of matching charging profiles within local capacity trendlines, e.g., capacity-based smart charging (Figure 5). Additionally, the operator can request the optimal EV charging energy demand, to prevent line or grid overloading.

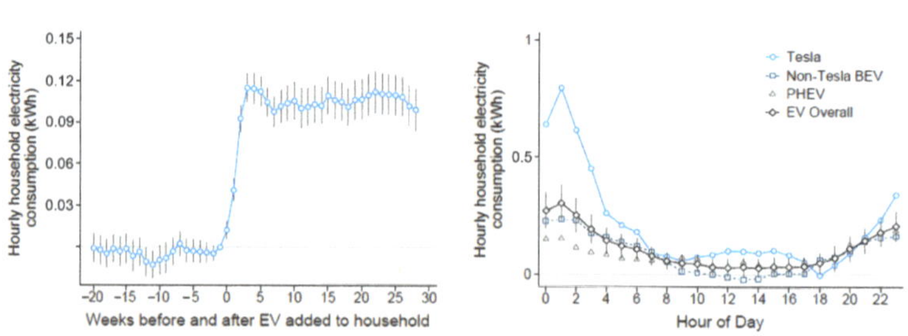

Figure 5. Impact of EV adoption on household electricity. **Left**: EV adoption (week 0) increases household electricity consumption by 0.12 kWh hourly or ca. 3 kWh per day. **Right**: effects are concentrated between 10 p.m. and 6 a.m., when vehicles are plugged in overnight to recharge. Reprinted from Ref. [18]. Services, such as capacity-based smart charging, help optimize energy consumption in the case of multi-dwelling unit applications, where multiple drivers might demand electricity at the same time.

3.2. Combining the OCPP with the ISO 15118 Standard Series for a Safe, Sustainable, and Automated Charging Network

The ISO 15118 "Road vehicles—vehicle to grid communication interface" is an international standard series, which contains specifications for the bidirectional communication between an electric vehicle and a charging station (CS) [10]. The ISO 15118 has been developed by the "International Organization for Standardization" (ISO) and represents a significant milestone in the advancement of electric vehicle technology. The standard addresses and solves the challenges that are associated with the interoperability and communication between EVs and the charging infrastructure, such as cybersecurity, ease of use for the driver, and smart dis-/charging technology. The series provides a comprehensive framework for the communication protocol between electric vehicles and chargers, converging seamless, green, and automated charging processes [10].

The ISO 15118 consists of multiple parts, each focusing on different aspects of the communication interface between electric vehicles and the charging infrastructure. Part 1, for example, serves as an introduction to the series, outlining general principles and defining use cases for vehicle-to-grid communication, such as the Plug and Charge use case. Part 2 of the series delves into the technical specifications of the network and application protocols. This includes the definition of the communication architecture, data formats, and security mechanisms, which are necessary for secure and reliable communication between electric vehicles and the charging infrastructure. The ISO 15118 standard series is designed to be scalable and adaptable to evolving technologies and industry requirements. Its modular structure allows for updates and additions to accommodate emerging features and advancements in electric vehicle technology.

The ISO 15118 addresses the security aspects of communication between electric vehicles and chargers, which is key to data security. It specifically outlines the security measures necessary to protect the communication interface from potential cyber threats. This includes authentication and authorization mechanisms, data integrity protection, and encryption techniques to ensure the confidentiality of the exchanged information. The Transport Layer Security (TLS v1.2) protocol is used to establish the encrypted communication session, while elliptic curve Diffie–Hellman (ECDH) is used to validate the process for one charging session [10]. AES-128-GCM (ISO 15118-20) is utilized to encrypt and decrypt instructions during a charging session using the TLS session key. The elliptic curve digital signature algorithm (ECDSA) will further verify the authenticity of the sender and the integrity of the received message (via SHA-256 as a cryptographic hash). These industry standard protocols ensure the charging process is secured, and minimize the risk of damaging the charger or vehicle from compromised devices [10,19].

Two major use cases of the ISO 15118 are Plug and Charge (PnC), i.e., automatic authorization and payment upon connecting an EVSE with the car, and vehicle-to-grid (V2G), i.e., a vehicle can supply energy back to the grid during down times. The PnC use case catalyzes the user experience, as a driver can simply plug the coupler into the vehicle and the necessary communication and initiation of the charging process occurs automatically, provisioning customer excellence without the necessity to rely on a secondary digital or physical payment option. The V2G functionality allows electric vehicles not only to receive power from the grid but also to feed stored energy back into the grid, contributing to grid stability and potentially creating new revenue streams for EV owners. This application is particularly important for fleet scenarios, which run on predicable schedules and thus can potentially support the demand during peak hours with the V2G technology.

3.2.1. The ISO 15118 Plug and Charge (PnC) Use Case

The ISO 15118 PnC use case provides an automated charging and payment process upon plugging the charger into the EV. The charging authentication and authorization is accomplished using digital certificates that are exchanged between the EV and the charger [10,19] (Table 1). No form of active consumer involvement is required, and the billing process happens in the back without any actions required by the driver (other than initially adding the payment to their platform).

Table 1. Basic certificate fields for a typical X.509v3 certificate, as used in ISO 15118 [10]. The EV's certificate is called the identity certificate, and is used to authenticate the EV to the charger. Similarly, the charger's digital contract certificate is used to authenticate itself to the EV.

Certificate Field	Description
Version	Version of certificate
Serial number	Unique number of certificate
Signature algorithm	Used signature algorithm
Issuer	Entity, who has issued and signed the certificate
Validity period	Time period, in which the certificate is valid
Subject	Entity, to which the certificate is issued
Public key	Public key corresponding to a private key
Issuer UID	Optional issuer unique identifier
Subject UID	Optional subject unique identifier
Extensions	Optional
Signature	Signature of the certificate generated by the issuer

The digital certificates are stored in the onboard system of the EV and then provided to the charger once plugged in. The certificates are signed by a third-party certificate authority (CA), and, in combination with encryption methods, the ISO 15118 ensures secure EV–charger communication and protected user contract data [20].

The vehicle's certificate is the identity certificate, which is used to authenticate the EV to the charging station. Similarly, the charger's digital contract certificate is used to authenticate the charger to the EV. By exchanging these two certificates through local charging cable communication, the EV and the charger can negotiate charging parameters, charging rates, and billing details agreed upon by the EV owner and the operator [10].

3.2.2. ISO 15118 Vehicle-to-Grid (V2G) Use Case

Electrical energy storage (EES) is one of the most effective support systems for balancing a green and dynamic DER grid [15]. EV traction batteries can be mobile resources, with typical capacities of 30–100 kWh of electrical energy [21]. For reference, an average household in the US consumes 30 kWh electrical energy per day [22]. Fleet applications, such as school buses, can be a predictable energy prosumer. The vehicles feed energy back to the grid during peak demand time when they are not in use and charge again off peak before they are required to dispatch. This technological milestone shall have a major positive impact on grid stabilization, while offsetting running costs, and help make electrification transformation sustainable and scalable. In a complete green energy cycle, the charger's central OCPP management system can be connected via OpenADR or OSCP to a capacity provider to receive dynamic updates on DER power availability.

The ISO 15118 provides the communication protocol between vehicle and charger. This communication solution together with further inverter requirements equips a charger with the potential to bring back green electrical energy from the vehicle's traction battery (originating from photovoltaics or wind power for example) to the grid (Figure 6) [10]. This energy can be used to power homes and businesses during peak demand periods, during emergencies, or when renewable energy sources are not active. In addition, EV owners can potentially generate an additional income stream by providing power to the grid, which reduces the cost of electrification [10].

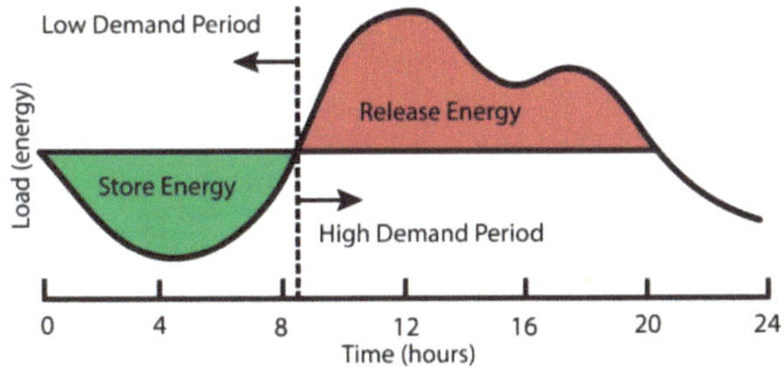

Figure 6. Electrical energy storage (EES) refers to the process of converting electrical energy into a stored form that can later be converted back into power when needed. Reprinted from Ref. [15]. The ISO 15118 V2G provides the communication basis between a vehicle and an EVSE to sell back (green or cheap) electrical energy from the EV's traction battery to the grid when demand is high.

In the ISO15118 V2G application, the EV and the charger authenticate each other by exchanging their identity and contract certificate, respectively. The charger must be able to support the bidirectional V2G data transfer and electricity flow between the EV and the grid. For data communication, the ISO 15118-20 can be implemented, which is currently in the process of being extended to AC bidirectional use cases. The technical specifications of the converter depend, among other things, on the on-board charger as well as the local grid requirements. We see V2G as a critical future use case to be established in the market and rolled out to a large customer base. As technical solutions are only at the beginning and the cost of investment is relatively high, the successful mass adoption of V2G will most likely be a few years from now. There is more work still to be done by authorities to establish a unified certification procedure for a bidirectional charging system.

In summary, the ISO 15118 standard series plays a crucial role in establishing a common and interoperable communication framework for electric vehicles and charging infrastructure (Figure 7). By addressing technical specifications, security considerations, and enabling advanced features like Plug and Charge and vehicle-to-grid capabilities, these series of standards contribute to the widespread adoption of electric vehicles and the development of a more efficient and sustainable transportation ecosystem.

Figure 7. The OCPP and ISO 15118—vehicle-to-grid (V2G)—turn an EV into a mobile and green EES that can contribute energy to the grid in times of high demand and generate an additional financial income stream.

3.3. *eRoaming: Charge Anywhere with a Single Mobile App*

The idea behind eRoaming is that a driver can "charge anywhere" at any destination charger using only one mobile app. As such, eRoaming contributes to a user-friendly driving experience and is a critical player in achieving EV mass adoption. It has been reported that the download of multiple apps for different destination chargers is one of the most dominant barriers that hinders a driver from purchasing an EV [1].

From a technical perspective, eRoaming requires the integration of a charger's cloud-based management system into an eRoaming hub or bilateral eRoaming platforms. These platforms store the needed EV certificates and payment options and allow the charger to validate the EV with their database. Globally leading eRoaming hubs include, for example, "e-clearing.net", "GIREVE", and "Hubject", while the "EVRoaming Foundation" supports bilateral webbing as well as the integration with hubs.

3.3.1. Hub-Based eRoaming

The largest hub-based eRoaming structure is Hubject [23]. The network originates from a joint venture between BMW, Bosch, EnBW, Enel X, Mercedes Benz, Innogy, Siemens, and Volkswagen, and is present around the globe, including US, Europe, and China. The roaming hub encompasses more than 300,000 charging stations, leading to a global user base of more than 10 M drivers. To connect a backend to the Hubject network the Open Inter Charge Protocol (OICP) or the Open Charge Point Interface (OCPI) are required.

GIREVE [24] and e-clearing.net [14] are two large European eRoaming hubs, maintaining their respective eRoaming networking protocol, the eMobility Interoperation Protocol (eMIP) and the Open Clearing House Protocol (OCHP), respectively. Both hubs also support the EV Roaming Foundation's protocol OCPI, which allows OCPI supporters peer-to-peer eRoaming networking as well as hub network relations to GIREVE and e-clearing.net.

3.3.2. Bilateral eRoaming

The non-profit EV Roaming Foundation maintains the free and independent Open Charge Point Interface (OCPI) protocol required to join its network [12]. Members of the global foundation are Google Maps, Last Mile Solutions, Freshmile, and more. The OCPI protocol supports bilateral as well as hub-based roaming. As such, the OCPI supports hybrid eRoaming network structures globally. Service functionalities of the OCPI protocol include authorization, reservation, tariff information, billing, real-time session information, etc. [12].

4. Practical Possibilities and Limitations of the Protocols

This review evaluates the possibilities and opportunities of combining the OCPP with open, cross-functional communication standards, such as the OpenADR, the ISO 15118 standard series, or eRoaming protocols. The goal of standardizing the communication between different players is to solve the major barriers to technical interoperability and capture the opportunities that come with a widespread EV adoption. Current examples of EV charging frustrations are chargers being too slow, too crowded, or not operable [1]. Governmental institutions and funding incentives, such as the National Electric Vehicle Infrastructure (NEVI) Program Formula by the U.S. Department of Transportation's (DOT) Federal Highway Administration (FHWA), strongly support a unified charging experience through their funding requirements [25].

Interoperability with the leading communication protocol for charger-to-backend communication, the OCPP by the Open Charge Alliance, and the ISO 15118 is a strong first step towards a user-friendly, consistent, and familiar charging experience. In a second phase, the integration of an EV infrastructure into a reliable, smart, and green DER/DR energy ecosystem can be realized by energy communication protocols, such as the OpenADR or OSCP [7,9,10,17].

The improvement of the OCPP relies on open-source development, so the protocol can be continuously updated as lessons are learned from real-world applications. Open-source application protocols have been proven to provide content that is more correct and reliable than proprietary implementations [16]. We believe the openness of a protocol, paired with the spirit of shared responsibility, will lead to a democratized and fair EV charging infrastructure characterized by high quality, convenience, and reliability. Limitations of the open protocol are the risk of technical interpretation between stakeholders, which have different technical solution approaches. This can lead to implementation friction and delayed roll outs of a standardized infrastructure platform. Testing tools, labs, and events are on the rise to speed up the interoperability process [26].

The OpenADR and OSCP protocols provide a standardized framework for communication between utilities and end users, ensuring interoperability across different systems and devices. This standardization streamlines the implementation of demand response programs. IP-based protocols encompass the capability to support large-scale deployments and the real-time feedback on currently and locally available grid capacity, which allows end nodes to quickly respond to changes in demand and grid conditions [10,16,17]. On the other hand, protocol implementations can be complex and require the orchestration of utilities, aggregators, and end users. Furthermore, the implementation of the protocol on older infrastructure may pose challenges.

Just like any IP-based protocol, the OCPP and the OpenADR rely on internet connectivity, which can be challenging for areas with unreliable or limited internet access. In addition, these protocols might raise security concerns regarding data privacy and network vulnerabilities. It is important to acknowledge that the protocols ensure high-level security mechanisms against cyber threats and are globally established for a safe widespread adoption [7,17].

In bilateral eRoaming agreements, such as the OCPI protocol, network providers and manufacturers sign peer-to-peer agreements to create a web of interoperable chargers. This process can be time- and resource-intensive, and the continuous maintenance of multiple bilateral agreements can introduce novel challenges. In addition, a bilateral roaming solution makes it harder for smaller players to enter the eRoaming market.

In central roaming hub solutions, eMSPs or operators can join an established network in the form of a hub organization. The hub director typically charges a fee for membership, which can potentially be re-directed to the end user. The hub-based eRoaming approach makes it easier for new and smaller players to enter the market against a fee without having to build, accumulate, and manage a large database of EVs and their payment preferences [12,14,23,24].

5. Outlook

Standardized communication in the EV industry enables long-term solutions, along with data sharing and diagnostics to enhance charger availability and uptime. Databases, e.g., the Alternative Fuels Data Center (DOE), display charger locations and availability across the US, fostering charger access and operability information [27]. While availability and downtime minimization are crucial, a standardized shared charger data forum also allows for the analysis of current charger status and consumer behavior, as well as energy usage forecasts, which are particularly important for the aim of fleet electrification (Figure 8).

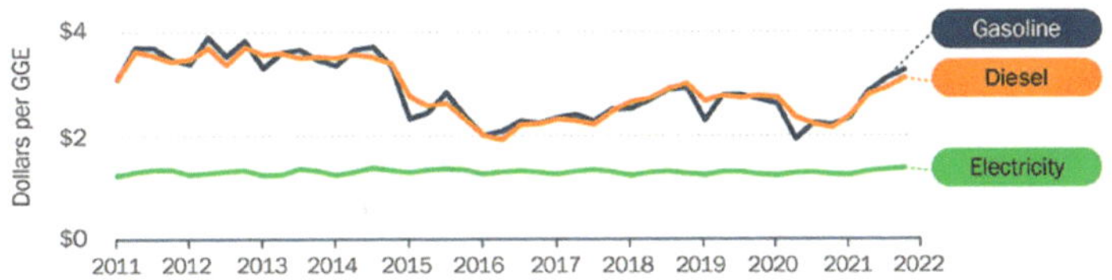

Figure 8. The comparison of electricity prices (USD) with gasoline or diesel shows that the price for electricity is lower and not as fluctuating, which makes it easier to predict costs over time. In addition, EVs offer high propulsion economy, which results in lower operating costs. Reprinted from Ref. [28].

Furthermore, the precise orchestration of energy consumption will be key to the success of mass electrification. Smart charging applications, such as described in the OCPP and OpenADR, allow for load balancing within an EVSE site to minimize energy consumption during peak hours or maximize usage within the limitations of a location's power systems (such as at an apartment building or place of business). Importantly, they can also assist with pushing greater capacity to charging systems when grid capacity is high or supported through active DER contribution, such as solar or wind.

Finally, standardized data sharing supports the development of new energy services and business models such as virtual power plants (VPP) and peer-to-peer energy trading forms.

Global Consortia of public and private EV infrastructure leaders, such as the Open Charge Alliance or the OpenADR Alliance, nurture the development, update, and adaption of international open communication protocols to standardize the EV charging industry and energy ecosystem. The success of a protocol is driven by market dynamics and stakeholder acceptance, together with regulated top-down decision by authorities.

Funding: This research received no external funding.

Acknowledgments: Thank you to my inspiring colleagues at Ampure. It is a true pleasure to be part of such a great team, driving tomorrow's green future together.

Conflicts of Interest: Silke R. Kirchner is an employee of Ampure. The paper reflects the views of the author and not the company.

References

1. FACT SHEET: Biden-Harris Administration Announces New Standards and Major Progress for a Made-in-America National Network of Electric Vehicle Chargers. Available online: https://www.whitehouse.gov/briefing-room/statements-releases/2023/02/15/fact-sheet-biden-harris-administration-announces-new-standards-and-major-progress-for-a-made-in-america-national-network-of-electric-vehicle-chargers/ (accessed on 11 February 2023).
2. The Expanding EV Market, Observations in a Year of Growth. Plug in America. Available online: https://www.atlasevhub.com/resource/the-expanding-ev-market-observations-in-a-year-of-growth/ (accessed on 15 February 2023).
3. What Is EV Charging Anxiety—And Is Range Anxiety a Thing of the Past? Available online: https://www.nationalgrid.com/group/what-ev-charging-anxiety-and-range-anxiety-thing-past#:~:text=Simply%20put,%20range%20anxiety%20is,be%20few%20and%20far%20between (accessed on 23 January 2023).
4. Kirchner, S.R.; Ruiz, H. OCPP Interoperability: Democratized Future of Charging. In Proceedings of the 36th International Electric Vehicle Symposium and Exhibition, Sacramento, CA, USA, 11–14 June 2023; pp. 1–12.
5. What Are LoRa and LoRaWAN? Available online: https://www.thethingsnetwork.org/docs/lorawan/what-is-lorawan/ (accessed on 13 February 2023).
6. What Are LoRa®and LoRaWAN®? Available online: https://lora-developers.semtech.com/documentation/tech-papers-and-guides/lora-and-lorawan/#:~:text=The%20name,%20LoRa,%20is%20a,areas%20(line%20of%20sight (accessed on 11 February 2023).
7. Open Charge Point Protocol 2.0.1. Available online: https://openchargealliance.org/protocols/open-charge-point-protocol/ (accessed on 23 January 2023).
8. OpenADR Alliance. DER Control and How to Implement Smart Inverter Management with OpenADR. Available online: https://www.openadr.org/assets/OpenADR%20for%20Smart%20Inverter%20Control_final.pdf (accessed on 15 February 2023).
9. Open Smart Charging Protocol. Available online: https://openchargealliance.org/protocols/open-smart-charging-protocol/ (accessed on 27 March 2023).
10. *ISO 15118*; Road Vehicles—Vehicle to Grid Communication Interface. International Organization for Standardization: Geneva, Switzerland, 2019.
11. Open Plug&Charge Protocol. Available online: https://github.com/hubject/opcp (accessed on 23 February 2023).
12. EVRoaming Foundation. Available online: https://evroaming.org/about-us/ (accessed on 1 March 2023).
13. Gireve, eMIP Protocol. Available online: https://www.gireve.com/wp-content/uploads/2022/09/Gireve_Tech_eMIP-V0.7.4_ProtocolDescription_1.0.14-en.pdf (accessed on 27 March 2023).
14. e-clearing.net. Available online: https://e-clearing.net/ (accessed on 2 February 2023).
15. U.S. Grid Energy Storage Factsheet. Available online: https://css.umich.edu/publications/factsheets/energy/us-grid-energy-storage-factsheet (accessed on 27 March 2023).
16. McParland, C. OpenADR Open Source Toolkit: Developing Open Source Software for the Smart Grid. In Proceedings of the 2011 IEEE Power and Energy Society General Meeting, Detroit, MI, USA, 24–28 July 2011.
17. OpenADR. Connecting Smart Energy to the Grid. Available online: https://www.openadr.org/ (accessed on 15 February 2023).

18. Burlig, F.; Bushnell, J.; Rapson, D.; Wolfram, C. Low Energy: Estimating Electric Vehicle Electricity Use. *AEA Pap. Proc.* **2021**, *11*, 430–435. [CrossRef]
19. Draz, M.; Kleeberg, T.; Vahrenholz, C.; Stichtenoth, F.; Deliverable 4.1 Plug & Charge Guidance Document. echarge4drivers. 2021. Available online: https://zenodo.org/records/6855650 (accessed on 27 March 2023).
20. Was Ist Die ISO 15118? Available online: https://www.tuev-nord.de/explore/de/erklaert/was-ist-die-iso-15118/#:~:text=Die%20ISO%2015118%20gew%C3%A4hrleistet%20dabei,Vertragsdaten%20der%20Nutzenden%20gesch%C3%BCtzt%20sind (accessed on 25 January 2023).
21. Electric Car Battery Basics: Capacity, Charging and Range. Available online: https://www.edmunds.com/car-technology/electric-car-battery-basics-capacity-charging-and-range.html (accessed on 24 January 2023).
22. What's the Average Home Energy Consumption in Your State? Available online: https://majorenergy.com/whats-the-average-home-energy-consumption-in-your-state/ (accessed on 11 February 2023).
23. Hubject. Available online: https://www.hubject.com/ (accessed on 20 January 2023).
24. Gireve Powering New Mobilities. Available online: https://www.gireve.com/ (accessed on 25 January 2023).
25. U.S. Department of Transportation. Available online: https://www.fhwa.dot.gov/bipartisan-infrastructure-law/nevi_formula_program.cfm/ (accessed on 19 April 2024).
26. Open Charge Alliance. Available online: https://openchargealliance.org/save-the-date-ocpp-plugfest-around-the-globe/ (accessed on 19 April 2024).
27. Alternative Fuels Data Center. Available online: https://afdc.energy.gov/ (accessed on 20 January 2023).
28. U.S. Department of Energy. Available online: https://afdc.energy.gov/vehicles/electric_fleets.html/ (accessed on 21 December 2023).

World Electric Vehicle Journal

MDPI

Article

Battery Research and Innovation—A Study of Patents and Papers

Hans Pohl [1,*] **and Måns Marklund** [2]

[1] Lindholmen Science Park, P.O. Box 8077, 402 78 Gothenburg, Sweden
[2] Cascelotte AB, 111 20 Stockholm, Sweden; mans.marklund@bergenstrahle.se
* Correspondence: hans.pohl@lindholmen.se

Abstract: This study of patent applications and scientific publications related to batteries is unique as it includes the volume of as well as qualitative indicators for both types of publications. Using carefully elaborated strategies to identify publications relating to batteries, this study provides data to discuss the critical balance to strike between investments in research and the more innovation-related aspects. The results show that China's dominance in publication volumes increases and that research with Chinese involvement is highly cited, whereas patent applications are slightly less valued than the world average. Quality-related indicators for Canada and the United States are very high for both scientific publications and patent applications. National differences in the proportions of patent applications and scientific publications are large, with Japan at one end with three patent applications per scientific paper and Canada at the other with almost seven scientific papers per patent application. On an actor level, data for Sweden indicate how the automotive industry started to file many patent applications in the decade starting in 2010. Finally, it is noted that this new approach to study a technological field appears promising as it gives new perspectives of relevance for policy actors and others.

Keywords: battery; research; patents; innovation; scientometrics

Citation: Pohl, H.; Marklund, M. Battery Research and Innovation—A Study of Patents and Papers. *World Electr. Veh. J.* **2024**, *15*, 193. https://doi.org/10.3390/wevj15050193

Academic Editors: Joeri Van Mierlo and Genevieve Cullen

Received: 19 March 2024
Revised: 7 April 2024
Accepted: 23 April 2024
Published: 29 April 2024

1. Introduction

Battery business is expanding rapidly. There is a global race to gain leadership along the whole battery value chain. Interestingly, even though production capacity is being scaled very rapidly, the investments in research are still also expanding dramatically. Many companies and countries are trying to gain market shares by developing competitive battery solutions. One decisive aspect is knowledge. With superior knowledge and associated intellectual property rights, the chances to gain and maintain a strong position increase.

The purpose of this study was to develop and test a method to analyze battery-related research and innovation. In this study, batteries include all types of electrochemical devices to store electrical energy, as well as super-capacitors. Through the use of two types of publications, patents and papers, this study addressed two steps in the value chain: research and innovation. By patents we mean patent applications as well as granted patents, and papers are here equal to articles, conference papers, books, book chapters, and reviews indexed in Scopus.

One challenge associated with investments in research and innovation is to find a balance between research-oriented more basic knowledge production and innovation-oriented activities leading to commercial development. Heavy investment in research but limited efforts to make use of the knowledge in new or improved products or services might lead to knowledge being wasted or exploited in other firms or countries. On the other hand, a limited involvement in research compared to subsequent steps toward the market might lead to a situation when the actors or the country is being surpassed by others working with superior technologies. In this study, we used scientific publications as a proxy for research and patent applications as an innovation indicator.

In addition to the development of a new method, we tested the method on data covering approximately two decades from 2000. For a selection of countries including Canada, China, France, Germany, Japan, Republic of Korea, Sweden and the United States, the following questions were studied:

- How do the volumes of patents and papers develop?
- How does the ratio of patents to papers develop?
- What is the share of academic–corporate papers?
- How is the quality of the patents and papers?

Moreover, but only for Sweden given the large efforts of manual work needed, it was also investigated on an actor and individual level to what extent patenting and academic publishing goes hand in hand.

Existing studies of patents and papers on batteries and vehicle electrification seldom combine and compare these two types of publications. One exception is [1], which used both types of publications to identify emerging trends. As highlighted by [2], actual patents form a small part of the total innovation activities, and by adding papers we cover a wider scope. The main contribution of our study is the combination and comparison of patent and paper data, including both volumes of publications and elaborated indicators related to their quality.

The methodology developed in this study is unique, at least in the context of batteries and vehicle electrification, and it delivers new insights relating to how different countries strike the balance between research and innovation.

The approach forwarded in this study can be used for any technology provided that it generates sufficient volumes of patents and papers. Given the broader perspective on innovation offered, it provides insights of relevance, not least for policy makers interested in the development of the innovation system.

The remainder of the paper is structured as follows. A review of previous literature follows; thereafter the methodology is described. The results section includes three sub-sections, with volumes and quality indicators on a national level in the first two sub-sections, followed by one sub-section on the actor level using data for Sweden. Finally, discussions and conclusions follow.

2. Quantitative Studies of Batteries and Vehicle Electrification

A background to the use of patent data and how it refers to papers is given in [3]. The study argues that patentometrics started to become important in the 1980s and that citations from patents to papers were used to better understand the links between science and technology.

In [1], the broader scope of energy storage was studied using both types of publications to identify emerging topics. Based on rather short search queries, publications were identified and clustered to find emerging topics. The number of citations was used to assess the relevance of each cluster, both for patents and papers. Papers were retrieved from Web of Science whereas patents were obtained from Derwent Innovation, a collection of patent data from 44 patent authorities. It can be noted that patent data for this study published in 2020 started to drop dramatically in 2016, as there is a time lag until patent applications are published. Within the battery domain, lithium–sulfur technologies were identified as emerging on the academic side, whereas multi-power systems where emerging within industry research.

Several papers use patent data for the study of batteries or their use in battery-electric vehicles. In [4], networks between organizations involved in electric and hybrid-electric vehicles were studied based on co-authorships of the patent applications. The search method was based on patent classes and patent data were from the European Patent Office's Global Patent Index Database. For the paper published in 2016, patent data until and including 2010 were used. One of their findings is that the networks toward the end of the period center around the large original equipment manufacturers, which could indicate that electric-vehicle technologies are maturing.

Using the same data source, another paper from the same year [5] addresses electric-vehicle technologies and presents which countries are most active in patenting. It also identifies some technological fields within electric-vehicle patenting. A small selection of patent classes was used to find relevant publications. Among the findings are that patenting activity within the electric-vehicle field has increased and that many innovations originate from Asian countries.

In an ambitious patent study of three energy-related technologies, li-ion batteries, hydrogen production and thermochemical conversion of biomass, the five largest countries in terms of patenting activity were covered [2]. Relevant patents were extracted using a combination of patent classes and keyword search from the European Patent Office's database. In the literature review, a good explanation of how patent applications relate to innovations is given, arguing that only a small part of all inventions is patented and thereof only a part is becoming innovation. A total of 5822 patents relating to li-ion batteries were found for the period 1995–2018. Japan dominated with approximately 50% of the total followed by China.

There have also been several studies that used scientific publications. One such study addressed the thermal management of li-ion batteries [6]. It used keywords to identify relevant publications in Scopus, but the development of the search string was not described. For the period 2000–2021, 983 papers were identified, and Chinese institutions dominated in terms of publication volumes, followed by a Canadian university. Volumes per country, institution, journal, and author were described, as well as total citation numbers per publication.

A study on a similar topic with the same approach had an explicit very short query to identify relevant papers in the period until 2018 [7]. It used clustering to identify trends and the distribution of keywords over time to study research trends, concluding among other conclusions that thermal management for li-ion batteries was a research gap. In terms of publication volumes, Chinese institutions dominated.

Another recent study used papers from Scopus to investigate electronic waste from electric vehicles [8]. A very short query was used to identify 593 publications during the period 2015–2023. These publications were then analyzed in different dimensions such as institutions, authors, collaborations, and networks. Batteries were among the most researched topics and Chinese institutions dominated in terms of volume. The citation count was used to investigate the importance of research.

Using Web of Science and a query with approximately 10 search terms, li-ion battery subfield fault diagnosis was studied [9]. The results indicate China's rapid growth in publication volumes since 2015, surpassing the United States to become clearly the largest producer of such papers in 2021. Vosviewer and other tools were used to analyze co-citations and collaboration networks.

A different approach to identify relevant papers is to use clusters generated based on citation relations [10]. A database with a Web of Science origin was used to study six sub-fields within battery research, as well as the whole scope of the European initiative Battery 2030+. The standing of Europe was compared with other countries or groups of countries in terms of volumes of papers as well as their field-normalized citation impact. It was noted that Europe was similar to China but well below North America in citation impact.

In a study of grid-connected Li-ion batteries, a five-step search strategy was deployed to discover the 100 most-cited papers in Scopus during the period 2010–2021 [11]. The study used search terms and the language of English and used subject filters as exclusion criteria. The United States had the highest number of publications in this top list followed by China.

A very short query "electric vehicle" was used to analyze relevant themes within battery-electric-vehicle research during the period 2000–2021 using data from Web of Science [12]. China was found leading in electric-vehicle research. In [13], a search query from a previous study from 2011 was reused to study li-ion battery research in India. Different types of electrified vehicles were studied using a query with search terms such as "hybrid NEAR/2 vehicle" and data from Web of Science [14]. Some patent data were also used, and whereas Japan was found to be most active in hybrid-electric-vehicle patenting, the United States led in paper volumes. Relating to electric vehicles, China produced the highest volume of papers.

Finally, to some extent representing the future of publication studies, a paper used text mining to shed some light on the content in battery-related publications [15]. This approach, which makes use of the full content in papers, is an interesting example of the opportunities and challenges with artificial intelligence tools.

This brief review of literature using patent and or paper data to study vehicle electrification and, in most cases, batteries, highlights from a methodological perspective that a combination and comparison of both types of publications is not common. None of the studies reviewed included attempts to analyze links between patents and papers, such as citations or having the same author. Moreover, more advanced quality-related indicators are not so common. If used, a direct citation count dominates, which has limitations, as the number of citations relates to the publication year as well as to the scientific field. Clustering and network analyses were often used to identify research and innovation trends.

Search strategies for patent studies were mainly based on patent classes, whereas papers typically were identified using queries. The development of the search strategy is not always explained, and the number of search terms is often limited.

Many of the papers reviewed include large sections with descriptive data covering the papers identified. In these sections, Chinese institutions often dominate, at least in terms of volume. A rapid growth starting in the period 2010–2015 is depicted, leading to China being the largest contributor of papers.

3. Methodology and Data

Critical for the study was to identify relevant publications. Patents were selected using patent classes, in line with a method described and used in an ambitious recent project led by the IEA [16]. The Swedish Intellectual Property Office was, in April 2022, commissioned to retrieve all battery patents from the global patent database DocDB, which then were further analyzed in a database for patent value assessment. Patent data until and including 2019 were considered sufficiently complete to be used in the analysis. This approach is in line with previous literature, which often uses patent classes to identify data and illustrates the significant lag between the year of study and the availability of complete patent data, c.f. [4,5].

Papers were selected using search terms in Scopus to be matched in the title or abstract of the paper. Scopus is the broadest abstract and citation database [17]. The query was developed in an iterative process, involving manual scrutiny of randomly selected papers to ensure that only relevant papers were selected. Papers from six productive battery researchers in Canada, the United States, Japan, and Sweden were used to test whether the query covered a sufficiently large share of these researchers' battery-related papers. The iterative process is described with some details in [18]. At the time of the study (June 2022), volume data for papers were almost complete until and including 2021.

The format of the query was: (A OR (B AND C)) AND NOT D, where

- A equals search terms specific for battery research, such as "electrochemical cell";
- B equals search terms often related to battery research, such as "battery";
- C equals a high number of search terms which in combination with B make it very likely that the publication deals with battery research, such as many different battery chemistries;
- D equals search terms in neighboring fields, such as "fuel cells", and words such as "batteryless".

The resulting query included hundreds of search terms. This approach led to an unexpected problem, as the standard query looks for matches in the title, abstract, and keywords. It was noted that the keywords include both the keywords given by the author(s) and other keywords, probably added by the journal. The latter keywords were in some cases broader, thus covering related fields not addressed in the paper. They did not work in combination with the "AND NOT" part of the query and thus a query only looking in the title and abstract was used.

In comparison to previous studies, the use of search terms to identify relevant publications is a dominating approach. However, three aspects differ. The first one is that previous studies do not always explain the strategy as to how the query was developed and how the precision of it was verified. A second difference is the use of only the title and abstract to search for relevant publications. Most other studies use the standard TITLE-ABS-KEY approach, which might work very well if "AND NOT" arguments are not used. Thirdly, the query developed differs substantially in size. Our query involved around 170 search terms plus the use of countries/regions and years to identify subsets. It is not always an advantage to use a very long query, but, for the purpose of this study, it was considered essential to ensure a reasonable coverage of all battery technologies over the 20-year period.

It is very difficult to capture all "battery-related" papers as blue-sky research, for example, does not always mention potential applications. Therefore, the resulting query underestimates the total volume and has a bias toward more applied battery research. A team of three battery experts from academy, business, and government supported in the development of the query.

The technical and economic value of patents was assessed using a composite index, the Technology Business Index (TBI), which combines several indicators, among them the patent's scope, family size, originality, generality, and backward and forward citations [19–21]. Percentiles were used to differentiate the patents, top 30% and top 10%.

We used a "full count" approach when a publication had several authors, both on individual and national levels. For example, this means that a publication with two authors, one from China and one from the United States, is counted fully for both countries. Various types of fractionalization constitute the main alternative, which, at least on the individual level, would have been rather confusing. Moreover, there are very few battery-related papers with many co-authors, which means that a full count approach does not lead to a severe bias in terms of volumes and citations.

For papers, standard citation indicators such as percentiles and the field-weighted citation impact, FWCI, were used. The latter is a normalized indicator based on the field, year, and type of publication. An average paper has FWCI 1.00 and if the paper has FWCI equaling 1.50, it is cited 50% more than the average publication.

This quality dimension was only used in a few previous studies and, typically, only with basic citation counts. To our knowledge, the quality indicator for patents has never been used in combination with different elaborated quality indicators for papers.

Given the sponsor of the project, the Swedish Energy Agency, the analysis had a focus on Sweden and the selection of countries for comparison was made from a Swedish perspective. In total, 11 countries were covered, some of which are not included in this paper, as they have relatively low patent volumes.

This study also included attempts to study institutions and individuals. For example, do researchers with many papers also have patents? This part of the study, which is unique in comparison to previous literature, was associated with a lot of manual work, and it was only carried out for Sweden. The main reason why this was laborious was the patent data quality, which made it difficult to identify people and institutions, as the names were indicated in many ways.

4. Results

4.1. National Level—Volumes of Patents and Papers

In Figure 1, the annual volumes of patent applications are indicated for all eight countries. Since 2011, China has had tremendous growth, becoming the largest patenting nation in 2014 and thereafter continued to increase the volume at the same pace. The dip in 2019 is probably due to incomplete data. Republic of Korea and Japan alternated as the number one until 2014 and thereafter as the number two. Since 2012, the United States has been in fourth place when it comes to battery-related patenting.

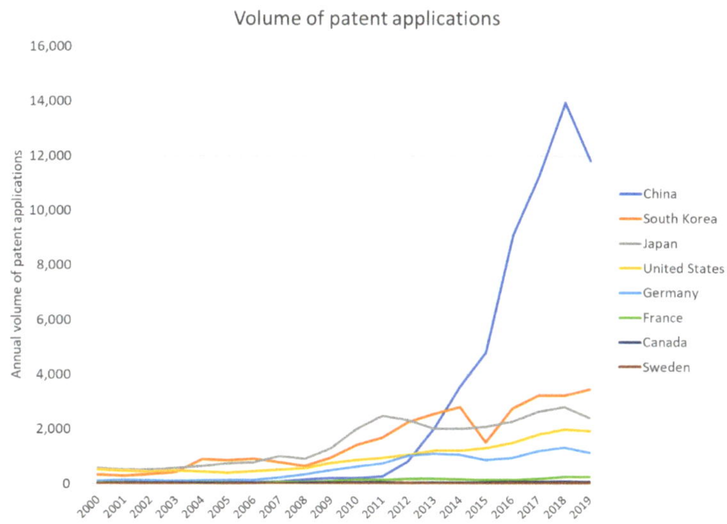

Figure 1. Development of patent volumes.

A closer look at the countries with lower volumes, see Figure 2, shows that Germany started patenting at an increasingly higher frequency in 2006, leaving the other countries included far behind. Sweden is clearly the country with the lowest volumes in the sample. Canada has, since 2012, developed to have approximately twice the annual volume compared to Sweden.

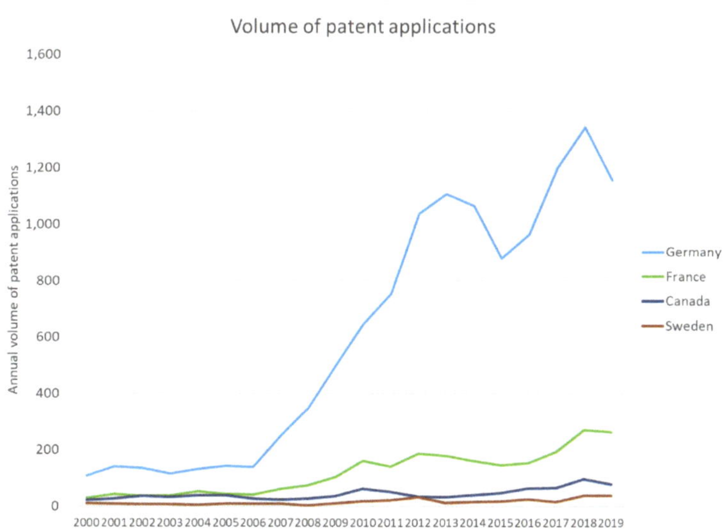

Figure 2. Development of patent volumes (countries with lower volumes).

On average there are approximately the same volumes of scientific publications relating to batteries as there are patent applications. When comparing Figure 1 with Figure 3 (below), it can be noted that China took the lead earlier in papers, in 2005, and that the United States since then has been the number two.

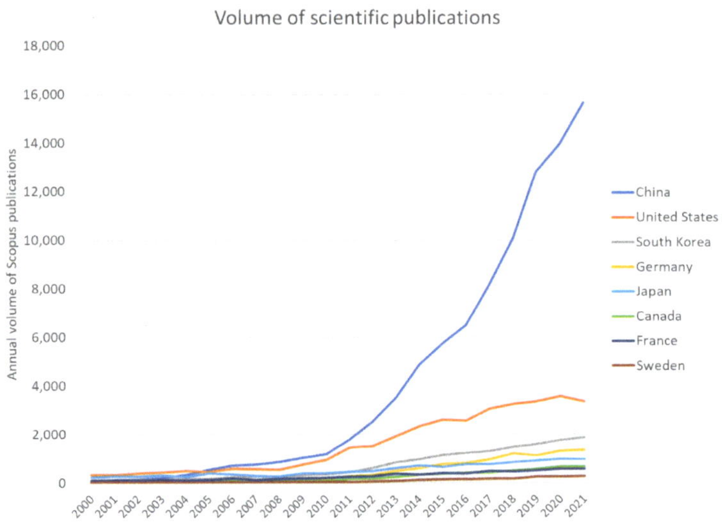

Figure 3. Development of paper volumes.

When China and the United States are removed, see Figure 4, the steep trajectory of Republic of Korea's papers becomes visible, overtaking Japan in 2011 and ten years later it had approximately twice the volume. A similar dramatic increase is also valid for Germany, which has more than quadrupled its paper volume in the last decade.

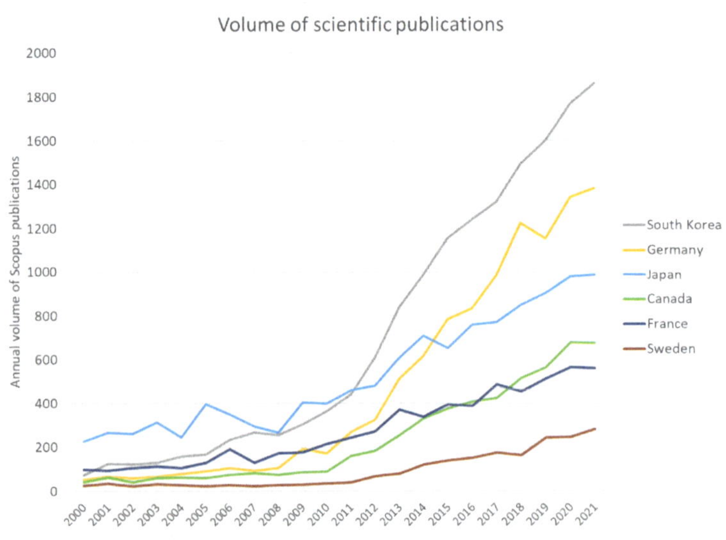

Figure 4. Development of paper volumes (excluding China and the United States).

A final times series is presented in Figure 5. Here the development of the volumes of patents and papers are possible to compare for China and the United States.

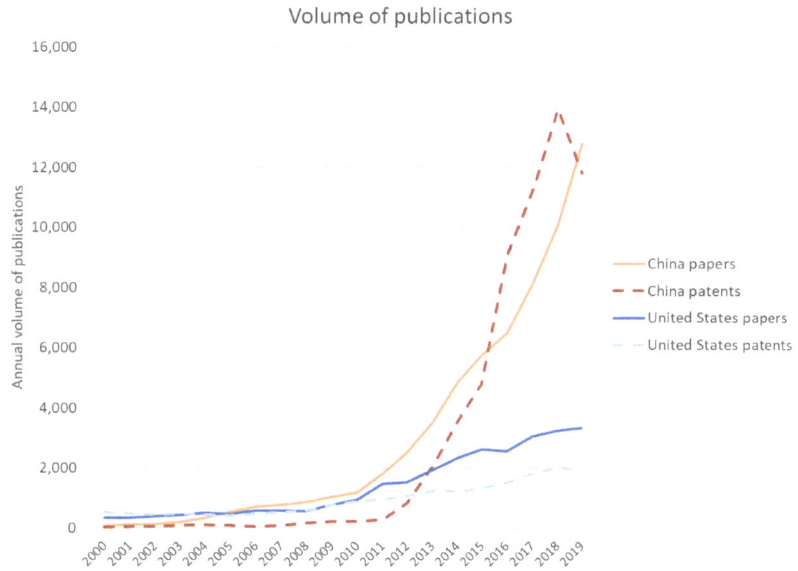

Figure 5. Development of patent and paper volumes for China and the United States.

The two countries show very different developments. China's paper volumes are much larger than the patent volumes until 2012, and thereafter the patent volumes after only a few years surpass the paper volumes. The United States had in the beginning of the period higher volumes of patents than papers. In 2010, the paper volumes started to increase more rapidly, and in the last period it clearly had higher volumes of papers. The dip in patent volume for China 2019 is probably due to incomplete data.

Three six-year periods were used to obtain a sufficient volume of patents for each period. In Table 1, the volumes of patents and papers for these three periods are presented.

Table 1. Comparison of paper and patent volumes.

	2002–2007			2008–2013			2014–2019		
	Paper	Patent	Paper/Patent	Paper	Patent	Paper/Patent	Paper	Patent	Paper/Patent
Canada	389	194	2.01	849	240	3.54	2619	390	6.72
China	2717	479	5.67	10,937	3,772	2.90	48,138	54,485	0.88
France	778	274	2.84	1452	845	1.72	2572	1188	2.16
Germany	499	915	0.55	1582	4391	0.36	5604	6608	0.85
Japan	1862	4349	0.43	2621	11,117	0.24	4643	14,300	0.32
Republic of Korea	1076	4267	0.25	2813	9590	0.29	7788	17,026	0.46
Sweden	159	43	3.70	284	97	2.93	988	149	6.63
United States	2984	2818	1.06	7182	5489	1.31	17,216	9796	1.76
World	13,775	14,939	0.92	33,831	38,541	0.88	102,132	111,518	0.92

Globally, the number of battery patents is slightly higher than the number of papers leading to a ratio around 0.9. A similar ratio applies for China in the last period included. In some countries, patent production dominates, among them Japan, Republic of Korea, and Germany. In others, the volumes of papers are clearly larger. Canada, Sweden, France, and the United States appear to focus more on research than patenting. For China, the share of patents per paper has increased over the periods, whereas, in Canada and the United States, the trend has been in the opposite direction. Globally, the ratio has been rather stable.

When looking at the period 2014–2019, Canada and Sweden are rather extreme with almost seven scientific papers per patent, whereas Japan is extreme in the other direction with approximately three patents per paper.

Another type of innovation indicator is academic–corporate co-publications, which are defined as scientific publications with at least two co-authors and at least one with an academic and one with a corporate affiliation. A high share of such publications is considered positive for innovations to materialize.

In Table 2, all countries except China have a higher share of academic–corporate papers within the battery field than the average for all papers in the country. In Canada, Germany, and Japan, the share is around twice as high.

Table 2. Academic–corporate collaboration (2014–2019).

Academic–Corporate Co-Publications (Share of)		
	Batteries	**All**
Canada	9.0%	4.3%
China	2.0%	2.7%
France	8.3%	6.3%
Germany	11.4%	6.5%
Japan	11.8%	6.4%
Republic of Korea	5.9%	4.9%
Sweden	9.3%	7.5%
United States	5.8%	4.7%

4.2. National Level—Quality-Related Indicators

In Table 3, two citation-based indicators for papers are presented, as well as TBI percentiles for patents. These indicators are explained above in the Section 3. Among the listed countries, battery papers are clearly more cited than all papers. The United States had the highest field-weighted citation impact, FWCI, as well as the highest share of papers in the top 10% citation percentile. Canada had the second highest FWCI and China the second highest share of papers in the top 10% percentile. Given China's dominance in paper production, it is interesting that the quantity does not come at the expense of quality, rather the opposite.

Table 3. Comparison of quality-related indicators for papers and patents (2014–2019).

	Paper Citation Data		**Patent TBI Value**	
	FWCI	**Top 10%**	**Top 10%**	**Top 30%**
Canada	2.47	43%	25%	54%
China	2.30	44%	7%	28%
France	1.89	32%	9%	24%
Germany	2.10	37%	6%	16%
Japan	1.60	29%	12%	34%
Republic of Korea	1.89	39%	8%	25%
Sweden	2.24	40%	16%	29%
United States	2.79	46%	23%	50%

Red (high value) to blue (low value).

The patent TBI values differ more between the countries than the citation impact indicators. Canada had the highest TBI values in both percentiles followed by the United States. Germany and China had the lowest TBI values. Japan, which was active in patenting, did so with a slightly better than average patent value.

A high share of academic–corporate papers is, as stated above, considered positive for innovation and it is also of interest to study whether the papers are cited. In Table 4, the citation impact for all battery papers and battery papers with academic–corporate collaboration are compared.

Table 4. Comparison of different types of battery papers.

Field-Weighted Citation Impact (2014–2018)	All	Academic–Corporate Collaboration
Canada	2.47	2.89
China	2.30	1.89
France	1.89	2.20
Germany	2.10	2.85
Japan	1.60	1.54
Republic of Korea	1.89	2.17
Sweden	2.24	1.49
United States	2.79	2.68

On a global level, academic–corporate co-publications are typically more cited [22]. In the battery field, this was also the case in four of the eight countries, with Germany exhibiting the largest positive difference. Sweden had a relatively large difference in the other direction; here, the academic–corporate collaboration clearly did not bring citation benefits.

4.3. Actor Level—Sweden

The number of patent applications and Scopus publications for the most recent period with reliable data is presented in Figure 6. The volumes vary between the years, but it is rather clear that both types of publications increase. The ratio between them is approximately 0.2, which means that for every patent application there are five scientific papers.

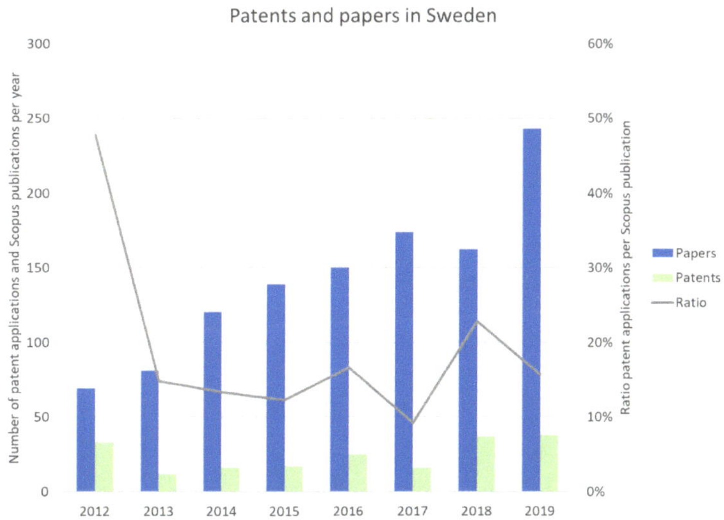

Figure 6. Development of patent and paper volumes for Sweden.

The gross list with patent applications for Sweden included more than 5000 items. It was associated with a lot of work to take care of all the name variants for people and organizations. Approximately 600 people and 124 companies had at least one patent application. Among them, 137 had both patent applications and scientific papers in the period 2000–2021. In Table 5, all people with at least 5 patent applications or 60 papers are included. Whereas all people with patent applications also have at least one paper in Scopus, the opposite is not always the case.

Table 5. Individuals with patent applications and/or papers.

Battery Researchers in Sweden with Minimum 5 Patents or 60 Papers 2000–2021			
	Battery Related		
Name	**Patents**	**Papers**	**Affiliation**
Legnedahl, Niklas	7	3	CEVT
Sturk, David	7	3	Autoliv
ASP, Leif	6	26	Chalmers University of Technology
Leijonmarck, Simon	6	13	KTH
Lindbergh, Goeran	5	116	KTH
Bryngelsson, Hanna	5	8	AB Volvo
Edstrom, Kristina	2	203	Uppsala University
Brandell, D.	0	150	Uppsala University
Johansson, P.	0	140	Chalmers University of Technology
Strömme, M.	0	73	Uppsala University
Younesi, Reza	2	62	Uppsala University
Matic, A.	0	62	Chalmers University of Technology

A long time series for companies is presented in Figure 7. During the oil crises in the 1970s, battery patenting was rather intensive. Since 2010, patenting activity has increased again.

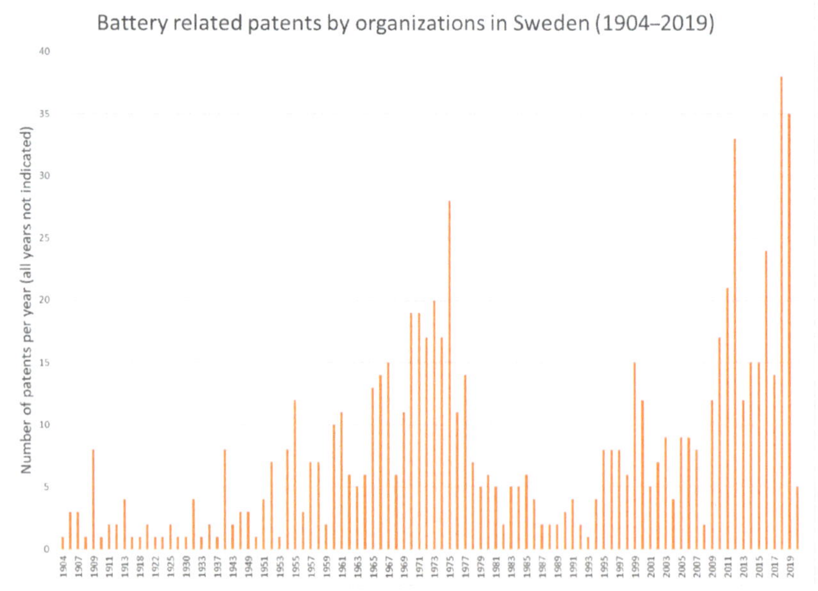

Figure 7. Development of patent applications in Sweden 1904–2019.

In Table 6, all companies with at least five patent applications during 2000–2020 are listed and divided into two time periods.

Table 6. Companies with patent applications in Sweden.

Battery Related Patents in Sweden (Minimum 5)		
Company	**2000–2010**	**2011–2020**
AB Volvo	5	34
Husqvarna	4	31
Scania CV	4	28
Nilar	16	5
Volvo Car Corporation	2	18
Ericsson Mobile Communications	7	7
Autoliv Development	3	8
Alelion Batteries	3	6
Effpower	5	1
Sony Mobile Communications	4	2
Lunalec	1	4

The automotive industry with AB Volvo, Scania CV, and Volvo Car Corporation appears to have increased its patenting activity substantially. Husqvarna, a company making garden equipment, has also intensified patenting. There are some battery manufacturers in Sweden, Northvolt probably being the most famous. Nilar, a battery company, filed for bankruptcy in December 2023. Another battery maker is Alelion Batteries, which also filed for bankruptcy in the autumn of 2023. Effpower terminated their operations in 2012.

Another very manual step in the analysis was to check which scientific publications were referenced in the patent applications. Slightly more than 100 papers could be identified in Scopus, where 92 were published in 1996 or later. In Table 7, the affiliations of the authors in these 92 papers are listed, including countries with at least 3 papers.

Table 7. Where scientific publications referenced in patent applications come from.

Papers	Country
32	United States
19	China
13	Sweden
10	Germany
8	United Kingdom
7	Australia
3	France
3	Italy
3	Taiwan

The United States dominates with one-third of the papers, followed by China and then Sweden. As the references are largely added by the reviewers of the patent applications, this reflects which literature they consider relevant.

Among the institutions affiliated in the papers, Linköping University in Sweden is included in seven papers, followed by institutions in the United States and the United Kingdom, see Table 8.

Table 8. Which institutions scientific publications referenced in patent applications come from.

Papers	Institution	Country
7	Linköping University	Sweden
4	Drexel University	United States
4	Imperial College London	United Kingdom
4	United States Department of Energy	United States
3	CNRS	France
3	RWTH Aachen University	Germany
3	University of New South Wales	Australia
3	University of Wollongong	Australia

5. Discussion—What Do the Numbers Say?

Quantitative studies have limitations and should be interpreted with care. It is often a good idea to use them as an input to generate an informed discussion among the actors in the field.

From a methodological perspective, the chosen approach appears promising. It is important to select a technological field that is large enough to result in reasonable volumes of publications. Analyses based on small numbers of publications seldom lead to solid results. One critical ingredient in the method was to involve experts in the battery field. The methodology for this study was developed in line with previous studies but added several unique or at least not so common features as explained in the Section 3. Among them, the combination of patents and papers to cover a broader range of innovation activities and the use of elaborated quality-related indicators are probably the most important ones.

The study confirms the massive development of Chinese patenting and research within the battery field. Whereas this has been indicated in several studies of papers, c.f. [6–9], the development in patenting activity is new to some extent. It is somewhat surprising how different the proportions of patents versus papers are in the countries studied and the diverging trends. The linear innovation model suggests a gradual development from research toward innovation, which in terms of patents and papers would mean that the ratio of patents per paper increases over time as the field matures. Data do not indicate such a trend, even though some countries, not least China, clearly had an increasing share of patents from 2002 to 2019. One possible interpretation is that the battery field is still developing rapidly with many new questions arising relating to everything from new chemistries to production methods.

The citation indicators and TBI percentiles highlight that the United States and Canada are strong in both patents and papers. China is stronger in papers, whereas Japan is somewhat stronger in patents. It should be noted that high quantity does not necessarily mean low quality. China, which made almost 50% of the global volume of battery papers in 2014–2019, did so with a high citation impact. Japan, which made three times more patents than papers in the same period also managed to achieve higher TBI values than the global average.

Academic–corporate collaboration is more frequent in the battery field than in general, at least when it comes to such co-publications. The associated citation impact varies between countries; some result in higher values and some in lower values than for all battery papers. As the citation impact is an important indicator for researchers, countries with a lower citation impact for academic–corporate papers might consider a closer study of how the collaborations are performing.

The actor level analysis focusing on Sweden provides interesting perspectives. Links between research and innovation are important and papers and patents provide data for a quantitative analysis of such links. It could be expected that a certain type of paper is more frequently referenced in patent applications. Potentially, it could be possible to trace an innovation from the original paper to one or several patent applications. In this study, we have not managed to make such a chronological ordering of the publications. But partly linked to the topic is the data for individuals with both patent applications and papers. Some individuals, but not many, carry out battery research resulting in many papers in combination with the writing of a few patent applications.

One policy implication of the study is that the battery field attracts large investments in knowledge production. Several countries show ambitions to secure a dominant position in the production of batteries for automotive and other applications. China dominates. Given the parallel investments in battery knowledge development and battery production, there appears to be an intricate balance between launching products onto the market and betting on the right technology. What if the massive investments in battery production become obsolete because they are not compatible with a new battery technology?

On a lower level, it appears rather easy to identify productive researchers. It might be relevant to nurture a dialogue with them to understand how research can be implemented. Even though there is a no right or wrong mix of patents and papers, a heavy focus on the latter indicates that there might be some missed opportunities.

In the case of Sweden, the battery companies with several patent applications since 2000 have not been successful. None of them were in operation in December 2023. This is worrying but should not be given too much emphasis. The battery industry is in a formative stage and a lot of changes are to be expected.

6. Conclusions

The purpose of this study was to develop and test a method to analyze the volumes of as well as qualitative aspects of patent applications and scientific publications. Battery development in several countries was used as a case. One conclusion is that this approach gives perspectives on battery research and innovation that are new and constitutes a valid starting point for further discussions on a policy level. For example, the substantial variations between countries in the volumes of papers versus patents triggers questions. What is the correct balance? How should we interpret China's rapidly increasing share of patents? By including the quality dimension for both types of publications, an estimation of whether it is only quantity or also quality is enabled. For clarity, we do not think that there is one ideal balance between the volumes of patents and papers. The balance depends on many factors, not least the speed of technology development. A publication study has many limitations, and one natural next step would be to discuss the findings with practitioners in the battery field. By doing so, the results can be scrutinized, and more nuances can be added. At the same time, the results have been communicated and potentially implemented to some extent. The results show that China during 2014–2019 dominated quantitatively and increasingly in both types of publications with a development toward a higher ratio of patent applications per scientific publications. The quality-related indicators show that the United States and Canada during the same period made highly cited scientific publications as well as patent applications with leading Technology Business Index values. On an actor level, the study illustrated how Swedish individuals and companies publish patents and papers. Automotive companies have recently started to file many patents relating to batteries.

Author Contributions: Conceptualization, H.P. and M.M.; methodology, H.P. and M.M.; software, H.P.; validation, H.P. and M.M.; formal analysis, H.P. and M.M.; investigation, H.P. and M.M.; resources, H.P.; data curation, H.P. and M.M.; writing—original draft preparation, H.P.; writing—review and editing, H.P.; visualization, H.P.; project administration, H.P. and M.M.; funding acquisition, H.P. and M.M. All authors have read and agreed to the published version of the manuscript.

Funding: This research was funded by the Swedish Energy Agency (Grant Number: P2022-01235).

Data Availability Statement: The original contributions presented in the study are included in the article, further inquiries can be directed to the corresponding author.

Conflicts of Interest: Hans Pohl is an employee of Lindholmen Science Park AB. Måns Marklund is an employee of Cascelotte AB. The paper reflects the views of the scientists, and not the companies.

References

1. Mejia, C.; Kajikawa, Y. Emerging topics in energy storage based on a large-scale analysis of academic articles and patents. *Appl. Energy* **2020**, *263*, 114625. [CrossRef]
2. Baumann, M.; Domnik, T.; Haase, M.; Wulf, C.; Emmerich, P.; Rösch, C.; Zapp, P.; Naegler, T.; Weil, M. Comparative patent analysis for the identification of global research trends for the case of battery storage, hydrogen and bioenergy. *Technol. Forecast. Soc. Change* **2021**, *165*, 120505. [CrossRef]
3. Hammarfelt, B. Linking science to technology: The "patent paper citation" and the rise of patentometrics in the 1980s. *J. Doc.* **2021**, *77*, 1413–1429. [CrossRef]
4. Crispeels, T.; Robert, D.; Verbeke, W.; Coosemans, T.; Van Mierlo, J. The Development of Hybrid and Electric Vehicles: Emergence and Development of the Patent Network. *World Electr. Veh. J.* **2016**, *8*, 611–622. [CrossRef]
5. Schmitt, G.; Scott, J.; Davis, A.; Utz, T. Patents and progress; intellectual property showing the future of electric vehicles. *World Electr. Veh. J.* **2016**, *8*, 635–645. [CrossRef]
6. Murugan, M.; Saravanan, A.; Elumalai, P.V.; Murali, G.; Dhineshbabu, N.R.; Kumar, P.; Afzal, A. Thermal management system of lithium-ion battery packs for electric vehicles: An insight based on bibliometric study. *J. Energy Storage* **2022**, *52*, 104723. [CrossRef]
7. Cabeza, L.F.; Frazzica, A.; Chafer, M.; Verez, D.; Palomba, V. Research trends and perspectives of thermal management of electric batteries: Bibliometric analysis. *J. Energy Storage* **2020**, *32*, 101976. [CrossRef]
8. Nurdini, A.; Nurcahyo, R.; Prabuwono, A.S. Waste from Electric Vehicle: A Bibliometric Analysis from 1995 to 2023. *World Electr. Veh. J.* **2023**, *14*, 300. [CrossRef]
9. Lan, J.; Wei, R.; Huang, S.; Li, D.; Zhao, C.; Yin, L.; Wang, J. In-depth bibliometric analysis on research trends in fault diagnosis of lithium-ion batteries. *J. Energy Storage* **2022**, *54*, 105275. [CrossRef]
10. Ahlgren, P.; Jeppsson, T.; Stenberg, E.; Berg, E.; Edström, K. A Bibliometric Analysis of Battery Research with the BATTERY 2030+ Roadmap as Point of Departure. 2022. Available online: https://www.diva-portal.org/smash/get/diva2:1754893/FULLTEXT01.pdf (accessed on 12 February 2024).
11. Wali, S.B.; Hannan, M.A.; Ker, P.J.; Abd Rahman, M.S.; Mansor, M.; Muttaqi, K.M.; Mahlia, T.M.; Begum, R.A. Grid-connected lithium-ion battery energy storage system: A bibliometric analysis for emerging future directions. *J. Clean. Prod.* **2022**, *334*, 130272. [CrossRef]
12. Barbosa, W.; Prado, T.; Batista, C.; Câmara, J.C.; Cerqueira, R.; Coelho, R.; Guarieiro, L. Electric Vehicles: Bibliometric Analysis of the Current State of the Art and Perspectives. *Energies* **2022**, *15*, 395. [CrossRef]
13. Dutta, B.; Kumarb, S. Scientometric study of lithium ion battery research in India during 1989 to 2020. *Ann. Libr. Inf. Stud.* **2021**, *68*, 430–441.
14. Wang, S.; Yu, J. A Bibliometric Research on Next-Generation Vehicles Using CiteSpace. *Recycling* **2021**, *6*, 14. [CrossRef]
15. El-Bousiydy, H.; Lombardo, T.; Primo, E.; Duquesnoy, M.; Morcrette, M.; Johansson, P.; Simon, P.; Grimaud, A.; Franco, A.A. What can text mining tell us about lithium-ion battery researchers' habits? *Batter. Supercaps* **2021**, *4*, 758–766. [CrossRef]
16. IEA; EPO. Innovation in Batteries and Electricity Storage: A Global Analysis Based on Patent Data. September 2020. Available online: https://www.iea.org/reports/innovation-in-batteries-and-electricity-storage (accessed on 13 March 2023).
17. Burnham, J.F. Scopus database: A review. *Biomed. Digit. Libr.* **2006**, *3*, 1. [CrossRef]
18. Pohl, H.; Karlström, M. Academic and Corporate Vehicle Electrification Research. *World Electr. Veh. J.* **2023**, *14*, 71. [CrossRef]
19. OECD. Enquiries Into Intellectual Property's Economic Impact 2015. Available online: https://one.oecd.org/document/DSTI/ICCP(2014)17/CHAP1/FINAL/En/pdf (accessed on 13 March 2023).
20. Donato, C.; Lo Giudice, P.; Marretta, R.; Ursino, D.; Virgili, L. A well-tailored centrality measure for evaluating patents and their citations. *J. Doc.* **2019**, *75*, 750–772. [CrossRef]
21. Williams, H. How Do Patents Affect Research Investments? Working Paper 23088, 2015. Available online: http://www.nber.org/papers/w23088 (accessed on 28 December 2023).
22. Pohl, H. Internationalisation, innovation and academic-corporate co-publications. *Scientometrics* **2021**, *126*, 1329–1358. [CrossRef]

World Electric Vehicle Journal

MDPI

Article

An Analysis of Vehicle-to-Grid in Sweden Using MATLAB/Simulink †

Jennifer Leijon [1,*], Jéssica Santos Döhler [1], Johannes Hjalmarsson [1], Daniel Brandell [2], Valeria Castellucci [1] and Cecilia Boström [1]

1 Division of Electricity, Department of Electrical Engineering, Ångström Laboratory, Uppsala University, 75237 Uppsala, Sweden

2 Department of Chemistry, Ångström Laboratory, Uppsala University, 75237 Uppsala, Sweden

* Correspondence: jennifer.leijon@angstrom.uu.se

† Presented at the 36th International Electric Vehicle Symposium and Exhibition (EVS36), Sacramento, CA, USA, 11–14 June 2023.

Abstract: With more electric vehicles introduced in society, there is a need for the further implementation of charging infrastructure. Innovation in electromobility may result in new charging and discharging strategies, including concepts such as smart charging and vehicle-to-grid. This article provides an overview of vehicle charging and discharging innovations with a cable connection. A MATLAB/Simulink model is developed to show the difference between an electric vehicle with and without the vehicle-to-grid capabilities for electricity grid prices estimated for Sweden for three different electric vehicle user profiles and four different electric vehicle models. The result includes the state-of-charge values and price estimations for the different vehicles charged with or without a bidirectional power flow to and from the electric grid. The results show that there is a greater difference in state-of-charge values over the day investigated for the electric vehicles with vehicle-to-grid capabilities than for vehicles without vehicle-to-grid capabilities. The results indicate potential economic revenues from using vehicle-to-grid if there is a significant variation in electricity prices during different hours. Therefore, the vehicle owner can potentially receive money from selling electricity to the grid while also supporting the electric grid. The study provides insights into utilizing vehicle-to-grid in society and taking steps towards its implementation.

Keywords: battery ageing; charging; simulation; smart charging; V2G (vehicle-to-grid); electric vehicle; infrastructure; electromobility; MATLAB/Simulink model

Citation: Leijon, J.; Santos Döhler, J.; Hjalmarsson, J.; Brandell, D.; Castellucci, V.; Boström, C. An Analysis of Vehicle-to-Grid in Sweden Using MATLAB/Simulink. *World Electr. Veh. J.* **2024**, *15*, 153. https://doi.org/10.3390/wevj15040153

Academic Editors: Joeri Van Mierlo and Genevieve Cullen

Received: 1 March 2024
Revised: 4 April 2024
Accepted: 5 April 2024
Published: 8 April 2024

1. Introduction

The increased number of electric vehicles (EVs) in society suggests the further development and implementation of new charging infrastructure and charging strategies, as well as the development of charging standards [1]. New ways of using and charging EVs may drive the transition towards electromobility. This article aims to provide an overview of the concepts of smart charging, vehicle-to-grid (V2G), vehicle-to-home (V2H), and vehicle-to-everything (V2X) [2]. An example of modeling EV charging and discharging to a grid in MATLAB/Simulink is presented. The analysis of the charging and discharging strategies includes potential pros and cons for different actors and what data could be of particular interest. The main goal is to contribute to the ongoing research discussion on future charging strategies of EVs. While there are many different types of charging strategies [3], this study is focused on cable charging (i.e., conductive charging), and it does not include an analysis of, e.g., wireless charging or battery swapping. While many previous studies focus on a larger international perspective on V2G, this study focuses on the modeling of V2G for the charging and discharging of new EVs to the Swedish electricity market as there has been a significant variation in electricity prices for the Swedish market recently, a strong interest in renewable energy sources (RES), as well as an ongoing trend toward introducing more EVs and implementing national charging infrastructure.

1.1. Charging Strategies for Electric Vehicles

Controlling the charging or discharging of EVs can potentially provide benefits in terms of, for example, lower charging costs, environmental aspects if charging occurs when there is a significant amount of RES feeding electricity to the grid, and providing grid balancing or ancillary services. However, a drawback of utilizing variable energy sources is a lower degree of utilization due to the intermittent nature of the RES, and therefore, energy storage could be used together with RES, as highlighted in, e.g., [4]. In [5], the different types of EV charging strategies are described as either uncontrolled or controlled charging and discharging strategies, where the controlled strategies are further divided into the subgroups: (i) indirect control, (ii) intelligent control, (iii) bidirectional control, and (iv) multistage control [5].

Controlled (unidirectional) charging, where the charging event is controlled in time, is often called smart charging. This is in contrast to uncontrolled charging, where the EV user charges whenever it is suitable based on the driving pattern [6]. In this context, the willingness of the driver to utilize, or not utilize, controlled charging is of interest [6]. One main benefit of smart charging for the EV owner is economical, which entails charging the vehicle when the cost of electricity is low rather than charging directly when the EV is connected to the charger. Smart charging could include starting and stopping the charging at certain times, or the power level of the charging is decreased or increased over the charging period. Depending on whether a customer has a variable or fixed-rate tariff, the cost optimization will look a bit different—but the fundamental purpose is the same: to minimize the total costs of charging. Smart charging could be beneficial for the grid owners too if the loads of the grid would be adapted to contribute to load balancing rather than charging all vehicles simultaneously. Controlled charging would therefore limit power peaks in the grid. This in turn could result in a limited need to reinforce the distribution grid and thus save money on installations and maintenance. However, controlled charging may affect the lifetime of the EV battery if the charging includes variable power levels. Smart charging may also include enhanced data-sharing. This suggests a concern for data protection and safety aspects in charging [7].

1.2. Vehicle-to-Grid

Bidirectional power flow between EVs and the distribution grid is often referred to as V2G. Certain EV models may provide discharging capabilities, such as the Nissan Leaf and Mitsubishi Outlander [8]. V2G operation is dependent on the inverter of the battery energy storage system (BESS), which has to be able to feed the current back to the grid [2]. There are several potential benefits of V2G considering the services provided to the grid [9]. V2G could potentially support the grid with ancillary services such as frequency regulation or peak shaving. The operation strategies may be differently applicable to various types of EVs, where V2G may be an interesting opportunity, e.g., for commercial EV fleets [10].

Recent research supports the fact that V2G technology has the potential to benefit electric utilities and microgrids, facilitating the integration of RES. Uncoordinated EV charging has a crucial impact on power systems [11], and extensive research has been conducted to analyze opportunities for the smart charging and discharging of EVs. V2G scenarios have been examined on a university campus [12], concluding that both V2G and stationary battery systems can be economical if the battery cost and electricity rates are considered.

From the user perspective, V2G functionality may contribute to extra revenue if the electricity can be sold back to the grid, especially during periods of electricity price peaks [13]. Looking at the techno-economic assessment of V2G in a microgrid, the authors in [14] highlight that several parameters impact the feasibility of V2G, e.g., the price of the chargers and the available capacity per car. Moreover, there are sociotechnical aspects of V2G that need to be further investigated, including, e.g., aspects regarding the motivation of the drivers in utilizing V2G and the driver's view on data sharing in the charging/discharging events [15]. For example, a comprehensive survey concluded that the income of users highly affects EV ownership and public interest in participating in V2G services [16]. The safety aspects of both charging and discharging to and from the grid are important to consider to ensure the protection of both the electric grid and the EV. Furthermore, the lack of concrete business models slows down V2G adoption [17]. V2G was investigated for the New York electricity market in [18] based on economic aspects and the availability of time for the charging.

When it comes to V2H, this concept enables house owners to utilize their EVs for energy storage at their homes. The EV would be both charged and discharged at the household. V2H could be an opportunity for EV owners to be more self-sustained, for example, when used together with photovoltaic (PV) systems on the roof of the house, and to ensure access to electricity even if the electric grid is not functioning properly. Vehicle-to-vehicle (V2V) enables charging and discharging between different EVs, whereas V2X is a broader concept including charging, discharging, and communication with EVs to the surrounding environment and society.

There are several safety and security aspects to consider for future EVs, especially if V2G is utilized. The risk of cyberattacks when utilizing EVs for load frequency control, and the need to detect and mitigate attacks, has been highlighted and modeled to support resilience [19,20]. Moreover, sensor attacks of the adaptive cruise control of vehicles could cause severe issues, as analyzed along with a proposed model in [19]. A recent review article discusses the benefits, challenges, and limitations of bidirectional charging and suggests research development directions [20]. In conclusion, the need for further research to address these challenges is compelling if the aim is to unleash the full potential of V2G. Technical aspects and also environmental, social, economic, and legal aspects need to be considered to make V2G a reality [21].

1.3. *Resilience of the Grid and Ancillary Services*

The main objective of vehicles is traditionally to transport people or goods if they are larger vehicles. Thus, EVs are typically treated as loads in power system analysis. Due to the possibility of utilizing EVs as mobile energy storage, it is of interest and relevance to investigate how EVs could increase resilience and manage distribution in grids and microgrids. Resilience includes the ability of a system to readapt after some disturbance [22]. In power systems, resilience is the capability of the system to prepare, adapt, withstand, and recover from any power outage [23]. In this regard, EVs can contribute to more reliable power systems by supporting the grid during typical outages and also support a more resilient power system that can sustain high-impact events [24]. These resilience-oriented events are generally known as low-probability high-impact events (for example, natural disasters and extreme weather events). Nowadays, such events may be increasing due to climate change, and the increase may be in both intensity and frequency, posing challenges to power systems. During certain events, the public might be evacuated. If so, EVs may not be available on-site, but EVs from nearby areas—not affected by the event—may be used [25]. During outages, EV batteries can be used as a backup resource, while after outages, EV batteries can be used to restore normal operation. The reliable charging of EVs during unusual events, including crises or natural disasters, could be analyzed more in future research. On a smaller scale, microgrids can be utilized for resilient power systems as long as they can survive critical loads and recover to normal operation after the events. Microgrids may provide good conditions for the development and implementation of solutions for grid resilience enhancement [26].

There are several interesting cases where EVs are used for grid applications. Firstly, EVs can be used efficiently for peak shaving and load to reduce the grid impact from a larger load. If the load has a high load factor originally, it could be reduced by implementing controlled charging and V2G. This is illustrated and analyzed in [27,28]. Secondly, EVs could also be used strategically in smaller energy systems consisting of one or a few loads together with local generation from, e.g., solar PV, potentially operating in island mode using the car batteries to balance the system. This would require an efficient and reliable control strategy for the inverters of the batteries in order to maintain the local grid's stability. Furthermore, if a large amount of EV chargers is operated by an aggregator, the cumulative capacity could be used strategically for more extensive grid applications. The capacity could be provided to the distribution system operator (DSO) or traded in available markets for ancillary services for power system stability or balancing purposes, e.g., frequency regulation services or balancing reserves.

It needs to be highlighted that EV batteries have a non-negligible cost. Also, EV owners may show a certain degree of skepticism in participating in ancillary services. EV batteries are considered degraded and not suitable for EVs when their capacity is reduced by 20–30%. However, the remaining 70–80% capacity can be used, after refurbishment, for less demanding purposes as a stationary storage system [29]. Thus, EV battery packs can contribute to grid resilience during their so-called second-life. Reusing batteries does not only enhance resilience, but it is also an environmentally friendly choice that can recover up to 20% of the initial battery cost [30]. However, there may be relevant aspects with regards to, e.g., the safety of the system or environmental aspects when refurbishing the batteries that need to be investigated further for second-life applications, and appropriate performance tests of retired batteries are important before reuse in other applications [31].

1.4. Vehicle Batteries and Stationary Batteries: Ageing

Providing V2G may result in the increased cycling of EV batteries depending on the use case, and this is relevant to estimate. However, if the vehicle battery is cycled more times, battery degradation becomes relevant to consider [13], and EV battery packs constitute a major part of the EV cost. Hence, it is important to evaluate battery aging when investigating V2G or V2X technology implementation. To ensure a long lifetime for the batteries and ensure safe usage, the state of charge (SOC) should be carefully estimated, as described, e.g., in [32], as well as the state of health (SOH) to better understand the aging process of the battery. The aging of the battery depends not only on how many times it has been charged and discharged but also under what circumstances it was charged and discharged (i.e., temperature, power level, etc.), and this relates to the cycle aging. When evaluating the aging of batteries in EVs due to increased cycling, it is often relevant to separate calendar aging from cycle aging. Gaining knowledge about the health of a used or retired EV battery can open up opportunities for the safe reuse of EV batteries in second-use applications [31]. In [33], empirical tests for capacity life loss were conducted on Li-ion cells for a selection of scenarios with varying C-rates, depth of discharge (DoD), and cell temperature. The results include several interesting aspects: first, the authors conclude that for lower C-rates the aging is not as dependent on the DoD effect as for higher C-rates. Second, capacity life loss models are presented for the chosen C-rates which can be implemented to estimate the capacity life loss for given conditions. According to [33], the capacity life loss can be estimated by Equation (1) when discharging with a current corresponding to C/2, that is

$$Q_{loss} = 30.330 \cdot \exp\left(\frac{-31500}{8.314 \cdot T}\right) A_h^{0.552} \tag{1}$$

In Equation (1), Q_{loss} is the estimated capacity loss (%), T is the absolute cell temperature, and A_h the energy throughput which is the product of the cycle number, DoD for the considered cycle (%), and the cell energy capacity (ampere hours). This is shown in Equation (2):

$$A_h = cycle\ number \cdot DoD \cdot Cell\ capacity. \tag{2}$$

Furthermore, when the current corresponds to C-rates higher than C/2, the capacity life loss model becomes more complex and can be described as

$$Q_{loss} = B \cdot \exp\left[\frac{-31700 + 370.3 \cdot C_{rate}}{R \cdot T}\right](A_h)^{0.55}. \quad (3)$$

In Equation (3), R is the gas constant, T is the absolute temperature, and B is a pre-exponential factor which decreases with increased C-rates and is determined in the fitting process of the capacity life loss model estimation. The values of B can be found in [33]. Another parameter relevant when modeling EV charging and discharging is the SOC in percentage. The method of SOC estimation includes the initial energy capacity $E_0(t)$ as the motor capacity of an EV is measured in kW. In this way, the relationship is given in [15] as

$$SOC(t) = SOC(t_0) - \frac{1}{E_0(0)} \cdot \int_0^t P_i(t)dt, \quad (4)$$

where $E_0(0)$ and $P_i(t)$ are the initial energy capacity and the instantaneous power fed from the battery into the load, respectively.

2. Methodology

The research design includes simulations in MATLAB/Simulink with study cases on the unidirectional charging and bidirectional charging of EVs in Sweden. The cases modeled in the simulations are based on the ongoing electrification in Sweden, where the electricity prices were significantly volatile at the end of 2022, resulting in higher household electricity costs during the winter months in 2022. Also, there is a significant number of EVs in Sweden. The novelty and the contribution of the model include an investigation of V2G charging based on estimated electricity prices in Sweden for several different EV owner user profiles in comparison to EV charging without V2G capabilities. The aim of the model is to gain knowledge on charging and discharging several EVs to the grid, with a control based on estimated local electricity prices. The objective of the V2G model is to investigate how V2G could function for a system of different vehicles, to compare the potential of EVs with and without V2G compatibilities, and to potentially enhance the economic revenue from charging and discharging EVs with V2G due to price variations.

2.1. *Modeling Charging and Discharging of Vehicles in MATLAB/Simulink*

The electric grid is represented in MATLAB/Simulink as a three-phase AC grid. The EV can be modeled in MATLAB/Simulink as a battery system. Therefore, the AC from the grid needs to be converted to DC for the battery, utilizing a converter. Thus, the converter system needs to be bidirectional to ensure a power flow in both directions. Input data to the model includes, for example, available data on EVs from [8]. To illustrate how EV chargers could operate dynamically by responding to an external control signal, the system is shown in Figure 1, where an aggregator plans the charging and discharging.

A set of chargers was modeled with varying characteristics. The EV charging was simulated for 24 h. The flowchart in Figure 2 presents the model with the different algorithms divided into different functions. An overview of the functionality of the proposed V2G model is also provided in Table 1.

Function 1 describes the initial conditions, including, e.g., the initial SOC value of each EV, as well as the electricity price set-point for when to sell or buy electricity (meaning when to charge or discharge the EV). The estimated electricity price in Sweden during a day with large hourly fluctuations was chosen to show how the chargers would operate during significantly different conditions. A price set-point was chosen in this simulation as 3 SEK/kWh (in Function 1), according to which the chargers would evaluate their operation mode. If the estimated electricity price exceeded the set-point, the chargers promoted the V2G mode, and for low prices, the chargers promoted the charging mode, as shown in Function 1 in the flowchart in Figure 2. Five of the vehicles in the simulation model could use the V2G mode (meaning that these vehicles could buy and sell electricity from or to the grid based on the estimated electricity price), and another five vehicles did not have this capability.

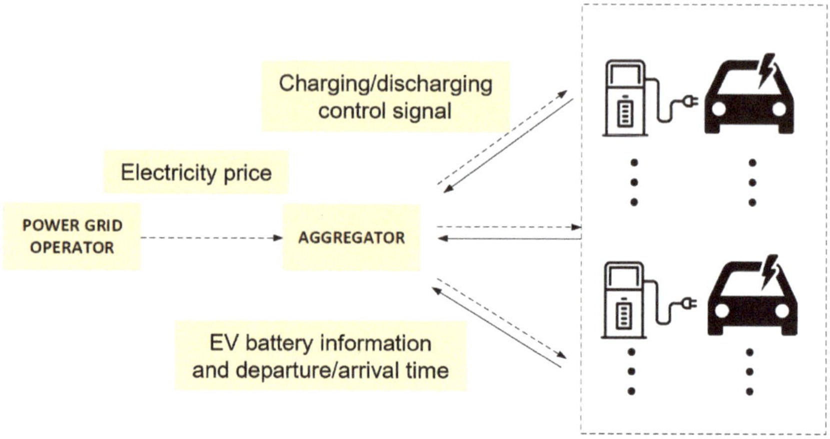

Figure 1. Parts of the model in MATLAB/Simulink, with several EVs, modeled as battery systems, connected to the three-phase grid. The different arrows show interconnection between the different parts of the figure and the dots represent several vehicles.

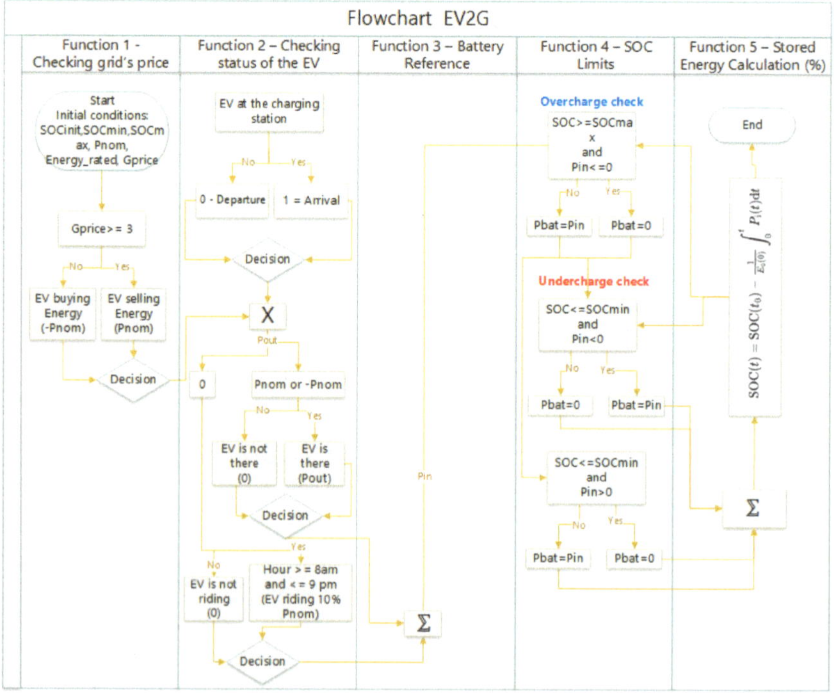

Figure 2. A sketch of a logic diagram of the reference to EV2G, providing an overview of the model for both charging and discharging.

Table 1. Overview of the functionality of the proposed V2G model.

Function	Functionality
Function 1	• Presents the initial parameters for the EVs, e.g., initial SOC, SOC limits, charging and discharging power, and electricity price set-point; • Evaluates current estimated electricity price to decide if the value is above or below the electricity price set-point.
Function 2	• Identifies if the EV is parked and connected to a charger or if the EV is being driven, with a limitation in time.
Function 3	• Summarizing the results from Functions 1 and 2 in order to set a power reference for the battery.
Function 4	• Identifies the current SOC value and evaluates this based on SOC limits; • Determines whether the EV should be charged or discharged.
Function 5	• A new SOC value is calculated.

In Function 2, it is determined whether the EV is parked and connected to a charger or not. If the EV is connected to a charger, there is an opportunity to use it for V2G with a set value of nominal power for charging and discharging, provided that the overall conditions (e.g., limits on SOC value, estimated electricity price, etc.) are fulfilled. If the EV is not parked, i.e., not connected to a charged, during the hours: 08:00 to 21:00, it is assumed that the EV is being driven and that the SOC value is dropping based on a set value.

In Function 3, the results from Functions 1 and 2 are summed up to decide whether the vehicle can be used for V2G or not.

In Function 4, the SOC of the EV is analyzed to find out if it is below or above the lower or upper SOC limits, set to 20% and 80%, respectively. The decision on whether to charge or discharge the EV depends also on the results from the previous functions.

Finally, in Function 5, the SOC of the EV battery is calculated based on Equation (4). The overall decision making for the vehicles in the model depends on all five functions described in the flowchart.

2.2. Input Data to the Case Study

The first version of the MATLAB/Simulink model for charging and discharging includes ten EVs, each with different estimated and assumed values regarding their battery systems (note that these values could vary), namely, three Nissan Leaf cars with BESS capacities of 40 kWh and an assumed 10 kW rated power for charging and discharging; two Mitsubishi Outlander, plug-in hybrids, with battery capacities of 13.8 kWh and a 3.7 kW rated power for charging and discharging; three Volvo cars, with a battery capacity of 69 kWh and an assumed 11 kW rated power for charging; and two Tesla cars with batteries of 57.5 kWh and an assumed 11 kW rated power for charging. The modeled EVs can be changed in user profiles (if the EVs are at home or away), initial SOC, maximum and minimum SOC, rated power for the charging and discharging, and battery capacity, to name a few configuration possibilities. This enables the modeling of different user profiles and different types of EVs. There is a trade-off between utilizing the car for personal transportation needs and enhancing the economic revenue from charging and discharging when there is a significant fluctuation in the electricity price.

The simulation model is a charging and discharging model of EVs based on a design approach, using MATLAB/Simulink in the phasor simulation type in 50 Hz for 24 h. This includes four different types of EVs, two of them in the charging and discharging mode (V2G) and the other two types in only the charging mode (EV). Three different user profiles are distributed among the vehicles. The model takes the estimated electricity price and user profile as input and generates the command to the vehicle. The case study focuses on the Swedish energy system. The electricity price over one day estimated for Sweden is used as input data to the model, and the estimation is shown in Figure 3. Electricity prices for different regions can be found on, e.g., Nord Pool [34] and in publications. The electricity price is based on the demand for electricity during the different hours. Based on Figure 3, the set-point of 3 SEK/kWh was chosen for this simulation (indicated by a dashed line). The high values of the estimated electricity price match a high national electricity demand, and the EVs could, at these moments, possibly contribute by selling additional electricity to the grid for support. Also, the opposite could occur, where the EV owner buys electricity from the grid when the estimated price is lower. The electricity price in Sweden varies with the days and seasons. The estimated price in Figure 3 varies over different days, where generally the prices are higher in the winter than in the summer. The electricity prices relate to the electricity production, with a significant amount of variable RES such as hydropower, wind power, and solar power in Sweden [35]. For a household with a PV system installed, the electricity purchased from the grid could be reduced in the summer, due to more sunlight and longer days, than in the winter.

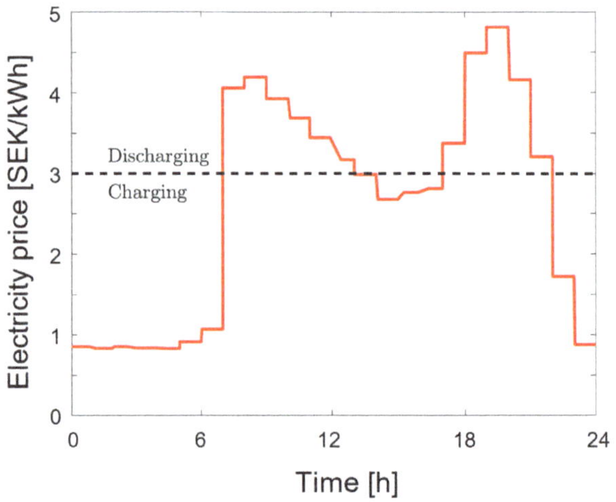

Figure 3. The electricity price estimated for a day in Swedish krona per kWh (SEK/kWh).

The user profile of the EV owner decides when the EV is at home, where an available bidirectional charger is assumed, or when the EV is not at home. The typical charging profiles of EVs vary (for example, if it is a weekend or weekday, the season, and type of life of the EV owner). Three user profiles utilized in this model are shown in Figure 4.

Considering the user profiles in Figure 4, the profiles follow patterns according to the following description; Profile 1: the car is parked for charging/discharging either at home or at work during some hours, and in between, it is driven a certain distance between the two locations. Profile 2: the car is parked at home in the morning and the evening, but during the daytime, there is no charging possibility at work. Profile 3: this profile corresponds to persons who work night shifts, where the car is parked at home during the daytime and parked at work during the nighttime with no charging possibilities. The three profiles are distributed among the ten EVs as follows: Profile 1 is added to Users 1, 6, 8, and 10; Profile 2 is added to Users 3, 7, and 9; and Profile 3 is added to Users 2, 4, and 5. The initial SOC is set to 50% for all ten cars, and the maximum and minimum SOC limits are set to 80% and 20%, respectively.

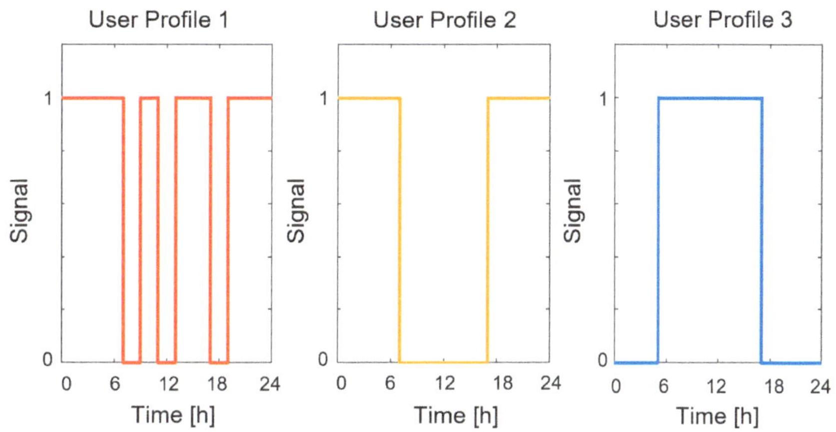

Figure 4. Three different user profiles of the EV owner, visualizing if the EV is parked or away. Signal 1 indicates that the car is at the charging station, whereas signal 0 means that it is not there.

3. Results and Discussion

The results from the simulations include, e.g., the SOC when using the EVs for bidirectional or unidirectional charging strategies, presented in Figures 5 and 6. The SOC values estimated for the EVs of the types Nissan Leaf and Mitsubishi Outlander, for different user profiles, are modeled and shown in Figure 5. These EVs are simulated to both charge from and discharge back to the grid, with V2G capabilities.

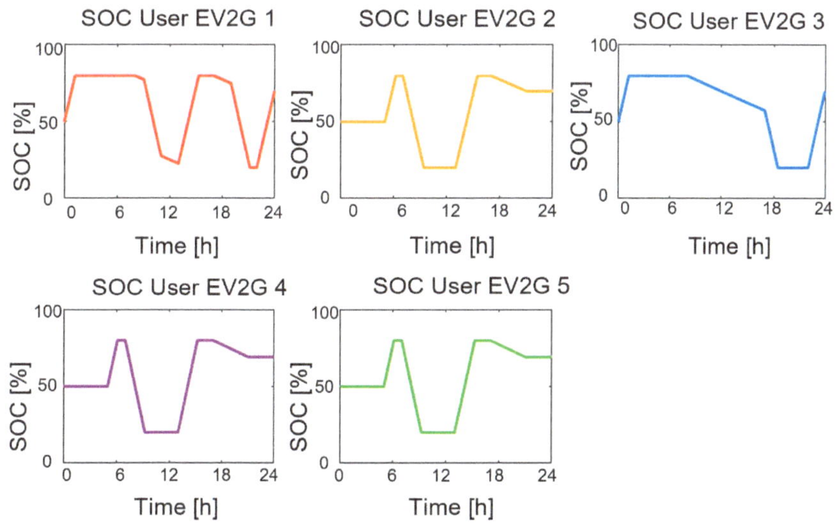

Figure 5. The state-of-charge value (%) over one day (hour) related to EV2G.

The SOC values estimated for the EVs of the types Volvo and Tesla are modeled and shown in Figure 6. It is noted that these EVs are only charged from the grid, with no V2G capabilities.

The different time periods of Figures 5 and 6 can be analyzed. According to Figure 3, before 07:00, the electricity price is below 3 SEK/kWh, allowing the vehicles only to charge until reaching the upper limit of the SOC. If the vehicle is not charging during this period, this is because of the user profile, meaning a car that is not connected to the charging station. This is the case for User 2, User 4, and User 5 (Profile 3 in Figure 4), presented in Figure 5.

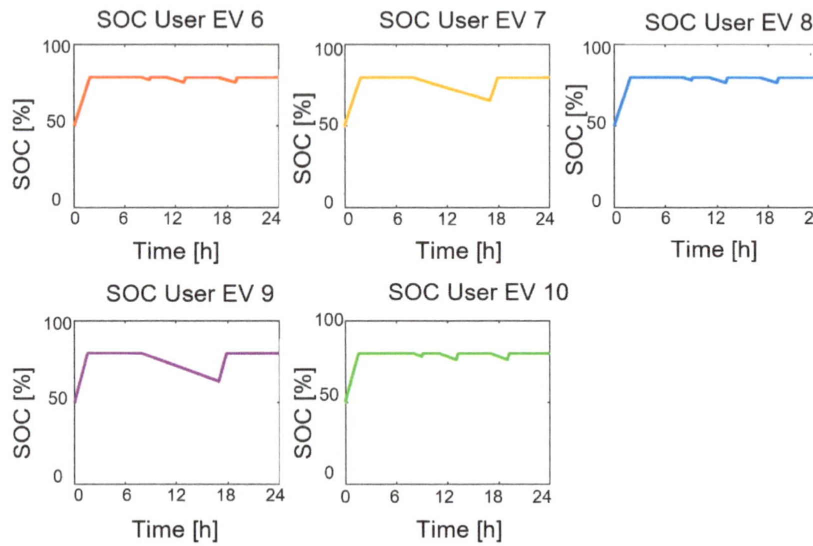

Figure 6. The state-of-charge value (%) over one day (hour) related to EV.

During the period 07:00 to 13:00, the vehicles only discharge as the electricity price is above 3 SEK/kWh, selling electricity to the grid until reaching the lower limit of SOC. However, it can be noticed that User 1, User 3, User 6, User 7, User 8, User 9, and User 10 in Figures 5 and 6 are discharging, even though the vehicles are not at the charging station. This occurs due to the vehicles traveling and discharging at an assumed 10% of the nominal power.

During the period 13:00 to 16:00, the vehicles only charge as the electricity price is below 3 SEK/kWh, buying the electricity from the grid until reaching the upper limit of SOC. However, it can be noticed that User 3, User 7, and User 9 (Profile 2 in Figure 4) are discharging because each vehicle is not connected to the charger and traveling, discharging at 10% of the nominal power.

From 16:00 to 21:00, the vehicles only discharge as the electricity price is above 3 SEK/kWh, selling the electricity to the grid until reaching the lower limit of SOC. From 21:00 to 07:00, the vehicles only charge as the electricity price is below 3 SEK/kWh, buying the electricity from the grid until reaching the upper limit of SOC. It can be noted that when the vehicle is not at the charging station, the SOC is constant, different from the other scenarios where the vehicles were traveling and discharging at 10% of the nominal power. This is due to the possibility of traveling during a certain period (from 08:00 to 21:00).

Aggregating the ten EVs to the grid, where half of the EVs provide V2G, the estimated cost of the charging or revenues from discharging and the power (kW) to and from the grid over one day are shown in Figure 7a,b.

From hour 00:00 to 07:00, in Figure 7a, the electricity price is below 3 SEK/kWh, and the EVs are charging (buying electricity from the grid). The negative signal represents that the grid is earning money from EV users. From hour 07:00 to 13:00, the electricity price is above 3 SEK/kWh, meaning that some of the EVs (i.e., the EVs with V2G capabilities parked at the charging station—User 1, 2, 3, 4, and 5) will sell electricity back to the grid, as can be indicated by the rise of the red curve in Figure 7a. But, the curve is still negative due to an imbalance between the electricity being sold to and bought from the electric grid. This variation (i.e., imbalance) in buying and selling electricity from and to the grid with all the EVs in the model is also shown in the final hours 13:00 to 24:00 in Figure 7a. Figure 7b shows the power to and from the electric grid based on the rated power for each EV for charging and discharging, presented in Section 2.2. The grid sells more electricity for charging the EVs than buys electricity from the EVs, which is reasonable since the grid buys only when the electricity price in this model is higher than 3 SEK/kWh, and only half of the EVs have V2G capabilities in the model. Charging and discharging based on the electricity price could be controlled by an aggregator to provide support to the electric grid. The charging and discharging of EVs will affect the power system. EV charging at high power levels, to provide a short charging time, can create power peaks in the electric grid. To use V2G on a large scale in society requires a robust electric grid, and V2G can also support the electric grid with balancing services and enhanced flexibility.

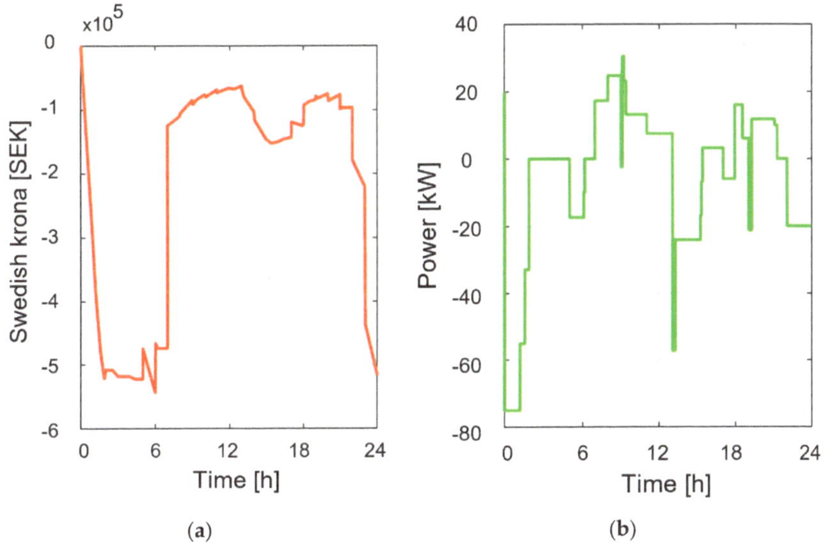

Figure 7. The price (SEK) in (**a**) and power (kW) to or from the grid in (**b**) to cover the charging and discharging of the 10 EVs over one day (hour).

While the main objective of this study is to investigate the potential economic revenues from V2G based on estimated local electricity price variations, future research could include an in-depth investigation of how the large-scale adoption of V2G may impact the electric grid (including, e.g., the load profiles of the grid). The benefits from utilizing V2G may not only be the economic revenues for each EV owner. V2G could potentially also support the local electric grid with grid balancing services, contribute with additional electricity at remote locations, or support the self-sufficiency of the EV owner if charged and discharged to a household.

However, the EV battery could be affected by this new bidirectional charging strategy. The lifetime of the EV batteries is affected by several factors, such as the ambient temperature and charging/discharging power levels. It is complex to estimate the SOH of an EV battery. Thus, the potential economic revenue from different charging strategies such as V2G is hard to estimate and varies from different specific cases. From the Swedish perspective, winters often provide negative ambient temperatures, and the charging or discharging of EVs outdoors may degrade the batteries faster, especially if the charging or discharging is carried out at high power levels (i.e., fast charging).

It is a challenge to propose a suitable economic compensation to an EV owner utilizing V2G as it should include both economic compensation for the electricity sold to the grid and for the potential EV battery wear. The opportunity to use V2G may also affect the warranty time of the EV, as well as the price of the EV on the second-hand market. The opportunities and challenges with implementing V2G in Sweden require further research, including both modeling and experimental work, to provide a deeper understanding of V2G from technical and economic perspectives.

As described in Section 2, the results presented are based on the charging and discharging of the ten vehicles during 24 h modeled in the MATLAB/Simulink simulation framework, with a phasor simulation type with 50 Hz. The model includes an algorithm for deciding when to charge or discharge the EV, including input data, e.g., SOC, estimated electricity price, EV type, and charging and discharging power levels. The simulation model is still at an early stage of development. In this first version of the model, all EVs have the same initial SOC, the analysis is only conducted for one day, the SOC lowered due to driving the EV is only roughly estimated, etc. This can be modified for in future versions, to simulate EVs driving a certain distance when it is not parked at home. Additional functionality can be included and added to the model to better simulate different types of EVs and V2G. Future simulations will be conducted with a real-time simulator, and so far, only first trials have been carried out to, for example, simulate transients. Choosing an appropriate control signal for V2G can sometimes result in a conflict of interest, e.g., if the estimated electricity price is low during local high-demand hours, which would suggest charging when the grid is already stressed. Therefore, it may be a good idea to prioritize the order of objectives if the chargers target both economic and technical objectives.

4. Conclusions

There are different charging and discharging strategies presented in the scientific literature, including smart charging strategies such as V2G, where the EV is not only charged from the electric grid but also discharged back to the grid. A simulation model of the charging and discharging of ten vehicles has been designed in MATLAB/Simulink. The model includes an algorithm for deciding when to charge or discharge the EV. The charging strategy for V2G capability in the model is related to the estimated electricity price, with the goal to charge the EV when the price is low and discharge when the price is high. The model also includes EVs with no V2G capabilities, meaning that these can only be charged from the grid. The results show how the SOC for different user profiles could vary over a day.

The maximum SOC value for using V2G was set to 80%, whereas the minimum value was 20%, and the starting value of each EV was 50%. The value 3 SEK/kWh was chosen as a set-point for when to charge (if the price was lower than 3 SEK/kWh) or when to discharge if V2G was an option (if the price was higher than 3 SEK/kWh). It was concluded that the model of V2G resulted in larger SOC differences (from 20% to 80% SOC) than if the V2G capability was not included. V2G can potentially support the power grid with grid balancing services.

The electricity price can vary with, e.g., different seasons and days due to the amount of RES connected to the electric grid. Therefore, the revenue from using the V2G will vary with different seasons and days. The electricity usage pattern may, however, be more or less similar for a workday in any season. If there are great variations in the electricity prices due to, for example, seasons with significant variations in electricity production from RES, the economic revenue from V2G will increase as the EV owner can buy electricity when the price is low and sell when the price is high. If the electricity price is more or less stable, which could be the case during some seasons, the financial incitements from using V2G will decrease. Generally, the electricity need in Sweden is greater in the winter than in the summer, and therefore, the V2G could be more important in the winter than in the summer.

While the estimated electricity price and variations over the day provide opportunities to create additional revenues, the battery system of the EV may be aged faster due to additional battery cycling. The results from the simulation show that the grid sells more electricity, due to EV charging, than buys electricity from the EVs due to V2G. There are limitations with this study, e.g., it is only based on MATLAB/Simulink simulations with no real experimental data from EV charging, and no experiments are included, and there are assumptions made on the type of EVs and the charging and discharging rates. Future research can include, e.g., improved estimations on discharging during driving, include a validation of simulated values in comparison to real-life data from EV charging, or include experiments on V2G in society. There are several barriers limiting the acceptance of V2G technologies, such as technical, economic, regulatory, social, political, and environmental challenges. Additionally, other important issues need to be addressed for the successful implementation of V2G, such as coordination among stakeholders, standardization, the deployment of charging stations, and the design of public policy incorporating EVs. This study investigates some of these aspects, bringing V2G technologies one step closer to more widespread implementation. This is the first step in modeling and understanding more about the opportunities and challenges with the charging and discharging of future EVs.

Author Contributions: Methodology, J.L., J.S.D. and J.H.; software, J.S.D.; validation, J.L., J.S.D. and J.H.; formal analysis, J.L., J.S.D., J.H., V.C., D.B. and C.B.; writing—original draft preparation, J.L, J.S.D., J.H. and V.C.; writing—review and editing, J.L., J.S.D., J.H., V.C., D.B. and C.B; visualization, J.S.D. and J.H.; supervision, C.B.; project administration, J.L. All authors have read and agreed to the published version of the manuscript.

Funding: This research was funded by Swedish Electromobility Centre (SEC), project: "Data exchange be-tween vehicle and power system for optimal charging", the Swedish Energy Agency project: "Re-source-efficient energy system solutions for airports with a high share of electric aviation (RES-Flyg)" grant number: P52433-1, the Resilient Competence Centre project: "Interregional Perspectives on Infrastructure Investment Optimization for Flight Electrification and Decarboni-zation—Uppsala University (iFED-UU)", SweGRIDS by the Swedish Energy Agency, Vattenfall AB, and E.ON, the Swedish Energy Agency project: "The impact of charging strategies on the electric car's battery system", grant number: P2022-01305, (in Swedish: "Laddningsstrategins in-verkan på elbilens batterisystem (LIEB)") and the SASUF project "Renewable energy sources for powering electric vehicles in South Africa". This project is conducted within the STandUP for En-ergy framework.

Data Availability Statement: The original contributions presented in the study are included in the article, further inquiries can be directed to the corresponding author.

Conflicts of Interest: The authors declare no conflicts of interest.

References

1. Das, H.S.; Rahman, M.M.; Li, S.; Tan, C.W. Electric Vehicles Standards, Charging Infrastructure, and Impact on Grid Integration: A Technological Review. *Renew. Sustain. Energy Rev.* **2020**, *120*, 109618. [CrossRef]
2. Islam, S.; Iqbal, A.; Marzband, M.; Khan, I.; Al-wahedi, A.M.A.B. State-of-the-Art Vehicle-to-Everything Mode of Operation of Electric Vehicles and Its Future Perspectives. *Renew. Sustain. Energy Rev.* **2022**, *166*, 112574. [CrossRef]
3. Leijon, J.; Boström, C. Charging Electric Vehicles Today and in the Future. *World Electr. Veh. J.* **2022**, *13*, 139. [CrossRef]
4. Sun, C.; Negro, E.; Vezzù, K.; Pagot, G.; Cavinato, G.; Nale, A.; Herve Bang, Y.; Di Noto, V. Hybrid Inorganic-Organic Proton-Conducting Membranes Based on SPEEK Doped with WO_3 Nanoparticles for Application in Vanadium Redox Flow Batteries. *Electrochim. Acta* **2019**, *309*, 311–325. [CrossRef]
5. Solanke, T.U.; Ramachandaramurthy, V.K.; Yong, J.Y.; Pasupuleti, J.; Kasinathan, P.; Rajagopalan, A. A Review of Strategic Charging–Discharging Control of Grid-Connected Electric Vehicles. *J. Energy Storage* **2020**, *28*, 101193. [CrossRef]
6. Kubli, M. EV Drivers' Willingness to Accept Smart Charging: Measuring Preferences of Potential Adopters. *Transp. Res. D Transp. Env.* **2022**, *109*, 103396. [CrossRef]
7. Johnson, J.; Berg, T.; Anderson, B.; Wright, B. Review of Electric Vehicle Charger Cybersecurity Vulnerabilities, Potential Impacts, and Defenses. *Energies* **2022**, *15*, 3931. [CrossRef]
8. Schram, W.; Brinkel, N.; Smink, G.; Van Wijk, T.; Van Sark, W. Empirical Evaluation of V2G Round-Trip Efficiency. In Proceedings of the SEST 2020—3rd International Conference on Smart Energy Systems and Technologies, Istanbul, Turkey, 7–9 September 2020. [CrossRef]
9. Tan, K.M.; Ramachandaramurthy, V.K.; Yong, J.Y. Integration of Electric Vehicles in Smart Grid: A Review on Vehicle to Grid Technologies and Optimization Techniques. *Renew. Sustain. Energy Rev.* **2016**, *53*, 720–732. [CrossRef]
10. Englberger, S.; Abo Gamra, K.; Tepe, B.; Schreiber, M.; Jossen, A.; Hesse, H. Electric Vehicle Multi-Use: Optimizing Multiple Value Streams Using Mobile Storage Systems in a Vehicle-to-Grid Context. *Appl. Energy* **2021**, *304*, 117862. [CrossRef]
11. Shariff, S.M.; Iqbal, D.; Saad Alam, M.; Ahmad, F. A State of the Art Review of Electric Vehicle to Grid (V2G) Technology. *IOP Conf. Ser. Mater. Sci. Eng.* **2019**, *561*, 012103. [CrossRef]

12. Clarke, A.D.; Makram, E.B. A Comprehensive Analysis of Plug in Hybrid Electric Vehicles to Commercial Campus (V2C). *J. Power Energy Eng.* **2015**, *3*, 24–36. [CrossRef]
13. Bhoir, S.; Caliandro, P.; Brivio, C. Impact of V2G Service Provision on Battery Life. *J. Energy Storage* **2021**, *44*, 103178. [CrossRef]
14. Van Kriekinge, G.; De Cauwer, C.; Van Mierlo, J.; Coosemans, T.; Messagie, M. Techno-Economical Assessment of Vehicle-to-Grid in a Microgrid: Case Study. In Proceedings of the 33nd Electric Vehicle Symposium (EVS33), Portland, OR, USA, 14–17 June 2020.
15. Bibak, B.; Tekiner-Moğulkoç, H. A Comprehensive Analysis of Vehicle to Grid (V2G) Systems and Scholarly Literature on the Application of Such Systems. *Renew. Energy Focus* **2021**, *36*, 1–20. [CrossRef]
16. Sovacool, B.K.; Kester, J.; Noel, L.; de Rubens, G.Z. Income, Political Affiliation, Urbanism and Geography in Stated Preferences for Electric Vehicles (EVs) and Vehicle-to-Grid (V2G) Technologies in Northern Europe. *J. Transp. Geogr.* **2019**, *78*, 214–229. [CrossRef]
17. Mojumder, M.R.H.; Ahmed Antara, F.; Hasanuzzaman, M.; Alamri, B.; Alsharef, M. Electric Vehicle-to-Grid (V2G) Technologies: Impact on the Power Grid and Battery. *Sustainability* **2022**, *14*, 13856. [CrossRef]
18. Zheng, Y.; Shao, Z.; Shang, Y.; Jian, L. Modeling the Temporal and Economic Feasibility of Electric Vehicles Providing Vehicle-to-Grid Services in the Electricity Market under Different Charging Scenarios. *J. Energy Storage* **2023**, *68*, 107579. [CrossRef]
19. Hu, Z.; Su, R.; Zhang, K.; Xu, Z.; Ma, R. Resilient Event-Triggered Model Predictive Control for Adaptive Cruise Control under Sensor Attacks. *IEEE/CAA J. Autom. Sin.* **2023**, *10*, 807–809. [CrossRef]
20. Zhou, Y.; Li, X. Vehicle to Grid Technology: A Review. In Proceedings of the 34th Chinese Control Conference, Hangzhou, China, 28–30 July 2015.
21. Taranto Rodrigues, L.; Gillott, M.; Waldron, J.; Naylor, S.; Rodrigues, L.; Shipman, R. Towards an Electric Revolution: A Review on Vehicle-to-Grid, Smart Charging and User Behaviour. In Proceedings of the 18th International Conference on Sustainable Energy Technologies—SET 2019, Kuala Lumpur, Malaysia, 20–22 August 2019.
22. Hussain, A.; Bui, V.H.; Kim, H.M. Microgrids as a Resilience Resource and Strategies Used by Microgrids for Enhancing Resilience. *Appl Energy* **2019**, *240*, 56–72. [CrossRef]
23. Panteli, M.; Trakas, D.N.; Mancarella, P.; Hatziargyriou, N.D. Power Systems Resilience Assessment: Hardening and Smart Operational Enhancement Strategies. *Proc. IEEE* **2017**, *105*, 1202–1213. [CrossRef]
24. Hussain, A.; Musilek, P. Resilience Enhancement Strategies For and Through Electric Vehicles. *Sustain. Cities Soc.* **2022**, *80*, 103788. [CrossRef]
25. Wang, Y.; Rousis, A.O.; Strbac, G. On Microgrids and Resilience: A Comprehensive Review on Modeling and Operational Strategies. *Renew. Sustain. Energy Rev.* **2020**, *134*, 110313. [CrossRef]
26. Castellucci, V.; Wallberg, A.; Flygare, C. Potential of Load Shifting in a Parking Garage with Electric Vehicle Chargers, Local Energy Production and Storage. *World Electr. Veh. J.* **2022**, *13*, 166. [CrossRef]
27. Wallberg, A.; Flygare, C.; Waters, R.; Castellucci, V. Peak Shaving for Electric Vehicle Charging Infrastructure—A Case Study in a Parking Garage in Uppsala, Sweden. *World Electr. Veh. J.* **2022**, *13*, 152. [CrossRef]
28. Flygare, C.; Wallberg, A.; Hjalmarsson, J.; Fjellstedt, C.; Aalhuizen, C.; Castellucci, V. The Potential Impact of a Mobility House on a Congested Distribution Grid—A Case Study in Uppsala, Sweden. In Proceedings of the CIRED Workshop on E-Mobility and Power Distribution Systems, Porto, Portugal, 2–3 June 2022.
29. Debnath, U.K.; Ahmad, I.; Habibi, D. Gridable Vehicles and Second Life Batteries for Generation Side Asset Management in the Smart Grid. *Int. J. Electr. Power Energy Syst.* **2016**, *82*, 114–123. [CrossRef]
30. Debnath, U.K.; Ahmad, I.; Habibi, D. Quantifying Economic Benefits of Second Life Batteries of Gridable Vehicles in the Smart Grid. *Int. J. Electr. Power Energy Syst.* **2014**, *63*, 577–587. [CrossRef]
31. Xu, J.; Sun, C.; Ni, Y.; Lyu, C.; Wu, C.; Zhang, H.; Yang, Q.; Feng, F. Fast Identification of Micro-Health Parameters for Retired Batteries Based on a Simplified P2D Model by Using Padé Approximation. *Batteries* **2023**, *9*, 64. [CrossRef]
32. Sun, J.; Jiang, T.; Yang, G.; Tang, Y.; Chen, S.; Qiu, S.; Song, K. A Novel Charging and Active Balancing System Based on Wireless Power Transfer for Lithium-Ion Battery Pack. *J. Energy Storage* **2022**, *55*, 105741. [CrossRef]
33. Wang, J.; Liu, P.; Hicks-Garner, J.; Sherman, E.; Soukiazian, S.; Verbrugge, M.; Tataria, H.; Musser, J.; Finamore, P. Cycle-Life Model for Graphite-$LiFePO_4$ Cells. *J. Power Sources* **2011**, *196*, 3942–3948. [CrossRef]
34. Nord Pool. Available online: https://www.nordpoolgroup.com/ (accessed on 4 April 2024).
35. Dong, S.; Li, H.; Wallin, F.; Avelin, A.; Zhang, Q.; Yu, Z. Volatility of Electricity Price in Denmark and Sweden. In *Proceedings of the Energy Procedia*; Elsevier Ltd.: Amsterdam, The Netherlands, 2019; Volume 158, pp. 4331–4337.

World Electric Vehicle Journal

MDPI

Article

Optimal Sizing of a Battery-Supported Electric Vehicle Charging Hub with a Limited-Capacity Grid Connection

Edward Heath *, Rick Wolbertus and Renée Heller

Faculty of Technology, Amsterdam University of Applied Sciences, 1091 GC Amsterdam, The Netherlands
* Correspondence: e.p.heath@hva.nl

Abstract: The ever-increasing electrification of society has been a cause of utility grid issues in many regions around the world. With the increased adoption of electric vehicles (EVs) in the Netherlands, many new charge points (CPs) are required. A common installation practice of CPs is to group multiple CPs together on a single grid connection, the so-called charging hub. To further ensure EVs are adequately charged, various control strategies can be employed, or a stationary battery can be connected to this network. A pilot project in Amsterdam was used as a case study to validate the Python model developed in this study using the measured data. This paper presents an optimisation of the battery energy storage capacity and the grid connection capacity for such a P&R-based charging hub with various load profiles and various battery system costs. A variety of battery control strategies were simulated using both the optimal system sizing and the case study sizing. A recommendation for a control strategy is proposed.

Keywords: electric vehicle; battery energy storage system; optimisation; genetic algorithm; charging hub

Citation: Heath, E.; Wolbertus, R.; Heller, R. Optimal Sizing of a Battery-Supported Electric Vehicle Charging Hub with a Limited-Capacity Grid Connection. *World Electr. Veh. J.* **2024**, *15*, 133. https://doi.org/10.3390/wevj15040133

Academic Editors: Joeri Van Mierlo and Genevieve Cullen

Received: 16 January 2024
Revised: 15 March 2024
Accepted: 17 March 2024
Published: 27 March 2024

1. Introduction

With the increased uptake of electric vehicles (EVs), the need for charging infrastructure is surging [1]. EVs are a part of a wider transition to electricity as an energy transmitter. Globally different economic sectors such as transportation, industry, and the heating of buildings are quickly being electrified [2]. Countries and cities that have historically relied on fossil fuels as the primary energy supply are facing issues as the electricity grid becomes congested, thus hampering the energy transition. For charging infrastructure, this implies that new grid connections or expansions are not available or a significant waiting list exists until grid expansion has been realised [3]. Charge point operators (CPOs) are looking for innovative ways to continue operations. These include battery-supported charging hubs.

Recently, a large number of sites have been installed with a battery energy storage system (BESS) at DC charging stations. Projects and studies with a BESS at large AC charging hubs have been missing. These projects are, however, more complex in terms of determining the optimal sizing of the system, as well as operating the system in the most efficient manner. These systems often require a lower, but more continuous, power than high-power DC systems. A temporary reduced power does not always have to be problematic. Inverter dimensioning and smart operation play a large role in the efficiency of the system. These unique features make dimensioning and the optimisation of these systems a different problem.

This paper is an expansion of the work presented at the EVS 36 conference [4].

1.1. Literature Review

There have been many studies that look into the use of batteries in combination with EV charging stations. Ref. [5] investigated the potential of a solar photovoltaic (PV) and BESS combination in a grid-tied, urban EV charging system. They optimised BESS and

PV array sizing, but failed to consider the space requirements for such a large PV array and the shading effects from surrounding buildings. They used hourly time resolution, which is not fine enough for such a volatile source as PV, nor for an EV charging station. Ref. [6] included a diesel generator as a back-up power source for peak times. Whilst this can work as a viable solution for grid-limited locations, it failed to consider the idea that many urban environments are aggressively cutting back on diesel engines to improve air quality and citizen health. The more suitable 15 min data resolution was used in this study, but failed to consider conversion losses for any source or the BESS and did not specify the power per charge point. Ref. [7] looked at the use of a PV + BESS combination to support a grid-connected EV fast-charging station under a few scenarios. Again, the size of a PV system, especially if it is to have any impact on a fast charger, would need to be very large; they stated in the paper that this would have to be between 0.5 MW_p–24 MW_p for different scenarios. A 0.5 MW_p PV array would require 1000 PV modules, and an array that is any larger would not be suitable for most urban environments.

The research on a BESS in combination with EV chargers has focused on fast chargers [8–10], and with good reason, given the grid volatility associated with the short duration and high power peaks from fast chargers. Again, refs. [8,9] did not use measured data and, instead, simulated EV loads; ref. [8] focused on plug-in hybrid EVs (PHEVs), a shortcoming given the prevalence of battery EVs (BEVs) nowadays, whilst [9] used four representational BEVs. Ref. [10] presented the optimal sizing of a BESS on a DC network with multiple fast chargers, a PV array, and industrial load. They used measured EV data; however, it is a general EV travel database and not specific to the case study. They do, however, investigate the charge scheduling of EVs and bi-directional charging to support the industrial load.

Access to reliable EV charging data is problematic. Refs. [5,8,9] simulated data using a mathematical formula. Ref. [6] derives data from another study. Ref. [7] uses data from a Dutch distribution network operator but included data from only two fast chargers and is, therefore, highly location specific.

Many other studies look into battery sizing optimisation in other applications, such as for prosumers in renewable energy communities [11], as neighbourhood-level storage at a low-voltage distribution level [12,13], and as storage in a microgrid setting [14–16].

Popular optimisation methods are the particle swarm optimisation algorithm, used in [5,7], and genetic algorithms (or variants thereof) such as in [11–13,17]. The objective function often seeks to minimise the annual investment cost, although other objective functions may include maximising self-consumption from PV production or minimising grid interaction.

1.2. Case Study

In a bid to further incentivise and provide for EVs whilst minimising the grid impact, Amsterdam, in partnership with local energy service provider Vattenfall and maintenance provider Heijmans, has installed a charging solution at a Park and Ride (P&R) facility. Thus, in order to be considered a small consumer, and therefore benefit from a faster connection time and reduced connection costs [18], a maximum grid connection capacity of 3 × 80 A was imposed. This pilot project serves as a learning opportunity for BESS-supported AC charging hubs. To further limit grid interaction, there was to be no energy flowing back to the grid from the BESS. This clause prevents profiting from energy arbitrage and limits the business case of this system.

Smart-charging strategies are often thought of as a first potential solution to grid-limited CPs, such as load shifting by suspending charge sessions or scaling current delivery with total CP power demand. In a P&R charging hub, smart charging can offer an improved charging efficacy both for the users and for the network operator. However, to ensure the user experience is not negatively impacted in the coming years due to increased EV penetration, this pilot project sought to investigate the use of a BESS.

1.3. Contribution

To the authors' knowledge, there have been no studies or projects besides the case study that supplied a charging hub using only a grid connection and BESS with the intention that the BESS limits grid loading during peak grid-load hours. This work uses measured data from the case study to address the oversized BESS and optimise the BESS and grid connection capacity using a variety of load profiles and 5 min time resolution. The model developed is a generalisable model of a BESS-supported type-2, level-2 charging hub, and is easily scalable for any number of CPs, grid connection capacity, BESS capacity, and load profile. The optimal system size was compared against the case study system for a number of BESS control strategies to form recommendations on sizing and control.

1.4. Structure

The paper is structured as follows: Section 2 introduces the optimisation problem, Section 3 describes the case study and details the model development, and Section 4 describes the chosen control strategies. Section 5 presents the optimal BESS and grid connection sizing for the case study and compares this against the currently installed case study sizing for a variety of control scenarios. For three different monthly loads, the change in the loss of load is plotted against the BESS capacity. Section 6 discusses the results in a broad and generalised frame, offering recommendations to future system developers and proposals for future work. Finally, Section 7 concludes the study.

2. Optimisation Problem

The sizing of a BESS in a grid-limited AC charging hub should be large enough to aid in supplying demand but not oversized such that there is an excess of capacity. A BESS remains an expensive investment so the intention is to keep the storage capacity minimal.

The Pymoo Python library [19] was used to form and solve the optimisation problem using a ($\mu + \lambda$) genetic algorithm. The optimisation problem considered mixed variables: integers over a range and the set of four discrete grid connection values. The Pymoo library only offers a genetic algorithm to solve such a mixed variable problem. Furthermore, future work will consider a multi-objective optimisation problem for which the NSGA-II algorithm can be used. The developers of the NSGA-II algorithm developed the Pymoo library, hence its choice. It is important to note that a genetic algorithm will find near-optimal solutions, and other algorithms may arrive at a more optimal solution.

The BESS control will have an effect on the optimisation process. In this optimisation problem, the standard 'base-case' control was used, which was derived from the case study. It is assumed that, in a newly installed and optimal system, the BESS would have individual phase power delivery, and, thus, phase imbalance and grid feedback would not be an issue. This means that the grid delivers up to the full connection capacity, and, if the BESS delivers power, it delivers only the remaining load above the maximum grid capacity.

The objective function is presented in Equation (1):

$$\min f(x) = \frac{C_B \cdot E_{B,max}}{L_{exp}} + C_G + 12 \cdot P_S \cdot \frac{LL}{\eta_{RT}} \tag{1}$$

where C_B is the cost of the BESS installation, currently approximately 650 €/kWh [20,21]. C_G is the cost of installation per grid connection capacity as presented in Equation (2) [18]. P_S is the profit from the electricity sale, assumed to be 0.1 €/kWh. The battery investment is annualised by dividing by the expected system lifetime L_{exp}, 10 years as per the battery supplier capacity warranty. Similarly, the monthly loss of load, LL, is annualised by multiplying by 12. η_{RT} is the mean battery round-trip efficiency. Operational and maintenance costs of the BESS are not included in this objective function since it is assumed that they would remain similar for a BESS regardless of its capacity.

$$C_G(P_G) = \begin{cases} 346\,€/year\ ,\ P_G = 3x25\ A \\ 1459\,€/year\ ,\ P_G = 3x35\ A \\ 2148\,€/year\ ,\ P_G = 3x50\ A \\ 3533\,€/year\ ,\ P_G = 3x80\ A \end{cases} \tag{2}$$

The case study system had a BESS power–energy ratio of 0.71. The constraint represented by Equation (3) was implemented to maintain a similar power–energy ratio. This allowed for some tolerance around the target value. To ensure a high quality of service, the constraint represented by Equation (4) was implemented, limiting the lost potential load, *LL*, to 100 kWh. This is equivalent to approximately 1.5% of the 7 MWh load profile used in the optimisation.

$$0.65{\cdot}E_{B,max} \leq P_{B,max} \leq 0.75{\cdot}E_{B,max} \tag{3}$$

$$LL = \sum\nolimits_{t_0}^{T} \frac{P_{EV,D}(t) + P_{Base}(t) + P_B(t) - P_G(t)}{12} \leq 100\text{ kWh} \tag{4}$$

where $P_{EV,D}(t)$ is the total EV power demand at time t, $P_{Base}(t)$ is the base load at time t, $P_B(t)$ is the power supplied by or delivered to the battery at time t, and $P_G(t)$ is the power supplied by the grid at time t. The battery current convention employed is a negative battery power for discharging. The sum of power over the entire time-series is then multiplied by 1/12 to convert from the 5 min time step to hours. T is the total time period of 1 month. All other constraints, Equations (5)–(8), were internal to the system model and were handled during simulation runtime. These included the power balancing, the battery state-of-charge (SOC) limits, and the battery charge/discharge power limits.

$$P_{Base}(t) + P_{EV,D}(t) + P_B(t) = P_B(t) + P_G(t),\ \forall t \in T \tag{5}$$

$$0.10{\cdot}E_{B,max} < E_B(t) < 0.95{\cdot}E_{B,max} \tag{6}$$

$$0 \leq P_{B,ch,t} \leq P_{B,ch,max} \tag{7}$$

$$0 \leq P_{B,dch,t} \leq P_{B,dch,max} \tag{8}$$

Due to the stochastic nature of the model, a single month-long load profile was formed and repeatedly used for the simulations in the optimisation process:

- A parent population of potential solutions was generated containing the decision variables $E_{B,max}$, $P_{B,max}$, and $I_{G,max}$;
- A simulation was performed for a single potential solution, and the outputs LL and η_{RT} were retrieved;
- Using these five decision variables, the objective function was evaluated and the results saved;
- This process was repeated for all possible solutions in the population of the current generation;
- A new parent population was created for the next generation, as the GA describes, allowing for crossover and mutation.

This study optimised the BESS capacity and grid connection capacity for a monthly load of 7 MWh (28 kWh/CP/day). Various BESS costs have been considered that span costs of up to the expected 2030 cost per kWh. The problem considered the base-case control strategy, defined in Section 4, and the discrete grid connection capacities of 3 × 25 A, 3 × 35 A, 3 × 50 A, and 3 × 80 A. The algorithm used a population size of 50 for 10 generations.

Additionally, the capacities were optimised for both a 5 MWh monthly load and a 6 MWh monthly load. All feasible solutions were plotted as the loss of potential load against the BESS capacity. These figures clearly illustrate the relationship between the BESS capacity and the loss of load.

3. Case Study

The main characteristics that define the BESS and CPs are presented in Figure 1 and Table 1. A 3 × 80 A grid connection is fed into the container housing the BESS, point *a* in Figure 1. The BESS is compiled from four battery stacks connected in parallel via four separate inverters, each fitted with a 100 A breaker. There is an air-conditioning unit within the BESS container to ensure a safe operating temperature is maintained. Leaving the container is the AC feeder line, point *b* in Figure 1, to which each of the eight dual-connector CPs are connected in parallel. Each dual-connection CP has had the phase connections rotated, as is standard [22]. In Case 2, single-phase EVs connect to the same CP, and phase rotation ensures they do not load the same phase. Each CP is fitted with a 35 A fuse per phase and each socket within the CP is fitted with a 20 A fuse per phase. This setup allows for the CPs to draw power from the grid, from the BESS, or a combination of the two. Similarly, the grid connection can feed power to both the CPs and the BESS given the available capacity. Conventional load sharing is applied when necessary [22]. The BESS is a commercially available system supplied by BECK [23]. The BESS uses lithium-ion technology, the common choice given its high cycle-life, high round-trip efficiency, and fast response time [24].

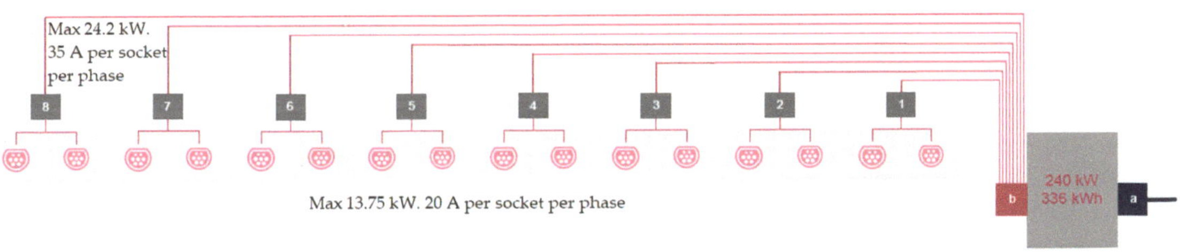

Figure 1. Schematic depicting the BESS and CPs. Point *a* is the grid connection rated at 3 × 80 A at 400 V, 55.4 kW. Point *b* is the output of the AC feeder line to the CPs from the BESS container.

Table 1. System components.

Component	Brand and Model	Specifications
Battery	BECK BESS Big Box	240 kW, 336 kWh
Charge points	Alphen Twin Public	3 × 35 A, 400 V, 24.2 kW per CP 3 × 20 A, 400 V, 13.75 kW per socket

The battery is considered to be oversized for the current operation. In the 6 months of data used to build the model, the BESS dropped below 50% SOC 15 times and the battery is cycled too frequently for low energies, as depicted in Figure 2. Frequent and small discharge/charge cycles are detrimental to battery lifetime [25,26]. Additionally, the battery is being charged and discharged at very low powers, resulting in low conversion efficiencies and high energy losses. Considering that the grid connection capacity could still be increased further from the programmed 3 × 25 A up to the physical connection of 3 × 80 A, which remains a cheaper option than the annualised battery system costs, the battery energy storage capacity could be reduced.

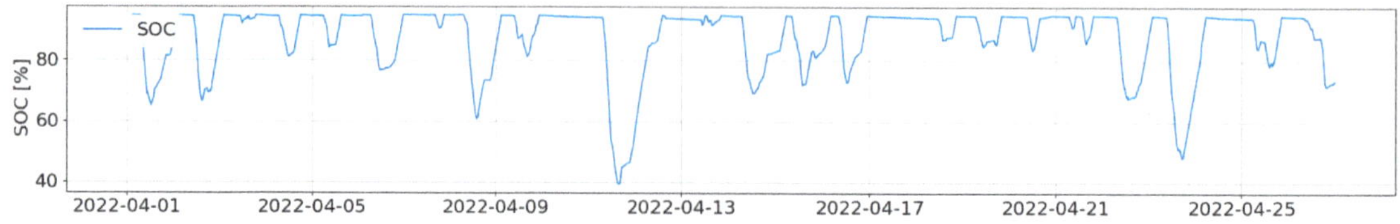

Figure 2. BESS state of charge over 4 weeks.

3.1. Model Development

The months of January–June 2022 were used to develop the model, with the month of July 2022 used to validate it. Individual charging sessions were identified and various session parameters determined, namely, the day of week, entry time, exit time, end of charging time, power delivery per time step, number of phases it is connected to, and the current per phase. The maximum charging power, total energy delivered, and connection and charging duration per charging event were deduced. The charging sessions were then filtered for charging duration and energy transfer, with limits of [0.5 h, 25 h) and [1 kWh, 80 kWh), respectively.

The entry events were separated by day since Friday, Saturday, and Sunday experienced different usage patterns than the working weekdays Monday–Thursday, as presented in Figure 3. This was to be expected, since, on Monday–Thursday, people generally follow similar work–life patterns. The peak connection time on weekdays is around 07:00–09:00, in line with commuter usage. There is also a second, smaller peak in the late afternoon from residents and evening visitors. It appears that a higher proportion of people arrive late on a Friday, to then park for the night or weekend. Saturday is a day in which people travel to the city for social/leisure purposes, whilst Sunday may still be regarded as a day of rest and, therefore, reduced P&R activity.

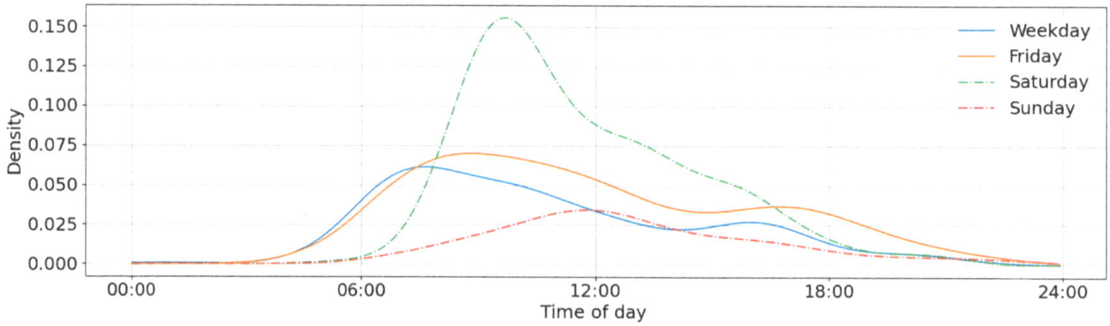

Figure 3. Probability density function of car arrival at a given time per day of week.

The charge sessions were then clustered into user groups using the Bayesian Gaussian Mixture Model (BGMM) from the Python scikit-learn library. When handling data with a distribution as visualised in Figure 4a, which plots the connection duration against the time of connection, a Gaussian Mixture Model (GMM) was found to be most suitable. BGMM performed better than a standard GMM due to the function's ability to minimise the number of clusters, resulting in two clusters per day type. The time of connection and connection duration were determined to be the best indicators of conscious user behaviour and are more strongly correlated than other variables. Including more than these two variables did not yield improved clustering performance.

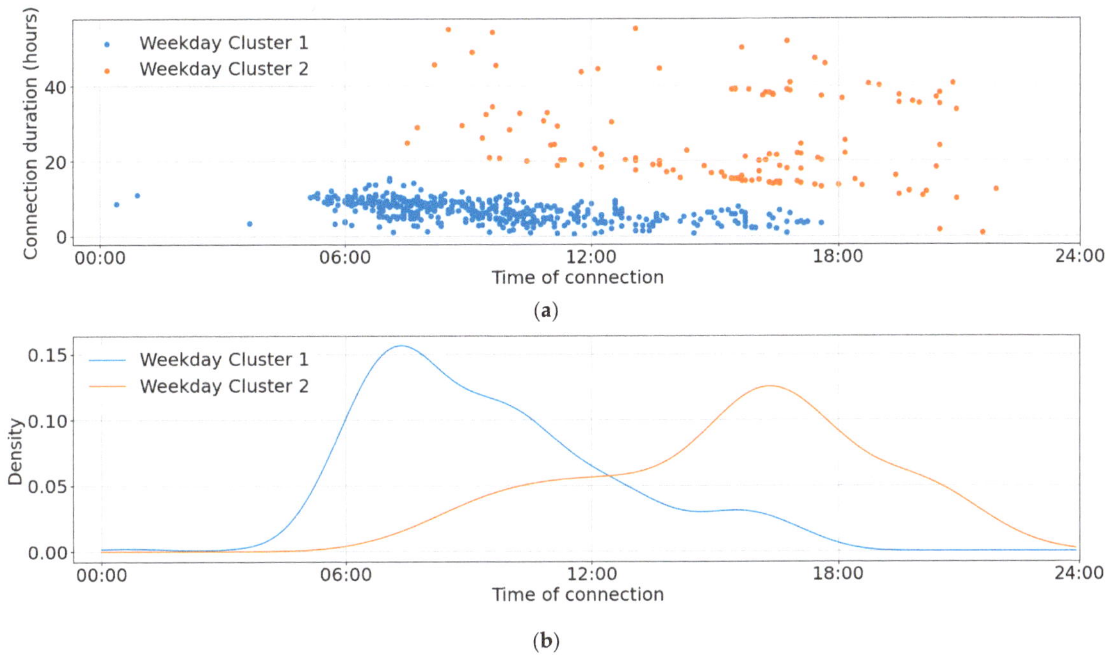

Figure 4. (**a**) Charge sessions clustered by time of connection and connection duration on weekdays. (**b**) Probability density function of time of connection per cluster on weekdays.

Figure 4a displays the clustered charge sessions on weekdays (Monday–Thursday). Cluster 1, in blue, shows the group of users dominated by commuters who typically arrive and depart on the same day. Cluster 2, in orange, represents a higher proportion of visitors who arrive in the late afternoon to evening, and park for longer durations. It is clear that the duration of connection has the largest influence on the clustering process.

The time of entry for the two clusters was plotted as a Kernel Density Estimation (KDE) curve with an independently normalised axis, as presented in Figure 4b for weekdays. A KDE plot introduces a normal Gaussian distribution per data point, and sums the curves to produce the density curve presented.

The model had a 5 min resolution. At each time step, the associated probability that a car would connect, as shown in Figure 3, was compared against a randomly generated number between 0–1. Upon connection, the charge session was assigned to a cluster by choosing between the normalised weighted probability of entry at that time step per cluster, as shown in Figure 4b. With the cluster determined, the nominal charging power was sampled, as was the energy demanded and connection duration. The number of phases it connected to was estimated from the nominal charging power. In this model, the line voltage was assumed to be constant at 230 V. Some noise was added to the charging power at each time step by sampling a normal distribution ($\sigma = 0.025$) to use as a multiplicative factor. A base load (system electronics and CP electronics), inverter self-consumption, and auxiliary battery load (air conditioning) were added. These were all sampled from distributions formed from the measured data. If the energy delivered in a charge session reached 75 kWh, then the charge session came to an end and the charge duration was updated. The maximum observed energy delivery in the measured data was 68.9 kWh.

The power was delivered by either the grid, the battery, or a combination of the two. In the case that the load was less than the maximum grid capacity, any residual grid capacity would charge the battery. If the load was greater than the maximum grid capacity and the battery was empty, conventional load sharing of the available grid capacity was employed, ensuring base loads were also provided for. Thus, assuming the base load was split evenly across the three phases, the power delivered to each EV at time t was calculated using Equation (9).

$$P_{EV,G,x}(t) = \frac{I_{Ph1,x}(t) + I_{Ph2,x}(t) + I_{Ph3,x}(t)}{I_{Ph1,T}(t) + I_{Ph2,T}(t) + I_{Ph3,T}(t)} \cdot (P_G(t) - P_{Base}(t)) \quad (9)$$

where $P_G(t)$ is available power from the grid at time t and $P_{Base}(t)$ is the total base power at time t. The numerator consists of the current draw per phase for EV x at time t, and the denominator consists of the total current demand per phase at time t.

For battery charging and discharging, the inverter efficiency was sampled from the efficiency curve depicted in Figure 5. The curve was fitted to the filtered data using Equation (10), and initial values $a = 0.95$, $b = -0.9$, and $c = 0.25$.

$$y = a + b \cdot e^{-c \cdot x} \quad (10)$$

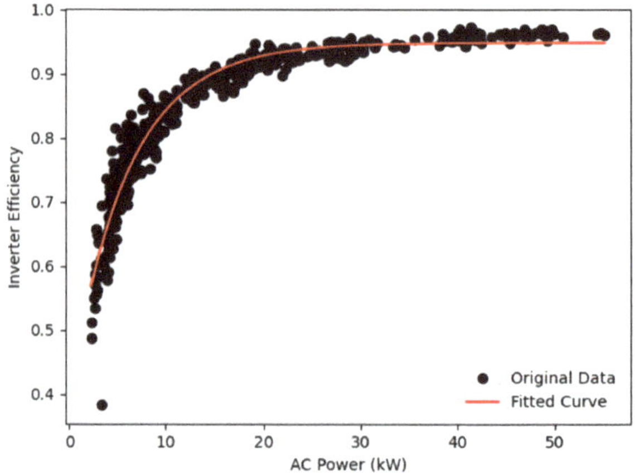

Figure 5. Inverter efficiency vs AC side power.

3.2. Model Validation

By comparing a 'base-case' simulation to the July 2022 data, the model was validated. The system usage in both the case study and simulation are presented in Figure 6a,b, respectively. A direct comparison of the system metrics is presented in Table 2, which shows that the fit of the model was in line with the measured data.

Table 2. Performance metrics of the case study and the simulation.

	Measured Data	Model Validation
Total load demand [kWh]	4869	4958
Number of charge events	219	221
Mean energy per charge event [kWh]	22.2	22.5
Mean charging power [kW]	7.8	7.7
Mean connection duration [hours]	12.2	14.9
Mean charging duration [hours]	3.9	3.7

The measured battery system round-trip efficiency was found to be 71.7%, and was influenced by the battery energy storage efficiency of 95.6% and the inverter efficiency for charging and discharging, as depicted in Figure 5.

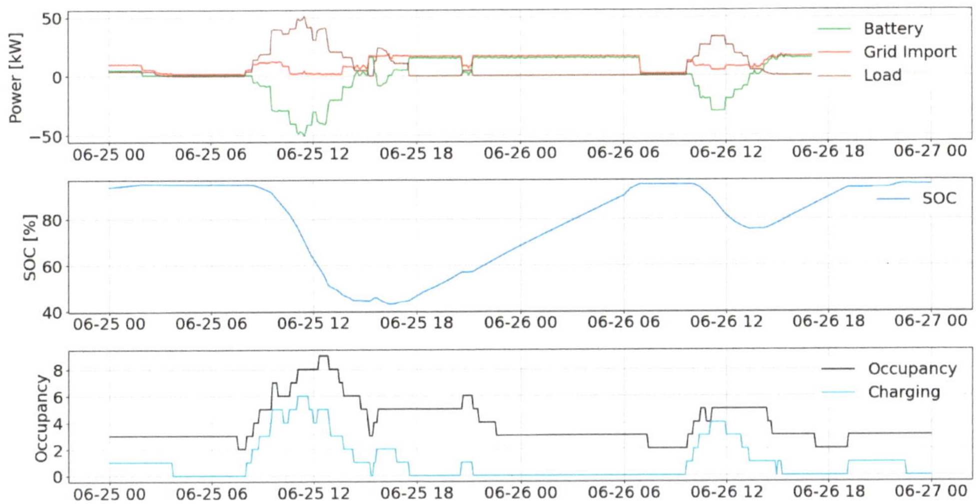

(**a**)

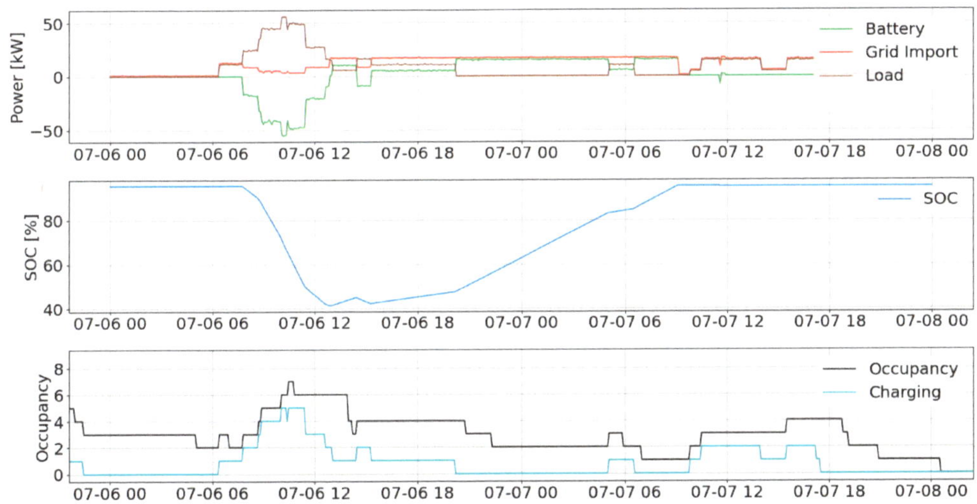

(**b**)

Figure 6. (**a**) Typical usage of the P&R system from measured data; (**b**) typical usage of the P&R system from the simulation.

4. Control Scenarios

The installed battery system round-trip efficiency was low due to the conversion losses across the inverter at low charging and discharging powers. Additionally, the battery made frequent and small discharge/charge cycles. Finally, there was no consideration for the power imported from the grid during peak grid-load hours. Given the current state of the power grid, it is logical to limit the power drawn from the grid during the peak grid-load hours of approximately 17:00 to 20:00. The following control scenarios were therefore decided upon:

1. The base case in which the grid supplies all load up to the maximum capacity. The battery supplies the remaining load above the maximum grid capacity.
2. During the peak grid-load hours of 17:00–20:00, all load is supplied by the battery. If there is no EV load, the grid will supply the base load. If the battery is drained, the grid will supply the load.
3. Charging/discharging the battery deadband of 10 kW and 15 kW. If the EV load is above this deadband, the battery supplies the entire load.
4. The combination of limited peak hour power draw and battery charge/discharge power deadband of both 10 kW and 15 kW.

Each scenario was simulated using the optimal sizing described in Section 2 and compared against the case study sizing. Due to the randomness in the stochastic model, each scenario was simulated five times and averaged. The chosen performance metrics were as follows:

- Potential load not delivered (lost load);
- BESS round-trip efficiency;
- Energy drawn from the grid during peak hours;
- Percentage of total load supplied by the BESS;
- Percentage of users still charging at the end of their session.

5. Results

5.1. Optimal System Sizing

Table 3 presents the optimal system sizing for three different BESS costs using the monthly load profile of 6899 kWh, which approximated the intended 7 MWh. The three costs span the current approximate cost of €750/kWh up to the expected cost of €250/kWh in 2030 [27]. Alongside this is the case study system sizing for which a simulation using the same load profile was performed, resulting in the higher monthly lost load and higher annual cost.

Table 3. Performance of the optimal sizing compared to the case study sizing, for a 7 MWh load profile.

	Case Study	Optimal System (€250/kWh)	Optimal System (€500/kWh)	Optimal System (€750/kWh)
Grid connection capacity, I_G	3 × 25 A	3 × 80 A	3 × 80 A	3 × 80 A
Battery energy storage capacity, E_B	336 kWh	100 kWh	69 kWh	49 kWh
Battery power capability, P_B	240 kW	71 kW	45 kW	34 kW
Annualised investment	€9518/year	€6115/year	€7101/year	€9340/year
Loss of potential load, LL	452 kWh	56 kWh	81 kWh	99 kWh

Clearly, a larger grid connection is preferable. Higher BESS costs result in an optimal sizing that favours a higher loss of load and smaller BESS capacity. However, the effect is minimal since the grid connection capacity cannot be increased further. If the LL constraint, Equation (4), were removed such that any LL was acceptable, then the BESS may not be included and LL may be much higher. However, such a system would be unsuitable as a charging hub. The optimal BESS sizing for a 7 MWh monthly load with at least one high-energy-demand day tends to become very small given that the average battery capacity of a newly available BEV is around 60 kWh [5].

The optimal system sizing, regardless of the BESS cost, delivers a much better quality of service to system users when compared to the case study system, for a reduced annual investment. For the highest BESS cost of €750/kWh, the loss of load was reduced from 6.5% of the total load to 1.5%, and the annual investment remained comparable. The predicted 2030 BESS cost of €250/kWh resulted in a loss of potential load of less than 1% of the total load, and the annual investment fell by 36%.

Figures 7–9 plot the BESS capacities against the loss of load for the three load profiles of 5 MWh, 6 MWh, and 7 MWh. They were each compiled from four optimisation calculations amounting to 200 feasible solutions. There are many more feasible solutions not shown.

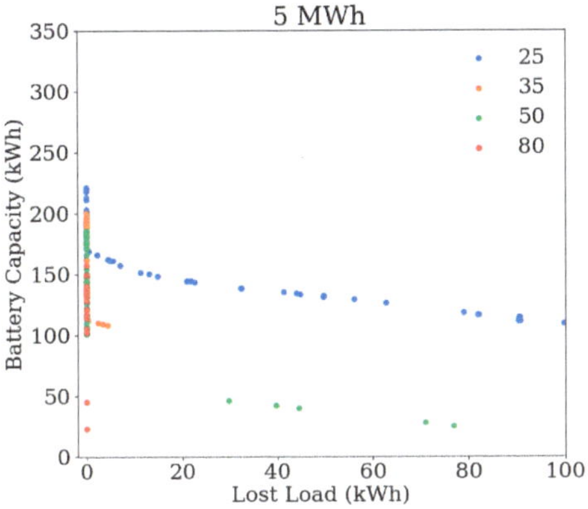

Figure 7. Feasible solutions to the optimisation problem considering a 5 MWh load profile.

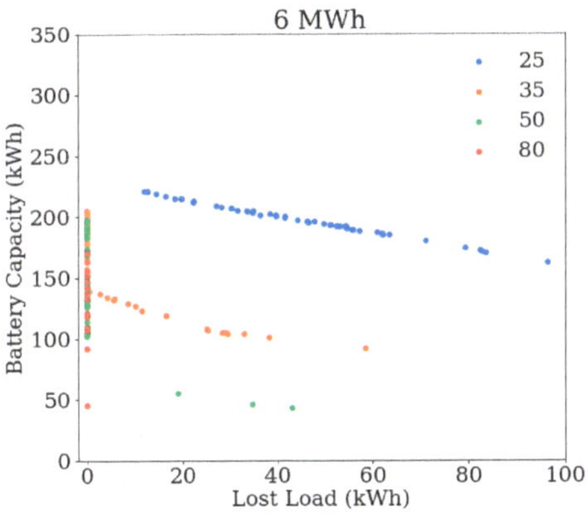

Figure 8. Feasible solutions to the optimisation problem considering a 6 MWh load profile.

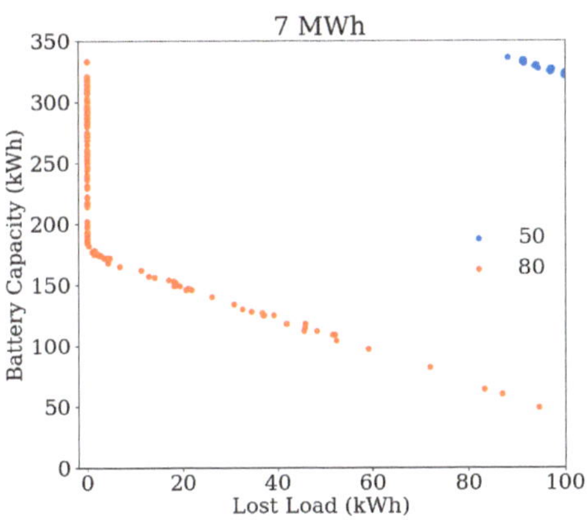

Figure 9. Feasible solutions to the optimisation problem considering a 7 MWh load profile.

The relationship between *LL* and annual investment is linear for all load profiles and grid connection capacities. The gradient differs for load profiles but appears to be consistent across grid connection capacities. These figures show the BESS capacity that is required for each grid connection capacity to ensure no loss of load.

For a 5 MWh monthly load and a 3 × 50 A grid connection, a BESS capacity of approximately 60 kWh is required to ensure no loss of load. For a 6 MWh monthly load and a 3 × 50 A grid connection, a BESS capacity of approximately 70 kWh is required to ensure no loss of load. For a 7 MWh monthly load and a 3 × 80 A grid connection, a BESS capacity of approximately 180 kWh is required to ensure no loss of load.

5.2. Comparative Analysis of Optimal Sizing and Case Study Sizing

The optimal system sizing calculated in Section 5.1 for a BESS cost of €500/kWh was used with a variety of control strategies, described in Section 4, for a monthly load profile of 5 MWh. These were then compared against the case study system sizing for the same scenarios. The abbreviations *BDB* and *PHBDB* refer to the control strategies Battery Deadband and Peak Hour Battery Deadband, respectively.

Figure 10 shows the loss of potential load. In every control strategy except the base case, the optimal sizing performed better than the case study sizing. All battery control strategies that imposed a battery charge/discharge deadband experienced a higher loss of potential load than the respective base case. This is because the battery, after supplying the full load for an extended period, will be drained, and, in some cases, the grid capacity is not enough to supply the full load.

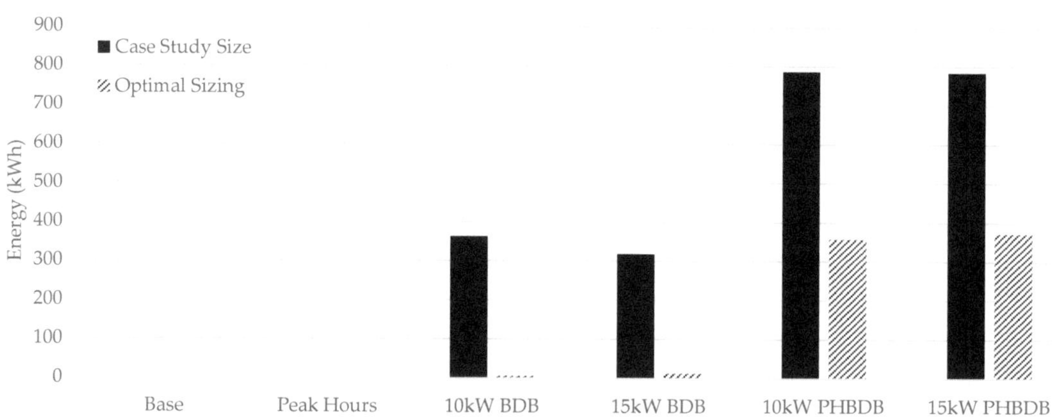

Figure 10. Potential load not supplied in different scenarios.

When using the optimal sizing, the battery system round-trip efficiency was increased for all control strategies with the exception of the base case, as depicted in Figure 11. The base case round-trip efficiency is lower with the optimal sizing because the battery would discharge at low powers; the maximum EV load was not much higher than the 55 kW grid connection. In fact, in some simulations, the battery would not be used at all.

A larger capacity grid connection can charge the BESS at relatively higher powers, resulting in a higher charging efficiency. The battery discharge deadband ensured the battery discharged at powers above the requirement, leading to an increased discharging efficiency. A higher charging/discharging battery power has a higher inverter conversion efficiency, as observed in Figure 5.

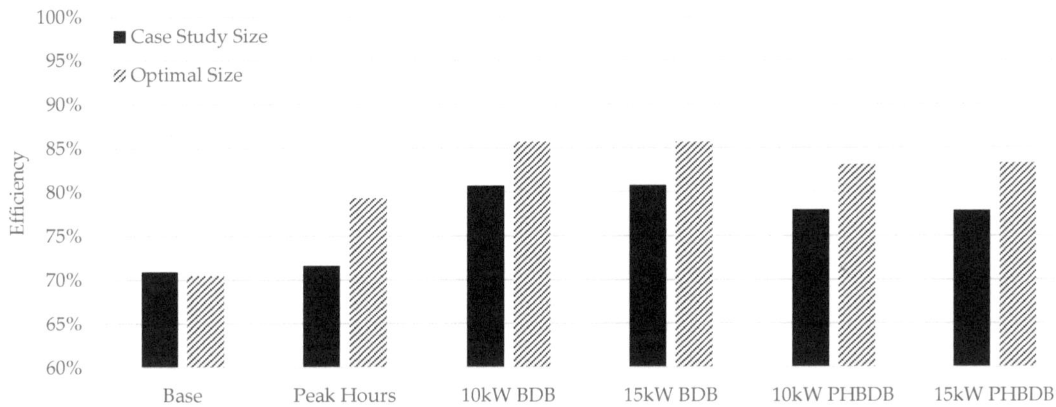

Figure 11. Battery system round-trip efficiency in different scenarios.

Figure 12 shows the volume of energy drawn from the grid during hours of peak grid load. Scenarios in which the grid exchange was limited still drew some power to cover base loads. This prevented the battery from discharging at low powers, thus maintaining a higher round-trip efficiency. By allowing the grid to supply the load during peak hours if the BESS was drained, the Peak Hours scenario had a negligible increase in energy drawn from the grid with respect to the 10 kWh PHBDB and 15 kWh PHBDB scenarios, from approximately 85 kWh to 110 kWh.

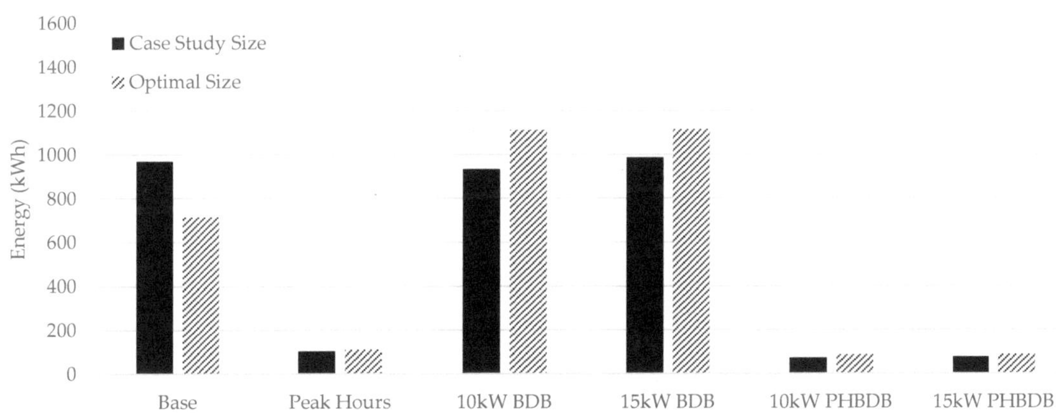

Figure 12. Energy imported from the grid during peak grid-load hours in different scenarios.

With the optimal sizing, an imposed battery deadband resulted in an increased grid import during peak hours, with respect to both the optimal base case and case study sizing. Above the deadband, the battery delivered the full load; therefore, at the end of the day, the battery was more depleted with respect to the base case. This is consistent with the P&R usage pattern which tends towards a high EV load in the morning and early afternoon due to commuters. With the optimal grid connection capacity, the high battery-charging power could fully recharge the battery in the three-hour window. Furthermore, the low battery utilisation in the optimal base case means the battery is not often recharged during these peak hours, hence the decrease with respect to the case study sizing.

The total load supplied by the BESS, displayed in Figure 13, is as expected. By enabling the battery to supply the full load during battery discharge periods, the battery will, of course, deliver more energy than the base case. Limiting the power draw during peak hours forces the battery to supply the load when otherwise it would not, namely, when the EV load is less than the grid capacity. In all cases, the optimal sizing resulted in a reduced battery utilisation with respect to the case study sizing.

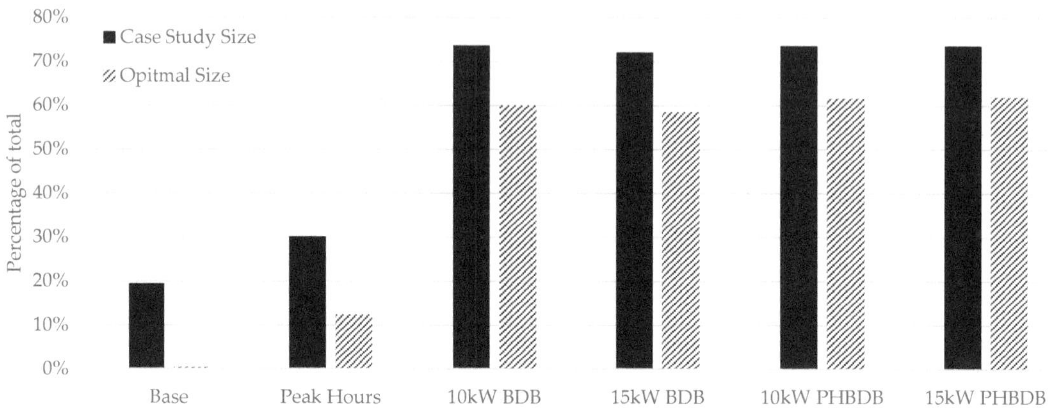

Figure 13. Total load supplied by the BESS in different scenarios.

The percentage of users that ended the charging session while the vehicle was still charging was fairly consistent across all scenarios and for both systems, as can be seen in Figure 14. Regardless of the system sizing, some users are simply not parked long enough to fully charge their cars. However, as shown in Figure 10, the optimal system sizing experienced less loss of load, meaning fewer times of insufficient capacity. Therefore, having a higher capacity grid connection and lower BESS capacity tends to result in the ability to deliver more energy and, therefore, generate higher revenue.

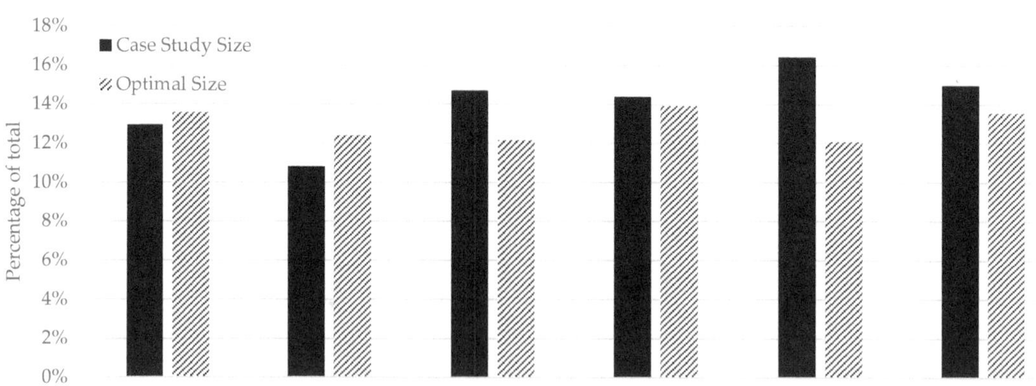

Figure 14. Percentage of users charging at the time of disconnection in different scenarios.

6. Discussion

The model presented in this study is for the charging of EVs in a charging hub with a stationary BESS and grid connection. The model is easily scalable for any number of CPs, BESS capacity, grid connection capacity, and load profile. The charge session data used in these simulations were measured at a P&R charging hub. Given the appropriate data, for example, from a workplace charging hub or shopping centre charging hub, the model is easily transferable.

The power grid difficulties faced across the Netherlands have the same basis—there is too little capacity to transmit and distribute power. The supply of power from distributed renewable energy resources to the grid during times of peak generation (high irradiance/high wind) is a problem in rural areas where such renewable energy farms are located. The installation of new wind and PV farms has been suspended due to a lack of capacity at peak power generation. Nationwide, in both rural and urban settings, the electrification and digitalisation of society has resulted in a rapid increase in the electric power demand. Stedin and Enexis, two other distribution network operators in the Netherlands, confirm this issue is present in other regions. For example, in the Province of Utrecht, 651 consumers are waiting to be connected to the grid with a total purchase capacity of 155 MW [28]. In Eindhoven Oost, there are 78 open connection requests with a total purchase capacity of 52.3 MW [29].

Therefore, one must consider what the goals of such a solution like a BESS are. A limited loss of load, limited grid interaction, and high BESS round-trip efficiency are all considered in this study.

The choice of load profile used in solving the optimisation problem had a large effect on the outcome. The simulated load profile used in the optimisation was chosen over other ~7 MWh profiles because it included a high demand day—a peak power demand of 81 kW which lasted over 3 h. This high demand day served to stress-test the sizing and ensures the optimal sizing is capable of serving future loads.

The control method used in solving the optimisation problem also had a large effect. For instance, if the system was optimised using the 15 kW PHBDB control strategy, the BESS would inevitably require a larger energy storage capacity to satisfy the constraint represented by Equation (4), the volume of potential load lost. Furthermore, these are only a selection of specific, yet limited, control strategies that were intended to address specific performance metrics. The optimal power dispatch and charge session scheduling which would result in an improved system performance were outside the scope of this study.

The control strategies investigated generally perform better with the optimal sizing rather than the case study sizing for the frequently observed 5 MWh monthly load. When the monthly load increases, the disparity between the performance of the optimal sizing and the case study sizing will increase. This is made clear in Table 3, where the loss of potential load was less than 1.5% for the optimal sizing and over 6.5% for the case study sizing.

The control strategy that limited grid interaction during peak grid-load hours yielded the most desirable results with the optimal sizing. The BESS round-trip efficiency was increased with respect to both the optimal sizing base case and the case study sizing, to 79%. Energy losses were kept low since the load was mostly supplied via the grid connection; the battery supplied only 12% of the load. The grid interaction during peak evening load hours was reduced to 110 kWh, compared to 713 kWh for the optimal sizing base case. Finally, there was no loss of potential load.

If the battery were to be used for grid ancillary services, such as frequency response and voltage control, then an additional revenue would be available for the battery, and the optimisation problem would be reformed. The optimal sizing would likely tend towards a larger battery to benefit from the ancillary service revenue whilst still maintaining the security of supply for the P&R users.

Dynamic charging tariffs are thought to be a good method for incentivising users to charge their EVs at low grid-load times and reduce the disruption to the power grid. This would have little effect in a P&R since the intended user groups associated with a P&R charging hub, namely, commuters and visitors, are not as flexible in their arrival and connection time as resident CP users.

Vehicle-to-Grid (V2G) is another rapidly progressing technology. During times of high electricity price, the EV can act as a battery and deliver power to a household when connected and laying idle on the driveway. In an urban neighbourhood that relies on public CPs, a fleet of EVs could be used to reduce evening peak residential loads behind the substation. V2G may be feasible in a P&R charging hub but only for specific users who meet certain criteria, such as commuters who park for the full working day. However, transferring energy from one commuter to another commuter may result in unsatisfied users. How V2G would be implemented in a P&R charging hub is yet unknown.

Whilst DC fast chargers are becoming more prevalent, their installation at a P&R is not necessary. They are suited for rapid turnover charge sessions, such as along motorways or in taxi ranks, or for high-battery-capacity vehicles, such as at bus depots or for heavy goods vehicles. Typical user connection durations are multiple hours at P&Rs. The measured data indicated the average connection duration to be 13 h. Therefore, level-2 charging will remain applicable for coming years.

Larger EV batteries are, of course, to be expected in the coming years; battery capacities greater than 100 kWh are already on the market. This study assumed a maximum charging demand of 75 kWh since the largest measured charge session was 68.9 kWh. It is hard to predict how larger EV battery capacities will affect charging behaviour since it is so highly dependent on social demographics, the availability of charging infrastructure, social and cultural norms, and personal preference. Considering the price of BEVs with large capacities and the rate at which EVs are penetrating the car fleet, it will be many years before such large-capacity BEVs are the norm.

Finally, it is clear that a multi-objective optimisation is required, in which grid interaction during peak hours is minimised, as well as the annual system cost. The intention of this system is to reduce grid loading for large charging hubs, especially during peak grid hours. Thus, the BESS should be adequately sized and appropriately controlled to service all EV users whilst maintaining a high round-trip efficiency and keeping grid interaction to a minimum. This could best be integrated using electricity pricing, such that the BESS is prioritised during times of high electricity price and the grid is prioritised during times of low electricity price.

Perhaps the most practical recommendation is that the battery be installed with individual phase control or to ensure an energy contract with the distribution network operator to allow power flow back to the grid. These design considerations will allow for power to be delivered individually and unevenly on separate phases.

7. Conclusions

This study used measured data from an installed EV charging hub with an on-site stationary battery (336 kWh/250 kW) and limited capacity grid connection (17.4 kW) to develop and validate a computer model in Python. A genetic algorithm was used to minimise the annual costs of the system by optimising the battery energy storage capacity and the grid connection capacity for a monthly load of 7 MWh. Three different battery costs were evaluated; the approximate current cost of €750/kWh, the expected 2030 cost of €250/kWh, and the middle €500/kWh. The optimal sizing, with the €500/kWh cost, a 55.4 kW grid connection, and a 69 kWh/45 kW battery, was then assessed using a variety of simple control strategies; namely, limiting grid power draw during peak evening grid-load hours, and implementing a battery charge/discharge deadband, and comparing this against the case study sizing. The limited peak hour grid interaction control strategy was determined to perform best with the optimal sizing.

The feasible solutions to the optimisation problem for three load profiles, 5 MWh, 6 MWh, and 7 MWh, were plotted as the battery capacity against the loss of potential load. These figures illustrated what battery capacity was required at each grid connection capacity to ensure no loss of potential load.

Finally, the limitations of this study were addressed and ideas for future work were presented.

Author Contributions: Conceptualisation, E.H. and R.W.; methodology, E.H. and R.W.; software, E.H.; validation, E.H.; formal analysis, E.H.; investigation, E.H.; resources, E.H. and R.W.; data curation, E.H.; writing—original draft preparation, E.H.; writing—review and editing, R.W. and R.H.; visualisation, E.H.; supervision, R.W. and R.H.; project administration, R.W. and E.H.; funding acquisition, R.W. and R.H. All authors have read and agreed to the published version of the manuscript.

Funding: This research is a part of the Future Charging project funded by SIA RAAK (NWO), project number: RAAK.PRO03.128.

Data Availability Statement: The data used in this project are not available for sharing and were shared with project members confidentially.

Conflicts of Interest: The authors declare no conflicts of interest.

References

1. Oude Groote Beverborg, B.; de Jager, A.; Heemskerk, J.; Fabius, B. *Laad Me: Strategisch Plan Laadinfrastructuur 2020–2030*; Techincal Report; Gemeente Amsterdam: Amsterdam, The Netherlands, 2020.
2. Nationale Agenda Laadinfrastructure. *The National Charging Infrastructure Agenda*; Technical Report; Ministry of Infrastructure and Water Management: The Hague, The Netherlands, 2021.
3. Liander. Available online: https://www.liander.nl/grootzakelijk/aansluiting/nieuw (accessed on 13 March 2023).
4. Heath, E.; Wolbertus, R.; Heller, R. Battery Supported Electric Vehicle Charging Plaza Using a Limited Capacity Grid Connection. In Proceedings of the Electric Vehicle Symposium 36, Sacramento, CA, USA, 11–14 June 2023.
5. Dai, Q.; Liu, J.; Wei, Q. Optimal photovoltaic/battery energy storage/electric vehicle charging station design based on multi-agent particle swarm optimisation algorithm. *Sustainability* **2019**, *11*, 1973. [CrossRef]
6. Mehrjerdi, H.; Hemmati, R. Electric vehicle charging station with multilevel charging infrastructure and hybrid solar-battery-diesel generation incorporating comfort of drivers. *J. Energy Storage* **2019**, *26*, 100924. [CrossRef]
7. Stecca, M.; Vermeer, W.; Soeiro, T.B.; Elizondo, L.R.; Bauer, P.; Palensky, P. Battery Storage Integration in EV Fast Charging Station for Increasing its Revenues and Reducing the Grid Impact. In Proceedings of the 2022 IEEE Transportation Electrification Conference & Expo (ITEC), Anaheim, CA, USA, 15–17 June 2022; pp. 109–113.
8. Negarestani, S.; Fotuhi-Firuzabad, M.; Rastegar, M.; Rajabi-Ghahnavieh, A. Optimal sizing of storage system in a fast charging station for plug-in hybrid electric vehicles. *IEEE Trans. Transp. Electrif.* **2016**, *2*, 443–453. [CrossRef]
9. Salapić, V.; Gržanić, M.; Capuder, T. Optimal sizing of battery storage units integrated into fast charging EV stations. In Proceedings of the 2018 IEEE International Energy Conference (ENERGYCON), Limassol, Cyprus, 3–7 June 2018; pp. 1–6.
10. Haupt, L.; Schöpf, M.; Wederhake, L.; Weibelzahl, M. The influence of electric vehicle charging strategies on the sizing of electrical energy storage systems in charging hub microgrids. *Appl. Energy* **2020**, *273*, 115231. [CrossRef]
11. Secchi, M.; Barchi, G.; Macii, D.; Moser, D.; Petri, D. Multi-objective battery sizing optimisation for renewable energy communities with distribution-level constraints: A prosumer-driven perspective. *Appl. Energy* **2021**, *297*, 117171. [CrossRef]
12. Mazza, A.; Mirtaheri, H.; Chicco, G.; Russo, A.; Fantino, M. Location and sizing of battery energy storage units in low voltage distribution networks. *Energies* **2019**, *13*, 52. [CrossRef]
13. Gu, T.; Wang, P.; Liang, F.; Xie, G.; Guo, L.; Zhang, X.P.; Shi, F. Placement and capacity selection of battery energy storage system in the distributed generation integrated distribution network based on improved NSGA-II optimisation. *J. Energy Storage* **2022**, *52*, 104716. [CrossRef]
14. Moghimi, M.; Garmabdari, R.; Stegen, S.; Lu, J. Battery energy storage cost and capacity optimisation for university research center. In Proceedings of the 2018 IEEE/IAS 54th Industrial and Commercial Power Systems Technical Conference (I&CPS), Niagara Falls, ON, Canada, 7–10 May 2018; pp. 1–8.
15. Khan, H.; Nizami, I.F.; Qaisar, S.M.; Waqar, A.; Krichen, M.; Almaktoom, A.T. Analyzing optimal battery sizing in microgrids based on the feature selection and machine learning approaches. *Energies* **2022**, *15*, 7865. [CrossRef]
16. Alawode, B.O.; Salman, U.T.; Khalid, M. A flexible operation and sizing of battery energy storage system based on butterfly optimisation algorithm. *Electronics* **2021**, *11*, 109. [CrossRef]
17. Koolman, G.; Stecca, M.; Bauer, P. Optimal battery energy storage system sizing for demand charge management in ev fast charging stations. In Proceedings of the 2021 IEEE Transportation Electrification Conference & Expo (ITEC), Chicago, IL, USA, 21–25 June 2021; pp. 588–594.
18. Liander. *Tarieven voor Aansluiting en Transport Elektriciteit: Voor Huishoudens en Zakelijke Klanten met een Kleinverbruikaansluiting per 1 Januari*; Technical Report; Liander: Arnhem, The Netherlands, 2023.
19. Blank, J.; Deb, K. Pymoo: Multi-objective optimisation in python. *IEEE Access* **2020**, *8*, 89497–89509. [CrossRef]
20. Ramasamy, V.; Zuboy, J.; O'Shaughnessy, E.; Feldman, D.; Desai, J.; Woodhouse, M.; Basore, P.; Margolis, R. *US Solar Photovoltaic System and Energy Storage Cost Benchmarks, with Minimum Sustainable Price Analysis: Q1 2022*; National Renewable Energy Lab (NREL): Golden, CO, USA, 2022.
21. Cole, W.; Frazier, A.W.; Augustine, C. *Cost Projections for Utility-Scale Battery Storage: 2021 Update*; National Renewable Energy Lab (NREL): Golden, CO, USA, 2021.
22. Alfen Charging Equipment. *Smart Charging Implementation Guide*; Technical Report; Alfen: Almere, The Netherlands, 2020.
23. BECK. Available online: https://www.team-elektro-beck.de/en/beck-automation/range-of-services/stationaeres-batteriespeichersystem.html (accessed on 5 February 2024).
24. Kim, D.K.; Susumu, Y.; Ali, Z.T.; Kim, Y.T. *Handbook on Battery Energy Storage System*; Asian Development Bank: Manila, Philippines, 2018; pp. 1–6.

25. Stecca, M.; Soeiro, T.B.; Elizondo, L.R.; Bauer, P.; Palensky, P. Lifetime estimation of grid-connected battery storage and power electronics inverter providing primary frequency regulation. *IEEE Open J. Ind. Electron. Soc.* **2021**, *2*, 240–251. [CrossRef]
26. Andrenacci, N.; Chiodo, E.; Lauria, D.; Mottola, F. Life cycle estimation of battery energy storage systems for primary frequency regulation. *Energies* **2018**, *11*, 3320. [CrossRef]
27. Ralon, P.; Taylor, M.; Ilas, A.; Diaz-Bone, H.; Kairies, K. *Electricity storage and renewables: Costs and markets to 2030*; International Renewable Energy Agency: Abu Dhabi, United Arab Emirates, 2017; p. 67.
28. Stedin. Available online: https://www.stedin.net/zakelijk/energietransitie/beschikbare-netcapaciteit/congestie-en-congestiemanagement/provincie-utrecht (accessed on 5 February 2024).
29. Enexis. Available online: https://indd.adobe.com/view/459b03c4-40b1-4cd8-bf9a-36b5aa4aad1b (accessed on 5 February 2024).

World Electric Vehicle Journal

MDPI

Article

Economic and Environmental Assessment of Technologies Optimizing the Execution of Long Trips for Electric Vehicles

Léa D'amore *, Daniele Costa and Maarten Messagie

MOBI Research Group, ETEC Department, Vrije Universiteit Brussel, Pleinlaan 2, 1050 Ixelles, Belgium; daniele.costa@vub.be (D.C.); maarten.messagie@vub.be (M.M.)
* Correspondence: lea.damore@vub.be

Abstract: Further advances in hardware and software features are needed to optimize battery and thermal management systems to allow for the execution of longer trips in electric vehicles. This paper assesses the economic and environmental impacts of the following features: eco-charging, eco-driving, smart fast charging, predictive thermal powertrain and cabin conditioning, and an advanced heat pump system. A Total Cost of Ownership (TCO) and externalities calculation is carried out on two passenger cars and one light commercial vehicle (LCV). The energy consumption data from the vehicles are based on experiments. The analysis shows more benefits for the LCV, while the smart fast-charging feature on the car shows a slight increase in TCO. However, negative results did not contribute significantly compared to the ability to install a smaller battery capacity for similar use.

Keywords: battery electric vehicle (BEV); driver experience; environment; extended range electric vehicle; energy consumption

Citation: D'amore, L.; Costa, D.; Messagie, M. Economic and Environmental Assessment of Technologies Optimizing the Execution of Long Trips for Electric Vehicles. *World Electr. Veh. J.* **2024**, *15*, 128. https://doi.org/10.3390/wevj15040128

Academic Editor: Ghanim A. Putrus

Received: 26 January 2024
Revised: 13 March 2024
Accepted: 20 March 2024
Published: 22 March 2024

1. Introduction

One of the challenges for battery electric vehicle (BEV) acceptance is autonomy for long trips, also known as "range anxiety". To tackle this issue, new hardware and software features providing strategies to enable the execution of long trips by BEVs were developed within the Connected Electric Vehicle Optimized for Life, Value, Efficiency and Range (CEVOLVER) project. More specifically, the project tackled the challenge of executing long trips in a reasonable time with a small battery capacity. This was achieved by using the features under study to try to increase battery autonomy and therefore optimize the execution of long trips without changing the battery itself. Such features are user oriented, such as eco-routing, eco-charging, and eco-driving. The project considered an approach based on users' experiences in different use cases to improve the comfort and usability of BEVs for long day trips. While it can be beneficial for reducing range anxiety, adding such features might have an impact on the overall cost of ownership and on the environmental performance of the vehicle. If not beneficial, especially in terms of cost, it could hinder the acceptance of BEVs with such solutions. This paper therefore focuses on the economic and environmental impacts of the features during the vehicle's ownership. The assessment includes the total cost of ownership (TCO) and external costs analysis regarding greenhouse gas emissions. The technological developments are compared to the baseline vehicles.

1.1. Range Anxiety and Technological Developments to Increase Battery Autonomy

While BEVs could help improve the environmental performances of the transport sector, their growth is facing some challenges. The main reasons hindering BEV acceptance from consumers' perspectives are range anxiety and the potential lack of charging infrastructure [1–7].

Range anxiety is a challenge that starts with its own definition, which can vary from one study to another, leading to different interpretations of how to tackle it. While

Liu et al. (2023) state that range anxiety refers more to "energy replenishment" anxiety and estimate that it is the main problem to solve [1], Rainieri et al. (2023) mention that one of the main sources of range anxiety is individual characteristics [5]. Regarding Franke et al. (2016) [2], the study defines range anxiety as "range stress", which is related to the fact that the resources to overcome the range are insufficient. However, these studies tend to agree on the fact that most BEVs available on the market can meet most consumers' travel needs [1,2,4,5,8]. Liu et al. (2023) go even a bit further by stating that ultra-long-range BEVs are actually not needed as they does not solve the problem of energy replenishment anxiety [1]. Furthermore, such cars raise the cost of BEVs due to high purchase and insurance costs, which can also hinder their acceptance. Using TCO and considering range anxiety, the study establishes that the optimal range would be 400 km. The study therefore states that the current BEV market could be sufficient for more than 98% of consumers' needs. Needell et al. (2016) also found that most existing and affordable vehicles can be sufficient to meet the energy needs of 87% of vehicle days in the United States [8]. Such findings are contradictory to the trend from the transport sector to produce BEVs with longer ranges [1]. Indeed, to face range anxiety issues, automotive companies are increasing the range of BEVs by increasing battery capacity and developing charging infrastructure, including fast charging. Those solutions come with some burden. Increasing battery capacity comes with different issues such as the rising cost of BEVs and also an increasing demand for critical materials such as cobalt, nickel, graphite, and lithium [1]. Regarding improving charging infrastructure, He et al. (2023) also pinpoints the fact that its growth depends on the adoption of EVs, as stakeholders are more reluctant to develop charging facilities without growing demand [3].

Several other solutions exist to tackle range anxiety challenges that do not necessarily involve changing the cars on the market. When range anxiety is defined by range stress or individual characteristics, the consensus is that learning experiences and range tolerance help to overcome the stress of not being able to reach a destination [2,3,6]. Other solutions are more technical and practical and are the focus of this paper. One main reason for range anxiety is the unreliability of autonomy and the variation of driving range throughout the usage of the vehicle [4,6]. Predictive models that can provide a more accurate range prediction for vehicles will help in that context. The accuracy is enhanced by collecting more parameters such as on-route data on traffic conditions and battery conditions [7,9,10]. In CEVOLVER, the feature that tackles a part of this issue is eco-charging, which uses real traffic conditions and is explained in more detail in Section 1.2. Another solution is to reduce the energy consumption of the vehicle. It can be achieved through thermal management systems that also help to enhance the life span of the battery. As assessed by Biswas (2020) [11], such systems generally include Heating, Ventilation and Air Conditioning (HVAC); Battery Management System (BMS); and Traction Cooling System (TCS). They ensure the optimal operating condition of the components based on their thermal efficiencies. Finally, eco-driving also helps reduce energy consumption for a certain trip [12]. It can be achieved through learning experiences and/or with advice while driving, such as suggested speed [13–15]. As for the driving range estimations, such add-on's accuracy benefit from on-route information and battery parameters [14,15]. Another possibility for enhancing eco-driving is vehicle platooning [16,17], but such technological advancement is still at an experimental stage.

When analyzed in the literature, the solutions' effectiveness in the studies is assessed through energy consumption gains, tested or simulated. It is not evaluated in terms of cost or environmental performances, which could be helpful to assess the effects on overall usage and to quantify possible burdens. When considering TCO and externalities analysis, the method is often used to compare costs of BEVs or alternative vehicles with equivalent Internal Combustion Engine Vehicles (ICEVs) [18–24]. However, some studies [1,25] quantified the economic performances to qualify the necessity of longer-range BEVs. As mentioned, Liu et al. (2023) calculated the TCO of BEVs with different ranges [1]. The study considers the battery replacement needs for a certain usage, which will differentiate between smaller and bigger EVs. The study shows that despite the battery replacement, the TCO is higher for higher electric range BEVs. Pfriem et al. (2013) found similar results for commercial fleet usage [25]. The TCO for the fleet is beneficial compared to commercial ICEVs when using small-range BEVs. Such studies used the TCO to promote the cost benefit of short-range BEVs and to question the actual need of long-range BEVs.

In this paper, the features under study are assessed in terms of economic and environmental aspects, also including the use of energy consumption data from testing under real driving conditions on open roads or test benches. This is because while the features might be successful in terms of executing longer trips without additional time, some burden in terms of costs or environmental performances might appear and hinder the application of such features. The quantification of the effect on costs will allow assessment of the significance of the potential burdens or benefits compared to the objectives of executing the longer trips on time. Furthermore, the emphasis on the cost and environmental potential benefit might help with the overall acceptance of BEVs with smaller battery sizes.

The next section will present the features and the system evaluated during the project.

1.2. System Description

The system includes three different parameters: the vehicle, the features tested and the use case. During the CEVOLVER project, six features were tested on three different vehicles in different use cases:

- One light commercial vehicle (LCV) with a 68 kWh battery;
- One passenger car with a 24 kWh battery (car 1);
- One passenger car with a 42 kWh battery (car 2).

The two passenger cars are identical except for the battery capacity. The baseline vehicle is defined as the vehicle without the CEVOLVER features switched on. Table 1 summarizes the systems considered for the experiments with the baseline vehicles, the corresponding use case, and the specific features switched on during testing. Each line of the table refers to one test that has been performed, once with the features not used and once with the features switched on. Thermal-related features have been tested on test benches and the others on open roads.

Table 1. Summary of baseline vehicles, the use cases and features. Legend: LCV—light commercial vehicle, NEDC—New European Driving Cycle.

Vehicle	Use Case	Feature
LCV	Parcel service daily job	Eco-charging
LCV	Parcel service daily job	Eco-driving
Car 1	Regular commute from home to work	Predictive thermal powertrain conditioning and predictive thermal cabin conditioning
Car 1	NEDC	Hardware changes in the heat pump
Car 2	Private visit of 350 km	Eco-charging
Car 2	Private visit of 350 km	Eco-charging and eco-driving
Car 2	Private visit of 350 km	Eco-charging and eco-driving and smart-fast charging

The use case describes the type of usage the vehicle faces and sets the boundaries of the experiments (i.e., the type of trips completed). The "parcel service daily job" means the vehicle is used for parcel delivery, mainly in urban areas. The charging of the vehicle is performed after returning to the distribution center. The "Regular travel to and from work" refers to a short-range trip from work to home, with a distance of 30 km. The charging is executed after arriving home at a charging station. The "private visit of 350 km" refers to occasional visits to relatives during the weekend or holiday trips. Since the trip is long, this use case assumes that one fast charging is required at a public charging station and one home charging during the visit.

As for the features, eco-charging determines the most energy- and time-efficient charging and routing strategy for the trip based on traffic conditions. Different parameters are considered, including traffic and weather conditions, which enhance the accuracy of such development. Still, the real value-add comes with the intelligent recommendation for fast charging that is optimized based on the assessment of the overall trip and not just the need to find the next charging station when the state of charge drops below a set value. The functionality of the feature is detailed in De Nunzio et al. (2020) [7]. Eco-driving ensures the speed recommendation to optimize energy consumption according to an analysis of the route and traffic conditions. The specificities are detailed in Ngo et al. (2021) [26]. In addition, smart fast charging conditions the battery before a fast charge to ensure the full charging power is available. It prevents the battery from overheating, which would lead to a longer charging time. The driving and charging conditions are based on the data gathered from the eco-charging features. The predictive thermal powertrain optimizes the use of the powertrain components based on their thermal efficiency, and the predictive thermal cabin conditioning ensures a comfortable cabin temperature while reducing the energy consumption from the climatization system. The software development is detailed in Wahl et al. (2022) [27] and in Chen et al. (2020) [28]. Finally, the advanced heat pump system developed in the project OPTEMUS allows the use of heat from electric components and batteries to warm up the cabin as described in the project website and in Ferraris et al. (2020) [29,30].

2. Materials and Methods

The assessment is based on the TCO and the assessment of external costs, focusing on greenhouse gas (GHG) emissions. TCO is a widely applied and accepted methodology to assess the economic impacts of a product. For all the vehicles, the TCO and external costs of the baseline vehicle will be compared to the TCO and external costs with the added developments. The geographic scope of the study is Belgium. However, Italy and Sweden conditions are also considered to cover different climate conditions for assessing the predictive thermal powertrain, cabin conditioning, and heat pump hardware changes since these goals are related to extreme weather conditions.

2.1. Total Cost of Ownership

The TCO methodology [31] compares the affordability of the vehicles by summing all costs that occur during the ownership of a vehicle. It can be defined as a tool to support understanding the actual cost of buying and using a particular good or service.

When calculating the TCO of a vehicle, there are two aspects to consider: Capital Expenditure (CAPEX), which are the one-time costs occurring to acquire fixed assets (e.g., the vehicle), and operating expenses (OPEX), which are the expenditures occurring during the operation of the vehicle at the present value (e.g., operational costs and non-operational costs). For vehicles, the TCO accounts for purchase costs, fuel, operating costs, and non-fuel operating costs. The TCO is based on the net present value of the vehicle's lifetime [32]. Therefore, Equation (1) is used for the one-time cost, and Equation (2) is used for recurring costs.

$$PV = A_t \frac{1}{(1+r)^t}, \tag{1}$$

$$PV = A_0 \times \frac{(1+r)^t - 1}{r \times (1+r)^t} \tag{2}$$

where:

PV is the present value given in EUR.
A_t is the one-time cost at time t.
A_0 is the annual recurring cost.
r is the real discount rate.
t is the time expressed as the number of years.

The real discount rate can be retrieved from the European Central Bank, considering the years 2011 to 2021. The critical assumptions for the TCO calculation are related to the vehicle's lifetime of ownership and are shown in Section 3.

2.2. Externalities

Externalities can be defined as uncompensated social or environmental effects due to social or economic activities [33]. In this study, the focus is on the climate change impact category. Therefore, externalities are based on the environmental impacts of the electricity consumed by the vehicles, which depends on the country-specific electricity production mix. The average carbon price for 2021, equivalent to 53.45 EUR/ton CO_2 [32,34], is considered. The carbon footprint is calculated with the Intergovernmental Panel on Climate Change (IPCC) characterization factors [35] with electricity mix data from the Ecoinvent 3.8 database [36]. Table 2 summarizes the carbon footprint and external costs for each country considered.

Table 2. Carbon footprint and external costs of electricity production per country.

Country	Carbon Footprint (kgCO_2 Eq/kWh)	External Cost (EUR/kWh)
Belgium	0.220	0.018
Sweden	0.022	0.001
Italy	0.395	0.021

2.3. Data Collection

The critical assumptions for the TCO calculation are related to the vehicle's lifetime of ownership (Table 3). The ownership of the vehicle is set to 10 years [37]. The discount rate is set to −3% [38]. The distance driven for the use cases does not necessarily cover the entire annual distance traveled with the vehicle. Therefore, additional kilometers are added to reach the average annual distance traveled in Belgium [39]. The impacts of the features are applied only to the distance the use case covers. This method allows an economic and environmental analysis of the developments per use case assessed during the experiments.

Table 3. Key assumptions for the vehicle life cycle. LCV—light commercial vehicle, NEDC—New European Driving Cycle.

Parameters	Value	Unit	Reference
Duration of ownership	10	Years	[37,39]
LCV annual distance	21,000	km	Aligned with the corresponding use case and the annual distance driven by a LCV in Belgium in 2019 [40]
Car 1 annual distance for the use case: regular travel to and from work	7500 (out of 15,000)	km	Aligned with the use case
Car 1 annual distance for the NEDC	15,000	km	Aligned with the distance driven per year in Belgium
Car 2 annual distance for the use case: private visit of 350 km	4200 (out of 15,000)	km	Aligned with the use case considering a once-a-month visit to relatives
Real discount rate	−3	%	[38]

2.3.1. Experimental Data

The energy consumption data presented (Figure 1) and duration of the trip are primary data obtained during the CEVOLVER experiments. Each feature was tested for the corresponding use case. First, the baseline vehicles were driven on a specific trip corresponding to the use case. Then, the same vehicles were driven using the additional project features.

Eco-charging, eco-driving, and smart fast charging were tested on open roads while the others were tested on test benches.

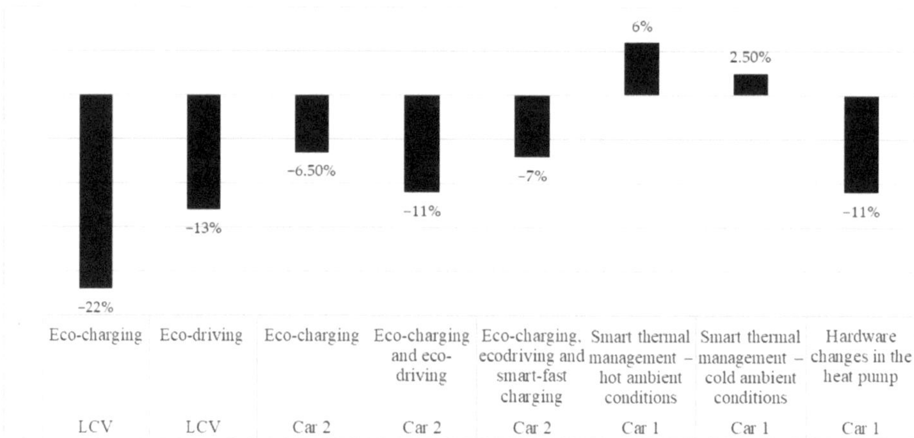

Figure 1. Summary of the changes in electricity consumption per vehicle from the baseline use case.

Regarding the LCV, eco-charging and eco-driving were tested separately, which means that the open road trips were slightly different for the two tests. For eco-charging, the trip with the baseline vehicle was the recommended one by the GPS to go from point A to point B. Then, eco-charging was used to define the most optimized road to take. Therefore, the two trips were not similar. For eco-driving testing, both trips were the same and corresponded to the optimized one provided by the eco-charging feature.

Specific Case of the Predictive Thermal Powertrain and Cabin Conditioning Features and the Hardware Changes in the Heat Pump

The experiments to reach the objectives set for thermal-related features are based on extreme weather scenarios. The features are indeed expected to help with energy consumption and the time to reach a certain temperature within the car during extreme temperature conditions. The provided data show the energy consumption for different trips for a certain ambient temperature (either −10 °C or +35 °C). Therefore, the extrapolation of these data and scenarios is slightly different than for the rest of the experiments.

As mentioned earlier, three countries with different temperature distributions are assessed for the boundaries: Belgium, Italy, and Sweden. Belgium is supposed to represent a middle-temperature condition, whereas Italy represents a hotter country and Sweden a colder one. The data were then adapted to each country. For Sweden, when necessary, the conversion rate taken is based on the one used by the European Alternative Fuels Observatory (EAFO), which is SEK 1 = EUR 0.097 [41].

The Open Power System Data provided an hourly temperature distribution for 2019 [42]. It helped to determine a percentage of time (Table 4) when the temperature is either above 28 °C or below 0 °C in a year. For the hardware changes, only the percentage related to cold temperatures is considered as only cold temperature conditions have been tested. This percentage is applied to the distance driven for the use case, and the changes in electricity consumption (Figure 1) are then applied to the corresponding kilometers. While it is difficult to predict the behavior of the developments and the savings potential under different temperature levels, the differences between the different predictions are expected to be marginal. Therefore, for simplification reasons, it is assumed that the potential saving from the experiment is the same for all ambient temperatures considered within "extreme weather conditions".

Table 4. Distribution of the temperature in a year for Belgium, Italy, and Sweden.

Country	Temperature below 0 °C (%)	Temperature over 28 °C (%)
Belgium (BE)	4.23	1.28
Italy (IT)	0.27	6.95
Sweden (SE)	18.46	0.01

2.3.2. Cost at Purchase Time

This section provides all purchase costs. It must be highlighted that the purchase price of a car in Table 5 can vary from region to region because of the choice of additional equipment consumers choose. These costs would affect the TCO by increasing or reducing the overall cost for both the use case with and without features. However, despite these price differences, the percentage changes between the two scenarios will stay the same.

Table 5. Summary of the costs at purchase time.

Costs		Value	Unit	Reference
LCV purchase cost		39,210	EUR	[43]
Car 1 purchase cost		29,424	EUR	[44]
Car 2 purchase cost		24,900	EUR	[45]
Registration costs	Belgium	0	EUR	Flanders, Belgium [46]
	Italy	150	EUR	Italy [47]
	Sweden	0	EUR	Sweden [46,47]
Features' cost	Eco-driving	0	EUR	Assumed to be included in the car purchase price.
	Eco-routing	0	EUR	
	Smart-fast charging	0	EUR	

2.3.3. Operational Costs

All operational costs, including electricity, are summarized in Tables 5–8. The cost of home charging is based on the average European price of electricity in the year 2019 [35], before the actual context of the energy crisis and geopolitical conflicts. This cost is considered constant for all ten years in this study. Given the actual context, the future and even actual electricity costs are very unstable and thus very difficult to predict. It will also impact the TCO. However, this TCO assessment focuses more on developing saving potential. Therefore, the results will still allow a first understanding of the economic impact of the features even without considering the situation at the time of writing.

Table 6. Operational costs for Belgium.

Operational Costs	Specificity	Value	Unit	Reference
Electricity cost at public charging	Chargers	0.32	EUR/kWh	[41]
	Fast chargers	0.60		
Electricity cost at home charging	all	0.22	EUR/kWh	[48]

Table 7. Operational costs for Italy.

Operational Costs	Specificity	Value	Unit	Reference
Electricity cost at public charging	Chargers	0.45	EUR/kWh	[41]
Electricity cost at home charging	all	0.22	EUR/kWh	[48]

Table 8. Operational costs for Sweden.

Operational Costs	Specificity	Value	Unit	Reference
Electricity cost at public charging	Chargers	0.29	EUR/kWh	[41]
Electricity cost at home charging	all	0.22	EUR/kWh	[48]

2.3.4. Non-Operational Costs

All non-operational costs, including road tax, insurance, maintenance, tire replacement, and technical control are summarized in Table 9. Maintenance, tire replacement, and technical control are vehicle specific. In addition, to estimate the real insurance cost in Belgium, a simulation was made for insurance costs with a specific person profile for both baseline vehicles. It is assumed that the person subscribed to the two types of insurance: the civil liability with basic protection rights and full omnium, which is a type of insurance in Belgium covering most car issues.

Table 9. Non-operational costs per vehicle.

Non-Operational Costs		Vehicle	Value	Unit	Reference
Small maintenance		Car 1 and 2	63	EUR/year	[49]
Large maintenance		Car 1 and 2	157.00	EUR/2 years	[49]
Maintenance before 5 years		LCV	185	EUR/year	[50]
Maintenance after 5 years		LCV	199	EUR/year	[50]
Tire replacements		LCV	591	EUR/40,000 km	[51–53]
		Car 1 and 2	234.44	EUR/40,000 km	[51–53]
Road tax	Belgium	All	0	EUR	[54]
	Italy	All	39.99	EUR/year after 4 years	[54]
	Sweden	All	35.55		[54]
Technical control	Belgium	LCV	59.80	EUR/year	[55]
		Car 1 and 2	45.10	EUR/year, after 4 years	[55]
	Italy	Car 1 and 2	79.02	EUR/every 2 years after 4 years	[56]
	Sweden	Car 1 and 2	58.20	EUR/year after 4 years	[57]
Insurance costs: civil liability	Belgium	LCV	655.96	EUR/year	[54]
		Car 1 and 2	248.19	EUR/year	[55]
	Italy	Car 1 and 2	344	EUR/year	[56]
	Sweden	Car 1 and 2	248.19	EUR/Once in year 3 Once in year 5 And once a year after	[57,58]

3. Results and Discussion

Figure 2 depicts the overall results from the TCO and externalities assessment comparing the use cases with or without the features. All developments considered resulted in rather small changes in the cost assessment. The changes ranged from −4% to +0.11%. The most significant and beneficial changes appeared for eco-charging and eco-driving with the LCV. These results are explained by the reduction in energy consumption. However, it is difficult to compare all vehicles and their respective results and confirm that the biggest changes would always be for vehicles like LCVs. Indeed, these differences may be due to one vehicle itself and the usage scenario differences that affect the direct extrapolation and boundaries of the TCO.

For the LCV, in both scenarios, the use of the developments reduces the TCO. A greater benefit is observed for the eco-charging features than for eco-driving; however, it is expected that the combination of the two would lead to an even bigger reduction of the energy consumption and therefore the overall TCO.

Cost savings were observed for cars but to a lesser extent. Regarding car 2, adding smart fast charging shows a slight reduction of benefit compared to the two other scenarios. It means that conditioning the battery to gain charging time also increases energy consumption for long-distance trips and, subsequently, the car TCO. Still, this negative environmental and cost effect is small and therefore remained less important than the fact that the driver can reach the destination on time for a long trip. The burden is also overcome using the other features. As expected, and shown by the results on car 1, the hardware changes are even more beneficial for countries with colder weather conditions. Small increases in energy consumption were observed for the thermal predictive conditioning features. Energy savings were actually shown regarding the powertrain energy efficiency, but it overlapped with other effects specific to the experiment. While the increase itself is also small, these additional costs are also neglectable as they depend on the experiment type and the drivers.

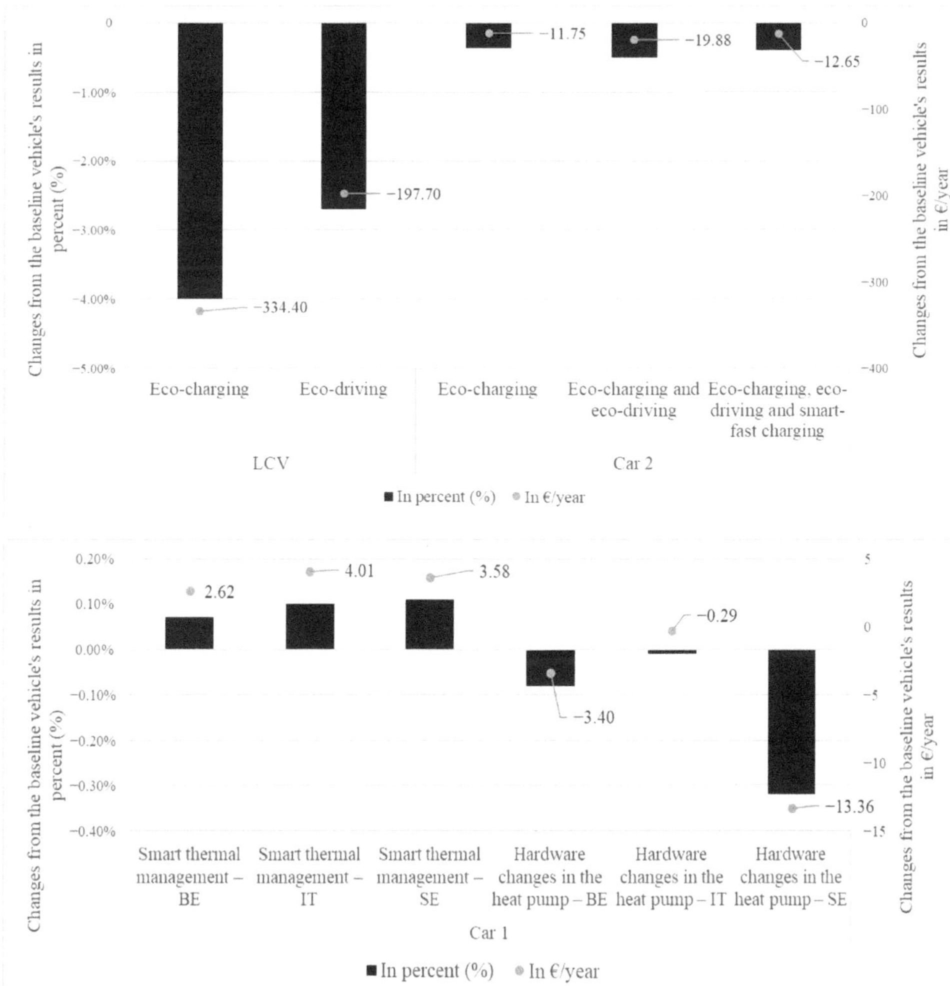

Figure 2. Overview of the changes in the TCO with the addition of the externalities' impacts. The bar graph is for the results in percent, and the point is for the results in EUR/year.

Overall, the results show that TCO and externalities reductions obtained by gains in energy efficiency are very small. This is because all demonstrator vehicles have the same battery capacity for the baseline tests and tests with enabled features. However, many features developed in CEVOLVER contribute to installing batteries with smaller capacities by ensuring outlier behavior (long trips, under severe ambient climate conditions), which typically determines the battery size and can be covered by smaller batteries with intelligent strategies, and as shown by the TCO and external costs, with no additional economic or environmental burden and even some small benefits. Therefore, greater differences are expected when comparing the TCO of the vehicle with CEVOLVER features to the TCO of a vehicle with the actual battery size required for similar usage.

4. Conclusions

This study examined the environmental and economic impacts of using features developed in the CEVOLVER project to reduce range anxiety in BEV drivers by improving the execution of long trips. While their effects are usually assessed in terms of energy consumption, this study took the approach to quantify the impacts in terms of costs for the overall usage. The features were of two kinds: hardware and software. They are related either to the driving and charging behavior or the thermal management system. A TCO and externalities approach has been carried out to understand their effects by comparing vehicles with and without the developed features. Several parameters including the duration of the trip or the energy consumption of the vehicles were retrieved from experiments performed during the project.

The use of the hardware and software features tested in CEVOLVER led to small environmental and economic impacts compared to the baseline vehicle. However, it proved that longer trips with the same vehicle are doable, with only a neglectable effect on TCO and no unexpected burden that could hinder their usage. The main advantage lies in the potential to reduce the vehicle's battery capacity for similar use. This would benefit the energy consumption in the use phase, costs, and also materials demand. Therefore, greater benefits are expected when considering the production phase in the externalities assessment. However, such benefits are not shown by the TCO and would require further research.

A limitation of the assessment was that extrapolating the experiments' results for the overall usage of the vehicle was in some cases not possible. As mentioned above, some use cases do not necessarily cover the entire usage of the vehicle. Further research to understand the effects on the additional distances could help show the full potential of the features on energy consumption reduction.

Author Contributions: Conceptualization, L.D., D.C. and M.M.; Data curation, L.D. and D.C.; Formal analysis, L.D.; Funding acquisition, M.M.; Investigation, L.D. and D.C.; Methodology, L.D., D.C. and M.M.; Project administration, D.C. and M.M.; Resources, D.C.; Supervision, D.C. and M.M.; Validation, L.D. and D.C.; Visualization, L.D.; Writing—original draft, L.D.; Writing—review & editing, L.D., D.C. and M.M. All authors have read and agreed to the published version of the manuscript.

Funding: This research received funding from CEVOLVER through grant agreements numbers 824295.

Institutional Review Board Statement: Not applicable.

Informed Consent Statement: Not applicable.

Data Availability Statement: The original contributions presented in the study are included in the article, further inquiries can be directed to the corresponding author.

Conflicts of Interest: The authors declare no conflicts of interest.

References

1. Liu, X.; Zhao, F.; Geng, J.; Hao, H.; Liu, Z. Comprehensive Assessment for Different Ranges of Battery Electric Vehicles: Is it Necessary to Develop a Battery Electric Vehicle with an Ultra-Long Range? *iScience* **2023**, *26*, 106654. [CrossRef]
2. Franke, T.; Rauh, N.; Günther, M.; Trantow, M.; Krems, J.F. Which Factors Can Protect Against Range Stress in Everyday Usage of Battery Electric Vehicles? Toward Enhancing Sustainability of Electric Mobility Systems. *Hum. Factors* **2016**, *58*, 13–26. [CrossRef]
3. He, X.; Hu, Y. Optimal Mileage of Electric Vehicles Considering Range Anxiety and Charging Times. *World Electr. Veh. J.* **2023**, *14*, 21. [CrossRef]
4. Rainieri, G.; Buizza, C.; Ghilardi, A. The psychological, human factors and socio-technical contribution: A systematic review towards range anxiety of battery electric vehicles' drivers. *Transp. Res. Part F Traffic Psychol. Behav.* **2023**, *99*, 52–70. [CrossRef]
5. Tamor, M.A. Examining the case for long-range battery electric vehicles with a generalized description of driving patterns. *Transp. Res. Part C Emerg. Technol.* **2019**, *108*, 1–11. [CrossRef]

6. Dong, J.; Xing, W.; Liu, C.; Lin, L.; Hu, L. The impact of reliable range estimation on battery electric vehicle feasibility. *Int. J. Sustain. Transp.* **2020**, *14*, 833–842. [CrossRef]
7. De Nunzio, G.; Thibault, L. Energy-optimal driving range prediction for electric vehicles. In Proceedings of the 2017 IEEE Intelligent Vehicles Symposium (IV), Los Angeles, CA, USA, 11–14 June 2017; pp. 1608–1613. [CrossRef]
8. Needell, Z.; McNerney, J.; Chang, M.; Trancik, J.E. Potential for widespread electrification of personal vehicle travel in the United States. *Nat. Energy* **2016**, *1*, 16112. [CrossRef]
9. Deepak, S.; Amarnath, A.; Krishnan, U.G.; Kochuvila, S. Survey on Range Prediction of Electric Vehicles. *i-PACT* **2019**, *1*, 1–7. [CrossRef]
10. Zhuo, S.; Li, H.; Bin Kaleem, M.; Peng, H.; Wu, Y. Digital Twin-Based Remaining Driving Range Prediction for Connected Electric Vehicles. *SAE Int. J. Elec. Veh.* **2024**, *13*, 23–26. [CrossRef]
11. Biswas, S. Thermal Management System and Performance Characteristics of Electric Vehicle. *SAE Tech. Pap.* **2020**, *28*, 22. [CrossRef]
12. Faria, R.; Marques, P.; Moura, P.; Freire, F.; Delgado, J.; de Almeida, A.T. Impact of the electricity mix and use profile in the life-cycle assessment of electric vehicles. *Renew. Sustain. Energy Rev.* **2013**, *24*, 271–287. [CrossRef]
13. Günther, M.; Rauh, N.; Krems, J.F. How driving experience and consumption related information influences eco-driving with battery electric vehicles—Results from a field study. *Transp. Res. Part F Traffic Psychol. Behav.* **2019**, *62*, 435–450. [CrossRef]
14. Naeem, H.M.Y.; Butt, Y.A.; Ahmed, Q.; Bhatti, A.I. Optimal-Control-Based Eco-Driving Solution for Connected Battery Electric Vehicle on a Signalized Route. *Automot. Innov.* **2023**, *6*, 586–596. [CrossRef]
15. Kim, Y.; Lee, I.; Kang, S. Eco Assist Techniques through Real-time Monitoring of BEV Energy Usage Efficiency. *Sensors* **2015**, *15*, 14946–14959. [CrossRef]
16. Liu, J.; Wang, Z.; Zhang, L. Efficient Eco-Driving Control for EV Platoons in Mixed Urban Traffic Scenarios Considering Regenerative Braking. *IEEE* **2023**, 1. [CrossRef]
17. Su, Z.; Chen, P. Eco-driving for Battery Electric Vehicles Using Traffic-aware Computationally Efficient Model Predictive Control. *IFAC* **2022**, *55*, 700–705. [CrossRef]
18. Cheon, S.; Lee, H.; Kim, A.; Choe, C.; Lim, H. Finding the most suitable vehicle type for projected years using analytic hierarchy process integrated with economic and environmental aspects. *J. Clean. Prod.* **2023**, *426*, 139075. [CrossRef]
19. Suttakul, P.; Wongsapai, W.; Fongsamootr, T.; Mona, Y.; Poolsawat, K. Total cost of ownership of internal combustion engine and electric vehicles: A real-world comparison for the case of Thailand. *Energy Rep.* **2022**, *8*, 545–553. [CrossRef]
20. Fulton, L. A Publicly Available Simulation of Battery Electric, Hybrid Electric, and Gas-Powered Vehicles. *Energies* **2020**, *13*, 2569. [CrossRef]
21. Morrison, G.; Stevens, J.; Joseck, F. Relative economic competitiveness of light-duty battery electric and fuel cell electric vehicles. *Transp. Res. Part C Emerg. Technol.* **2018**, *87*, 183–196. [CrossRef]
22. Kumar, D.; Kalghatgi, G.; Agarwal, A.K. Comparison of Economic Viability of Electric and Internal Combustion Engine Vehicles Based on Total Cost of Ownership Analysis. In *Transportation Systems Technology and Integrated Management*; Energy, Environment, and Sustainability; Springer: Berlin/Heidelberg, Germany, 2023.
23. Di Vece, G.; Di Nunno, D.; Bilancia, M.; Verdino, V. Development of a Total Cost of Ownership Model to Compare BEVs, FCEVs and Diesel Powertrains on Bus Applications. *SAE Tech. Pap 2022-37-0030.* **2022**, 13. [CrossRef]
24. Flaris, K.; Mitropoulos, L.; Kepaptsoglou, K.; Kouretas, K.; Vlahogianni, E. Analysis of total cost of ownership for conventional and alternative vehicle technologies: Evidence from France. *Adv. Transp. Stud.* **2021**, *55*, 153–166. [CrossRef]
25. Pfriem, M.; Gauterin, F. Less range as a possible solution for the market success of electric vehicles in commercial fleets. In Proceedings of the 2013 World Electric Vehicle Symposium and Exhibition (EVS27), Barcelona, Spain, 17–20 November 2023; pp. 1–8. [CrossRef]
26. Ngo, C.; Solano-Araque, E.; Aguado-Rojas, M.; Sciarretta, A.; Chen, B.; Baghdadi, M.E. Real-time eco-driving for connected electric vehicles. *IFAC-PapersOnLine* **2021**, *54*, 126–131. [CrossRef]
27. Wahl, A.; Wellmann, C.; Krautwig, B.; Manns, P.; Chen, B.; Schernus, C.; Andert, J. Efficiency Increase through Model Predictive Thermal Control of Electric Vehicle Powertrains. *Energies* **2022**, *15*, 1476. [CrossRef]
28. Chen, B.; Wulff, C.; Etzold, K.; Manns, P.; Birmes, G.; Andert, J.; Pischinger, S. A Comprehensive Thermal Model for System-Level Electric Drivetrain Simulation with Respect to Heat Exchange Between Components. In Proceedings of the 2020 19th IEEE Intersociety Conference on Thermal and Thermomechanical Phenomena in Electronic Systems (ITherm), Orlando, FL, USA, 21–23 July 2020; pp. 558–567. [CrossRef]
29. Optimised Energy Management and Use (OPTEMUS) Concept. Available online: http://www.optemus.eu/ (accessed on 24 March 2023).
30. Ferraris, W.; Bettoja, F.; Casella, M.; Rostagno, M.; Tancredi, A. Heat Pump for BEVs: Architectures and Performance Analysis. In Proceedings of the 2020 CO2 Reduction for Transportation Systems Conference, Turin, Italy, 14 June 2020. [CrossRef]
31. Brown, R.J. A new marketing tool: Life-cycle costing. *Ind. Mark. Manag.* **1979**, *8*, 109–113. [CrossRef]
32. EU Carbon Price Tracker. Available online: https://ember-climate.org/data/data-tools/carbon-price-viewer/ (accessed on 21 September 2022).
33. Lebeau, K.; Lebeau, P.; Macharis, C.; Van Mierlo, J. How expensive are electric vehicles? A total cost of ownership analysis. *World Electr. Veh. J.* **2013**, *6*, 996–1007. [CrossRef]

34. European Comission; Directorate-General for Research and Innovation; Bickel, P.; Friedrich, R. *ExternE Externalities of Energy Methodology 2005 Update*; European Commission: Luxembourg, 2005.
35. Pachauri, R.K.; Meyer, L.A. *Climate Change 2014: Synthesis Report. Contribution of Working Groups I, II and III to the Fifth Assessment Report of the Intergovernmental Panel on Climate Change*; IPCC: Geneva, Switzerland, 2014.
36. Wernet, G.; Bauer, C.; Steubing, B.; Reinhard, J.; Moreno-Ruiz, E.; Weidema, B. The ecoinvent database version 3 (part I): Overview and methodology. *Int. J. Life Cycle. Assess* **2016**, *21*, 1218–1230. [CrossRef]
37. Average Age of the EU Vehicle Fleet, by Country. Available online: https://www.acea.auto/figure/average-age-of-eu-vehicle-fleet-by-country/ (accessed on 21 September 2022).
38. European Central Bank. Available online: https://data.ecb.europa.eu/ (accessed on 21 September 2022).
39. Blackley, J. How Long Do People Keep Their Cars? *iSeeCars*. 2022. Available online: https://www.iseecars.com/ (accessed on 30 September 2022).
40. Morlet, N. 25% de kilomètres parcourus en moins en 2020 en Belgique. *DH les sportsI*, 7 April 2021.
41. AVERE. Pricing of Electric Vehicle Recharging in Europe. Available online: www.eafo.eu (accessed on 30 September 2022).
42. Open Power System Data. Data Package Weather Data. Version 16 September 2020. 2020. Available online: https://data.open-power-system-data.org/weather_data/2020-09-16 (accessed on 30 September 2022).
43. Ford. Available online: https://www.ford.com/commercial-trucks/e-transit/ (accessed on 21 September 2022).
44. Fiat 500 USA. Available online: http://www.fiat500usa.com/2013/04/fiat-500e-pricing-revealed.html (accessed on 21 September 2022).
45. L'argus. Available online: https://www.largus.fr/fiche-technique/Fiat/500/Ii/2022/Berline+3+Portes/E+95ch+Action-2322775.html (accessed on 21 September 2022).
46. ACEA. *Electric Vehicles: Tax Benefits Purchase Incentives*. 2021. Available online: https://www.acea.auto/fact/overview-electric-vehicles-tax-benefits-purchase-incentives-in-the-european-union-2022/ (accessed on 30 September 2022).
47. ACEA. *ACEA Tax Guide*. 2021. Available online: https://www.acea.auto/publication/acea-tax-guide-2022/ (accessed on 30 September 2022).
48. Eurostat. Available online: https://ec.europa.eu/eurostat/databrowser/view/NRG_PC_204/default/table?lang=en&category=nrg.nrg_price.nrg_pc (accessed on 21 September 2022).
49. Fiat. Available online: https://fiat-garage.be/fr (accessed on 21 September 2022).
50. Ford. Available online: https://www.fr.ford.be/apres-vente/service-entretien/ford-economy-service/entretien (accessed on 21 September 2022).
51. Midas. Available online: https://www.midas.be/fr/nos-prestations/pneus (accessed on 21 September 2022).
52. Pirelli. Available online: https://www.pirelli.com/global/en-ww/road/how-many-kilometres-can-a-tyre-cover (accessed on 21 September 2022).
53. Michelin. Available online: https://www.michelin.be/nl/auto/home-auto (accessed on 21 September 2022).
54. Autosécurité. Available online: https://www.autosecurite.be/wp-content/uploads/2022/08/FM0860-Bareme-des-redevances-CT-2022-FR.pdf (accessed on 21 September 2022).
55. Yuzzu. Calculons Ensemble Votre Assurance Auto Personnalisée. 2022. Available online: https://www.yuzzu.be/fr/assurance-auto/simulation (accessed on 21 September 2022).
56. Statista Research Department. Average Annual Cost of Vehicle Insurance Policies in Italy in Selected Months of 2019 (in Euros). 2022. Available online: https://www.statista.com/statistics/871878/average-price-of-vehicle-insurance-in-italy/ (accessed on 21 September 2022).
57. Bilprovningen. Available online: https://www.bilprovningen.se/ (accessed on 21 September 2022).
58. Transport Styrelsen. Available online: https://www.transportstyrelsen.se/en/road/Vehicles/motor-vehicle-inspection2/motor-vehicle-inspection-of-passenger-cars-and-lorries-not-exceeding-3500-kg-in-total-weight/ (accessed on 21 September 2022).

World Electric Vehicle Journal

Article

A Digitalized Methodology for Co-Design Structural and Performance Optimization of Battery Modules †

Theodoros Kalogiannis *, Md Sazzad Hosen, Joeri Van Mierlo, Peter Van Den Bossche and Maitane Berecibar

Battery Innovation Centre, MOBI Research Group, Vrije Universiteit Brussel (VUB), Pleinlaan 2, 1050 Brussel, Belgium; md.sazzad.hosen@vub.be (M.S.H.); joeri.van.mierlo@vub.be (J.V.M.); peter.van.den.bossche@vub.be (P.V.D.B.); maitane.berecibar@vub.be (M.B.)

* Correspondence: theodoros.kalogiannis@vub.be

† Paper Presented at EVS36 the 36th International Electric Vehicle Symposium and Exhibition, Sacramento, CA, USA, 11–14 June 2023.

Abstract: In this study, we present an innovative, fully automated, and digitalized methodology to optimize the energy efficiency and cost effectiveness of Li-ion battery modules. Advancing on from today's optimization schemes that rely on user experience and other limitations, the mechanical and thermal designs are optimized simultaneously in this study by coupling 3D multi-physical behavior models to multi-objective heuristic optimization algorithms. Heat generation at various loading and ambient conditions are estimated with a physics-based, fractional-order battery cell-level model, which is extrapolated to a module that further accounts for the interconnected cells' heat transfer phenomena. Several key performance indicators such as the surface temperature increase, the temperature variations on the cells, and heat uniformity within the module are recorded. For the air-cooled study case, the proposed coupled framework performs more than 250 module evaluations in a relatively short time for the whole available electro-thermal-mechanical design space, thereby ensuring global optimal results in the final design. The optimal module design proposed by this method is built in this work, and it is experimentally evaluated with a module composed of 12 series-connected Li-ion NMC/C 43Ah prismatic battery cells. The performance is validated at various conditions, which is achieved by accounting the thermal efficiency and pressure drop with regard to power consumption improvements. The validations presented in this study verify the applicability and overall efficiency of the proposed methodology, as well as paves the way toward better energy and cost-efficient battery systems.

Keywords: lithium-ion; electro-thermal model; battery thermal management; multi-physics and multi-objective optimization; particle swarm optimization; energy storage; structural design; battery module

Citation: Kalogiannis, T.; Hosen, M.S.; Van Mierlo, J.; Van Den Bossche, P.; Berecibar, M. A Digitalized Methodology for Co-Design Structural and Performance Optimization of Battery Modules. *World Electr. Veh. J.* **2024**, *15*, 115. https://doi.org/10.3390/wevj15030115

Academic Editor: Michael Fowler

Received: 15 December 2023
Revised: 1 March 2024
Accepted: 13 March 2024
Published: 16 March 2024

1. Introduction

Currently, lithium-ion batteries (Li-ion) are utilized in many electro-motive applications and in grid support due to their good operating efficiency and lifetimes [1,2], and they are also used in attempts to reduce greenhouse emissions, as well as in attempts to transition from the fossil fuel era [3,4]. Nonetheless, to obtain the optimal performance of Li-ion cells, they have to be monitored and preserved within a safe operating area (SoA) [5]. The thermal, safe-operating window is defined by the manufacturer according to the cell's chemistry and shape. Moreover, it depends on the current rate and it is usually between 20 °C to 40 °C [6]. For multi-cell designs, the cell-to-cell interactions, enclosure shields, and high-power demands (e.g., fast charge) can increase the cells' temperature or create heat non-uniformities with severe safety and performance implications. Hence, a battery thermal management system (BTMS) is always utilized to maintain a predefined SoA [7,8].

Various approaches for a BTMS can be found in the literature and in real-life applications, and they are categorized based on thermal resistance (direct/indirect cooling/heating inter-phase) or the cooling mediums (water-based/dielectric/air/phase-change, etc.) that are implemented in electric vehicles to preserve the temperature. In addition, they are typically tailor based, and they are used to account for low costs, good performance, scalability, modularity, etc. [9,10]. Many studies have been carried out that have evaluated and optimized air-based cooling BTMSs [11]. The air cooling module is the simplest in terms of mechanical and thermal design, and it was selected in this study for a proof-of-concept of our proposed methodology; however, it was also used without limiting the proposed method to air-based applications.

In [12], the optimal configuration of a battery pack composed of cylindrical Li-ion cells was proposed, whereby the authors showed that a cell arrangement with a small length-width ratio, in addition to an inlet/outlet configuration that facilitates the air flow over most of cells, can significantly improve cooling efficiency. Park [13] evaluated several air-cooled BTMS designs for Li-ion cells by theoretical investigations on a proposed numerical model. The aforementioned author showed that BTMS efficiency is highly dependent on the uniform distribution of the air passage, which can be achieved by adding a tapered manifold and pressure relief ventilation to the BTMS.

Chen et al. [14] proposed a flow resistance network model to capture the friction pressure loss along the BTMS channel, which was seen as the frictions between air and channel wall in air-based BTMSs. By these means, they calculated the velocities of the cooling channels and modeled the heat transfer and temperature distribution of the battery cells. The coupling of these models showed improvements in the thermal management efficiency of the BTMSs. Optimization of the air-cooled modules was also performed in [15], where tje authors investigated the influence of the air inlet/outlet angles and the width of the air flow channels between the battery cells. Similar assessments were performed in [16] for both a U-type and Z-type air-based BTMS, where it was also concluded that the cooling efficiency and power consumption can be improved by optimizing the aforementioned parameters. Moreover, experimental and numerical evaluations on air-cooled BTMSs were performed in [17]. Those authors investigated various performance parameters, such as the channel size, channel locations, mass flow rates, and temperature influence, and they calculated the pressure drop during constant current operation. They proposed a J-type BTMS by integrating the Z- and U-type designs, as well as by means of surrogate modeling, and through this approach they optimized the heat distribution in the battery module.

In order to estimate the optimal channel position on air-cooled BTMSs, the authors in [18] investigated several BTMS types with different input and output channels topologies, albeit with the same design parameters such as cell spacing, channel size, air flow rate, and temperature. They performed a numerical study on various designs and concluded that the cooling efficiency was improved if the channel regions were in the middle of the plenums. Li et al. [19] evaluated the effect of the manifold size and mass flow rate on a U-type Li-ion module that was composed of 36 battery cells. The authors of the current study concluded that increasing the channel size of the mass flow rate can deteriorate the temperature uniformity of the cells; in addition, an optimization was suggested to balance the air flow density and rate with the cell-to-cell variations and energy cost.

Among the research that has been already performed in the field, to the best of our knowledge, there has been limited information presented on performing co-design optimizations with a combination of multi-objective and multi-physics models.

Hence, in this study, we investigated the modeling capabilities of a physics-based, fractional-order electro-thermal model that was coupled to a 3D multi-physics module model. This model was used to evaluate the cooling performance of a proposed optimized BTMS, which was composed of 12 Li-ion prismatic cells. The methodology was initially presented in our previous study [20], where the optimized battery module was compared to the baseline model. The selected solutions of the co-design optimization model were thoroughly presented and evaluated, and this was achieved by taking into account the multi-objective optimization methods presented here [21] without the performance of laboratory assessments on the developed battery modules.

In this work, the design space from the optimization was built and evaluated to bring validity to our proposed methodology. The findings from [20] indicated that, over a broad design space with various parameters to account for (i.e., geometrical and performance indicators), a unique battery module design can be selected. Meanwhile, from over 250 structural and performance investigations, the algorithm was able to propose a single battery module that simultaneously minimizes all the selected objectives.

In this paper, we built the proposed battery module in our labs, and we tested it with various current profiles for functional assessments. A detailed comparison between the experimental and modeling performance was further performed to assess and validate the proposed design optimization methodology. Moreover, we performed a second step optimization study to evaluate and maximize the cost effectiveness of the proposed methodology. In this work, we evaluated the performance with both static and dynamic loading profiles to generate various heat profiles on the battery cells and modules.

Three cases were utilized to assess the applicability of the method, i.e., high heat-generation static currents, a conventional discharge/charge cycle, and a dynamic loading profile, according to the worldwide harmonized light vehicle test procedure (WLTP). Once the single cell and the module models were validated, the proposed BTMS was strategically optimized to balance between the temperature management and pressure drop, which was achieved by exploring various mass flow rates, inlet/outlet channel sizes, and geometrical parameters.

The introduction is followed by Section 2, in which the experimental methodology is described. Thereafter, we have Section 3, which details the single cell physics-based model and multi-cell module model development. In Section 4, the electro-thermal experimental validation is performed at three loading conditions. Section 5 details the numerical optimization that was conducted on the proposed air-cooled BTMS. Lastly, in Section 7, the paper is concluded and a discussion on future steps is conducted.

2. Experimental Setup

2.1. Li-Ion Cell Properties

A prismatic NMC/C lithium-ion cell was used in this study, as shown in Figure 1. The cell was designed for high-energy applications with a nominal capacity at 43 Ah, and it is capable of a current rate (C-rate or C) of up to a 2C constant loading.

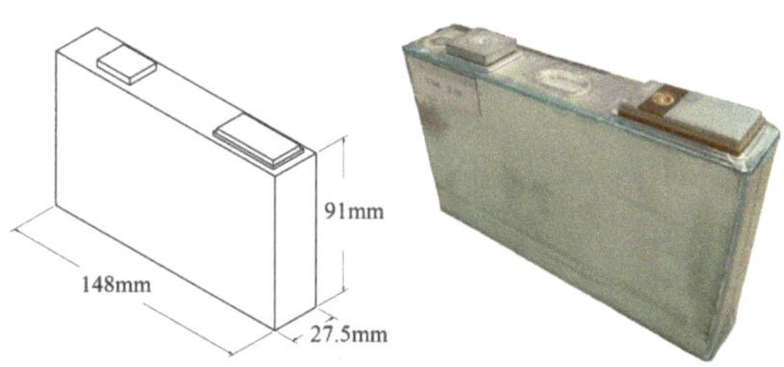

(a) (b)

Figure 1. (**a**) CAD cell design. (**b**) The actual NMC/C prismatic cell.

The main electrical and mechanical parameters of the Li-ion prismatic cell used in this study are displayed in Table 1.

Table 1. Electro-mechanical properties of the Li-ion cell.

Main Characteristics	Value	Unit
Chemistry	NMC/C	(-)
Shape	Prismatic	(-)
Nominal voltage	3.65	(V)
Nominal capacity *	43	(Ah)

Table 1. *Cont.*

Main Characteristics	Value	Unit
End-of-charge maximum voltage	4.2	(V)
End-of-discharge cut-off voltage	3	(V)
Volumetric energy density *	424	(Wh/L)
Specific energy density *	186.8	(Wh/kg)
Specific power *	>1200	(W/kg)
AC impedance (1 kHz)	<1	(mOhms)
Recommended charge current rate (continuous) *	1 C	(-)
Maximum charge C-rate	2 C	(-)
Cell dimensions	148 × 91 × 27.5	(mm)
Positive tab	6 × 18 × 22	(mm)
Negative tab	6 × 18 × 38	(mm)
Weight	0.840	(kg)

* At 1 C and 25 °C.

2.2. BTMS Configuration

A Z-type BTMS with parallel air flow to the cells is proposed in this study. Figure 2 shows the design configuration of the proposed air-cooled BTMS.

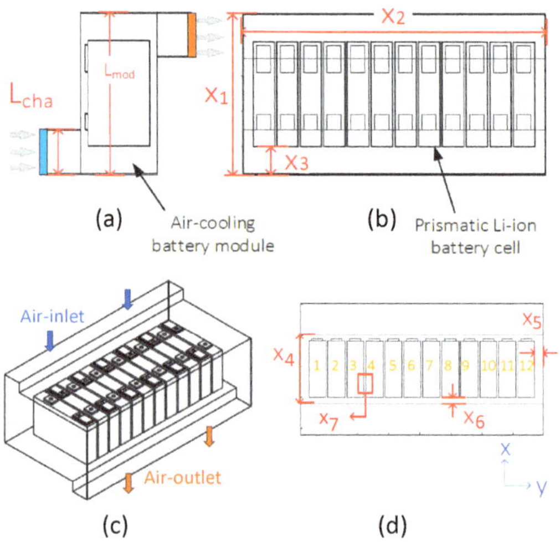

Figure 2. (**a**) Side view. (**b**) Top view. (**c**) The proposed BTMS inlet/outlet channels. (**d**) The BTMS cell configurations with the cells placed in the module (cell 1 to cell 12).

In Figure 2a, the inlet/outlet orientation and size is shown; in this study, this is denoted as L_{cha}, and it is approximately equal to a third over the total channel's area, which is denoted as L_{mod}. The inlet and outlet channels are, respectively, represented by blue and orange colors. In Figure 2b, the top view of the BTMS is shown, with the BTMS total length and width dimensions being denoted as x_1 and x_2, respectively. Meanwhile, x_3 represents the space from the cells to the BTMS's side wall. In Figure 2c, the proposed Z-type parallel-flow BTMS is shown with air inlet and outlet channels, whereas Figure 2d shows the internal configuration of the cells with respect to the BTMS boundaries, where x_4 is the height of the BTMS; x_5 and x_6 are the distances between the outer cells to the x- and y-axis side walls, respectively; and x_7 is the cell-to-cell distance. Lastly, it should be noted that the symmetries for all sides were considered. The proposed dimensions for the implementation of the BTMS were gathered and are shown in Table 2.

Table 2. Mechanical parameters of the proposed air-based BTMS.

Main Parameter	Implemented Value	Unit
x_1	201.2	(mm)
x_2	412.9	(mm)
x_3	26.6	(mm)
x_4	152.6	(mm)
x_5	20	(mm)
x_6	30.8	(mm)
x_7	3.9	(mm)

2.3. Test Bench

For the implementation of the BTMS, we initially electro-thermally characterized the prismatic cell. The electrical process included the capacity test that was conducted to obtain the actual cell's value at various temperatures and current rates. Also, it included impedance measurements, which were performed with the hybrid pulse power test (HPPC), as well as with measurements, to obtain the open-circuit voltage of the cells at different states of charges (SoCs). The raw data obtained from the tests were used to map the proposed modeling behavior. A particle swarm optimization (PSO) was utilized to extract the model parameters, the detailed experimental characterization process of which can be found in [22]. The cooling efficiency of a BTMS is often described with the maximum temperature (T_{max}), the temperature difference among the cells (DT), the temperature difference on the cell surface ($Cell_T$), and the total volume (V) that the battery module occupies. The primal objective, serving as an attempt through which to improve the volumetric and gravimetric energy densities of the modules, is to keep the cells within a safe operation temperature area but simultaneously minimize the volume or the weight. The significance of these KPIs is related to temperature management, which plays a crucial role in assessing the safety, reliability, and efficiency of Li-ion batteries. The maximum temperature rise observed on the outer casing or surface of the various form factors or formats during its operation is referred to as surface temperature increase (T_{max}). Elevated levels may signal operational inefficiencies, including internal resistance rise, poor power capabilities, overcharging, or excessive current consumption. These inefficiencies could serve as early warning signs of thermal runaway, which is a dangerous scenario characterized by a rapid escalation in battery temperature that could potentially result in the venting, fire, or explosion of the cells. The temperature difference (DT) among the cells is also referred as thermal uniformity in a multi-cell topology. Non-uniform temperature distribution can lead to localized hotspots, which accelerate degradation and reduce the overall lifespan of the battery. Monitoring temperature variations helps in identifying potential thermal management issues, such as poor heat dissipation, uneven cooling, or cell-to-cell variations in performance. Last but not least, the temperature difference on the cell surface ($Cell_T$) indicates the heat generation from the core of the cells to the outer casing, which can lead to uneven heat distribution and to local hot spots that affect the efficiency and reliability of Li-ion batteries (as it induces thermal stress on individual cells and increases the risks of thermal runaway).

In this method, the objective of the algorithm was defined so as to minimize these four parameters, and this was achieved by taking into account the complex multi-variable and multi-physical requirements of the design. The multi-objective particle swarm optimization algorithm (MOPSO) was developed in MATLAB (and was published in our previous work [21]). It is linked here to the 3D model created in Comsol to solve the multidisciplinary costs, which can mathematically be expressed as in Equation (1):

$$\begin{aligned} \min_i F(x_i) &= \left[T^i_{max}, DT^i, Cell^i_T, V^i\right], \forall\, i = 1 \leq x_9 \leq 9 \\ s.t. &= \left[x_1, x_2, x_3, x_4, x_5, x_6, x_7, x_8\right] \\ at &\left[v, C_r, T_{amb}, v_{temp}\right]. \end{aligned} \tag{1}$$

The mass flow rate v will be further examined in this work as a second-optimization step that accounts for cost efficiency.

The thermal properties of the cell were found with a micro-pulse heat profile that was injected into the cell, whereby the same number of Ah was extracted and injected in order to be SoC-independent, which could result in influencing electrical and thermal parameters such as the cells' resistances. The extracted specific heat and the convection transfer coefficient were found in our previous work [23]. Once the cells were characterized, the module development took place, which was based on the optimized selected geometry presented in Table 2. For the 12S1P module implementation, an electrical characterization of more than 100 cells was performed to select the ones with the closest values in a Ragone plot in terms of capacity and impedance. A variation of up to 3% and 11% was exhibited at the beginning of life (BoL) and at the same testing conditions, respectively. A total of 12 cells were selected that showed similar performance in terms of cycle capacity and impedance at the various conditions, whereas the BTMS design that was implemented was based on BTMS 5, which showed—in simulation—a potential 15% decrease in the maximum temperature increase on the cells, a 5% decrease in the required volume, a 70% decrease in the temperature uniformity among the cells, and approximately a 40% decrease in the better heat distribution on a cell [20]. The selected battery module was constructed with a plexi glass that iwass precisely manufactured to follow the optimized dimensions. In addition, three fans were placed in the proposed inlet location with a variable mass flow rate and power consumption. The battery module was placed in a room with a control ambient temperature at 25 °C and a specific convection rate.

The hardware setup for the testing included a climate chamber to preserve the cells' surface temperatures, which were close to the various environmental conditions. The single cells were connected to a PEC ACT0550 tester, (PEC, Leuven, Belgium) which is capable of up to DC 5 V measurements. In addition, the module was connected to a PEC SBT8050, which is capable of up to 80 V DC. Both of the testers had a ±0.005% voltage accuracy. The module was connected to a pre-charge circuit and a fuse relay safety box, and it was monitored by a commercially available battery management system (BMS). An anemometer and a temperature data logger were used to measure the mass flow rates and to obtain the cells' surface temperatures in the module. The overall test bench is shown in Figure 3, where the green and blue paths indicate the cell and module level testing, respectively.

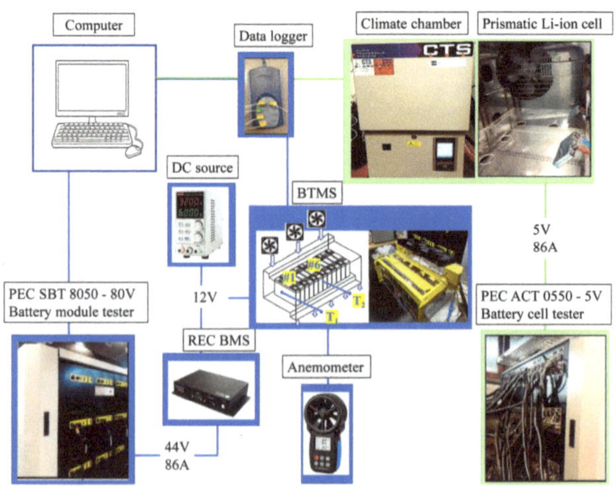

Figure 3. Experimental setup during the cell- (green) and module-level (blue) characterizations and validations.

3. Model Development

Physically meaningful battery models can be employed by means of differential equations and fractional order calculus (FOC), where the lumped capacitor of the ECM [22] is replaced with constant phase elements (CPE), which can greatly improve the modeling accuracy [24,25]. A Thevenin FOC model was presented in [26] that can achieve significant improvements on its electrical behavior compared to an empirical model. The modeling impedance was correlated to the actual cells by way of comparisons to the electrochemical impedance spectroscopy (EIS) results at various cell aging states. Several FOC models can be found in the literature. However, the Thevenin equivalent with a current dependency on a charge transfer resistance that is based on the Butler–Volmer approximation (BVE) [27] combined with a diffusion element (Warburg) [28] has shown a good performance and balance between computational burden and accuracy. An R-(BVE//CPE)-W FOC coupled to a thermal model is thus proposed in this study.

3.1. The Physics-Based FOC Electro-Thermal Model

3.1.1. Electrical Part

The FOC model describes the impedance behavior according to the CPE elements, as expressed in Equation (2):

$$Z_{CPE}(f) = \frac{1}{Q(j2\pi f)^{\alpha}}, (0 < \alpha < 1), \tag{2}$$

where j is the imaginary part, and α and Q are the variables used to obtain the CPE. An ideal capacitor can be represented in case α is equal to 1; when α is equal to 0, it is instead treated as a pure resistor. Moreover, α can be obtained directly from the EIS measurements [29], or it can be numerically calculated with FOC calculus, where the real-order differential operator can be defined as in Equation (3):

$$_{\alpha}\mathcal{D}_t^{\alpha} = \begin{cases} \frac{d^{\alpha}}{dt^{\alpha}}, & \alpha > 0 \\ 1, & \alpha = 0 \\ \int\limits_{\alpha}^{t} (d\tau)^{\alpha}, & \alpha < 0, \end{cases} \tag{3}$$

where the fractional order in a real domain is denoted as α. The discrete form of the FOC can be numerically solved according to the Grünwald–Letnikov (GL) definition in fractional calculus [30]:

$$\mathcal{D}^{\alpha} f(t) = \lim_{\Delta t \to 0} \frac{1}{(\Delta t)^{\alpha}} \sum_{j=0}^{[t/\Delta t]} (-1)^j \binom{\alpha}{j} f(t - j\Delta t), \tag{4}$$

where the sampling time is represented with Δt; the user-defined memory length is denoted as $[t/\Delta t]$; and $\binom{\alpha}{j}$ is the Newton binomial coefficient, which is calculated as in Equation (5):

$$\binom{\alpha}{j} = \begin{cases} 1, & j = 0 \\ \frac{\alpha!}{j!(\alpha-j)!} = \frac{\Gamma(\alpha+1)}{\Gamma(j+1)\Gamma(\alpha-j+1)}, & j > 0. \end{cases} \tag{5}$$

The proposed FOC model was depicted based on Kirchhoff's voltage law in accordance with the following expressions:

$$\begin{cases} U_{batt} & = V_{OC} - \eta_{Ohmic} - \eta_{ct} - \eta_{diff} \\ U_{batt} & = f(C_r, SoC, \theta) \\ \mathcal{D}^{\alpha}\eta_{ct} & = \frac{I_{batt}}{Q_{ct}} - \frac{\eta_{ct}}{Q_{ct}R_{BVE}} \\ \mathcal{D}^{\beta}\eta_{diff} & = \frac{1}{Q_{diff}}\eta_{diff}, \end{cases} \tag{6}$$

where V_{OC} represents the open-circuit voltage of the cells; η_{Ohmic} is the voltage drop of the Ohmic resistance; and η_{ct} and η_{diff} represent the voltage across the CPE and diffusion elements, respectively. The terminal voltage of the cell is denoted as U_{batt} and is dependent on the SoC, current rates (C_r), and temperature (θ). The fractional orders α and β were assigned to the CPE elements for the polarization effects with factors Q_{ct} and Q_{diff}, respectively. R_{BVE} shows the activation polarization resistance, which was based on BVE approximation [31,32].

$$R_{BVE,s} = \alpha_{0,s} \cdot \frac{ln\left[\frac{1}{\alpha_{1,s}} I_{batt,s} + \sqrt{\left(\frac{1}{\alpha_{1,s}} I_{batt,s}\right)^2 + 1}\right]}{\frac{1}{\alpha_{1,s}} I_{batt,s}} \tag{7}$$

where s is the SoC counterl $\alpha_{0,s}$ and $\alpha_{1,s}$ are fitting parameters; and $I_{batt,s}$ is the current passing through the cell at every sample. In using the GL from Equations (3) and (4), we obtain the following transformations (Equations (8) and (9)):

$$\mathcal{D}^{\alpha}\eta_{ct} = \frac{1}{T_s^{\alpha}} \sum_{j=0}^{[k]} (-1)^j \binom{\alpha}{j} \eta_{ct,k-j} = \frac{I_{batt,k}}{Q_{ct,k}} - \frac{\eta_{ct,k}}{Q_{ct,k} R_{BVE,k}}, \tag{8}$$

$$\mathcal{D}^{\beta}\eta_{diff} = \frac{1}{T_s^{\beta}} \sum_{j=0}^{[k]} (-1)^j \binom{\beta}{j} \eta_{diff,k-j} = \frac{1}{Q_{diff,k}} \eta_{diff,k}, \tag{9}$$

where k is the time step. The solutions of these two formulas were derived from the battery voltage U_{batt} at the discrete time domain, as in Equation (10):

$$U_{batt,k} = V_{OC,k} - \eta_{Ohmic,k} - \eta_{ct,k} - \eta_{diff,k}, \tag{10}$$

where the corresponding voltage drops are calculated as (Equations (11)–(13))

$$\eta_{Ohmic,k} = I_{batt,k-1} R_{Ohmic}, \tag{11}$$

$$\begin{aligned} \eta_{ct,k} = & \left(\alpha - \frac{T_s^{\alpha}}{Q_{ct} R_{BVE}}\right) \eta_{ct,k-1} \\ & + \frac{T_s^{\alpha} I_{batt,k-1}}{Q_{ct}} - \sum_{j=2}^{[N_s]} (-1)^j \binom{\alpha}{j} \eta_{ct,k-j}, \end{aligned} \tag{12}$$

$$\eta_{diff,k} = \frac{T_s^{\beta} I_{batt,k-1}}{Q_{diff}} - \frac{1}{T_s^{\beta}} \sum_{j=0}^{[N_s]} (-1)^j \binom{\beta}{j} \eta_{diff,k-j}. \tag{13}$$

3.1.2. Thermal Part

For the thermal branch, and in order to obtain the heat dissipation rates, the temperature gradient was calculated as in Equation (14) [33]:

$$\begin{cases} \frac{dU_{cell}}{dt} & = Q_{gen}(t) - Q_{loss}(t) = m\, C_p\, \frac{dT}{dt} \\ Q_{loss}(t) & = Q_{conv}(t), \end{cases} \tag{14}$$

where U_{cell} represents the internal energy of the Li-ion and Q_{gen} is the generated heat rate based on the cell's Joule losses. Q_{loss} represents the heat loss expressed by the convective heat transfer to the environment. Also, m is the cell mass, C_p is the heat capacity, and T is the surface temperature. The heat transfer to the ambient temperature was calculated as shown in Equation (15):

$$\begin{cases} Q_{gen}(C_r, SoC, \theta) & = R_{Ohmic}(C_r, SoC, \theta)\, I_{batt}^2 \\ & + R_{BVE}(C_r, SoC, \theta)\, I_{ct}^2 \\ Q_{loss}(C_r, SoC, \theta) & = h_{conv} S (T_{amb} - T_{cell}), \end{cases} \tag{15}$$

3.1.3. Coupled Electro-Thermal Model

Where h_{conv} is the cell-level convection heat transfer coefficient and S is the cell cross-section area. The thermal properties of the model were calculated in our previous work [23]. Figure 4 shows the coupled 1D electro-thermal model.

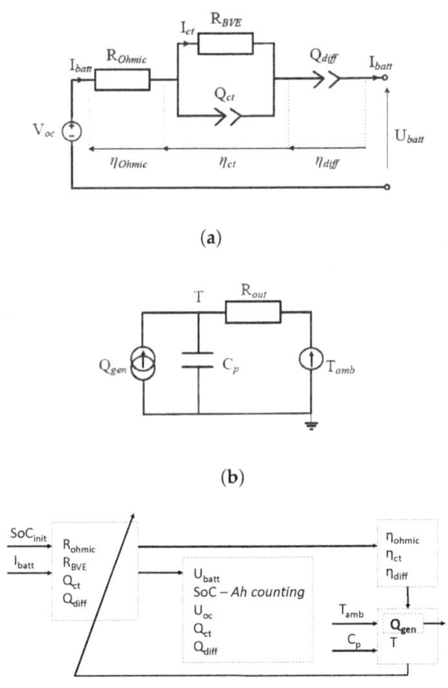

Figure 4. (**a**) The electrical FOC model. (**b**) The thermal branch (**c**) The coupling of the electrical FOC with the thermal 1D model.

3.2. 3D Numerical Model

The proposed 1D electro-thermal model derived the heat losses of the cell over time, which were fed to the 3D model as an input to further evaluate the temperature gradients at the module level. The 3D models' thermal properties, such as conductivity, cell density, and specific heat, were obtained from a characterization process that was presented in our previous work [23], whereas the rest domain parameters were found in the literature from the study of [34], and they are gathered in Table 3.

Table 3. Input parameters of the multi-physics model.

Main Parameter	Air [34]	NMC/C [23]	Aluminum Tabs [34]
Density ρ (kg/m^3)	1.165	2268	2700
Specific heat C_p (J/kg·K)	1005	933.7	900
Thermal conductivity λ (W/m·K)	0.0267	0.82 (λ_x) 4.43 (λ_y) 2.72 (λ_w)	238
Electrical resistance R' (Ω)			9.97×10^{-6}

The model was developed with a COMSOL computational fluid dynamics (CFDs) simulation tool, whereas the temperature and fluid fields were solved with a finite element method. We used a turbulent, single-phase, and incompressible fluid, the mass, momentum, and energy conservation of which were described according to the Reynolds average Navier–Strokes equations with a k-ε turbulence model [34,35] for the air flow area (Equations (16)–(18)) and the battery cells (Equation (19)):

$$\frac{\vartheta \rho_f}{\vartheta t} + \nabla \cdot \left(\rho_f \bar{v}\right) = 0, \tag{16}$$

$$\rho_f \frac{\vartheta \bar{v}}{\vartheta t} + \rho_f (\overrightarrow{v} \cdot \nabla) \bar{v} = \nabla \bar{p} + [\nabla \cdot (\mu \nabla \bar{v}) - \phi], \tag{17}$$

$$\rho_f C_p \frac{\vartheta T_f}{\vartheta t} + \left(\rho_f C_p \overrightarrow{v}\right) \nabla T_f = \nabla \cdot \left[\left(\lambda_a + \frac{\mu_t}{\sigma_t}\right) \nabla T_f\right], \tag{18}$$

$$\rho_c C_{p}, c \frac{\vartheta T_c}{\vartheta t} = \nabla \cdot \left[\overrightarrow{\lambda_c} \nabla T_c\right] + \dot{Q}_{gen} + \dot{Q}_{tab}, \tag{19}$$

where ρ_f represents the fluid density; $\bar{v}$ shows the average velocity; and the viscosity, pressure, and Reynolds stress are denoted by μ, p, and ϕ, respectively. Also, the time-dependent heat dissipation created an unsteady temperature in the airfield region, which is described by Equation (18). In this case, the fluid temperature is denoted as T_f, the thermal conductivity of air is λ_a, and μ_t is the turbulent dynamic viscosity. The temperature equation for the battery module is described in Equation (19), with ρ_c, C_p, c, and T_c the cell's density, specific heat, and surface temperature, respectively, and where $\overrightarrow{\lambda_c}$ is the thermal conductivity at each direction.

The heat generation included the losses generated from the aluminum cell's tabs in accordance with the following formula:

$$Q_{tab} = \frac{R' \cdot I_{batt}^2}{V l_{tab}} * N, \tag{20}$$

where N is the 24 accounted tabs in a 12S1P topology. The boundary conditions for a cell were between the cell's surface and the ambient temperature, as calculated in Equation (15). For the module level, the heat convection transfer coefficient varied during testing time, and it was calculated by the software based on the air properties at the selected design (mass-flow speed, channel sizes, temperature, etc.). The model was solved with COMSOL software v5.5 using the MUMPS solver with a default physics-controlled unstructured tetrahedral mesh, as well as with the non-slip boundary conditions being imposed to the walls and the initial temperature being set at the ambient temperature i.e., 25 °C.

4. Experimental and Numerical Studies

The BTMS evaluations were performed at a 25 °C ambient temperature for a constant current discharge, as well as with maximum allowed C-rates (2 C), a discharge/charge cycle at the recommended rates (1 C), and a dynamic loading profile (WLTC). For each case, the single cell electrical validation is shown with the modeling voltage in comparison with the experimentally obtained results. In addition, their relative error V_{err} is used as a model accuracy indicator as follows:

$$V_{err} = \frac{V_{exp} - V_{batt}}{V_{exp}} * 100\%. \tag{21}$$

Furthermore, the single-cell temperature behavior was evaluated and the corresponding heat generation was obtained based on Equation (15). The analysis was continued for the multi-cell model by comparing the modeling to the experimental temperature behaviors, which was measured at two different locations. The first thermocouple was placed in the outer area of the first cell (noted as T_1) and the second (noted as T_2) between the fifth and sixth cells of the 12S1P module. Both were placed in the center of the cells, and, by these means, the thermal uniformity among the cells could be monitored and assessed.

For the module assessments, the initial SoC of the cells was 85% and this decreased by up to approximately 20% in order to bypass any of the balancing processes from the BMS. During the experimental implementation, a velocity of 1 m/s was measured via an anemometer, which corresponded to an approximately 0.012 m^3/s mass flow rate for a channel size L_{cha} that was set at 60 mm.

4.1. Maximum Static Discharge

A 2 C constant current profile over the whole available SoC was loaded to the battery for single-level electrical validation.

4.1.1. Cell-Level Static Evaluation

The single-cell model voltage behavior is shown against the experimental behavior along with their relative errors in Figure 5. It was observed that the proposed model can map a high accuracy with the voltage behavior over the whole SoC range. Only a slight deviation occurs during the last steps of the discharge, and this is possibly due to the unexplored lithium diffusion processes that occur simultaneously in the cells [36], which could have been better captured with extra CPEs in the model. Nonetheless, for the proposed model, the relative error stayed below 4% over the whole experiment.

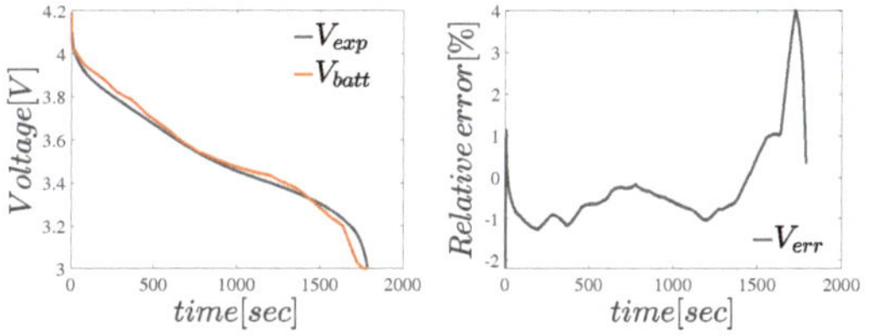

(a) (b)

Figure 5. (**a**) Static single cell voltage behavior. (**b**) Relative error.

The thermal validation of the model can be seen through a comparison of the experimental with the simulation results. The temperature behavior for this case is shown in Figure 6 with the corresponding heat generation. To accord with the real conditions of the air flow in the thermal chamber, in a single-cell model, the heat transfer coefficient was set to 10 $W/(m^2{\cdot}K)$.

The abovementioned diffusion effect did not highly affect the temperature accuracy in the simulation. A good agreement was achieved over the whole experiment, which helped with verifying the proposed process and allowed it to proceed with multi-cell assessments.

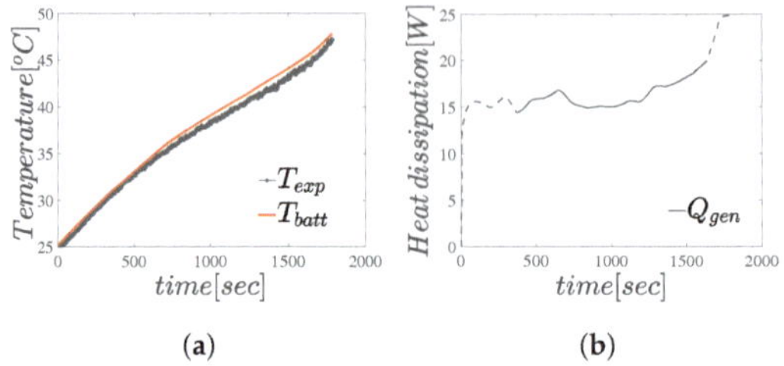

Figure 6. (**a**) Single cell temperature behavior at a constant current profile. (**b**) Heat generation.

4.1.2. Module-Level Static Validation

The module was validated for a static current with an initial SoC at approximately 85%, which was then discharged by up to 20%. Figure 6b shows, with solid lines, the heat generation input to the CFD model. The experimental and numerical results are shown at the locations of T_1 and T_2 in Figure 7. In the same figure, the blue line shows the behavior of a natural convention model (NC BTMS) without any cooling, and it was obtained from the numerical solution.

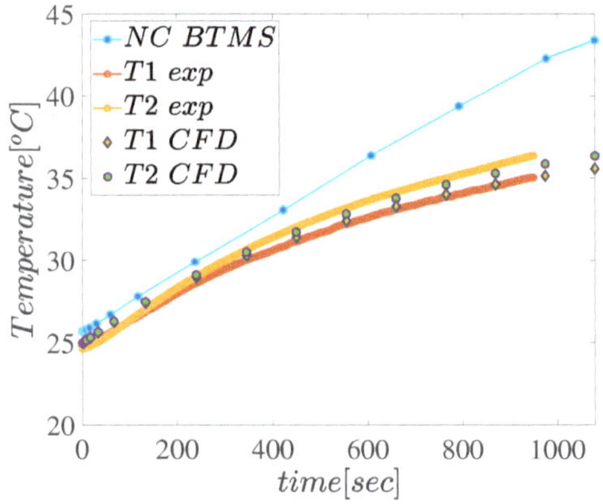

Figure 7. Constant current BTMS validation.

It was observed that the proposed BTMS model fit, with a good agreement, the experimental data and could keep the maximum temperature lower than 36 °C, which is approximately 9 °C lower than what was encountered with the NC-BTMS. Also, the heat uniformity was below 3 °C, which signifies the efficiency of the proposed air-based BTMS under this demanding loading. The selected battery module presented in Figure 8a is a result of the digitalized methodology over the various available designs with respect to inlet/outlet topology, cell spacing, etc. More details can be found in our previous publication [20]. The heat distribution at the end of the test profile is shown in Figure 8 for the proposed thermo-mechanical design (a), as well as the solution and natural convection (b), at the end (1200 s) of the discharge. The maximum temperature evolution was improved by approximately 10 °C, while the thermal uniformity among the cells was improved by approximately 4 °C with a maximum deviation of 1 °C (as can be further investigated here [20]).

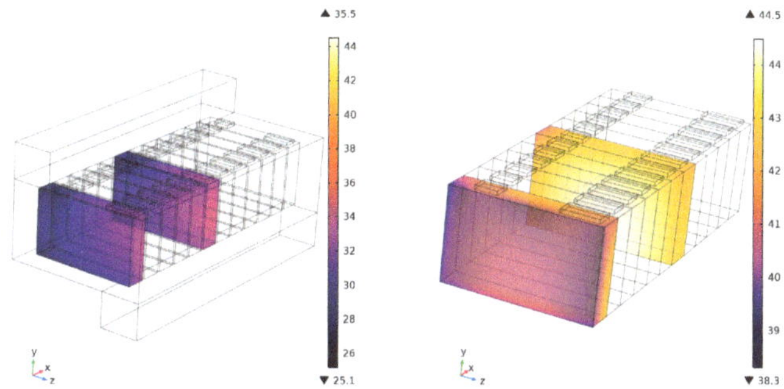

(**a**) (**b**)

Figure 8. (**a**) The proposed BTMS. (**b**) The natural convection heat distribution.

4.2. Discharge–Charge Cycle

The current profile of a constant discharge–charge cycle that was applied on both topologies is shown in Figure 9.

The cells were charged to an approximate 85% SoC, and they were then discharged with 1 C for 2000 s. They experienced an immediate charge with the same C-rate until the SoC had reached the initial level.

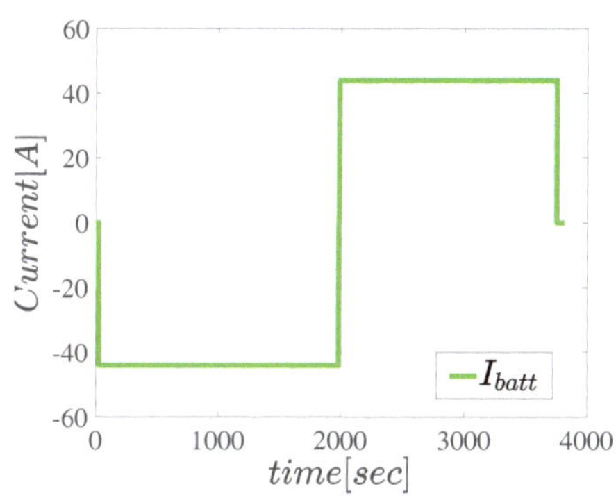

Figure 9. The discharge-charge cycle with a 1 C rate.

4.2.1. Cell-Level Cycle Evaluation

The experimental and modeling voltages are shown with the respective relative errors in Figure 10.

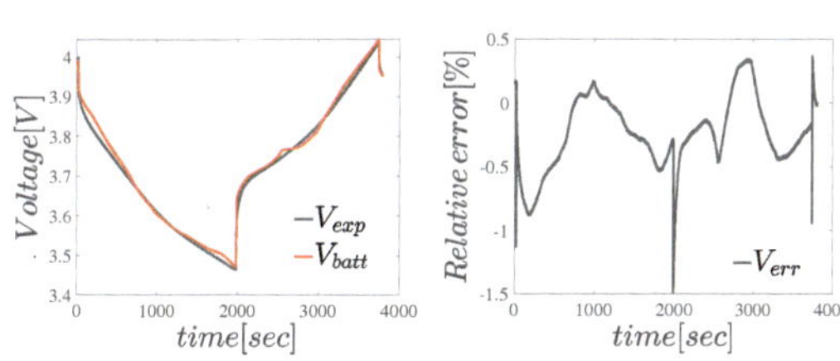

(**a**) (**b**)

Figure 10. (**a**) The discharge-charge single cell voltage behavior. (**b**) The relative error.

The activation and concentration polarization effects are shown with higher and lower time constants in the model. They can be tracked with high accuracy, as indicated by the relative error that stayed below 1.5% for the whole test. The corresponding temperature behavior, validation, and heat generation are illustrated in Figure 11.

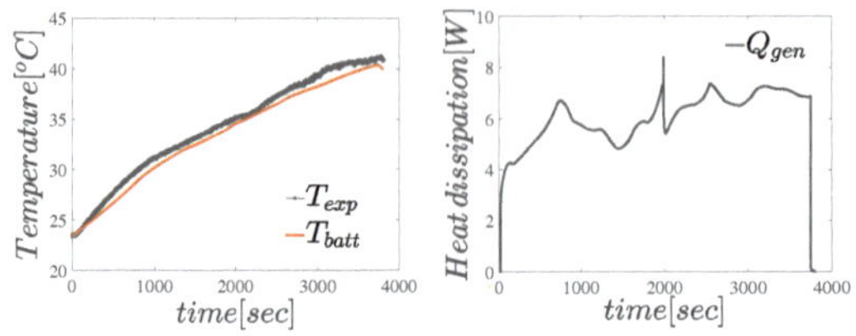

(**a**) (**b**)

Figure 11. (**a**) The single cell temperature behavior with a current cycle. (**b**) Heat generation.

Compared to the previous case, it was observed that the voltage could be realized with a single diffusion element. Hence, a good thermal agreement was achieved, whereas it was also observed that the charge resistance was slightly higher than discharge, as indicated by the heat dissipation curve.

4.2.2. Module-Level Cycle Validation

A time-dependent heat generation was supplied to the CFD model. Even though, for the two current pulses, the test time was higher and the heat generation was based on the impedance data from both charge and discharge profiles, the temperature behavior can be tracked with a high accuracy, as shown in Figure 12.

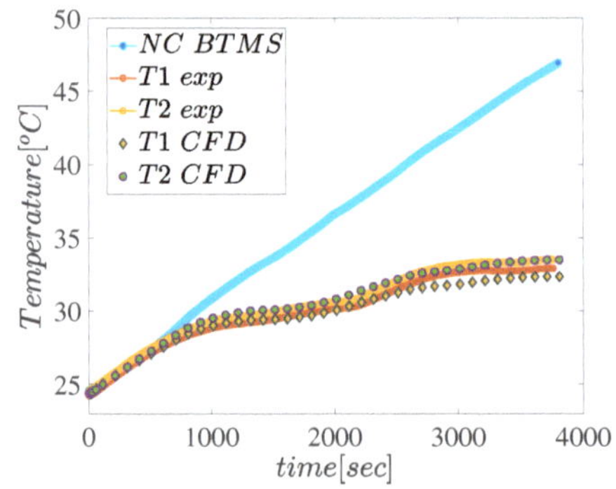

Figure 12. The discharge–charge current BTMS validation.

Also, the proposed module was capable of keeping the maximum temperature below 35 °C when compared to the NC-BTMS, while the heat uniformity was established with thermocouples T_1 and T_2, as well as by the good agreement between the experimental and CFD values.

4.3. Dynamic Loading—WLTC

In the last study case, the WLTC dynamic current profile, as shown in Figure 13, was applied to the cells. It was based on a mixture of low and more demanding current pulses with the maximum C-rate being limited at 1.5 C (60 A).

The applied profile was composed of four consecutive WLTC cycles that discharged the cells for an approximate 60% SoC. Similar to previous cases, to avoid the balancing effects of the BMS, the initial SoC was set at around 85% and the test was conducted in 25 °C ambient conditions.

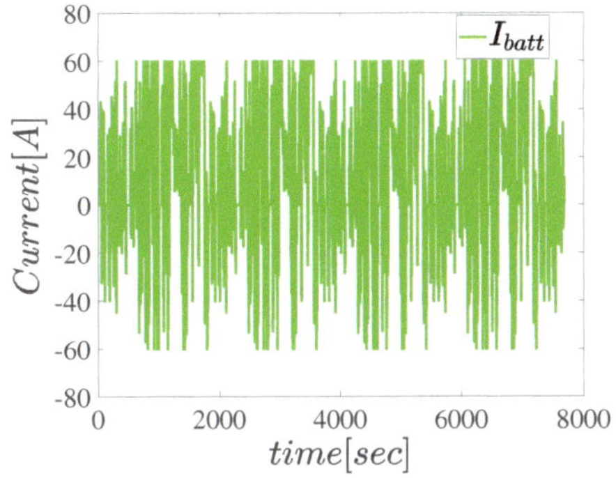

Figure 13. The dynamic cycle based on the WLTC profile.

4.3.1. Cell-Level WLTC Evaluation

The R-(BVE//CPE) branch was found to be very efficient for the fast dynamics that were mainly accounted for in this profile. Throughout the whole test, the relative error did not exceed 0.5% of the actual voltage value, as shown in Figure 14.

Hereafter, the proposed electro-thermal model captured the temperature behavior with high accuracy, the corresponding heat generation of which is shown in Figure 15.

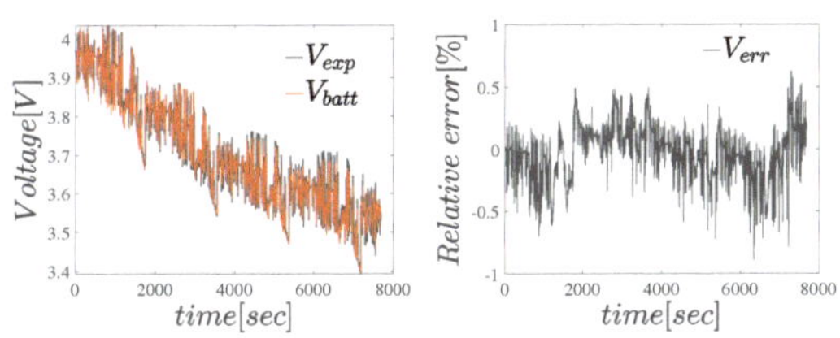

(**a**) (**b**)

Figure 14. (**a**) The dynamic profile single cell voltage behavior. (**b**) The relative error.

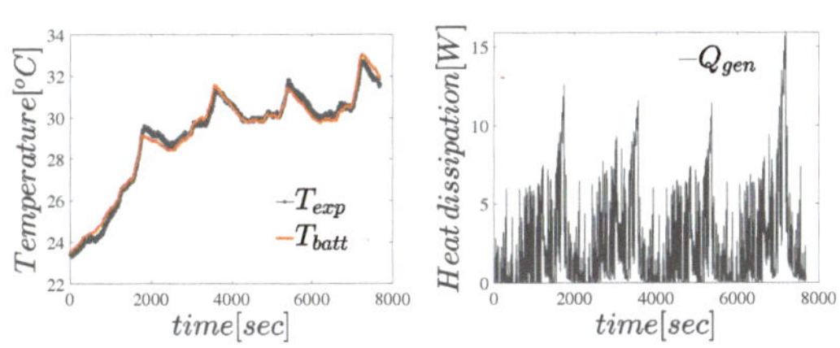

(**a**) (**b**)

Figure 15. (**a**) The single cell temperature behavior with a WLTC profile. (**b**) The heat generation.

4.3.2. Module-Level WLTC Validation

The dynamic profile was applied to the 12S1P air-based module, and the experiment data when compared against the CFD are shown in Figure 16.

A good agreement was achieved both on the maximum temperature and heat distribution for the proposed module. It was observed that the consecutive current cycles could elevate the maximum temperature at the end of the experiment (approx. 20% SoC) to 38 °C. Based on the NC-BTMS model, which is indicated with the blue line, the temperature had an increasing rate of change. This could lead to surpassing the safety window if no cooling solution is applied. However, the proposed BTMS showed a stable thermal trend without exceeding 28 °C.

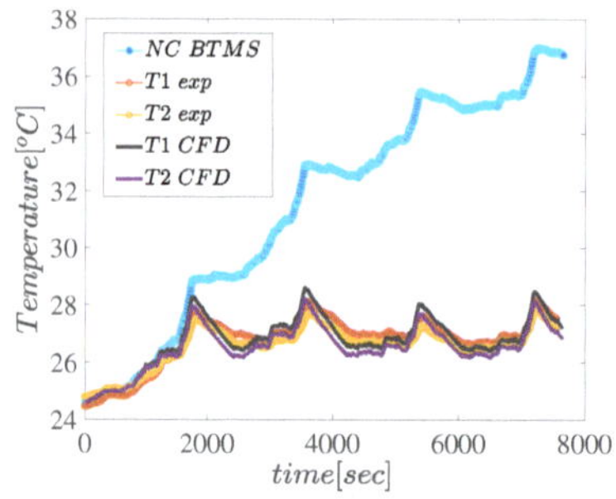

Figure 16. The dynamic current BTMS validation.

5. BTMS Cost-Effectiveness Study

The proposed module was strategically optimized, as detailed in this section. For this purpose, the validated numerical model was used to help visualize and predict the temperature behavior when using various structural designs.

5.1. The Inlet Coolant Flow Rate

Different inlet coolant flow rates were evaluated to help determine the influence on the temperature evolution of the proposed BTMS. The coolant was set at the same temperature as the ambient, i.e., 25 °C, whereas, for the evaluation and validation steps, a 0.012 m^3/s flow speed was selected. Figure 17a demonstrates the temperature contours of 0.01 m^3/s to 0.04 m^3/s. It was observed that the temperature declined when increasing the flow rate. With the coolant rate increasing from 0.01 m^3/s to 0.04 m^3/s, the maximum temperature dropped to 8 °C at the end of a 2 C static discharge loading, and this was as a result of the enhanced convective heat transfer between the cells and the air coolant. However, the increased flow rate could proportionally increase the hot spots on the cells and deteriorate the uniformity in the module. The local hotter zones could create a higher localized negative aging effect on the cells as a faster degradation was expected at those points. Operations outside the SoA can jeopardize the safety and performance of the module. To find the optimal flow rate, the pressure drop was plotted with the temperature behavior, as shown in Figure 17b. It was evident that the higher the flow rate, the steeper the temperature drop and the cooling speed. The results depicted in the figure verified the fact that a non-linear relationship was obtained between the coolant rate and the pressure drop. To balance between the pressure loss and the thermal management, a mass flow rate at the intersection of those curves, at approximately 0.023 m^3/s, was selected.

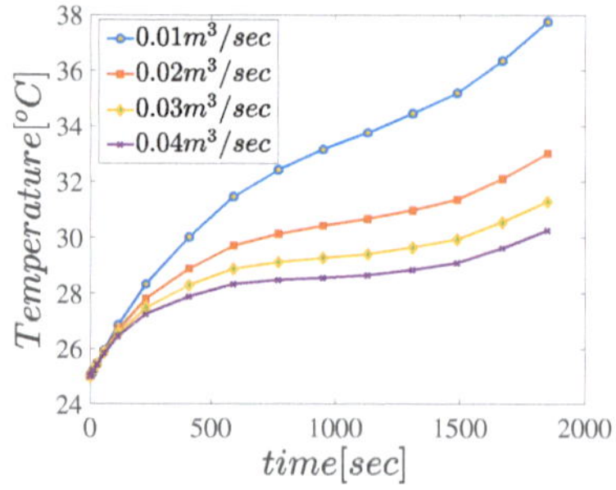

(a)

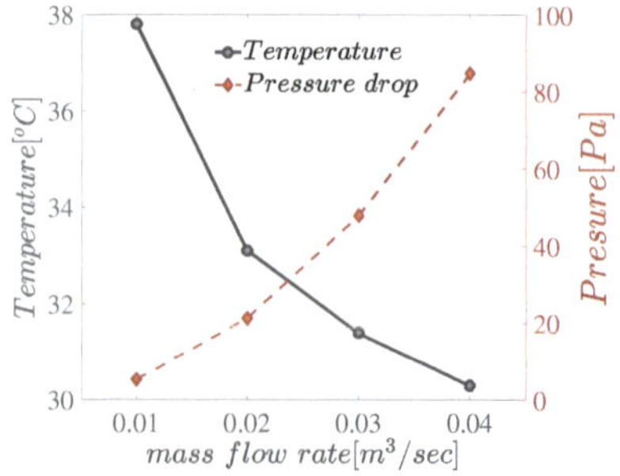

(b)

Figure 17. (**a**) Temperature behaviors. (**b**) The pressure drop at various mass flow rates.

5.2. Channel Size

The design of the channels' input and output played an important role in managing the temperature and pressure losses of the proposed BTMS. In this part, the evaluation was performed with channel sizes of 20 mm to 60 mm. The symmetries of the input and output channels were recorded, whereas, for the main case study and experimental validation, the channels were designed with an approximately 55 mm size. It is shown in Figure 18 that, from increments in the channel size, the monitored maximum temperature decreased by up to 2 °C. On the other hand, the pressure drop showed a sharp increase that was up to four times higher as the channel size increased. The study was performed with the previously optimized mass flow rate (0.023 m^3/s).

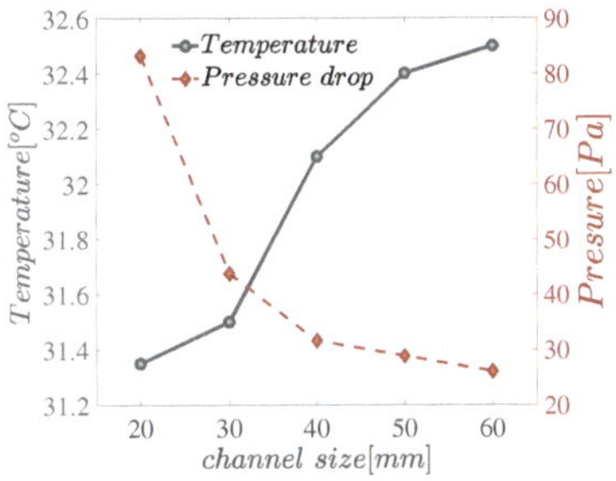

Figure 18. The pressure drop and maximum temperature at various channel sizes.

Hence, the optimized BTMS was to be implemented with a trade-off value, for which the temperature and the cost could be balanced, of approximately 30 mm for the channel size.

5.3. Cell-to-Cell Space

The last key parameter to be optimized based on the pressure drop and the heat distribution was the cell-to-cell distance. The optimal flow rate and channel size obtained in previous sections were now accounted for. In the following Figure 19, it is clearly shown that, when increasing the cell's intermediate distances, the cell's maximum temperature at the end of a 2 C discharge process was decreased by up to approximately 4 °C. It was also observed that the drop decreased from 75 Pa to 45 Pa. This meant that, by increasing the cell-to-cell distance, not only did the temperature decline, but a pressure drop reduction was also observed, as illustrated in Figure 19b. Nonetheless, increasing the distance by more than 4 mm could lead to extra costs when the temperature is not proportionally dropped. An approximate 4 mm distance was hence selected as the optimized value for the proposed BTMS.

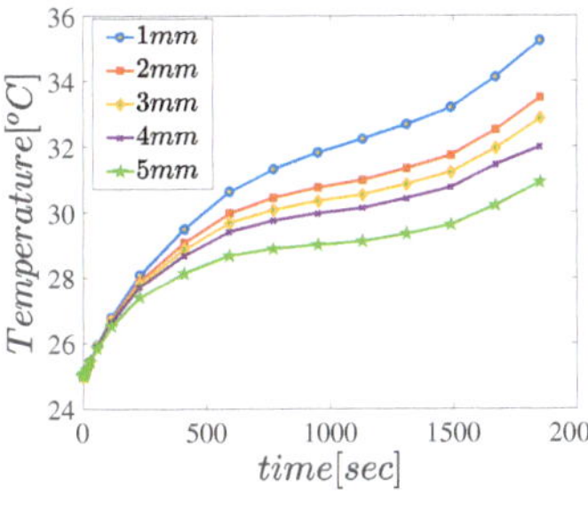

(**a**)

Figure 19. *Cont.*

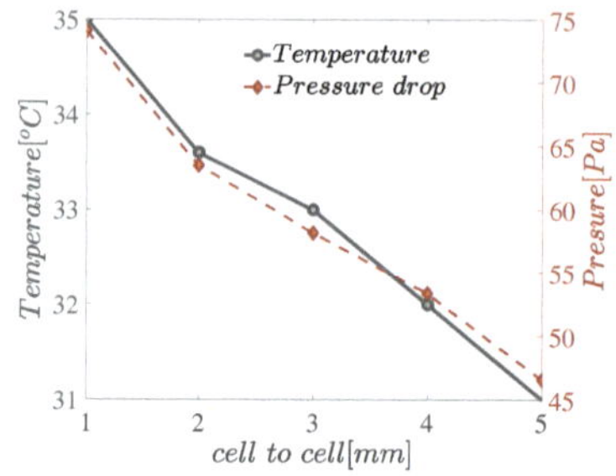

(b)

Figure 19. (a) Temperature behaviors. (b) The pressure drop at various cell-to-cell distances.

6. Discussion

While the design optimization method presented in this work can significantly enhance the performance, efficiency, and safety of Li-ion battery systems, several constraints may affect its broader adoption across diverse applications. In this section, we try to elaborate on the challenges that might arise during the design optimization method for the purposes of future consideration and wider adoption. The first challenge was computational power and efficiency. We demonstrated that our method required a non-stop optimization process of approximately 5 days [20]. Hence, the multi-objective design optimization methods implemented in 3D environments involved complex mathematical models, simulations, and algorithms, which were used to analyze various design parameters and optimize performance criteria, as was presented in our work. Implementing these methods requires substantial computational resources, including high-performance computing systems, advanced software tools, and expertise in the numerical methods. With the trade-off that can be made on the various modeling approaches, the modeling accuracy of the battery systems can vary significantly. An accurate battery behavior is essential for effective design optimization. However, developing accurate battery models that can capture the internal electrochemical processes, the thermal behavior at various conditions, and mechanical interactions can be challenging. Furthermore, the validation of these models against experimental data is crucial but often time-consuming and costly. Such model inaccuracies or uncertainties can lead to suboptimal designs or unexpected performance issues, thereby undermining the reliability and effectiveness of design optimization methods. This is one of the main purposes that validations were presented after the construction of the proposed module design, as well as after further analyses, in the current work. Moreover, a global implementation of the proposed method should concern the numerous interdependent parameters, including the cell chemistry, geometry, materials, thermal management strategies, and operating conditions of the various cell chemistries and formats. Exploring this vast design space to identify optimal solutions while considering trade-offs between conflicting objectives might be challenging. An extra important constraint that one should consider is that the optimized designs generated through computational methods may not always be readily manufacturable or scalable to large-scale production. Design constraints imposed by manufacturing processes, material availability, supply chain logistics, and cost considerations may limit the practical feasibility of implementing optimized designs in real-world applications. Bridging the gap between design optimization and manufacturability is essential for ensuring the broader adoption of the proposed method.

7. Conclusions and Future Work

In this work, the performance of an air-cooled, battery thermal management system (BTMS) was studied for a battery module composed of 12 high-energy, prismatic Li-ion cells connected in series.

A physically meaningful electrical model was built based on fractional-order calculus, which helped to map the impedance of the cells with high accuracy. The single cell electrical model was coupled to a thermal branch, and it was evaluated with three different current profiles, a maximum static discharge current, a discharge–charge cycle, and a consecutive dynamic profile. A good agreement between the modeling and the experimental values was achieved, and this was underlined with the low relative errors obtained in each study case. Hence, the heat generation was derived and supplied to an efficient three-dimensional model, which was also validated against the experimental results obtained from various current profiles. The roles of natural convection (NC-BTMS) and forced convection were studied separately for the proposed BTMS under intense static and dynamic loads.

By optimizing the key performance parameters, such as the mass flow velocity and the channel size, one can conclude that their increment leads to a maximum temperature reduction. Also, the cell-to-cell distance increase had a reverse impact on the temperature and the pressure drop. By these means, the thermal management of the proposed architecture was enhanced while the pressure drop was kept at a minimum range.

A significant improvement can be concluded from this study—one that is related to the overall physics-based structural optimization of the battery modules. The coupling of high-fidelity models with global optima multi-objective algorithms could consider the following: (1) a wider design space with many objectives (which is proven in this study, i.e., that co-design electrical-thermal and mechanical objectives can be solved at the same time); (2) the optimization steps that are evaluated can be significantly increased when compared to user-based or single-objective solutions (i.e., the sampling time related to the computational time and step-time related to the amount of derived solutions that could be set accordingly); and (3) the optimization results presented in this paper are accurate versus the real-life experiments for various loading scenarios.

The method we have presented here was studied with a air-cooled battery module design. To optimize this design, we used a multi-objective optimization that used four objectives and several constraints, as presented in Equation (1). By changing the cost functions and the constraints, the use case could be adapted to other designs and battery configurations, such as when investigating the liquid channel sizes, mass flows, and pressure drops, in order to provide the most suitable solutions. We are currently working on such investigations, and future publications will present the performance of the proposed methods to a different design set. Forthcoming works might include a study on the proposed BTMS when it is applied to a higher energy and power battery pack, as well as an evaluation of different thermal management solutions for different objectives and constraints.

Author Contributions: Conceptualization, T.K.; methodology, T.K.; software, T.K.; validation, T.K. and M.S.H.; formal analysis, T.K.; investigation, T.K.; resources, M.B. and J.V.M.; data curation, T.K.; writing—original draft preparation, T.K. and M.S.H.; writing—review and editing, T.K. visualization, T.K.; supervision, P.V.D.B. and M.B.; project administration, J.V.M.; funding acquisition, M.B. All authors have read and agreed to the published version of the manuscript.

Funding: This research was made possible thanks to the research project GEIRI. This research was part of the the GEIRI project (funding number: SGRIKXJSKF[2017]632).

Data Availability Statement: The datasets presented in this article are not readily available because the data are part of an ongoing study.

Conflicts of Interest: The authors declare no conflicts of interest.

References

1. Deng, J.; Bae, C.; Denlinger, A.; Miller, T. Electric Vehicles Batteries: Requirements and Challenges. *Joule* **2020**, *4*, 511–515. [CrossRef]
2. Kebede, A.A.; Hosen, M.S.; Kalogiannis, T.; Behabtu, H.A.; Assefa, M.Z.; Jemal, T.; Ramayya, V.; Van Mierlo, J.; Coosemans, T.; Berecibar, M. Optimal sizing and lifetime investigation of second life lithium-ion battery for grid-scale stationary application. *J. Energy Storage* **2023**, *72*, 108541. [CrossRef]

3. Peters, J.F.; Baumann, M.; Zimmermann, B.; Braun, J.; Weil, M. The environmental impact of Li-Ion batteries and the role of key parameters—A review. *Renew. Sustain. Energy Rev.* **2017**, *67*, 491–506.
4. Balali, Y.; Stegen, S. Review of energy storage systems for vehicles based on technology, environmental impacts, and costs. *Renew. Sustain. Energy Rev.* **2021**, *135*, 110185. [CrossRef]
5. Gandoman, F.H.; Jaguemont, J.; Goutam, S.; Gopalakrishnan, R.; Firouz, Y.; Kalogiannis, T.; Omar, N.; Van Mierlo, J. Concept of reliability and safety assessment of lithium-ion batteries in electric vehicles: Basics, progress, and challenges. *Appl. Energy* **2019**, *251*, 113343.
6. Rezvanizaniani, S.M.; Liu, Z.; Chen, Y.; Lee, J. Review and recent advances in battery health monitoring and prognostics technologies for electric vehicle (EV) safety and mobility. *J. Power Sources* **2014**, *256*, 110–124. [CrossRef]
7. Song, W.; Chen, M.; Bai, F.; Lin, S.; Chen, Y.; Feng, Z. Non-uniform effect on the thermal/aging performance of Lithium-ion pouch battery. *Appl. Therm. Eng.* **2018**, *128*, 1165–1174. [CrossRef]
8. Iraola, U.; Aizpuru, I.; Gorrotxategi, L.; Segade, J.M.C.; Larrazabal, A.E.; Gil, I. Influence of voltage balancing on the temperature distribution of a Li-ion battery module. *IEEE Trans. Energy Convers.* **2015**, *30*, 507–514. [CrossRef]
9. Arora, S. Selection of thermal management system for modular battery packs of electric vehicles: A review of existing and emerging technologies. *J. Power Sources* **2018**, *400*, 621–640. [CrossRef]
10. Xie, P.; Jin, L.; Qiao, G.; Lin, C.; Barreneche, C.; Ding, Y. Thermal energy storage for electric vehicles at low temperatures: Concepts, systems, devices and materials. *Renew. Sustain. Energy Rev.* **2022**, *160*, 112263. [CrossRef]
11. Akinlabi, A.A.; Solyali, D. Configuration, design, and optimization of air-cooled battery thermal management system for electric vehicles: A review. *Renew. Sustain. Energy Rev.* **2020**, *125*, 109815. [CrossRef]
12. Peng, X.; Cui, X.; Liao, X.; Garg, A. A thermal investigation and optimization of an air-cooled lithium-ion battery pack. *Energies* **2020**, *13*, 2956. [CrossRef]
13. Park, H. A design of air flow configuration for cooling lithium ion battery in hybrid electric vehicles. *J. Power Sources* **2013**, *239*, 30–36. [CrossRef]
14. Chen, K.; Wang, S.; Song, M.; Chen, L. Configuration optimization of battery pack in parallel air-cooled battery thermal management system using an optimization strategy. *Appl. Therm. Eng.* **2017**, *123*, 177–186. [CrossRef]
15. Xie, J.; Ge, Z.; Zang, M.; Wang, S. Structural optimization of lithium-ion battery pack with forced air cooling system. *Appl. Therm. Eng.* **2017**, *126*, 583–593. [CrossRef]
16. Chen, K.; Song, M.; Wei, W.; Wang, S. Structure optimization of parallel air-cooled battery thermal management system with U-type fl ow for cooling ef fi ciency improvement. *Energy* **2018**, *145*, 603–613. [CrossRef]
17. Liu, Y.; Zhang, J. Design a J-type air-based battery thermal management system through surrogate-based optimization. *Appl. Energy* **2019**, *252*, 1–13. [CrossRef]
18. Chen, K.; Wu, W.; Yuan, F.; Chen, L.; Wang, S. Cooling efficiency improvement of air-cooled battery thermal management system through designing the flow pattern. *Energy* **2019**, *167*, 781–790. [CrossRef]
19. Li, M.; Liu, Y.; Wang, X.; Zhang, J. Modeling and optimization of an enhanced battery thermal management system in electric vehicles. *Front. Mech. Eng.* **2019**, *14*, 65–75. [CrossRef]
20. Kalogiannis, T.; Akbarzadeh, M.; Hosen, M.S.; Behi, H.; De Sutter, L.; Jin, L.; Jaguemont, J.; Van Mierlo, J.; Berecibar, M. Effects analysis on energy density optimization and thermal efficiency enhancement of the air-cooled Li-ion battery modules. *J. Energy Storage* **2022**, *48*, 103847. [CrossRef]
21. Kalogiannis, T.; Hosen, M.S.; Gandoman, F.H.; Sokkeh, M.A.; Jaguemont, J.; Berecibar, M.; Van Mierlo, J. Multi-objective particle swarm optimization and training of datasheet-based load dependent lithium-ion voltage models. *Int. J. Electr. Power Energy Syst.* **2021**, *133*, 107312. [CrossRef]
22. Barai, A.; Uddin, K.; Dubarry, M.; Somerville, L.; McGordon, A.; Jennings, P.; Bloom, I. A comparison of methodologies for the non-invasive characterisation of commercial Li-ion cells. *Prog. Energy Combust. Sci.* **2019**, *72*, 1–31.
23. Akbarzadeh, M.; Kalogiannis, T.; Jaguemont, J.; He, J.; Jin, L.; Berecibar, M.; Van Mierlo, J. Thermal modeling of a high-energy prismatic lithium-ion battery cell and module based on a new thermal characterization methodology. *J. Energy Storage* **2020**, *32*, 101707. [CrossRef]
24. Wang, B.; Li, S.E.; Peng, H.; Liu, Z. Fractional-order modeling and parameter identification for lithium-ion batteries. *J. Power Sources* **2015**, *293*, 151–161. [CrossRef]
25. Brivio, C.; Musolino, V.; Merlo, M.; Ballif, C. A Physically-Based Electrical Model for Lithium-Ion Cells. *IEEE Trans. Energy Convers.* **2019**, *34*, 594–603. [CrossRef]
26. Wang, Y.N.; Chen, Y.Q.; Liao, X.Z. State-of-art survey of fractional order modeling and estimation methods for lithium-ion batteries. *Fract. Calc. Appl. Anal.* **2020**, *22*, 1449–1479. [CrossRef]
27. Xiong, R.; Tian, J.; Shen, W.; Sun, F. A Novel Fractional Order Model for State of Charge Estimation in Lithium Ion Batteries. *IEEE Trans. Veh. Technol.* **2019**, *68*, 4130–4139. [CrossRef]
28. Xiong, R.; Tian, J. A comparative study on fractional order models for voltage simulation of lithium ion batteries. In Proceedings of the 2019 IEEE 89th Vehicular Technology Conference, Kuala Lumpur, Malaysia, 28 April–1 May 2019; pp. 1–5. [CrossRef]
29. Mawonou, K.S.; Eddahech, A.; Dumur, D.; Beauvois, D.; Godoy, E. Improved state of charge estimation for Li-ion batteries using fractional order extended Kalman filter. *J. Power Sources* **2019**, *435*, 226710. [CrossRef]

30. Petras, I. *Fractional-Order Nonlinear Systems. Modeling, Analysis and Simulation*; Higher Education Press: Beijing China; Springer: Berlin/Heidelberg, Germany, 2011.
31. Zhu, J.; Sun, Z.; Wei, X.; Dai, H. Studies on the medium-frequency impedance arc for Lithium-ion batteries considering various alternating current amplitudes. *J. Appl. Electrochem.* **2016**, *46*, 157–167. [CrossRef]
32. Farmann, A.; Sauer, D.U. Comparative study of reduced order equivalent circuit models for on-board state-of-available-power prediction of lithium-ion batteries in electric vehicles. *Appl. Energy* **2018**, *225*, 1102–1122. [CrossRef]
33. Kalogiannis, T.; Jaguemont, J.; Omar, N.; Van Mierlo, J.; Van den Bossche, P. A comparison of internal and external preheat methods for NMC batteries. *World Electr. Veh. J.* **2019**, *10*, 18. [CrossRef]
34. Chen, K.; Chen, Y.; Li, Z.; Yuan, F.; Wang, S. Design of the cell spacings of battery pack in parallel air-cooled battery thermal management system. *Int. J. Heat Mass Transf.* **2018**, *127*, 393–401. [CrossRef]
35. Akbarzadeh, M.; Jaguemont, J.; Kalogiannis, T.; Karimi, D.; He, J.; Jin, L.; Xie, P.; Van Mierlo, J.; Berecibar, M. A novel liquid cooling plate concept for thermal management of lithium-ion batteries in electric vehicles. *Energy Convers. Manag.* **2021**, *231*, 113862. [CrossRef]
36. Gantenbein, S.; Weiss, M.; Ivers-Tiffée, E. Impedance based time-domain modeling of lithium-ion batteries: Part I. *J. Power Sources* **2018**, *379*, 317–327. [CrossRef]

World Electric Vehicle Journal

MDPI

Article

Simulation-Based Assessment of Energy Consumption of Alternative Powertrains in Agricultural Tractors

Antti Lajunen [1,*], Klaus Kivekäs [1], Vincent Freyermuth [2], Ram Vijayagopal [2] and Namdoo Kim [2]

[1] Department of Agricultural Sciences, University of Helsinki, 00790 Helsinki, Finland; klaus.kivekas@helsinki.fi

[2] Argonne National Laboratory, Lemont, IL 60439, USA; nkim@anl.gov (N.K.)

* Correspondence: antti.lajunen@helsinki.fi

Abstract: The objectives of this research were to develop simulation models for agricultural tractors with different powertrain technologies and evaluate the energy consumption in typical agricultural operations. Simulation models were developed for conventional, parallel hybrid electric, series hybrid electric, fuel cell hybrid, and battery electric powertrains. Autonomie vehicle simulation software (version 2022) was used for the simulations and the tractor models were simulated in two tilling cycles and in a road transport cycle with a trailer. The alternative powertrains were configured to have at least the same tractive performance as the conventional, diesel engine-powered tractor model. The simulation results showed that the potential of the parallel and series hybrid powertrains to improve energy efficiency depends heavily on the tractor size and the operating cycle conditions. The fuel cell hybrid and battery electric powertrains have a higher potential to reduce energy consumption and emissions but still have inherent technical challenges for practical operation. The battery-powered electric tractor would require improvements in the storage energy density to have a comparable operational performance in comparison to other powertrains. The fuel cell hybrid tractor already provided an adequate operating performance but the availability of hydrogen and refueling infrastructure could be challenging to resolve in the farming context.

Keywords: energy consumption; agricultural tractor; modeling; alternative powertrain; simulation

Citation: Lajunen, A.; Kivekäs, K.; Freyermuth, V.; Vijayagopal, R.; Kim, N. Simulation-Based Assessment of Energy Consumption of Alternative Powertrains in Agricultural Tractors. *World Electr. Veh. J.* **2024**, *15*, 86. https://doi.org/10.3390/wevj15030086

Academic Editors: Joeri Van Mierlo and Genevieve Cullen

Received: 12 January 2024
Revised: 8 February 2024
Accepted: 21 February 2024
Published: 27 February 2024

1. Introduction

Alternative powertrains have been increasingly implemented in different types of on-road vehicles for increasing energy efficiency and reducing emissions [1,2] and electrification is also on the way for off-road machinery [3,4]. The recent technological developments in powertrain electrification [5] and increased fossil fuel prices are also starting to make alternative powertrains and fuels relevant options for agricultural tractors. Unlike passenger vehicles, agricultural tractors have not yet been the most interesting application for powertrain electrification. The uncertainties about future developments regarding fossil fuels, environmental legislation, and emission standards have increased interest in the development of hybrid electric, fully electric, and fuel cell hybrid powertrain solutions [6]. Therefore, it is reasonable to believe that powertrain electrification will also be one of the major technology trends for agricultural tractors in the coming years. Recent scientific research results indicate that there could be a significant potential to increase energy efficiency with alternative powertrains [7]. The main architectures for suitable alternative electrified powertrains have been studied and the benefits of using electric power for numerous agricultural implements have been well recognized [8,9]; however, most of the existing research studies evaluating alternative powertrains for agricultural tractors focus only on single powertrain options and, therefore, a balanced comparison between the different technologies is required. This research presents a comparison—in terms of energy consumption and operational performance—by taking into account the most relevant alternative powertrain options for agricultural tractors. This article is a revised and expanded version of a paper entitled "Simulation of Alternative Powertrains in Agricultural Tractors" [7], which was presented at EVS36 in Sacramento on 12 June 2023.

Some agricultural tractor manufacturers have introduced new concepts for alternative powertrains and have launched prototype tractor models; they are even starting to produce versions of hybrid electric powertrains, but large-scale electrification still has many challenges to overcome. Several companies and research institutions are working on prototype battery electric tractors to reduce greenhouse gas emissions and dependence on fossil fuels in agriculture. John Deere has planned to launch an electric tractor by 2026, the small electric tractor by Fendt (e100 V Vario) has already been launched, and CNH Industrial is developing the New Holland T4 Electric Power and Farmall 75C Electric, which are both lithium-ion battery-powered all-electric utility tractors. Research has been ongoing to improve battery technologies for electric tractors. Increased energy density, longer battery life, and faster charging times are crucial aspects of the success of electric agricultural tractors [10]. The integration of electrified tractors with precision agriculture technologies is also a growing area of interest for manufacturers. Some governments have been offering incentives and subsidies to encourage the adoption of electric vehicles in agriculture. These policies aim to reduce emissions, promote sustainable farming, and support the transition to cleaner energy sources.

Powertrain electrification has spread steadily from passenger cars to utility vehicles and, today, even to heavy on-road vehicles [11,12]. There is also increasing development for off-road machinery, especially since 2021 as energy costs have increased exceptionally and there are many uncertainties surrounding the use of fossil fuels in the future. Higher technology costs can be a major barrier to using alternative powertrains in agricultural tractors, although previous research on heavy vehicles and off-road machinery suggests that the higher development and component costs can be paid off with benefits when assessing the cost on a lifecycle basis [13,14]. The electrification of farm vehicles started with small-sized machines, for example, there are already electrified versions of telehandlers and small loaders available for purchase [15]. Because modern agricultural tractors are used for a wide variety of field operations, road transport, and other supporting work such as front-end loading or mixer wagon operation, there are several different variants of basic agricultural tractors. However, from very small-sized tractors (engine power < 50 kW) up to very large tractors (engine power up to 300 kW), conventional agricultural tractors have quite a similar powertrain topology [16]. This similarity might limit the opportunities to introduce new powertrain designs and favor the minimal modification of the existing layout to avoid too many modifications in the production lines. This is also the case due to the multipurpose aspect of agricultural tractors, providing a universal operator for a vast variety of farm purposes.

Over the last few years, research studies have been carried out to estimate the benefits and feasibility of hybrid electric powertrains in agricultural tractors. For many reasons, compact and medium-sized tractors (engine power between 50 and 100 kW) have often been the baseline for hybridization studies. Troncon et al. (2019) studied the feasibility of hybridization for specialized orchard and vineyard tractors using a mild parallel-hybrid system [17,18]. The challenge was to fit the electric system in a rather limited space and deliver an adequate performance. Their simulated research results indicated that fuel consumption would be 15–35% lower depending on the duty cycle operation. In another study, an ICE-based platform was converted to a parallel hybrid powertrain with a downsized engine and electric motor [19]. The downsizing was about 29% (from 77 kW to 55 kW of engine power), the electric motor maximum power was 60 kW, and the battery size was 25 kWh. The fuel economy savings were evaluated using simulations of high and low power duty cycles, which clearly showed that hybridization had only a marginal benefit on high power cycles (on average about a 5% reduction) and a significant benefit on low power cycles, having a reduction of over 30% on average [19]. Mendes et al. (2019)

investigated the hybridization of a tractor backhoe loader by focusing on using electrical power produced by a generator for the hydraulic system with supercapacitors as the energy storage [20]. Simulations on real-world recorded duty cycles indicated over a 50% reduction in fuel consumption. Mocera and Martini (2022) proposed a hybrid eCVT power-split hybridization for a specialized agricultural tractor [21]. Their performance simulations showed that the hybrid tractor would have a comparable performance in typical use of the tractor and fuel savings of 10–20%.

Alternative fuels, such as biodiesel, biogas, e-fuels, or hydrogen for internal combustion engines, have the potential to lower greenhouse gas emissions compared to traditional fossil fuels. This can contribute to mitigating environmental impacts associated with agricultural activities. Some alternative fuels are derived from renewable sources, offering the advantage of sustainability. For instance, biofuels can be produced from crop residue or organic waste, providing a renewable and potentially carbon-neutral energy source. Certain alternative fuels, like biogas, can be produced locally, promoting regional economic development. The adoption of alternative fuels is hindered by the lack of widespread infrastructure for production, distribution, and refueling [22]. Establishing a robust infrastructure is crucial for the successful integration of alternative fuels into agricultural practices. Some alternative fuels have a lower energy density than traditional fossil fuels, which can impact the overall range and efficiency of agricultural tractors [23]. This challenge requires advancements in fuel storage and utilization technologies.

Considering off-road vehicles and machinery in general, agricultural tractors differ from other machinery because they are often used for various purposes and many different types of field operations. Therefore, it is important to develop methods that provide the tools for evaluating the benefits of powertrain electrification of agricultural tractors [24]. Vehicle modeling and simulation methods are a practical and rather fast way of analyzing and comparing different powertrain solutions. Different from many other vehicles, agricultural tractors are used on different types of field surfaces and in different conditions, which creates specific challenges for modeling [25]. Reliably and accurately simulating tire–soil interactions need high-fidelity models, e.g., FEM—(Finite Element Method) or DEM—(Discrete Element Method) based models, that need laborious development and require significant amounts of computational capacity [26,27]. In addition, acquiring reliable validation data for high-fidelity tire–soil interaction models from field operations can be rather challenging [28]. For effectively comparing and evaluating the performance of alternative powertrains, less computationally intensive models are typically used, such as numerical simulation.

This research presents a numerical modeling and simulation approach for evaluating alternative powertrains in agricultural tractors using Autonomie vehicle simulation software [29]. Off-road vehicles and machinery are typically simulated in a different way to on-road vehicles because they usually perform repetitive tasks and do not have a traditional speed profile to follow. Instead, agricultural tractors are simulated based on distance, by giving a target speed based on the distance traveled. Also, as these types of machines often do heavy work, the resistance force from implements must be integrated into the model by, for example, simulating agricultural field work such as plowing or harrowing. Naturally, in typical field work, like field cultivation, the power requirement can consist of a passive draft force or an active power that uses the power take-off (PTO) or hydraulic power in an implement. For evaluating alternative powertrains in agricultural tractors, numerical modeling and simulation provide an effective approach to generating different simulation cases, comparing component sizing, and then evaluating the benefits in several use cases.

In this research, conventional, parallel hybrid electric, series hybrid electric, fuel cell hybrid, and battery electric powertrains were modeled and simulated in dedicated operating cycles. The powertrain models were parametrized based on the performance of a conventional tractor with a diesel engine and dual-clutch transmission. The operating cycles were generated based on field measurements carried out in the Viikki Research Farm at the University of Helsinki, Finland. According to the simulation results, the benefits of hybridization and electrification were evaluated and the operating performance was analyzed.

2. Materials and Methods

2.1. Simulation Model Development

Autonomie software (version 2022) [29] was used for the tractor model development and for running the simulations in multiple cycles. This software has been developed by the Argonne National Laboratory (ANL), to be used as a vehicle system simulation tool for assessing the energy consumption, performance, and cost of advanced vehicle powertrain technologies in various types of vehicles [12]. The simulations and model configurations can be executed in a dedicated interface called AMBER, which has been developed as a universal graphical user interface for multiple simulation applications and allows workflows to be run with different software developed by ANL [30]. All the simulations were carried out by using AMBER and, thus, the model development was performed in Autonomie, and configuration and parametrization were performed in AMBER. Autonomie was originally designed for on-road vehicle simulations and, therefore, implementing off-road simulation models with distance-based cycles required some modifications to the driver and vehicle control systems. Otherwise, the software is well suited to off-road vehicle simulation as long as a representative operating cycle can be generated. The first versions of the agricultural tractor models with a time-based simulation approach were developed during previous research, which focused on conventional tractor model development and the electrification of agricultural tractors [31]. The previously developed simulation models were updated by modifying them to be suitable for distance-based cycle simulations. Also, more representative operating cycles were developed based on the measurements carried out in an agricultural field environment and during road transport tractor tests.

The modeled powertrain options included diesel-powered conventional, parallel hybrid electric, series hybrid electric, fuel cell hybrid, and battery electric powertrains. The conventional and parallel hybrid models have a diesel engine as a power source and a dual-clutch transmission; dual-clutch transmission was chosen for its high energy efficiency [32]. The parallel hybrid has a pre-transmission layout with an electric drive and uses a battery pack for electrical energy storage. The series hybrid tractor model also has a conventional diesel engine attached to a generator, one electric drive motor, and a three-speed gearbox. The fuel cell hybrid and electric tractor models have a fully electric powertrain consisting of a battery, one electric drive motor, and a three-speed gearbox. The fuel cell hybrid model has a fuel cell stack as the primary power source and a small battery pack for power load leveling. The electric tractor has a large energy-type battery pack for energy storage. A lithium-ion battery model was used for energy storage in all of the electrified simulation models. Figures 1 and 2 present the powertrain layouts of the different tractor models in the Autonomie software. The vehicle dynamics block is illustrated in Figure 2 and includes a transfer case, front and rear final drives, wheels, and chassis model. The transfer case splits the driving power between the front and rear axles. All the tractor models have the same driver model, which determines the speed and acceleration demand. The external loads generated by an implement or trailer are taken into account in the chassis block of the models. The hybrid powertrains have dedicated energy management strategies (EMSs) for ensuring driving performance and minimizing energy consumption when possible. Power-following and charge-sustaining EMSs were used for all the hybrid powertrains to ensure performance and keep the battery state of charge within predetermined limits.

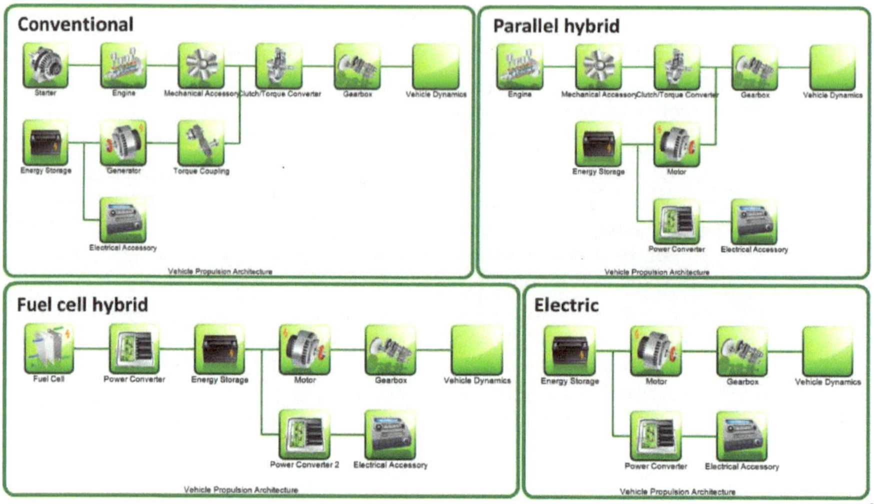

Figure 1. Powertrain layouts of the conventional, parallel hybrid, fuel cell hybrid, and electric tractor models in the Autonomie software.

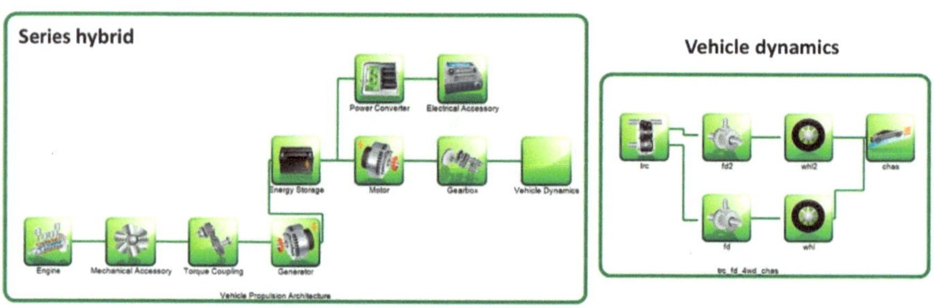

Figure 2. Series hybrid powertrain layouts and vehicle dynamics block layouts of the tractor models.

2.2. Model Parameters

Two different baseline tractor sizes were chosen for the conventional tractor models—a medium size with an engine-rated power of 112 kW, and a large size with an engine-rated power of 225 kW. The tractor models were configured using the Autonomie libraries that provide component initialization data for a wide range of components used in light- and heavy-duty vehicles. The powertrain component sizing was determined in a way that the alternative powertrains had at least the same tractive performance in comparison to the conventional, diesel-engine-powered tractor models. The total weight of each powertrain was estimated based on the main component weights, and the results indicated that no major differences in total weight needed to be considered. Therefore, all the models were simulated with the same total weights of 5000 kg (medium size) and 10,000 kg (large size). The size of the battery in the electric tractor model was limited to less than 200 kWh for the large-sized tractor and 100 kWh for the medium size tractor to not exceed the total tractor weight. Table 1 presents the general technical specifications of the conventional tractor models and, thus, the engine and transmission parameters. The general technical specifications include the front and rear axle gear reductions, tire sizes, and total weights, which were the same for all the tractor models.

Table 1. Conventional tractor powertrain and general technical specifications.

Component	Medium-Sized Tractor	Large-Sized Tractor
Diesel engine	maximum power 112 kW, maximum torque 580 Nm	maximum power 225 kW, maximum torque 1154 Nm
Transmission	eight-speed dual-clutch transmission (DCT) with three ranges	eight-speed dual-clutch transmission (DCT) with three ranges
Rear axle [1]	bevel set ratio of 2.93:1 and planetary gear ratio of 6:1	bevel set ratio of 3.28:1 and planetary gear ratio of 6:1
Front axle [1]	bevel set ratio of 2.30:1 and planetary gear ratio of 6:1	bevel set ratio of 2.48:1 and planetary gear ratio of 6:1
Tires [1]	front: 380/85R28, rear: 460/85R38	front: 540/65R30, rear: 650/65R42
Weight [1]	5000 kg	10,000 kg

[1] Same parameters for all tractor models.

Tables 2 and 3 show the powertrain specifications for the parallel hybrid, series hybrid, fuel cell hybrid, and battery electric tractor models. Parallel and series hybrid types have a downsized diesel engine. The parallel hybrid has a similar dual-clutch transmission to the conventional tractor but needs only two ranges for the same driving performance. All the hybrid models have a rather small battery pack because this is mostly used for peak power shaving and storing regenerated braking energy. Based on the evaluation of the typical field and road operations, it was determined that a three-speed gearbox is sufficient to cover the typical agricultural tractor operations and provide high energy efficiency. The electric driving motor was dimensioned based on the performance requirement set by the baseline conventional tractor. Depending on the different duty cycles and operations, an optimization study could be carried out to evaluate the influence of the component sizes on the operating performance. However, this would be more interesting if the design and operating costs were included in the analysis.

Table 2. Specifications for the hybrid and electric powertrains of the medium-sized tractor models.

Component	Parallel Hybrid	Series Hybrid	Fuel Cell Hybrid	Electric
Diesel engine/Fuel cell stack	Diesel engine: power 90 kW, torque 466 Nm	Diesel engine: power 92.5 kW, torque 480 Nm	Fuel cell stack: max power 80 kW	---
Transmission	Eight-speed (DCT) with two ranges	Three-speed gearbox	Three-speed gearbox	Three-speed gearbox
Battery configuration	6 Ah cell, 180 cells in series in a pack, 648 V, 3.9 kWh	6 Ah cell, 180 cells in series in a pack, 648 V, 3.9 kWh	6 Ah cell, 180 cells in series in a pack, 648 V, 3.9 kWh	33 Ah cell, four packs in parallel, 192 cells in series in a pack, 720 V, 95 kWh
Electric motor	max power 50 kW, max torque 201 Nm, max speed 4400 rpm	max power 112 kW, max torque 304 Nm, max speed 8000 rpm	max power 112 kW, max torque 304 Nm, max speed 8000 rpm	max power 112 kW, max torque 304 Nm, max speed 8000 rpm

Table 3. Specifications of the hybrid and electric powertrains of the large-sized tractor models.

Component	Parallel Hybrid	Series Hybrid	Fuel Cell Hybrid	Electric
Diesel engine/Fuel cell stack	Diesel engine: power 175 kW, torque 898 Nm	Diesel engine: power 185 kW, torque 949 Nm	Fuel cell stack: max power 160 kW	---
Transmission	Eight-speed (DCT) with two ranges	Three-speed gearbox	Three-speed gearbox	Three-speed gearbox

Table 3. *Cont.*

Component	Parallel Hybrid	Series Hybrid	Fuel Cell Hybrid	Electric
Battery configuration	6 Ah cell, 180 cells in series in a pack, 648 V, 3.9 kWh	6 Ah cell, 180 cells in series in a pack, 648 V, 3.9 kWh	6 Ah cell, 180 cells in series in a pack, 648 V, 3.9 kWh	33 Ah cell, eight packs in parallel, 192 cells in series in a pack, 720 V, 190 kWh
Electric motor	max power 100 kW, max torque 542 Nm, max speed 4400 rpm	max power 225 kW, max torque 611 Nm, max speed 8000 rpm	max power 225 kW, max torque 611 Nm, max speed 8000 rpm	max power 225 kW, max torque 611 Nm, max speed 8000 rpm

2.3. Operating Cycles

Experimental measurements were carried out in the Viikki Research Farm at the University of Helsinki using a typical agricultural tractor (Valtra N141) and a chisel plow to acquire data to evaluate different levels of load resistances for the operating cycles. The measurements were made in October 2022 in a stubbled field, as presented in Figure 3. The tractor data were measured from the CAN bus by a developed data logger consisting of a mini-computer, CAN shield logger, and a GPS module. Location data were logged using a u-blox ZED-F9P GPS module, which was connected to a u-blox ANN-MB-00 GPS antenna. The operational data that were recorded from the CAN bus included, among other things, engine data, vehicle speed, and the linkage draft force.

Figure 3. Field measurements with a tractor and chisel plow.

Two tillage cycles were generated with target speeds of 8 and 12 km/h. For both cycles, three levels of load resistance were defined—light, medium, and high. The resistance load was applied only when the implement was in use during operation. The tillage cycles are illustrated in Figure 4 with the target speed and the different levels of force as load resistances for the large-sized tractor. For the medium-sized tractor, the target speed and the lowest load resistance were the same as for the large-sized tractor. The higher loads were gradually lowered, being approximately 50% of the load resistance in comparison to the large-sized tractor cycles.

Figure 5 shows the 27 km long measured road cycle with the elevation profile and the 20 km long generated road cycle. The measured road cycle corresponds to a typical road transport operation performed with agricultural tractors with a trailer between fields and a farm. The 27 km roundtrip cycle was measured from the route that has been used for tractor comparison tests by a Finnish magazine. The road cycle data included multiple tractor test recordings containing tractor operational data. The large-sized tractor was simulated in the measured cycle and the medium-sized tractor in the generated cycle, which has a lower top speed and elevation. Simulations were carried out with a trailer, having total weights of 10,000 kg and 15,500 kg for the large-sized tractor, and 6400 kg and 10,000 kg for the medium-sized tractor. These loads correspond to payloads of 30% and 60% for 18 t (large tractor) and 12 t (medium tractor) trailers.

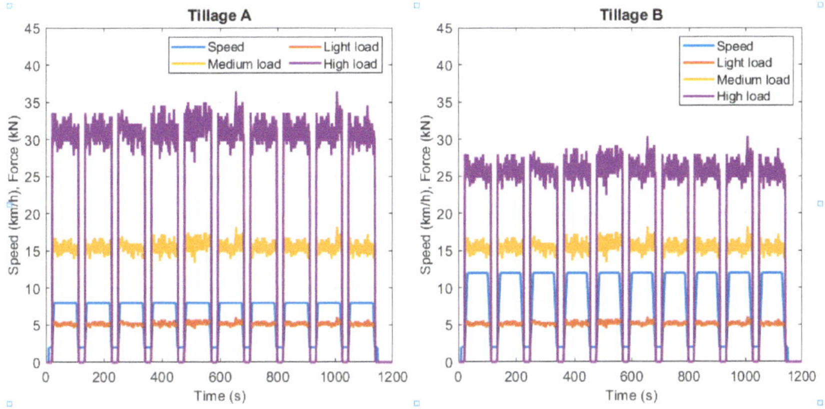

Figure 4. Generated tillage cycles (Tillage A and B) with three defined load resistance profiles for the large-sized tractor.

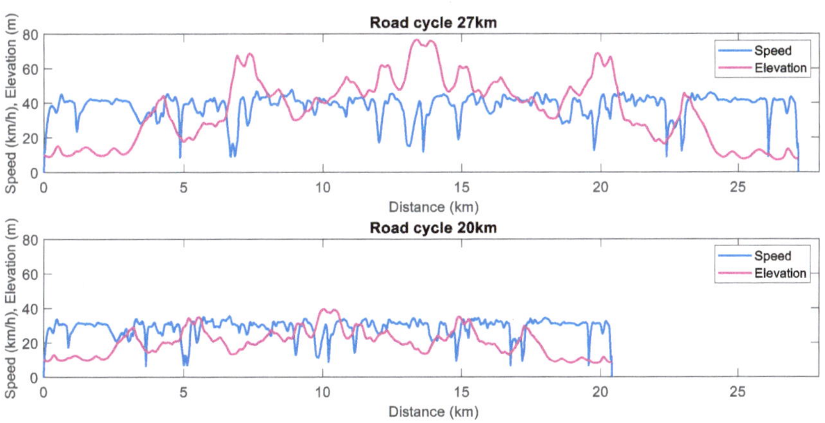

Figure 5. Measured and generated road cycles for tractor–trailer simulations.

3. Results

3.1. Driving Performance

MATLAB software (version 2021b) was used for analyzing the simulation results. Overall, all the simulations were successfully carried out and it was concluded that all the models were operating correctly. It was observed that the target speed was followed quite well, without any major deviations in all cycles, although the electrified powertrains did have more precise control for following the target speed, especially during slow-speed driving that did not need gear changes. The speed traces in the Tillage A cycle for all the large tractor models are illustrated in Figure 6. The conventional tractor did not follow the lower target speed very closely, but at higher speeds, the speed control worked fine. Also, the load resistance in the tillage cycles had some influence on the driving dynamics, and this will be a focus point in future research when developing more advanced driver models for agricultural tractors. In the road cycles, there was very little difference in the driving speeds between the tractor models due to the more dynamic nature of the cycle. Only the hard acceleration phases generated some lagging for the conventional and parallel hybrid tractor models, because of the consecutive gear shifting.

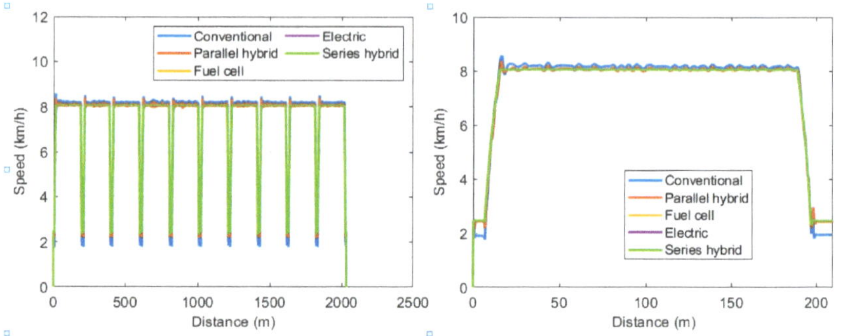

Figure 6. Speed traces of large tractor models in Tillage A cycle.

3.2. Energy Consumption

Energy consumption was calculated as on-board energy use and, therefore, no charging losses were considered. Figures 7 and 8 present the cumulative energy consumption for the simulated tractor models in the Tillage A and road cycles. The results correspond to the medium workload for the Tillage A cycle and the higher payload for the road cycle. The cumulative energy consumption illustrates that there was a gradual energy-saving potential along the tillage cycle due to the higher powertrain efficiency. Only the series hybrid powertrain was less efficient than the conventional powertrain under the higher load situations. In the road cycle, the advantage of regenerating braking energy increased energy savings, especially for tractor models that had the fully electric powertrain. The alternative powertrains showed better performance (in terms of energy consumption) for the medium-sized tractor compared to the large-sized tractor.

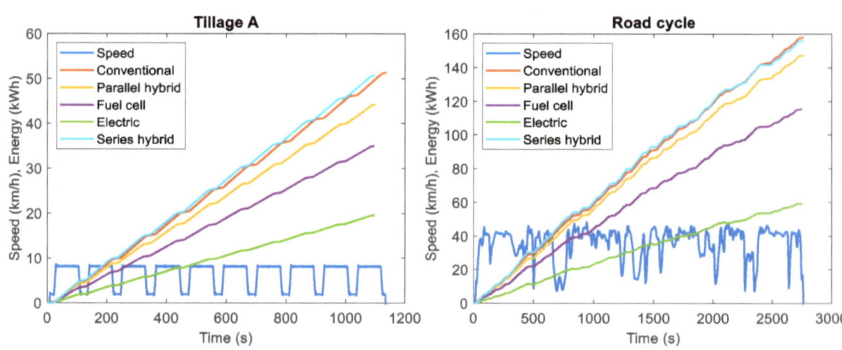

Figure 7. Cumulative energy consumption in the Tillage A and Road cycles for the large-sized tractor models.

A comparison of the energy consumption between the different cycles was made using the units of kWh/km. These units are not necessarily useful in terms of agricultural work but allow a comparison of the results obtained from different simulations. Figures 9 and 10 show the calculated energy consumption results for all the simulated cycles. The highest consumption was obtained in the Tillage A cycle with the high workload. The consumption increased quite rapidly in the function of the workload in both tillage cycles. Only for the electric tractor model was the increase less strong. Distance-based energy consumption was much lower in the Road cycle, which is due to the much higher driving speed. The payload increase had less influence on the energy consumption in the road cycle than the increase in the workload in the tillage cycles.

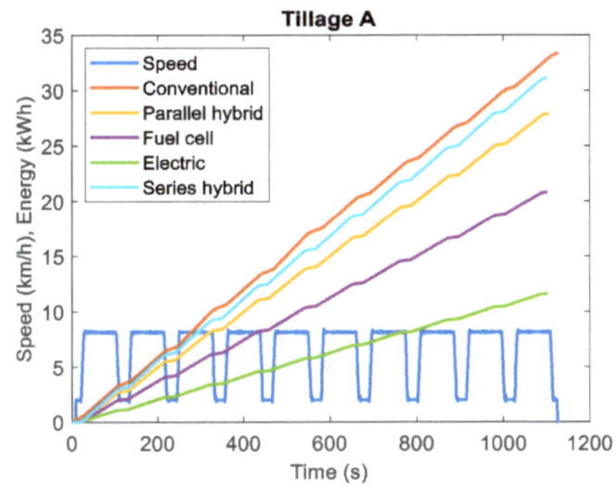

Figure 8. Cumulative energy consumption in the Tillage A and Road cycles for the medium-sized tractor models.

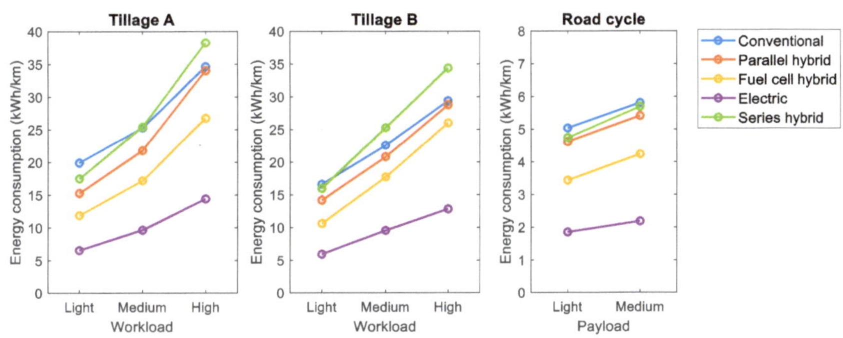

Figure 9. Energy consumption in kWh/km for all simulated cycles for the large-sized tractor.

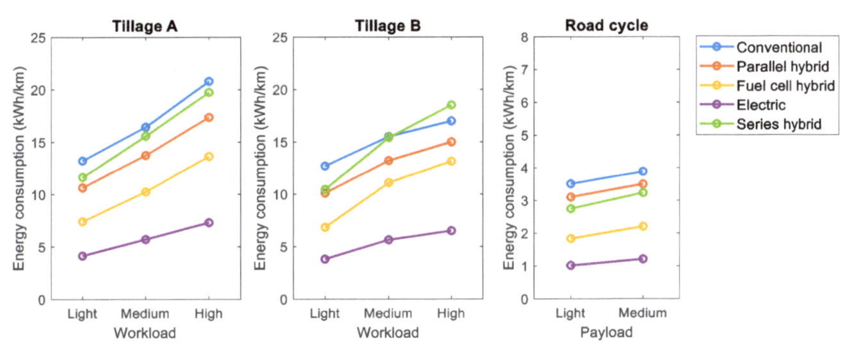

Figure 10. Energy consumption in kWh/km for all simulated cycles for the medium-sized tractor.

Depending on the workload, the fuel consumption of the large conventional tractor model was 12.7–23.0 L per hour (L/h) in the Tillage A cycle, 14.9–27.1 L/h in the Tillage B cycle, and 18.2–20.9 L/h in the Road cycle. For the medium-sized tractor, the fuel consumption values were 8.6–13.8 L/h (Tillage A), 11.6–15.7 L/h (Tillage B), and 9.5–10.5 L/h (Road cycle). These values correspond to typical the fuel consumption of diesel-powered tractors in tillage operations. The hydrogen consumption of the large fuel cell hybrid tractor model was 2.4–5.4 kg per hour (kg/h) in the Tillage A cycle, 3.0–7.3 kg/h in the Tillage B cycle, and 3.7–4.5 kg/h in the Road cycle. For the medium-sized fuel cell hybrid tractor, the hydrogen consumption was 1.5–2. 7 kg/h (Tillage A), 1.9–3.7 kg/h (Tillage B), and 1.5–1.8 kg/h (Road cycle).

The reduction potential in the energy consumption of the alternative powertrains is shown in Figures 11 and 12. These results clearly show that there is a significant potential for reducing energy consumption with the battery electric powertrain. The average reduction potential was 60–70% in the tillage and road cycle. The potential to reduce energy consumption with the fuel cell hybrid varied from 20% to 30% for the large tractor and from 35% to 45% for the medium-sized tractor. The parallel hybrid had on average 10–15% higher energy efficiency than the conventional tractor, but the gain was reduced with higher load resistance so that the variation in the potential was due to the operating conditions; thus, less reduction can be achieved with higher workload cycles with the hybrid powertrains. The series hybrid powertrain has a much higher potential to reduce energy consumption with the medium-sized tractor than with the large-sized tractor; however, not all the electrification benefits can be demonstrated with the passive duty cycle and, therefore, the powertrain benefits should also be evaluated in different types of operating cycles. There were no major differences in simulation results between the two tillage cycles, with the Tillage B cycle being slightly more demanding due to the 50% higher target speed.

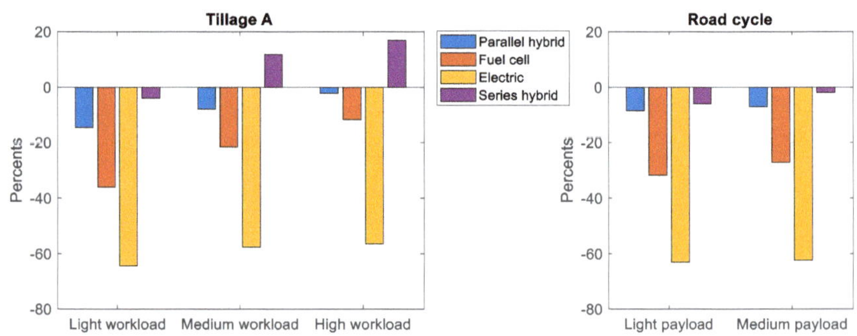

Figure 11. Potential for reducing energy consumption in the Tillage A and Road cycles of the large-sized tractor.

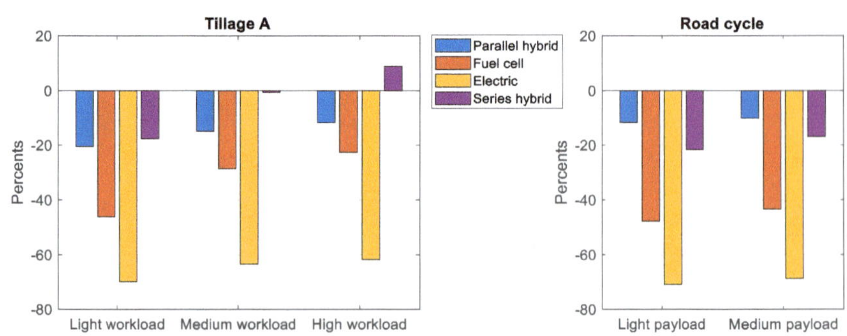

Figure 12. Potential for reducing energy consumption in the Tillage A and Road cycles of the medium-sized tractor.

3.3. Distribution of Losses

From the simulation results, the breakdown of powertrain losses was calculated for all simulations in order to evaluate the energy losses between the different powertrains. Figures 13 and 14 present the distribution of the powertrain losses of the large-sized tractor models in the Tillage A and Road cycles. The presented bar diagrams illustrate the total energy consumption in units of kWh. For the conventional, parallel hybrid, and series hybrid tractors, the major energy losses were generated by the heat losses of the power source (PS), i.e., the diesel engine. Depending on the cycle and workload, the energy loss portion of the power source was 65–70% for the conventional, parallel hybrid, and series hybrid tractor, 44–48% for the fuel cell hybrid, and 7–10% for the electric tractor. It is important to notice that the portions of the auxiliary loads in the energy losses were significant, especially when compared to the transmission losses. The increase in workload in the tillage cycles significantly increased the overall energy consumption. The highest increase occurred in the work and power source losses, especially for the conventional, parallel hybrid, and series hybrid tractor models. The increase in the payload in the Road cycle had much less of an influence on the overall energy consumption than the increase in the workload in the tillage cycles.

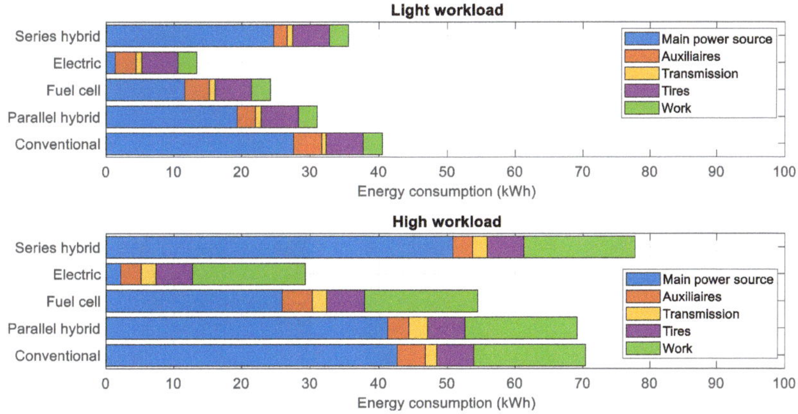

Figure 13. Distribution of energy losses for the different large tractor models in the Tillage A cycle with two workloads.

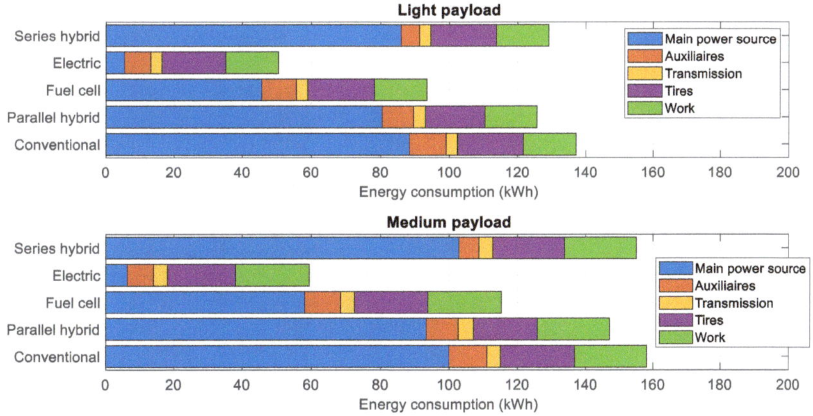

Figure 14. Distribution of energy losses for the different large tractor models in the Road cycle with two payloads.

3.4. Operating Time

The operating performance was evaluated based on the calculated operating times in the simulated cycles. The fuel tank size for the conventional large-sized tractor was 500 L and 350 L for the parallel and series hybrid. The hydrogen storage was assumed to be 36 kg of compressed hydrogen at 700 bars. This is comparable to the amount of hydrogen storage capacity in the fuel cells of hybrid city buses. The on-board energy capacities were 50%

less for the medium-sized tractor model. The operating time variations in the simulated cycles are presented in Figures 15 and 16. It can be observed that there were no major differences between the cycles but very significant differences between the tractor models. The conventional, parallel, and series hybrid tractors had very long operation times without refueling, which is typical nowadays for agricultural tractors. The fuel cell hybrid already offers quite a reasonable operating time without refueling, from 5 h up to 15 h. The major challenge for battery electric tractors is the low energy density of the energy storage and, therefore, the operating time remained very low in comparison to the other tractor models. The operating time could be increased by adding battery capacity, but this is challenging in terms of weight and available volume. Another solution could be fast charging, but access to high-power charging in the farming context could prove difficult.

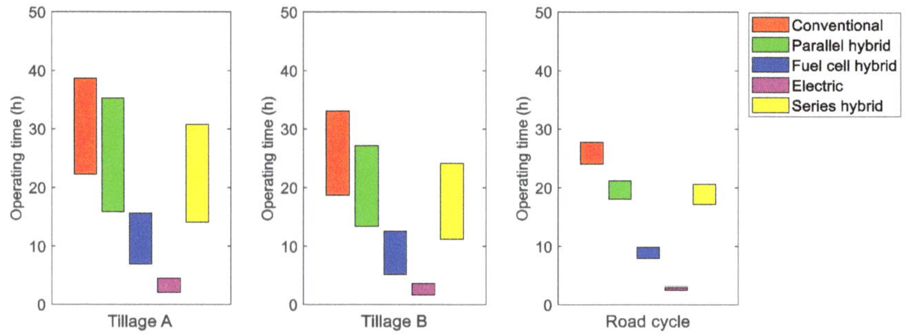

Figure 15. Calculated variations in operating times for different cycles for the large-sized tractor.

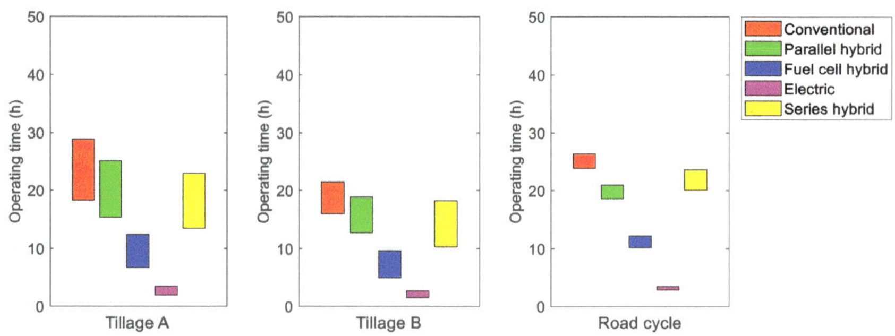

Figure 16. Calculated variations in operating times for different cycles for the medium-sized tractor.

4. Discussion

The research results clearly indicate the significant potential to reduce the energy consumption of agricultural tractors by powertrain electrification. Over the years, different electrified powertrain topologies have been proposed for vehicles, and as with other types of vehicles, such as city buses [14], the benefit of electrified powertrains for agricultural tractors is typically dependent on the duty cycle or work task carried out with the vehicle. In many scientific and practical research studies [17–19,33,34], parallel hybrid electric powertrain topology has been recognized as being suitable for agricultural tractors. One of the main reasons for this could be that it would not need any major modifications to conventional tractor designs but, instead, could be implemented by adding a motor/generator in the place of the flywheel along with a small battery pack or even supercapacitors [20]. The results indicate that the parallel hybrid electric powertrain would provide meaningful energy savings for medium-sized tractors and, when operating with lighter loads, large-sized tractors. Similar conclusions were drawn in recent research by Beligoj et al. (2022), who evaluated the feasibility and life-cycle cost of a parallel hybrid powertrain for different sizes of agricultural tractors [35]. They concluded that very little energy consumption reduction or cost saving would be attained with large-sized tractor (engine power of 210 kW) hybridization, but small-sized orchard tractors and medium–large-sized tractors with medium workloads would provide considerable savings in life-cycle costs. The lower fuel consumption would offer reductions in operational costs and decrease the carbon footprint of these tractors.

The series hybrid electric powertrain has shown to be less interesting for vehicle applications that do not have very repeatable duty cycles or for which there are several use cases. This is the case for passenger vehicles and for agricultural tractors because these are used for a wide variety of purposes by different types of professional and individual users. The simulation results showed the variable potential of a series hybrid electric powertrain, including notable benefits for the medium-sized tractor but less encouraging results in the case of the large-sized tractor. Nevertheless, more detailed simulations should be performed to evaluate the potential of the series hybrid powertrain for different types of agricultural tasks. In comparison to parallel hybrid powertrains, the series hybrid powertrain could provide the possibility of using the electric power take-off (ePTO) and electrified implements, which would be much harder to accomplish with the parallel hybrid due to the limited amount of on-board electric power [35].

Hydrogen as a vehicle fuel is gaining more and more interest as a method for reducing the use of fossil fuels and harmful emissions. Fuel cell systems have been used as the main power sources in vehicles for a long time, but the technological cost and lack of fueling infrastructure are still barriers that have not been fully resolved. Even though fuel cells can be considered as a mature technology, it is not technologically easy to design an agricultural tractor with a fuel cell system because of the spatial requirements for the stack, hydrogen storage, and auxiliary systems. Recent research by Ahluwalia et al. (2022) concluded that the fuel cell system could be cost competitive for agricultural tractors if the targeted improvements to the cost, performance, and durability of the technology could be achieved [36]. Much more research is needed to find the best solutions for alternative fuels for use in agricultural vehicles. For example, methane or methanol might be preferred over hydrogen because of its low volumetric energy density and adapted infrastructure requirements [37]. As a potential fuel for internal combustion engines, burning hydrogen in an engine also has some challenges in terms of NOx emissions and engine knocking [38], and the storage challenge would remain the same as for the fuel cell systems.

Adopting alternative fuels allows for a diversification of energy sources in agriculture. This reduces dependency on a single energy resource by enhancing energy security and resilience in the face of changing market conditions. Using alternative fuels may reduce reliance on imported fossil fuels, providing a pathway towards greater energy independence for agricultural operations. However, implementing alternative fuel technologies in agricultural tractors may require substantial upfront investments. Farmers may be hesitant to adopt these technologies due to concerns about costs and the need for specialized equipment. The compatibility of alternative fuels with existing tractor engines and performance characteristics is a critical consideration. Adapting engines to run efficiently on alternative fuels without compromising power output and durability is still a technical challenge.

The batteries in electric vehicles have seen tremendous technological development and market success, essentially in all on-road vehicle categories; even for 40 ton heavy-duty trucks and battery-powered tractors have been designed and manufactured. Hence, battery and power electronics technology is certainly mature enough even for heavy-duty machinery. The simulation results show that energy consumption could be reduced by up to 70%, which comes from a much higher powertrain efficiency. However, this higher powertrain efficiency does not mitigate the fact that many agricultural field operations require high power or high workload operation. This ultimately leads to high energy requirements and, therefore, the focus must be on the total amount of required on-board energy. The simulations in this research were performed with the consideration that all the tractor models have the same performance and, therefore, the total weight was limited. It could be said that a higher battery capacity than was used in this research could be installed into battery-powered electric tractors [39]. In this case, the tractor weight would increase, which would have some influence on performance and energy consumption, but this influence would need to be evaluated with more detailed simulations. Another challenge that remains to be resolved is battery charging; it is not clear whether every farm could have access to high-power fast charging. Thus, preliminary studies on the fully electrification of agricultural tractors have concluded that it would be more profitable to have a battery exchange system rather than high-power charging systems [40].

Overall, electrification is being applied to agricultural tractors and there are more possibilities than challenges. More research is needed to evaluate the different use cases, namely duty cycle operations, and, especially, life-cycle energy consumption, emissions, and cost [41]. Available electric power would allow the electrification of many auxiliary devices that could lead to additional savings by reducing the idling losses that are quite important for agricultural tractors [42]; Molari et al. (2019) stated that agricultural tractors may remain idle from 10% to 43% of their entire operating time [43]. This would provide additional savings with electrification because unnecessary idling could be easily avoided by shutting down the engine.

5. Conclusions

Simulation models for conventional, parallel hybrid electric, series hybrid electric, fuel cell electric, and battery electric agricultural tractors were developed using Autonomie software. Simulations of three different work cycles were carried out with different workloads to evaluate energy consumption and operating performance. The results show that the battery electric powertrain provided the most energy-efficient powertrain option for agricultural tractors. However, the operating performance was relatively poor because the energy density of lithium-ion batteries does not provide a long enough operating time without fast charging. Furthermore, providing fast charging in agricultural contexts could prove challenging. The simulation results indicate that fuel cell hybrid tractors could provide substantial energy savings in comparison to the diesel-powered, conventional powertrain. The major advantage is the much higher efficiency of the fuel cell system compared to diesel engines. A reasonable amount of hydrogen storage would provide an adequate operating performance of more than 10 h of operation without refueling. It remains to be validated whether the fuel cell system with storage tanks would be a feasible solution, especially for larger-sized tractors. The parallel hybrid powertrain does not provide significant energy savings with high workloads, but medium-sized parallel hybrid tractor models show relatively good performance in terms of energy consumption and operating time.

Author Contributions: Conceptualization, A.L. and N.K.; methodology, A.L.; software, A.L. and R.V.; validation, A.L. and N.K.; formal analysis, K.K.; investigation, A.L. and K.K.; resources, V.F., R.V. and N.K.; data curation, A.L.; writing—original draft preparation, A.L.; writing—review and editing, A.L., K.K. and N.K.; visualization, A.L. and N.K.; supervision, V.F., R.V. and N.K.; project administration, N.K. and V.F.; funding acquisition, N.K. All authors have read and agreed to the published version of the manuscript.

Funding: The authors would like to acknowledge the financial support of Gurpreet Singh (Vehicle Technologies Office, U.S. Department of Energy). The submitted manuscript was created by the UChicago Argonne, LLC, Operator of Argonne National Laboratory ("Argonne"). Argonne, a U.S. Department of Energy Office of Science laboratory, is operated under Contract No. DE-AC02-06CH11357. The U.S. Government retains for itself, and others acting on its behalf, a paid-up nonexclusive, irrevocable worldwide license in said article to reproduce, prepare derivative works, distribute copies to the public, and perform publicly and display publicly, by or on behalf of the Government. Open access funding provided by University of Helsinki.

Data Availability Statement: The original contributions presented in the study are included in the article, further inquiries can be directed to the corresponding author.

Conflicts of Interest: The authors declare no conflicts of interest.

References

1. Ajanovic, A.; Haas, R.; Schrödl, M. On the Historical Development and Future Prospects of Various Types of Electric Mobility. *Energies* **2021**, *14*, 1070. [CrossRef]
2. Balazadeh Meresht, N.; Moghadasi, S.; Munshi, S.; Shahbakhti, M.; McTaggart-Cowan, G. Advances in Vehicle and Powertrain Efficiency of Long-Haul Commercial Vehicles: A Review. *Energies* **2023**, *16*, 6809. [CrossRef]
3. Hegazy, O.; Barrero, R.; Van den Bossche, P.; El Baghdadi, M.; Smekens, J.; Van Mierlo, J.; Vriens, W.; Bogaerts, B. Modeling, analysis and feasibility study of new drivetrain architectures for off-highway vehicles. *Energy* **2016**, *109*, 1056–1074. [CrossRef]
4. Lin, T.; Lin, Y.; Ren, H.; Chen, H.; Chen, Q.; Li, Z. Development and key technologies of pure electric construction machinery. *Renew. Sustain. Energy Rev.* **2020**, *132*, 110080. [CrossRef]
5. Bilgin, B.; Magne, P.; Malysz, P.; Yang, Y.; Pantelic, V.; Preindl, M.; Korobkine, A.; Jiang, W.; Lawford, M.; Emadi, A. Making the Case for Electrified Transportation. *IEEE Trans. Transp. Electrif.* **2015**, *1*, 4–17. [CrossRef]
6. Khan, A.U.; Huang, L. Toward Zero Emission Construction: A Comparative Life Cycle Impact Assessment of Diesel, Hybrid, and Electric Excavators. *Energies* **2023**, *16*, 6025. [CrossRef]
7. Lajunen, A.; Kivekäs, K.; Freyermut, V.; Vijayagopal, R.; Kim, N. Simulation of Alternative Powertrains in Agricultural Tractors. In Proceedings of the International Electric Vehicle Symposium and Exhibition (EVS36), Sacramento, CA, USA, 11–14 June 2023.
8. Scolaro, E.; Beligoj, M.; Perez Estevez, M.; Alberti, L.; Renzi, M.; Mattetti, M. Electrification of Agricultural Machinery: A Review. *IEEE Access* **2021**, *9*, 164520–164541. [CrossRef]
9. Tetzlaff, S. System-wide electrification and appropriate functions of tractor and implement. *Landtechnik* **2015**, *70*, 203–216.
10. Nizam Uddin Khan, F.M.; Rasul, M.G.; Sayem, A.S.M.; Mandal, N. Maximizing energy density of lithium-ion batteries for electric vehicles: A critical review. *Energy Rep.* **2023**, *9* (Suppl. S11), 11–21. [CrossRef]
11. Martinez-Boggio, S.; Monsalve-Serrano, J.; García, A.; Curto-Risso, P. High Degree of Electrification in Heavy-Duty Vehicles. *Energies* **2023**, *16*, 3565. [CrossRef]
12. Vijayagopal, R.; Rousseau, A. Benefits of Electrified Powertrains in Medium- and Heavy-Duty Vehicles. *World Electr. Veh. J.* **2020**, *11*, 12. [CrossRef]
13. Lajunen, A. *Energy Efficiency of Conventional, Hybrid Electric, and Fuel Cell Hybrid Powertrains in Heavy Machinery (2015-01-2829)*; SAE Technical Paper; SAE: Warrendale, PA, USA, 2015.
14. Lajunen, A.; Lipman, T. Lifecycle cost assessment and carbon dioxide emissions of diesel, natural gas, hybrid electric, fuel cell hybrid and electric transit buses. *Energy* **2016**, *106*, 329–342. [CrossRef]
15. Beltrami, D.; Iora, P.; Tribioli, L.; Uberti, S. Electrification of Compact Off-Highway Vehicles—Overview of the Current State of the Art and Trends. *Energies* **2021**, *14*, 5565. [CrossRef]
16. Renius, K.T. *Fundamentals of Tractor Design*; Springer Nature: Baldham, Germany, 2020.
17. Troncon, D.; Alberti, L.; Mattetti, M. A feasibility study for agriculture tractors electrification: Duty cycles simulation and consumption comparison. In Proceedings of the IEEE Transportation Electrification Conference and Expo (ITEC), Detroit, MI, USA, 19–21 June 2019.
18. Troncon, D.; Alberti, L.; Bolognani, S.; Bettella, F.; Gatto, A. Electrification of agricultural machinery: A feasibility evaluation. In Proceedings of the International Conference on Ecological Vehicles and Renewable Energies (EVER), Monte-Carlo, Monaco, 8–10 May 2019.
19. Dalboni, M.; Santarelli, P.; Patroncini, P.; Soldati, A.; Concari, C.; Lusignani, D. Electrification of a Compact Agricultural Tractor: A Successful Case Study. In Proceedings of the IEEE Transportation Electrification Conference and Expo (ITEC), Detroit, MI, USA, 19–21 June 2019.
20. Mendes, F.E.G.; Brandao, D.I.; Maia, T.; Braz de Filho, J.C. Off-Road Vehicle Hybridization Methodology Applied to a Tractor Backhoe Loader. In Proceedings of the IEEE Transportation Electrification Conference and Expo (ITEC), Detroit, MI, USA, 19–21 June 2019.
21. Mocera, F.; Martini, V. Numerical Performance Investigation of a Hybrid eCVT Specialized Agricultural Tractor. *Appl. Sci.* **2022**, *12*, 2438. [CrossRef]
22. Nevzorova, T.; Kutcherov, V. Barriers to the wider implementation of biogas as a source of energy: A state-of-the-art review. *Energy Strategy Rev.* **2019**, *26*, 00414. [CrossRef]
23. Simikic, M.; Tomic, M.; Savin, L.; Micic, R.; Ivanisevic, I.; Ivanisevic, M. Influence of biodiesel on the performances of farm tractors: Experimental testing in stationary and non-stationary conditions. *Renew. Energy* **2018**, *121*, 677–687. [CrossRef]
24. Briggs, I.; Murtagh, M.; Kee, R.; McCulloug, G.; Douglas, R. Sustainable non-automotive vehicles: The simulation challenges. *Renew. Sustain. Energy Rev.* **2017**, *68*, 840–851. [CrossRef]
25. Birkmann, C.; Fedde, T.; Frerichs, L. Drivetrain, Chassis and Tire-Soil Contact Influence on Power Shift Operations in Standard Tractors. *Landtechnik* **2018**, *73*, 146–160.

26. Witzel, P. The Hohenheim Tyre Model: A validated approach for the simulation of high volume tyres–Part I: Model structure and parameterisation. *J. Terramech.* **2018**, *75*, 3–14. [CrossRef]
27. Battiato, A.; Diserens, E. Tractor traction performance simulation on differently textured soils and validation: A basic study to make traction and energy requirements accessible to the practice. *Soil Tillage Res.* **2017**, *166*, 18–32. [CrossRef]
28. Witzel, P. The Hohenheim Tyre Model: A validated approach for the simulation of high volume tyres–Part II: Validation. *J. Terramech.* **2018**, *75*, 15–24. [CrossRef]
29. Vijayagopal, R.; Rousseau, A. *System Analysis of Multiple Expert Tools (2011-01-0754)*; SAE Technical Paper; SAE: Warrendale, PA, USA, 2011.
30. AMBER. Argonne National Laboratory. Available online: https://amber.anl.gov/ (accessed on 31 December 2023).
31. Lajunen, A. Simulation of energy efficiency and performance of electrified powertrains in agricultural tractors. In Proceedings of the IEEE Vehicle Power and Propulsion Conference (VPPC), Merced, CA, USA, 1–4 November 2022.
32. Seeger, J. New Dual Clutch Transmission for Tractors. *ATZ Offhighway* **2012**, *5*, 58–67. [CrossRef]
33. Tebaldi, D.; Zanasi, R. Modeling Control and Simulation of a Parallel Hybrid Agricultural Tractor. In Proceedings of the Mediterranean Conference on Control and Automation (MED), Puglia, Italy, 22–25 June 2021.
34. Zahidi, Y.; El Moufid, M.; Benhadou, S.; Medromi, H. An Assessment of Low-Cost Tractor Motorization with Main Farming Implements. *World Electr. Veh. J.* **2020**, *11*, 74. [CrossRef]
35. Beligoj, M.; Scolaro, E.; Alberti, L.; Renzi, M.; Mattetti, M. Feasibility Evaluation of Hybrid Electric Agricultural Tractors Based on Life Cycle Cost Analysis. *IEEE Access* **2022**, *10*, 28853–28867. [CrossRef]
36. Ahluwalia, R.K.; Wang, X.; Star, A.G.; Papadias, D.D. Performance and cost of fuel cells for off-road heavy-duty vehicles. *Int. J. Hydrog.* **2022**, *47*, 10990–11006. [CrossRef]
37. Ahlgren, S.; Baky, A.; Bernesson, S.; Nordberg, Å.; Norén, O.; Hansson, P.A. Tractive power in organic farming based on fuel cell technology–Energy balance and environmental load. *Agric. Syst.* **2009**, *102*, 67–76. [CrossRef]
38. Hosseini, S.H.; Tsolakis, A.; Alagumalai, A.; Mahian, O.; Lam, S.S.; Pan, J.; Peng, W.; Tabatabaei, M.; Aghbashlo, M. Use of hydrogen in dual-fuel diesel engines. *Prog. Energy Combust. Sci.* **2023**, *98*, 101100. [CrossRef]
39. Brenna, M.; Foiadelli, F.; Leone, C.; Longo, M.; Zaninelli, D. Feasibility Proposal for Heavy Duty Farm Tractor. In Proceedings of the International Conference of Electrical and Electronic Technologies for Automotive, Milan, Italy, 9–11 July 2018.
40. Lagnelöv, O. Electric Autonomous Tractors in Swedish Agriculture: A Systems Analysis of Economic, Environmental and Performance Effects. Ph.D. Thesis, Swedish University of Agricultural Sciences, Uppsala, Sweden, 2023.
41. Martelli, S.; Mocera, F.; Somà, A. Carbon Footprint of an Orchard Tractor through a Life-Cycle Assessment Approach. *Agriculture* **2023**, *13*, 1210. [CrossRef]
42. Saetti, M.; Mattetti, M.; Varani, M.; Lenzini, N.; Molari, G. On the power demands of accessories on an agricultural tractor. *Biosyst. Eng.* **2021**, *206*, 109–122. [CrossRef]
43. Molari, G.; Mattetti, M.; Lenzini, N.; Fiorati, S. An updated methodology to analyse the idling of agricultural tractors. *Biosyst. Eng.* **2019**, *187*, 160–170. [CrossRef]

World Electric Vehicle Journal

MDPI

Article

Multi-Use Optimization of a Depot for Battery-Electric Heavy-Duty Trucks

Florian Biedenbach [1,*] and Kai Strunz [2]

[1] Forschungsstelle für Energiewirtschaft (FfE) e.V, 80995 Munich, Germany
[2] Chair of Sustainable Electric Networks and Sources of Energy (SENSE), School of Electrical Engineering and Computer Science, Technische Universität Berlin, 10587 Berlin, Germany
* Correspondence: fbiedenbach@ffe.de

Abstract: Battery-electric trucks offer a high battery capacity and good predictability, making them attractive for the implementation of bidirectional charging strategies. Nevertheless, most of the previous charging strategy studies focus on electric passenger cars. These charging strategies are usually formulated as separate use cases like tariff-optimized charging, arbitrage trading, peak shaving, and self-consumption optimization. By combining different use cases, their economic potential can be increased. In this paper, we introduce a model to optimize charging processes in depots for electric vehicles considering the combination of different use cases. This model is applied to a depot for battery-electric trucks. The savings obtained through optimized bidirectional charging highlight the enormous potential of this technology for the future, especially in the heavy-duty sector.

Keywords: bidirectional charging; smart charging; vehicle to grid; modeling; optimization; electric vehicles; battery electric trucks

Citation: Biedenbach, F.; Strunz, K. Multi-Use Optimization of a Depot for Battery-Electric Heavy-Duty Trucks. *World Electr. Veh. J.* **2024**, *15*, 84. https://doi.org/10.3390/wevj15030084

Academic Editors: Joeri Van Mierlo and Genevieve Cullen

Received: 18 January 2024
Revised: 19 February 2024
Accepted: 21 February 2024
Published: 26 February 2024

1. Introduction

Controlled and bidirectional charging has recently become an extensively discussed topic. A variety of publications that deal with this technology predict its high relevance in the near future [1,2]. The Original Equipment Manufacturers (OEMs) have discovered its importance as well, and the first bidirectional vehicles are on the market [3]. Nevertheless, past considerations have mostly revolved around battery-electric vehicle (EV) passenger cars and not focused on battery-electric heavy-duty trucks (BETs). However, taking into account that heavy-duty and bus traffic is responsible for 6% of all European greenhouse gas emissions, a major wave of electrification in this area is necessary [4]. Registration statistics show that this area is still dominated by diesel vehicles, while BETs represent only 1.5% of the current truck market [5]. Various challenges impede the market roll-out of BETs. These challenges include the high acquisition costs and the limited availability of grid connection capacity in depots [6]. The use of controlled and bidirectional charging can address these challenges by reducing operating costs and the required grid connection capacity [7]. When evaluating the requirements for controlled and bidirectional charging, BETs offer several advantages over passenger cars. Due to the higher charging power and the bundling of many vehicles in one depot, a high marketable capacity can quickly be achieved at one location. The use of bidirectional charging in BET depots can exploit these advantages and support the roll-out of BETs, making the research topic of this paper highly relevant.

Previous work on BETs typically covered a comparison of the technology with diesel or hydrogen-based power trains in terms of CO_2 emissions, cost, and technical feasibility [8–10]. Those studies usually acknowledge the advantages of BETs in tons of emission reductions and cost, but the availability of BETs with sufficient battery capacity for long-haul transport is noted as an issue [9,11]. Apart from limited real-world observations [10], most of those studies are assumption-based and use synthetic driving profiles. When it comes to the optimization of charging processes for BETs, there is a lack of existing knowledge.

A study that already examines the optimization of charging processes for BETs is [12]. While the routes of the BETs are optimized, variable prices are not considered. In [13], the charging processes for trucks in a depot are optimized based on charging costs, and bidirectional charging is included. However, the authors used assumed driving profiles and price time series. Furthermore, the considered optimization period is only two days. These simplifications eliminate high price fluctuations, which are particularly important to determine the revenue of arbitrage trading [14].

Even though most of the prior work on the topic of charging management and optimization of charging strategies excludes BETs, there is a large amount of literature focusing on passenger cars. Related prior work distinguished between different use cases of controlled and bidirectional charging. These use cases have mostly been considered separately in previous studies [15]. For the use case self-consumption optimization, the self-consumption rate is maximized by shifting charging processes to times of PV generation [2,16]. Minimizing the peak load at a grid connection point is the objective of the use case peak shaving [7]. The optimization of charging with a variable electricity tariff, where charging processes are shifted to times of low prices, can be referred to as tariff-optimized charging [17]. The batteries are charged at times when electricity prices are low, and they feed the electric energy back into the grid at times when electricity prices are high, such as in the use case arbitrage trading [14]. To increase the economic efficiency, use cases can also be combined as a so-called multi-use objective, which has already been investigated for stationary storage facilities [18,19]. Apart from [15], this methodology has not yet been applied to EVs.

There already exist various publications on the optimization of charging processes in bus depots. The use cases defined above can also be found in these publications. The authors of [20,21] include the peak-shaving use case in their optimization. Tariff-optimized charging is examined in [21–23]. A few publications have already addressed the combination of different use cases for charging optimization in bus depots. The authors of [21] combine peak shaving with tariff-optimized charging and additionally include timetable shifting in the optimization problem. In [24], a depot is integrated into a virtual power plant, and arbitrage trading is combined with the provision of power system services. However, even for bus depots, the combination of all four use cases introduced above has not yet been examined.

While the optimization strategies developed for passenger cars or buses can be adapted for BETs, the results are expected to differ due to changes in battery capacities, charging power, and parking duration. Therefore, we see a need for further research in optimizing charging processes for BETs, especially with regard to multi-use optimization. In this paper, we tackle this gap by developing a model that supports the combination of the use cases self-consumption optimization, peak shaving, tariff-optimized charging, and arbitrage trading within a multi-use optimization. This model is then applied to a real depot for BETs. The combination of different use cases and their application to BETs is the novelty value of this work. The developed model and input data are described in Section 2. By using the model, possible savings from bidirectional charging of the BETs are determined and presented and discussed in Section 3. The final conclusion and an outlook are given in Section 4. The results of this study can be used by freight forwarders and OEMs as an orientation for expectable savings and for the prioritization of charging strategies. A preliminary version of this study has already been presented in [25].

2. Materials and Methods

2.1. Optimization Model

The optimization model eFlame was primarily developed to optimize several use cases for bidirectional charging separately. In [14,26], the use cases arbitrage trading and self-consumption optimization are elaborated upon. The use case peak shaving is dealt with in [7]. The novelty of the present paper is the combination of the use cases in the context of a multi-use optimization that was not implemented before. Figure 1 shows all power flows relevant to the optimization. At this point, we describe the optimization problem covering decision variables, objective function, and constraints. Linear optimization is a method that can be used to solve problems where the objective function and constraints are linear functions of the decision variables. The constraints can be formulated as equalities or inequalities. A mixed-integer linear program (MILP) problem includes integer decision variables. A comprehensive introduction to linear optimization is given in [27]. Since the model was primarily developed for optimizing battery-electric cars, the vehicles are generally referred to as EVs in the following model description.

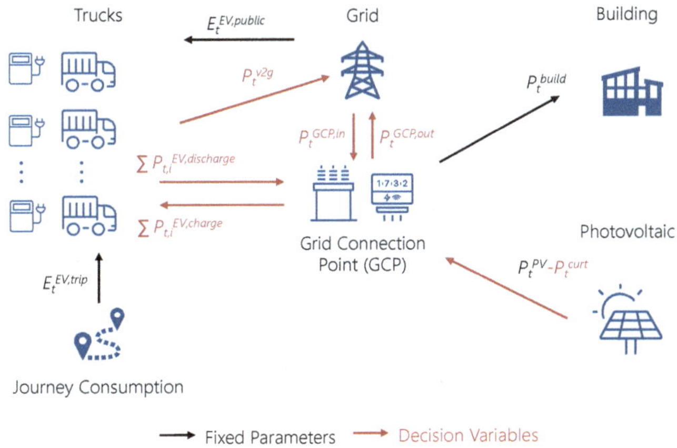

Figure 1. Schematic representation of interrelations in the optimization model.

For all decision variables, the non-negativity constraint applies. The constraint is exemplarily defined in Equation (1) for the received power $P_t^{GCP,in}$, and the feed-in power $P_t^{GCP,out}$ at the Grid Connection Point (GCP), but it can be applied to the remaining decision variables. The total number of time steps t in the observation horizon is represented by n.

$$P_t^{GCP,in} \geqslant 0, \qquad P_t^{GCP,out} \geqslant 0 \qquad \forall\, t \in T = \{1, ..., n\} \tag{1}$$

The photovoltaic (PV) generation is not a decision variable, but it may be influenced during the optimization via the curtailment P_t^{curt}. With this optimization variable, the generation of the PV system can be reduced, e.g., to prevent feed-in at negative prices. Using the decision variable $P_t^{GCP,peak}$, the maximum power at the grid connection point is determined. The charging power P_t^{charge} and discharging power $P_t^{discharge}$ and the energy capacity of the battery E_t^{EV} are further decision variables that are related to the EVs. Furthermore, there is the decision variable P_t^{v2g}, which is used to observe how much energy from the vehicles is fed back into the grid. The remaining decision variables b_t^{charge}, $b_t^{discharge}$, b_t^{out}, and b_t^{in} are boolean variables, which are used to ensure that the power flow exchanges with the vehicles and the grid connection point are only in one direction at any time instant.

The objective of the optimization model is to maximize the revenue. The established objective function shown in Equation (2) consists of four terms: the cashflow from arbitrage trading at the spotmarket CF^{spot}, costs through levies C^{levies}, costs through grid fees $C^{gridfee}$, and a term that evaluates the opportunity costs due to battery degradation $C^{bat,deg}$. The different terms are defined in the following.

$$\max\left(CF^{spot} - C^{levies} - C^{gridfee} - C^{bat,deg}\right) \tag{2}$$

The cash flow, the difference between cash in- and outflows, from arbitrage trading at the spot market CF^{spot} is calculated in Equation (3). Different market data can be selected for the price time series p_t^{in} and p_t^{out}, but constant values may also be used.

$$CF^{spot} = \sum_{t=1}^{n} \left(P_t^{GCP,in} \cdot p_t^{in} \cdot \Delta t - P_t^{GCP,out} \cdot p_t^{out} \cdot \Delta t \right) \qquad \forall\, t \in T \tag{3}$$

Consumers have to pay a gridfee $C^{gridfee}$ to the Distibution System Operator (DSO) for the use of the grid infrastructure. In Germany, the grid fee for commercial customers is divided into a usage price p^{usage} and a capacity price p^{cap}. The usage price depends on the energy consumed, whereas the capacity price depends on the annual peak power. The grid fee is included in the objective function of Equation (4).

$$C^{gridfee} = \sum_{t=1}^{n} P_t^{GCP,in} \cdot p^{usage} \cdot \Delta t + P_{t=n}^{GCP,peak} \cdot p^{cap} \qquad \forall\, t \in T \tag{4}$$

Additionally, various taxes and levies are charged on electricity, and those are summarized through C^{levies} and shown in Equation (5). Stationary battery storage may be partially exempt from levies, and such an exemption is also being discussed for bidirectional vehicles. The problem is to determine how much energy is actually fed back into the grid. This is especially problematic in combination with PV systems. Via the subtrahend of Equation (5), a partial exemption from the levies on energy fed back into the grid is implemented. The decision variable P_t^{v2g} represents the power the vehicles feed into the grid and is introduced later in (17) and (18). It is an auxiliary variable calculated from the other power variables and is therefore not directly included in the power balance following in Equation (7). A partial exemption may be dynamically parameterized via the levies on V2G $p^{levies,v2g}$ that are still charged even if the energy is fed back into the grid. If $p^{levies,v2g}$ is set equal to p^{levies}, no exemption occurs. A full exemption can be achieved by setting $p^{levies,v2g}$ equal to zero.

$$C^{levies} = \sum_{t=1}^{n} P_t^{GCP,in} \cdot p^{levies} \cdot \Delta t - \sum_{t=1}^{n} P_t^{v2g} \cdot \left(p^{levies} - p^{levies,v2g} \right) \cdot \Delta t \qquad \forall\, t \in T \tag{5}$$

The opportunity costs from battery degradation $C_{bat,deg}$ are included in the optimization problem by using Equation (6) based on [28]. The calculation of the degradation costs $C_{bat,deg}$ is based on the use of the battery and determined by the decrease of the available capacity C^{loss} from a cycling aging model. The costs result primarily from the total charge quantity throughput, which is defined by the charging and discharging power. The price of the battery is represented by $c_{bat,buy}$, and $E^{EV,max}$ is the capacity of the battery. The used model assumes the end of life of the battery at a loss of 20% of the initial capacity.

$$C_{bat,deg} = \frac{c_{bat,buy} \cdot E^{EV,max}}{20\%} C_{loss}(P_t^{EV,charge}, P_t^{EV,discharge}) \qquad \forall\, t \in T \tag{6}$$

The optimization model is restricted by several constraints concerning the GCP and the EVs. We start by introducing the boundary conditions of the GCP. According to the law of conservation of energy, the incoming power flows at the GCP must be equal to the outgoing power flows. This is ensured by Equation (7). The load profile of the building P_t^{build} is integrated into the optimization as a static time series.

$$\begin{aligned} P_t^{GCP,in} + \sum_{i=1}^{n_{EV}} P_t^{EV,discharge} + P_t^{PV} = \\ P_t^{GCP,out} + \sum_{i=1}^{n_{EV}} P_t^{EV,charge} + P_t^{curt} + P_t^{build} \qquad \forall\, t \in T \end{aligned} \tag{7}$$

For the determination of the grid fee $C^{gridfee}$ in Equation (4), the annual peak power at the GCP $P_t^{GCP,peak}$ is required. Using Equation (8), the power peak is updated continuously during the optimization. Thus, the last time step n contains the annual power peak.

$$P_t^{GCP,peak} \geqslant P_t^{GCP,in}, \qquad P_t^{GCP,peak} \geqslant P_{t-1}^{GCP,peak} \qquad \forall\, t \in T \tag{8}$$

Equations (9) and (10) are introduced to prevent energy from being purchased and fed in simultaneously at the GCP. In consequence, the boolean decision variables b_t^{in} and b_t^{out} are used. $P^{GCP,max}$ describes the maximum grid connection capacity, which results from the transformer and structural conditions at the grid connection point. The combination of Equations (8) and (9) ensures the grid connection capacity $P^{GCP,max}$ is always greater than or equal to the annual power peak $P_t^{GCP,peak}$.

$$P^{GCP,max} \cdot b_t^{in} \geqslant P_t^{GCP,in}, \qquad P^{GCP,max} \cdot b_t^{out} \geqslant P_t^{GCP,out} \qquad \forall\, t \in T \tag{9}$$

$$b_t^{out} + b_t^{in} \leqslant 1 \qquad \forall t \in T \tag{10}$$

The following constraints are related to the EVs and apply separately for each EV. The energy balance of the vehicle battery must be maintained to preserve the physical consistency of the EVs. The energy stored in the EV battery in the first time step is defined by the constraint Equation (11). For the first time step, this equation defines the stored energy as equal to the initial stored energy plus the charged energy at the GCP minus the discharged energy and the energy consumed during trips $E_t^{EV,trip}$ plus the energy charged at public stations $E_{t=1}^{EV,public}$. Constant efficiencies for charging $\eta^{EV,charge}$ and discharging $\eta^{EV,discharge}$ are considered.

$$\begin{aligned} E_{t=1}^{EV} &= SOC_{t=1}^{EV} \cdot E^{EV,max} + P_{t=1}^{EV,charge} \cdot \eta^{EV,charge} \cdot \Delta t \\ &- P_{t=1}^{EV,discharge} \cdot \eta^{EV,discharge} \cdot \Delta t - E_{t=1}^{EV,trip} + E_{t=1}^{EV,public} \end{aligned} \tag{11}$$

For the remaining time steps, Equation (12) applies, where the initially stored energy is replaced by the stored energy of the previous time step.

$$\begin{aligned} E_t^{EV} &= E_{t-1}^{EV} + P_t^{EV,charge} \cdot \eta^{EV,charge} \cdot \Delta t \\ -P_t^{EV,discharge} \cdot \eta^{EV,discharge} \cdot \Delta t &- E_t^{EV,trip} + E_t^{EV,public} \qquad \forall\, t \in \{2, \ldots, n\} \end{aligned} \tag{12}$$

Equation (13) ensures that the vehicles are always charged with a minimum State of Charge $SOC^{EV,dep,min}$ at departure. The condition is only valid for the time steps in which a vehicle departs, as indicated by the boolean variable $b_t^{EV,dep}$. This variable is determined before the optimization based on the driving profiles and is only equal to one if the vehicle departs. To ensure that the condition can also be met if the vehicle is only plugged in for a short time and thus the minimum SOC cannot be reached, a buffer E_t^{buffer} is integrated into the condition. This buffer is also determined before the optimization.

$$E_t^{EV} + E_t^{buffer} = SOC^{EV,dep,min} \cdot E^{EV,max} \cdot b_t^{EV,dep} \qquad \forall\, t \in T \tag{13}$$

Apart from public charging, each EV can only be charged or discharged if it is connected to a charging point at the GCP, and this is ensured by Equations (14) and (15). The boolean variable b_t^{EV} is determined before the optimization based on the driving profiles. If the vehicle is plugged in, the variable is one, and otherwise it is zero. We assume that each vehicle has its own charging point. To prevent the EVs from charging and discharging at the same time, the decision variables b_t^{charge} and $b_t^{discharge}$ are added to Equations (14) and (15). Equation (16) prevents both variables from being equal to one simultaneously. If only unidirectional charging is considered, the Equations (15) and (16) are omitted, and $P_t^{EV,discharge}$ is set to zero via a further boundary condition.

$$b_t^{EV} \cdot b_t^{charge} \cdot P^{EV,charge,max} \geqslant P_t^{EV,charge} \qquad \forall\, t \in T \tag{14}$$

$$b_t^{EV} \cdot b_t^{discharge} \cdot P^{EV,discharge,max} \geqslant P_t^{EV,discharge} \qquad \forall\, t \in T \tag{15}$$

$$b_t^{charge} + b_t^{discharge} \leqslant 1 \qquad \forall\, t \in T \tag{16}$$

Finally, boundary conditions are required to determine the power fed back from the EVs into the grid P_t^{v2g}. This variable is necessary to calculate the exemption from levies in Equation (5). Therefore, we choose a power balance based approach and rearrange Equation (7) according to the discharged energy. Since power can only be fed into the grid if no energy is purchased, $P_t^{GCP,in}$ is set to zero. The discharged power is replaced by the introduced decision variable P_t^{v2g}, resulting in Equation (17). The boundary condition in Equation (18) ensures that P_t^{v2g} cannot become greater than the feed-in power.

$$P_t^{v2g} \leqslant P_t^{GCP,out} - P_t^{PV} + P_t^{curt} + P_t^{build} + \sum_{i=1}^{n_{EV}} P_t^{EV,charge} \qquad \forall\, t \in T \tag{17}$$

$$P_t^{v2g} \leqslant P_t^{GCP,out} \qquad \forall\, t \in T \tag{18}$$

Since the model is intended to examine entire years and since the use of boolean variables makes it a mixed-integer optimization problem, the computational effort required to solve the problem is rather high. In order to be able to solve it with a reasonable computational effort, the model is computed as a rolling optimization. The determination of the annual power peak is a special aspect of the rolling optimization, which will be explained in the following using the schematic diagram in Figure 2. For rolling optimization, the whole optimization period is divided into m smaller optimization time periods of uniform size. In individual optimization steps, each of the smaller optimization periods is optimized one after the other. The results of an optimization step are passed as start values to the next step. By using an overlapping period, we increase the prediction horizon for the optimization. After the m-th step, the first run of the optimization is finished. According to Equation (8), the power peak is continuously updated as shown in Figure 2 below. As can be seen in the figure, the first optimization steps are limited by a lower power peak compared to the later steps. Therefore, in a second optimization run, the affected steps before the occurrence of the annual power peak are optimized again with the updated power peak.

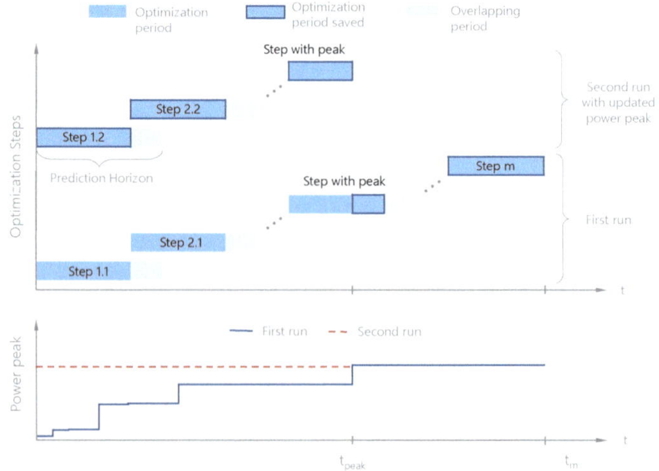

Figure 2. Schematic diagram explaining the used rolling optimization process.

The sequence of the used optimization model eFlame is illustrated schematically in Figure 3. After importing the input parameters and input data described below, the optimization problem is set up. The optimization problem is solved sequentially considering the charging strategies: uncontrolled charging (ref), unidirectional charging (uni), and bidirectional charging (bidi). The results are examined separately for each charging strategy.

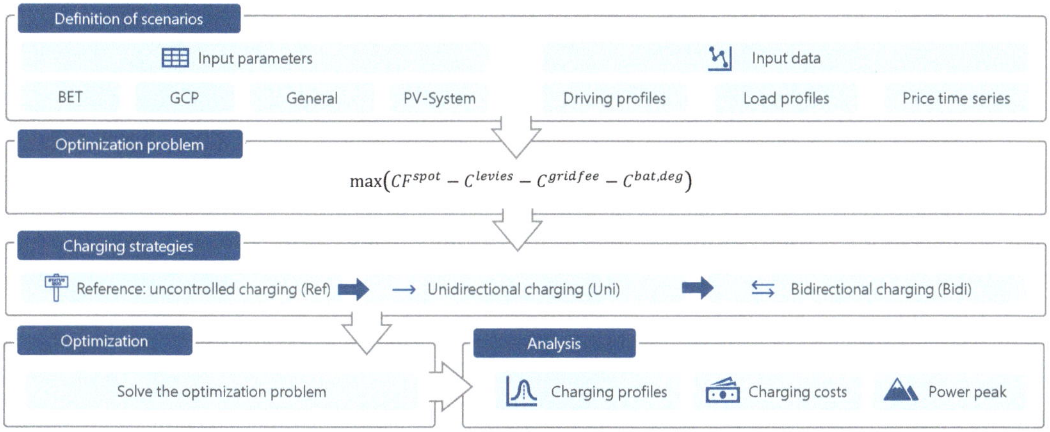

Figure 3. Schematic diagram explaining the used methodology.

2.2. Input Data

As mentioned in Section 1, prior research on the topic of BETs has relied on assumptions regarding driving profiles. In this paper, we had the opportunity to use real-life data from a depot of a freight forwarding company in Germany. The company primarily operates in the short-haul segment. The data were provided within the framework of the project NEFTON in which partners from industry and science jointly develop a Megawatt Charging System (MCS) for BETs. Mobility data of the company's trucks, historical load profiles of its buildings, and information about the PV system are included in the data. The selected depot can serve as a real-life example.

In the project NEFTON, driving data from several fleets of German fleet operators were recorded using high-resolution GPS data loggers. The recorded dataset includes 1.26 million km of driving data and is openly available in anonymized form in [29]. Only the driving data of the depot under consideration were extracted from this dataset. Since the data were recorded for trucks using diesel fuel, our investigation builds on the observation that the company desires to keep its services in the same way with electric trucks. The data are available for different lengths of time and were extended to uniform periods using a Markov process. To avoid oversizing the vehicle batteries, the missions in the dataset are divided into two clusters depending on the distance traveled. Missions with a distance of more than 200 km are grouped into the cluster regional transport and those with less than 200 km into the cluster local transport, which is similar to the classification of [30]. The annual driving profiles are taken as given and are presented in the following. Figure 4 shows the average percentage of vehicles in different locations for the two clusters. It can be seen that especially the mobility profiles from the Local Transport cluster have very high idle times at the depot and that at least 50% of the BETs are always present at the depot. On weekends and at night, most of the vehicles are located at the depot. The driving profiles of the cluster Regional Transport show significantly lower idle times at the depot. During daytime on weekdays, 80% of the vehicles are absent. On weekends, almost 40% are not at the depot. In addition, the driving profiles of the Regional Transport cluster show high parking durations in industrial areas and other locations. The difference between the two clusters is also evident from the characteristic values included in Table 1. The annual kilometrage of the Local Transport cluster is about 14,000 km. This is significantly lower than the kilometrage of the Regional Transport cluster of about 66,000 km. The electrical energy consumption for the driving profiles is determined using the model from [31]. The average annual consumption determined in this way is also included in Table 1. The variables $b_t^{EV,dep}$, E_t^{buffer}, b_t^{EV}, $E_t^{EV,trip}$ and $E_t^{EV,public}$ are determined based on the driving profiles and serve as inputs for the optimization model.

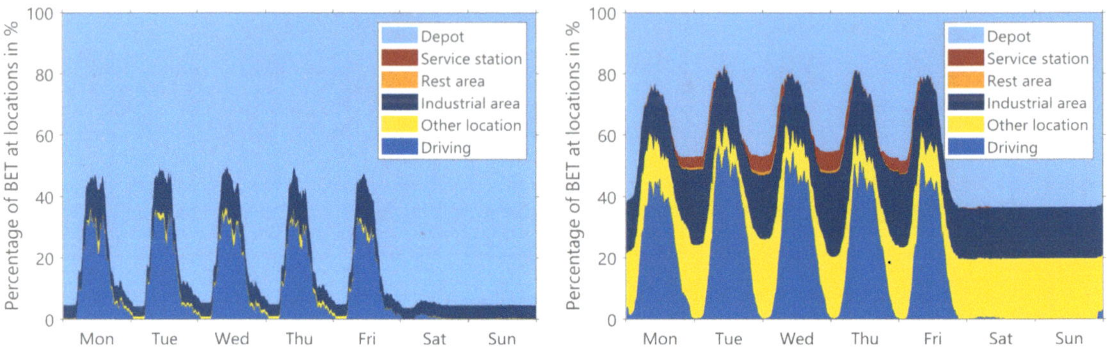

Figure 4. Layered percentage of BETs at different locations over the week. (**a**) Local Transport cluster, (**b**) Regional Transport cluster.

Table 1. Characteristics of the used driving profiles.

Characteristics	Local Transport	Regional Transport
Daily kilometrage (Weekdays/Weekends)	53.8 km/0.75 km	250 km/4.2 km
Percentage at depot (Weekdays/Weekends)	78.20%/95.19%	37.80%/63.40%
Annual kilometrage	14.382 km	65.750 km
Average consumption per km	1.1 kWh/km	1.26 kWh/km
Annual energy consumption	14.9 MWh	83.4 MWh

In addition to the driving profiles, the load profile of the building of the depot P_t^{build} is another important input for the optimization. The used load profile shown in Figure 5b for an average week relies on real data of the depot. From the annual time series, the average was determined for each quarter-hour of the week as well as the ranges in which 80% and 100% of the values lie. The plot shows that there are significant load peaks in the evening hours on weekdays, indicating suitability for peak shaving. The load is significantly lower at weekends and at night than it is during the day on weekdays.

We assume that the depot pays variable electricity prices based on the prices of the electricity exchange. Therefore, we used the intraday auction prices as electricity prices p_t^{in} and p_t^{out} for the optimization. In Europe, there are various short-term markets on the power exchange. One of those markets is the intraday auction. Due to the shorter time slices of quarter hours compared to the day-ahead market, in which hourly products are traded, this market offers higher price spreads. Thus, the revenue opportunities for flexibilities like bidirectional EVs are increased. The development of the prices of the intraday auction from the beginning of 2019 to the end of 2022 is shown in Figure 5a. As a consequence of the energy crisis, the price has risen from around 4 ct/kWh to a maximum of over 70 ct/kWh, and also the price spreads increased significantly.

The PV generation is determined as a time series depending on the historical irradiation data on CAMS level as a function of the orientation of the PV plant and its peak power [32]. The irradiation data are used for the location of the depot for the weather year 2012. The weather year is chosen based on the recommendation in [33].

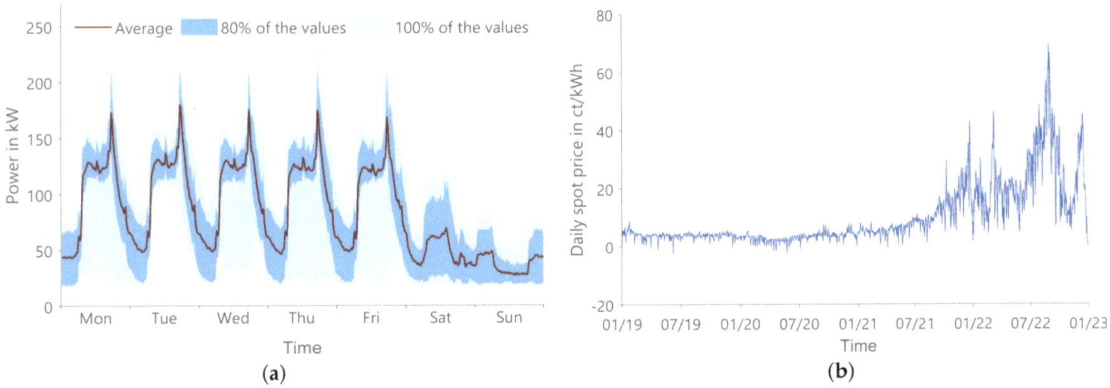

Figure 5. Visualization of input data: (**a**) electric load profile of the depot building, (**b**) daily average price intraday auction.

2.3. Input Parameters

After introducing the data source and the model in the previous sections, the input parameters are presented in the following. For this purpose, we define a base scenario for which the input parameters are listed in Table 2. By varying different parameters of this base scenario, various sensitivities are examined. For the sensitivity analysis, one parameter of the base scenario is changed, while the rest of the parameters are left unchanged. The varied parameters of the sensitivity analysis are also included in the table. The base year is 2021, and the optimization is performed at a time step size of 15 min. As Figure 2 illustrates, we use a rolling approach and divide the examined years into 61 optimization steps. The observation period of each step is seven days, consisting of the optimization period of six days and one day of overlap. In contrast to real-world charging management systems that apply forecasts, we assume perfect foresight for each optimization step. In the base scenario, no exemption of levies on energy fed back into the grid is assumed. Therefore, p^{levies} is set to be equal to $p^{levies,v2g}$. However, the exemption is considered in the sensitivity analysis. In the base scenario, no limitation of the grid connection capacity is considered. Thus, $P^{GCP,max}$ is set to the oversized value of 5 MW. A limitation of $P^{GCP,max}$ is examined in the sensitivity analysis. The grid connection capacity is minimally limited to 700 kW, since a lower capacity would result in the curtailment of the PV system in times of high irradiation. The feed-in tariff of 0.06 EUR/kWh is an assumed value suitable for Germany and is only used in the reference simulation as p_t^{out}. It is also assumed that 30 BETs of the depot are electrified. The number of electric vehicles is one of the sensitivities examined. According to the distribution from the dataset, 30% of the vehicles are used for regional traffic and 70% are used for local traffic. The appropriate driving profiles are divided among the BETs according to the distribution, and a battery capacity of 250 kWh for local and 500 kWh for regional traffic is assumed. Based on [34], the price of the vehicle battery $c_{bat,buy}$ is set to 139 EUR/kWh. The parameters of the PV system are selected according to the system of the real depot.

Table 2. Parameters of the base scenario and sensitivities.

Category	Parameter	Symbol	Unit	Value	Sensitivities
General	Year			2021	2019, 2020, 2022
	Time step size	t	h	0.25	
	Optimization period		h	168	
	Overlapping period		h	24	
GCP	Levies on V2G	$p^{levies,v2g}$	EUR/kWh	p^{levies}	0.02, 0
	Price for public charging	p^{public}	EUR/kWh	0.50	
	Max. grid connection capacity	$P^{GCP,max}$	MW	5	0.7, 1, 1.5, 2
	Feed-in remuneration PV (ref)		EUR/kWh	0.06	
BETs	Number of vehicles	n_{EV}		30	20, 40, 50
	Efficiency of charging	$\eta^{EV,charge}$		0.926	
	Efficiency of discharging	$\eta^{EV,discharge}$		0.921	
	Capacity of vehicle battery	$E^{EV,max}$	kWh	250/500	
	Minimum SOC at departure	$SOC^{EV,dep,min}$		1	
	Maximum charging/ discharging power	$P^{EV,max}$	kW	100	50, 200, 300
	Price of battery	$c_{bat,buy}$	EUR/kWh	139	
PV system	Peak power		kW	1000	0, 2000
	Azimuth angle		°	0	
	Tilt angle		°	35	

In addition to the year 2021 of the base scenario, the years 2019, 2020, and 2022 are also examined. For the optimization of the different years, several parameters have to be varied. In contrast, only one parameter is changed at a time in the sensitivity analysis presented below. The other parameters remain unchanged. In consequence, these year-dependent parameters are separated in Table 3. For the reference simulation, a constant price based on the average day-ahead price is assumed for p_t^{in} [35]. For the levies, the real historical values for Germany from [36] are used. The prices for the grid fees are also based on historical values of the grid operator Netze BW, where the depot under consideration is located [37]. We use the prices for medium voltage networks and consider an annual usage time of less than 2500 h.

Table 3. Year-dependent parameters.

Year	p_t^{ref} (EUR/kWh)	p^{levies} (EUR/kWh)	p_{usage} (EUR/kWh)	p_{cap} (EUR/kW)
2019	0.038	0.131	0.047	16.37
2020	0.030	0.135	0.052	18.36
2021	0.097	0.133	0.054	18.65
2022	0.245	0.495	0.056	19.20

3. Results and Discussion

In order to better understand the results presented in the following, we first look at a single example day. A sunny weekday in August from the base scenario in 2021 is chosen. Figure 6 is intended to explain the charging strategies and shows the important time series from the optimization results for the example day. The results for the reference with uncontrolled charging are shown on the left, and those for bidirectional charging are shown on the right. In the upper diagram, the power of the different components is plotted as a stacked area diagram. The resulting power at the grid connection point $P_t^{GCP} = P_t^{GCP,in} + P_t^{GCP,out}$ is shown as a black line. The center diagram illustrates for each time step how many vehicles are attendant and how many of them are charging or discharging. The given prizes are shown in the lower diagram. Levies and grid fees are not included in the prices.

With uncontrolled reference charging, the vehicles are charged immediately when they arrive at the depot. Even though some vehicles arrive and charge at midday, this leads to charging processes in the evening and at night where the power of the PV system is unavailable. The unused energy from the PV system is fed into the grid for the low feed-in tariff, and more expensive energy is purchased from the grid in the evening hours. The situation is different with the bidirectional charging strategy. According to the optimization problem presented in Section 2.1, the objective of the optimization is to maximize the revenue. One way to achieve this is to shift the charging process to times when PV power is available, since this power is not priced in the optimization problem. This shifting is clearly visible in the diagram because the area of the BET charging matches the PV generation. Energy can also be fed into the grid to maximize the revenue. Such a feed-in takes place on the example day from around 6 PM, when many vehicles are available and high energy prices are reached. Due to the oversized grid connection capacity of 5 MW, a large number of BETs discharge at the same time, resulting in a high feed-in power of over 2 MW. Because of the power price integrated in Equation (8), the annual power peak of the reference of 1.3 MW is lowered in the optimization to 0.4 MW. The power price only affects the purchased power, which allows the feed-in with a higher power. Figure 6 also clearly shows that outside the times with PV generation, the vehicles supply each other and also the building with energy.

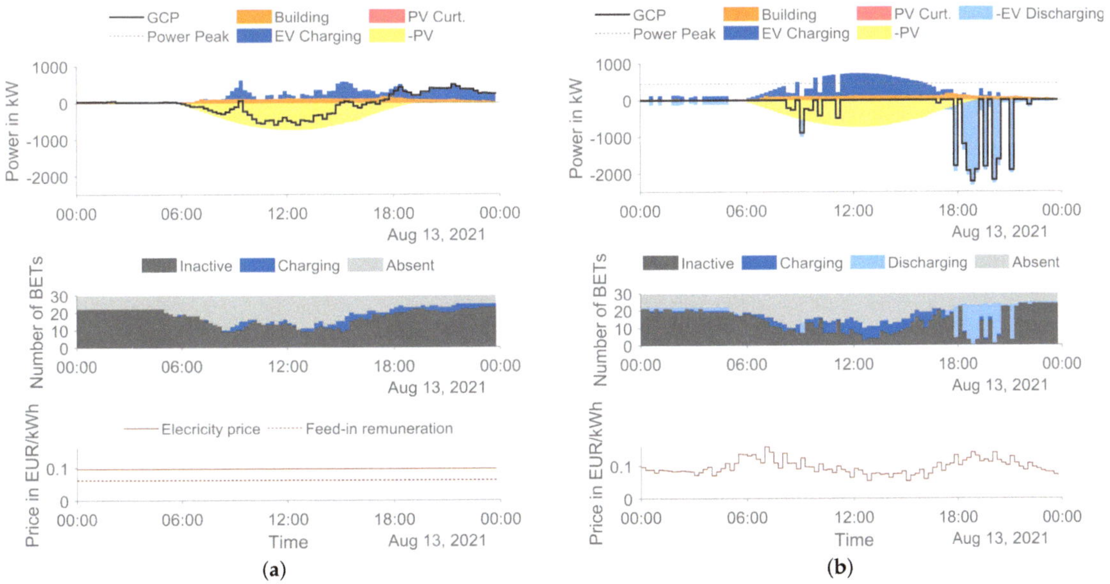

Figure 6. Results for an example day for different charging strategies.: (**a**) reference, (**b**) bidirectional.

The results of the base scenario are compared with those of the other examined years in Figure 7. Figure 7a shows the annual savings for the optimization with unidirectional (uni) and bidirectional (bidi) BETs. The savings are calculated from the difference between the costs in the reference simulation and the respective charging strategy and are normalized per vehicle. Before 2021, the savings are modest at about 2000 EUR/BET even with bidirectional vehicles. As energy prices rise from 2021 (cf. Figure 5), savings also increase significantly. Thus, almost 3300 EUR/BET can be achieved in 2021 with the bidirectional and 1500 EUR with the unidirectional charging strategy. In 2022, the savings skyrocket up to more than 10,000 EUR/BET. On the one hand, this can be explained by the fact that the reference costs in 2021 and 2022 rise due to the higher prices. On the other hand, the increasing price spreads and falling levies are responsible for the high savings, as this makes arbitrage trading significantly more attractive. In comparison, the authors of [13] estimate lower savings of 1515 EUR/BET. The deviation mainly results from the simplified price time series they use, which does not adequately reflect realistic price spreads of the spot market.

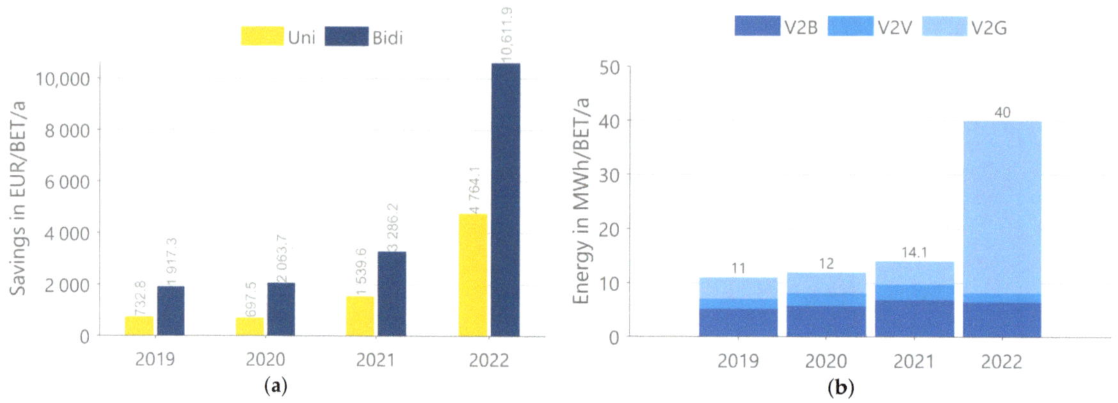

Figure 7. Results of the analyzed years: (**a**) annual savings, (**b**) discharged energy.

A corresponding observation can be made in Figure 7b, where the average discharged energy per BET and year is illustrated. It only represents the results from the bidirectional charging strategy, because only here can discharging occur. The diagram contains information on how much energy is fed back to the building (V2B), to other vehicles (V2V), or to the grid (V2G). Discharging into the grid takes place in order to generate revenues. V2G dominates the discharged energy in 2022. This explains the high revenues discussed above. Since the load of the building cannot flexibly respond to prices and PV generation, the BETs can, through discharging, supply the building with cheaper energy from the PV system or the grid in time steps with high electricity prices. Furthermore, V2B can serve to reduce the annual power peak. The same applies for V2V, where vehicles with high parking duration can supply frequently driving vehicles with cheap energy. V2V is thus another way to reduce charging cost. The share of V2B is relatively similar in all years and is slightly higher in 2021 and 2022 than in previous years. V2V takes the smallest share of the discharged energy in all years. In the reference scenario, the self-consumption rate is around 50% in all the examined years. The optimization increases this ratio to almost 65% with unidirectional BETs and 95% with bidirectional BETs.

To examine the influence of individual parameters on the results, the results of the sensitivity analysis for the bidirectional charging strategy are shown in Figure 8. In the sensitivity analysis, we varied various parameters that could impact the savings. Depending on the results, the values of these sensitivity parameters are selected iteratively. The diagram shows the percentage deviation of the sensitivity parameters from the parameters of the base scenario on the x-axis and the annual savings on the y-axis. The absolute values of the sensitivity parameters are provided in the last column of Table 2. The point in the diagram where the parameter variation is zero contains the savings of the base scenario already shown in Figure 7 (2021). The reduction of the grid connection capacity $P^{GCP,max}$ has the least impact on the savings. With the limitation of 700 kW (parameter variation = 86%), the grid connection capacity is still large enough and the savings decrease only minimally. A reduction of $P^{GCP,max}$ below 700 kW would reduce the savings more significantly, but then the PV system (1 MW peak) has to be curtailed. More points are calculated for this parameter to determine the boundary where no curtailment occurs. The savings decrease without a PV system but increase with a larger PV system. With higher charging and discharging power, savings can be increased. In the analysis, the charging power of 300 kW (parameter variation = 200%) leads to increased savings above 3700 EUR per BET. A larger number of vehicles reduces the savings per vehicle. The parameter with the strongest impact on the savings is the levies on the energy fed back. In the base scenario, we assume the worst case for V2G without any exemption from levies for energy fed back into the grid. Therefore, $p^{levies,v2g}$ is set equal to p^{levies}. With full exemption ($p^{levies,v2g} = 0$), the annual savings per BET strongly increase to 5300 EUR. However, these savings are only possible with a significantly higher, discharged energy from the BETs to the grid.

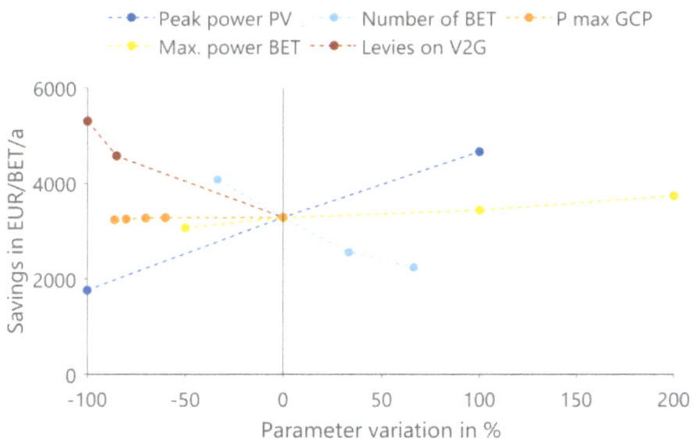

Figure 8. Results of the sensitivity analysis for the bidirectional charging strategy.

4. Conclusions and Outlook

This study presents a model to optimize charging processes in depots for EVs. In addition to the chargers, the depots may be equipped with a PV system and an inflexible load, e.g., from a building. The objective of the optimization is to minimize the charging cost. The cost reduction is achieved by increasing the self-consumption rate, reducing the annual peak load, shifting charging processes to time steps with low energy prices, and arbitrage trading. Through this combination of different use cases, a multi-use optimization is implemented. The optimization is implemented on a rolling basis. Despite a higher computing effort, even large depots with hundreds of vehicles can be optimized using this method. The economic benefits of V2G can be compromised by levies on purchased energy fed back into the grid. A full or partial exemption from levies for bidirectional EVs could solve this problem in the future. The implementation of this exemption is difficult, as it may only apply to the energy fed back into the grid. Energy consumed while driving must be taxed. If EVs feed into the grid and are charged by energy from the grid plus a PV system, then no exemption may apply to fed-in energy provided by the PV system. A partial exemption from levies is therefore implemented in the presented model. The amount of the exemption from levies can be chosen freely, and the approach even works with PV systems.

The model presented is used to optimize a depot for BETs. The study shows that the examined depot is very well suited for implementing bidirectional charging strategies. The operator of the depot can benefit monetarily from it. Due to the large PV system and the long duration of attendance of the BETs, the depot under consideration offers excellent conditions for optimization. In the base scenario, the bidirectional charging strategy can save 3300 EUR per vehicle and year compared to uncontrolled charging. A self-consumption rate of 95% can be achieved and the peak load can be significantly reduced. Arbitrage trading is only worthwhile when price spreads are high like in the examined years 2021 and 2022. Levies on fed-back energy impede arbitrage trading. According to the results of the sensitivity analysis, the exemption from levies can significantly increase savings. We examined the exemption from levies within a sensitivity analysis. At least a partial exemption from levies would be a precondition for the successful operation of V2G.

For further research, we propose a three-step strategy: Firstly, the model can be readily adapted to include additional use cases, e.g., providing frequency control. Secondly, instead of solely decreasing charging costs, it could be further developed to minimize the total cost of ownership (TCO) of the depot. Thirdly, the method can be applied to examine other depots by simply exchanging the database. The model is not limited to depots for BETs and can also be used to optimize depots for passenger cars or buses. This paper is therefore a basis for further research on the topic of bidirectional charging in depots for EVs.

Author Contributions: Conceptualization, F.B. and K.S.; methodology, F.B.; software, F.B.; validation, F.B. and K.S.; formal analysis, F.B.; investigation, F.B.; resources, F.B.; data curation, F.B.; writing—original draft preparation, F.B.; writing—review and editing, F.B. and K.S; visualization, F.B.; supervision, K.S.; project administration, F.B.; funding acquisition, F.B. All authors have read and agreed to the published version of the manuscript.

Funding: This research was conducted as part of the activities of the FfE in the project NEFTON funded by the Federal Ministry of Economics and Climate Action (BMWK) (funding code: 01MV21004E).

Institutional Review Board Statement: Not applicable.

Informed Consent Statement: Not applicable.

Data Availability Statement: The data used for the modeling are explained and cited in Section 2.2. The data of the considered freight forwarding company are confidential and cannot be shared publicly. The simulation model used is not available open access.

Acknowledgments: Special thanks go to Yannic Blume, Julius Engasser, Christian Peteranderl, and Maximilian Zähringer for their support during the interpretation of the results. We would also like to thank Georg Balke for providing and processing the mobility data.

Conflicts of Interest: The authors declare no conflicts of interest.

Abbreviations

The following abbreviations are used in this manuscript:

BET	battery-electric truck
bidi	bidirectional
CF	cash flow
EV	electric vehicle
GCP	Grid Connection Point
MCS	Megawatt Charging System
OEM	Original Equipment Manufacturer
PV	photovoltaic
ref	reference
SOC	State of Charge
TCO	total cost of ownership
uni	unidirectional
V2B	vehicle to building
V2G	vehicle to grid
V2V	vehicle to vehicle

References

1. Høj, J.; Juhl, L.; Lindegaard, S. V2G—An Economic Gamechanger in E-Mobility? *WEVJ* **2018**, 9, 35. [CrossRef]
2. Kern, T.; Kigle, S. Modeling and evaluating bidirectionally chargeable electric vehicles in the future European energy system. *Energy Rep.* **2022**, 8, 694–708. [CrossRef]
3. Nissan Approves First Bi-Directional Charger for Use With Nissan LEAF in the U.S. 2022. Available online: https://usa.nissannews.com/en-US/releases/nissan-approves-first-bi-directional-charger-for-use-with-nissan-leaf-in-the-us (accessed on 17 February 2024).
4. CO_2 Emissions from Heavy-Duty Vehicles Preliminary CO_2 Baseline (Q3–Q4 2019) Estimate. 2022. Available online: https://www.acea.auto/files/ACEA_preliminary_CO2_baseline_heavy-duty_vehicles.pdf (accessed on 17 February 2024).
5. New Commercial Vehicle Registrations: Vans +14.6%, Trucks +16.3%, Buses +19.4% in 2023. 2024. Available online: https://www.acea.auto/cv-registrations/new-commercial-vehicle-registrations-vans-14-6-trucks-16-3-buses-19-4-in-2023/ (accessed on 17 February 2024).

6. Earl, T.; Mathieu, L.; Cornelis, S.; Kenny, S.; Ambel, C.C.; Nix, J.C. Analysis of long haul battery electric trucks in EU Marketplace and technology, economic, environmental, and policy perspectives. In Proceedings of the 8th Commercial Vehicle Workshop, Graz, Austria, 17–18 May 2018.
7. Kern, T.; Bukhari, B. Peak Shaving—A cost-benefit analysis for different industries. In Proceedings of the 12. Internationale Energiewirtschaftstagung an der TU Wien, Wien, Austria, 8–10 September 2021.
8. Cunanan, C.; Tran, M.K.; Lee, Y.; Kwok, S.; Leung, V.; Fowler, M. A Review of Heavy-Duty Vehicle Powertrain Technologies: Diesel Engine Vehicles, Battery Electric Vehicles, and Hydrogen Fuel Cell Electric Vehicles. *Clean Technol.* **2021**, *3*, 474–489. [CrossRef]
9. Nykvist, B.; Olsson, O. The feasibility of heavy battery electric trucks. *Joule* **2021**, *5*, 901–913. [CrossRef]
10. Link, S.; Plötz, P. Technical Feasibility of Heavy-Duty Battery-Electric Trucks for Urban and Regional Delivery in Germany—A Real-World Case Study. *WEVJ* **2022**, *13*, 161. [CrossRef]
11. Liimatainen, H.; van Vliet, O.; Aplyn, D. The potential of electric trucks—An international commodity-level analysis. *Appl. Energy* **2019**, *236*, 804–814. [CrossRef]
12. Zähringer, M.; Wolff, S.; Schneider, J.; Balke, G.; Lienkamp, M. Time vs. Capacity—The Potential of Optimal Charging Stop Strategies for Battery Electric Trucks. *Energies* **2022**, *15*, 7137. [CrossRef]
13. Razi, R.; Hajar, K.; Hably, A.; Mehrasa, M.; Bacha, S.; Labonne, A. Assessment of predictive smart charging for electric trucks: A case study in fast private charging stations. In Proceedings of the 2022 IEEE International Conference on Electrical Sciences and Technologies in Maghreb (CISTEM), Tunis, Tunisia, 26–28 October 2022; pp. 1–6. [CrossRef]
14. Kern, T.; Dossow, P.; von Roon, S. Integrating Bidirectionally Chargeable Electric Vehicles into the Electricity Markets. *Energies* **2020**, *13*, 5812. [CrossRef]
15. Englberger, S.; Abo Gamra, K.; Tepe, B.; Schreiber, M.; Jossen, A.; Hesse, H. Electric vehicle multi-use: Optimizing multiple value streams using mobile storage systems in a vehicle-to-grid context. *Appl. Energy* **2021**, *304*, 117862. [CrossRef]
16. Roselli, C.; Sasso, M. Integration between electric vehicle charging and PV system to increase self-consumption of an office application. *Energy Convers. Manag.* **2016**, *130*, 130–140. [CrossRef]
17. Biedenbach, F.; Valerie, Z. Opportunity or Risk? Model-Based Optimization of Electric Vehicle Charging Costs for Different Types of Variable Tariffs and Regulatory Scenarios from a Consumer Perspective. In Proceedings of the CIRED Porto Workshop 2022: E-mobility and Power Distribution Systems, Porto, Portugal, 2–3 June 2022. [CrossRef]
18. Battke, B.; Schmidt, T.S. Cost-efficient demand-pull policies for multi-purpose technologies—The case of stationary electricity storage. *Appl. Energy* **2015**, *155*, 334–348. [CrossRef]
19. Parra, D.; Patel, M.K. The nature of combining energy storage applications for residential battery technology. *Appl. Energy* **2019**, *239*, 1343–1355. [CrossRef]
20. Jahic, A.; Eskander, M.; Schulz, D. Charging Schedule for Load Peak Minimization on Large-Scale Electric Bus Depots. *Appl. Sci.* **2019**, *9*, 1748. [CrossRef]
21. Duan, M.; Liao, F.; Qi, G.; Guan, W. Integrated optimization of electric bus scheduling and charging planning incorporating flexible charging and timetable shifting strategies. *Transp. Res. Part C Emerg. Technol.* **2023**, *152*, 104175. [CrossRef]
22. Verbrugge, B.; Rauf, A.M.; Rasool, H.; Abdel-Monem, M.; Geury, T.; El Baghdadi, M.; Hegazy, O. Real-Time Charging Scheduling and Optimization of Electric Buses in a Depot. *Energies* **2022**, *15*, 23. [CrossRef]
23. Houbbadi, A.; Trigui, R.; Pelissier, S.; Bouton, T.; Redondo-Iglesias, E. Multi-Objective Optimisation of the Management of Electric Bus Fleet Charging. In Proceedings of the 2017 IEEE Vehicle Power and Propulsion Conference (VPPC), Belfort, France, 11–14 December 2017; pp. 1–6. [CrossRef]
24. Raab, A.F.; Lauth, E.; Strunz, K.; Göhlich, D. Implementation Schemes for Electric Bus Fleets at Depots with Optimized Energy Procurements in Virtual Power Plant Operations. *World Electr. Veh. J.* **2019**, *10*, 5. [CrossRef]
25. Biedenbach, F.; Blume, Y. Size matters: Multi-use Optimization of a Depot for Battery Electric Heavy-Duty Trucks. In Proceedings of the EVS36, Sacramento, CA, USA, 11–14 June 2023.
26. Kern, T.; Dossow, P.; Morlock, E. Revenue opportunities by integrating combined vehicle-to-home and vehicle-to-grid applications in smart homes. *Appl. Energy* **2022**, *307*, 118187. [CrossRef]
27. Ploskas, N.; Samaras, N. *Linear Programming Using MATLAB*, 1st ed.; Springer International Publishing AG: Cham, Switzerland, 2017.
28. Preis, V.; Biedenbach, F. Assessing the incorporation of battery degradation in vehicle-to-grid optimization models. *Energy Inform.* **2023**, *6*, 33. [CrossRef]
29. Balke, G.; Adenaw, L. Heavy commercial vehicles' mobility: Dataset of trucks' anonymized recorded driving and operation (DT-CARGO). *Data in Brief.* **2023**, *48*, 109246. [CrossRef]
30. Borlaug, B.; Moniot, M.; Birky, A.; Alexander, M.; Muratori, M. Charging needs for electric semi-trailer trucks. *Renew. Sustain. Energy Transit.* **2022**, *2*, 100038. [CrossRef]
31. Sripad, S.; Viswanathan, V. Performance Metrics Required of Next-Generation Batteries to Make a Practical Electric Semi Truck. *ACS Energy Lett.* **2017**, *2*, 1669–1673. [CrossRef]
32. Schroedter-Homscheidt, M.; Hoyer-Klick, C.; Killius, N.; Lefèvre, M.; Wald, L.; Wey, E.; Saboret, L. *User's Guide to the CAMS Radiation Service—Status December 2017*; ECMWF: Reading, UK, 2017.

33. Guminski, A.; Fiedler, C.; Kigle, S.; Pellinger, C.; Dossow, P.; Ganz, K.; Jetter, F.; Kern, T.; Limmer, T.; Murmann, A.; et al. *eXtremOS Summary Report*; Forschungsstelle für Energiewirtschaft e. V.: Munich, Germany, 2021. [CrossRef]
34. Veronika Henze. Lithium-Ion Battery Pack Prices Rise for First Time to an Average of $151/kWh. 2022. Available online: https://about.bnef.com/blog/lithium-ion-battery-pack-prices-rise-for-first-time-to-an-average-of-151-kwh/ (accessed on 17 February 2024).
35. EPEX-Spot SE. *Market Data of EPEX-Spot SE*; EPEX-Spot SE: Paris, France, 2023.
36. BDEW. BDEW Electricity Price Analysis Beginning of 2023. 2023. Available online: https://www.bdew.de/service/daten-und-grafiken/bdew-strompreisanalyse/ (accessed on 17 February 2024).
37. Netze BW GmbH. *Prices for the Use of the Electricity Distribution Grid of Netze BW GmbH*; Netze BW GmbH: Stuttgart, Germany, 2023.

World Electric Vehicle Journal

MDPI

Article

BANULA—A Novel DLT-Based Approach for EV Charging with High Level of User Comfort and Role-Specific Data Transparency for All Parties Involved

Daniel Stetter [1,*], Tobias Höpfer [1], Marc Schmid [1], Ines Sturz [2], Simon Falkenberger [2] and Nadja Knoll [2]

[1] Fraunhofer-Institute for Industrial Engineering (IAO), Nobelstrasse 12, 70569 Stuttgart, Germany; tobias.hoepfer@iao.fraunhofer.de (T.H.); marc.schmid@iao.fraunhofer.de (M.S.)
[2] Institute of Human Factors and Technology Management (IAT), University of Stuttgart, Allmandring 35, 70569 Stuttgart, Germany; ines.sturz@iat.uni-stuttgart.de (I.S.); simon.falkenberger@iat.uni-stuttgart.de (S.F.); nadja.knoll@iat.uni-stuttgart.de (N.K.)
* Correspondence: daniel.stetter@iao.fraunhofer.de

Abstract: The core goal of the BANULA research project is to combine customer-oriented and grid-compatible charging of electric vehicles. It addresses the current challenges of the e-mobility ecosystem from the perspective of grid operators and charging infrastructure users and creates added value for every mass market role involved. In the project, the idea of a virtual balancing group based on blockchain technology is implemented. Thereby, it enables extended data acquisition, a real-time data exchange between grid and market participants, proper balancing and grid node-specific load flow determination and, thus, load management.

Keywords: charging; power supply infrastructure; mass market; data acquisition; load management

Citation: Stetter, D.; Höpfer, T.; Schmid, M.; Sturz, I.; Falkenberger, S.; Knoll, N. BANULA—A Novel DLT-Based Approach for EV Charging with High Level of User Comfort and Role-Specific Data Transparency for All Parties Involved. *World Electr. Veh. J.* **2024**, *15*, 79. https://doi.org/10.3390/wevj15030079

Academic Editors: Joeri Van Mierlo and Genevieve Cullen

Received: 15 December 2023
Revised: 30 January 2024
Accepted: 17 February 2024
Published: 21 February 2024

1. Introduction

One of the major challenges on the pathway to a high penetration of electric vehicles (EV) is the ramp-up of a widespread and reliably available charging infrastructure. Due to the need to significantly expand the number of charging stations, the construction of public charging stations has been strongly promoted in Germany and other European countries over the course of recent years. In this context, various publicly funded programs, such as "Publicly accessible charging infrastructure for electric vehicles in Germany" [1], have been established to create incentives for the erection of charging infrastructure. As a result, the number of charging points in Germany has steadily and significantly increased. Quantitatively, the number of public charging points has increased from 17,108 in 2018 to 77,191 in 2022. In 2022 alone, over 28,000 charging points were installed. The strong expansion of public charging infrastructure affects both AC charging as well as DC fast-charging stations. Some 20% of the expansion of charging stations in 2022 in Germany were DC fast-charging stations [2].

Despite this rapid growth in the number of public charging points, great efforts still have to be made to further increase the number of accessible charging points in order to meet an ever-increasing demand from the sharply rising number of electric vehicles in the German market. Estimates for the required number of public charging points range from 350,000 [3] to one million [4] in 2030. Similar developments with respect to the need for charging stations and growth of the latter, as well as the EV market, can be observed in other European countries, such as France and the Netherlands [5,6]. The US is also planning a significant expansion of charging infrastructure in the upcoming years [7] and, hence, has also allocated a significant amount of government funds.

In addition to a high number of charging points, easy and, most of all, reliable access to charging infrastructure for end customers is an indispensable requisite for the success of e-mobility as a whole. In order to use public charging stations, consumers currently conclude charging contracts with e-mobility providers (EMPs), thereby gaining access to all available charging stations within their direct charging network. Most providers offer roaming contracts so that their customers are capable of charging their electric vehicles using the infrastructure of other third-party providers [8]. Moreover, the EMP is responsible for pricing and billing of charging processes as a service towards end users as well as third party CPOs. These CPOs are responsible for operating their charging stations and able to exert influence on the operation, taking legal, economic and factual circumstances into account. Both market players, therefore, enable users to charge electric vehicles as a combined service.

However, there is significant potential to improve the current charging and billing processes between the market participants involved. Users often lack certainty about the exact roaming and billing conditions for charging stations of third-party providers in general and whether they can use their charging contract with a specific charging station at all. In addition, users are forced to conclude specific charging contracts—just like gasoline charge cards nowadays– and cannot use their household electricity contract. Moreover, the lack of usage of forecast data places a disproportionate burden on CPOs regarding the acquisition of the correct amount of energy for their charging stations—unless they are eligible for balancing using synthetic profiles. This, however, is not a feasible pattern for a full-scale market penetration of EVs for various reasons.

Finally, distribution grid operators (DSO) need full transparency in terms of location, power and energy load with regard to occurring charging processes in their grids, all of which are not available at the time being. This will dramatically gain importance, as millions of charging points are to be accommodated by power grids.

This paper proposes a solution to the issues raised. We present a block-chain approach, building upon distributed ledger technology and yielding solutions to the challenges of all players involved: costumers, charging point operators, distribution as well as transmission system operators, e-mobility providers and balancing group managers.

In June 2023, parts of this research were presented at the EVS36 Symposium in Sacramento, USA.

Past and Ongoing Research Activities

Within the BANULA project, the authors are particularly addressing the development and applicability of a novel and innovative e-mobility ecosystem from the perspective of a multitude of stakeholders. It also focuses on the applicability of communication and control of charging points of the current system in the BANULA ecosystem via blockchain. From today's perspective, the charging electricity is allocated on the balance sheet of the supplier of the charging infrastructure operator. In this model, the charging station of the charging infrastructure operator fills the role of the final consumer. However, completely removing the e-mobility provider from responsibility for the balancing and forecasting of charging processes is neither goal-driven for a correct balancing group management nor does it reflect reality. The first, as yet imperfect, approaches to a solution are offered by the German E-Mobility network usage contract [9]. In essence, this involves the allocation of the electricity quantities drawn from the grid to the balancing group of the respective e-mobility provider rather than to the balancing group of the charging infrastructure operator in accordance with MaBiS (Market rules for the execution of balancing group settlement electricity). In addition to the costs of the charging current, this also affects the network charges, in particular the provision of corresponding power, as well as the costs of construction, maintenance and upkeep of the actual infrastructure. Our research includes the creation of transparency for the availability of measurement and billing data in real time, the reduction of contractual complexity with roaming providers or in the context of roaming, the simple routing of costs for the construction and operation of the infrastructure and network charges, and the allocation of costs according to the source.

Many of these issues have been addressed in recent years, but the challenge of a comprehensive and user-friendly charging infrastructure has so far only been inadequately solved (Triebke et al. [10]). Kihm and Trommer [11] model the future market for electric vehicles as well as the associated substitution of conventional energy sources with respect to the use of electric vehicles. The authors have emphasized that both the charging infrastructure and a coherent regulatory framework for corporate customers are important elements for the diffusion of electromobility. However, the study merely considers financial aspects from the user's perspective and, thus, does not address either local or system-allocated effects of the diffusion of electric vehicles. In general, electric vehicles for load management are widely viewed positively. Lopez et al. [12] simulate the load shifting of individual electric vehicles and the resulting load-smoothing possibilities incentivized by minimizing the purchase costs for electricity. By doing so, they demonstrate the suitability of electric vehicles for load management. Babrowski et al. [13] analyze the load-shifting potential of electric vehicles and additional implications due to the availability of charging infrastructure at the workplace. The authors conclude that load-shifting potential exists and discuss the consequences of load management deployment of electric vehicles on electricity generation. However, specific system-wide impacts are not analyzed nor quantified in detail. Also, the applicability and potential for load shifting using decentralized technologies are neither considered nor compared to centralized technologies.

Other projects focus on a large variety of aspects with regard to charging station rollout (LamA—Laden am Arbeitsplatz [14]), charging pattern optimization (eFlotten- und Lademanagement [15], Shared E-Fleet [16], eMobility-Scout [17], ChargeLounge [18], InFlott—Integriertes Flottenladen [19]), inclusion of smart meter gateways into the charging IT landscape (LamA-connect [20]), various boundary conditions of charging (C/sells [21], SPARCS [22], IMEI—Erforschung integrierter Mobilitäts- und Energieinfrastrukturen [23], GeMo—Gemeinschaftliche Mobilität [24]) and charging infrastructure as a fundamental pillar of a smart grid (Charge@Work [25]).

In the referenced projects, the existing roles, involved parties and systems of electromobility have been used and the functions have been embedded within the framework of the current ecosystem. This existing ecosystem is to be expanded and combined with the ecosystems of the energy grids and markets. For this purpose, new roles, processes, responsibilities and systems are to be defined and developed, and a new approach—based on blockchain technology—is tested to carry out charging processes. The topics of balancing group management (both technical and legal aspects) and the consideration of flexibility have also not yet been integrated into electromobility ecosystems nor implemented in blockchain approaches so far. Finally, regulation has to be adapted, very much the way the German regulator has recently put thoughts into this process of "Netzzugangsregeln zur Ermöglichung einer ladevorgangscharfen bilanziellen Energiemengenzuordnung für Elektromobilität" [26]. All these aspects are taken into consideration within the work presented as follows.

2. Materials and Methods

One objective of the project is to implement the BANULA concept in the current energy and electromobility market. Therefore, it is necessary to adapt to the methods, processes, software systems and regulations in the German energy market. In 2020, the German federal network agency (Bundesnetzagentur), as the responsible regulation authority, passed a regulation—BK6-20-160 [27]—to improve and enhance access to the electrical grid.

For electromobility, as one of the biggest new use cases in the electrical market, the NZR-EMob [26] is part of this framework. It contains cornerstones for energy quantity balancing for specific charging processes and the associated and necessary gird access rules. These new grid access rules for electromobility enable an alternative settlement model for energy quantity allocation in the balance sheet compared to the approach used today.

The BDEW-German Association of Energy and Water Industries (Bundesverband der Energie- und Wasserwirtschaft) is the largest energy industry association in Germany and represents the interests of the electricity and energy sector. The BDEW published applications rules [28] for the implementation of the above-mentioned NZR-EMob in the

German energy sector. These application rules are a current vehicle for the BANULA ecosystem and will be laid out below. The BDEW has formulated a process description entitled "Model 2 for balancing energy quantity allocation options for specific charging processes". Within the BDEW application rule, Model 1 (Figure 1) describes the current balancing model and Model 2 (Figure 2) describes the options under the NZR-EMob.

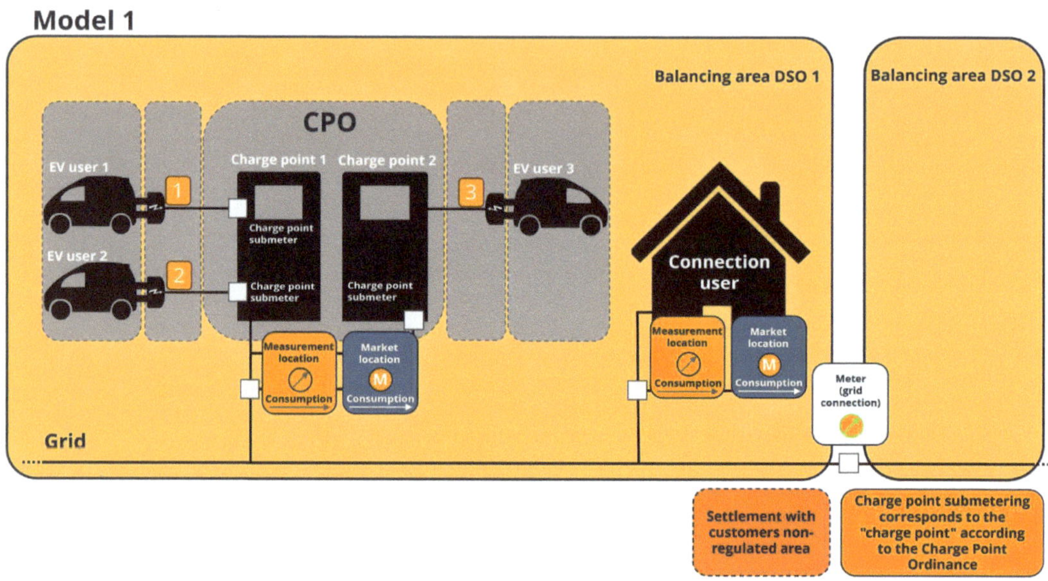

Figure 1. Model 1 is based on NZR-EMob and within the BDEW application rule, Model 1 describes the current balancing model in the actual energy market where the charge point is handled as a usual consumer. Two charging processes (1, 2) are carried out at charging station 1 and one charging process at charging station 2 (3). Each charging process is measured by the charge point register and the complete energy is measured by the measurement location. The energy of each charging process is balanced by the charge point owner via standard load profile [28].

As a basis for the description of Model 2, the BDEW used the already established market communication processes for business processes for the supply of electricity to customers (GPKE) [29], switching processes in electricity metering (WiM Strom) [30] and the market rules for the implementation of balancing group billing for electricity (MaBiS) [31], which were adapted in relation to electromobility. Only a few new market communication processes were introduced for Model 2. It also only describes the processes of the energy industry, not specifically related to the electromobility sector.

While the energy quantity of the market location is balanced in the already established settlement model, Model 2 provides for a charging process-specific energy quantity allocation on the balance sheet. In Model 2, the energy quantity of the market location is no longer balanced, but treated like a grid time series in terms of balancing and a charging process-specific balancing energy quantity allocation takes place in the balancing area of the charging point operator. This means that the charging point operator must ensure a balancing group allocation for each charging process in its balancing area. The distribution system operator no longer has balancing responsibility for the market location registered in Model 2. The charging point operator is, therefore, obliged to register a balancing area in the corresponding control area with the balancing coordinator. The balancing area of the charge point operator is not limited to the grid area of a distribution grid operator. Balance discrepancies (delta quantities) must be borne by the balancing group manager of the balancing area and market locations that are to be settled in accordance with Model 2 must be reported to the respective distribution system operator.

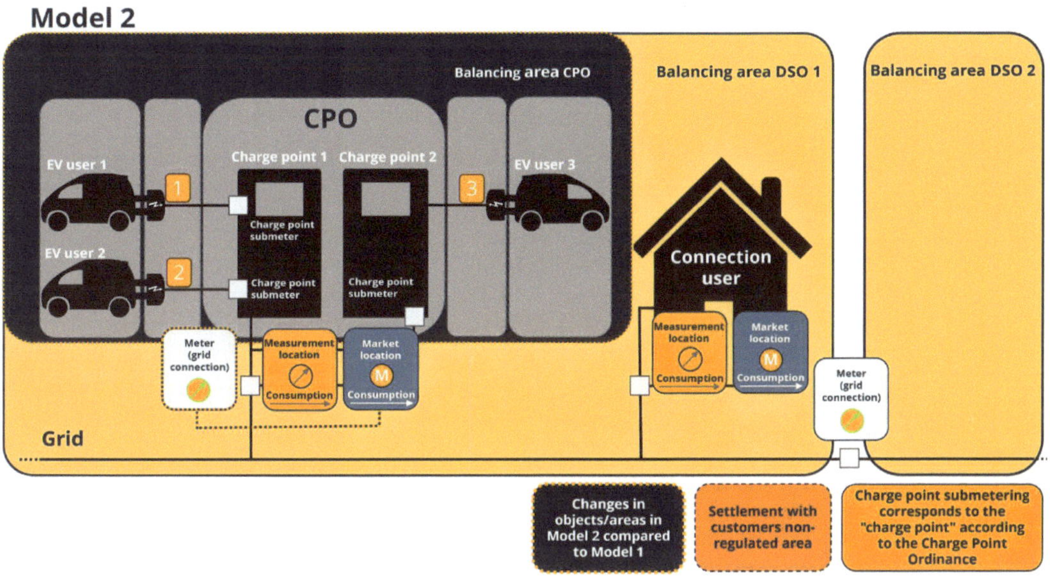

Figure 2. Model 2 describes the options under the NZR-EMob, which are balancing energy quantity allocation options for specific charging processes. The charge points are removed from the balancing area of the DSO 1 and added to the balancing area of the CPO. Two charging processes (1, 2) are carried out at charging station 1 and one charging process at charging station 2 (3). Each charging process is measured by the charge point register and the complete energy is measured by the measurement location. The energy of each charging process is balanced by the charge point operator, based on the measurements of the new meter [28].

General rules for both models:

- A market location can only be assigned to one model at a time, either Model 1 or Model 2.
- The following applies to the commissioning of a market location with the consumption type "e-mobility charging station".
- The nonstandardized "new installation" process is carried out as for any other market location according to the principles of the respective distribution system operator.
- In the new installation process, the market location is initially assigned to Model 1 by the DSO.
- To participate in accordance with Model 2, the charging point operator must apply for and use the IDs relevant to the corresponding role (market partner ID and electricity grid operator number) from BDEW.
- The ID of the market location and the ID of the metering location remain unchanged regardless of the model assignment.

Additional regulations for Model 2

- Balancing of the energy quantity of the market location in the balancing area of the distribution system operator does not take place. Instead, the energy quantity of the market location is balanced via a grid time series between the balancing areas of the distribution system operator and the charging point operator.
- The distribution grid operator is the grid operator responsible for the grid time series. The charging point operator is the neighbouring grid operator.

- The energy quantities of the charging processes in the balancing area of the charging point operator are balanced.
- The charging point operator has the aggregation responsibility for the energy quantity of the charging processes in the charging point operator's balancing area.
- The energy quantity of a charging process can only be allocated to one charging point operator's balancing area.
- The change of supplier within Model 2 is currently not procedurally structured and must be carried out bilaterally.

As the BANULA concept deals with personal data between multiple parties that may not trust each other, a distributed and trustworthy technology is to be used for data access and proof of validity: Distributed ledger technology (DLT). DLT is a concept that aims to store and manage data in a decentralized manner. Unlike traditional centralized systems, where a central authority or institution has utter control over the data, DLT enables distributed storage and processing of information across a network of participants. This is achieved through the use of cryptography and consensus-based mechanisms that allow participants to agree on a common data consensus without the need for a central authority. As a result, DLT offers a high level of transparency, security and resistance to failure and tampering.

A well-known example of DLT is blockchain technology, which was first introduced in Satoshi Nakamoto's white paper "Bitcoin: A Peer-to-Peer Electronic Cash System" [32]. In addition to the public blockchain, there is also the so-called permissioned blockchain, in which network participants must be approved in advance. This ensures a certain degree of privacy and control over the network. Permissioned blockchains are used in sectors such as financial services and corporate environments where special compliance requirements and data protection regulations apply.

3. BANULA's Concept

The fundamental concept of the BANULA ecosystem is a holistic approach to combine energy economic processes and energy balancing, on the one hand, with the commercial processes in the e-mobility framework on the other hand to make them both more efficient in favor of the end customers and the electricity market roles. BANULA provides correct accounting between all parties involved as they have to implement a new common communication network.

Within this new ecosystem, charging point operators provide their infrastructure to e-mobility service providers and do not need to procure the charging electricity. This may sound arbitrary at first; however, the consequences are profound. The e-mobility providers have to procure the necessary charging electricity for their own customers while at the same time benefiting the system as a whole, they are able to create much better energy procurement forecasts than charging station providers. This decreases the overall energy grid imbalances as the correct amount of energy can be purchased. Distribution system operators can use this approach to gain full transparency of the charging loads within their grid, reduce imbalances within balancing groups and improve overall energy grid stability. Moreover, balancing errors are no longer to be covered by grid operators. In short, errors that are the responsibility of the e-mobility service providers are to be met physically or financially, the latter imposing a strong incentive for correct balancing.

Operationally, in order to meet current regulations, within BANULA, a virtual grid area is implemented in which all charging points relevant operated through an EMP are aggregated by the charging processes of its customers.

BANULA acts as the operator of this virtual grid area and is in direct exchange with the adjacent physical distribution system operators in order to coordinate grid operator processes directly with each other. The management of the charging energy (in terms of energy balancing groups) is carried out by any number of EMPs and not by a single supplier who supplies the physical grid connection point to which the charging infrastructure is connected (An example from the concrete application would be a charging process at a

LamA charging point on the campus of a Fraunhofer institute by any EMP. This EMP now manages this charging point in the virtual grid for the charging time of its customer, even if the charging point is located on the Fraunhofer campus). By enabling a mechanism to decouple the supplier of the charging station and the supplier of a specific charging process, customers can choose a specific EMP for a charging process, completely independent of the CPO of the charging infrastructure and EVSE they want to use. Thus, an EMP always balances the charging current withdrawals of its customers based on the authentication at the charging station (e.g., by means of RFID) in terms of a different accounting allocation. Against the background of this approach, it becomes necessary for the EMP to ensure correct balancing group management—with regard to the charging processes of its customers. This is a major contribution to ensuring system stability while further increasing the charging capacities of electric vehicles through a proper allocation of balancing responsibility. As stated earlier, by shifting the balancing responsibility to the EMP, the costs of balancing errors are reduced for the grid operator. In this way, costs and risks can be allocated appropriately. The EMP's designated balancing group is assigned the withdrawals of a large number of charging points depending on the usage behaviour of its customers.

To implement BANULA's novel ecosystem, blockchain technology is used to provide a data architecture that all participants in the ecosystem can use and build upon (For more properties that distinguish a blockchain, see Section 4). This corresponds to a back-end system of market communication in order to be able to allocate charging energy quantities to the supplier or suppliers of charging electricity within the 15 min period relevant for balancing. Blockchain technology, as a decentralized medium, manages and regulates the interaction of the different parties involved. It enables a timely, accurate, tamper-proof and transparent allocation of the

- charged energy quantities per charging pole,
- customers to the balancing groups,
- balancing areas,
- duration of use as well as data necessary for the billing of the grid usage.

It also offers the opportunity to integrate information about the network status into the charging management of the EMPs.

For grid operations, it offers opportunities to balance the provision of flexibility on a plant-by-plant basis, to assign these to corresponding market roles, and to assign the intended use of flexibility usage. The coupling of the grid (transmission system operator, distribution system operator) and the market (electric mobility provider, balancing group coordinator), thus, provides a data and information interface to communicate grid events and restrictions directly to the market in accordance with German regulation (§ 13(2) EnWG). The overall system with the interfaces and information to be exchanged is shown in Figure 3.

As far as the issue regarding customers of an arbitrary EMP charging at any charging infrastructure of a given CPO is concerned, in the novel BANULA ecosystem, a charging process works as follows:

1. A given client establishes a contract with fixed terms (i.e., cost per kWh) with an arbitrary EMP. The client's authentication dataset is assigned to its respective EMP in a decentralized DLT/blockchain network. The client is now capable and eligible to use any charging point, which is part of the decentralized virtual grid area.
2. In order to start a charging process, the client carries out the authentication process either via presenting an RFID, registering in a mobile app or through "plug and charge" building upon an implementation of the ISO 15118-1:2019 standard [33]. Through the DLT network, the client's authentication and eligibility to be granted access to the charging point in question are verified. If all criteria are met, the charging process is enabled, and the respective kWh are assigned to the EMP's energy economic balancing sheet.

3. The charging process starts, in parallel, the BANULA-DLT network aggregates all relevant data, e.g., typical charge detail records for other participants in the ecosystem, such as charged energy, time stamps, etc. The respective EMP as well as the distribution system operator (DSO) in whose grid area the event takes place are both provided with the time series data of the charging process. The CPO whose charging point was used is provided with the data needed in order to bill the use of the charging infrastructure, i.e., the contribution margin to its fixed costs.
4. Following this process, the DSO has complete knowledge of all charging processes within its grid. Moreover, this information is available for each charging point in real time, which in turn enables the DSO to gain a better understanding of load flows in the grid.
5. In order for this concept to work, all charging points are to be balanced in a so-called virtual grid, although physically, they are clearly and obviously part of a DSO's real distribution grid.
6. In complete analogy to existing processes in Germany's energy economic regulatory framework, the virtual system operator establishes a summarized load time series on a monthly basis for each EMP in a temporal resolution of 15 min. This time series data are used for an exact ex post balancing of the charged kWh for each EMP. Where applicable, upstream and downstream grid operators can tap the same kind of summarized load time series data for their balancing purposes. If applicable, EMPs are subject to financial punishment for any physical deviations compared to their balancing sheets.
7. Because real-time charging information is available, grid congestion can be determined in time and further developments using i.e., artificial intelligence farther down the road will allow for predictive grid management e.g., by establishing incentives to charge at different times or locations.

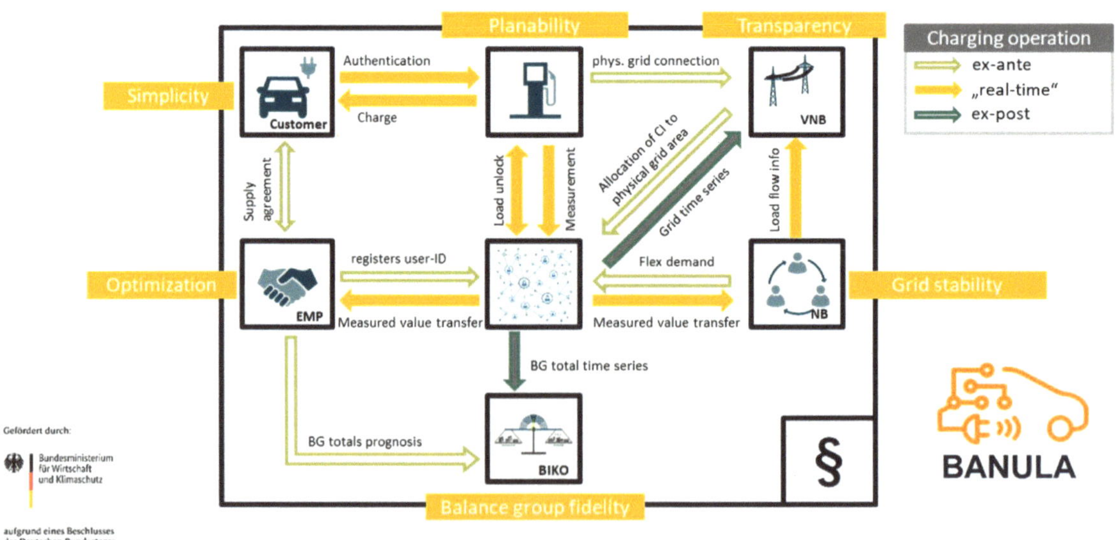

Figure 3. Brief overview of the EV charging ecosystem and the operational steps carried out during charging processes. Within the context of the BANULA ecosystem, these processes are significantly enhanced using a blockchain-based approach employing distributed ledger technology for the benefit of all participants of the system.

Figure A1 displays the proposed charging progress in the new ecosystem as a swimlane process diagram (Business Process Model and Notation 2.0/BPMN 2.0, https://www.

omg.org/spec/BPMN/, accessed on 23 March 2023). The Figure A1 shows the market participants involved and their interactions during the charging process. As indicated, data are exchanged between market participants of the energy market and the electromobility ecosystem during the charging process.

3.1. *Respective Perspectives of Each Market Participant*

In the current energy market and electromobility ecosystem, each market participant has its own defined role with its respective tasks, advantages and disadvantages—for its role and the entire system. The roles in the current ecosystem are defined for the time being and the companies fill these roles as their business cases. As the BANULA ecosystem aims to reduce the disadvantages of the system, it must also maintain and improve the advantages. Based on the respective role of the market participant, different perspectives need to be considered to leverage the new ecosystem in the current market.

3.1.1. End Customer

Currently, customers are free to choose any EMP, but they cannot charge with certainty after finding a charging point at a previously agreed-upon price; even at the same charging point, the costs of a comparable charging process can differ significantly depending on the EMP. In practice, users would have to check before each charging process whether they want to accept the price or continue the quest for the next charging point. If the user has explicitly concluded a contract with an EMP that includes, for example, 100% green power for charging, the guarantee of this power quality cannot, under the current regime, be mapped independently in the roaming case. In addition, it is currently not possible for customers to reliably use all available charging stations with just one charging contract.

Thus, the addressed needs for action from the customer's perspective are:

- Removal of access barriers and creation of price transparency;
- Sourcing of advertised and purchased "quality" charging power, e.g., regional, green, etc.;
- Reliable access to all available charging stations with just one charging contract,
- Usage of their own energy provider at every charging station: charge your own PV electricity—even on the road.

3.1.2. E-Mobility Provider

EMPs that enable their customers to charge on the basis of a peer-to-peer contract with a CPO or in the context of roaming have been insufficiently involved to date in the correct balancing of charging processes. Particularly in the case of roaming, there is no need for the EMP that enables its customer to charge at charging infrastructure, to make an accurate forecast under the current regime. Suppliers who provide electricity to charging stations of CPOs balance for annual withdrawals of up to 100,000 kWh using synthetic load profiles (SLP). However, proper SLPs that sufficiently consider the various use cases of the charging infrastructure and, in particular, the frequently spontaneous charging do not yet exist; even if they did, they would very likely reflect reality poorly.

Thus, the addressed needs for action from the EMP's perspective are:

- Enable access to any charging infrastructure under transparent and simple conditions;
- Establish a system that solves the access deficits of today's roaming;
- Introduce a central energy balancing group forecast of all customers across Germany or a control zone by the EMP (establishment of reliable forecasts of charging energy to be procured).

3.1.3. Charge Point Operator

The charging energy is currently assigned to the supplier of the CPO in the balance sheet. In this picture, the charging station of the CPO fills the role of the final consumer. However, completely removing the EMP from responsibility for the balancing and forecasting of charging processes, as has been the case to date, is not expedient and is the subject of heated debate within the industry. The first, as yet imperfect, approach to a solution is offered by the E-Mob network usage contract [9]. In essence, this involves the allocation of the electricity quantities drawn from the grid on a balancing group basis according to the MaBiSF, not to the balancing group of the CPO, but to balancing groups designated by the respective EMP. However, neither essential details have been specified yet, nor are technical solutions available. In addition to the costs of the charging current, this also affects the network charges, in particular the provision of corresponding power, as well as the costs of the construction, maintenance and repair of the actual infrastructure.

Thus, the addressed needs for action from the CPO's perspective are:

- Create transparency for the availability of measurement and billing data in real time;
- Reduce contract complexity with roaming providers;
- Wire through simply the costs for infrastructure construction, operation as well as network charges;
- Allocate costs according to the originator;
- Procuring and balancing of charging electricity in line with the polluter-pays principle.

3.1.4. Distribution System Operator

DSOs bear a significant balancing group deviation risk of their network groups due to the current balance sheet mapping of charging processes and have to expect high consumption peaks in distribution network strands that are currently merely inadequately measured. Furthermore, distribution grid operators do not know the charging load at certain grid points and install sensor technology to operate the grid safely.

Thus, addressed needs for action from the DSO's perspective are:

- Exploiting synergies and creating transparency: what happens where in the grid, in real time (so that grid stability measures can be initiated to minimize balancing group deviations and the necessary risk capital);
- Form appropriate aggregation points that can be forecasted and managed;
- Create incentives for EMPs to make predictable withdrawals and avoid power peaks.

3.1.5. Transmission System Operator

If the existing system is continued, the TSOs will also be increasingly exposed to uncontrolled and hard-to-predict use of balancing energy in the physical balancing of their networks. This would result as a direct consequence of schedule deviations in the downstream distribution networks.

Thus, addressed needs for action from the TSO's perspective are:

- Increase in balancing group reliability;
- Support system security through the systemic use of flexibility by managing the load in the distribution network and its IT-based proof of delivery.

4. Implementational Questions

The BANULA project initiative develops a blockchain-based data platform that enables a tamper-proofed and German-regulated billing of charging processes for all market participants. The overall purpose is to establish a new ecosystem, which will be of benefit to all players involved. The question as to who will eventually operate the system is yet to be resolved.

The main objective of the project is to make public charging points accessible to all end consumers in the most transparent terms and to best prepare all parties involved for the mass market penetration of electric cars. For this purpose, the ecosystem proposed aggregates charging points of a charging station operator in a specific grid area into a virtual charging point network. It integrates all involved market roles and enables trustworthy data exchange. Blockchain technology manages and regulates the interaction of the various players. In addition, EMPs, CPOs, DSOs and TSOs gain full transparency as to occurring charging events. This provides an accurate, tamper-proof and transparent allocation of the charged energy per charging point, per customer and balancing group, per usage period as well as the data required for the grid fees. Within our contribution to EVS 37, we would like to further discuss this approach with specialists from around the world. Within the project team, we raise the following five questions during the project duration and create guidelines for the technical implementation.

4.1. Reasoning for Blockchain-Based Approach

As an underlying communication platform, the blockchain as a distributed ledger offers security, full transparency and auditable traceability over all interactions of the participants. In addition, the blockchain is not operated by a single party, but rather operates in the form of a distributed network that "belongs" equally to all stakeholders and to rules that all stakeholders have jointly defined (governance model). New participants (market parties) can join at any time, but they can only use the system if they submit to the common set of rules (based on the rights defined by the stakeholders, e.g., by means of previously defined rights for the individual roles). By introducing digital market roles (identities), blockchain technology can be used to include market parties in their role deposited by an authority in an automated as well as standardized manner via an "authority" model. Based on the best fitting governance model for the ecosystem—which is also part of our research—a blockchain-based approach delivers the technological aspects for each role to interact with each other.

4.2. Is a Blockchain Performing Well Enough to Deliver Real-Time Results Even with a High Number of Participants?

Depending on the use case, different blockchain technologies and concepts are available. Permissioned Blockchains are only accessible to consortia and also offer transaction times for high-performance requirements that enable almost real-time processing (a few milliseconds) compared to previous matching mechanisms (approx. 15 min). However, depending on the project requirements, the use of public blockchains may also be appropriate, e.g., to ensure easy access by the public (the end users). Deciding which blockchain concepts (or combinations of them) are suitable for operational use is also part of our research. As for now, the current plan is to implement a blockchain-based approach where the blockchain itself only holds a limited set of data but offers an up-to-date lookup table for each role of the system. The blockchain—let it be called BANULA Data Hub (DHB) for now—knows all the application programming interfaces (API) for each party in the ecosystem and also the necessary rights to interact with this party, as indicated in Figure 4 below.

4.3. Can All Data in the Blockchain Be Viewed by All Actors and How Do We Ensure Data Protection and Privacy?

Depending on the blockchain technology used, there are various options for protecting sensitive data. Following the governance model mentioned above and the associated different roles, access to the data available in the BDH network and its processing options can be comprehensively regulated. For example, in Hyperledger Fabric, it is possible to separate different parts of the distributed ledger network for different use cases. In addition, encryption can be used to secure the transmission of data within these subareas. The project also investigates the possibilities for controlling access to the protection and its suitability for the different use cases. In any case, it is important that market players or all participants are only allowed to see the data in plain text for which they have authorization (e.g., EMP A only sees the measurement time series of its assigned customers and EMP B does not see these data in plain text, but only as an encrypted value for consensus building, which cannot be deciphered by EMP B). As shown in Figure 4, the blockchain works as a gatekeeper and permission management system with suitable smart contracts programmed by the government organization in the ecosystem.

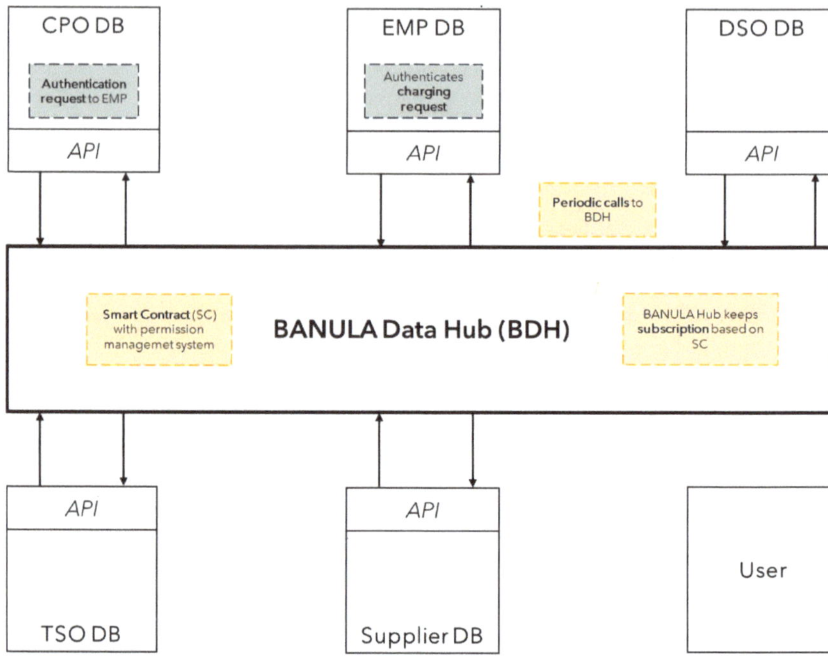

Figure 4. A schematic description of the blockchain-based approach for a DLT-based platform as lookup table and gatekeeper.

4.4. What Is the Strategy for Existing Hard- and Software Systems?

The project will develop an integration concept for existing charging points as well as new charging hardware to be set up by the Fraunhofer charging network "LamA" and the Lidl/Schwarz Group as a blueprint for the scalability of the project architecture solutions. In total, the participating project partners can so far integrate up to 10,600 charging points into the BANULA network. The solution explicitly aims to make a further inventory (outside the partners' charging networks) integrable. Overall, there are two starting points here.

Top-down: On the one hand, regulatory requirements are needed for later implementation, e.g., on the part of the regulation authorities of the German federal network agency (Bundesnetzagentur) with regard to virtual network areas concerning the network connection of charging stations. Discussions are planned here on the part of the transmission system operators. Furthermore, the topic is to be discussed within the framework of the European TSO-DSO cooperation. In the next few years, the network code flexibility is to be developed at the European level. Basic elements of our question can be directly incorporated here.

Bottom-up: Parking operators and large employers can already participate in the network today in order to achieve fair and equitable treatment of the workforce with regard to mobility subsidies (no free fuel for e-drivers) and the charging of guests.

4.5. Transfer Strategy for Europe

A decentralized solution (like DLT) also offers more flexibility than a centralized platform for onboarding additional regions (or environments). A DLT network is not subject to the sovereignty of a single provider and, due to its decentralized orientation, by definition, offers more openness for the onboarding of new stakeholders. Due to the (within the framework of the project initially Germany-wide) implementation on a national level, overarching solutions for simple loading in the virtual grid area must already be developed within the project for the four German TSO control areas. These processes are scalable across Europe, since the European TSOs are galvanically as well as information-technically coupled and the concept of the balancing group management is analogously structured. The model requires an expansion of the German (and potentially also the European) regulatory framework by extending and thus improving the existing balancing system for charging processes. The virtual grid approach is easy to integrate into the existing balancing system, as it is based upon it—comparable to the, e.g., traction current system or balancing between two grid operators. However, it is also easily transferable to all other EU countries, as they have a balancing system that is comparable to Germany's in the main points. In this respect, the EU makes hardly any particular specifications. A European rollout is, therefore, possible. Specifically, a coupling or interoperability of the platforms of the company Elia (EnergyBlocks) and BANULA is already being considered. Elia is very interested in the results of the BANULA project since both concepts are based on a virtual grid area. The company Elektromaps from Spain, which provides information about charging points throughout Europe, has expressed interest in participating in the project.

In a further step, BANULA will be presented by TransnetBW and 50Hertz in entso-e in the context of the established TSO-DSO cooperation. Currently, the managing director of TransnetBW is chairman of the working group for coordinated cooperation with the new European DSO association "EUDE". This provides the best conditions for coordinating processes between the two associations throughout Europe with regard to a virtual grid area. Against this backdrop, on the one hand, we carry the concepts to Europe, and on the other hand, we also want to actively accompany European developments (e.g., FlexHub, Equigy) in order to derive possible synergies for BANULA.

In summary, the technology is thus transferable to other countries. On the operator side, the Schwarz Group is a project partner that operates many hundreds of charging points not only in Germany, but across many other European countries. It is interested in a solution for all of its charging points, so it also has a great intrinsic interest in developing an international solution.

5. Results and Practical Applications

To prove and validate the project's underlying concepts, a pilot of the entire system and all players involved is implemented. For each role mentioned above, there is at least one organization necessary to adapt and implement the idea to make the ecosystem applicable in the current German energy market. Therefore, the project is composed of a multitude of entities in order to be capable of proving the concept end-to-end. As the approach of the system shall eventually be rolled out in the real world's energy market, the pilot is applied to the infrastructure of two large German charging networks and not only in a laboratory setup. For this reason, the new ecosystem is based on the current German and European regulations of the energy and e-mobility market. In the German energy market, a voluntary regulation system for CPOs to create a transaction-based energy balancing group was implemented in 2020 (BK6-20-160 [27]); however, it is not widely used by CPOs in Germany due to the lack of adequate incentives. For the pilot project, this regulation system is analyzed, the flaws are exposed and practicable improvements are made. Hence, the pilot project focuses especially on improvements, which create benefits for the CPOs to participate in this voluntary regulation system but also integrate into the current technical market solutions. Furthermore, the new ecosystem shall include the application of bidirectional charging and decrease the administrative burden as charge points—in Germany—so far are not considered as energy market locations but as energy metering locations.

To fulfil the ecosystem's transparency objectives, the pilot project evaluates the usage of distributed ledger technologies and implements the best fitting solution into the pilot implementation. Therefore, the pilot will be split into multiple implementation and testing stages. In the first stage a prototype—the minimal viable product (MVP)—is implemented and tested. Therefore, the technical feasibility of the project is shown by implementing the necessary roles for a limited set of use cases and charging stations. For this approach only the least features of the roadmap are implemented and tested in a friendly environment with known drivers, CPOs, EMPs, grid operators and—if necessary—shadow balancing. Thus, we can implement and test the MVP, even if the current regulations are not fully satisfied. The roadshow across Germany—from Freiburg to Berlin—shall demonstrate the crossregional approach across the four balancing zones operated by the German TSOs.

The use cases to be demonstrated in this roadshow—shown in Figure 5—cover the most important roles for the combination of the e-mobility ecosystem and the energy market ecosystem. The goal of the roadshow is to enable system- and grid-compatible charging at any charging point. In the roadshow, two electric vehicles with charging cards from two different EMPs will start the journey from Freiburg to Berlin. They will charge simultaneously, also at charging points of one and the same charging station, but authenticated through the different EMPs. The charging quantities will be allocated to the respective balancing group of the respective EMP and will appear in the correct balancing group bookings. The supplier of the physical grid connection point (to which the charging infrastructure is connected) will not change—however, the two selected EMPs will deliver the energy for the charging processes. After the MVP is realized and tested successfully, the next stage of the project with more users, charging stations and use cases will be implemented. For testing the ecosystem under real conditions in Germany, a large number of charging stations will be integrated into the BANULA ecosystem and the functionality will be proven and validated within a one-year fleet trial. These vehicles will be handed over to different groups of test subjects who differ in terms of their usage behaviour and the provided charging options. The results of the testing groups will form the basis for evaluating the application of the ecosystem to end users and improving the pilot project. In addition, a roadshow across Germany—from Freiburg to Berlin—will demonstrate the crossregional approach in the four balancing zones operated by the German TSOs.

During the lifetime of the pilot, a series of workshops with experts in the field of energy and mobility will be held and a new expert community will be founded. The feedback of the expert rounds will be integrated into the pilot and the interim results will be published to the community.

The entire pilot of the BANULA ecosystem is divided into multiple MVPs, the roadshow being one of them. Because the German energy market is highly regulated and the BANULA ecosystem demands that many players in the energy market adapt their software systems and processes, many perspectives need to be considered in the implementation of the entire pilot. Adapting the current German law [26] and following the BDEWs application documents [27] is not easy, as most of the current software systems and processes do not support them. In order to develop a functional pilot, all use cases of the BANULA ecosystem have been broken down into technical use cases—there are currently 24 of them—which were, in turn, grouped into several MVPs. Therefore, each MVP is a standalone part of the ecosystem, which addresses different technical and functional parts of the BANULA ecosystem.

The first MVP of the entire pilot and the basis of the entire ecosystem—the ability to charge with a different energy supplier than the energy supplier of the charging point and account for the energy amount to the correct energy balancing group—is implemented and tested at one location in one distribution grid in one transmission network. Simultaneously, the same processes and implementations are prepared at four other locations. All of these four locations are on different properties, while three of them are in new distribution grids and two of them are in new transmission networks.

Based on the experiences of implementing the BANULA ecosystem on different premises in different distribution networks in different transmission networks, each case has its own tasks to handle. Even as the law supports an innovative concept as the BANULA ecosystem, the software systems and processes of the current energy market players do not support a smooth implementation at every location at present.

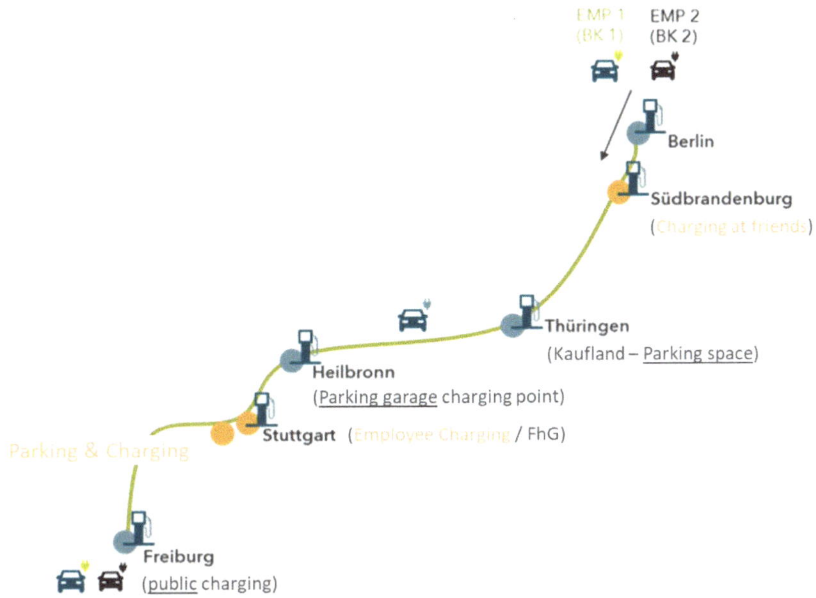

Figure 5. Overview of the planned roadshow across Germany—from Freiburg to Berlin—to demonstrate the MVPs important use case.

6. Discussion

The content of the BANULA concept is sufficiently complex and innovative that certain reflections and assumptions are certainly open to discussion. As discussed above, the added values for each role are, therefore, preliminary considerations and may change depending on what applies in practice. Questions such as the following arise: Does the transfer of responsibility for energy procurement from the role of CPO to the role of EMP actually lead to more reliable consumption forecasts and an associated increase in balancing group loyalty? Do users appreciate the new added value offered by the BANULA ecosystem to a sufficient extent to achieve market penetration? Is a new business model for CPOs that explicitly excludes energy procurement and sales attractive enough to prevail over the existing model? The topics of grid fees and ad hoc charging and how these can be integrated into BANULA are of particular interest. It is crucial that BANULA achieves a significant market share with CPOs and EMPs in order to realize the added values of the individual market roles, e.g., transparency regarding usability and prices (from the customer's perspective).

Of course, the ecosystem described in this document harbours risks as well as far-reaching opportunities, which can be outlined but not conclusively recorded. In any case, the use of the BANULA platform allows further added value to be realized beyond the basic idea of BANULA—think of dynamic prices or bidirectional billing, for example. However, the transaction costs are decisive for the profitability of these offers. For the BANULA platform to thrive as a vehicle for ecosystem transformation, proper governance is required. Therefore, the corporate structure must be well thought out, especially in view of the fact that the platform is a decentralized system.

7. Conclusions

With the concept of BANULA, a new innovative ecosystem for the operation and billing of charging processes is developed. The concept of BANULA offers added value for all market roles in the ecosystem and brings together energy economic processes, balancing charging energy quantities and the processes in electromobility. The benefits for selected stakeholders are as follows: End customers can choose their electricity supplier at BANULA charging points and are not bound to the supplier associated with the charging point; EMPs offer their electricity at all BANULA charging stations and develop new business and tariff models for their customers; CPOs (Charge Point Operators) can focus on operating the charging stations and delegate the procurement of the corresponding electricity quantities to the EMP and its suppliers; and distribution and transmission grid operators benefit from increased transparency in their respective electricity grids.

To achieve these objectives, BANULA defines new processes and roles, reassigns responsibilities and develops a technical backbone layer. The latter is based on blockchain technology (distributed ledger technology) and offers security, full transparency and auditable traceability. BANULA adopts a blockchain-based approach where the blockchain itself stores only a restricted amount of data. However, it serves as a real-time lookup table for each role within the system. To ensure data integrity and security, a comprehensive concept for roles and permissions will be developed to prevent data misuse. The described model is technically and procedurally operational within the current German legal framework. The feasibility will be demonstrated at selected charging stations.

An example of this is charging electric vehicles at public charging stations using the electricity generated from one's own photovoltaic system. To leverage added value at the European level, there are plans to transfer and adapt the entire system at the EU level.

Author Contributions: Conceptualization, T.H.; Methodology, D.S., T.H. and S.F.; Validation, T.H., M.S., I.S. and N.K.; Formal analysis, M.S., I.S., S.F. and N.K.; Resources, M.S.; Data curation, M.S., I.S., S.F. and N.K.; Writing—original draft, D.S., T.H., M.S., I.S., S.F. and N.K.; Writing—review & editing, D.S., T.H., M.S., I.S., S.F. and N.K.; Visualization, I.S. and S.F.; Supervision, D.S. and T.H.; Project administration, D.S., T.H. and S.F.; Funding acquisition, D.S. All authors have read and agreed to the published version of the manuscript.

Funding: This research was funded by the German Federal Ministry of Economics and Climate (BMWK) under the funding number 01MV21016A with administrative support by German Aerospace Center (DLR-PT).

Data Availability Statement: No new data were created or analyzed in this study. Data sharing is not applicable to this article.

Acknowledgments: We gratefully acknowledge the financial support of the German Federal Ministry of Economics and Climate (BMWK) as well as administrative support by German Aerospace Center (DLR-PT). Furthermore we thank the entire project consortium for their cooperation: Our full partners Badenova AG & Co. KG, Becker Büttner Held PartGmbB, OLI Systems GmbH, smartlab Innovationsgesellschaft mbH, TransnetBW GmbH, Universität Stuttgart (IAT and IEH) as well as our associated partners Schwarz Immobilien Service GmbH & Co. KG, Netze BW GmbH, 50 Hertz GmbH, Amprion GmbH, ZSW e.V., Stadtwerke Stadtroda GmbH, Stadtwerk am See GmbH & Co. KG and e-clearing.net GmbH.

Conflicts of Interest: The authors declare no conflicts of interest.

Appendix A

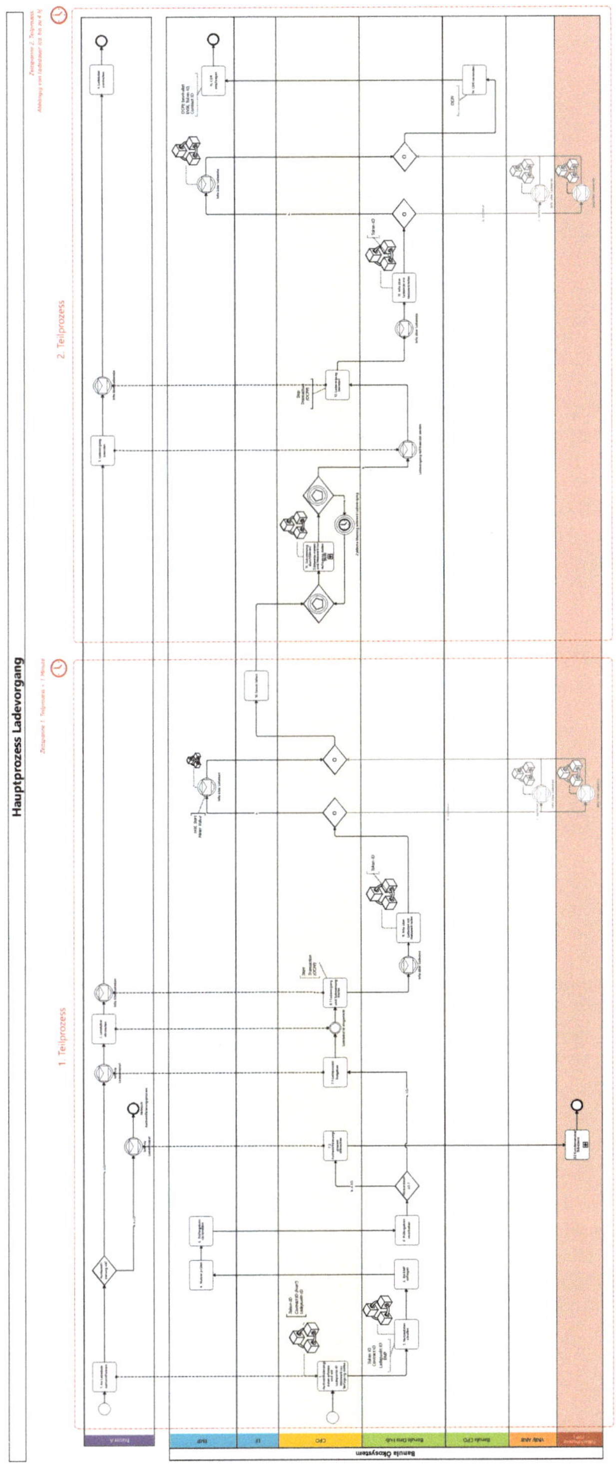

Figure A1. Display of the charging progress in the BANULA ecosystem as a swim-lane process diagram (BPMN 2.0).

References

1. Bundesministerium für Verkehr und digitale Infrastruktur. *Bekanntmachung der Förderrichtlinie "Öffentlich zugängliche Ladeinfrastruktur für Elektrofahrzeuge in Deutschland"*; Bundesministerium für Verkehr und digitale Infrastruktur: Berlin/Heidelberg, Germany, 2021.
2. Nationale Leitstelle Ladeinfrastruktur. *Report der Nationalen Leitstelle Ladeinfrastruktur—Öffentliche Ladeinfrastruktur Dezember 2022*; Nationale Leitstelle Ladeinfrastruktur: Berlin/Heidelberg, Germany, 2023.
3. BDEW. Available online: https://www.bdew.de/presse/presseinformationen/kapferer-zum-eine-million-ladesaeulenprogramm/ (accessed on 23 March 2023).
4. Bundesministerium für Digitales und Verkehr (BMDV). *Masterplan Ladeinfrastruktur II der Bundesregierung*; Bundesministerium für Digitales und Verkehr (BMDV): Berlin/Heidelberg, Germany, 2022.
5. L'Avere-France. *Baromètre National des Infrastructures de Recharge Ouvertes au Public*; L'Avere-France: Paris, France, 2023.
6. Dutch National Charging Infrastructure Agenda, Dutch National Charging Infrastructure Agenda Brochure. 2022. Available online: https://www.agendalaadinfrastructuur.nl/PageByID.aspx?sectionID=208529&contentPageID=1767173 (accessed on 23 March 2023).
7. The White House. Available online: https://www.whitehouse.gov/briefing-room/statements-releases/2021/12/13/fact-sheet-the-biden-harris-electric-vehicle-charging-action-plan/ (accessed on 23 March 2023).
8. USEF Foundation. *USEF Position Paper Electric Mobility*; USEF Foundation: Arnhem, The Netherlands, 2015.
9. BDEW. Available online: https://www.bdew.de/service/anwendungshilfen/zusatzrahmenvereinbarung-zum-netznutzungsvertrag-strom-netzzugangsregeln-zur-ermoeglichung-einer-ladevorgangscharfen-bilanziellen-energiemengenzuordnung-fuer-elektromobilitaet/ (accessed on 23 March 2023).
10. Triebke, H.; Göhler, G.; Wagner, S. Data analysis of PEV charging events in rural and business environments—A load behaviour comparison. In Proceedings of the 13th Symposium of Hybrid and Electric Vehicles, Braunschweig, Germany, 23–24 February 2016; University Stuttgart and Fraunhofer IAO: Stuttgart, Germany, 2016.
11. Kihm, A.; Trommer, S. The new car market for electric vehicles and the potential for fuel substitution. *Energy Policy* **2014**, *73*, 147–157. [CrossRef]
12. López, M.A.; La Torre, S.D.; Martín, S.; Aguado, J.A. Demand-side management in smart grid operation considering electric vehicles load shifting and vehicle-to-grid support. *Int. J. Electr. Power Energy Syst.* **2015**, *64*, 689–698. [CrossRef]
13. Babrowski, S.; Heinrichs, H.; Jochem, P.; Fichtner, W. Load shift potential of electric vehicles in Europe. *J. Power Sources* **2014**, *255*, 283–293. [CrossRef]
14. LamA. Available online: https://www.lama.zone/ (accessed on 23 March 2023).
15. eFlotte. Available online: https://www.keim.iao.fraunhofer.de/de/projekte/eflotte.html (accessed on 14 December 2023).
16. Shared E-Fleet. Available online: https://www.digital.iao.fraunhofer.de/de/leistungen/Energiesysteme/Shared_E-Fleet.html (accessed on 14 December 2023).
17. eMobility-Scout. Available online: http://www.emobilityscout.de/ (accessed on 23 March 2023).
18. Chargelounge. Available online: https://idw-online.de/de/news664310 (accessed on 14 December 2023).
19. Inflott. Available online: https://www.keim.iao.fraunhofer.de/de/projekte/inflott.html (accessed on 23 March 2023).
20. LamA-Connect. Available online: https://www.lama-connect.de/ (accessed on 23 March 2023).
21. C/SELLS. Available online: https://smartgrids-bw.net/projekte/c-sells-das-schaufenster-fuer-intelligente-energie/ (accessed on 14 December 2023).
22. SPARCS. Available online: https://sparcs.info/ (accessed on 23 March 2023).
23. IMEI. Available online: https://www.keim.iao.fraunhofer.de/de/projekte/imei.html (accessed on 23 March 2023).
24. GeMo. Available online: https://www.gemo.fraunhofer.de/ (accessed on 23 March 2023).
25. Charge@Work. Available online: https://www.muse.iao.fraunhofer.de/de/projekte/charge-at-work.html (accessed on 23 March 2023).
26. Bundesnetzagentur. *Netzzugangsregeln zur Ermöglichung einer ladevorgangscharfen bilanziellen Energiemengenzuordnung für Elektromobilität (NZR-EMob)*; Bundesnetzagentur: Bonn, Germany, 2020.
27. Bundesnetzagentur. Available online: https://www.bundesnetzagentur.de/DE/Beschlusskammern/1_GZ/BK6-GZ/2020/BK6-20-160/BK6-20-160_beschluss_db.html (accessed on 23 March 2023).
28. BDEW. Available online: https://www.bdew.de/service/anwendungshilfen/ergaenzung-der-marktregeln-fuer-die-durchfuehrung-der-bilanzkreisabrechnung-strom-mabis/ (accessed on 23 March 2023).
29. Bundesnetzagentur. Available online: https://www.bundesnetzagentur.de/DE/Beschlusskammern/BK06/BK6_83_Zug_Mess/831_gpke/gpke_node.html (accessed on 21 March 2023).
30. Bundesnetzagentur. Available online: https://www.bundesnetzagentur.de/DE/Beschlusskammern/1_GZ/BK6-GZ/2018/BK6-18-032/BK6-18-032_anlage_2_wim.pdf (accessed on 28 March 2023).
31. Bundesnetzagentur. Available online: https://www.bundesnetzagentur.de/DE/Service-Funktionen/Beschlusskammern/BK06/BK6_83_Zug_Mess/833_mabis/mabis_node.html (accessed on 7 March 2023).

32. Nakamoto, S. *Bitcoin: A Peer-to-Peer Electronic Cash System*; Decentralized Business Review; 2008. Available online: https://bitcoin.org/bitcoin.pdf (accessed on 5 March 2023).
33. International Organization for Standardization, ISO 15118-1:2019. Available online: https://www.iso.org/standard/69113.html (accessed on 7 March 2023).

Disclaimer/Publisher's Note: The statements, opinions and data contained in all publications are solely those of the individual author(s) and contributor(s) and not of MDPI and/or the editor(s). MDPI and/or the editor(s) disclaim responsibility for any injury to people or property resulting from any ideas, methods, instructions or products referred to in the content.

World Electric Vehicle Journal

MDPI

Article

Optimization of H_2 Supply to the Refuelling Infrastructure for Long-Haul Trucks: Centralized versus Local H_2 Production, and Using Transportation by Tanker Truck or Pipeline †

Nafisa Mahbub * and Hajo Ribberink

CanmetENERGY Research Centre, Natural Resources Canada, 1 Haanel Drive, Ottawa, ON K1A 1M1, Canada; hajo.ribberink@nrcan-rncan.gc.ca

* Correspondence: nafisa.mahbub@nrcan-rncan.gc.ca

† This article is a revised and expanded version of a paper entitled Optimization of H2 Supply and Refuelling Infrastructure for Long Haul Trucks, which was presented at 36th International Electric Vehicle Symposium and Exhibition (EVS36), Sacramento CA, USA, 14–17 June 2023.

Abstract: In a simulation study, it was investigated how the costs of supplying H2 for the refuelling of long-haul trucks along highways in Canada can be minimized by optimizing the design of the refuelling infrastructure. Scenarios using local or centralized blue H2 production were evaluated using two different modes of H2 transportation (liquid H2 tanker trucks and pipelines). For each scenario, the average H2 supply costs were determined considering H2 production costs from facilities of different sizes and transportation costs for H2 that was not produced locally. Average H2 supply costs were 2.83 CAD/kg H2 for the scenario with local H2 production at each refuelling site, 3.22–3.27 CAD/kg H2 for scenarios using centralized H2 production and tanker truck transportation, and 2.92–2.96 CAD/kg H2 for centralized H2 production scenarios with pipeline transportation. Optimized scenarios using the cheaper transportation option (tanker truck or pipeline) for each highway segment had average H2 supply costs (2.82–2.88 CAD/kg H2) similar to those of using only local H2 production, with slightly lower costs for the scenario using the largest H2 production volumes. Follow-on research is recommended to include the impact of CO2 transportation (from blue H2 production) on the design of an optimum supply infrastructure.

Keywords: long-haul trucks; hydrogen; infrastructure; simulation; tanker truck; pipeline; transportation cost

Citation: Mahbub, N.; Ribberink, H. Optimization of H_2 Supply to the Refuelling Infrastructure for Long-Haul Trucks: Centralized Versus Local H_2 Production, and Using Transportation by Tanker Truck or Pipeline. *World Electr. Veh. J.* **2024**, *15*, 57. https://doi.org/10.3390/wevj15020057

Academic Editors: Joeri Van Mierlo and Genevieve Cullen

Received: 21 December 2023
Revised: 25 January 2024
Accepted: 31 January 2024
Published: 8 February 2024

1. Introduction

Around 27% of global CO_2 emissions come from the transportation sector, which makes it the second largest emitter [1]. Over the last decade, significant momentum has built for the electrification of passenger vehicles mainly using battery electric technology. However, the use of hydrogen (H_2) as a transportation fuel has significant potential to support the transition towards a low carbon economy since it does not emit carbon at the end-use point of combustion, has good storage life, and it can be transported by roads, ships, or pipelines in gaseous or liquid forms [2].

To drastically reduce emissions and eliminate the use of diesel in the long-haul trucking sector, it is expected that H_2 fuel cell vehicles will be needed as battery electric vehicles cannot provide the same utility as diesel vehicles. Significant studies have been conducted on different aspects of using hydrogen as a clean fuel for long-distance transportation.

- Kumar et al. developed a framework to analyze the supply chain cost of low-carbon hydrogen exports from Alberta, Canada, to several viable destinations in North America, the Asia–Pacific, and Europe [3]. The supply chain includes all unit operations ranging from hydrogen production with carbon capture and storage, hydrogen pipeline transportation, liquefaction, shipping, and regasification at the destinations. A techno-economic assessment has been conducted to estimate the supply chain cost of different viable pathways considering the energy, material, and capacity. Results show that within North America, transporting hydrogen blended with natural gas using the existing natural gas pipelines could reduce the price by 17%. Further cost savings around 28% were achieved while transporting ammonia to the Asia–Pacific in comparison to shipping liquified hydrogen. An analysis of overseas H_2 supply chains has been conducted by Lim et al. considering the economic (unit H_2 cost), environmental (carbon footprint), and technological aspects [4]. The supply chains include all the operations ranging from H_2 production, ship transportation, to inland distribution. Several supply chains were compared varying the economies of scale, amount of H_2, and distance. Results show that the use of liquid organic hydrogen carrier, liquid hydrogen, and ammonia are the most potentially feasible options for H_2 carriers considering these criteria.

- Stolen et al. developed a well-to-tank analysis to calculate the costs, energy consumption, and greenhouse gas (GHG) emissions for supplying hydrogen to fuel cell electric vehicles (FCEVs) [5]. The study followed a holistic approach considering the whole supply chain that includes the storage and transportation of hydrogen. The study discussed different hydrogen infrastructure technologies from ecological and economic points of view. Compression and liquefaction have been mentioned as the state-of-the-art H_2 technologies, whereas liquid organic hydrogen carrier (LOHC) has been identified as the most promising H_2 technology for the near future from an economic perspective. However, further research is needed regarding the system design of the LOHC-supplied refuelling stations and the heat source for dehydrogenation.
- Barbir et al. considered a wide range of hydrogen refuelling station (HRS) capacities and configurations [6]. For example, locating the hydrogen production and refuelling station within an existing wind farm in Croatia or nearby the end users, or site the hydrogen production within the wind farm and install the refuelling station nearby the users, etc. The study assumed that hydrogen is delivered to the refuelling station with a tube trailer and when hydrogen was produced within the wind farm, a mobile refuelling station was used for consumers in different locations. The techno-economic analysis of each hydrogen refuelling station configuration was conducted to estimate the levelized cost of hydrogen production—the capital, operational, and maintenance costs. The study results show that, since the capacity and location of the hydrogen refuelling stations depend on the users, it was difficult to identify the optimum configuration without the hydrogen infrastructure development in Croatia. However, the authors mentioned that the results could play a significant role in the implementation of hydrogen infrastructure in Croatia in the near future.
- Hurskainen and Ihonen conducted a techno-economic assessment for point-to-point large-scale road transportation of hydrogen [7]. The researchers compared liquid organic hydrogen carriers (LOHC), compressed H_2 gas delivery by trucks, and on-site production of hydrogen using water electrolysis. Results show that the LOHC supply chain was the most economic option for long-distance hydrogen transportation by road. However, to achieve economic feasibility, the heat supply method for releasing hydrogen at the end-user site and the investment costs were found as the most critical parameters to consider.
- Qing et al. assessed four possible low-carbon hydrogen supply chains for a hydrogen refuelling station located in Shanghai [8]. The study analyzed the feasibility of using renewable hydrogen as a transportation fuel for fuel cell vehicles. Two routes considered on-site hydrogen production powered by a stand-alone or grid-connected photovoltaic (PV)–wind generation system separately, whereas the other two routes considered off-site hydrogen supply. The off-site hydrogen is also produced by a stand-alone or grid-connected PV–wind generation system located in the Qinghai Province, since it is a rich renewable energy area. The H_2 is then delivered to Shanghai

by liquid hydrogen tanker trucks. The study found the off-site production supply chains as feasible options. The study mentioned transporting liquid hydrogen for long distance using trucks is more economical compared to transporting compressed gaseous hydrogen due to its higher energy density. Although this study was focused on H_2 supply for passenger vehicles, the results would also apply to H_2 supply for long-haul trucks.

- Kumar et al. conducted a process-based techno-economic assessment of hydrogen transportation pathways [9], Including, for example, pure hydrogen (hydrogen pipeline and truck transport of gaseous and liquified hydrogen), hydrogen–natural gas blends (pipeline), ammonia (pipeline), and liquid organic hydrogen carriers (pipeline and rail). The authors estimated the costs and GHG emissions for high-capacity long distance H_2 transportation, such as 1000 km, 3000 km. Kumar et al. identified the hydrogen pipelines and hythane (hydrogen and natural gas blends) as the least expensive H_2 transportation pathways for long distances [9]. The ammonia, liquid organic hydrogen carrier, and truck transportation pathways were found to be more than 1.5 times expensive than the pure H_2 pipelines.
- The International Energy Agency (IEA) assessed the opportunities and issues related to different alternative H_2 transportation options [10]. Pipeline and shipping options were discussed for long-distance H_2 transport such as 1000, 2000, and 3000 km. The report also elaborated on local H_2 distribution options such as trucks carrying gaseous H_2, liquid H_2, ammonia, and LOHC and pipelines with 100 tonnes per day and 500 tonnes per day capacities. H_2 conversion and reconversion technologies are also considered in the study. The study identified pipeline H_2 transportation to be the cheapest option for less than 1500 km distances compared to ammonia and LOHC, which were more cost effective for overseas H_2 transportation over longer distances. For local distribution, pipelines with high capacity were referred to as more cost effective than tanker trucks for H_2 transportation over longer distances [10].

The above-mentioned studies focused on different aspects of H_2 used as a transportation fuel technology such as H_2 production technologies, techno-economic assessments of low-cost hydrogen transportation, H_2 export by overseas or inland routes, assessments of low carbon hydrogen exports from cheaper production locations like Canada, economic, technological, and environmental impacts of H_2 supply chains, and on different storage technologies of H_2. However, very few studies have been addressing the significance of the H_2 refuelling infrastructure on the overall H_2 supply chain from a cost perspective. Current H_2 prices are far above the level needed for cost-effective operation of H_2-based transportation. Cost reductions in every part of the H_2 supply chain will be needed to realize a sustainable cost level.

This study investigated how costs in the supply of H_2 for the refuelling of long-haul trucks along major highways in Canada can be minimized by optimizing the design of the supply infrastructure. Given the high cost of green hydrogen, this study focussed on the use of blue hydrogen, produced from natural gas with carbon capture. Scenarios of local and centralized blue H_2 production were evaluated to investigate whether cost savings from centralized H_2 production on a larger scale would outweigh the additional costs of H_2 transportation to refuelling sites without H_2 production. The study analyzed two pathways for H_2 transportation: tanker trucks carrying liquid H_2 and pipelines transporting gaseous H_2. The study used state-of-the-art H_2 transportation and refuelling infrastructure design parameters both for pipelines and tanker trucks.

First, scenarios that used one of the two transportation methods for all refuelling sites in the total network were analyzed. Then the relationship between H_2 volume, transportation cost, and transportation distance were investigated, enabling the evaluation of optimized scenarios, which used the least cost transportation option (either tanker truck or pipeline, based on local conditions) for different refuelling sites along the highways.

The paper is divided in the following sections: Section 2 discusses the methodology; description of the H_2 supply pathways, scenarios, modes of transportation, and costs considered. Section 3 presents the results and discussions, followed by conclusions in Section 4.

2. Methodology

A simulation model was developed to estimate the H_2 demand for heavy-duty long-haul trucks along major highways in Canada. Over 11,000 km of Canadian major highways were considered (see Figure 1). Data from provincial transportation authorities (like the Ontario Ministry of Transport) were used to estimate the annual average daily truck traffic (AADTT) flow for each of the highway segments, resulting in a large range in AADTT values from 500 trucks per day on more quiet highways to 17,000 trucks per day on the busiest segments. The model assumed that each highway segment should provide the H_2 needed for all trucks that drive on that highway segment and for the total length of that segment.

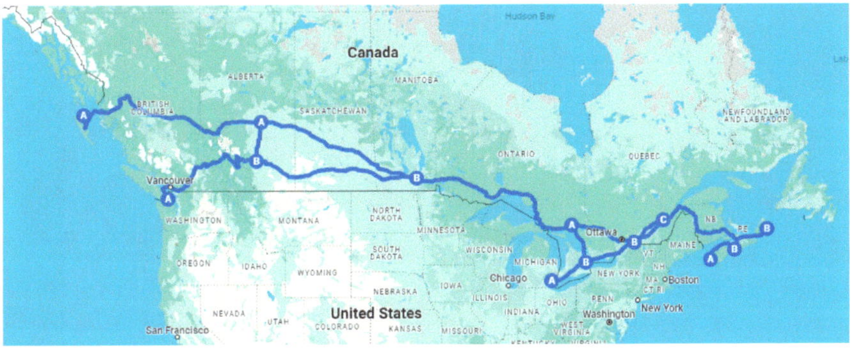

Figure 1. Overview of major highways in Canada.

The required H_2 production per highway segment was then calculated from the truck traffic kilometers driven on that segment and a fixed H_2 consumption of 10 kg/100 km.

2.1. Hydrogen Refuelling Sites, Production Sites, and Pumps

To determine the number of H_2 refuelling sites per highway, the highways were segmented based on the truck traffic flow. To minimize H_2 transportation costs, production sites were co-located with refuelling sites. The standard distance between refuelling sites was assumed to be 100 km. However, the actual distance between refuelling sites on some highway sections was a bit shorter if the segment length was not exactly a multiple of 100 km. Given space constraints on truck rest stops, a maximum of 20 H_2 pumps per refuelling site were allowed, requiring refuelling stations to be placed closer together on the busiest highway sections.

The number of refuelling sites for highway segment i, $N_{R,i}$, was calculated using Equation (1):

$$N_{R,i} = L_i / D_i \tag{1}$$

in which L_i is the length of highway segment i and D_i the distance between H_2 refuelling sites on that highway segment. Similarly, the number of H_2 production sites for highway segment i, $N_{P,i}$, was calculated:

$$N_{P,i} = N_{R,i} / R_j \tag{2}$$

with R_j being the number of refuelling sites supplied by a single production site in scenario j.

2.2. Scenarios and Transportation Options

Different scenarios were developed to investigate the impact of the size of the H_2 production facility on the amount of hydrogen that would need to be transported, the transportation distance, and on the costs to supply H_2 to refuelling sites. Larger production sites were placed further apart, effectively increasing the number of refuelling sites that would be supplied by one production facility. Four scenarios were evaluated with a focus on identifying the potential to optimize the H_2 supply to the refuelling infrastructure for long haul trucks:

- Scenario 1: Each refuelling site had its own production facility; hence, no transportation was required between production and refuelling stations. Production sites were about 100 km from each other.
- Scenario 2: Each production site supplied two refuelling sites (one of them being the co-located refuelling site), and the distance between production sites was around 200 km. Under this scenario, there were two times as many refuelling sites as production sites, requiring half of the total amount of H_2 produced to be transported to a neighbouring refuelling site over 100 km distance (one-way) away.
- Scenario 3: Each production site supplied H_2 to three refuelling sites, and there was generally one production site per 300 km of highway. With three times as many refuelling sites as production sites, two thirds of all refuelling sites needed to have their H_2 transported from the neighbouring production site, which was placed at the middle location of the three refuelling sites. The H_2 transportation distance was 100 km (one-way).
- Scenario 4: Each production site supplied five different refuelling sites, because production sites were placed about 500 km apart. Under this scenario, there were five times as many refuelling sites as production sites, requiring four fifths of all the H_2 to be transported to neighbouring refuelling sites from the central H_2 production station. On average, the H_2 was transported over 150 km (one-way), as the distance to the closest refuelling sites was 100 km, and 200 km to the outer refuelling sites.

For each scenario of supplying H_2 to a network of refuelling sites under Scenarios 2–4, two transportation variants were evaluated: tanker trucks and pipelines. The unit operations considered in the tanker truck pathway included production, liquefaction, transportation, and regasification, while the pipeline pathway considered production and transportation (see Figure 2). The unit operations are explained in detail in Sections 2.3.1 and 2.3.2. For clarity, Scenario 1 does not use either of these pathways, because all H_2 is produced locally at each refuelling site.

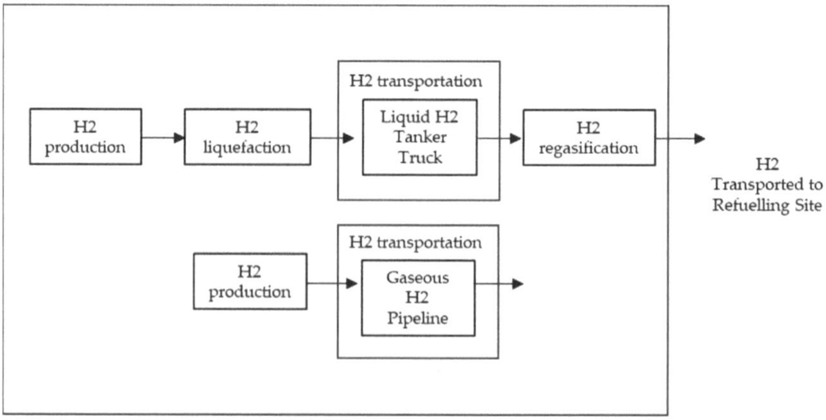

Figure 2. Unit operations for two pathways of H_2 supplying H_2 to refuelling sites using different transportation methods: tanker trucks (top) and pipelines (bottom).

2.3. *Hydrogen Production Cost and Transportation Costs*

In the economic analysis, the total costs to supply H_2 to the refuelling sites were calculated as the sum of the production costs and the transportation costs.

$$TC_H = C_P + C_T \tag{3}$$

Here, TC_H, C_P, and C_T are the total H_2 supply cost, the H_2 production costs, and the costs for transportation of H_2 from production to the refuelling sites, respectively.

For the H_2 production costs, a correlation between the costs of production and the production volume was used, based upon the results for blue H_2 production by steam methane reforming (SMR) from [11]. The SMR cost curve was extrapolated to determine values for the smaller production volumes (between 5 and 100 tonnes/day) needed in this study (see Figure 3). The hydrogen production cost numbers developed by [11] include the cost to capture the associated CO_2, but they exclude the costs to transport CO_2 to storage sites.

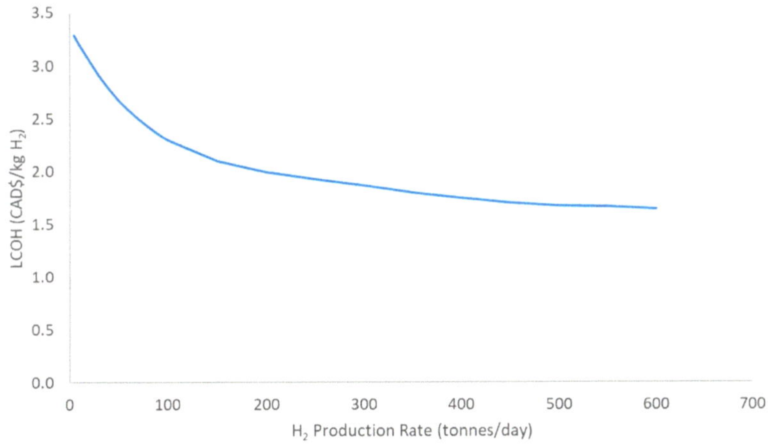

Figure 3. Levelized costs of (blue) H_2 production (LCOH), based on [11].

It should be noted that the overall cost numbers shown in the results of this study are the average costs over the total highway network and for all H_2 supplied (a total of 2978 tonnes per day), considering the H_2 production costs from facilities of different sizes, and distributing the total H_2 transportation costs (for transportation over different distances) over all H_2 supplied (i.e., the total of H_2 that needed transportation and the H_2 that was produced on-site and did not need transportation).

Sections 2.3.1 and 2.3.2 discuss the details of the H_2 transportation costs by liquid H_2 tanker truck and by pipeline, respectively.

2.3.1. H_2 Transportation by Liquid H_2 Tanker Truck

The calculation of the costs to transport H_2 by tanker truck included the capital cost for the liquid H_2 tanker truck, the driver salary, diesel fuel costs, and maintenance and repair costs (including tires). Table 1 presents the technical, operational, and economic data for H_2 transport by tanker truck used in this study. All costs are given in 2020 Canadian dollars (CAD).

Table 1. Data and assumptions for H_2 transportation by liquid tanker truck.

	Values	Units	Reference
Capacity of liquid H_2 tanker truck	4000	kg	[12]
Lifetime of liquid H_2 tanker trailer	20	years	[12]
Lifetime of truck cab	5	years	[12]
Capital costs of liquid H_2 tanker trailer	$852,000	CAD	[12]
Capital costs of truck cab	$108,000	CAD	[12]
Diesel price per litre	1.59	CAD	[13]
Diesel consumption per km	0.39	litre	[13]
Diesel costs per km	0.83	CAD	Calculated
Truck driver salary per hour	26.42	CAD	[14]
Average speed truck	50	km/h	[12]
Total time needed for loading and unloading	3	H	[12]
H_2 liquefaction costs	1.34	CAD/kg	[10]
Conversion rate USD/CAD	1.3415		[15]
Adjusted inflation factor (2008 to 2020)	1.2		[16]
Driving days per year	300	days	Assumed
Shifts (round trips) per day	3		[12]
Operational cost (repair, maintenance, tires)	0.159	CAD/km	[17]

The truck capital cost includes the costs for tanker, undercarriage, and cab. The liquid H_2 tanker can transport 10 times more hydrogen than a tube trailer for compressed H_2 gas [12], hence, it was decided to use the liquid H_2 tanker truck.

A diesel price of CAD 1.59 per liter was used in the tanker truck scenarios, based on the average diesel retail price in 17 Canadian cities over the period of May 2020–November 2023 [13].

The total salary costs per roundtrip to deliver H_2 to a refuelling site was calculated based on the total time spent on the job, which included driving time and loading and unloading time.

While the costs of liquefying hydrogen were substantial (1.34 CAD/kg), the costs for regasification at the refuelling sites to convert it back to gaseous H_2 were found to be negligible [10], and, hence, they were not included in this study.

The capital costs of the tanker trailer and cab were calculated using a net present value (NPV) approach. A 10% minimum acceptable rate of return (MARR) was used in the NPV calculation of the truck transportation infrastructure for a 20-year lifespan. The capital recovery factor (CRF) given by Equation (4) was used to obtain the annual capital cost.

$$CRF = \{I^*(1 + i)^n/((1 + i)^n - 1)\} \quad (4)$$

Here, i is the MARR (10%) and n is the project lifespan of 20 years.

2.3.2. H_2 Transportation by Pipeline

For the calculation of the costs to transport H_2 by pipeline, capital, operational, and maintenance costs were considered for the two main components, namely, the pipeline and the H_2 compressor.

The pure hydrogen transportation cost equations from [9,18] were used to estimate the pipeline capital cost and compressor capital cost, which included the equipment purchase and installation cost.

The pipeline capital cost was assessed using the following equation from [9,18].

$$C_{pipe} = (\{1171^*(D/25.4)^2 + 15{,}251^*(D/25.4) + 329{,}705\}^*L + 767{,}845)^*AF \quad (5)$$

C_{pipe} is the H_2 pipeline capital cost (CAD), D is the pipeline diameter (mm), and L is the pipeline length (km). A cost factor (Alberta factor 1.15) has been considered in this equation to reflect a 15% cost increase from the average North American value [9,19].

The internal pipeline diameter for the pipelines of the network used in this study was based on the maximum daily amount of H_2 to be supplied to a refuelling site, which was 77 tonnes. In [9], a 16-inch (406.4 mm) pipeline was utilized for transporting 607 tonnes of H_2 per day. Using a simplified scaling method, which assumes that pipeline throughput is linear with the area of the pipe opening, a 6-inch (152.4 mm) diameter was found to be sufficient for transporting up to 85 tonnes per day. It was decided to use this diameter for all pipelines in the network, because of the limited cost savings from using pipelines with smaller diameters (a pipeline with half the diameter (3-inch/76.2 mm) has a 75% lower capacity than a 6-inch pipeline, but costs only 16% less) in comparison to the opportunity to use the pipeline network for additional clients in the future.

The compressor capital cost was estimated using the factorial method. Applying this method, the purchase equipment cost of an equipment was first estimated, then the cost was multiplied by the factors (installation costs, offsite costs, design and engineering, and contingencies) to obtain the capital installation cost [9]. The purchase equipment cost (C_{PEC}) and the installed capital costs (C_{ICC}) were determined using Equations (6) and (7) from [9,11].

$$C_{PEC} = 30{,}746 * P^{0.6089} \tag{6}$$

$$C_{ICC} = (C_{PEC} * \sum C_{IF} + OSBL + D\&E + Contingency) * AF \tag{7}$$

Here, P is power consumption by the compressor motor (kW). The factors used in the C_{ICC} equation, namely, material and labour installation factor (C_{IF}), off-site battery limit cost (OSBL), the design and engineering cost (D&E), and the contingency charge (Contingency) have values of 1.385, 0.3, 0.3, and 0.1 respectively.

The operating costs and maintenance costs were also considered to calculate the overall cost for a hydrogen pipeline transportation infrastructure. The operating and maintenance costs for the pipeline were assumed to be 1.5% and 3% of the pipeline capital cost, respectively [9]. The compressor operating cost was not considered in this study and would have been negligeable in comparison to the much higher costs to operate the pipeline.

The pipeline and compressor have an assumed lifetime of 25 and 10 years, respectively. A 10% minimum acceptable rate of return (MARR) was considered to calculate the net present value (NPV) of the pipeline infrastructure for a 25-year lifespan. The capital recovery factor (CRF) given by Equation (4) was used to obtain the annual capital cost. All costs for the pipeline scenario were adjusted to 2020 CAD.

The annualized capital, operating, and maintenance costs for all pipelines were added together into the total annual costs for H_2 transportation. These costs were distributed over all H_2 produced/supplied, to calculate the contribution of transportation to the average supply costs per kg of H_2 for the pipeline scenarios.

It was assumed pipelines can be buried and they can follow the same trajectory as truck routes for H_2 transportation. Hence, the distance between the H_2 refuelling stations and the H_2 production stations were kept the same as for truck transportation cases.

3. Results and Discussion

3.1. Hydrogen Production and Refuelling Sites

Different scenarios were evaluated for local and centralized H_2 production (as explained in Section 2.2), varying the size of the production facility and the associated distance between production sites, and the number of refuelling sites that were supplied by one production site (see Table 2). For Scenario 1, where the H_2 required for each refuelling site is produced locally, there is no need for any H_2 transportation. For the other scenarios, the total H_2 transportation distances were determined based upon the number of H_2 refuelling sites that did not have on-site production and their distances to the nearest H_2 production site.

Table 2. Number of production and refuelling sites for the scenarios evaluated.

	Approximate Distance between Refuelling Sites (km)	Number of Refuelling Sites	Number of Pumps	Number of Refuelling Sites per Production Site	Approximate Distance between Production Sites (km)	Number of Production Sites
Scenario 1	100	125	843	1	100	125
Scenario 2	100	125	843	2	200	67
Scenario 3	100	125	843	3	300	51
Scenario 4	100	125	843	5	500	38

The results in Table 2 show a clear reduction in the number of production sites for scenarios in which one production site would supply an increasing number of refuelling sites. However, this reduction is not linear, because the number of required H_2 production sites was evaluated per highway segment. If a highway segment was smaller than the standard distance between two production sites under a certain scenario, it would still have its own production facility. Similarly, if the length of the highway segment was 1.5 times the standard distance between production sites, the segment would have two production sites.

3.2. *H_2 Supply Costs and Potential Cost Savings*

The total cost per kg of hydrogen supplied to the refuelling sites for each of the scenarios are presented in Tables 3 and 4. The cost numbers shown here are *average* costs over the total highway network for all H_2 supplied, 2978 tonnes H_2 per day, considering the full range of H_2 costs from production facilities of different sizes, and distributing the H_2 transportation costs over all H_2 supplied.

Table 3. Total H_2 cost per kg supplied and H_2 costs savings for different scenarios using liquid H_2 tanker trucks for transportation.

Scenario	Number of Production Sites	Average H_2 Production Volume (Tonnes/Day)	Average H_2 Production Costs (CAD/kg)	Average H_2 Transport. Costs (CAD/kg)	Average Cost of H_2 Supplied to Refuelling Sites (CAD/kg)	H_2 Costs Savings (%)
Scenario 1	125	23.8	2.83	0.00	2.83	
Scenario 2	67	44.5	2.55	0.66	3.22	−13.6%
Scenario 3	51	58.4	2.45	0.81	3.26	−15.1%
Scenario 4	38	78.4	2.31	0.96	3.27	−15.5%

Table 4. Total H_2 cost per kg supplied and H_2 costs savings for different scenarios using pipelines for transportation.

Scenario	Number of Production Sites	Average H_2 Production Volume (Tonnes/Day)	Average H_2 Production Costs (CAD/kg)	Average H_2 Transport Costs (CAD/kg)	Average Cost of H_2 Supplied to Refuelling Sites (CAD/kg)	H_2 Costs Savings (%)
Scenario 1	125	23.8	2.83	0.00	2.83	
Scenario 2	67	44.5	2.55	0.40	2.95	−4.2%
Scenario 3	51	58.4	2.45	0.51	2.96	−4.7%
Scenario 4	38	78.4	2.31	0.62	2.92	−3.3%

H_2 production costs varied significantly across the evaluated scenarios. The average H_2 production cost ranged from 2.31 CAD/kg to 2.83 CAD/kg, with the lowest cost for Scenario 4, which had, on average, the highest production volume per production facility.

For the truck transport scenarios (see Table 3), the costs of the H_2 liquefaction had a major impact on the overall H_2 transportation costs (i.e., the average of the costs for H_2 that needed transportation and for H_2 that did not need transportation). For the scenarios using centralized H_2 production, more refuelling sites need transportation when moving from Scenario 2 to Scenario 4, hence, the results in Table 3 show an increase in the average transportation costs. Due to the high costs for liquefaction, the scenario needing no transportation due to local H_2 production at every refuelling site (Scenario 1) was overall more cost-effective in supplying H_2 than the scenarios using centralized H_2 production.

For the pipeline scenarios, H_2 was transported in gaseous form, avoiding the need for costly liquefaction. Similar to the scenarios using tanker trucks, the transportation cost increased when moving from Scenario 2 to Scenario 4 to more centralized H_2 production (see Table 4), with each scenario needing more H_2 to be transported. The overall costs to supply H_2 by pipeline were fairly similar for all scenarios with centralized H_2 production.

The results in Tables 3 and 4 show that while pipeline transport results in lower H_2 supply costs than truck transport, all scenarios involving H_2 transportation are still more expensive than the scenario that has on-site H_2 production at all refuelling sites and does not need any H_2 transportation (Scenario 1).

The evaluated scenarios for H_2 supply using tanker truck or pipeline transportation, however, may not be optimized scenarios, because they used the same transportation method for the total network of refuelling sites. In the next section, first the relationship between the transportation costs and transportation distance is investigated for tanker truck transportation and for pipelines, after which results for a mixed scenario using both truck transport and pipelines are presented.

3.2.1. Comparison of H_2 Transportation Costs between Tanker Truck and Pipelines

Figure 4 compares the per kg H_2 transportation cost for different transportation methods and for a range in daily H_2 demand (volume), focusing on transportation between locations that are 100 km and 200 km apart. For the tanker truck scenarios, the roundtrip driving distance per delivery was twice the distance between locations, i.e., 200 km and 400 km, respectively, for the examples illustrated in Figure 4.

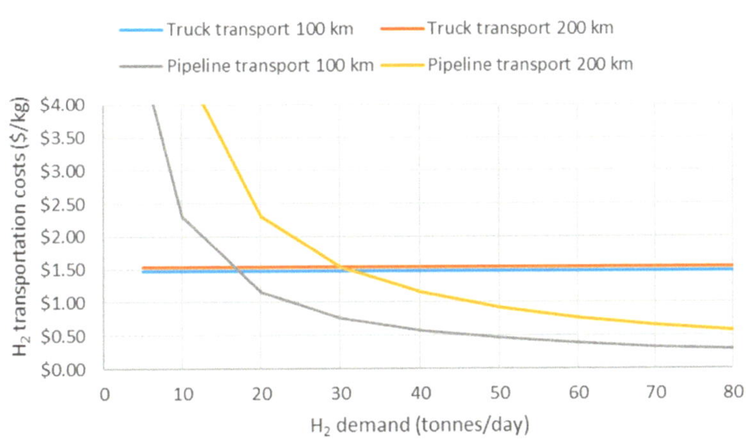

Figure 4. Comparison of transportation costs per kg of H_2 between liquid H_2 tanker trucks and pipelines.

From Figure 4, it is observed that the pipeline transportation costs change with hydrogen demand whereas the truck transportation costs per kg of H_2 remain constant. This is caused by the different characteristics of the two transportation methods. Since the tanker truck capacity is fixed (4000 kg of hydrogen per tanker truck), the per kg H_2 transportation costs are also constant with demand, because if more H_2 will need to be transported, this will be achieved by a larger number of tanker trucks, each having the same H_2 transportation costs per kg of H_2 transported. For pipelines, the overall transportation costs are dominated by capital costs, which are fixed once a network with a certain capacity has been constructed. Using this pipeline network for different H_2 demands will then lead to differences in H_2 transportation costs. If the H_2 throughput of the pipeline (the H_2 demand) was increased, the transportation cost would decrease and vice versa.

Figure 4 shows that truck transportation is more cost-effective for lower H_2 demands (i.e., for refuelling sites along quiet highways) and that pipeline transportation is favored for higher H_2 demands (refuelling sites along busy highways). For a distance of 100 km between production and refuelling sites, pipelines have lower costs for H_2 demands starting around 17 tonnes per day, while for a 200 km transportation distance, pipelines will result in lower costs for demands over 30 tonnes per day.

3.2.2. Optimized Scenarios Using a Mix of H_2 Transportation Methods

The results from Figure 4 were used to create optimized scenarios for Scenarios 2 to 5 by selecting the lowest costs transportation method (tanker truck or pipeline) for each highway segment. Thus, the optimized scenario represents a combination of transportation modes to supply H_2 to the total network of refuelling sites across highways in Canada.

Table 5 presents details on the optimized scenarios and compares the average cost to supply H_2 from centralized production to the costs for when using local H_2 production (Scenario 1). It was observed that for each scenario, almost half of the refuelling sites used tanker truck transportation, while the other sites utilized pipelines. The optimized scenarios had average H_2 supply costs that were further reduced in comparison to the scenarios that used either tanker truck or pipeline transport and were similar to those of Scenario 1.

Table 5. Total H_2 cost per kg supplied and H_2 costs savings for optimized scenarios using a mix of tanker trucks and pipelines for H_2 transportation.

Scenario	Number of Refuelling Sites Needing Transportation	Number of Refuelling Sites Using Transportation by Truck	Number of Refuelling Sites Using Transportation by Pipeline	Average Cost of H_2 Supplied to Refuelling Sites (CAD/kg)	H_2 Cost Savings (%)
Scenario 1	N/A	N/A	N/A	2.83	
Scenario 2	58	27	31	2.88	−1.9%
Scenario 3	74	36	38	2.87	−1.5%
Scenario 4	87	43	44	2.82	0.3%

Although the average H_2 supply costs for the optimized scenarios utilizing centralized H_2 production were comparable to those for the scenario using decentralized H_2 production, there is one additional aspect that is outside the scope of the current study, but that will need to be taken into account for a full view on the optimum H_2 supply infrastructure for refuelling sites for long-haul trucks along highways in Canada: The study used cost information for H_2 production from natural gas, which included the costs of CO_2 removal, but not the costs to transport CO_2 to the storage site. These costs may be substantial, because for a large fraction of the production sites along the Canadian highway network it is expected that CO_2 may need to be transported over long distances. This may have a significant impact on the overall H_2 supply costs and may influence whether centralized or decentralized H_2 production would be most cost-effective. It is, therefore, recommended that the transportation of CO_2 from the H_2 production sites to the storage location(s) will be included in follow-on research on an optimized H_2 supply infrastructure.

4. Conclusions

In this study, it was analyzed how the costs of supplying H_2 to refuelling sites for long-haul trucks along major highways in Canada can be minimized by optimizing the design of the refuelling infrastructure. Scenarios using local or centralized blue H_2 production were evaluated using two different modes of H_2 transportation (liquid H_2 tanker trucks and pipelines).

For each scenario, the average H_2 supply costs were determined considering H_2 production costs from facilities of different sizes and transportation costs for H_2 that was not produced locally. Average H_2 supply costs were 2.83 CAD/kg H_2 for the scenario with local H_2 production at each refuelling site. Scenarios with centralized H_2 production had average H_2 supply costs of 3.22–3.27 CAD/kg H_2 when using tanker trucks for H_2 transportation and of 2.92–2.96 CAD/kg H_2 when utilizing pipeline transportation. The high costs for H_2 liquefaction were a major factor in the higher supply costs for the tanker truck transportation scenarios.

The different characteristics of the two H_2 transportation methods allowed for the creation of optimized scenarios, which utilized each transportation mode when it would be the cheaper option (tanker trucks for refuelling sites with lower H_2 demand along quiet highways, and pipelines for sites along busy highways that had higher H_2 demand). The average H_2 supply costs for the optimized scenarios of centralized H_2 production (2.82–2.88 CAD/kg H_2) were similar to the average supply costs when using local H_2 production at each refuelling site (2.83 CAD/kg H_2), with slightly lower costs for the scenario using the largest H_2 production volumes.

While the results of this study seem to indicate that there is little difference in the results between scenarios using local or centralized H_2 production, the transport of CO_2 from the production site of the blue H_2 to a storage location were out of scope for this study. Follow-on research is recommended to investigate how the CO_2 transportation costs will impact the H_2 supply costs and may influence the design of an optimum supply infrastructure for a network of refuelling sites for long-haul trucks.

It is recommended that in future research, the results obtained from blue hydrogen pathways will be updated to include the costs for CO_2 transportation and storage, and to compare them to those of green hydrogen pathways to identify the optimum solution.

Author Contributions: Conceptualization, H.R. and N.M.; formal analysis, H.R. and N.M.; investigation, H.R. and N.M.; writing—original draft preparation, N.M.; writing—review and editing, H.R. and N.M.; supervision, H.R.; project administration, H.R.; funding acquisition, H.R. All authors have read and agreed to the published version of the manuscript.

Funding: Funding for this work was provided by Natural Resources Canada through the Carbon Capture, Utilization and Storage (CCUS) Program and the Program of Energy Research and Development (PERD).

Data Availability Statement: Data sharing is not possible for this article due to confidentiality issues with the source data.

Conflicts of Interest: Nafisa Mahbub and Hajo Ribberink are employees of CanmetENERGY Research Centre. The paper reflects the views of the scientists, and not the company.

References

1. International Energy Agency. CO2 Emissions from Fuel Combustion: Overview. Available online: https://www.iea.org/reports/CO2-emissions-from-fuel-combustion-overview (accessed on 10 December 2022).
2. Di Lullo, G.; Giwa, T.; Okunlola, A.; Davis, M.; Mehedi, T.; Oni, A.O.; Kumar, A. Blending blue hydrogen with natural gas for direct consumption: Examining the effect of hydrogen concentration on transportation and well-to combustion greenhouse gas emissions. *Int. J. Hydrogen Energy* **2021**, *46*, 19202–19216. [CrossRef]
3. Okunlola, A.; Giwa, T.; Di Lullo, G.; Davis, M.; Gemechu, E.; Kumar, A. Techno-economic assessment of low-carbon hydrogen export from Western Canada to Eastern Canada, the USA, the Asia-Pacific, and Europe. *Int. J. Hydrogen Energy* **2022**, *47*, 6453–6477. [CrossRef]
4. Kim, A.; Yoo, Y.; Kim, S.; Lim, H. Comprehensive analysis of overall H_2 supply for different H_2 carriers from overseas production to inland distribution with respect to economic, environmental, and technological aspects. *Renew. Energy* **2021**, *177*, 422–432. [CrossRef]
5. Reuß, M.; Grube, T.; Robinius, M.; Preuster, P.; Wasserscheid, P.; Stolten, D. Seasonal storage and alternative carriers: A flexible hydrogen supply chain model. *Appl. Energy* **2017**, *200*, 290–302. [CrossRef]
6. Simunovi, J.; Pivac, I.; Barbir, F. Techno-economic assessment of hydrogen refueling station: A case study in Croatia. *Int. J. Hydrogen Energy* **2022**, *47*, 24155–24168. [CrossRef]

7. Hurskainen, M.; Ihonen, J. Techno-economic feasibility of road transport of hydrogen using liquid organic hydrogen carriers. *Int. J. Hydrogen Energy* **2020**, *45*, 32098–32112. [CrossRef]
8. Chen, Q.; Gu, Y.; Tang, Z.; Wang, D.; Wu, Q. Optimal design and techno-economic assessment of low-carbon hydrogen supply pathways for a refueling station located in Shanghai. *Energy J.* **2021**, *237*, 121584. [CrossRef]
9. Di Lullo, G.; Giwa, T.; Okunlola, A.; Davis, M.; Mehedi, T.; Oni, A.O.; Kumar, A. Large-scale long-distance land-based hydrogen transportation systems: A comparative techno-economic and greenhouse gas emission assessment. *Int. J. Hydrogen Energy* **2022**, *47*, 35293–35319. [CrossRef]
10. International Energy Agency. The Future of Hydrogen. Available online: https://www.iea.org/reports/the-future-of-hydrogen (accessed on 16 October 2023).
11. Elnigoumi, A. Investigating Cost Effective Pathways for Blue Hydrogen Production in Alberta. Unpublished. Master's Thesis, University of Calgary, Calgary, AB, Canada, 2021. Available online: http://hdl.handle.net/1880/113957 (accessed on 5 November 2023).
12. Yang, C.; Ogden, J.M. *Determining the Lowest-Cost Hydrogen Delivery Mode*; Institute of Transportation Studies: UC Davis, CA, USA, 2008. Available online: https://escholarship.org/uc/item/0st9s56s (accessed on 5 November 2023).
13. Monthly Average Retail Prices for Gasoline and Fuel Oil, by Geography. Available online: https://www150.statcan.gc.ca/t1/tbl1/en/tv.action?pid=1810000101 (accessed on 20 December 2023).
14. Truck Driver Salary in Canada. Available online: https://ca.indeed.com/career/truck-driver/salaries (accessed on 15 December 2023).
15. Annual Exchange Rates. Available online: https://www.bankofcanada.ca/rates/exchange/annual-average-exchange-rates/ (accessed on 10 November 2023).
16. CPI Inflation Calculator. Available online: https://www.in2013dollars.com/us/inflation/2008?endYear=2020&amount=1 (accessed on 11 November 2023).
17. Leslie, A.; Murray, D. *An Analysis of the Operational Costs of Trucking: 2023 Update June 2023*; American Transportation Research Institute: Arlington, VA, USA, 2023.
18. Parker, N. Using Natural Gas Transmission Pipeline Costs to Estimate Hydrogen Pipeline Costs. Available online: https://escholarship.org/content/qt2gk0j8kq/qt2gk0j8kq_noSplash_cfbe115e54fba9e62c107c7ac2f3ef17.pdf (accessed on 12 November 2023).
19. Olateju, B.; Kumar, A. Techno-economic assessment of hydrogen production from underground coal gasification (UCG) in Western Canada with carbon capture and sequestration (CCS) for upgrading bitumen from oil sands. *Appl. Energy* **2023**, *111*, 428–440. [CrossRef]

World Electric Vehicle Journal

Article

Reducing the Environmental Impact of Large Battery Systems with Conductive Electric Road Systems—A Technical Overview

David Wenander * and Mats Alaküla

Division of Industrial Electrical Engineering and Automation, LTH, Faculty of Engineering, Lund University, P.O. Box 117, SE-221 00 Lund, Sweden; mats.alakula@iea.lth.se
* Correspondence: david.wenander@iea.lth.se

Abstract: A radical transformation of the transport industry is required in order to achieve a fossil-fuel-free vehicle fleet and reach the greenhouse gas emissions goals. Electrification plays a crucial role in this radical process. An electric road system (ERS) is a road that supplies power to electric vehicles as they drive on it, offering numerous advantages. These include an extended driving range, decreased reliance on batteries, and increased flexibility and convenience for drivers, eliminating the need to stop for recharging. This paper highlights the transformative potential of ERS in revolutionizing the land transport sector. Through thorough testing with a conductive ERS demonstrator, the viability of the presented technology is validated. Essential aspects like power transfer, efficiency, safety, and environmental impact showcase ERS's adaptability and scalability across diverse vehicle types. This study recommends widespread ERS support for battery electric vehicles, emphasizing the route toward a sustainable future.

Keywords: electric road system (ERS); dynamic charging; charging infrastructure; battery electric vehicle (BEV); charging; infrastructure

Citation: Wenander, D.; Alaküla, M. Reducing the Environmental Impact of Large Battery Systems with Conductive Electric Road Systems—A Technical Overview. *World Electr. Veh. J.* **2024**, *15*, 59. https://doi.org/10.3390/wevj15020059

Academic Editor: Joeri Van Mierlo and Genevieve Cullen

Received: 15 December 2023
Revised: 1 February 2024
Accepted: 5 February 2024
Published: 8 February 2024

1. Introduction

The decarbonization of road transport is a fundamental step toward significant reductions in global CO_2 emissions. The electrification of road transport is a promising path toward CO_2 reduction [1]. This electrification of road vehicles is however challenging from several viewpoints. The electric traction system (electric machine and corresponding inverter, excluding the battery) is an improvement compared to combustion-based propulsion, since an electric traction system is significantly smaller, lighter, and more efficient than the corresponding combustion-based traction system. The challenge lies in how the energy is transferred to and stored onboard a battery electric vehicle (BEV).

The electric energy can be transferred in three different ways:

1. The onboard battery energy storage can be filled via the direct transfer of electric energy, called charging. The charging rate is constrained by the capacity of the battery to receive power. Modern full-electric-vehicle batteries are designed to store large amounts of energy, enabling faster charging. Fast charging, exemplified by the Kia EV6 [2], can achieve partial recharging from 10% to 80% in about 20 min. Extended charging times, such as during night-time, result in lower charging rates.
2. An alternative is to replace an empty battery with a full one, called battery swapping. There are several automotive OEMs for both commercial vehicles and private cars, like the Chinese car OEM NIO [3], that are designed for systematic battery replacement in just a few minutes in dedicated battery-swapping stations. The empty battery is then recharged in the battery swapping station at a lower charge rate than demanded at a fast charging station.
3. A continuous supply of electric energy can be provided at least for parts of a travelled distance on a public road. This kind of energy transfer can be referred to as an "electric

road system" (ERS), which partly replaces the role of fast charging. While on an ERS, energy can be provided for both propulsion of the vehicle and charging an onboard battery. The need for onboard energy storage is reduced. As presented in [4], even with fractional deployments of an ERS on a main road system, the battery capacity installed in the majority of BEVs can be reduced by 50–70%.

1.1. Electric Roads

An electric road refers to a road that includes some kind of technology that facilitates electric energy transfer from the road to the vehicles driving on it. The energy transfer can be made in a conductive way, meaning that some kind of sliding contacts are used, as used in trains, trams, and trolleybuses. Energy transfer can also occur inductively utilizing a high-frequency magnetic field connecting a transmitting component, typically located just beneath the road surface, and a receiving component installed in vehicles, usually beneath the vehicle body. The company Electreon [5] manufactures an example of this technology. This paper focuses on ERSs in the form of conductive ERSs.

Conductive ERSs exist in at least three different forms:

1. Siemens eHighway [6] is an example of conductive transfer from above the vehicle via catenary lines. This technology is derived from train technology and is only designed to supply power to electric trucks. The catenary lines are located on a highway in Germany and was first commissioned using an electric truck in 2017.
2. Honda has introduced a conductive transfer system from the side of the road [7]. With this technology, the vehicle connects to a continuous, parallel, two-pole supply on the side of the road via an arm extended from the side of the vehicle. Honda has developed a demonstrator ERS based on this technology located on a racing track. The system has demonstrated exceptional performance, effectively transferring power exceeding 450 kW at speeds of 150 km/h.
3. There are multiple companies that have built ERS demonstrators using conductive transfer from the road surface, for instance, Alstom [8], Elways [9], and Elonroad [10]. For this type of conductive ERS technology, a mechanical arm or mechatronic device establishes a connection with a continuous, parallel, two-pole supply located in the road. Both Alstom and Elways have tested their own versions of this technology with electric trucks in enclosed environments. Elonroad has tested this technology on a public road using an electric bus, a passenger car, and a resistive load.

To validate the viability of conductive ERSs, this paper reports on experiences with commissioning an ERS demonstrator with different road vehicles drawing power from it.

1.2. Aspects of Different Charging Technologies

As outlined in [1], a fast-charging-based road transport system requires a ratio of 1:100 or denser of fast charging stations for BEV passenger cars. In Sweden, with about 5 million cars [11], this corresponds to about 50,000 fast charging spots distributed along approximately 15,000 km of national and European roads. This results in at least 100 fast chargers every 30 km on average across Sweden. Along traffic-intense roads, where an even greater demand is expected, even more fast chargers will be needed.

Meeting the increasing demand for a high ratio of fast chargers to BEVs, coupled with a lower overall vehicle count, presents a similar challenge in allocating fast chargers for commercial vehicles. Modeling indicates that with the implementation of an ERS for road transport, approximately 3000 to 4000 km out of the total 15,000 km of national and European roads would require ERS coverage [12]. This coverage would facilitate "non-stop" travel across the country, with vehicles needing 50–70% less battery capacity than a fast-charger-based system.

These conclusions highlight two challenges with electromobility: (I) a huge number of fast charging stations is needed in a fast-charging-based system; (II) by deploying ERS technology on a wide scale, the abundant requirement of fast chargers can be diminished and the amount and size of BEV batteries can be reduced significantly.

1.3. Related CO_2 Emissions

Figure 1 illustrates the CO_2 emissions originating from both the manufacturing and driving phases of passenger cars, local distribution trucks, and long-haul trucks in Sweden. The analysis considers a fully electrified vehicle fleet, comparing a fast-charger-based system (left bar) with an ERS-based system (right bar) in terms of required battery amount. Note that the CO_2 emissions from driving various vehicle types (depicted in the yellow field) remain constant for both types of charging infrastructure. The CO_2 emissions from driving were calculated using the CO_2 emissions related to electricity production from [13] and consumption per vehicle from [4].

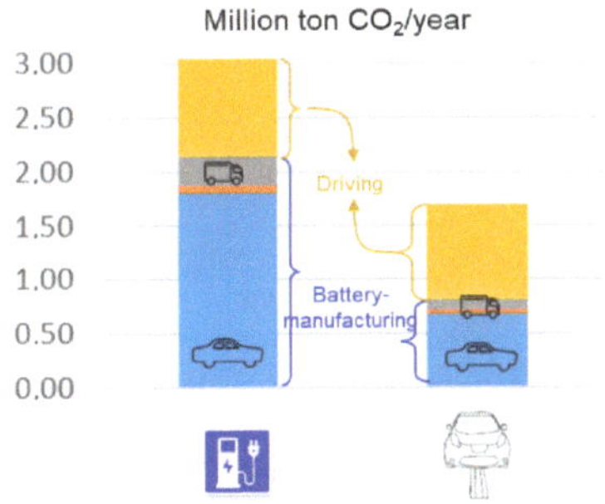

Figure 1. CO_2 emissions from BEVs with a full electrification of cars and two types of trucks in Sweden using either a fast-charger based system (left bar) or an ERS-based system (right bar).

However, the CO_2 emissions related to battery manufacturing (indicated by the blue field for passenger cars, gray for long-haul trucks, and orange for local distribution trucks) are non-negligible compared to the emissions from driving, regardless of the charging infrastructure. Moreover, employing an ERS significantly reduces the CO_2 emissions from battery manufacturing. The calculation of battery manufacturing emissions involved several steps:

1. Determination of the distribution of vehicle types and their quantities in Sweden based on data from [11].
2. Assumption of battery sizes: 30 kWh for passenger cars, 200 kWh for local distribution trucks, and 250 kWh for long-haul trucks [1].
3. Computation of battery amounts for a fast-charger-based system compared to an ERS-based system, relying on [4].
4. Adoption of an assumed emissions factor of 60 kg CO_2/kWh related to battery production [14].

In conclusion, the figure highlights that the CO_2 emissions from battery manufacturing are dominated by passenger car batteries and that the selection of charging technology significantly impacts CO_2 emissions, indirectly via the reduced battery need.

2. Materials and Methods

This paper presents a conductive electric road demonstrator built by the company Elonroad [10], located on a public road in an urban environment in Lund, Sweden, that has been used to assess the potential of the presented electric road technology. The project related to the demonstrator is known as the Evolution Road project [15], which consists of partners from industry, academia, as well as the public sector in Sweden, and is funded by the Swedish Transport Administration. Since the project started in 2019, a number of aspects related to the technology has been tested encompassing the electric power transfer to the installation and build procedures. In this paper, results and experiences from the project concerning the following areas are presented:

- Electric power transfer;
- Efficiency calculations;
- Electrical safety;
- Mechanical safety *;
- Electromagnetic compatibility (EMC) *;
- Acoustic tests *;
- Environmental perspective *.

Areas labeled with an asterisk provide a concise overview of the gained experiences and results and are presented in Sections 2.3–2.5.

2.1. The Electric Road Demonstrator

The conductive electric road demonstrator consists of a transformer, a rectifier, and a segmented conductive electric rail with a total length of 850 m embedded in the top layer of the road surface, as presented in Figure 2. In the following subsection, each subsystem of the demonstrator is presented.

Figure 2. The ERS demonstrator.

2.1.1. Alternating Short-Segmented Electric Road

The electric road presented in this paper is a form of conductive electric road known as an alternating short-segmented electric road (ASSE). This means that the short segments (1 m long) are arranged so that, as illustrated in Figure 3, every other short segment (dark grey) is permanently connected to 0 V DC, and the remaining short segments can be connected to either 0 V DC or 650 V DC by means of solid-state switches integrated in the road. To mitigate the risks associated with hazardous voltage levels, a short segment is activated with 650 V DC only when a vehicle is positioned on top of it. As a convenience and safety measure, each short segment is equipped with LEDs to indicate the activation status of 650 V DC. The activation of short segments is facilitated through wireless communication between the vehicle and ERS.

The ASSE is rectangular in cross-section and built in 10 m sections that are integrated in a groove in the road and secured by a bitumen mass, as presented in Figure 4. A 40 cm wide and 6 cm deep groove is milled in the top layer of the road. The 10 m sections are then placed in the groove hanging on bars, thus are level with the surrounding asphalt. The remaining space is filled with a bitumen mass that fixes the rail to the road and the bars are removed. The whole process is fast and has no negative impact on the underlying layers of the road body. Once installed, the visual impact of the road is very limited, as shown in Figure 2. However, more knowledge and research are required regarding the rectifier station installations and cable installation between the rectifier station and the electric road, as these installations are expected to be more time-consuming and labor-intensive than the installation of the electric road.

The presented ERS technology does not only supply vehicles with power but also offers control over both charge access as well as billing data for the charging vehicles. In addition, the LEDs in the ASSE offer additional safety services such as alerts of traffic accidents and risks of traffic congestion.

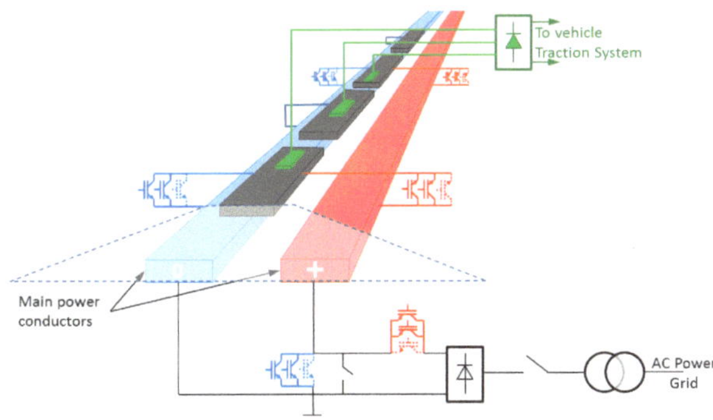

Figure 3. Overview of the basic principle of a conductive ERS taken from [16]. All elements in green are placed onboard the vehicle and represent the interface toward the electric road. The rest of the vehicle is not shown for clarity.

Figure 4. (**Left**): Installation of a 10 m section of ASSE into a groove. (**Right**): Moulding a 10 m section of ASSE into a groove in the road.

2.1.2. The Rectifier Station

A 6-pulse passive rectifier (bottom right in Figure 3) is used to provide 650 V DC to the two internal main power conductors (illustrated in blue and red) in the ASSE. The demonstrator is connected to a 400 V grid; therefore, a 400 kVA 400/450 V transformer is used to achieve a mean three-phase rectified voltage of 650 V DC. In addition to the rectifier and the transformer, the rectifier station also contains solid-state switches (used to connect and disconnect the electric road) and computers that can handle billing and data as well as survey and control the short segments and their corresponding LEDs in the road.

2.1.3. The Power Receiver in the Vehicle—The "Pick-Up"

In order to ensure a smooth power transfer between the ASSE and the vehicle, at least three sliding contacts are required (illustrated in green in Figure 3). The device containing the sliding contacts, presented in Figure 5, is referred to as the "pick-up". The pick-up is a mechatronic device equipped with sensors that automatically ensure contact with the ASSE as it has the capability to lower, raise, and move the sliding contacts laterally. Additionally, it can exert a specific contact force on the contacts. As the pick-up adjusts its position automatically, limited input or attention is required from the driver.

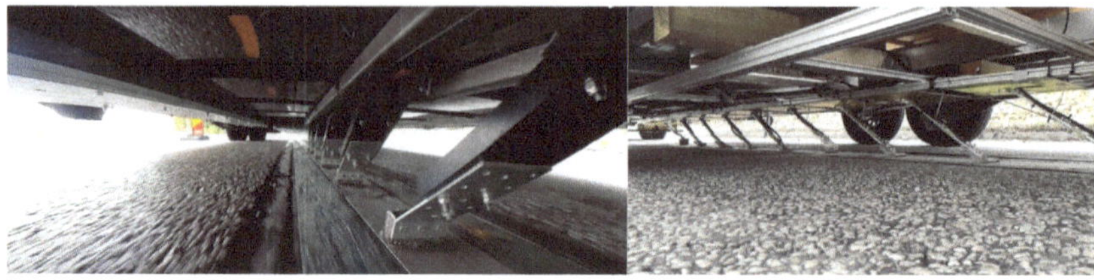

Figure 5. Two different types of pick-ups mounted on different vehicles providing contact between ASSE and vehicle.

As a vehicle drives along the ERS and draws power, the sliding contact points alternate between short segments that are connected to either 650 V DC or 0 V DC. As a consequence, an onboard rectifier is required as the voltage perceived by the vehicle has an alternating voltage polarity from the pick-up. In addition to the onboard rectifier, a DC/DC converter is required onboard the vehicle for two reasons: (I) to match the voltage level of the high-voltage battery with the ERS supply and (II) to provide galvanic isolation between the ERS supply and the traction voltage system (TVS) of the BEV.

As roads are subjected to harsh weather conditions, the contact surface can be wet, icy, or even covered in snow. In addition, small pebbles as well as sand can accumulate on the contact surface. To mitigate the risk of poor contact during cold conditions, the project has successfully used a snow plough to clear the contact surface. Small objects, such as sand or pebbles, are cleared by the pickup due to the nature of the sliding contact. However, small objects have the potential to lift the sliding contact, posing a risk of breaking the current path to the rectifier and potentially causing an arc.

To address this issue concerning poor contact, two measures have proved effective during the project: (I) dividing one sliding contact into several smaller contacts arranged in parallel and (II) incorporating more than three sliding contacts per vehicle. Both of these strategies provide alternative paths for the pick-up currents, mitigating the risk of arcing. Although the pick-up has performed well during the project, with an efficiency well over 97% (as shown in Section 3.2), little is known of its performance in terms of losses at higher speeds. In addition, service intervals and maintenance costs related to the pick-up are hitherto unknown.

2.2. Electric Power Transfer and Efficiency Calculations

In this project, three different vehicles draw power from the electric road: an electric bus, a passenger car, and a resistive load mounted on a trailer. In Figure 6, the interfaces where voltage and current are measured in the demonstrator are depicted as follows: A—grid, B—after the transformer, C—after the rectifier in the rectifier station, and E—after the rectifier in the vehicle. These measurements serve as the basis for calculating power, losses, and efficiency at these interfaces. A full description of the measurement system is presented in [17].

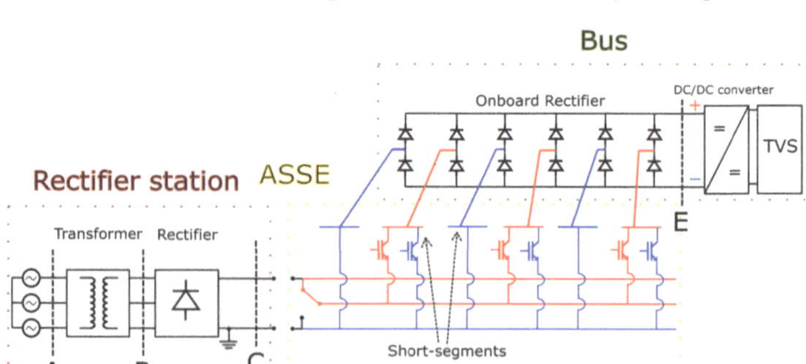

Figure 6. An overview of the measurement interfaces of the ERS demonstrator and the electric bus. Red elements represent 650 V DC and blue represent 0 V DC.

2.3. Safety

There are many safety aspects to consider related to conductive ERS technology. The following issues are considered in this paper:

- The two main areas of *electrical safety* in a conductive ERS are related to the risk of touch events related to hazardous voltages. This means that there should be no risk related to touching the electric road or a vehicle that is connected to an electric road. Firstly, touch protection of the ASSE is accomplished by the design, as the outer exterior of the ASSE is grounded to 0 V potential in the rectifier station. In addition, short segments are only activated with hazardous voltage levels of 650 V DC when a vehicle is located over it; otherwise, they remain at safe ground potential of 0 V, as presented in Section 2.1.1.
  Secondly, the high-voltage battery, ERS supply voltage, and corresponding TVS in the vehicle are galvanically isolated from the chassis to prevent hazardous touch currents from flowing through a person's body (standing on conductive ground) when touching the vehicle's chassis. Figure 7 presents the basic concept of an isolation fault between the positive high-voltage pole of the ERS supply and the vehicle's chassis while a human body is in contact with the vehicle's chassis and ground. This event causes current to flow through the human body to ground and back to the rectifier station.
  However, to mitigate this risk, the vehicle is equipped with an isolation fault monitoring system. The isolation fault monitoring system is designed to ensure that the impedance between the high-voltage poles in the TVS and the chassis is unaltered and kept in the magnitude of MΩ. If an isolation fault occurs (meaning that the impedance is instantly drastically reduced), the intended isolation fault monitoring system is triggered and disconnects the high-voltage battery from the TVS as well as the vehicle from the ERS supply.
  In order to gain a greater understanding of this phenomenon, measurements of touch currents were obtained when a human model, defined according to standard [18], is connected to the electric bus chassis while the bus was drawing power from the ERS. No isolation faults were present in the bus or the ERS demonstrator during these measurements. A 4444 Picoscope [19] was used to measure the voltage over the human model in order to assess the magnitude of touch current, as outlined in the related standards [18,20]. The results of these measurements are presented in Section 3.3 in this paper.

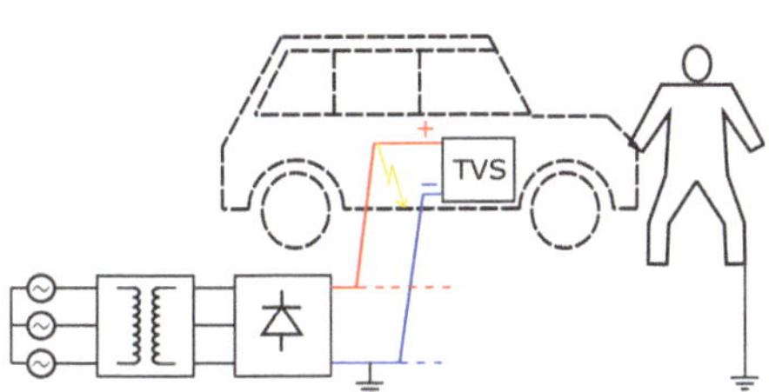

Figure 7. Conceptual overview of the electric safety issue of touch current when a vehicle is drawing power from the ERS and a human body is connected to the vehicle's chassis. Red elements represent 650 V DC and blue represent 0 V DC.

- *Mechanical safety* refers to the road friction on the short segments of an ASSE, which, as a conductive ground-based technology, can cause poor friction compared to the adjacent friction on the original road. To mitigate this risk, a pattern is engraved in the short segments that is small enough to not interfere with the sliding contacts but provide enough grip to the rubber in the wheels. Throughout the project, a number of friction test have been conducted by two different parties, Ramboll consultancy and The Swedish National Road and Transport Research Institute (VTI). All tests concluded that the ASSE fulfilled the friction requirements for public roads in Sweden.

- *Electromagnetic compatibility (EMC)* is briefly mentioned in this paper as the ERS must comply with standards concerning radiated emissions. However, as there are no specific standards concerning ERSs, measurements were conducted based on best practices concerning EMC as well as standards addressing similar infrastructure, for instance, railways, tramways, and trolleybuses. Three different organizations conducted measurements at different occasions: Research Institute of Sweden (RISE), Swedish Defence Materiel Administration (FMV), and Department of Biomedical Engineering, LTH, Faculty of Engineering. Measurements of radiated emissions, while the bus is drawing power from the demonstrator ERS, identified peaks across a broad spectrum of frequencies. To investigate this issue, further measurements are planned to be conducted in a laboratory environment, aiming to mitigate the uncertainties associated with background emissions.

2.4. Acoustic Tests

There are many aspects to consider in an ERS, for instance, the issue of noise pollution. In order to assess the impact of the sliding contacts concerning noise, acoustic tests were performed with the ERS demonstrator. The tests were performed according to ECE-R51 [21] with the electric bus on the ERS demonstrator. The tests were performed at night time in order to reduce the influence of background noise. The bus drove at different speeds with the pick-up lowered and raised for each speed as to determine the impact of the sliding contact. Initial results showed no significant increase in noise emissions with the pick-up connected to the ASSE.

2.5. Environmental Perspective

As ERSs have the benefit of reducing CO_2 emissions by allowing BEVs to have smaller batteries, this paper also briefly addresses the environmental perspective of recycling the ERS. As the ERS demonstrator is still in commission, the process of recycling is in a preliminary phase. However, initial measures of recycling subsystems and parts of the ERS have thus far produced promising results. The project is presently assessing the economic benefits of recycling, with more detailed information forthcoming. However, it is evident that recycling the entire system post use poses no practical challenges, and the metal value of the components far surpasses the recycling costs.

A primary motivation behind the project is the need to lower the overall need for batteries in a future, mostly for BEVs. A compelling argument for ERS-based infrastructure capable of serving all BEV types lies in the fact that approximately 90% of the batteries in existing and future vehicles are and will be in smaller vehicles, with about 10% in heavy duty trucks (HDTs) [1]. Infrastructure that only services HDTs would have a comparatively minor impact on the environmental footprint of battery production and the complexity of recycling batteries. In conclusion, from an environmental standpoint, comprehensive support for all BEV types is recommended.

3. Results

3.1. Electric Power Transfer

Two types of electric power transfer tests are presented in this paper: (I) two vehicles (a passenger car and the electric bus) are drawing power from the ERS demonstrator simultaneously, see Figure 8 (left), and (II) a resistive load mounted on a trailer, see Figure 8 (right), draws a power of 280 kW at a speed of 80 km/h.

The two-vehicle test showed that the road was able to provide individual supply to several vehicles. In Figure 9, the total drawn power in interface C (interfaces defined in Section 2.2) is presented while a passenger car (a Nissan Leaf) and an electric bus (Solaris Trollino 15) draw power from 140 m of electric road. The measurements were sampled at 200 kHz using LEM sensors. The full measurement system is presented in [17].

Figure 8. (**Left**): An electric bus and a passenger car drawing power from the ERS demonstrator simultaneously. (**Right**): The trailer designed to draw 300 kW.

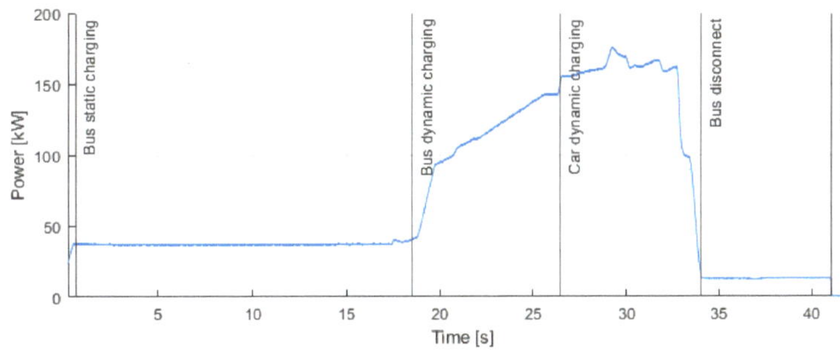

Figure 9. Bus and passenger car drawing power from the ERS demonstrator.

At 0.5 s, static charging is initiated, and the bus is standing still while charging from the ERS demonstrator and drew a power of 37 kW. During static charging, the drawn power is limited to approximately 40 kW to avoid the risk of overheating the contact points in the pick-up. As the bus starts to drive, at 18.5 s, the drawn power to the battery increases from 37 kW to roughly 80 kW. As dynamic charging commences, the bus accelerates, and the total drawn power in interface C increases. While the bus is accelerating, the passenger car starts to draw 13 kW dynamically (located behind the electric bus) at 26.5 s; consequently, the total drawn power is instantly increased from 142 kW to 155 kW. At 32.5 s, the bus reaches a top speed of approximately 20km/h, and, at 33 s, the bus starts to brake until it disconnects from the electric road at 34 s. At 41 s, the passenger car drives off the electric road and disconnects.

Throughout this charging event, the onboard DC/DC converter in the passenger car is configured to draw 13 kW. Due to the low state of charge (SoC) in the car's high-voltage battery, the entire 13 kW is allocated to the battery circuit, where it is distributed between the battery and the electric traction system. Since the DC/DC converter sets a limit on the drawn power from the ERS demonstrator, the drawn power remains constant regardless of whether the passenger car is charging statically or dynamically.

In Figure 10, measurements of the drawn power and speed from the ERS demonstrator are presented from interface C (power on left y–axis, and speed on right y–axis) with the trailer acting as the vehicle load. These measurements were obtained using the Elonroads internal measurement system sampled at approximately 1 kHz. After 0.9 s, the trailer connects to the first ASSE section and draws a power of 280 kW. As different versions of the 10 m sections of ASSE were used during this measuring event, the first ASSE 10 m sections that the trailer connected to were not equipped with sensors to measure speed. Consequently, the speed of the trailer was unknown until 1.3 s when the trailer made contact with a 10 m section of ASSE that could provide speed measurements. When speed measurements were available at 1.3 s, the trailer had a speed of 75 km/h and reached a speed of 80 km/h at 2.5 s. After 3 s, the car that was pulling the trailer started to break and reached a speed of 13 km/h after 6.5 s. The trailer disconnected from the ASSE at 8.1 s, and the power was 0 kW at this time. The trailer came to a complete stop after 8.6 s.

As presented in Figure 10, the ERS demonstrator is able to provide a continuous power of 280 kW with no disruption between 0.9 and 8.1 s at speeds up to 80 km/h. This verifies that the presented ERS technology is capable of providing high power levels at high speeds resembling highway conditions for trucks. This high power level of 280 kW is mainly intended for heavy trucks and not expected to be drawn by cars that are expected to draw up to 50 kW from the ERS. Although the ERS demonstrator is designed to provide 300 kW of power, tests have shown that the installed transformer in the rectifier station is slightly undersized. Consequently, the transformer restricts the trailer from drawing power at these levels.

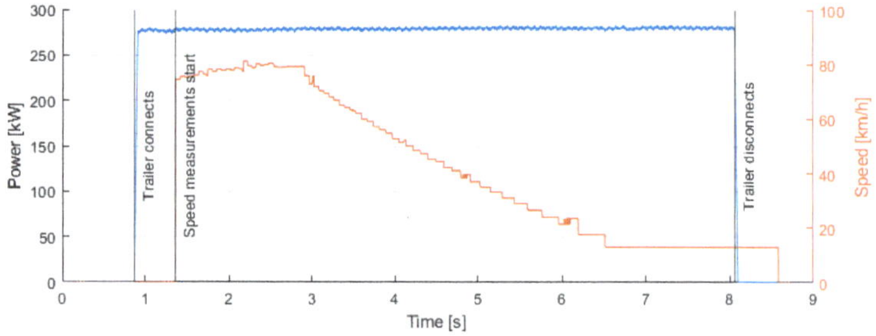

Figure 10. Trailer drawing power from the ERS demonstrator. The left y–axis shows the drawn power (blue line), and right y–axis shows the speed of the trailer (orange line).

3.2. Efficiency Calculations Based on Tests

To evaluate the efficiency between the grid and the vehicle for the showcased ERS technology, a charging event was conducted wherein the electric bus draws power from the ERS demonstrator. Figure 6, previously introduced in Section 2.2, presents an overview of the ERS demonstrator and the electric bus with the corresponding measurement interfaces (A–E), where current and voltage are used to calculate the losses and efficiency for the ERS demonstrator. Interface A–C represents the losses in the transformer and the rectifier, and interface C–E represents the losses in the underground cable to the ASSE, the ASSE, the sliding contact, and the onboard rectifier.

In Figure 11, a charging event with 140 m of electric road and the electric bus drawing power from it is presented. In the upper plot, losses (left y–axis) and drawn power (right y–axis) at interfaces A–C, C–E, and A–E are presented with a moving average filter with a time constant of 50 ms; in the lower plot, the efficiency (left y–axis) and drawn power (right y–axis) at interfaces A–C, C–E, and A–E are presented with a moving average filter with a time constant of 250 ms. Losses and efficiencies were calculated based on measurements of voltage and current using the measurement system presented in [17].

At 2 s, the bus is at standstill and draws 39 kW from the electric road. The drawn power, supplied to the battery during static charging, is limited to 39 kW in order to avoid overheating the contact points in the pick-up. After 18 s, the bus starts to accelerate and dynamic charging is initiated. Consequently, the previous power limit to the battery increases from 39 kW to 79 kW at 19 s. Simultaneously, the drawn power required for propulsion increases as the bus accelerates, reaching a peak power of 187 kW at 22.5 s. By 25 s, the bus has attained its final speed of 40 km/h, and at 27 s, the drawn power reduces as the bus starts to cruise at 40 km/h.

During static charging, the loss in the ERS demonstrator from grid to vehicle (interface A–E, upper plot) is 2.1 kW, which corresponds to an efficiency of 94% (interface A–E, lower plot). A total of 1.7 kW of these losses occur in the rectifier station (interface A–C, upper plot), which results in a efficiency of 95% for the rectifier station. The corresponding loss in the ASSE and up to the BEV DC/DC converter (interface C–E, upper plot) is 0.4 kW, which results in a efficiency of 98.7%. During dynamic charging, the total losses in the ERS demonstrator (interface A–E, upper plot) at peak power (187 kW at 22.5 s) is 9 kW, which results in a total efficiency of 95.4% (interface A–E, lower plot). The losses in the rectifier station are now 3.6 kW (interface A–C, upper plot) and that in the ASSE up to the DC/DC converter in the BEV (interface C–E, upper plot) reaches 5.4 kW. Although the losses increase at peak power, and generally during dynamic charging, the efficiency for the rectifier station (interface A–C, lower plot) increases to 98.1%, and the efficiency for the ASSE and up to DC/DC converter in the BEV decreases to 97.2%. After peak power at 22.5 s, the efficiency for the ASSE and the vehicle (interface C–E) increases to 97.4%.

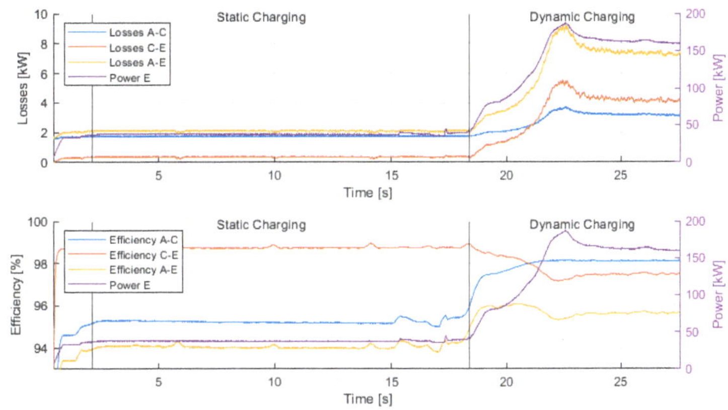

Figure 11. Power (right y–axis), losses (upper plot, left y–axis), and efficiency (lower plot, left y–axis) when the bus is drawing power from the ERS demonstrator.

To conclude, from Figure 11, it is clear that the presented conductive ERS technology offers great efficiency performance. For the ASSE and the vehicle (interface C–E), an efficiency between 97.2 and 98.7% shows great promise for the presented ERS technology as high efficiency is crucial for ERS deployment on a wide scale.

However, it is also evident that the efficiency performance is dependent on the power drawn by the vehicles. During static charging, the efficiency for the rectifier station is lower than during dynamic charging. This is because the relatively constant magnetization losses cause the efficiency of the 400 kVA transformer in the rectifier station to increase with increasing load. At lower loads (around 40 kW in this case), the no-load losses in the transformer are high compared to the load-losses, which results in a low efficiency for the transformer. This indicates that the transformer choice in conjunction with drawn power load from the ERS are important factors for the efficiency of the presented ERS technology.

During static charging, the losses in the ASSE and in the vehicle are smaller than during dynamic charging, which is due two main reasons: (I) The resistive losses in the system increase with the square of the drawn current. Hence, a higher drawn power results in higher losses, as seen in Figure 11 when dynamic charging commences. (II) During dynamic charging, the contact resistance in the sliding contact between the pick-up and short segment increases. As a result, the resistive losses in the system at interface C-E increase as both the resistance and the drawn current increase, which leads to an overall reduction in efficiency.

3.3. Electrical Safety Test—Touch Current

For a conductive ERS, the issue of electrical safety is critical. One of the most important safety aspects is the issue of touch current related to the vehicle's chassis, meaning that there should be no safety risks of touching an electric vehicle's chassis while it is charging on an electric road. In order to assess the electrical safety aspect of touch current, measurements were conducted on the ERS demonstrator with a model of a human body connected to electric bus chassis while the bus was drawing power from the demonstrator.

In Figure 12, the measurement setup for the touch current tests is presented, where a human model is connected between a BEV chassis and ground while the BEV is drawing power from an electric road. The human model (defined in [18]) comprised a resistor R_s (1.5 kΩ), connected in parallel with a capacitor C_s (0.22 nF), which was connected in series with the resistor R_b (0.5 kΩ). Resistor R_s and C_s correspond to the impedance of a human body's skin, while resistor R_b corresponds to the impedance of a human body excluding the skin. The voltage V_b over resistor R_b was measured in order to assess the current that flows through the human model when connected to the vehicle's chassis.

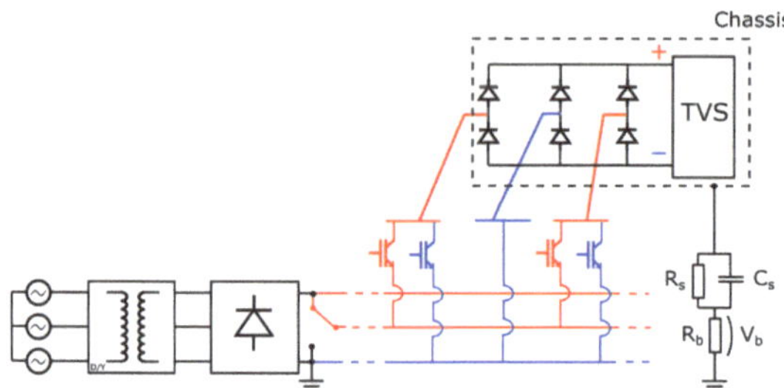

Figure 12. Overview of the measurement setup related to touch current when a human model is connected between a BEV chassis and the ground while the BEV is drawing power from an ERS. Red elements represent 650 V DC and blue represent 0 V DC.

Figure 13 presents measurements of the voltage V_b and current through the human model I_b when the electric bus is charging on 80 m of electric road with a human model connected between its chassis and ground. The upper plot shows the voltage V_b, and the lower plot shows the current I_b. In both plots, the unfiltered value (left y–axis) as well as a moving root mean square (RMS) value (right y–axis) with a time constant of 100 ms are shown. The relevant levels of touch current are presented as RMS values in the relevant standards [18,20]. Although the high-voltage system of the bus is isolated from the chassis, there is a current that flows through the human body to the ground when the bus is drawing power from the ERS demonstrator. This is caused by inadvertent capacitive coupling (also known as-parasitic capacitance) between the high-voltage poles of the TVS and the bus chassis. As presented in [22], the greatest impact on the voltage V_b is the parasitic capacitance between the output of the converters in the TVS and the bus chassis, for instance, the output of the onboard high voltage battery charger and the traction inverter, see Figure 14. Therefore, the magnitude of touch current is greatly related to which subsystems and corresponding converters are active in the TVS during charging.

In Figure 13, between 0.5 s and 3.5 s, the onboard charger starts its sequence to initiate static charging from the ERS demonstrator. After 3.5 s, the bus draws 38 kW of power from the electric road statically, and the RMS of V_b and I_b reaches values of 7.7 V and 15.6 mA, respectively. As dynamic charging starts at 5.5 s, the traction inverter starts to draw power from the ERS, and the current through the capacitive coupling between the output of the traction inverter and bus's chassis, see Figure 14, increases the RMS of V_b to 10.6 V and of I_b to 21.2 mA in Figure 13.

The presented measurements show that the BEVs that charge from an ERS must account for this parasitic capacitance in their onboard converters. Despite the observed touch current values exceeding the recommendations specified by relevant standards [18,20], these findings do not implicate an inherent fault in the presented ERS technology. Two primary factors account for this discrepancy. Firstly, the particular electric bus that was used for the measurements lacked complete double isolation between the TVS and chassis, a feature now considered standard in modern BEVs. Single isolation elevates the risk of isolation faults and the occurrence of parasitic capacitance. Secondly, the electric bus does not have a fully functioning isolation fault monitoring system that is adapted for the presented conductive ERS technology. Finally, it is probable that mitigating the touch current issues associated with this phenomenon could be achieved through measures such as (I) using an isolated DC/DC converter or (II) minimizing the parasitic capacitances during the vehicle's design phase.

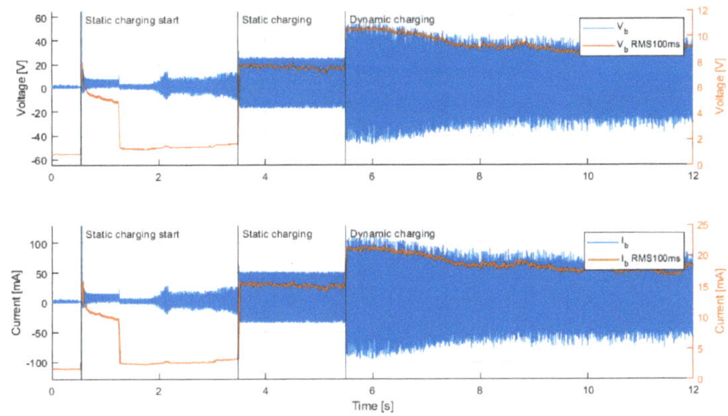

Figure 13. Measurements of the human model connected between the bus chassis and the ground. Upper plot: Blue lines represents the unfiltered voltage V_b (**left** y–axis) and red (**right** y–axis) with a moving RMS filter of 100 ms. Lower plot: Blue lines represents the unfiltered current I_b (**left** y–axis) and red (**right** y–axis) with a moving RMS filter of 100 ms.

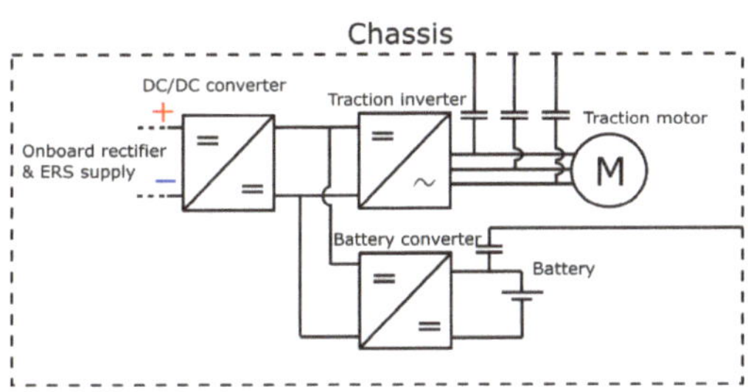

Figure 14. An overview of the bus's TVS that illustrates the inadvertent capacitive coupling between the output of the onboard converters and the bus's chassis.

4. Discussion

Electrifying road transport demands strategic choices, where ERS emerges as a transformative solution, mitigating both infrastructure demands and environmental footprint. The showcased ERS demonstrator, along with its corresponding installation, commissioning, and comprehensive testing within the Evolution Road project, not only validates and establishes the viability of the presented conductive ERS technology but also confirms the feasibility of the overall concept of ERS. This paper explores various aspects and presents tests essential for the widespread deployment of ERS, including electric power transfer, efficiency evaluation, electrical and mechanical safety, acoustic emissions, EMC, and environmental considerations.

The electric power transfer tests showed the presented ERS technology's capability of supplying power to two vehicles simultaneously and showcasing its suitability for highway implementation by using a trailer equipped with a resistive load. While the drawn power was constrained to 280 kW, the limitations were not attributed to the design of the ERS demonstrator but rather to the transformer in the rectifier station. Also, power levels exceeding 300 kW are only considered for HDTs, whereas passenger cars are expected to draw power in the range of 50 kW. Given the aforementioned distribution between passenger cars and HDTs, these expected power levels pose no threat to an ERS's power capabilities. These results reinforce the ERS technology's adaptability and scalability for various vehicle types and power requirements with no implications.

The ERS demonstrator's performance in terms of efficiency further showcases the potential of this ERS technology for wide deployment. An efficiency of over 97% at power levels of over 150 kW during dynamic charging between rectifier station and onboard DC/DC converter in the vehicle, see Figure 11 interface C–E, is a noteworthy achievement. Although these results are impressive, the efficiency performance of the presented technology under high-speed conditions, involving more challenging contact scenarios on rough terrain with dirt, leaves, and snow, is hitherto unknown. At higher speeds, it is anticipated that the pick-up may introduce risks in terms of reliable functionality due to poor contact performance. However, the pick-up design is still in its early stages, and preliminary designs and tests have indicated substantial potential for improvement.

Ensuring electrical safety is paramount for conductive ERSs. This paper presents the critical concern of touch current related to the vehicle's chassis during charging. Touch current measurements, conducted with a human model connected to an electric bus's chassis drawing power from the ERS demonstrator, revealed inadvertent capacitive coupling issues. The voltage and current through the human model were assessed during charging, highlighting the impact of parasitic capacitance between the output of the onboard converters in the bus TVS and the bus's chassis.

Despite touch current values exceeding recommended standards, it was clarified that this does not inherently condemn ERS technology. Limitations in the tested electric bus, lacking complete double isolation between the TVS and chassis, and the absence of a fully functioning isolation fault monitoring system contribute to the observed touch current values. Mitigation strategies, such as employing an isolated DC/DC converter or minimizing the parasitic capacitances between the output of the onboard converters and chassis during vehicle design, are proposed to address touch current concerns effectively.

Although the suggested solutions will be the subject of further work and their effectiveness is unknown, their impact and cost on vehicle design is expected to be minor. While the phenomenon of capacitive coupling between the output of the onboard converters and chassis is a novel consideration for conductive ERS safety, capacitive coupling between the chassis and TVS is a well-known phenomenon, as it is considered in various standards [23]. Given the automotive industry's existing familiarity with and management of such issues, the incorporation of these mitigating strategies is not anticipated to yield significant implications for overall vehicle design and cost.

Despite the promising potential of ERS as a charging infrastructure, three significant challenges remain unresolved: Firstly, for widespread deployment of ERSs, it is imperative to establish common regulations and standardization on an international scale, ensuring the implementation of compatible systems that facilitate international transport. The second challenge concerns financing and business models during both deployment and operation. The responsibility for conducting and financing the deployment and operation of ERS remains uncertain, whether it should be undertaken by governments, companies, or private vehicle owners. Thirdly, in connection with operational concerns, the matter of maintenance and its associated costs remains unknown.

Throughout the year, the demonstrator has remained active and operational, enduring various harsh weather conditions, but not without maintenance work. Although maintenance is expected to be required for ERSs, accurately estimating the actual maintenance needs proves challenging due to the experiences gained from the demonstrator. This difficulty arises partly from reduced traffic volumes and partly from the experimental nature of the installation. Given the absence of a fully operational conductive ERS on public roads, additional knowledge regarding maintenance and associated costs is essential.

However, as this paper focused on technical issues related to ERSs, these three challenges are beyond the scope of this paper. Nevertheless, addressing these challenges is pivotal for achieving widespread deployment of ERSs.

5. Conclusions

The Evolution Road project illuminates the transformative potential of ERS in reshaping the land transport sector. Extensive testing has revealed ERS as a solution that effectively balances charging infrastructure needs with environmental concerns. While challenges related to regulation, operation, maintenance, financing, and business models remain, the adaptability of the ERS across various vehicles highlights its viability and scalability. To ensure an environmentally sustainable future, it is imperative to advocate for the comprehensive support of all types of BEVs through ERS integration. Future endeavors involve refining efficiency, addressing touch current concerns, and advancing societal integration for widespread ERS adoption.

Author Contributions: Conceptualization, M.A. and D.W.; methodology, M.A. and D.W.; software, D.W.; validation, D.W.; formal analysis, D.W.; investigation, D.W.; resources, M.A.; data curation, D.W.; writing, D.W.; writing—review and editing, D.W. and M.A.; supervision, M.A.; project administration, D.W. and M.A.; funding acquisition, M.A. All authors have read and agreed to the published version of the manuscript.

Funding: This research and the Evolution Road project were funded by the Swedish Transport Administration.

Institutional Review Board Statement: Not applicable.

Informed Consent Statement: Not applicable.

Data Availability Statement: Data available on request due to restrictions. The data presented in this study are available on request from the corresponding author. The data are not publicly available due to nondisclosure agreements within the Evolution Road project.

Acknowledgments: The authors express gratitude for the assistance and direction provided by various project members, notably Anna Wilkens at Innovation Skåne for her involvement and contribution to the first draft and outline of this paper. Special acknowledgment is extended to Johan Gran from the Department of Biomedical Engineering, LTH, Faculty of Engineering at Lund University for his valuable feedback on the EMC results.

Conflicts of Interest: The authors declare no conflicts of interest.

References

1. IEA—International Energy Agency. "Global EV Outlook 2023". Available online: https://www.iea.org/reports/global-ev-outlook-2023 (accessed on 1 February 2024).
2. KIA. Available online: https://www.kia.com/ (accessed on 1 February 2024).
3. NIO. Available online: https://www.nio.com/ (accessed on 1 February 2024).
4. Márquez-Fernández, F.J.; Bischoff, J.; Domingues-Olavarría, G.; Alaküla, M. Assessment of future EV charging infrastructure scenarios for long-distance transport in Sweden. *IEEE Trans. Transp. Electrif.* **2021**, *8*, 615–626. [CrossRef].
5. Electreon Website. Available online: https://www.electreon.com/ (accessed on 1 February 2024).
6. Boltze, M. eHighway—An Infrastructure for Sustainable Road Freight Transport. In *Proceedings of the CIGOS 2019, Innovation for Sustainable Infrastructure*; Springer: Singapore, 2020; pp. 35–44.
7. Tajima, T. Study of 450-kW Conductive ERS at 150km/h. In Proceedings of the 3rd Electric Road Systems Conference, Frankfurt am Main, Germany, 7–8 May 2019; pp. 1–7.
8. Veyrunes, P.; Duprat, P.; Hourtane, J.L. Ground-level feeding systems from rail to road. In Proceedings of the 2017 IEEE Transportation Electrification Conference and Expo, Asia-Pacific (ITEC Asia-Pacific), Harbin, China, 7–10 August 2017; pp. 1–4. [CrossRef].
9. Elways AB Website. Available online: http://elways.se/ (accessed on 1 February 2024).
10. Elonroad AB Website. Available online: http://elonroad.com/ (accessed on 1 February 2024).
11. The Swedish Transport Agency. Available online: https://www.transportstyrelsen.se/sv/vagtrafik/statistik/Fordonsstatistik/ (accessed on 1 February 2024). (In Swedish)
12. Rogstadius, J. *Interaction Effects between Battery Electric Trucks, Electric Road Systems and Static Charging Infrastructure*; Technical Report; RISE Research Institutes of Sweden AB: Gothenburg, Sweden, 2022. Available online: http://www.diva-portal.org/smash/get/diva2:1712747/FULLTEXT03.pdf (accessed on 1 February 2024). (In Swedish)
13. Statista. Available online: https://www.statista.com/statistics/1290491/carbon-intensity-power-sector-sweden/ (accessed on 1 February 2024).

14. MIT Climate Portal. Available online: https://climate.mit.edu/ask-mit/how-much-co2-emitted-manufacturing-batteries (accessed on 1 February 2024).
15. The Evolution Road Project Website. Available online: https://www.evolutionroad.se/en/ (accessed on 1 February 2024).
16. Abrahamsson, P. Thermal Management of Conductive Electric Road Systems. Ph.D. Thesis, Division of Industrial Electrical Engineering and Automation, Lund University, Lund, Sweden, 2020.
17. Wenander, D.; Abrahamsson, P.; Márquez-Fernández, F.J.; Alaküla, M. Measuring electric properties of a conductive electric road. In Proceedings of the 2021 AEIT International Conference on Electrical and Electronic Technologies for Automotive (AEIT AUTOMOTIVE), Torino, Italy, 17–19 November 2021; pp. 1–6. [CrossRef].
18. *IEC 60990*; Methods of Measurement of Touch Current and Protective Conductor Current. IEC: Geneva, Switzerland, 2016.
19. Picotech Website. Available online: https://www.picotech.com/download/datasheets/picoscope-4444-data-sheet.pdf (accessed on 1 February 2024).
20. *IEC 60479*; Effects of Current on Human Beings and Livestock. IEC: Geneva, Switzerland, 2018.
21. Regulation No 244 51 of the Economic Commission for Europe of the United Nations (ECE-R51). Available online: https://eur-lex.europa.eu/legal-content/EN/TXT/PDF/?uri=CELEX:42018X0798&rid=8#:~:text=This%20Regulation%20contains%20provisions%20on,normal%20driving%20in%20urban%20traffic (accessed on 1 February 2024).
22. Wenander, D.; Márquez-Fernández, F.J.; Alaküla, M. Modelling Electric Transients in a Conductive Electric Road System. In Proceedings of the 2023 AEIT International Annual Conference (AEIT), Rome, Italy, 5–7 October 2023; pp. 1–6. [CrossRef]
23. *ISO 6469-3:2021*; Electrically Propelled Road Vehicles—Safety Specifications—Part 3: Electrical Safety. ISO: Geneva, Switzerland, 2021.

World Electric Vehicle Journal

MDPI

Communication

Flexibility Potential of Smart Charging Electric Trucks and Buses †

Christian Will [1,*] and Fabian Ocker [2]

1 Daimler Truck AG, Fasanenweg 10, 70771 Leinfelden-Echterdingen, Germany
2 TenneT TSO GmbH, Bernecker Straße 70, 95448 Bayreuth, Germany
* Correspondence: christian.will@partner.kit.edu

† This article is a revised and expanded version of a paper entitled "Flexibility potential of smart charging of electric trucks and buses", which was presented at 36th Electric Vehicle Symposium & Exposition, 11–14 June 2023, Sacramento, CA, USA.

Abstract: In addition to passenger vehicles, battery-electric trucks and buses could offer substantial flexibility to the energy system. Using a Bass diffusion model, we extrapolated the unidirectional charging needs and availability of trucks in five of eleven typical applications, as well as city buses, for Germany until 2040. Combined, these heavy-duty vehicles could provide up to 23 GW of down-regulating flexibility potential (i.e., in case of excess power supply) in 2040. The resulting revenues could contribute to reducing electricity costs for depot operators. These results illustrate the need to provide easy and automated market access to heavy-duty vehicle fleets.

Keywords: heavy-duty electric vehicles; electric trucks; electric buses; smart charging; flexibility potential

Citation: Will, C.; Ocker, F. Flexibility Potential of Smart Charging Electric Trucks and Buses. *World Electr. Veh. J.* **2024**, *15*, 56. https://doi.org/10.3390/wevj15020056

Academic Editors: Joeri Van Mierlo and Genevieve Cullen

Received: 22 December 2023
Revised: 13 January 2024
Accepted: 30 January 2024
Published: 7 February 2024

1. Introduction

The European electricity grid is maintained and operated by unbundled grid operators for ultra-high and high voltage levels by so-called transmission system operators (TSOs). TSOs co-create and partly operate markets to solve physical challenges such as frequency deviations or bottlenecks in the grid (i.e., congestions). These are referred to as ancillary services [1] and can be divided into four flexibility segments: two ancillary services, balancing power and congestion management, as well as congestion alleviation and the wholesale market. These flexibility segments consider regulatory, technological, and economic framework conditions, as well as the involvement of key stakeholders. Due to the increasing share of electricity generation from renewable sources, as well as the increasing electrification of heating and transport sectors, more flexibility will be needed in the future, in particular on the demand side. The first two segments are the most promising for the integration of demand-side flexibility from electric vehicles. They are briefly introduced in the following discussion, and the temporal order of market closures in Germany is provided in Figure 1.

- **Balancing power** provides upward regulation (supplying additional energy to the grid) and downward regulation (drawing excess energy from the grid) to guarantee a constant equilibrium between electricity generation and consumption, and thus maintain a stable system frequency of 50 Hertz at any time. In particular, the uncertainty of wind and solar generation forecasts is an important driver for the increasing need for flexibility to keep the system in balance. German TSO TenneT expects the need for flexibility to grow by up to 3 GW by 2030. Balancing power is procured in three "qualities" representing different speeds and durations of intervention, namely, frequency containment reserve (FCR), automatic frequency restoration reserve (aFRR), and manual frequency restoration reserve (mFRR). All three are procured through auctions until a certain time on the previous day (D-1).
- **Congestion management** aims to solve an energy transmission (or distribution) problem by making use of remedial actions, such as redispatch and feed-in management.

The task is to match market outcomes with the physical restrictions of the grid during real-time operation. Locational shifts in generation (wind and solar), increasing peak supply, and new demand centers increase needs in this segment. TenneT expects additional flexibility needs in this segment of up to 9 GW by 2030.

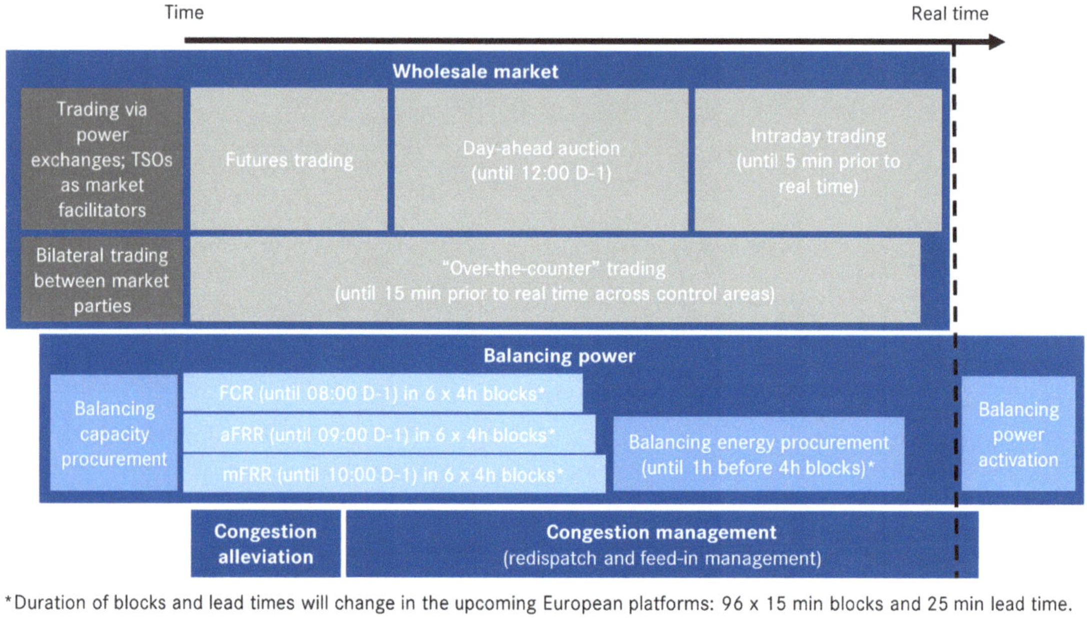

Figure 1. Temporal sequence of market closures for flexibility segments (dark blue) in Germany.

Although the use of battery-electric vehicle (BEV) passenger cars to provide flexibility to the power grid has been investigated extensively (e.g., [2–5]), the body of research regarding the flexibility potential of battery-electric trucks and buses is much smaller. Although the technical implementation of smart charging should be similar for all vehicle sizes [6], the impact of electric truck and bus charging on electricity grids appears significant due to larger batteries, longer distances travelled, and larger charging powers [7,8]. Although buses have well-planned routes with high temporal synchronization [9], truck use cases are diverse (cf. Section 2). Initially, Borlaug et al. found that the early ramp-up of short-haul, predictable truck use cases can likely be accommodated with existing infrastructure in the US [10]. However, this may change with increasing penetration of the technology to more demanding use cases [11]. As a first step, minimizing depot peak-load already strains the grid less and lowers electricity costs for truck fleets due to reduced demand charges [12]. Taljegard et al. [13] showed that a completely electrified transport sector using bidirectional charging, including trucks and buses, could reduce necessary investments in the energy system to meet peak-power by 50% in Sweden, Germany, the UK, and Spain.

In contrast, we aimed to investigate in detail the flexibility and remuneration potential on a per-depot level, focusing on comparing different vehicle use cases. We considered unidirectional conductive direct current (DC) charging using the CCS2 charging standard (combined charging system). As a refined market framework is currently in place only for balancing power, our quantitative analysis focused on this flexibility segment rather than congestion management.

This feasibility study examined how electrified medium- and heavy-duty trucks and city buses can provide flexibility to the energy system by investigating key economic, regulatory, legal, and technical aspects. This paper is structured as follows. Section 2 describes our methodological approach and use case assumptions. Results for initial considerations, technical flexibility, and remuneration potential are discussed in Section 3. Section 4 presents our conclusions and discusses future work.

2. Materials and Methods

The approach taken in this study was twofold (cf. Figure 2). First, expert workshops with representatives from Daimler Truck and TenneT were held. Second, flexibility and marketing potential were derived for a range of use cases and extrapolated over exemplary market ramp-ups.

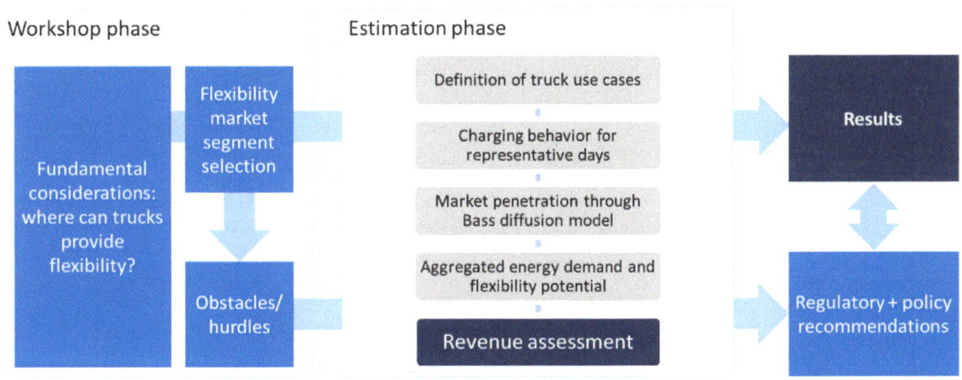

Figure 2. Graphical representation of methodology.

The goals of the workshops were to establish a common understanding of the subject matter between two vastly different industries, further focus our approach, and coordinate the quantification methodology. In total, four online workshops were conducted during the time span of 12 April 2021 until 18 May 2021. The number of participants in these workshops ranged between 10 and 14 employees of TenneT and Daimler Truck. The experience level ranged from expert level (technical, regulatory, or economic) to project leads (of other related projects of TenneT and Daimler Truck) and manager level. The workshops were structured as follows: focus presentations by participants on specific topics, participants were split in multiple groups for deep dives, and participants worked together using a prepared digital whiteboard.

The following three tables describe the parameters used to describe a city bus use case (Table 1) and major truck use cases (Tables 2 and 3). The city bus use case was based on a large, electrified depot in a major German city. Unlike in truck use cases, columns in Table 1 describe spectrums for various parameters rather than specific routes or use cases.

Table 1. Parameters for the "city bus" use case.

Available battery capacity	kWh	350	
Max. available charging power	kW	80	
Energy demand per day	kWh	Min 200	Max 550
Time departure 1	h	Earliest 05:30	Latest 08:30
Time arrival 1	h	Earliest 11:00	Latest 15:00
Time departure 2	h	None, or earliest 13:30	None, or latest 17:00
Time arrival 2	h	None, or earliest 19:00	None, or latest 24:00
Vehicles in example depot		149	

Table 2. Parameters for "line haul" (LH 1–3) and "retail/distribution" (R/D 4–6) use cases.

		LH 1	LH 2	LH 3	R/D 4	R/D 5	R/D 6
Available battery capacity	kWh	600	600	600	600	400	400
Max. available charging power	kW	300	300	50	50	150	150
Energy demand per day	kWh	650	600	350	575	350	400
Time departure 1	h	05:30	06:00	07:00	08:00	05:00	05:00
Time arrival 1	h	17:00	16:00	15:00	16:00	13:00	13:00
Time departure 2	h	-	-	-	-	14:00	14:00
Time arrival 2	h	-	-	-	-	20:00	20:00
Variability of departure		avg.	avg.	large	low	low	low
Vehicles per example depot		50	50	45	20	30	30

Table 3. Parameters for "construction" (Con 7–9) and "waste" (Wa 10–11) use cases.

		Con 7	Con 8	Con 9	Wa 10	Wa 11
Available battery capacity	kWh	600	400	400	400	400
Max. available charging power	kW	150	50	50	50	50
Energy demand per day	kWh	475	300	275	375	300
Time departure 1	h	08:00	08:00	08:00	07:30	07:00
Time arrival 1	h	12:00	16:00	16:00	15:30	15:00
Time departure 2	h	13:00	-	-	-	-
Time arrival 2	h	16:00	-	-	-	-
Variability of departure		average	average	average	low	very low
Vehicles per example depot		10	10	10	15	30

Line haul segments (LH 1–3) summarize a wide variety of long, medium, and short haul applications, transporting all types of goods either on demand or on daily return trips. Retail and distribution routes (R/D 4–6) are usually shorter but more plannable (cf. "variability of departure"), and often include multiple trips per day to retail locations, supermarkets, or distribution locations.

Construction uses cases (Con 7–9) include transportation of building material or equipment to and from construction sites as well as haulage within the site. Waste collection in urban environments and transport between collection and deposition/incineration sites are further prime uses cases for electrification (Wa 10–11).

Although stylized, these parameters allowed detailed modelling of flexibility potentials for exemplary depots for every use case. Flexibility potential is a function of battery state-of-charge and charging power; i.e., the energy volume that can be made available for flexibility marketing. We assumed minimizing peak load as the default charging strategy and as the baseline for the assessment of flexibility potential. Within the limits of ensuring that vehicles are fully charged for their next route, the vehicles' state-of-charge, and the available charging power, charging load could deviate from the minimum depot load schedule, and this flexibility could be offered to the energy market. In an extensive Excel tool, this calculation was conducted for every example depot for the average weekday.

For the flexibility calculation, we assumed that a sufficiently sized grid connection existed or would be built at the depot to enable installed chargers to be simultaneously used at maximum capacity. In combination with over-night idle times, these assumptions allowed for the deterministic calculation of positive (delayed charging processes) and negative (accelerated charging processes) flexibility potential in MW per depot. The potential was assumed to be equal for every day of the week; weekends and bank holidays were not modelled.

In the next step, we created a ramp-up scenario for every use case for Germany using a Bass diffusion model [14] as applied by Ensslen et al. [15] for passenger BEV. Innovation coefficients were used to calculate the share of diesel vehicles being replaced by BEVs over time. The scenario was based on expert assessments (the vehicle ramp-up at the basis of this analysis represents a potential scenario and does not represent a sales prognosis of Daimler

Truck AG), market data [16], and an external source for the bus use case [17]. Furthermore, each use case had a cap on its electrification potential at full diffusion due to the limitations of BEVs, e.g., regarding the range, cargo load, or power demand of ancillary consumers, which was accounted for in the scenario. Looking only at use cases most relevant for flexibility marketing (i.e., with sufficient idle time and early electrification potential), we focused the discussion on five of eleven truck use cases and the city bus use case. Their scenario ramp-up numbers are listed in Table 4.

Table 4. Ramp-up approximation of number of vehicles on the road in Germany.

Use Case	2025	2030	2035	2040
Line haul 2	1200	9300	29,000	37,000
Line haul 3	8300	31,300	68,000	94,000
Retail 5	5000	22,800	58,000	86,000
Construction 7	200	2300	13,000	22,000
Waste 11	1500	6500	13,000	16,000
Total of all use cases	**30,900**	**151,700**	**411,000**	**606,000**
City bus	6900	20,300	31,000	36,000

The flexibility potential per depot were then scaled to the entirety of Germany and aggregated for flexibility marketing. Revenue calculations were based on market data for 2020 and 2021 from the German balancing market platform regelleistung.net [18], and considered both theoretical revenues from the power bid as well a conservative energy bid. Note that we did not model costs, and therefore did not make any claims regarding profitability. Likely cost components, e.g., include increased grid fees, software licenses, prequalification, and market access fees.

3. Results

3.1. Expert Workshops

As the workshops brought together a mix of participants from different levels of expertise across a range of topics from two different industries, opinions, and therefore results, were faceted and diverse. Nevertheless, the workshop series yielded three key take-aways:

1. Logistics businesses will not use electrified vehicles if there is no positive business case, based, e.g., on vehicle price, electricity costs, or incentives for earning additional revenue by providing flexibility services.
2. Promising flexibility segments include balancing power and congestion management (i.e., redispatch).
    a. Although for balancing power the asset location (e.g., the depot) is less important, it is crucial for congestion management because spatial bottlenecks in the electricity network need to be solved.
    b. Technically, trucks and buses can participate in all three balancing types: FCR, aFRR, and mFRR. However, the "higher quality" balancing types (FCR and aFRR) are most suitable because the charging of batteries can be adjusted quickly, and trucks and buses have enough capacity that can be shifted.
    c. In Germany, the regulatory framework for loads and storages under "Redispatch 3.0" is yet to be shaped, while in the Netherlands the so-called GOPACS platform already offers market-based remuneration. Depot operators only provide the redispatch service if they reduce their electricity costs from a market-based remuneration. Therefore, it was decided to focus the following quantification on balancing power within the currently available market framework.
3. The crowd balancing platform "Equigy" enables a more efficient provision of balancing power and congestion management from decentralized, distributed flexibility sources.

a. The crowd balancing platform is not a marketplace, but creates the framework conditions for decentralized prequalification and efficient accounting for the increasing amount of small and distributed assets. This ultimately lowers market entry barriers.

Beyond these key takeaways, many other topics were discussed. Opinions diverged regarding the following points:

4. Not all participants in the workshops agreed that marketing flexibility potential on the wholesale power market should be out-of-scope due to the wholesale markets' strong liquidity and ease of use.
5. The focus on solely Germany was discussed across several workshops. The reason for this discussion was that markets for balancing power are largely integrated in Europe; thus, changes to integrate electrified busses and trucks often requires European regulatory changes.
6. Regarding technical challenges to the integration of electrified busses and trucks, there are differences between countries in which Equigy operates. For example, the Netherlands already uses a practical implementation in which EVs can provide aFRR, but this is not yet the case for Germany.

3.2. Flexibility and Revenue

Positive and negative flexibility potentials (in MW) for grid operation are illustrated in Table 5. Technical flexibility potential is substantial for line haul and retail truck use cases, and large bus depots also play a substantial role in the early morning hours. With a theoretical potential for over 4 GW of positive and negative flexibility from 4 pm to 4 am (peaking at over 23 GW of negative flexibility in the 20:00–24:00 4 h block and at over 7 GW of positive flexibility in the 00:00–04:00 block), all examined use cases combined could have a significant impact on, for example, the balancing power market in 2040. For context, the current demand in 2022 for positive and negative balancing power in Germany is approximately 7.1 GW.

Table 5. Maximum positive and negative (−) flexibility potential for Germany in 2025, 2030, and 2040 [MW].

	00:00–04:00	04:00–08:00	08:00–12:00	12:00–16:00	16:00–20:00	20:00–24:00
2025	+529 −1146	+13 −26	+4 −13	0 −47	+266 −659	+354 −1048
2030	+2210 −5960	+46 −77	+13 −39	0 −138	+1238 −3981	+1613 −5765
2040	+7066 −22,593	+154 −137	+23 −70	0 −245	+4183 −16,095	+5542 −23,113

Figure 3 illustrates the potential revenue from flexibility provision, and therefore the reduction potential for the total cost of ownership [EURct/kWh] for truck customers. In practice, depot operators may have electricity contracts with flexibility aggregators who grant remuneration or rebates on electricity prices in exchange for flexibility. The revenue potential is larger in the aFRR market, and the largest revenue results for truck use cases were line haul 2 and waste 11, while the bus use case and truck use case retail 5 had the lowest potential. For aFRR, the revenue potential could be significant, given that average electricity prices for German industry are approximately 20 EURct/kWh. If transport companies could facilitate flexibility marketing reliably, significant rebates on their electricity costs would be possible.

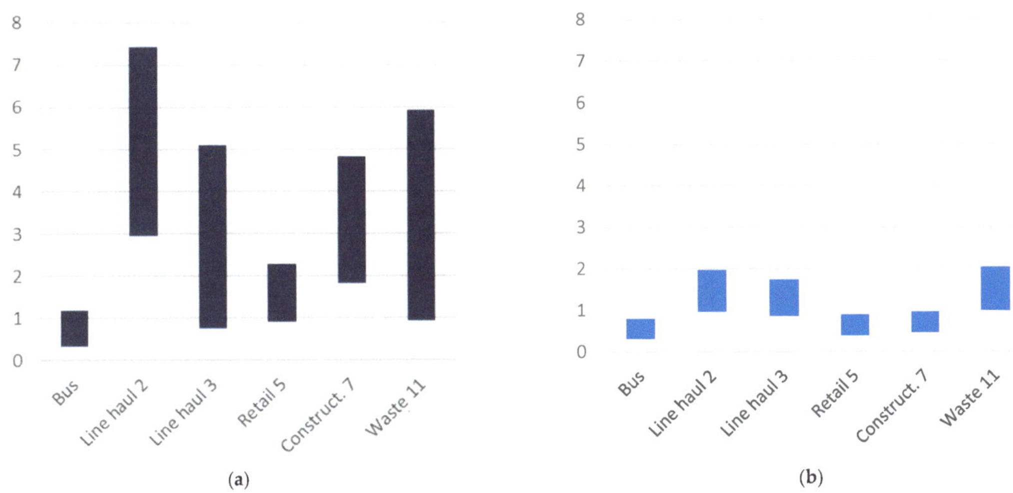

Figure 3. Range of maximum possible revenue per consumed kWh from (**a**) aFRR (capacity and energy) and (**b**) FCR in EURct/kWh (minimum revenue with 2020 prices, maximum with 2021 prices).

There are several limitations to these findings. First, the analysis did not allow for profitability conclusions because only the revenue side was presented (i.e., costs were not included). Second, the flexibility potential assumed that it could be offered over the entire bid timeframe, which is not possible in practice because actual flexibility delivery can considerably reduce the potential. Furthermore, flexibility potentials were based on only a selection of bus and truck use cases (six of twelve) and considered only weekdays (neither weekends nor bank holidays). Finally, we used market data from 2020 and 2021 to illustrate revenue ranges; predictions of future prices require further analysis.

4. Discussion

This study laid the foundation for a mutual understanding of the interaction of energy and transport sectors by assessing the flexibility and revenue potentials of electrified trucks and buses. We showed the significant technical potential of shifting charging times of specific truck and bus use cases to offer balancing power. Furthermore, this offering could lead to notable revenues that should be used to compensate depot operators for the flexibility provided.

Policy recommendations for balancing power include prequalification criteria, which should avoid redundancy and minimize costs for balancing service providers (e.g., by establishing largely automated prequalification processes). Furthermore, vehicle operators' risk of insufficient state-of-charge must be nullified through smart IT solutions. Due to a current lack of marketability, we excluded congestion management from the quantification analysis of this study, despite the expected impact of truck and bus charging on distribution grids [5,6]. A market-based approach should complement the existing cost-based provision of redispatch services and address these decentralized generation or consumption assets, for which there is no mandatory participation in the current redispatch regime. This means that an attractive market solution is needed to allow voluntary participation by consumers and businesses, rather than mandatory load reductions.

5. Conclusions

The electricity system is changing: a growing share of volatile renewable production meets higher and more dynamic loads on all consumption levels. This study investigated how demand–response in the form of battery-electric trucks and buses could offer substantial flexibility to the energy system. In a small series of expert workshops, we built a common understanding of the key aspects of this topic and aligned a research approach. Consequently, we used a Bass diffusion model to extrapolate the unidirectional charging needs and availability of trucks in five of eleven typical applications, as well as city buses, for Germany until 2040. Combined, these heavy-duty vehicles could provide up to 23 GW of down-regulating flexibility potential (i.e., in case of excess power supply) in 2040. The resulting revenues could contribute to reducing electricity costs for fleet operators, thereby improving the attractiveness of zero-emission technologies. These results illustrate the need to provide easy and automated market access to heavy-duty vehicle fleets.

A full economic examination regarding the profitability potential is advisable in future work. This includes, in particular, a quantitative assessment of the cost side and of the effects of delivering balancing energy on the flexibility potential. Further research is needed to quantitatively compare other marketing options, e.g., congestion management, intraday arbitrage trading, or even pure behind-the-meter cost minimization using on-site solar generation. A logical expansion of the model could integrate bidirectional charging, which should further increase flexibility potentials, especially when considering weekends and public holidays. Furthermore, a technical pilot could be informative regarding open topics in standardization or availability of equipment.

Author Contributions: Conceptualization, C.W. and F.O.; methodology, C.W. and F.O.; software, C.W. and F.O.; validation, C.W. and F.O.; formal analysis, C.W. and F.O.; investigation, C.W. and F.O.; resources, C.W. and F.O.; data curation, C.W. and F.O.; writing—original draft preparation, C.W. and F.O.; writing—review and editing, C.W. and F.O.; visualization, C.W. and F.O.; supervision, C.W. and F.O.; project administration, C.W. and F.O.; funding acquisition, C.W. and F.O. All authors have read and agreed to the published version of the manuscript.

Funding: This research received no external funding.

Data Availability Statement: The original contributions presented in the study are included in the article, further inquiries can be directed to the corresponding author.

Acknowledgments: We would like to thank the co-authors and participants from both companies for their invaluable input and support. This study did not benefit from public funding.

Conflicts of Interest: Christian Will is employee of Daimler Truck AG and Fabian Ocker is employee of TenneT TSO GmbH. The paper reflects the views of the scientists, and not the company.

References

1. Dena. Dena-Studie Systemdienstleistungen 2030. 2022. Available online: https://www.dena.de/themen-projekte/projekte/energiesysteme/dena-studie-systemdienstleistungen-2030/ (accessed on 24 March 2023).
2. García-Villalobos, J.; Zamora, I.; San Martín, J.I.; Asensio, F.J.; Aperribay, V. Plug-in electric vehicles in electric distribution networks: A review of smart charging approaches. *Renew. Sustain. Energy Rev.* **2014**, *38*, 717–731. [CrossRef]
3. Mwasilu, F.; Justo, J.J.; Kim, E.-K.; Do, T.D.; Jung, J.W. Electric vehicles and smart grid interaction: A review on vehicle to grid and renewable energy sources integration. *Renew. Sustain. Energy Rev.* **2014**, *34*, 501–516. [CrossRef]
4. Habib, S.; Kamran, M.; Rashid, U. Impact analysis of vehicle-to-grid technology and charging strategies of electric vehicles on distribution networks—A review. *J. Power Sources* **2015**, *277*, 205–214. [CrossRef]
5. Hildermeier, J.; Kolokathis, C.; Rosenow, J.; Hogan, M.; Wiese, C.; Jahn, A. Smart EV Charging: A Global Review of Promising Practices. *World Electr. Veh. J.* **2019**, *10*, 80. [CrossRef]
6. Johnsen, D.; Ostendorf, L.; Bechberger, M.; Strommenger, D. Review on Smart Charging of Electric Vehicles via Market-Based Incentives, Grid-Friendly and Grid-Compatible Measures. *World Electr. Veh. J.* **2023**, *14*, 25. [CrossRef]
7. Liimatainen, H.; van Vliet, O.; Aplyn, D. The potential of electric trucks—An international commodity-level analysis. *Appl. Energy* **2019**, *236*, 804–814. [CrossRef]
8. Walz, K.; Rudion, K.; Moraw, C.-M.; Eilers, M. Probabilistic impact assessment of electric truck charging on a medium voltage grid. In Proceedings of the 2022 IEEE PES Innovative Smart Grid Technologies Conference Europe (ISGT-Europe), Novi Sad, Serbia, 10–12 October 2022; pp. 1–6. [CrossRef]
9. Rodrigues, A.; Seixas, S. Battery-electric buses and their implementation barriers: Analysis and prospects for sustainability. *Sustain. Energy Technol. Assess.* **2022**, *51*, 101896. [CrossRef]
10. Borlaug, B.; Muratori, M.; Gilleran, M.; Woody, D.; Muston, W.; Canada, T.; Ingram, A.; Gresham, H.; McQueen, C. Heavy-duty truck electrification and the impacts of depot charging on electricity distribution systems. *Nat. Energy* **2021**, *6*, 673–682. [CrossRef]

11. Bernard, M.R.; Tankou, A.; Cui, H.; Ragon, P.-L. Charging Solutions for Battery-Electric Trucks. White Paper by ICCT. 2022. Available online: https://zevalliance.org/wp-content/uploads/2022/12/charging-infrastructure-trucks-zeva-dec22.pdf (accessed on 12 January 2024).
12. Al-Hanahi, B.; Ahmad, I.; Habibi, D.; Masoum, M.A.S. Smart charging strategies for heavy electric vehicles. *eTransportation* **2022**, *13*, 100182. [CrossRef]
13. Taljegard, M.; Göransson, L.; Odenberger, M.; Johnsson, F. Electric vehicles as flexibility management strategy for the electricity system—A comparison between different regions of Europe. *Energies* **2019**, *12*, 2597. [CrossRef]
14. Bass, F. A new product growth for model consumer durables. *Manag. Sci.* **1969**, *15*, 215–227. [CrossRef]
15. Ensslen, A.; Will, C.; Jochem, P. Simulating electric vehicle diffusion and charging activities in France and Germany. *World Electr. Veh. J.* **2019**, *10*, 73. [CrossRef]
16. MAPIS. Markant Price Monitor (MPM). 2022. Available online: https://www.markant.com/en/our-services/market-price-analysis/mapis (accessed on 24 March 2023).
17. pwc. E-Bus-Radar: Wie Elektrisch Ist der Öffentliche Nahverkehr in Deutschland? 2022. Available online: https://www.pwc.de/e-bus-radar (accessed on 24 March 2023).
18. Regelleistung.net. Datacenter. 2021. Available online: https://www.regelleistung.net/apps/datacenter/tenders/?productTypes=PRL,SRL,MRL&markets=BALANCING_CAPACITY,BALANCING_ENERGY&date=2022-11-17&tenderTab=PRL$CAPACITY$1 (accessed on 24 March 2023).

World Electric Vehicle Journal

MDPI

Article

Series-Hybrid Powertrains: Advancing Mobility Control in Electric Tracked Vehicle Technology

Dersu Çeliksöz * and Varlık Kılıç

FNSS Savunma Sistemleri A.Ş., 06830 Ankara, Turkey
* Correspondence: dersu.celiksoz@fnss.com.tr

Abstract: This work focuses on developing a mobility control system for high-speed series-hybrid electric tracked vehicles, which operate with independent traction motors for each track. The scope of this research includes modeling a series-hybrid powertrain specific to military tracked vehicles and conducting an in-depth analysis of its dynamic behavior. Subsequently, this study conducts a critical review of mobility control approaches sourced from the literature, identifying key techniques relevant to high-inertia vehicular applications. Building on foundational models, this study proposes a robust closed-loop mobility control system aimed at ensuring precise and stable off-road vehicle operations. The system's resilience and adaptability to a variety of driving conditions are emphasized, with a particular focus on handling maneuvers such as steering and pivoting, which are challenging operations for tracked vehicle agility. The performance of the proposed mobility control system is tested through a series of simulations, covering a spectrum of operational scenarios. These tests are conducted in both offline simulation settings, which permit meticulous fine-tuning of system parameters, and real-time environments that replicate actual field conditions. The simulation results demonstrate the system's capacity to improve the vehicular response and highlight its potential impact on future designs of mobility control systems for the heavy-duty vehicle sector, particularly in defense applications.

Keywords: hybrid electric tracked vehicles; hybrid electric military vehicles; vehicle control; mobility control system; series-hybrid electric powerpack; tracked vehicle dynamics; steering maneuver; torque management; robust control; terrain adaptability

Citation: Çeliksöz, D.; Kılıç, V. Series-Hybrid Powertrains: Advancing Mobility Control in Electric Tracked Vehicle Technology. *World Electr. Veh. J.* **2024**, *15*, 47. https://doi.org/10.3390/wevj15020047

Academic Editors: Joeri Van Mierlo and Genevieve Cullen

Received: 29 December 2023
Revised: 24 January 2024
Accepted: 31 January 2024
Published: 2 February 2024

1. Introduction

Tracked vehicles are essential for various sectors, including the automotive industry, defense, construction, and agriculture, due to their superior off-road capabilities. The recent shift towards hybrid electric drive systems, similar to those in wheeled vehicles, has gained momentum thanks to the advantages they offer. Sivakumar's study states that the hybridization of military vehicles offers significant benefits, including improved fuel efficiency, enhanced drivability, and silent running, yet faces considerable development challenges due to the demand for robust and environment-resistant components [1]. Many studies agree that adding electric power to military vehicles could make them quieter and work better, while also providing extra electric power when needed, as summarized in a study that introduces a new hybrid power system for these vehicles that aims to cut down on weight and use less fuel without compromising on how well the vehicle performs [2]. Furthermore, the adaptation of series-hybrid drives to tracked vehicles implies a need for distinct controller requirements: a power management algorithm for effective power distribution among the power sources (battery and generator set) and a mobility control algorithm for independent motor operation to achieve the desired motion control variables, including sprocket velocity and track speed differential during maneuvers. It is also crucial to operate these systems within a region that ensures robustness and optimal performance.

FNSS, a global leader in the land systems sector, is at the forefront of this innovation, not only producing wheeled and tracked armored combat vehicles, turrets, and engineering vehicles, but also exploring hybrid powertrain technologies [3]. The focus of this study revolves around the hybridization of an armored tracked FNSS vehicle named KAPLAN HYBRID, as shown in Figure 1. This work was presented by Çeliksöz at the EVS36 symposium in Sacramento, USA, in June 2023, together with the publication of a conference paper in which the development of a control system for the vehicle is focused on [4]. This paper builds upon the conference paper, providing an extension that details the development of a mobility control system specifically tailored for FNSS's series-hybrid tracked vehicles. The approach commences with the modeling of vehicle dynamics and a hybrid powertrain using MATLAB Simulink, followed by a review of existing mobility control strategies detailed in the literature. Subsequently, in a simulated environment, a robust closed-loop mobility control system is designed and tested, while the power management system is analyzed as a black-box model that sets instantaneous traction power limits to ensure clarity and maintain focus on mobility controller development. The system's robustness is validated by simulating the vehicle model against diverse dynamic traction power constraints in various scenarios.

Figure 1. KAPLAN HYBRID, developed by FNSS [3].

Before delving into the details of this work, it is essential to note that the current literature documents a range of control methodologies for hybrid systems. In terms of dual independent motor-driven tracked electric vehicles, Zhang et al. studied an improved steering control utilizing a model predictive control approach, leading to efficient adherence to specified speed and turn rates [5]. Their research demonstrates the potential to enhance vehicle stability under varied driving conditions through an intelligent application of predictive control algorithms. Zeyu et al. implemented a neural network combined with a PID algorithm to refine the vehicle's steering control [6]. Even though the traction motor torques are initially defined by the neural network, the introduction of PID adjustments accommodates variable terrain coefficients. Hu et al. discussed a dual-motor drive strategy that inherently computes torques for a predetermined target turning radius, employing a PID controller to achieve the radius more rapidly [7]. Additionally, Pei et al. proposed a torque fuzzy compensation control strategy to enhance steering execution, where torque distribution is regulated by applying direct multipliers derived from fuzzy logic [8]. Summary of the corresponding study is presented in Figure 2.

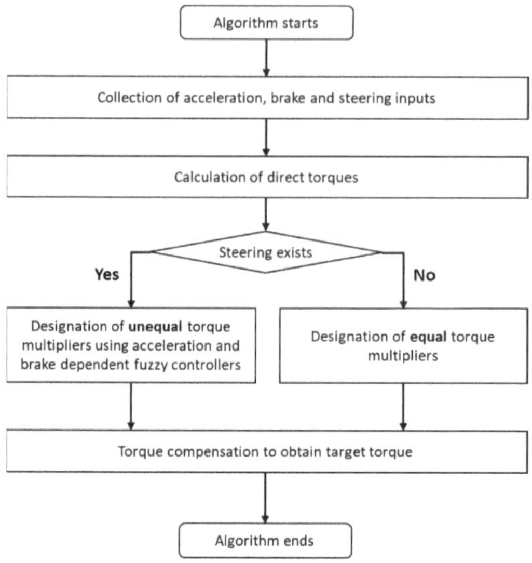

Figure 2. Summary of torque fuzzy compensation control strategy by Pei et al. [8].

In their review of electric tracked vehicle mobility, Kotiev et al. address the complexities of achieving balanced torque in high-speed scenarios, particularly through individual drives on each sprocket [9]. Traditional methodologies are determined insufficient for accurate analysis, causing the authors to incorporate neural networks and simulation tools to better predict the thrust requirements for various steering conditions. Their work emphasizes sophisticated design and testing protocols in the realm of advanced electric vehicle systems.

Alongside independent dual-motor drive studies, researchers in the tracked EV community have also suggested power coupling drive solutions, in which mobility is improved using a transmission between left and right traction units. Zhai et al. introduced a steering stability control methodology that significantly enhances handling stability and safety for high-speed tracked vehicles equipped with a four-motor distributed drive system [10]. This method utilizes a steering coupling system and a direct yaw torque control strategy to optimize torque distribution, ensuring improved steering performance under various conditions. In their subsequent work, Zhai and co-authors address a pivotal aspect of electric tracked vehicle design, focusing on handling behavior during dynamic steering of a dual motor drive system [11]. Their research articulates the requirements of dual motors for torque and power for effective maneuverability, leading to the conceptualization of a new steering system. An investigation by Huang et al. into the dynamics of dual-motor drive tracked vehicles during high-speed steering proposes a novel electromechanical coupling device coupled with an optimized torque distribution control approach, resulting in a significant improvement in power output and steering performance [12]. Zhai and his team, furthermore, have devised a power coupling steering system that effectively counters insufficiencies in motor torque and power during differentiated steering demands [13]. The system's capability has been rigorously validated through simulations, demonstrating its effectiveness in maintaining vehicle stability and enhancing maneuverability, particularly in steering scenarios that historically placed substantial demands on engine torque and power.

Overall, a review of the literature indicates that steering ability could potentially be synthesized and augmented through the application of either advanced closed-loop control software or simpler software assisted by an optimized transmission mechanism. This work will focus on the development of control software using the former solution for dual independent electric drive.

Research extends the domain of knowledge by developing a special mobility control system for FNSS's specific series-hybrid electric tracked vehicle. Emphasizing technological innovations in hybrid drive systems for armored combat vehicles, this work contributes insights that improve the understanding and application of these systems, setting a foundational model for further study and practical deployment in the sector. Our principal contributions and conclusions underscore technology's progression in hybrid drive systems and exemplify their potential for robust, real-world functionality in military vehicle applications. Moreover, this study provides a comprehensive evaluation of the system's performance under a variety of combat scenarios, ensuring that vehicles are equipped to handle diverse and challenging terrains. The discussions included are not limited to theoretical assessments; they also present insights from extensive field tests that enhance the practical feasibility and dominance of the hybrid system.

2. Materials and Methods

Plant and control system models are deliberately designed to further research initiatives and assess an algorithm's performance with a variety of control inputs and driving cycles. Plant modeling incorporates power sources, the traction system, and the vehicle's framework, whereas the control system is explored in terms of power management and mobility control systems. These modeling activities are carried out distinctly for plant and control systems, and this division is to enable the examination of diverse vehicle configurations and powertrain systems. After each section is individually shaped, they are then unified to construct an integrated model depicted in Figure 3. A more detailed explanation of each model is provided in the following sections, offering a deeper insight into how each component functions within the entire system.

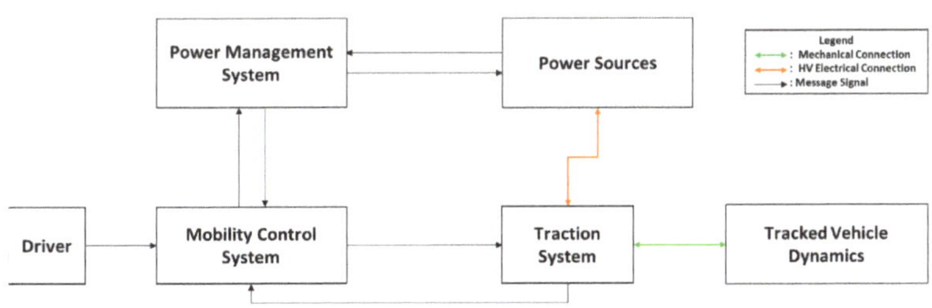

Figure 3. Block diagram of the hybrid tracked vehicle model and controller.

2.1. Modeling the Plant: Hybrid Electric Tracked Vehicle

In this section, attention is given to the modeling process for the plant of the hybrid electric tracked vehicle. The focus is on the representation of hybrid electric propulsion systems specific to tracked vehicles and on modeling techniques that capture the dynamic relationships between the traction system's components and the vehicle's overall architecture. Furthermore, power sources are briefly defined, even though the scope of this work is powertrain and mobility of the vehicle rather than energy management.

2.1.1. Electric Traction System

The electric traction system, which is presented in a simplified schematic in Figure 4, is modeled through a composite approach that brings together electric motors, gearboxes, and friction brakes connected to left and right sprockets. This configuration is carefully constructed to accurately reflect the system's mechanical and electrical interactions, ensuring that the model provides a realistic representation of vehicle dynamics.

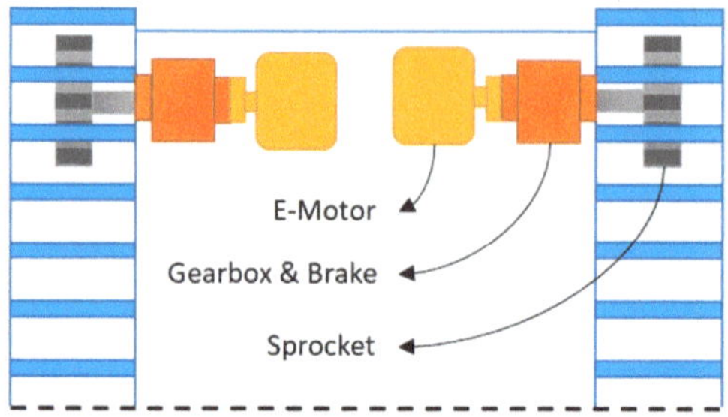

Figure 4. Schematic of electric traction system.

In this configuration, e-motors are modeled as a source of torque featuring a unity transfer function, which means an absence of delay in torque delivery upon request, as long as it does not exceed the motor's torque reserve. To preserve this characteristic, the system utilizes the smallest value between the requested torque and the available torque at the current shaft speed, as illustrated in Equation (1). Subsequently, the torques generated by the motors are scaled by the gear ratio and modified for gearbox efficiency.

$$T_{out,Motor} = \min\{T_{avail,Motor}(w), T_{Request}\} \tag{1}$$

In Equation (1); $T_{out,Motor}$, $T_{avail,Motor}(w)$, and $T_{Request}$ represent motor torque output, speed-dependent available motor torque, and requested torque via mobility control system, respectively. Speed-dependent available motor torque is obtained from the full load curve, which represents the speed–torque characteristics of the electric motor. Even though the full load curve is distinct in electric motors, it displays a standard trend: a constant torque at lower speeds transitioning to a constant power at higher speeds. This pattern is elaborated on in Aiso's research [14], as exemplified in Figure 5.

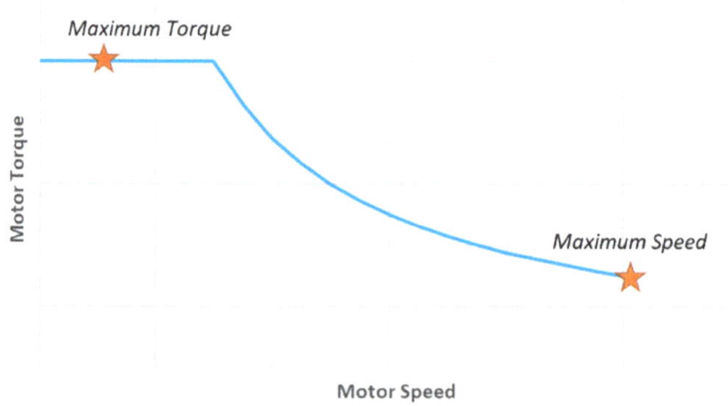

Figure 5. Full load characteristic curve for an electric motor.

Prior to transitioning to vehicle dynamics calculations, the torques from the friction brakes are added to those from the gearbox output, as shown in Equation (2). The resulting torque is then supplied to the subsystem governing the dynamics of the tracked vehicle.

$$T_{Sprocket} = i_{GB}\,\eta_{PT}\,T_{out,Motor} - T_{Brake} \tag{2}$$

In Equation (2), $T_{Sprocket}$, i_{GB}, η_{PT}, and T_{Brake} represent sprocket torque output, gear ratio, powertrain efficiency, and applied brake torque, respectively.

2.1.2. Dynamics of the Tracked Vehicle

The dynamics of the tracked vehicle are captured using a 3-DOF (degrees of freedom) vehicle model that considers the equations of motion along the longitudinal (x), lateral (y), and yaw (z) axes, which are detailed in Figure 6 for a left maneuver. These equations, detailed in Figure 6 for a left maneuver, are represented by (3), (4), and (5). They are derived for this maneuver assuming a center of gravity in the middle of the lateral axis, rather than the longitudinal axis.

$$m_{Vhc}\, a_x = (F_{Traction,left} - F_{Resistance,left}) + (F_{Traction,right} - F_{Resistance,right}) = \Sigma F_{x,i} \quad (3)$$

$$m_{Vhc}\, a_y = (Q_{Lateral,left} + Q_{Lateral,right})\, x_{center,rear} - (Q_{Lateral,left} + Q_{Lateral,right})\, x_{center,front} = \Sigma F_{y,i} \quad (4)$$

$$I_{Vhc,zz}\, \Phi' = \Sigma F_{x,i}\, y_{Resultant} - \Sigma F_{y,i}\, x_{Resultant} \quad (5)$$

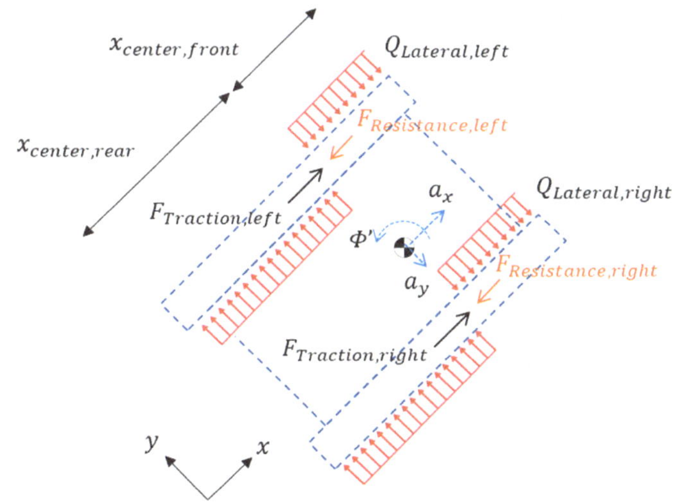

Figure 6. Loads on a tracked vehicle during left maneuver.

In the above expressions, the terms m_{Vhc}, $I_{Vhc,zz}$, a_x, a_y, and Φ' denote the mass of the vehicle, inertia about the vertical axis at the vehicle's center of gravity, accelerations in the longitudinal and lateral directions, and the rate of yaw, respectively. The forces $F_{Traction,left}$, $F_{Traction,right}$, $F_{Resistance,left}$, and $F_{Resistance,right}$ are indicative of the traction and longitudinal resistance forces acting on the vehicle for left and right tracks, respectively. While $Q_{Lateral,left}$ and $Q_{Lateral,right}$ characterize distributed side frictional forces per length interacting with the left and right tracks, lateral distance between the vehicle's cog (center of gravity) and the vehicle's front and rear end are denoted by $x_{center,front}$ and $x_{center,rear}$. Furthermore, the net forces along the longitudinal and lateral axes are given by $\Sigma F_{x,i}$ and $\Sigma F_{y,i}$, with the corresponding distances from these net forces to the vehicle's center of gravity being represented by $x_{Resultant}$ and $y_{Resultant}$, correspondingly.

Dynamic calculations are followed by integrations of computed accelerations to obtain speed components of the vehicle in the longitudinal, lateral, and yaw axes.

2.1.3. Power Sources

Power sources for the hybrid vehicle comprise a combination of a battery and a generator set involving a diesel engine and an electric generator. In this work, it is assumed that the available power and the summation of the power sources' reserve remains constant for a clear investigation of the mobility control system.

2.2. *Mobility Control Method*

- Control Theory and Big Picture

Based on the research and findings presented in previous sections, it is found that the vehicle in question, a high-speed off-road military vehicle, requires a strong and effective closed-loop control system to achieve the targeted maneuverability at high speeds across various terrains. It has been determined that while producing overall torque in response to the driver's input is correlated with the accelerator pedal's position, the distribution of torque should be adjusted based on feedback from the speed difference between the sprockets. This differential is related to the angle of the steering wheel set by the driver. In other words, a certain speed differential between the electric motors is decided, corresponding to the given steering wheel angle, through the employment of the closed-loop controller. The strategy for mobility control can be seen in Figure 7, demonstrated via block diagrams. There are four primary subsystems, each with special roles. The Driver block is designed to feed specific driver commands for varying test runs and is separate from the onboard vehicle control system. The Driver command preprocessor and the closed-loop controller blocks are critical to the control system, handling the computational side of mobility control and transforming driver instructions into specific torque demands for the left and right motors. Lastly, the traction system represents the plant, including mathematical models of the electric motor and gearbox. The outputs from the traction system are the torques delivered to the left and right sprockets, which are the main inputs of the vehicle's dynamic model.

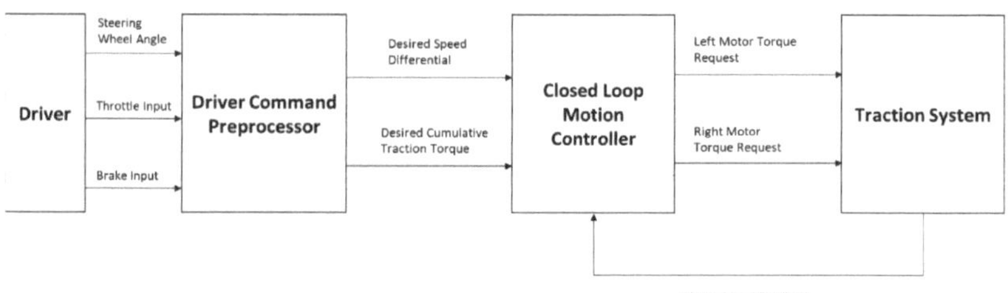

Figure 7. Big picture of mobility control system.

- Preprocessing and Shaping Control Commands

Within the Driver Command Preprocessor section, throttle and steering inputs are preprocessed depending on the selected gear and current vehicle speed. The shaping of the inputs is performed by three main functions: steering shaping, pivot shaping, and throttle shaping.

The steering shaping function aims to transform steering commands into differences in motor velocities through a suitable shaping strategy. FNSS Savunma Sistemleri A.Ş., operating out of its Ankara, Turkey location, has compiled test records from standard tracked vehicles for calibration purposes. Upon analyzing the change in speed at maximum steering, a pattern is noted where the difference in speed across the motors at full steer changes with vehicle velocity. However, this increase is interrupted by sudden changes at certain velocities, making the pattern non-linear. A closer inspection shows that these dips coincide with the gear shifting points of traditional gearboxes. For consistent steering behavior, a smooth curve is mapped over the test data, excluding these dips. The normalized version of these test data together with the smoothing curve is depicted in Figure 8, ensuring confidentiality.

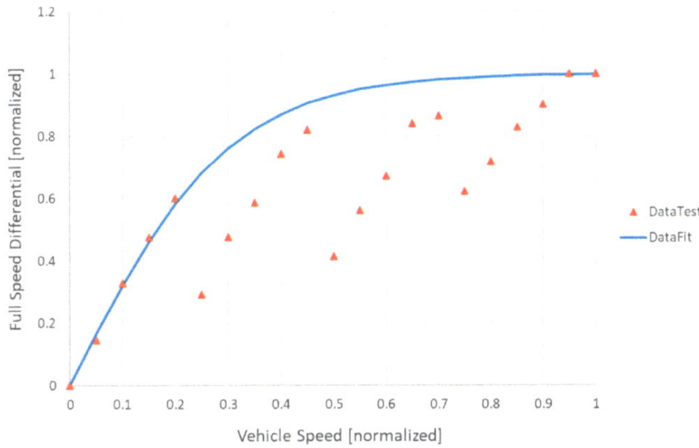

Figure 8. Steering shaping strategy in normalized form.

To further refine the effectiveness of the steering shaping function, FNSS has incorporated adaptive elements into the function to cater to a range of variables, such as terrain type, vehicle mode, and the selected gear. This adaptive approach ensures that steering inputs are not only translated into consistent motor velocity differences across the usual operational spectrum but also provides compensation for less predictable conditions that could affect handling, such as a scenario in which one of the tracks is in contact with a more slippery surface. By integrating a dynamic feedback loop, steering inputs result in real-time adjustments to motor outputs, thus delivering an enhanced and more responsive driving experience for operators.

The pivot maneuver is executed through a specifically designed shaping function. Initially, the maximum speed range is established. Subsequently, the position of the accelerator pedal is correlated to this range to achieve the required speed differential. In addition, the angle of the steering wheel is utilized to decide the pivot turn's direction, allowing the driver to command a counterclockwise (CCW) or clockwise (CW) rotation by steering, regardless of the actual degree of the wheel's turn. The plot for this pivot shaping strategy can be found in Figure 9, providing a visual representation of the maneuvering process.

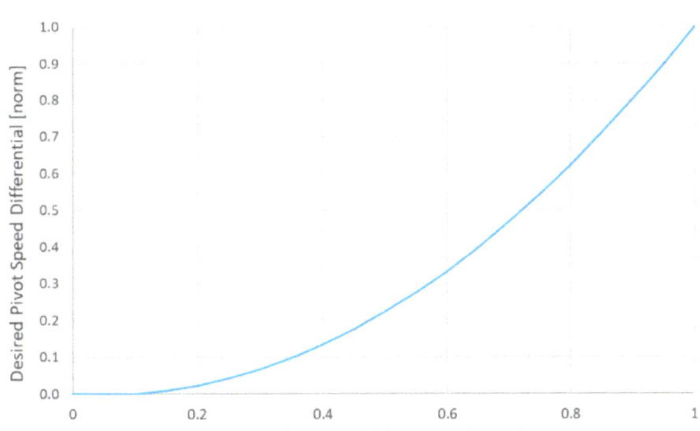

Figure 9. Pivot shaping strategy in normalized form.

Enhancing the pivot shaping function further, FNSS has considered vehicle load dynamics to ensure stability during pivot turns. The shaping strategy is thus calibrated to maintain the vehicle's balance by investigating vehicle dynamics in offline simulation environments and real-time tests for different conditions. This calibration is particularly critical when operating with varying loads or executing pivots on steep gradients.

The throttle shaping function is developed to adjust the input received from the accelerator pedal based upon the selected vehicle mobility mode, with the objective of improving the driver's experience. This is achieved by mapping the position of the pedal to specific throttle values, which correspond to a range of distinct operational modes, as depicted in Figure 10.

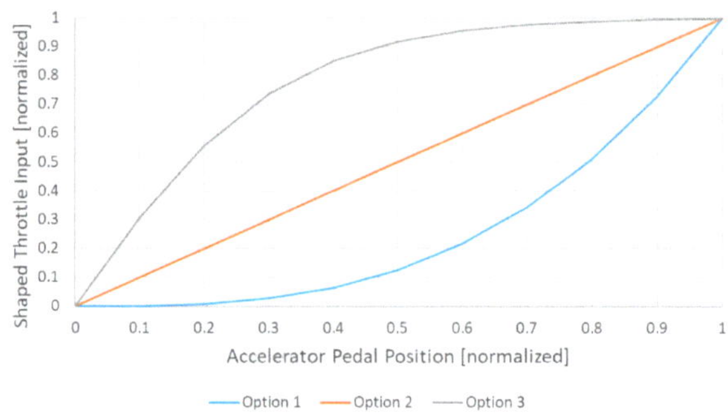

Figure 10. Throttle shaping options.

In scenarios where safety is significant, such as in the preliminary testing phase of the algorithm, a conservative throttle response is preferred. This scenario is optimally supported by a shaping function like option 1 in Figure 10, which is characterized by a more gradual and controlled acceleration curve. On the other hand, for circumstances that demand a more robust and dynamic performance, such as during an aggressive driving test, the shaping curve should approach option 3 in Figure 10. This latter option is fine-tuned to yield a sharper and more immediate increase in throttle response, reflecting the vehicle's need for rapid acceleration. Throughout many testing protocols and simulation exercises, option 2, which represents linear throttle shaping, is the preferred choice. This option is beneficial because it provides a straightforward correlation between pedal input and throttle output. Such predictability enables the control engineer to isolate and evaluate the performance characteristics of each subsystem without the added complexity that non-linear shaping options might introduce.

By employing these shaping functions, total cumulative torque demand and desired speed difference variables are designated. Based on these desired inputs, a closed-loop motion controller is operated, and the torque demands of left and right traction motors are determined.

2.3. *Power Management Method*

The mobility control system serves a dual purpose within the vehicle's control architecture. Primarily, it is responsible for directly actuating the traction motors. However, the scope of the mobility control system's functionality extends to playing a key role in the vehicle's overall power management by continuously monitoring the instantaneous power requirements of the traction system.

As the vehicle operates, the mobility control system calculates the immediate power demands needed for traction by multiplication of torque demand, measured speed, and corresponding traction efficiency. Once the power calculations are performed, the mobility control system communicates a traction power request to the power management algorithm, which is another significant system of the hybrid tracked vehicle in question, as explained in the work of Akar et al. [15]. This request for power is carefully evaluated to control how the power management system should regulate the generator set's output. By providing this traction power request, the mobility control system ensures that the power management system can adjust the generator's output dynamically, matching the generated power with the traction system's demands. This synchronous operation is crucial for ensuring that the difference between generated and demanded power does not exceed the limits of electric battery.

Another aspect influenced by traction power is the sizing of the EV battery and configuration of the generator set, dictating the selection of an internal combustion engine and electric generator of optimal size. This choice becomes even more complex as power requirements directly impact other considerations for battery and generator selection. Choosing the appropriate battery for a specific application relies on assessing key battery properties and finding the right balance among them. For electric and hybrid vehicles, crucial factors include the necessary power and range, which dictate battery pack design within the constraints of available space. While lithium-ion (Li-ion) batteries are prevalent due to their high energy density, significant voltage output, and minimal maintenance, no single battery type is universally optimal for all uses. After the assessment of the environmental conditions, a Li-ion-type battery is used in this vehicle and the control parameters are tuned accordingly.

3. Simulations and Results

Using the modeling approach and equations mentioned previously, a mobility control system was developed and simulated using MATLAB Simulink 2021b. An overview of the Simulink model is illustrated in Figure 11.

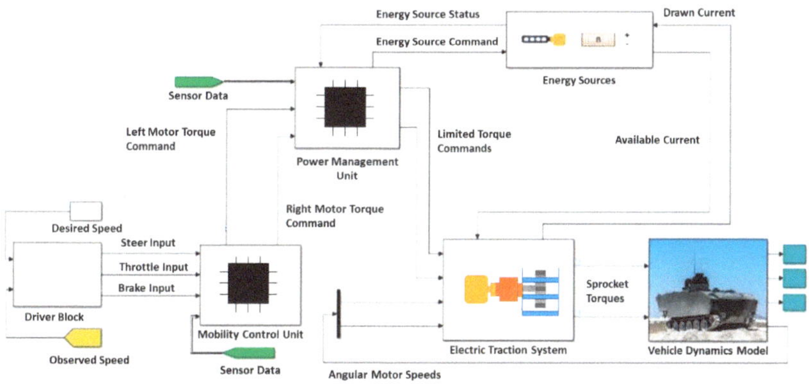

Figure 11. Simulink model of the hybrid tracked vehicle.

Functional scenarios, including pivot maneuvers, steering actions, and forward and reverse movements, were studied by extensive simulations to analyze the vehicle's behavior under various operational conditions. To analyze the results, a simulation postprocessing interface was developed, which permits the in-depth observation of the vehicle's motion dynamics and the associated responses from the powertrain, including variables like sprocket torques and track speeds.

An example of this simulation platform can be seen in Figure 12, which specifically illustrates the interface during a scenario involving a forward steering motion at a predetermined longitudinal speed of the vehicle. In this simulation, it is observed that upon reaching the 50th second of the simulation time, the driver introduces a steering command, inputting a 40% steering angle into the system. Due to this action, the motor torques experience a substantial and rapid increase. After this initial jump in torque caused by the need to adjust the vehicle's trajectory according to the steering input, the motor's torque levels exhibit a stabilization, converging to a steady-state value. The steady-state torques involved in steering are now balanced, and the vehicle maintains the new desired directional path at a constant speed.

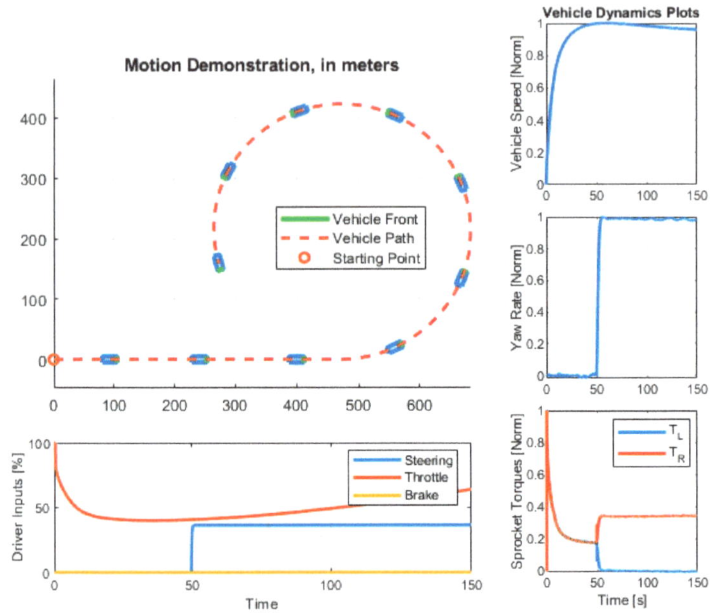

Figure 12. Simulation interface for a cornering maneuver.

Additionally, Figure 13 indicates that the speed differential between the left and right electric motors reaches the preset target specified by the steering shaping function. An upward trajectory in speed difference is observed as the steering input rises, a response predicted from the closed-loop control system. The presence of noise in the actual speed differential originates from the characteristics of the electric motors' encoders. It is assumed that the encoders are influenced by white noise to test the controller's capability to cope with this type of disturbance.

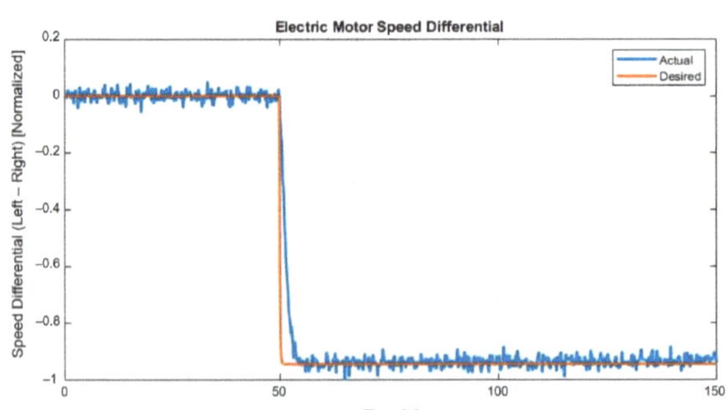

Figure 13. Speed differential during cornering.

Observations of the vehicle's longitudinal response are demonstrated in Figure 14 as well. It is noted that there is a proportional increase in the vehicle's total torque, correlating with the rise in the accelerator pedal position, regardless of the steering input. This effect is achieved by applying a linear throttle shaping function, as explained in the mobility control system section.

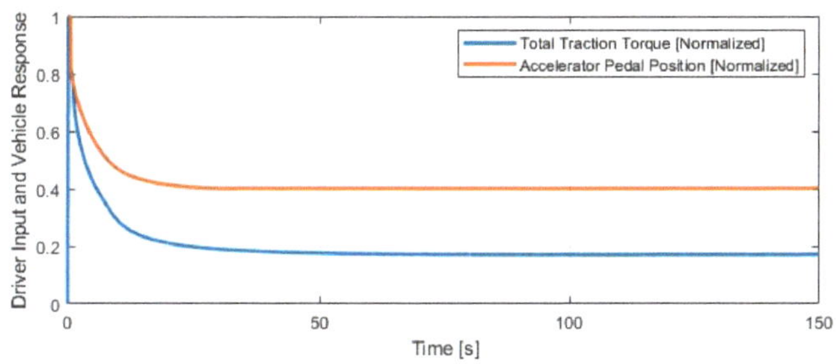

Figure 14. Total traction torque of the vehicle in response to accelerator pedal position.

The simulations successfully confirmed the effectiveness of the control methods discussed earlier in shaping vehicle dynamics, ensuring that the system can fully manage both the total traction force and the vehicle's turning radius based on driver commands. This controllability enables consistent maneuvering at fixed speeds in straight lines as well as during complex directional changes where high maneuverability is required. The ability to precisely adjust the speeds of individual sprockets strengthens the vehicle's flexibility to cope with different terrain types and driving conditions while enhancing the system's overall handling and dependability.

4. Discussion and Conclusions

This paper investigates the mobility control system of a series-hybrid electric tracked vehicle. The objective of the system is to control electric motor torques to produce the entire torque demanded by the throttle, while achieving the necessary speed differential when steering inputs are made. Traditional testing methods on the electrified tracked vehicle helped establish the speed differential target.

Results from practical demonstrations indicate that the intended motion is achievable with the deployed mobility control system. The findings, thus, suggest that similar architectures can be applied generically across the domain of electric propulsion systems in heavy-duty vehicles. Moreover, the speed differential methodology enhances system adaptability across varying terrains. Even though different friction coefficients may necessitate varied sprocket torques for steering actions, the employment of a closed-loop speed differential strategy yields the correct sprocket torques. This validates the significance of torque management and its responsiveness to external conditions, which holds universality for other applications. Variations in radius due to changes in slip characteristics are compensated for by the driver's input, effectively incorporating human interaction as part of the loop.

The integration of regenerative braking into the series-hybrid system enables transmission reversibility and further demonstrates the potential for increased efficiency and reduced wear on braking components. By converting kinetic energy into electrical energy during deceleration phases, energy is fed back into the battery, thus extending the operational range of the vehicle. Initial tests exhibit promising results, with noticeable energy recovery without compromising the stability or control of the vehicle. The ability to regenerate is critically important for hybrid systems since the vehicle in question is a high-weight off-road piece of machinery operating on inclines, where reversibility becomes crucial for boosting efficiency. In such challenging environments, the system's ability to convert kinetic energy into electrical energy for storage in the battery significantly enhances the operational range and energy conservation, underscoring its value for vehicles navigating steep terrain. The interaction between regenerative braking and torque distribution algorithms is crucial; it maintains driving dynamics that are consistent with driver expectations. This addition to the mobility control system complements the existing architecture, presenting a beneficial approach to vehicle energy management and efficiency optimization.

Our mobility control system represents a significant advancement in the management of series-hybrid electric tracked vehicles, and future refinements could be made to fine-tune the relationship between driver inputs and vehicle performance, supporting the human–machine interface for optimal control. Furthermore, current studies show that the transition from engine-driven auxiliary systems to electric-powered ones is accelerating [16]. This transition not only streamlines the management of auxiliary systems but also reinforces the essential connection between vehicle efficiency and advanced control technologies, setting a benchmark for future EV system innovations. Illustrative of this trend is a study by Pugi et al. (2021), which highlights the significant enhancements made in electric directional drilling machines, marking them as exemplary cases of electrification in auxiliary elements and contributing to the sustainable progress of urban infrastructure development [17].

In conclusion, the findings presented herein emphasize the generic value of this research in shaping the future of vehicle control systems. Particularly, the control of electric motor torques within a high-inertia platform of armored tracked vehicles demonstrates the agility that can be attained through advanced engineering. This closed-loop speed differential control system offers a new paradigm in mobility, where precision and adaptability are improved together by promoting a high level of responsiveness relative to conventional tracked vehicles. Mobility control enhances vehicle management and performance, while efforts to electrify auxiliary systems contribute to the overall efficiency and evolution of electric vehicle technologies. Finally, the navigation strategies developed here could serve as a basis for autonomous or semi-autonomous control systems in similar military vehicles, significantly impacting the field of vehicle automation.

Author Contributions: Conceptualization, D.Ç. and V.K.; methodology, D.Ç. and V.K.; software, D.Ç.; validation, D.Ç. and V.K.; formal analysis, D.Ç. and V.K.; investigation, D.Ç. and V.K.; resources, D.Ç. and V.K.; data curation, D.Ç. and V.K.; writing—original draft preparation, D.Ç.; writing—review and editing, D.Ç. and V.K.; visualization, D.Ç. and V.K.; supervision, D.Ç. and V.K.; project administration, D.Ç. and V.K.; funding acquisition, D.Ç. and V.K. All authors have read and agreed to the published version of the manuscript.

Funding: This work was carried out as part of the Hybrid Electric Tracked Vehicle Development project of FNSS Savunma Sistemleri A.Ş. and was fully funded by FNSS Savunma Sistemleri A.Ş.

Data Availability Statement: The data and software materials used in this study are owned by FNSS Savunma Sistemleri A.Ş. and cannot be shared due to confidentiality policies of the company.

Acknowledgments: Our thanks go to the R&D Team at FNSS Savunma Sistemleri A.Ş., whose assistance was crucial in completing this work.

Conflicts of Interest: Dersu Çeliksöz and Varlık Kılıç are employees of FNSS Savunma Sistemleri A.Ş. The paper reflects the views of the scientists, and not the company.

References

1. Sivakumar, P.; Reginald, R.; Viswanath, G.; Viswanath, H.; Selvathai, T. Configuration Study of Hybrid Electric Power Pack for Tracked Combat Vehicles. *Def. Sci. J.* **2017**, *67*, 354–359. [CrossRef]
2. Randive, V.; Subramanian, S.C.; Thondiyath, A. Design and analysis of a hybrid electric powertrain for military tracked vehicles. *Energy* **2021**, *229*, 120768. [CrossRef]
3. FNSS. Available online: https://www.fnss.com.tr/en (accessed on 6 November 2023).
4. Çeliksöz, D.; Kılıç, V. Motion Control System Development of Serial Hybrid Electric Tracked Vehicles. In Proceedings of the 36th International Electric Vehicle Symposium and Exhibition (EVS36), Sacramento, CA, USA, 11–14 June 2023.
5. Zhang, J.; Ma, X.; Yuan, D.; Zhang, P. Steering Control of Dual Electric Drive Tracked Vehicle Based on Model Predictive Control. In Proceedings of the IOP Conference Series: Materials Science and Engineering 782, Qingdao, China, 28–29 December 2019. [CrossRef]
6. Chen, Z.-Y.; Zhang, C.-N. Control strategy based on BP neutral network plus PID algorithm for dual electric tracked vehicle steering. In Proceedings of the 2nd International Conference on Advanced Computer Control, Shenyang, China, 27–29 March 2010. [CrossRef]
7. Hu, J.; Tao, J.; Zhao, W.; Han, Y. Modeling and simulation of steering control strategy for dual motor coupling drive tracked vehicle. *J. Braz. Soc. Mech. Sci. Eng.* **2019**, *41*, 190. [CrossRef]

8. Li, P.; Yan, J.; Tu, Q.; Pan, M.; Jiang, C. A Steering Control Strategy Based on Torque Fuzzy Compensation for Dual Electric Tracked Vehicle. *Filomat* **2018**, *32*, 1953–1963. [CrossRef]
9. Kotiev, G.; Padalkin, B.; Miroshnichenko, A.; Stadukhin, A.; Kositsyn, B. A Theoretical study on the high-speed electric tracked vehicle mobility. In Proceedings of the IOP Conference Series: Materials Science and Engineering 820, Moscow, Russia, 1–2 October 2019. [CrossRef]
10. Zhai, L.; Zhang, X.; Wang, Z.; Mok, Y.M.; Hou, R.; Hou, Y. Steering Stability Control for Four-Motor Distributed Drive High-Speed Tracked Vehicles. *IEEE Access* **2020**, *8*, 94968–94983. [CrossRef]
11. Zhai, L.; Sun, T.M.; Wang, Q.N.; Wang, J. Lateral stability control of dynamic steering for dual motor drive high-speed tracked vehicle. *Int. J. Automot. Technol.* **2016**, *17*, 1079–1090. [CrossRef]
12. Huang, H.; Zhai, L. A Power Coupling System for Electric Tracked Vehicles during High-Speed Steering with Optimization-Based Torque Distribution Control. *Energies* **2018**, *11*, 1538. [CrossRef]
13. Zhai, L.; Huang, H.; Kavuma, S. Investigation on a Power Coupling Steering System for Dual-Motor Drive Tracked Vehicles Based on Speed Control. *Energies* **2017**, *10*, 1118. [CrossRef]
14. Aiso, K.; Akatsu, K. Performance Comparison of High-Speed Motors for Electric Vehicle. *World Electr. Veh. J.* **2022**, *13*, 57. [CrossRef]
15. Akar, A.; Çalık, G.; Çeliksöz, D. Battery Sizing & EV Mode PMU Algorithm in Military Armored Hybrid Electric Vehicles. In Proceedings of the 36th International Electric Vehicle Symposium and Exhibition (EVS36), Sacramento, CA, USA, 11–14 June 2023.
16. Charged EVs. Electrifying Auxiliaries: A Flexible System Approach for Multiple Auxiliaries. Available online: https://chargedevs.com/newswire/electrifying-auxiliaries-a-flexible-system-approach-for-multiple-auxiliaries (accessed on 22 January 2024).
17. Pugi, L.; Berzi, L.; Grasso, F.; Savi, R.; Vita, V. Design and Simulation of an Electrified Directional Drilling Machine. *Int. J. Mech. Control* **2021**, *22*, 17–29.

World Electric Vehicle Journal

MDPI

Article

An Empirical Study of the Policy Processes behind Norway's BEV-Olution

Erik Figenbaum

Institute of Transport Economics, Gaustadalleen 21, N-0349 Oslo, Norway; efi@toi.no

Abstract: Norway's large battery electric vehicle (BEV) market and fleet are not the result of a comprehensive policy plan. Using the multiple streams (MS) framework and document analysis, it was identified that the most important Norwegian BEV policy decisions were made using inadequate policy processes that fall outside of traditional politics. This is contrary to the MS framework postulate that three independent streams of problems, policy solutions, and politics must align to pave the way for new policies. Politicians had limited information about the effects of policies they introduced in this "learning by doing process". Impact assessments were rarely made. The decision rationale was often not documented. The future market expectation and thus the national budget consequences were low when important policy decisions were made, whereas the political gain was high. The processes were more aligned with traditional politics after 2014. The ambitious ZE vehicle targets for 2025 and the climate policy targets for 2030 locked in incentives, despite rising tax losses. In sum, these developments created the world's largest per-capita BEV market. To avoid negative issues and keep the BEV policies' potential to support the BEV transition, politicians should ensure that sufficient knowledge is available when making decisions about future policies. Such decisions should be taken transparently within traditional politics, be properly assessed as with EU policy processes, and regularly reviewed as with the California ZEV mandate process. The required knowledge should be developed in open-access research.

Keywords: policy; incentive; strategy; passenger car; government

Citation: Figenbaum, E. An Empirical Study of the Policy Processes behind Norway's BEV-Olution. *World Electr. Veh. J.* **2024**, *15*, 37. https://doi.org/10.3390/wevj15020037

Academic Editors: Joeri Van Mierlo and Genevieve Cullen

Received: 18 November 2023
Revised: 5 January 2024
Accepted: 17 January 2024
Published: 24 January 2024

1. Introduction

Norway is the world leader in per capita BEV diffusion (battery electric vehicle) [1]. In total, 690,000 BEVs were on the road at the end of 2023, which was 24% of the passenger car fleet. Another 7% were plug-in hybrid vehicles (PHEVs) [2]. The BEV market share passed 79% in 2022, with PHEVs accounting for another 8.5% [2], as seen in Figure 1. In 2010, there were only 3000 BEVs in Norway [3]. BEVs are now pursued as a measure to reduce transport GHG emissions, and the current target is to only sell ZEVs from 2025, which is the world's most ambitious timeline. The consensus among most automakers is that BEVs will become the dominant technology for passenger vehicles. PHEVs will not count towards the EU's new vehicle 0 g CO_2/km 2035 target, and the sale of hydrogen fuel cell vehicles (FCEV) is miniscule, despite having the same incentives as BEVs. FCEVs were discredited after a filling station explosion in 2019 and subsequent filling station closures. Only two models have been available in the market, the Toyota Mirai and the Hyundai NEXO.

The incentives for BEVs include the exemption from registration tax from 1990, the 1996–2003 exemption from annual tax and the reduction from 2004, the zero rate value-added tax (VAT) on BEV purchases from 2001 and BEV leasing from 2015, free parking from 1999 to 2017 and a parking fee reduction from 2018, free road tolls from 1997 to 2017 and a road toll reduction from 2018, reduced ferry rates from 2009, and access to bus lanes in the Oslo area from 2003 and nationally from 2005, with some rush hour limitations from 2015. Finally, there has been a re-registration tax exemption from 2018 and there was a reduction in 2022, as well as a reduced benefit tax on company cars from 2000 to 2022.

Support schemes for normal chargers have been in place in Oslo since 2008 and nationally since 2009. Support for fast chargers was introduced in 2011 and scaled to cover all main roads between 2015 and 2020. Most fast chargers are now deployed on commercial terms.

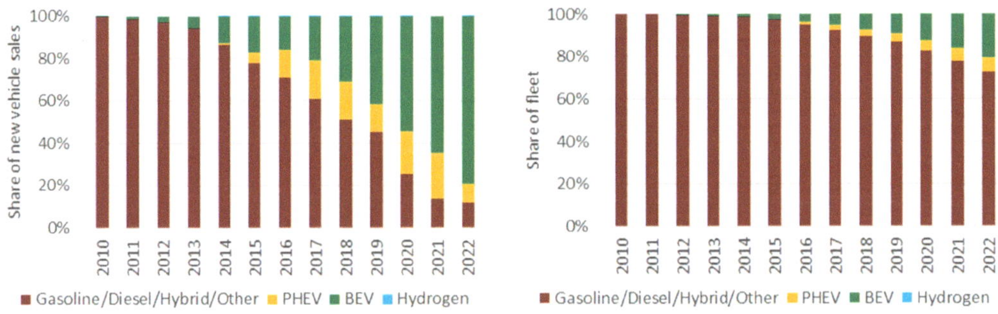

Figure 1. Sales shares and fleet shares for ICEVs, PHEVs, BEVs, and hydrogen FCEVs from 2010 to 2021. Source: Statistics Norway.

Norway established through these incentives a market where BEV producers competed on equal terms and where BEVs became competitive with ICEVs earlier than elsewhere. This gives rise to the research question of this article: How and why were the BEV policies and incentives established and what did the politicians know about BEVs when the actual decisions were made? Understanding how these policies came about can be used to improve the policy processes ahead and help other countries seeking to accelerate the transition to ZEVs. They need to understand how a country like Norway, without an automotive industry, could become such a leader in the ZEV transition in order to develop efficient policies.

Section 2 of this article includes a description of the methods and materials used in this study. The results are in Section 3, the discussion is in Section 4, and the conclusion is in Section 5.

2. Materials and Methods

2.1. Literature Review

The rapid market introduction of BEVs in Norway is the result of large incentives that were introduced in 1990 [3–5]. The first incentives were intended to enable market experiments and establish knowledge about BEVs' technical potential in Norway's demanding climatic conditions [3,4]. Further incentives were introduced [3–5] to build a niche market and establish a BEV industry between 1999 and 2002 and 2007 and 2010. Norway had then a world-leading BEV developer, THINK [5], which Ford owned from 1999 to 2002. Ford needed low-cost BEVs for the California ZEV mandate. The market remained small up until 2011. It was limited by the high cost and limited supply and quality of BEVs [3,5]. Norwegian BEV industrialisation ended in 2010 due to a lack of funding for THINK and other entrepreneurs in the wake of the global financial crisis. [3,4]. The market took off in 2011 with the availability of high-volume OEM BEVs that were competitive with ICEVs due to the Norwegian incentives [6]. The OEMs expanded their offerings further in 2016, which allowed a BEV regime to gradually emerge [5] and compete with the existing ICEV regime. A long-term policy framework was a prerequisite for success [7], and the user value of the incentives was high, according to user surveys [8–10]. BEV sales expanded in 2020 with the availability of longer-range models [11,12]. An expanding charging infrastructure ecosystem supported long-distance travel and single-vehicle ownership [13], but user-friendliness was lacking due to a myriad of different suppliers with different apps and payment systems [14]. The policy focus since 2021 has been a controlled downscaling of incentives, as signalled by the publishing of the principles for the future of vehicle taxation in Norway in the 2021 National Budget [15]: "A sustainable vehicle taxation system must include technology neutral equitable taxes on purchase and use of vehicles and take into account the transition to ZEVs towards 2025".

The Norwegian 1990 to 2022 societal BEV development process has previously been analysed using the technology innovation system framework [3,4] and the multi-level perspective (MLP) [5,16]. These analyses found that an alignment of the factors required to achieve rapid BEV diffusion occurred after 2010. The costs, effectiveness, and impacts of policies [17–22], including the total cost of ownership [6], have been analysed, with the conclusion that the incentives have been vital in the development of the market. Downscaling the incentives while keeping sales up may be possible according to the latest research [23], although user surveys show that most of the incentives have been and are still important [8–10]. The knowledge available to politicians when introducing BEV policies has been limited [24]. Cities had an important role in BEV policy development [5,25] due to local incentives such as free parking and support schemes for chargers. Local assets such as clean electricity and policy learnings over time have also been important. [7,26]. Some criticisms of the policies do exist, especially on the combined size of the incentives [27,28], but most research has focused on the positive or factual aspects of the electromobility transition and how to reach national targets. Politicians have focused on making BEVs a story, as seen in the Appendix A overview of the suggested policies in party programmes and government declarations from 1990 to 2023.

Norway became the world leader in BEV adoption without anyone having analysed in detail exactly how and why the Norwegian BEV policies and incentives were established or what politicians knew about BEVs and the impacts of the policies when the actual decisions were made. This article aims to fill this knowledge gap and improve the understanding of policy processes. This knowledge can aid policy development for the electrification of light commercial vehicles and trucks and should be of interest to other countries seeking to accelerate their transition to ZEVs. This understanding may also be relevant for other policy areas.

Understanding the BEV policy development processes has also not been an important research theme in other countries. There are singular examples for Sweden and Denmark [29], Germany [30], France [31], and the UK [30], and for regions such as the EU [32], the Nordic countries [33], and California [31,34]. California and the EU conduct large, transparent, and publicly available impact assessments when introducing new ZEV and vehicle CO_2 policies [35,36] so there is less need to study how these processes proceeded. Cross-country analyses have provided additional information about the efficiency of BEV policies in different contexts [26,29–31,33,37,38] and, sometimes, on how they came about [26]. The conclusion is, however, that BEV policy development processes are understudied in general. Yet, this topic is of special interest in the Norwegian case as the policies led to market shares above 80%. This achievement came at a considerable tax loss cost but without much resistance. There is thus a need to increase the understanding of the overall process in Norway.

2.2. *A Brief Overview of Norwegian Politics and Policy Processes*

Norway has a tradition of technology-neutral politics developed in thoroughly documented processes defined by the "Instructions for official studies and reports" ("Utredningsinstruksen") [39–41]. A strong social economics bureaucracy in the Ministry of Finance oversees the national budget process and has written procedures and methods for how policy changes should be evaluated [42–45]. The essence of these requirements is that all relevant aspects of all types of governments that internally or externally develop policy proposals should be thoroughly evaluated using a specific method that captures the economic impacts.

Large policy changes are normally introduced in gradual policy processes, as illustrated in Figure 2. They start with a public report evaluating policy change needs and implications for the national budget, the public, businesses, and stakeholders. The government develops the suggested changes into a proposal in the annual national budget. A debate in parliament on the budget and policy changes follows. A recommendation from one of the parliament committees is made before making a decision in plenum. Large policy changes are often anchored as broad political agreements between the parties in parliament for stability reasons. New governments build politics from the existing situation and rarely reverse recent reforms.

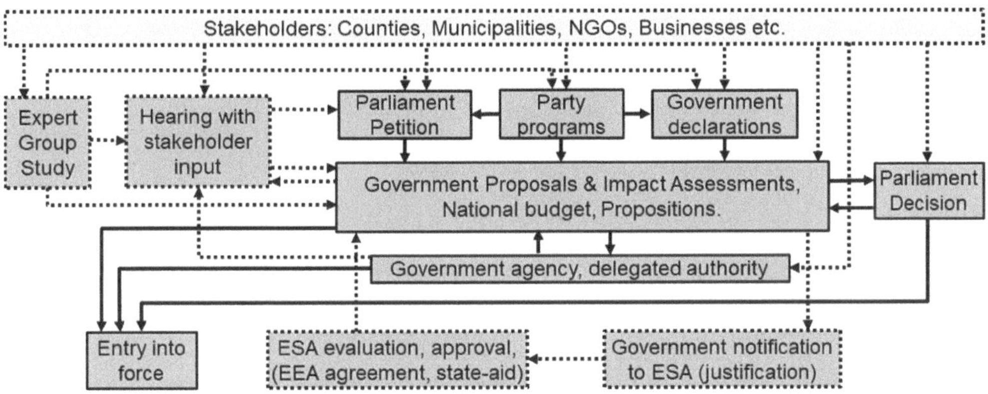

Figure 2. Generic BEV policy development process flow. Source: Author.

Taxes are adjusted in small steps in a government's internal annual national budget process to provide stability for market actors. Vehicle importers have, for instance, already pre-ordered vehicles for the following year when a national budget has been made public in early October. Large, unexpected tax changes can influence the value of a dealer's stock of new and second-hand vehicles and can thus be a challenge.

Policies that require a law change and affect businesses, consumers, or other governance levels go through a structured process with a public hearing after parliament or the government has proposed a law change. Potentially unwanted effects can be identified, and adjustments can be made before the law enters into force. Law changes that only affect the national governance level can be made directly by parliament. Parliament can also petition the government to introduce specific policies. The government responds with an analysis of the impacts in the next national budget documents or propositions to parliament.

Norwegian politics is, however, less stable than before as cross-party coalitions have become the rule. This leads to very detailed government declarations that regulate the policies that the government will pursue up until the next election, including vehicle taxation and BEV incentives. These declarations are the result of long negotiations in which party programmes are the starting point. This means that decisions can have been made about politics even before any impact assessments have been made about their effectiveness, costs, or other impacts. Small pro-BEV parties can in this way have a high impact on BEV policies.

The Norwegian relationship with the EU is regulated via the EEA agreement, which essentially means that the four freedoms of the EU—the free movement of goods, capital, services, and people—applies also to Norway. The EFTA Surveillance Authority (ESA) has the role that the EU court has within the EU, i.e., to verify the legality of the policies proposed in terms of state aid and EU regulations related to the four freedoms.

2.3. Method

This study of Norwegian BEV policy development processes was based on a systematic document analysis. This method was chosen because all the relevant facts about the development of the large and costly Norwegian BEV incentives should have been properly documented in publicly available documents if the structured Norwegian policy processes depicted in Figure 2 were followed. A second reason is that documents are the only reliable source of information that span the entire 1990 to 2023 timeframe of this study. A third reason was to avoid memory bias.

The first target of the document study was to identify the level of knowledge of BEVs and the expectations for future developments at the time when important BEV policy decisions were made. The second target of the document study was to identify the degree to which structured Norwegian political processes were followed and, specifically, the instructions for official studies and reports and the guidelines for policy analysis. Combined, these two approaches make it possible to answer this article's research question. The actors that were involved in the decisions would not have had access to the full information gathered retrospectively in this article. They may also have acted on biased information from market actors and stakeholders. Neutral information on BEV usability in Norwegian conditions was hardly available up until 2010.

The 1990–2023 BEV policy development process has been split into seven periods and evaluated against the structured Norwegian governance processes using Kingdon's [46] multiple streams (MS) framework. This framework is appropriate for the study of policy development processes. Kingdon states that policy agendas are set by the dynamics of three "streams" of processes that are essentially independent of each other: a stream of problems, a stream of policies, and a stream of politics—the 3Ps [34]. When these three streams align, a policy window is created that provides opportunities for policy actors to push their views on policy problems and solutions and set the policy agenda, i.e., pave the way for BEV support policies. An agenda is defined by Kingdon [46] as "the list of subjects or problems to which government officials, and people outside of government closely associated with those officials, are paying some serious attention at any given time". Collantes and Sperling [46] found the framework useful for EV policy analysis but questioned if these three streams are independent of each other.

2.4. Materials

Great effort was put into identifying all the relevant documents that deal with different aspects of BEV development since 1990. The materials included 261 articles, reports, books, and other documents from research, government and civil services, consultants, NGOs, and industry, as shown in the overview in Table 1. In addition to these documents, the analysis draws on information from the annual national budget documents and protocols of policy decisions and debates in parliament. Documents with relevance to the policy development processes were subsequently included in the analysis and complemented by press articles identified through the Norwegian Retriever news archive service. Many of the reviewed documents prior to 2000 are not publicly available now but were disseminated to the public and policy makers when they were published. They fill a void in the knowledge of the early development and come from the author's archive. Most documents up to 2010 and most of the press articles are in Norwegian language.

Table 1. Overview of the documents analysed.

	Electromobility Norway	Theory/International	Total
Peer-reviewed research articles	22	18	40
Editor-reviewed research articles	2	1	3
Monographs (scientific book, PhD thesis)	3	2	5
Book chapter in scientific book	3	3	6
Scientific research paper	3	3	6
Reports—research/scientific	28	3	31
Reports—authorities using scientific approach	2	0	2
Reports—consultants	10	0	10
Reports—organisations	2	1	3
Popular science book	2	0	2
Press articles	66	9	75

Table 1. *Cont.*

	Electromobility Norway	Theory/International	Total
Other news articles, websites	14	3	17
Private actor documents	17	6	23
Public actor documents	10	1	11
Political documents	18	0	18
Law texts	1	2	3
Other references	4	2	6
Total	207	54	261

3. Results

Sections 3.1–3.7 analyse the detailed policy processes behind the introduction and revision of each policy and incentive—split into seven periods—before assessing the overall process in Section 3.8. Each subsection starts with a brief overview of the main activities of the period and contains a flowchart that shows a chronological chain of the market, policy, and knowledge development.

3.1. 1990–1997—Policies Enabled Market Experiments to Verify BEVs' Potential

In 1990 when the first was imported, BEVs were unknown [3]. A vehicle registration code for BEVs did not even exist. The gasoline three-way catalyst became obligatory in 1989, but local pollution was still a problem in cities. Politicians saw BEVs as a local pollution reduction measure, whereas the energy sector saw BEVs as a new electricity market [3]. The first incentives—the registration tax (1990) and the annual tax exemption (1996)—enabled market experimentation. PIVCO saw potential for producing BEVs, being inspired by California's ZEV mandate requirements for BEV sales from 1998 and French and Swiss BEV activities. PIVCO (THINK) tested BEV prototypes 3–4 years before starting industrialisation. The target was to produce 5000–10,000 BEVs/year [3]. PIVCO BEVs were advanced for their time, having Ni–Cd batteries, an ABS, a driver airbag, and a 50–80 km range [3]. The competitor Kewet had BEVs with lead–acid batteries and a 30–40 km range. A small number of Peugeot, Citroën, and Renault Ni–Cd-battery BEVs were also available. The National Institute of Technology tested BEVs' capability [3], with inconclusive results. BEVs were seen as a positive concept with good potential, but they were small, had low top speeds, and the build quality and durability were poor. They were also too expensive, even with incentives [6]. The Electric Vehicle Association (EVA, Norstart) was established to increase BEV interest, improve incentives, and support industrial development [47]. Free road tolls were introduced in 1997 to make BEVs more competitive, after NGO lobbying [3,47,48]. This incentive became important in later years with toll roads everywhere. At the end of this period, the BEV fleet counted 105 and two models were sold. They had a range of 30–60 km. The timeline is shown in Figure 3.

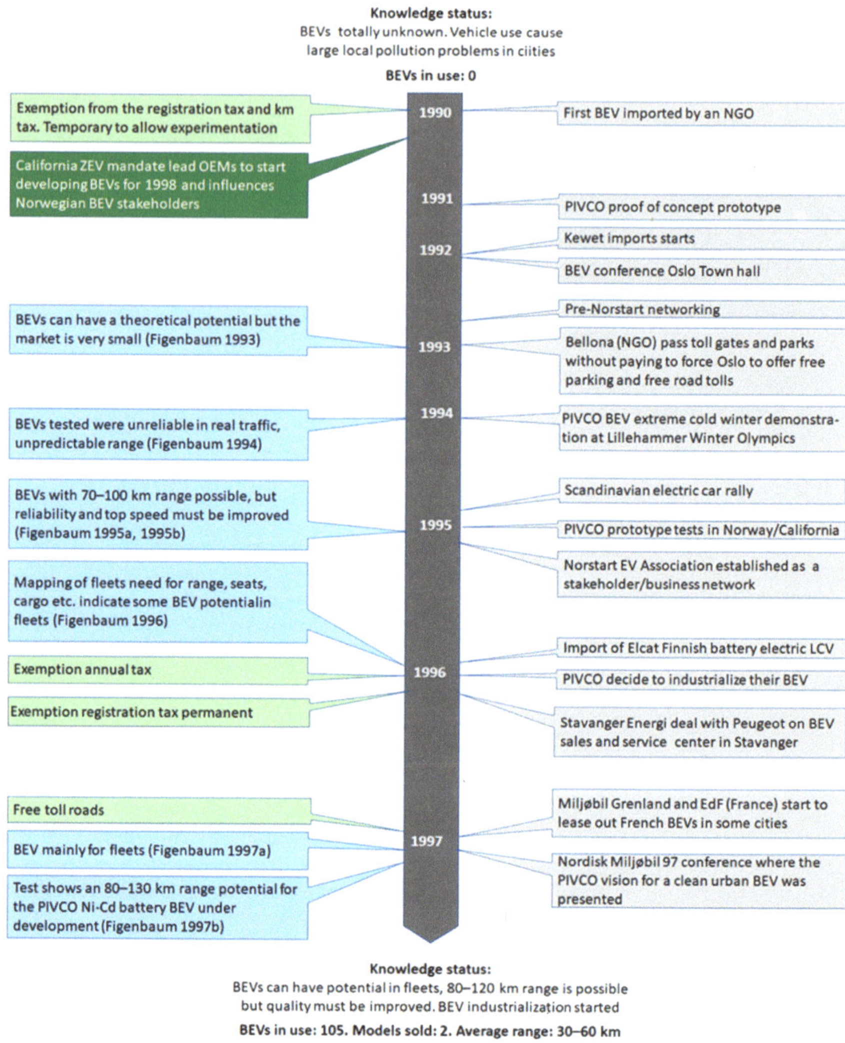

Figure 3. Timeline of policies, market activities, and research publications 1990–1997 [49–55]. Light green: Norwegian policies. Dark green: international policies. Blue: research results. Grey: market activities. Source: Author.

Exemption from registration tax and km tax (1990, 1996)

High vehicle registration taxes generate government income. Because taxes doubled 1990 BEV prices [6], NGO and industry actors lobbied for an exemption [47] to enable market tests. The finance minister was positive [56]. A temporary exemption in the 1990 National Budget [57] was endorsed by parliament [58]. BEVs were also exempted from km tax. No impact assessments were made. The tax loss was negligible, with five BEVs in the fleet [3]. The exemption became permanent when Parliament adopted the 1996 National Budget vehicle tax reform [59,60]. This was again without an impact assessment, contrasting the well-prepared reform itself [59,61]. The exemption lasted until 2023 when a weight tax was introduced for all vehicles [62].

Exemption and reduction from annual tax (1996)

The annual tax exemption was decided in the parliament's 1996 National Budget vehicle taxation reform debate and was documented in a sentence in the minutes [60]. There was no impact assessment as it had not been proposed in the reform [59]. With 50 BEVs in the fleet [3], the budget impact was negligible. BEV owners had to pay [63] a traffic accident tax (EUR ~40) from 2004 after it became part of the annual tax. A 2015 vehicle taxation policy settlement [64] decision to introduce half annual tax in 2018 was broken in 2016. Parliament instead decided on a full tax exemption from 2018 [65] during a process to change the annual tax to a vehicle insurance tax [66]. In the national budget process for 2021, parliament endorsed the government's 70% ICEV rate proposal [15]. The incentive was removed in April 2022 [67].

Free road tolls (1997)

An environmental NGO and the pop group A-ha, supported by the EV Association and Oslo Energi (DSO), wanted an exemption from road tolls and parking fees for BEVs in Oslo in the early 1990s. They thus refused to pay [47]. The pressure [3,5,24,47] made Oslo decide to offer free toll roads in 1995 [47,68] and free parking from 1997 [69]. Oslo wanted to reduce pollution and have PIVCO/THINKs BEV factories in Oslo [70]. National laws inhibited the introduction. The Norwegian Public Roads Administration (NPRA) stated that road tolls could by law only be used to build roads. Parliament changed the law in 1997 without any impact assessment, stating that BEVs were environmentally comparable to the already exempted buses [71]. The Minister of Transport stated that toll road companies were not impacted, but longer payment periods or higher rates could be required. In the 2017 National Budget, parliament decided that BEVs could pay maximum 50% of the rate for ICEVs for parking and road tolls, to be decided by local authorities [65], which was changed to 70% in 2023 [62].

3.2. 1998–2002—Policies Supported BEV Industrialisation

This period started optimistically. PIVCO changed its name to THINK and industrialised a city BEV that was launched at the 1998 Brussels World Electric Vehicle Symposium. Lotus Engineering (UK) improved the quality of the product and aided the production start-up. The 1998 Asian crisis hit the Norwegian economy with falling oil prices. THINK lost capital and went bankrupt in 1998 [3]. Ford bought THINK in 1999 to obtain a low-cost BEV to meet California's ZEV mandate [3]. Production started in late 1999 after product improvements and the introduction of a better-quality assurance system. Sales started in Norway and some European markets. A model for California was developed [47]. THINK reached the global BEV forefront. Small numbers of Kewet and French BEVs were imported and sold to fleets, enthusiasts, and free-road-toll beneficiaries. Politicians became BEV proponents with Ford owning THINK and introduced free parking and a zero-rate VAT on BEVs. Ford sold THINK in 2002 after it became clear that the 2003 California ZEV mandate no longer required BEVs as they were seen as technically immature by the legislators. THINK also had technical problems with the California model, and Ford had economic problems and had to save costs [3]. THINK was sold to an Indian investor. The period ended pessimistically with a global downturn. BEVs were seen as not being market-ready. The fleet had, however, grown to 871 by the end of 2002 [3]. The complete timeline is shown in Figure 4.

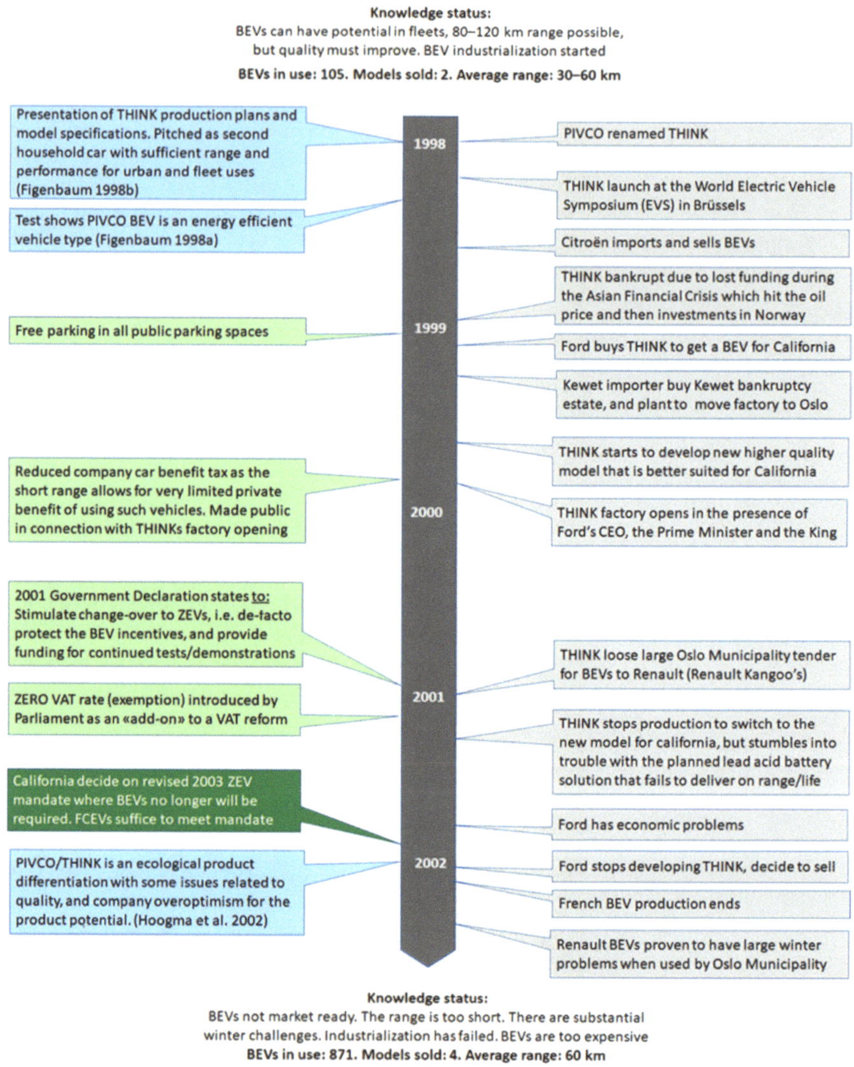

Figure 4. Timeline of policies, market activities, and research publications 1998–2002 [72–74]. Light green: Norwegian policies. Dark green: international policies. Blue: research results. Grey: market activities. Source: Author.

Free public parking (1999)

Free public parking (see also Section 3.1) came [75,76] after a Ministry of Transport parking law revision [77] and law change process [78]. The fee losses were negligible, with only 285 BEVs in the fleet [3]. In 2016, parliament followed up on the 2015 vehicle tax policy settlement [64] and decided to let local authorities decide on BEV parking fees from 2017. During the 2017 National Budget debate, parliament decided on a maximum of a 50% ICEV rate, to be decided by local authorities [65], but it was never implemented.

Reduced company car benefit tax (2000)

The prime minister introduced—without any impact assessment—a reduced company car benefit tax at the 1999 THINK factory opening to support industrialisation [79,80]. The rationale was that BEVs had lower private benefits due to their short range and long charge times. BEV company car sales were low as the zero-rate VAT did not apply to leased BEVs until 2015. A tax revision from 2005 set BEVs' value to 75% of their list price before calculating tax [81]. It was set to 50% in 2009 [82], 60% in 2018 [66,83], 80% in 2022 [84,85], and 100% in 2023 [62].

Zero-rate VAT (Value-Added Tax) (2001)

THINK BEVs were too expensive [3], but Ford, preferring mandatory public fleet targets [3] as in the US, was inactive on tax incentives. So, EVA and Bellona (NGOs) lobbied [47] for a VAT exemption during the 2001 National Budget VAT reform process. This reform, which was based on a VAT expert group report [86], but a BEV exemption was not part of the proposed reform. The EVA and Bellona told parliament that a VAT exemption was needed to support THINK. Politicians were positive according to an EVA document [87]. Parliament decided [88] on a zero-rate VAT to make BEVs more economical to buy as part of the reform [89]. A NOK 10 million tax loss for 250 BEVs sold was estimated for 2001 [3]. There was no impact assessment in the national budget to support the decision [90]. A gradual VAT re-introduction from 2018 was proposed in the 2015 Revised National Budget [91], but parliament decided to keep it in place through 2017 after EFTA Surveillance Authority (ESA) approval (see Section 3.5) [92]. It was later extended through 2020 [93] and to 2022 [94]. In the 2022 Revised National Budget [95], it was proposed that the zero VAT rate be replaced by a support scheme covering the VAT up to a price of NOK 500,000. Parliament decided on a full exemption up to NOK 500,000 and full VAT on the part of the price above NOK 500,000. This scheme, formalised in the 2023 National Budget [62], will last until 2025 [96].

3.3. 2003–2006—Policies Remained in Place as Global BEV Markets Collapsed

A four-year global BEV downturn followed the previous periods' optimism [3,5]. BEVs were not considered market-ready, Ford had left THINK, and the French activity ended. Norwegian activity also plummeted. The EV Association lost most of its members and barely survived [3]. THINK went bankrupt again, in spite of BEVs obtaining access to bus lanes in the Oslo area from 2003 and nationally from 2005. THINK was bought by Norwegian investors who saw the potential for BEVs in the increased global interest in GHG emission reduction [3] measures. ElbilNorge had bought the bankrupt Kewet in 1999 and in 2005 established a small production of the 4-wheel MC (L7e registration) Buddy based on the Kewet model. New actors imported used French BEVs [3,5]. The 2001 government declaration contained a sentence on keeping the incentives in place, which was important for future developments. The government-appointed Low Emission Committee [97] found BEVs to be vital for Norway to become a low-emission society by 2050 and suggested supportive policies. A slow market continued to develop through this difficult period. The BEV fleet now counted 1656 [3]. Several models were imported in this period. In the end, two were available. The timeline is shown in Figure 5.

Access to bus lanes (2003/2005)

In 2001, the NPRA planned to ban minibuses from bus lanes [98]. Consumers used them to avoid rush hour queues. This would [5] thus make room for environmentally friendly BEVs, and lobbyists [99] also hoped to turn around Ford's decision to sell THINK [47] with such an incentive. The Minister of Transport [100,101] agreed to test it out in the Oslo area in 2003 to see if buses were delayed by the (then) slow BEVs [3]. Buses were not delayed, BEV demand increased [102,103], and the incentive became permanent and nationwide in 2005. Minibuses were thrown out in 2009 [104]. The test replaced the impact assessment normally required for policy changes. The motivation was to reduce pollution and support the market [105,106]. It had no budget impacts and was not state aid [92]. A passenger has in some places been required in rush hour in some places since 2015 [107–109] due to increased bus lane congestion.

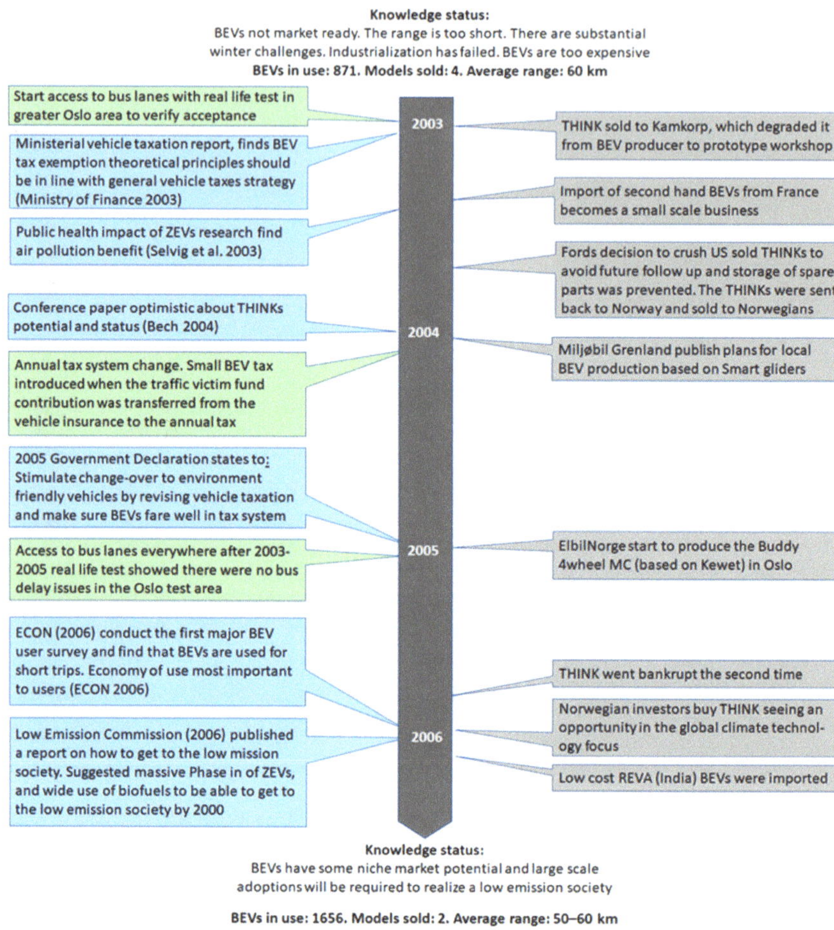

Figure 5. Timeline of policies, market activities, and research publications 2003–2006 [97,103,110–112]. Light green: Norwegian policies. Dark green: international policies. Blue: research results. Grey: market activities. Source: Author.

3.4. 2007–2010—Policies Supported BEV Industrialisation in the Global Climate Policy Spur

THINK expanded with international investors, hired previous staff, and, in 2008 at the Geneva Auto show, launched the model developed under Ford for the Californian market [3]. It had double the range of other models (130 km) but used expensive Li–Ion and Ni–$NiCl_2$ batteries. ElbilNorge improved the Buddy and increased production. The incentives were still in place and a new one was added—the reduced ferry rate from 2009 [5]. A new vehicle GHG emission reduction target was introduced. The first public charging networks were put in place in the Oslo municipality in 2008 and across Norway from 2010 with support from Transnova, a new government agency. The Electrification Resource Group appointed by the Ministry of Transport saw great BEV potential leading up to 2020 and suggested new incentives [113]. The global financial crisis hit THINK and ElbilNorge hard. Both went bankrupt in 2010/2011 [3] when the BEV breakthrough started with sales of OEM BEVs. The fleet had increased to 3360 and six models were sold [3]. The timeline is shown in Figure 6.

Average new vehicle CO_2 emissions target of 120 g/km by 2012 (2007)

Norway's first average new vehicle CO_2 emissions target of 120 g/km by 2012 came during a 2007 government press conference [114]. The 2007 Climate Policy Bill [115] had no specific vehicle target, only a sentence about phasing in ZEVs. The vehicle importers had lobbied for the 120 g/km 2012 target to favour diesel ICEVs, which, in theory, reduce CO_2 emissions by 20–25% compared to gasoline ICEVs. The target was to be achieved by tuning the CO_2 element of the registration tax. No impact assessment was published. The parliament majority formalised the target as part of a climate policy settlement [116]. It was more ambitious than the EU's voluntary 130 g/km target for 2015, which became an EU regulation in 2009 [117].

Increased vehicle allowance for business trips (2008)

A higher km allowance for government employees' use of private BEVs for business trips was introduced, without impact assessment, by the government in 2008 [118]. It was introduced as a measure to support THINK's reopened factory and because BEVs' total cost of ownership was higher than that for ICEVs.

Reduced ferry rates (2009)

In 2009, the Minister of Transport introduced [119] reduced ferry rates. This was based on a voter's idea [120]. The BEV was free of charge, but the driver paid the regular fee. No expert group study or impact assessment was made. No stakeholders were involved. The idea came from a voter and the rationale was to support industry, reduce energy consumption and environmental impacts, and spread BEVs to coastal areas [120]. The NPRA had the delegated power to implement the change. With 2424 BEVs in the fleet in 2008 [3], mainly in cities, the incentive would not strain ferry operators' budgets. Ferry operators could from 2018 charge BEVs 50% of the ICEV rates [65].

Transnova funding agency (2009) transport GHG emission reduction measures, first charger support programme

In parliament's climate policy settlement [116] for the 2007 Climate Policy Bill [115], it was decided that a new funding agency, Transnova, should fund clean transport projects. To battle the 2009 financial crisis, the government decided to support the installation of chargers across Norway [121]. Transnova thus established [122] a national NOK 50 million, first-come-first-serve support scheme for normal public chargers [123]. Standard Schuko household-type outdoor sockets were chosen due to a lack of standards and 2500 chargers were supported. Transnova supported the first 24 fast chargers in 2011–2012 [124] with leftover funds.

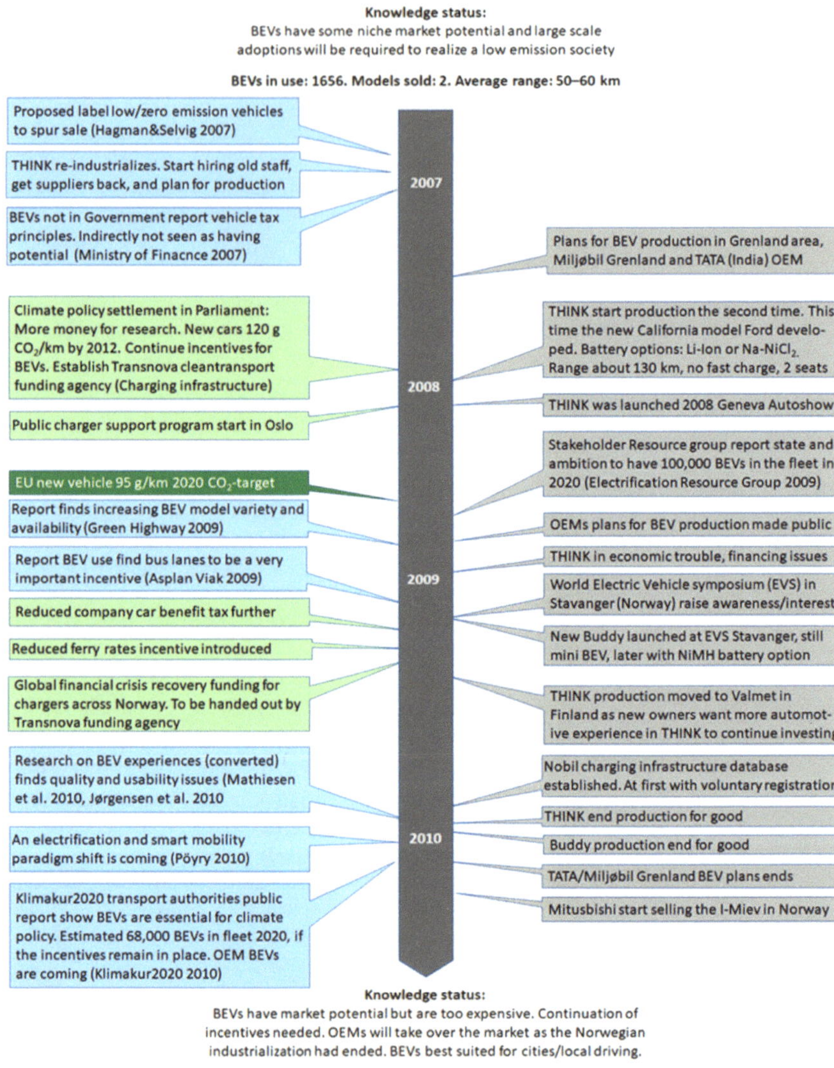

Figure 6. Timeline of policies, market activities, and research publications 2007–2010 [125–132]. Light green: Norwegian policies. Dark green: international policies. Blue: research results. Grey: market activities. Source: Author.

3.5. 2011–2015—Policies Supported the Roll-Out of Increasing Numbers of OEM BEVs

BEVs from the Mitsubishi, Citroën, Peugeot, and Nissan OEMs sold well. Buyers saved time using bus lanes and saved money on road tolls and parking fees. More models came on the market when other OEMs started production. Existing outdoor sockets were used for charging, but "wall-box" installations expanded after dealers bundled them with BEV purchases. Improved Li–Ion batteries enabled longer ranges at a decreasing cost. Public chargers and fast chargers supported the market. A national fast charger infrastructure connected southern Norway's cities by 2015. Tesla developed the first long-range-capable BEV, supported by their growing network of superchargers. Dealers gave buyers a one-year-free EV Association membership. The EVA became a large consumer NGO supporting

BEV owners across Norway and influencing policy processes. The policy processes became more complex as sales increased and the impact on tax revenues became noticeable. The new average vehicle CO_2 emissions of 85 g CO_2/km by 2020 target (introduced in 2012) meant that BEVs, PHEVs, or FCEVs had to be sold [133]. Politicians simultaneously decided to continue the incentives until 2015 or until 50,000 BEVs had been sold. In 2015, Norway stated its intention in the Paris Agreement to reduce GHG emissions by 40% compared to 1990 levels. The average range of the BEVs sold more than doubled during this period, so BEV user appeal and sales increased substantially. The fleet reached 69,134 at the end of 2015, with 14 models sold. The average range was 120–176 km. The timeline is shown in Figure 7.

Average new vehicle CO_2 emissions of 85 g/km by 2020 (2012)

The 2012 Climate Policy Bill [134] proposed a reduction in average new vehicle CO_2 emissions to 85 g/km by 2020, which would de facto require the sales of BEVs, PHEVs, or FCEVs [134]. The Klimakur 2020 expert study [132] provided a knowledge base for CO_2 reduction measures in 2020, without proposing this specific target. The measure was inspired by the EU's voluntary 95 g/km by 2020 target from a 2009 EU regulation [117], which became a firm policy in a 2014 regulation [135].

Keep incentives in place until there are 50,000 BEVs in the fleet or through 2015 (2012)

Parliament decided in the 2012 climate policy settlement [136] to keep incentives in place until 50,000 BEVs were in the fleet or through 2015. This reduced uncertainty about the incentives, although, in 2011, the government stated that it had no plans to change them [137]. Parliament thus linked for the first time the level and timeframe of BEV incentives to both a long-term vehicle (85 g CO_2/km by 2020) and climate policy targets. No impact assessment supported the decision, apart from the Klimakur 2020 report [132].

Keep incentives in place until the end of 2017 (2013)

The 2013 government declaration [138] specified that the BEV incentives should last through 2017, regardless of sales. The small party Venstre was a strong BEV policy proponent and had this included in the declaration as a condition for supporting the new government. Impact assessments were partly found in the 2012 Climate Policy Bill [134] and the Klimakur 2020 [132] report.

Zero-rate VAT for BEV leasing and battery replacement (2015)

In the 2014 National Budget process, parliament petitioned [139] the government to introduce zero-rate VAT on BEV leasing and batteries [140] based on an NGO/auto-sector report [141]. The rationale was to treat leasing as equal to purchase and address battery replacement cost concerns [8,142]. It had been discovered that the EFTA Surveillance Authority (ESA), which supervises the European Economic Area agreement with the EU, had to be notified [92] to evaluate the impacts on the trade agreement between the EU and Norway. Notifications should have been sent also for other BEV incentives. The government proposed the incentive in a 2015 Revised National Budget (RNB) document [91] after notifying the ESA [143]. The ESA confirmed compliance through 2017 [92], including also the zero rate for BEV purchases. A formal decision was made during the RNB 2015 debate [91] over VAT law changes [144]. It entered into force in 2015 with a NOK 40 million first-year estimated tax loss [91]. This process followed the political tradition in Norway because the ESA notification required a proper rationale and impact assessment. The ESA found that the bus lane access was not state aid and that the registration tax exemption and free parking were in place before the 1994 EEA agreement, so these incentives were acceptable. The remaining BEV incentives were found to be proportional to targets.

Vehicle taxation policy settlement (2015)

The RNB 2015 document [91] stated that BEVs should fare better than ICEVs in the tax system to support the 2020 and 2030 climate policy targets and the 85 g CO_2/km target for 2020. Yet, several policy changes were proposed to limit the increasing tax losses from BEV sales. The incentives were expected to last through 2017 and the registration tax exemption was expected to last through 2020, but the zero-rate VAT was to be replaced by a support scheme after 2017, initially set as equal to the zero-rate VAT, as proposed by the Green Tax Committee [145]. This support was to be reduced as the technology improved and sales increased. A re-introduction of annual tax from 2018 and a removal of the company car tax advantage was also proposed. Parliament agreed [146] to this in a settlement with the government. It was also decided that local authorities should define parking fees and allow access to bus lanes and that the government should develop an environmental tariff differentiation system for toll roads and ferries. In the 2017 National Budget, parliament dismissed [65] the VAT support scheme and annual tax re-introduction. The 2025 target to only sell ZEVs introduced in 2017 was more important.

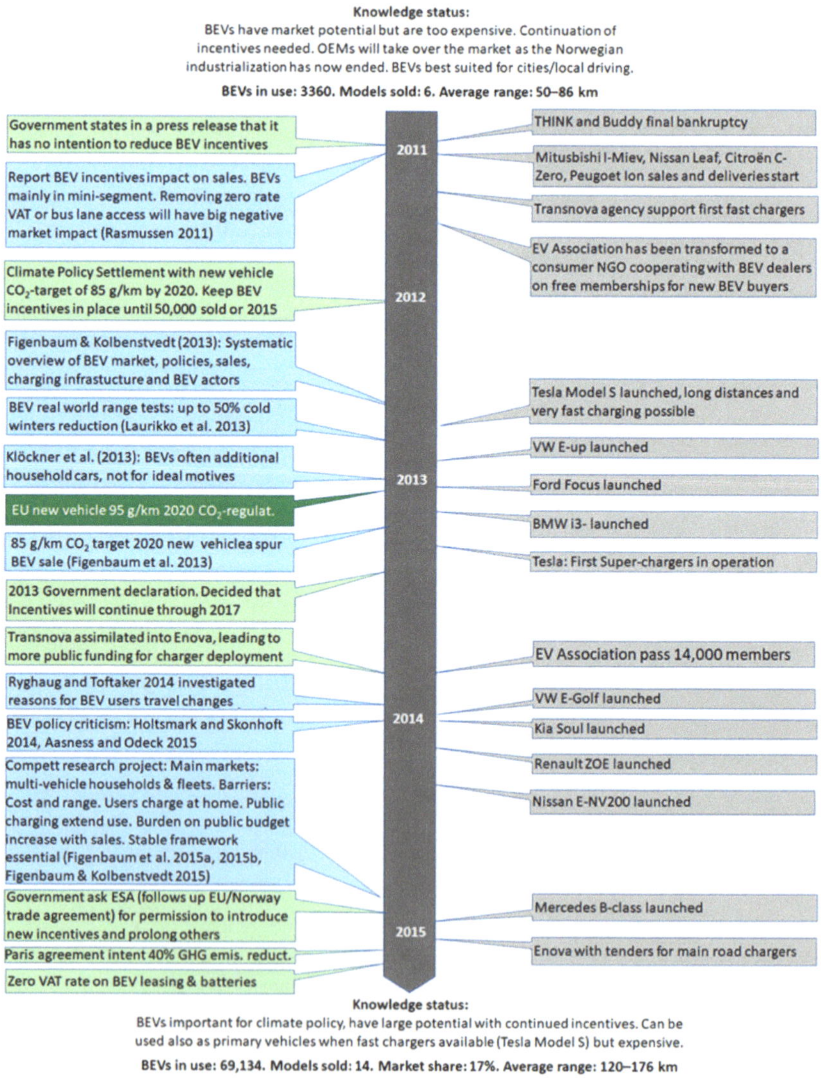

Figure 7. Timeline of policies, market activities, and research publications 2011–2015 [7,27,28,37,133, 147–152]. Light green: Norwegian policies. Dark green: international policies. Blue: research results. Grey: market activities. Source: Author.

3.6. 2016–2020—Policies Supported the Mass Market to Achieve GHG Emission Reductions

BEVs now had a good foothold in the market, although the market share in 2016 was 1% lower than in 2015. The first year without growth. The reason for this was that buyers were waiting for longer-range BEVs that had been announced. When these longer-range and lower-cost BEVs became available from traditional and new Chinese OEMs, the market expanded rapidly. BEVs average range doubled, and the model variety expanded. BEVs became an alternative for all households. Charging networks supported long-distance driving across Norway. The national target that only BEVs were to be sold from 2025 was, however, so ambitious that the politicians kept the incentives in place, despite growing tax losses and criticism. The EU 2019 CO_2 targets for 2025 and 2030 de-facto required European ZEV shares, and access to BEVs improved further and costs continued to decrease. The electromobility transition accelerated. The Klimakur2030 public report by the environment, transport, and energy authorities [153] saw BEVs as a top priority for reducing national GHG emissions. Yet, a Ministry of Transport toll road expert group suggested road tolls for BEVs [154]. This was because the purpose of these tolls is to finance road and transport reduction measures and public transport in cities [154]. Norway's Paris Agreement NDC to reduce GHG emissions by 40% by 2030 compared to 1990 entered into force in 2016 and was increased to 50% to 55% in 2020. The fleet had now reached 339,912 (12% of the total fleet) and 43 models were sold [3]. The average range increased from 209 to 393 km [3]. There was a broad understanding among stakeholders and politicians that BEVs were the future. Figure 8 shows the timeline.

Keep incentives in place through 2020 (2016/2017)

In the 2017 National Budget negotiations, parliament petitioned the government [65] to continue the zero VAT rates until 2020, introduce an annual tax exemption from 2018, and ask for ESA notification [155]. The rationale was to keep up the momentum towards the 2025 100% ZEV target. No impact assessment was made apart from in the "after the fact" ESA notification. ESA gave approval through 2020 [93].

Only sell ZEVs from 2025 (2017)

The 2016 National Transport Plan (NTP) suggested a target [156] of only selling ZEVs from 2025 and increasing biofuel use to reduce transport GHG emissions by 50% by 2030 compared to 1990. This target was derived from the national 2030 40% GHG emission reduction commitment of the Paris Agreement [157]. Insights came from an Environment Agency report [158]. Parliament approved the NTP and thus the ZEV target [159,160] in 2017. The incentives remained mostly unchanged until 2022 following this decision.

Exemption from re-registration tax (2018)

In 2014, the EV Association had proposed [161] an exemption from re-registration. In the negotiations over the national budget for 2018, parliament decided to ask the government to obtain ESA approval for the exemption [155], which the ESA approved until the end of 2020 [93]. It was adopted as part of the national budget for 2018 [155]. No expert group report supported the decision, but the ESA notification and the 2018 National Budget did contain an impact assessment. The rationale was to reduce the transaction cost and value loss of BEVs. In 2022, the tax was set to 25% of ICEVs [67] and, from 2023, it was set back to 100% [62].

Right to charge for flat owners in joint properties (2018)

The right to charge for flat owners in joint properties came about in a 2017 law change process [162], following a petition from parliament to the government, which was preceded by EV Association pressure. The law stated, "A section owner may, with the consent of the board, construct a charging point for electric cars in connection with a parking space available to the section, or other places designated by the board. The board can only refuse to consent if there is a valid reason". Later, a new sentence was added: "A section owner who has the right to park on the owner section association's property, but without disposing of his own space, may demand that a charging point be set up for an electric car. The board shall comply with the claim unless there is objective reason to refuse. The board decides where to set up the charging point" [163].

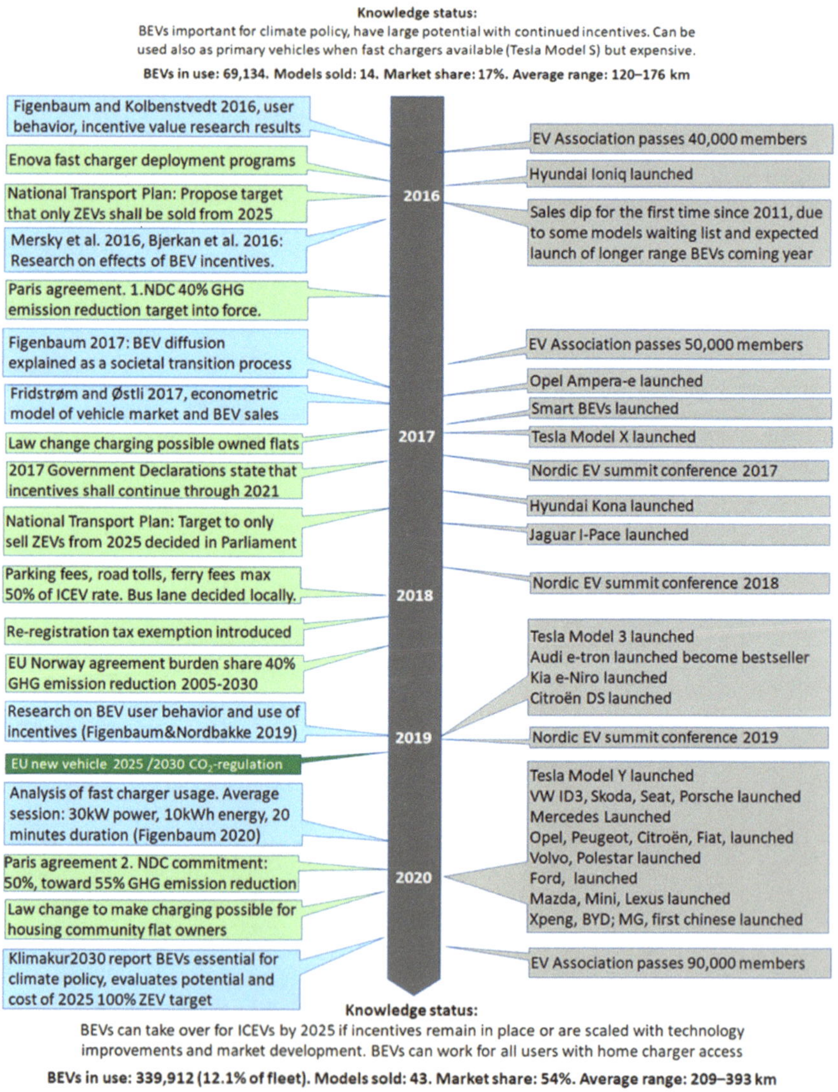

Figure 8. Timeline of policies, market activities, and research publications 2016–2020 [5,9,10,13,18–20,153]. Light green: Norwegian policies. Dark green: international policies. Blue: research results. Grey: market activities. Source: Author.

The 50% rule for road tolls, parking fees, and ferry tickets, and acknowledging local authority co-decisions (2018)

In the national budget negotiations for 2017, parliament decided [65] that ZEVs should pay a maximum of 50% of the ICEV rates for toll roads, parking, and ferries to reduce the income losses associated with the exemptions while keeping some ZEV incentives. Municipalities were allowed to make decisions within this limit. Changes to the toll road tariff system [164] and the NPRAs National Ferry Tariffs [165] for national main roads followed. This was thus a combination of a major national law change and minor administrative changes. No impact assessments were made. The 50% rate for parking was never implemented [166]. The 2017 parking regulation revision had replaced the obligatory exemption, with a possibility to offer exemption [167]. Municipalities could thus in practise charge BEVs 100% of the rate of ICEVs, despite the parliament's decision.

Action plan for infrastructure for alternative fuels in transport (2019)

The Oslo municipality funded the first large deployment of chargers in 2008 [168]. A national scheme followed in 2009 with financial crisis funding [121,122]. Normal and fast chargers had since then been supported by the Transnova [169] and Enova [170] funding agencies, counties [171], and municipalities [172,173], without a coherent national plan. The government's alternative fuel infrastructure plan published in 2019 [174] targeted a coherent alternative fuel infrastructure deployment and support for the implementation of the EU Dir. 2014/94/EU on infrastructure for alternative fuels [175]. The plan was presented in an expert group report followed by a public hearing. A final plan has yet to be adopted.

Keep incentives in place through 2021 (2020)

A decision to keep the zero VAT rates and registration tax exemption in place until 2021 came after the 2018 government declaration [176] had stated this intention. An ESA notification [177] was sent asking to extend the incentives through 2022, which the ESA approved [94]. No formal impact assessment was made but both the notification and the ESA decision contained a thorough evaluation of impacts and a justification.

Strategy for post-2025 vehicle taxation (2020)

Vehicle taxes provide a large portion of government income and are normally adjusted in small annual steps in the national budget process to avoid market distortion. The post-2025 general vehicle taxation principles were presented in the national budget documents for 2021 [15] and in the Climate Policy Bill to parliament [178] to provide market actors with a long-term perspective on vehicle taxes. The main principles stated were as follows: "A sustainable car tax system has a stable tax base, put a price on the external costs of vehicle use, taxes purchase and ownership of vehicles technology neutral, and takes care of distributional effects".

Right to charge for flat owners in housing communities (2020)

The right to access to charging infrastructure for flat owners was expanded to housing communities in 2020 following a thorough law change process. The results were included in a 2020 bill to parliament [179], which then made the formal decision. The law on housing communities was updated accordingly [163].

3.7. 2021–2023—Policies Downscaled to Preserve Government Income but Still Meet Targets

In 2023, Norway increased its Paris Agreement GHG emission reduction obligation to 55% [180]. The BEV market share passed 80% and the fleet share reached 24% at the end of 2023. The BEV fleet reached 690,000 [2]. Politics was focusing on reaching the target of only selling ZEVs, i.e., BEVs, from 2025. A gradual incentive downscaling was, however, initiated to preserve government tax income and because of diminishing user barriers. Home charging access in dense cities and for flat owners was still a barrier. The remaining 75–80% of Norwegians live in detached, semi-detached, and row houses where charging is easily accessible. Fast charging networks now covered all of Norway and were mainly built on commercial terms without support. There were, however, increasing charge queues at peak travel times as the building of chargers was outpaced by the expanding fleet and because more users drove long distances with BEVs. Long-range BEVs were available in all sizes and segments from traditional and Chinese OEMs, but some use areas, such as heavy, long-distance towing, could still not be covered satisfactorily. The timeline is shown in Figure 9.

Charging infrastructure strategy proposal (2022)

The development of a charging infrastructure strategy was started in 2022 after a parliament petition that the government should secure the building of a comprehensive charging infrastructure [181], which researchers, the EVA, and other NGOs saw as a major barrier to meeting the 2025 ZEV target [182]. The petition was sent during a parliament debate and public hearing [183] over the climate policy bill [178]. Another petition asked the government to develop a national charging infrastructure strategy to secure coordination between public authorities and develop more user-friendly charging infrastructure [184]. A charging expert group report was published in March 2022 [185], and stakeholders were invited to comment on it [186].

Proposal of the removal of zero-rate VAT, to be replaced by a support scheme (2023). VAT to be introduced on the part of the purchase price exceeding NOK 500,000

An expert group report [145] and a previous government [91] had proposed replacing zero-rate VAT with a support scheme. In the revised national budget for 2022 [95], the government proposed a scheme equal to 25% VAT up to NOK 500,000, i.e., capped at NOK 125,000. The incentive would move from the national budget income side, which is balanced by oil sector income, to the expense side, balanced against all other spending. It was stated to be a more equitable system for the future. Parliament decided, however, to keep the VAT exemption in place for a price up to NOK 500,000 and introduce VAT on the part of the purchase price exceeding that sum from 2023 [187,188], and to keep this scheme until 2025.

New weight tax on all vehicles (2023)

This tax on all new vehicles above 500 kg came as a big surprise in the 2023 National Budget [62]. BEVs, due to their heavier weight, had a higher tax than ICEVs. No impact assessment was published.

Removal of reduced re-registration tax incentive (2023)

The re-registration tax incentive was removed in the 2023 National Budget [62] proposal, which was endorsed by parliament [189]. No impact assessment was published.

Removal of reduced company car benefit tax (2023)

The reduced company car benefit tax was removed in the 2023 National Budget [62] proposal and endorsed by parliament as the budget proposal was not changed [189]. No impact assessment was published.

The 70% rule for toll roads (2023)

It was decided during the national budget process for 2023 that BEVs can from 2023 be charged up to 70% of the toll road rate charged for ICEVs [62,190].

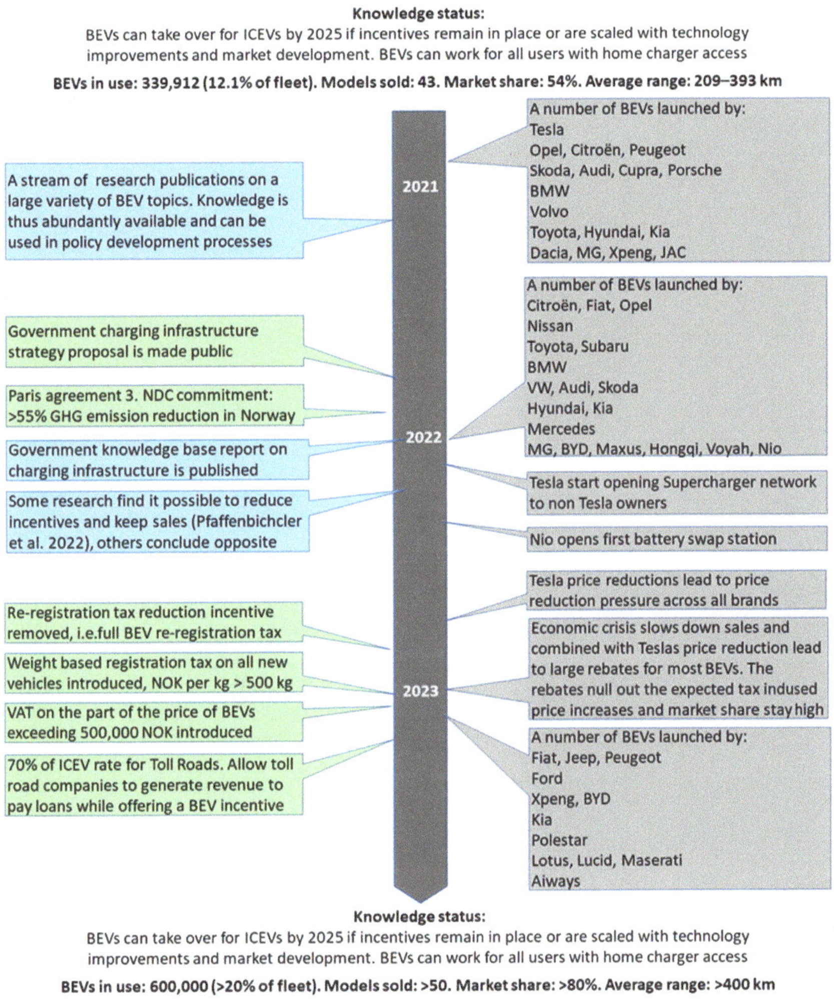

Figure 9. Timeline of policies, market activities, and research publications 2021–2023 [23,191]. Light green: Norwegian policies. Dark green: international policies. Blue: research results. Grey: market activities. Source: Author.

3.8. 1990–2023—The Policy Processes from Infancy to Mass Market and Beyond

The 33-year-long time horizon of the Norwegian BEV policy framework stands out. Large incentives covering many aspects of BEV purchase and ownership remained in place for a long time after their introduction, as seen in Table 2. The incentives came about in a learning-by-doing process where politicians introduced BEV-friendly policies based on stakeholder input and pressure. Lobbyism is easier in a small country like Norway with good access to politicians compared to large countries. BEV interest thus developed broadly, and the policies were adopted into party programmes and government declarations over time, as seen in Appendix A.

Table 2. Timeline of the main incentives and their 2023 status. Source: Author.

Incentive	Introduction	1st Major Revision	2nd Major Revision	3rd Major Revision	4th Major Revision	5th Major Revision	Status 2023
Registration tax exemption	1990, temporary	1996, permanent	2023, weight tax element introduced				Weight tax as for ICEVs, other parts exempted
Annual tax exemption	1996	2004, partial reduction	2018, BEVs fully exempted, changed to tax on insurance	2021, partial reduction	2022, full tax as for ICEVs		Full tax as for ICEVs
Road toll exemption	1997	2018, max 50% of ICEVs, local decision	2023, max 70% of ICEVs, local decision				Max 70% of ICEVs, local decision
Parking fee exemption	1999	2017, local authorities can decide	2018, BEVs 50% of ICEVs				50% rate still not implemented
Reduced company car benefit tax	2000	2005, new tax system, BEVs 75% of ICEVs	2009, 50% of ICEVs	2018, 60% of ICEV	2022, 80% of ICEV	2023, full tax as for ICEVs	Full tax as for ICEVs
Zero-rate VAT purchases	2001	2023, full VAT on price above NOK 500,000					Full VAT on price above NOK 500,000
Reduced ferry rates	2009, national car ferries	2018, max 50% of ICEVs, ferry operator to decide, includes county ferries					Max 50% of ICEVs, ferry operator to decide, includes county ferries
Zero-rate VAT leasing	2015	2023, full VAT on price above NOK 500,000					Full VAT on price above NOK 500,000
Re-registration tax exemption	2018	2022, 25% of ICEV rate	2023, full tax as for ICEVs				Full tax as for ICEVs
Access to bus lanes	2003, Oslo area test	2005, access to all bus lanes in Norway	2015, passenger in the car in rush hour, local authority decides				2015, passenger in the car rush hour, local authority decides

About 20% of the policy processes were improper from a traditional politics point of view, as seen in Table 3 and the flowcharts in Figure 10. Another 28% were inadequate. Some were parliamentary add-ons to traditional political processes. One example is the zero-rate VAT that was added by parliament to an otherwise well-prepared VAT reform.

Another is the 2012 decision to keep the BEV incentives in place until 2015 or when 50,000 BEVs were sold, which came during a parliamentary climate policy debate. Several incentives came during late-night parliament national budget negotiations (Table 3, policy nos. 4, 8, 18). Others originated from parliament (Table 3, policy nos. 5, 27) and were thus not "prepared" by the government. No impact assessments, therefore, supported these decisions, but policy no. 27 went through an after-the-fact process. In some cases, only a sentence shows that the decision was made. The temporary registration tax exemption became de facto permanent when a vehicle taxation reform was passed by parliament, without even being mentioned in the reform documents. Incentives affecting consumer-oriented laws have been carried out as proper law change processes (Table 3, policy nos. 6, 28, 30). Most processes since 2014 have been proper (Table 3, policy nos. 22–25, 27, 29–31, 33–35) apart from the last year's incentive downscaling (Table 3, policy nos. 26, 32, 37–40).

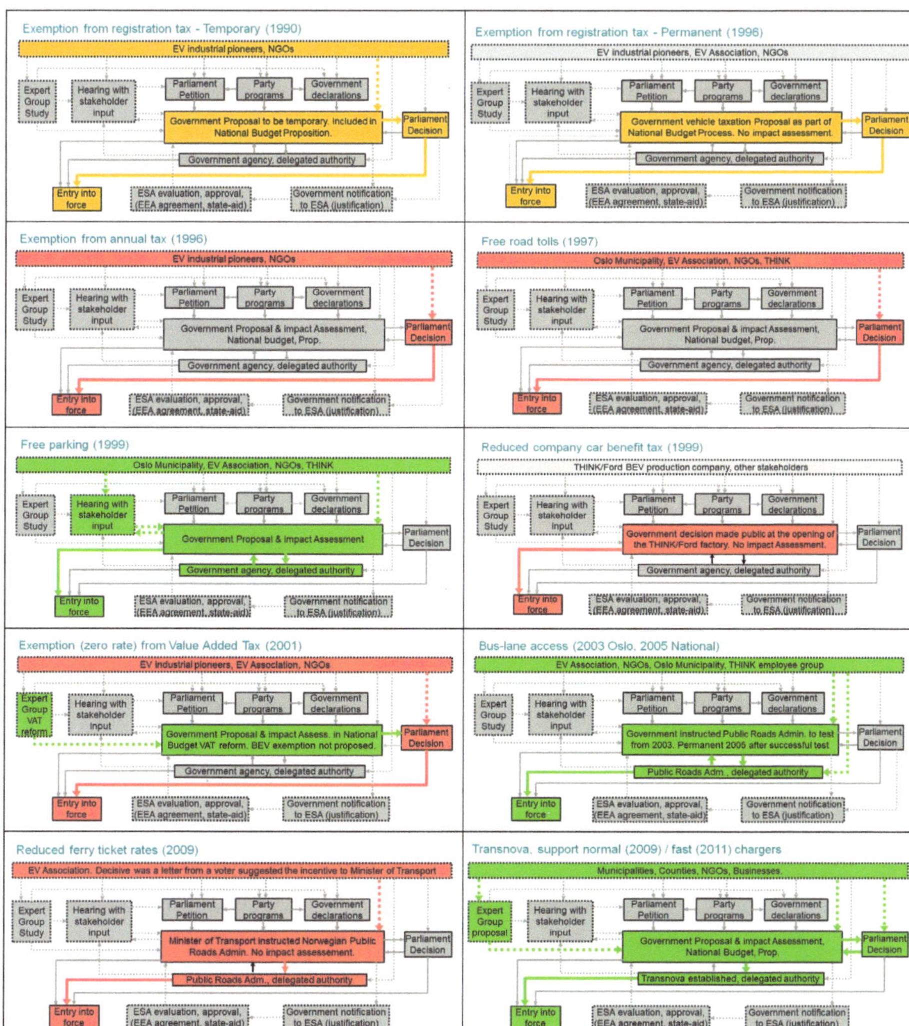

Figure 10. *Cont.*

Figure 10. BEV policy development process flows. Green = OK, orange = inadequate, red = improper. Grey: Elements of the traditional policy processes, as seen in Figure 2. The thick arrows shows the actual policy process flow for each policy. Source: Author.

Table 3. BEV policy and incentive process summary. Green = normal, yellow = deviations, red = improper. Source: Author.

	No.	BEV Policy	Type	Year Decided	Year Initiated	Effect	Market Impact	Impact Assessm.	Process	New BEVs Sold	Market Share	BEV Fleet (Incl. Used)	Average Range km	Public Chargers	Fast Chargers	Press Articles	Reports Articles
1990–1997	1	Registra tax exemption, temporary	Tax	1989	1990	Market pull	High	No	Nationa budget doc.	5	0.0%	5	30	0	0	-	1
	2	Km tax exemption	Tax	1989	1990	Market pull	Low	No	Nationa budget doc.	5	0.0%	5	30	0	0	-	1
	3	Registra tax exemption, permanent	Tax	1995	1996	Market pull	High	No	Gov. prop. to parliament	10	0.0%	50	60	11	0	27	1
	4	Annual tax exemption	Tax	1995	1996	Market pull	Medium	No	Nationa budget debate	10	0.0%	50	60	11	0	27	1
	5	Toll road exemption	Fee/Lav	1997	1997	Market pull	High	No	Parliame law change	42	0.0%	147	60	30	0	49	0
	6	Parking fee exemption	Fee/Lav	1999	1999	Market pull	Medium	Yes	Law change w. hearing	101	0.0%	285	60	No data	0	89	0
	7	Reducec company car benefit tax	Tax	1999	2000	Market pull	Low	No	Nationa budget doc.	101	0.1%	285	60	No data	0	89	0

Table 3. *Cont.*

	No.	BEV Policy	Type	Year Decided	Year Initiated	Effect	Market Impact	Impact Assessm.	Process	New BEVs Sold	Market Share	BEV Fleet (Incl. Used)	Average Range km	Public Chargers	Fast Chargers	Press Articles	Reports Articles
1998–2002	8	Zero-rate VAT BEV purchase	Tax/Lav	2000	2001	Market pull	High	No	Nationa budget process	207	0.2%	468	60	No data	0	71	0
	9	Bus lanes Oslo area (test 2003–2005)	Adm.	2003	2003	Market pull	High	No	Real-life test	15	0.0%	1081	60	No data	0	208	0
2003 –2006	10	Bus lanes Norway (Oslo test ok)	Adm.	2005	2005	Market pull	High	No	Experier from test	26	0.0%	1320	50	No data	0	285	1
	11	New car average CO_2 emis. <120 g/km	Target	2007	2012	Support target	Low	No	Governr process	240	0.0%	1903	50	No data	0	1276	2
	12	Increase car allowance for business trips	Adm.	2008	2008	Cost compensation	Low	No	Governr process	443	0.1%	2424	77	No data	0	2002	1

Table 3. *Cont.*

	No.	BEV Policy	Type	Year Decided	Year Initiated	Effect	Market Impact	Impact Assessm.	Process	New BEVs Sold	Market Share	BEV Fleet (Incl. Used)	Average Range km	Public Chargers	Fast Chargers	Press Articles	Reports Articles
2007–2010	13	Ferry ticket price reduction	Adm.	2008	2009	Market pull	Medium	No	Governr process	443	0.1%	2424	77	No data	0	2002	1
	14	TRANSI funding agency start	Adm.	2009	2010	Barrier reduction	Medium	Yes	Nationa budget doc.	295	0.1%	2753	56	No data	0	4482	5
	15	Transno support for normal chargers	Adm.	2009	2010	Barrier reduction	Low	No	Parliame propositon	295	0.1%	2753	56	No data	0	4482	5
	16	Transno support for fast chargers	Adm.	2010	2011	Barrier reduction	Low	No	Transno decision	599	0.2%	3360	86	1163	0	4041	5
	17	New car average CO_2 emis. <85 g/km	Target	2012	2020	Support target	Medium	No	Governr process	3950	2.9%	9581	131	3433	58	4215	11
	18	Keep incentives to 2015/50 BEVs	Decisior	2012	2015	Market stability	High	No	Parliame agreement	3950	2.9%	9581	131	3433	58	4215	11

Table 3. *Cont.*

	No.	BEV Policy	Type	Year Decided	Year Initiated	Effect	Market Impact	Impact Assessm.	Process	New BEVs Sold	Market Share	BEV Fleet (Incl. Used)	Average Range km	Public Chargers	Fast Chargers	Press Articles	Reports Articles
	19	Keep incentives in place through 2017	Decisior	2013	2017	Market stability	High	No	Governr declaration	7888	5.6%	18,916	185	4538	131	6680	19
	20	TRANSI merged into ENOVA	Adm.	2014	2015	More resources	High	Partial	Governr process	16,830	13%	38,652	185	5744	270	10,389	13
	21	ENOVA strategy for fast chargers	Adm.	2015	2016	Barrier reduction	High	No	ENOVA internal process	25,785	17%	69,134	176	6550	449	10,539	14
	22	ENOVA support for fast chargers 2015–2022	Adm.	2015	2016	Barrier reduction	High	No	Delegate authority	25,785	17%	69,134	176	6550	449	10,539	14
2011–2015	23	Zero-rate VAT BEV leasing/batt	Tax/Lav	2015	2015	Market pull	Low	Yes	Parliame petition	25,785	17%	69,134	176	6550	449	10,539	14

Table 3. *Cont.*

No.	BEV Policy	Type	Year Decided	Year Initiated	Effect	Market Impact	Impact Assessm.	Process	New BEVs Sold	Market Share	BEV Fleet (Incl. Used)	Average Range km	Public Chargers	Fast Chargers	Press Articles	Reports Articles
24	Only sell ZEVs from 2025—proposa	Target	2016	2025	Propose target	High	No	Nationa Transport Plan	24,222	16%	97,532	209	7830	757	9196	24
25	Only sell ZEVs from 2025—decision	Target	2017	2025	Support target	High	No	Nationa Transport Plan	33,025	21%	138,983	301	6858	1211	11,876	17
26	Keep incentives in place through 2020	Decisior	2017	2020	Market stability	High	No	Parliame petition	33,025	21%	138,983	301	6858	1211	11,876	17
27	Re-registrat tax exemption	Tax	2017	2018	Market pull	Low	Yes	Parliame petition	33,025	21%	138,983	301	6858	1211	11,876	17
28	50% of ICEV rate parking/roa toll/ferr	Fee/Lav	2017	2018	Incentiv reduct.	Medium	No	Nationa budget agreement	33,025	21%	138,983	301	6858	1211	11,876	17

Table 3. *Cont.*

	No.	BEV Policy	Type	Year Decided	Year Initiated	Effect	Market Impact	Impact Assessm.	Process	New BEVs Sold	Market Share	BEV Fleet (Incl. Used)	Average Range km	Public Chargers	Fast Chargers	Press Articles	Reports Articles
	29	Right to charge, flats/joi prop-erties	Law	2017	2018	Barrier reduction	Low	Yes	Parliame petition/law revision	33,025	21%	138,983	301	6858	1211	11,876	17
	30	Action plan for alternative fuels infrastr.	Strategy	2019	n/a	Barrier reduction	Low	No	Governr strategy	60,316	42%	260,692	397	12,132	2399	18,316	44
	31	Right to charge, flats/ho communities	Law	2020	2021	Barrier reduction	Low	Yes	Governr law revision	76,804	54%	339,912	393	14,073	3390	15,161	50
	32	Keep incentives in place through 2021	Decisior	2020	2021	Market stability	High	Yes	Governr declaration	76,804	54%	339,912	393	14,073	3390	15,161	50
2016–2020	33	Policy strategy for post-2025 vehicle taxes	Strategy	2020	2025	Market stability	Low	Yes	Nationa budget doc.	76,804	54%	339,912	393	12,962	3390	15,161	50

Table 3. *Cont.*

No.	BEV Policy	Type	Year Decided	Year Initiated	Effect	Market Impact	Impact Assessm.	Process	New BEVs Sold	Market Share	BEV Fleet (Incl. Used)	Average Range km	Public Chargers	Fast Chargers	Press Articles	Reports Articles
34	Partial re-introduc re-registrat tax	Tax	2021	2022	Incentiv reduction	Low	Yes	Nationa budget doc.	122,539	66%	461,661	>400	12,962	4035	19,390	No data
35	Chargin infrastructure strategy proposal	Strategy	2022	n/a	Barrier reduction	Low	Yes	Governr strategy	152,707	78%	600,464	>400	17,558	5667	18,738	No data
36	Re-introduc VAT price > NOK 500 k	Tax/Lav	2022	2023	Incentiv reduction	Medium	Yes	Nationa budget doc.	152,707	78%	600,464	>400	17,558	5667	18,738	No data
37	Full re-registrat tax	Tax	2022	2023	Incentiv removal	Medium	Yes	Nationa budget doc.	152,707	78%	600,464	>400	17,558	5667	18,738	No data
38	Weight-based registration tax	Tax	2022	2023	Incentiv reduction	Low	Yes	Nationa budget doc.	152,707	78%	600,464	>400	17,558	5667	18,738	No data

Table 3. *Cont.*

	No.	BEV Policy	Type	Year Decided	Year Initiated	Effect	Market Impact	Impact Assessm.	Process	New BEVs Sold	Market Share	BEV Fleet (Incl. Used)	Average Range km	Public Chargers	Fast Chargers	Press Articles	Reports Articles
2021–2023	39	Reintroc of full company car tax	Tax	2022	2023	Incentiv removal	Low	No	Nationa budget doc.	152,707	78%	600,464	>400	17,558	5667	18,738	No data
	40	Road tolls can be up to 70% of ICEVs rates	Fee/Lav	2022	2023	Incentiv reduction	Medium	No	Nationa budget doc.	152,707	78%	600,464	>400	17,558	5667	18,738	No data

The reasons for the lack of proper policy processes up to 2010 could be the large political interest in BEVs, a lack of knowledge [24], a sense of urgency as BEVs were uncompetitive without incentives, the need for transport sector GHG emission reductions [6], a willingness to support BEV industrialisation, and that the expected tax losses were low for the first few years after each incentive was introduced.

4. Discussion

Development in the problems stream: Air pollution was a major issue in Norwegian cities in the 1990s. The three-way catalyst became obligatory in 1989, but the slow fleet turnover caused cities to look to BEVs to reduce air pollution. Energy companies needed a new electricity revenue stream after an energy market reform. The BEV developer PIVCO saw an opportunity to produce a city BEV using a low-volume production process. This market was uninteresting for the OEMs. They saw BEVs as California-ZEV-mandate-compliance cars. Norwegian actors were inspired by French and Swiss BEV developments and the California ZEV mandate. Market experiments started after the costly registration tax was exempted in 1990. The actors established the EVA to improve the policy framework. Research found the early BEVs to be of poor quality and in need of improvements to be marketable. PIVCO planned to solve these issues through industrialisation. The clean air motivation had been reduced by 1998–2000. The focus shifted to industrialisation when Ford owned THINK (PIVCO) from 1999 to 2002 and in 2010 when other investors had taken over. The total cost of ownership was almost competitive with ICEVs for users, with free parking and free road tolls when the zero-rate VAT was introduced. A GHG emission reduction focus emerged when it became clear that the Kyoto Agreement GHG emission trading system did not work, and national policies would be required. The obligations of the Paris GHG emission reduction agreement would not be possible without BEVs as other transport measures had low potential and would be unpopular [192]. Local pollution came back on the agenda with rising diesel shares and the EU diesel emission regulation scandal [193]. Research showed that BEVs became multi-vehicle households' "workhorse" after OEM BEVs became available in 2011 [8–10]. The limited range was not an issue as they also owned an ICEV. The range of the latest generation of BEVs of all sizes was also sufficient for single-vehicle households. The market share reached 17% in 2015, 54% in 2020, and 80% in 2022. The EU's 2020–2030 vehicle CO_2 regulations de facto require ZEVs to be sold and show that Norway is on the right track, but ahead of other European countries.

Development in the policy solutions stream: BEVs have never been mandated in Norway. The market is too small for automakers to develop specific vehicles. Market pull incentives were used at first to allow for experiments and reduce local pollution, and later to support the build-up of a Norwegian BEV industry. High vehicle taxes since the 1960s made it possible to support BEVs through large tax exemptions. Politicians and municipalities were pressurised by NGOs to introduce incentives such as free parking and free road tolls. The policy effects were not well understood but the tax losses were initially low. The Norwegian BEV industry was globally leading at a time when OEMs saw BEVs as California "compliance" cars. Politicians failed, however, to support the BEV industry through the 2009 financial crisis. A government fund invested a small amount in THINK but demanded experienced auto industry involvement, which led to a production relocation to Valmet in Finland. All Norwegian BEV industry activity had ended by 2011. OEMs developed BEVs to meet the EU's vehicle CO_2 regulations and the policy focus shifted to support climate policy by replacing ICEVs with BEVs when they became available. The market took off. BEVs contributed significantly to GHG emission reduction 2020 targets. By backcasting the 2030 Paris Agreement GHG emission obligation, it became clear that the national target had to be to only sell BEVs from 2025. This was so ambitious that the incentives remained in place likely longer than they would have without the target. Increased research on user needs and the effects of policy changes supported policy development through the 2010s. The increasing vehicle tax losses were masked out by the increased oil sector income. BEV policies thus did not get in the way of other priorities. The BEV tax loss was about 5% of the oil income that balanced the 2020 National Budget, as seen in Figure 11. The incentive reductions from 2023 had little impact as Tesla lowered their prices and forced other OEMs to do the same.

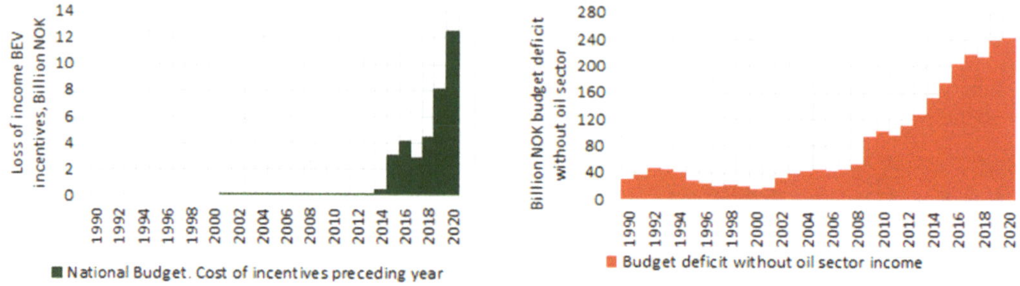

Figure 11. National budget, cost of BEV incentives for preceding years, and budget deficit without the oil and gas sector. Billion NOK. Source: author, based on data from the 1990–2020 national budget documents.

Developments in the political stream: The structured development processes of Norwegian politics were not followed for important BEV policies up until 2014. Decisions were made in poorly documented ad-hoc processes in parliament. The incentives would likely not have fared well in regular political processes with rigorous cost-benefit analyses. There was a lack of information about BEVs' potential and effects on government income in the early years. Such decisions were to some extent anchored in the party programmes, as seen in the overview in Appendix A. The party programmes became more positive to ZEVs over time. The government's reliance on small, pro-environment parties as coalition partners and the 2025 ZEV target caused BEV policies to continue despite increasing tax losses. They became "protected" by government declarations. Politicians were disproportionally receptive to BEV policies suggested by advocacy groups and entrepreneurs. Industrialisation, a policy area where political and financial risks seem to be more accepted, was a strong policy driver for the 1997 to 2010 developments. The most important incentives were agreed across parties in parliament to ensure stability. BEV policies found little opposition as they consisted of positive measures, and Norway did not have to worry about an incumbent ICEV production industry as other countries do. Some precedence for substantial vehicle tax exemptions existed before BEVs for, for instance, airbags and three-way catalysts. BEV policies contributed to Norway's international environmental standing. The discovery that BEV politics violated the EEA agreement with the EU led to an alignment of BEV politics with traditional politics.

Politicians did not want to make the best-informed decisions. When the two large studies [113,126] of climate policy measure options were carried out by energy, transport, and environment authorities in 2010 and 2020, the government's mandate precluded recommending packages of policy measures based on the best available knowledge. The 2020 mandate reads as follows: "A specialist group is established to carry out an investigation of possible measures and means of implementation of climate policy targets in 2030 but shall not make recommendations". The 2010 mandate was similar. It seems that the politicians did not want expert advice so that they could cherry-pick options matching party programmes. These authorities are, however, underlying government ministries. This may have led to the conclusion that they could only present possibilities, not policy suggestions. The actual targets and measures that were decided upon by politicians were not the same as those evaluated in these studies, and the impact assessments were thus not representative. This lack of competence-based politics was also seen in the National Transport Plan process where politicians often prioritised uneconomic projects over good projects [194].

The politics stream was side-lined. The main incentives were developed in the policy solutions stream outside of traditional politics as a response to issues in the problem stream.

These issues were put on the political agenda by lobbyists, i.e., industrial entrepreneurs, NGOs, and local authorities, and later by traditional vehicle importers. The Norwegian BEV policy process thus did not follow the multiple streams (MS) framework postulate that the 3Ps—the problems, policy solutions, and politics streams—must align to pave the way for new policies.

The lack of knowledge within all three streams was an issue through the 1990s and 2000s. Politicians lacked knowledge about how BEVs could function in Norway and solve issues in the problem stream. BEVs' potential to reduce pollution was, for instance, vastly overestimated, given THINK's low level of planning for BEV production. Research on the potential of BEVs was therefore initiated. The earliest incentives were intended to be temporal to build knowledge to enable decision making within traditional politics. Later policies supported industrial development, a policy area with high risk and lower documentation requirements, so the incentives were decided ad-hoc, without impact assessments. The policy results were inconclusive. The market remained slow, and more incentives were added with unclear effects but high political visibility at a low initial cost. The incentives were on the less-visible-income side of the national budget, and they did not compete against policies on the expense side of the budget, where competition for funding is hard.

The feedback from the problem stream was not clear. The early buyers were fleets seeing a marketing advantage and "irrational" enthusiasts. They made large sacrifices in terms of comfort, vehicle size, usability, quality, and reliability. This may have led to a misconception in the politics stream of BEVs' potential to solve issues in the problems stream. The user base expanded with the zero-rate VAT from 2001 and the bus lane access from 2003, but sales were hampered by the low access to BEVs, again leading to mixed signals to the other streams. The market did not respond until the OEMs took over in 2011. Norway's demand-side measures have from then on perfectly matched the EU's supply-side measures.

Researchers have supplied increasingly enhanced knowledge to all three streams since 2010. User behaviour and needs have been analysed and statistical models of future demand have been established since 2010. This knowledge has been used to investigate how to reach the increasingly ambitious ZEV targets. It could have been used to develop impact assessments that would have aligned BEV politics with traditional politics earlier. This did not happen until the government had to assess the impacts and justifications in the notifications sent to ESA in 2014.

It is unlikely that the large package of BEV incentives could have been established within traditional politics. Traditional politics requires sufficient information to be able to write impact assessments and make well-documented decisions. BEV policies before 2014 were mainly decided outside traditional politics in a poorly documented "learning-by-doing" process, not following the instructions of official studies and reports ("Utredningsinstruksen") [39–41], nor the Ministry of Finance procedures for evaluating policies [42–45]. The reasons for this could be a lack of knowledge, strong political interests, and minority governments' reliance on small, BEV-friendly parties. Economists' first best solution for reducing GHG emissions, the Kyoto global trading system, failed, so national policies became the focus. Finally, there was an increasing GHG emission reduction urgency through the 2010s. The three streams of problems, policies, and politics were not fully aligned until 2017 when parliament endorsed the target to only sell ZEVs from 2025. The three streams came close to being aligned two years earlier when the government found out that it had to consult the ESA about the legality of the policies. The first notifications to the ESA did contain impact assessments in line with traditional Norwegian politics, but they were written after the decision had been made.

The tax income losses may have been higher than necessary, and the industrial support may have been too low. It is not certain that the entire incentive package is really needed to be able to reach the target of only selling ZEVs from 2025. Some incentives could potentially have been gradually scaled back earlier. The target itself may also not be optimal. The costs could have been lower if the target had been 80% or 90% or if the 100% target had been phased in over a few years. Another issue is that the large incentives failed to build a Norwegian BEV industry. Norway could potentially have had a BEV industry today had some incentives been refocused to industrial support, especially during the 1998 and 2009 financial crises when private funding became unavailable to BEV producers.

The effectiveness of the BEV incentives should have been monitored, given the high cost of the incentives. Norway spent, according to the 2024 National Budget [195], NOK 28.6 billion on tax incentives for BEVs in 2023 and NOK 39 billion in 2022. Compared to the tax income on ICEVs, the tax incentive costs were even higher due to the loss of fuel taxes (electricity tax is lower) and because registration tax is partially based on a vehicle's CO_2 emissions. Given these high costs, governments should have invested more in research on the effectiveness of BEV policies.

The transition to BEVs continues and deepens. Norway will in the coming years spend large resources to transform the transport sector into a zero-emission sector mainly powered by renewable electricity. All new city buses shall be zero-emission from 2025. Every new, small LCV shall be zero-emission by 2025 and every new, large LCV shall be zero-emission by 2030. The transition will spread to the trucking sector, which must transition much faster than BEVs to reach the National Transport Plan target that 50% of trucks sold should be zero-emission in 2030, as few were in the 2023 fleet [196]. Parliament changed the 2030 truck target to 100% but included biogas during the national budget negotiations for 2024 [197]. The main recommendation for the truck segment is to link incentives to a long-term plan, with regular public reviews of the progress and the need for policy changes. These plans and reviews should, due to the large resources that will have to be used to speed up the truck transition, be supported by policy cost-effectiveness research.

The Norwegian BEV policy processes deviate from those of other countries and regions. The oil income made it easy to continue the incentives when the market took off in 2011. Other countries must evaluate BEV policy expenses against other policy needs or use budget-neutral measures such as ZEV mandates or bonus/malus systems. Norway has no ICEV producers, whereas some countries must consider the effects on their vehicle producers. Norwegian electricity is almost 100% hydro-electric based and without GHG emissions, and most Norwegians have or can obtain access to home charging. The stable Norwegian tax exemptions are on the less-visible-income side of the national budget. The support schemes in other countries are on the expense side, are affected by frequent policy changes, and often run out of money mid-year. The large user privileges were enabled by the spare capacity in the bus lanes and the toll roads spread across the country. The ad-hoc policy process differs from other countries' structured processes.

5. Conclusions

The Norwegian BEV policy process was investigated using a combination of document analysis and the multi-stream (MS) policy analysis framework. The MS framework states that policies can gain traction when the three streams of problems, policy solutions, and politics align. The analysis reveals that this prerequisite was not fulfilled when powerful BEV policies were decided in Norway, not as part of a comprehensive plan but in an ad-hoc "learning-by-doing" process outside traditional politics. The early BEV policy decision processes were also inadequate in terms of traceability and the documentation of their impacts and rationale. Politicians and other actors had until 2010 no or inadequate information about the effects of the BEV policies they introduced. They were disproportionally receptive to arguments from the problems stream about the need for BEV support policies.

At first, the stated target in the problem stream was to improve air quality; then, industrialisation became the focus, before it shifted to GHG emission reduction. Incentives that addressed these problems were developed in the policy solutions stream outside traditional politics. Given the market status and expectations for the future when the most important policy decisions were made, the immediate consequences, budget impacts, and risks were seen as low, and the political gain was seen as high. The post-2010 processes when the market share increased from <1% to above 80% have been more in line with traditional politics after politicians discovered that they had to notify the ESA. ESA notifications contain proper justifications and impact assessments, as is expected for decisions made in the politics stream. The 3Ps of the MS framework aligned, although some of the processes were still inadequate in terms of transparency. The ambitious target to only sell ZEVs from 2025 and the ambitious climate policy targets for 2030 led to a lock-in that protected the incentives from down-scaling. These developments created the world's largest per-capita BEV market.

To avoid negative issues but keep the potential to support the transition to ZE vehicles, politicians should ensure that sufficient knowledge about status and uncertainties is available when decisions about BEV policies are made. Decisions should be properly assessed within traditional politics as with EU policy processes and regularly reviewed as with the California ZEV mandate mid-term reviews. The required knowledge for decision making should be developed in continuous open-research activities and in other publicly available documents.

New insights into BEV policy development processes can be gained by comparing the Norwegian process with that of other countries. This would be of particular interest to countries that aim to expand their BEV market. The analysis of policy processes should also be expanded to heavy-duty truck electrification while it is still in an early stage. The knowledge of the policy processes for passenger vehicle electrification can then be used to devise better policy processes related to the target of only selling ZE trucks in Norway by 2030. More research is also required on policies to improve the knowledge of charging infrastructure deployment strategies and policies that improve usability.

Funding: This study was funded by the Research Council of Norway under contract no. 267848, with a contribution from Mobility Zero Emission Energy Systems, Norwegian Centre for Environmentally Friendly Energy Research, co-sponsored by the Research Council of Norway (FME MoZEES, p-no. 257653) and 40 partners from research, industry, and public sectors.

Data Availability Statement: Data are contained within the article.

Conflicts of Interest: The author declares no conflict of interests.

Appendix A

Table A1. BEV policies in Partyk programmes per parliament period. Grey color: The ruling parties over the full or main part of the period.

	Socialist		Labour	Centre				Conservative	
	Rødt	**SV**	**AP**	**SP**	**MDG**	**V**	**KRF**	**H**	**FRP**
	The Reds (Socialist)	The Socialist Party	Labour Party Sociodemocrat	Centre Parti (rural/farmers)	The Green Party	The Liberal Party	The Christian Democratic Party	Conservative Party	Progress Party (populist)
1989–1993		Stricter emission regulations. Use of natural gas in the transport sector.	Less vehicle use in cities, use road tolls Emission reductions for diesel cars, BEVs/ZEVs were not mentioned.	Use best available emission reduction technology for all vehicle types.	Not available	Use fuel/other taxes to stimulate a switch over to gas and electricity. No tax on safety/environment equipment.	Differentiate taxes based on emissions. No tax on safety/environment equipment.	Reduce tax on environment equipment (i.e., catalytic converters).	Proposes strong reduction in vehicle taxes to enable people to buy safe and less polluting cars (i.e., new cars).
1993–1997		Favourable conditions for BEVs, low fuel consumption vehicles, and biofuels.	BEVs or ZEVs not mentioned. Transfer from vehicles to pub. Transport. New tech. mentioned to reduce pollution but no details.	Favourable conditions for BEVs. Move taxes from purchase to use and exempt BEVS. Work to reduce vehicle use in cities.	Reduce transport as much as possible, prioritise electricity-based transport. Avoid fossil-fuel transport. Introduce restrictions on ICEVs and later bans.	Move taxes from purchase to use, use gas as the main alternative energy carrier. Adjust purchase tax to enable installation of emission-reducing equipment.	No mentions apart from requiring more stringent emission limits.	Stimulate change over to vehicles using less fuel.	Chapter on motor vehicles. Proposes reduction in vehicle taxes. Tax income from transport to be transfer-ed back to the sector.

Table A1. *Cont.*

	Socialist		Labour	Centre				Conservative	
	Rødt	**SV**	**AP**	**SP**	**MDG**	**V**	**KRF**	**H**	**FRP**
1997–2001		Support use of BEVs to improve air-quality in cities. Increased use of biodiesel, car sharing support, testing of hydrogen.	Stimulate increased testing and change to gas, electricity, or H_2 for transport, using vehicle tax system. Increase diesel tax.	BEVs to be 100% exempted from taxes. Reduce vehicle use in cities and increase public transport use.	Reduce vehicle-based transport. Vehicles should be powered by clean electricity and biogas. Develop car sharing with cleaner vehicles. Ban ICEVs in cities in 10 years.	Support increased use of electric vehicles. Move taxes to vehicle usage. Differentiate tax based on fuel consumption.	Norway a front-runner for more environ-mentally friendly transport. Vehicle users pay real societal costs. Tax system to stimulate BEVs and other low-emission vehicles	No mention of vehicles in particular. A general text on how taxes shall reflect environmentall costs.	Chapter on motor vehicles. Propose strong reduction in vehicle taxes. Tax income from transport to be transferred back to the sector.
2001–2005		Use taxes to support low energy use/alternative fuels. Support testing and increase adoption in public fleets. Rebates for car sharing. Less traffic volume with city road tax.	Same as 1997 apart from diesel tax not mentioned. New in 2001: Action plan for large cities that target increased use of ZEVs.	Full-tax-exempt BEVs (VAT, reg. tax, etc.). Continue local incentives (road, toll parking). Support BEV demo projects, H_2 in transport. 10% of the fleet to be emission-free by 2005.	Reduction in car-based transport. Vehicles should be powered by clean electricity and biogas. Nat. gas preferred over other fossil options. Car sharing with cleaner vehicles. Higher taxes on ICEVs, later bans.	Stimulate buying, testing, and use of ZEVs. Make use of vehicles in cities more expensive. Move taxes from purchase to vehicle usage.	Norway a front-runner for more environ-mentally friendly transport. Vehicle users pay real societal costs. Tax system to stimulate BEVs and other low-emission vehicles. Natural gas as alt. fuel. Support H_2.	Stimulate use of BEVs and other low- and zero-emission vehicles. Reduce total taxes on vehicles.	Chapter on motor vehicles. Proposes strong reduction in vehicle taxes as measure to renew fleet and reduce emissions. Increased BEV adoption, reduced city pollution.

Table A1. *Cont.*

	Socialist		Labour	Centre				Conservative	
	Rødt	**SV**	**AP**	**SP**	**MDG**	**V**	**KRF**	**H**	**FRP**
2005–2009		No mention of policies for vehicles other than support the opposite, i.e., public transport.	H_2 cars same incentive as BEVs, indirectly support BEV incentives. No mention of new BEV policies. Build H_2 fuel stations.	Support increased use of biofuels incl. sales obligation. Support ZEVs and LEVs through the tax system. Keep BEV incentives in place.	Not available.	Focused moved to a hydrogen society as the vision of the future with same tax advantages for hydrogen as BEVs. Biofuels also in focus.	Norway a front-runner for environmentally friendly transport. More focus on use of and research on H_2 and other ZEVs. Hydrogen tax exempt. Support biofuel use/prod.	Reduce vehicle taxes to make it easier to buy safe and more environmentally friendly vehicles.	Expand use of NG in transport sector by building infrastructure. Increased use of BEVs/HEVs and fleet renewal to reduce city pollution.
2009–2013	Focus on public transport measures and policies for reduction of vehicle-based transport.	2015 ban on car only using fossil fuel. Plan scaling up sales of ZEVs, incl. importer obligations. Public ZEV procurement. Plan for charging infrastructure deployment.	Reward ZEVs and LEVs in the tax system so that they take over as soon as possible. Also use biogas for transport.	Taxes support ZEVs, PHEVs, LEVs, HEVs, and biofuels. Minimum 20% ZEVs sold 2020, rest LEVs. Build charging infrastructure. H_2 available. Scrappage bonus of NOK 40,000 for buying environmentally friendly car.	Ban on gasoline and diesel cars from 2013. Certified biofuels. Registration tax based on CO_2 emissions and immediate ban on high emitters. Vehicles OK where public transport does not suffice, but new technology should be used.	Vehicle taxes stim-ulate use of LEVs, BEVs, and FCEVs. Adjust annual (CO_2) emission targets for new cars from 2015, ban sale of fossil cars. Public fleets should buy LEVs/ZEVs. Minimum 10% share of biofuel and H_2 by 2013. Build more infrastructure for ZEV, H_2, and biofuel.	Increased use of hydrogen, HEVs, PHEVs, and BEVs. Increased use of CO_2-neutral fuels (biofuels). Public fleets to only procure LEVS or ZEVs. Norway should push for an end to global ICEV production by 2020.	Reduce vehicle taxes to make it easier to buy safe and more environmentally friendly vehicles. Remove taxes on ZEVs. Build alternative fuel and charging infrastructure.	Reduce vehicle taxes.

Table A1. *Cont.*

	Socialist		Labour	Centre				Conservative	
	Rødt	**SV**	**AP**	**SP**	**MDG**	**V**	**KRF**	**H**	**FRP**
2013–2017	BEVs or vehicles not mentioned. Focus on public transport, vehicle-based transport reduction policies.	BEVs compete with ICEVs. Keep incentives. Support other emission-free options and infrastructure, i.e., chargers. 50% of public fleets shall be BEVs/PHEVs.	Continue to use the tax system to reduce emissions from transport.	Reduce GHG emissions from transport, support 2nd-gen. biofuels. Phase in new and environmentally friendly vehicle technology.	All new vehicles BEVs or HEVs, but highest priority is public transport. Use road pricing to curb city traffic. Stimulate car sharing solutions Remove VAT for BEV lease/batteries. Keep/expand BEV incentives to 2020.	Target world's most environmentally friendly transport. Keep BEV incentives. Build fast chargers between cities. Remove VAT for BEV leasing/batteries. Support 2nd-gen. biofuel development. Strenghten Trans-nova and BEV, H_2, and biofuel infrastructure. Expand biofuel and H_2.	Increased use of H_2, HEVs, PHEVs, BEVs, and biofuels. Public fleets only LEVS or ZEVs. Norway push end of global ICEV production by 2020. Build charging stations. Keep incentives until 10% PEVs on road or 2020.	Continue tax exemption for ZEVs. Build infrastructure for BEVs and hydrogen, use public procurement. Develop biofuel strategy.	Strong reduction in vehicle taxes and increase in scrappage bonus to renew vehicle fleet to make it safer and more environmentally friendly.

Table A1. *Cont.*

	Socialist		Labour	Centre				Conservative	
	Rødt	**SV**	**AP**	**SP**	**MDG**	**V**	**KRF**	**H**	**FRP**
2017–2021	Build good, fast charger capacity in all munici-pal centres. Focus on public trans-port mea-sures and vehicle-based trans-port reduc-tion policies.	Keep BEV incentives in place through 2021. Max half rate of ICEVs for ZEVs for road tolls, parking, and ferries. Biofuel production to be developed. Use 2nd-gen. biofuels in intermedium term.	Pursue ZEVs and biofuels. Build biofuel fueling stations and charging infrastructure.	Continue ZEV support. Keep purchase incentives, slowly phase out local incentives by 2030. Use taxes to get 100% ZEV share by 2025. Stronger focus on biofuels. Intensify building of charging and energy stations. Strong support for H_2 use.	Phase out sales of ICEVs by 2020. Public fleets must buy ZEVs. Higher ICEV taxes. Build energy stations to support ZEVs everywhere. Always be cheaper to select a ZEV. Less traffic in cities and support ZEVs in districts. Support charging station building and BEV leasing. Continue ZEV advantages until competitive.	Keep incentives until ZEVs competitive by themselves, at least until 2025. ZEVs have lasting advantage of half price of ICEVs for road tolls/ferries. Ensure good infrastructure for fast/normal charging across the country. Cooperate w. companies on nationwide energy stations with chargers and biofuel and H_2 dispensers. Public fleets buy ZEVs (not police).	Emission-free sector by 2030. Increase BEVs, H_2, HEVs, PHEVs, and biofuel. Public fleets to only procure LEVs or ZEVs. Norway push end to global ICEV production by 2020. Build charging, hydrogen, and biofuel stations faster. Keep incentives until 110% PEVs on road or 2020.	Zero emission vision for transport. Shall be worthwhile to go for ZEVs. Support for infrastructure.	Proposes strong reduction in vehicle taxes and increase in scrappage bonus to renew vehicle fleet to make it safer and more envi-ronmentally friendly.

Table A1. *Cont.*

	Socialist		Labour	Centre				Conservative	
	Rødt	**SV**	**AP**	**SP**	**MDG**	**V**	**KRF**	**H**	**FRP**
2021–2025	References NTP target. Build BEV fast-charging capacity in municipal centres. Maximum-limit BEV subsidies. Tax above limit. Increase H_2 efforts.	Reach 2025 target 2 years earlier/2023. Economic to buy a ZEV. Increase BEV tax followed by a larger ICEV tax. VAT price > NOK 600,000. Investigate an ICEV ban and BEV sharing. Tighten ICEV vs. BEV loop-hole leasing.	Reach the 2025 ZEV target. Reduce tax incentives on expensive BEVs and increase CO2 tax on ICEVs. VAT on prices exceeding NOK 600,000. Increase ICEV taxes. Build fast chargers.	Reach the 2025 ZEV target for new vehicles. Avoid single BEV focus due to blackout risk. VAT on price > NOK 600,000. Propose building of 10,000 fast chargers and national plan for charging infrastructure. Gradual reduction in user advantages.	Reach the 2025 ZEV target for new vehicles 2 years earlier/2023. Support rural environmentally friendly transport and BEV leasing and large-scale building of chargers, incl. for flat owners. Stimulate car sharing. Incentives until ZEVs competitive. Increase taxes on ICEVs and CO_2.	Reach the 2025 ZEV target for new vehicles. We are the BEV advocate and will ensure BEV advantages through 2025. The advantage must remain until BEVs reach a competitive price. Secure charging infrastructure is available in the whole country.	Reach the 2025 ZEV target for new vehicles. Build charging infrastructure (housing communities and common garages mentioned specifically). Establish a sustainable taxation system ensuring it is economical to buy a BEV over an ICEV.	Follow up the 2025 target for new vehicles. Build fast and ultra-fast chargers incl. for flat owners. Gradually step down ZEV incentives, starting with the most expensive, but always more economic to buy a BEV.	Proposes strong reduction in vehicle taxes in general, BEVs only mentioned as an example.

Table A2. Overview of BEV content of government declarations and BEV incentives/targets governments have introduced.

GovernmenPeriod		Parties	Support in Parliament	Government Declaration	BEV-Related Topics in Declaration	BEV Incentives and Targets Introduced		
						Year Decided	**Intro Year**	**Description**
Brundtland	9 May 1986–16 October 1989	AP	Minority	Brundt-land 1986 "Speech"	No mentions of climate or CO_2, no measures within transport related to greenhouse gas emissions. Note that the Brundtland commision (UN) "Our common future" report came in 1987, leading to a higher political focus on environment issues, including climate, following years.	1989	1990	Temp. registration tax exemption to allow BEV experiments.
						1989	1990	Km tax exemption to allow BEV experiments.

Table A2. *Cont.*

Governmen	Period	Parties	Support in Parliament	Government Declaration	BEV-Related Topics in Declaration	BEV Incentives and Targets Introduced		
						Year Decided	**Intro Year**	**Description**
Syse	16 October 1989–3 November 1990	H, KrF, SP	Minority	Lysebu 1989	None, CO_2 emission reduction of high priority, reforestation to reduce CO_2. Reduce local pollution from transport with 3-way catalysts and traffic measures			Carried through the decicions in parliament in late 1989 to provide exemptions from registration and km tax.
Brundtland 3	3 November 1990–25 October 1996	AP	Minority	Brundt-land 1990 "Speech"	None, prioritise global climate policy agreement, focus on sector overarching environmental policies	1995	1996	Permanent registration tax exemption.
						1995	1996	Annual tax exemption.
Jagland	25 October 1996–17 October 1997	AP	Minority	Jagland 1996	None, but talks about an ecologically sustainable society.	1996	1997	Road toll exemption.
Bondevik 1	17 October 1997–17 March 2000	KrF, SP, V	Minority	Voksenåsen 1997	None, high priority to reduce greenhouse gases, focus on Kyoto negotations and reaching an agreement. Transportation: Focus on reducing and supporting public transport.	1997	1999	Free parking.
						1999	2000	Reduced imposed benefit tax on disposing a company car, 50% reduction.
Stoltenberg 1	17 March 2000–19 October 2001	AP	Minority	Stoltenberg 2000, "Speech from the throne"	None specific, mentions Kyoto as a breakthrough, Norway being a frontrunner on environmental issues.	2000	2001	VAT exemption.
Bondevik 2	19 October 2001–17 October 2005	KrF, H, V	Minority	Sem 2001	None, high priority to reduce greenhouse gases, focus on Kyoto negotiations and reaching a global agreement. Transportation: Focus on reducing and supporting public transport.	2002	2003	Bus lane access test—Greater Oslo.
							2004	Traffic insurance tax moved to annual tax, BEV owners had to pay that tax.
						2005	2005	Bus lane access permanent and national.

Table A2. *Cont.*

Governmen	Period	Parties	Support in Parliament	Government Declaration	BEV-Related Topics in Declaration	BEV Incentives and Targets Introduced		
						Year Decided	Intro Year	Description
Stoltenberg 2	17 October 2005–7 October 2009	AP, SV, SP	Majority	Soria Moria 1 2005	Follow up Kyoto, work for more ambitious global climate policy agreement, strive for increased use of environmentally friendly vehicles, make it economical to buy low-emission vehicles, biofuel focus.	2008	2009	Reduced ferry rates.
						2008	2008	Increased km allowance for electric car use on business trips.
						2008	2012	Average new vehicle CO_2 emissions below 120 g/km by 2012. Broad agreement in parliament.
						2009	2009	Creation of support agency Transnova.
						2009	2010	Financial crisis support programme for chargers.
	7 October 2009–16 October 2013	AP, SV, SP	Majority	Soria Moria 2 2009	Work for a strong international climate agreement, exceed Kyoto obligations by 10%, transport policy shall support climate policy, action plan for ZEV/LEV introduction, biofuels support, charging stations to be built.	2011	2011	First support programme for fast chargers.
						2012	2020	Average new vehicle CO_2 emissions below 85 g/km by 2020.
						2012	2015	Keep incentives in place until the end of 2015 or 50,000 BEVs are in the fleet. Broad agreement in parliament.

Table A2. *Cont.*

GovernmenPeriod		Parties	Support in Parliament	Government Declaration	BEV-Related Topics in Declaration	BEV Incentives and Targets Introduced		
						Year Decided	Intro Year	Description
Solberg	16 October 2013–17 January 2018	H, FrP	Minority, supported by V, KrF	Sundvolden 2013	Continue BEV tax regime until 2017, go through tax policy, follow-up on 2012 climate policy settlement in Stortinget.	2013	2017	Keep purchase incentives in place until the end of 2017.
						2013	2015	Zero VAT rate for leasing and battery replacement.
						2014	2015	Restriction on bus lane access on west corridor to Oslo in rush hour introduced.
						2014	2015	Transnova assimilated into Enova.
Solberg contin-ued						2015	2015–18	Support programme for fast chargers along major roads.
						2016	2016	Restriction on bus lane access on southeast corridor to Oslo in rush hour introduced.
						2016	2017	Only sell ZEVs by 2025. Broad agreement in parliament.
						2016	2017	Re-registration tax exemption.

Table A2. *Cont.*

GovernmenPeriod	Parties	Support in Parliament	Government Declaration	BEV-Related Topics in Declaration	BEV Incentives and Targets Introduced		
					Year Decided	Intro Year	Description
Solberg contin-ued						2018	Law change parking: Parking facilities and public parking can charge full rate for BEVs, and must offer up to 6% of spaces with charging access.
					2016	2018	Parliament decide rule that maximum rate for toll roads, parking, and ferries shall be 50% of the rate of ICEVs, local authorities to decide on the level up to the maximum.
					2017	2017	Law change for condominiums to regulate access to charging in common parking facilities.
					2017	2018	Full exemption from annual tax, or rather the tax on insurance that replaced the annual tax.

Table A2. *Cont.*

Governmen	Period	Parties	Support in Parliament	Government Declaration	BEV-Related Topics in Declaration	BEV Incentives and Targets Introduced		
						Year Decided	Intro Year	Description
Solberg contin-ued	17 January 2018–22 January 2019	H, FrP, V	Minority, KrF support	Jeløya 2018	40% greenhouse gas emission reduction by 2030 over 1990, suggest new ambitious target to Paris Agreement. NTP vehicle targets basis for policy (only ZEVs sold from 2025).	2018	2021	
	22 January 2019–13 January 2020	H, FrP, V, KrF	Majority	Grana-volden 2019	More ambitious climate policy target, 50% reduction in transport sector by 2030, NTP vehicle targets (only ZEVs sold from 2025).	2019	2021	Keep purchase incentives in place until end of 2021.
						2019		Action plan for alternative fuel infrastructure.
	13 January 2020–end 2021	H, V, KrF	Minority	Grana-volden 2019	See above	2020	2020	Law change for housing communities to regulate access to charging in common parking facilities.
						2020	2021–22	Keep zero-rate VAT through 2022.
						2020	2021	Reduced-rate insurance tax (30% below ICEV rate).
						2020	2025	Future principles for vehicle taxation post-2025 transition period 2022–2025.

Table A2. *Cont.*

GovernmenPeriod		Parties	Support in Parliament	Government Declaration	BEV-Related Topics in Declaration	BEV Incentives and Targets Introduced		
						Year Decided	Intro Year	Description
Støre	14 October 2021-This day	A, SP	Minority, with SV primary supporter	Hurdals-platt-formen	More ambitious climate policy target, -55% reduction of Norwegian emissions by 2030 compared to 1990.			
					Reduce GHG emissions from transport and contribute to meeting national climate policy goals.			
					Make it attractive to select LEVs and ZEVs, 100% of new vehicles fossil-fuel-free from 2025, contribute to ZEVs keeping their competitive advantage vs. ICEVs.	2022		Strategy process for charging infrastructure initiated.
					Arrange the tax system so that it is fair and contributes to cuts in greenhouse gas emissions.	2022	2023-	Proposal to replace zero-rate VAT with support scheme.

References

1. Figenbaum, E. Norway the World Leader in BEV Adoption. In *Who's Driving Electric Cars*; Understanding Consumer Adoption and Use of Plug-in Electric Cars; Springer Nature: Berlin, Germany, 2020. Available online: https://link.springer.com/book/10.1007/978-3-030-38382-4#about (accessed on 16 January 2024).
2. *Norwegian Vehicle Register*; Norwegian Public Roads Administration: Oslo, Norway, 2024.
3. Figenbaum, E. The 1990 to 2020 Technology Innovation System (TIS) Supporting Norway's Bev Revolution. *SSRN* **2022**. Available online: https://ssrn.com/abstract=4061401 (accessed on 20 December 2023). [CrossRef]
4. Langeland, O.; George, C.; Figenbaum, E. Technological Innovation System and Transport Innovations: Understanding Vehicle Electrification in Norway. In *Innovations in Transport: Success, Failure and Societal Impacts*; Edward Elgar: Cheltenham, UK, 2022. Available online: https://www.elgaronline.com/edcollchap-oa/book/9781800373372/book-part-9781800373372-14.xml (accessed on 16 January 2024).
5. Figenbaum, E. Perspectives On Norway's Supercharged Electric Vehicle Policy. *Environ. Innov. Soc. Transitions* **2017**, *25*, 14–34. [CrossRef]
6. Figenbaum, E. Retrospective Total Cost of Ownership Analysis of Battery Electric Vehicles in Norway. *Transp. Res. Part D Transp. Environ.* **2022**, *105*, 103246. [CrossRef]
7. Figenbaum, E.; Kolbenstvedt, M. *Competitive Electric Town Transport: Main Results from COMPETT—An Electromobility + Project*; Report 1422/2015; Institute of Transport Economics: Oslo, Norway, 2015. Available online: https://www.toi.no/publications/competitive-electric-town-transport-main-results-from-compett-and-electromobility-project-article33368-29.html (accessed on 16 January 2024).
8. Figenbaum, E.; Kolbenstvedt, M.; Elvebakk, B. *Electric Vehicles–Environmental, Economic and Practical Aspects: As Seen by Current and Potential Users*; Compett and TØI report 1329/2014; Institute of Transport Economics: Oslo, Norway, 2014. Available online: https://www.toi.no/publications/electric-vehicles-environmental-economic-and-practical-aspects-as-seen-by-current-and-potential-users-article32644-29.html (accessed on 16 January 2024).
9. Figenbaum, E.; Nordbakke, S. *Battery Electric Vehicle User Experiences in Norway's Maturing Market*; TØI report 1719/2019; Institute of Transport Economics: Oslo, Norway, 2019. Available online: https://www.toi.no/publications/battery-electric-vehicle-user-experiences-in-norway-s-maturing-market-article35709-29.html?deviceAdjustmentDone=1 (accessed on 16 January 2024).
10. Figenbaum, E.; Kolbenstvedt, M. *Learning from Norwegian Battery Electric and Plug-In Hybrid Vehicle Users—Results from a Survey of Vehicle Owners*; TØI report 1492/2016; Institute of Transport Economics: Oslo, Norway, 2016. Available online: https://www.toi.no/publications/learning-from-norwegian-battery-electric-and-plug-in-hybrid-vehicle-users-results-from-a-survey-of-vehicle-owners-article33869-29.html (accessed on 16 January 2024).
11. Figenbaum, E.; Thorne, R.J.; Amundsen, A.H.; Pinchasik, D.R.; Fridstrøm, L. *From Market Penetration to Vehicle Scrappage—The Movement of Li-Ion Batteries through the Norwegian Transport Sector*; TOI report 1756/2020; Institute of Transport Economics: Oslo, Norway, 2020. Available online: https://www.toi.no/publications/from-market-penetration-to-vehicle-scrappage-the-movement-of-li-ion-batteries-through-the-norwegian-transport-sector-article36209-29.html (accessed on 16 January 2024).
12. Transport & Environment. *Electric Surge: Carmakers' BEV Plans across Europe 2019–2025*; Transport & Environment: Brûssel, Belgium, 2019. Available online: https://www.transportenvironment.org/publications/electric-surge-carmakers-electric-car-plans-across-europe-2019-2025 (accessed on 16 January 2024).
13. Figenbaum, E. Battery Electric Vehicle Fast Charging—Evidence from the Norwegian Market. *World Electr. Veh. J.* **2020**, *11*, 38. [CrossRef]
14. Figenbaum, E.; Wangsness, P.B.; Amundsen, A.H.; Milch, V. Empirical Analysis of the User Needs and the Business Models in the Norwegian Charging Infrastructure Ecosystem. *World Electr. Veh. J.* **2022**, *13*, 185. [CrossRef]
15. Ministry of Finance. National Budget Documents 2021. 2020. Available online: https://www.regjeringen.no/no/statsbudsjett/2021/id2741050/ (accessed on 16 January 2024).
16. Ryghaug, M.; Skjølsvold, T.M. How policies and actor strategies affect electric vehicle diffusion and wider sustainability transitions. *Proc. Natl. Acad. Sci. USA* **2023**, *120*, e2207888119. [CrossRef]
17. Fevang, E.; Figenbaum, E.; Fridstrøm, L.; Halse, A.H.; Hauge, K.E.; Johansen, B.J.; Raaumm, O. Who goes electric? The anatomy of electric car ownership in Norway. *Transp. Res. Part D* **2021**, *92*, 102727. [CrossRef]
18. Bjerkan, K.Y.; Nørbech, T.E.; Nordtømme, M.E. Incentives for promoting Battery Electric Vehicle (BEV) adoption in Norway. *Transp. Res. Part D Transp. Environ.* **2016**, *43*, 169–180. [CrossRef]
19. Mersky, A.C.; Sprei, F.; Samaras, C.; Qian, Z. Effectiveness of incentives on electric vehicle adoption in Norway. *Transp. Res. Part D* **2016**, *46*, 56–68. [CrossRef]
20. Fridstrøm, L.; Østli, V. The vehicle purchase tax as a climate policy instrument. *Transp. Res. Part A* **2017**, *96*, 168–189. [CrossRef]
21. Fearnley, N.; Pfaffenbichler, P.; Figenbaum, E.; Jellinek, R. *E-Vehicle Policies and Incentives—Assessment and Recommendations*; TØI-rapport 1421/2015; Transportøkonomisk Institutt: Oslo, Norway, 2015. Available online: https://www.toi.no/publications/e-vehicle-policies-and-incentives-assessment-and-recommendations-article33367-29.html (accessed on 16 January 2024).
22. Aurland-Bredesen, K.J. Too green to be good: The efficiency loss of the Norwegian electric vehicle policy. *J. Environ. Econ. Policy* **2017**, *6*, 404–414. [CrossRef]
23. Pfaffenbichler, P.; Emberger, E.; Fearnley, N.; Figenbaum, E. Simulating the effects of tax exemptions for plug-in electric vehicles in norway. In Proceedings of the European Transport Conference, Milan, Italy, 7–9 September 2022.

24. Figenbaum, E. The contribution of research and knowledge accumulation in the development of the Norwegian battery electric vehicle market. *Transp. Res. Procedia* **2023**, *72*, 4127–4134. [CrossRef]
25. Bjerkan, K.Y.; Bjørge, N.M.; Babri, S. Transforming socio-technical configurations through creative destruction: Local policy, electric vehicle diffusion, and city governance in Norway. *Energy Res. Soc. Sci.* **2021**, *82*, 102294. [CrossRef]
26. Lemphers, N.; Bernstein, S.; Hoffman, M.; Wolfe, D.A. Rooted in place: Regional innovation, assets, and the politics of electric vehicle leadership in California, Norway, and Quebec. *Energy Res. Soc. Sci.* **2022**, *87*, 102462. [CrossRef]
27. Holtsmark, B.; Skonhoft, A. The Norwegian support and subsidy policy of electric cars. Should it be adopted by other countries? *Environ. Sci. Policy* **2014**, *42*, 160–168. [CrossRef]
28. Aasness, M.A.; Odeck, J. The increase of electric vehicle usage in Norway—Incentives and adverse effects. *Eur. Transp. Res. Rev.* **2015**, *7*, 34. [CrossRef]
29. Haustein, S.; Jensen, A.F.; Cherchi, E. Battery electric vehicle adoption in Denmark and Sweden: Recent changes, related factors and policy implications. *Energy Policy* **2021**, *149*, 112096. [CrossRef]
30. Mazur, C.; Contestabile, M.; Offer, G.J.; Brandon, N.P. Assessing and comparing German and UK transition policies for electric mobility. *Environ. Innov. Soc. Transit.* **2015**, *14*, 84–100. [CrossRef]
31. Calef, D.; Goble, R. The allure of technology: How France and California promoted electric and hybrid vehicles to reduce urban air pollution. *Policy Sci.* **2007**, *40*, 1–34. [CrossRef]
32. Haas, T.; Sander, H. Decarbonizing Transport in the European Union: Emission Performance Standards and the Perspectives for a European Green Deal. *Sustainability* **2020**, *12*, 8381. [CrossRef]
33. Kester, J.; Noel, L.; Zarazua de Rubens, G.; Sovacool, B.K. Policy mechanisms to accelerate electric vehicle adoption: A qualitative review from the Nordic region. *Renew. Sustain. Energy Rev.* **2018**, *94*, 719–731. [CrossRef]
34. Collantes, G.; Sperling, D. The origin of California's zero emission vehicle mandate. *Transp. Res. Part A Policy Pract.* **2008**, *42*, 1302–1313. [CrossRef]
35. California Air Resources Board. Midterm Review Report. Advanced Clean Cars. 2017. Available online: https://ww2.arb.ca.gov/resources/documents/2017-midterm-review-report (accessed on 16 January 2024).
36. Commission Staff Working Document Impact Assessment Part 1 Accompanying the Document Proposal for a Regulation of the European Parliament and of the Council amending Regulation (EU) 2019/631 as Regards Strengthening the CO_2 Emission Performance Standards for New Passenger Cars and New Light Commercial Vehicles in Line with the Union's Increased Climate Ambition. Document 52021SC0613. SWS/2021/613 Final. Available online: https://eur-lex.europa.eu/legal-content/EN/TXT/?uri=celex:52021SC0613 (accessed on 16 January 2024).
37. Figenbaum, E.; Fearnley, N.; Pfaffenbichler, P.; Hjorthol, R.; Kolbenstvedt, M.; Emmerling, B.; Jellinek, F.; Bonnema, G.M.; Ramjerdi, F.; Iversen, L.M. Increasing competitiveness of e-vehicles in Europe. *Eur. Transp. Res. Rev.* **2015**, *7*, 28. Available online: http://link.springer.com/article/10.1007/s12544-015-0177-1 (accessed on 16 January 2024). [CrossRef]
38. Münzel, C.; Plötz, P.; Sprei, F.; Gnann, T. How large is the effect of financial incentives on electric vehicle sales?—A global review and European analysis. *Energy Econ.* **2019**, *84*, 104493. Available online: https://www.sciencedirect.com/science/article/pii/S0140988319302749 (accessed on 16 January 2024). [CrossRef]
39. Lovdata. Instruks Om Utredning Av Konsekvenser, Foreleggelse Og Høring Ved Arbeidet Med Offentlige Utredninger, Forskrifter, Proposisjoner Og Meldinger Til Stortinget. For-2000-02-18-108. 2000, Revised 2005 and 2016. Available online: https://lovdata.no/dokument/LTI/forskrift/2000-02-18-108 (accessed on 16 January 2024).
40. Lovdata. Instruks Om Utredning Av Konsekvenser, Foreleggelse Og Høring Ved Arbeidet Med Offentlige Utredninger, For-Skrifter, Proposisjoner Og Meldinger Til Stortinget. For-1994-12-16-4062. 1994. Available online: https://lovdata.no/dokument/INSO/forskrift/1994-12-16-4062 (accessed on 16 January 2024).
41. Lovdata. Regelverksinstruksen. Bestemmelsen Om Arbeidet Med Offentlige Utredninger, Lover, Forskrifter, Stortingsmeldinger Og-Proposisjoner. For-1985-08-30-9952. 1985. Available online: https://lovdata.no/dokument/INSO/forskrift/1985-08-30-9952 (accessed on 16 January 2024).
42. Ministry of Finance. Rundskriv R. Prinsipper Og Krav Ved Utarbeidelse Av Samfunnsøkonomiske Analyser. 2021. Available online: https://www.regjeringen.no/globalassets/upload/fin/vedlegg/okstyring/rundskriv/faste/r_109_2021.pdf (accessed on 16 January 2024).
43. Ministry of Finance. Prinsipper Og Krav Ved Utarbeidelse Av Samfunnsøkonomiske Analyser MV. R-109/14. Available online: https://www.regjeringen.no/globalassets/upload/fin/vedlegg/okstyring/rundskriv/faste/r_109_2014.pdf?id=2220435 (accessed on 16 January 2024).
44. Ministry of Finance. Behandling Av Kalkulasjonsrente, Risiko, Kalkulasjonspriser Og Skattekostnad I Samfunnsøkonomiske Analyser. R-109/2005. Available online: https://www.regjeringen.no/globalassets/upload/fin/vedlegg/okstyring/rundskriv/faste/r_109_2005.pdf (accessed on 16 January 2024).
45. Ministry of Finance. Behandling Av Diskonteringsrente, Risiko, Kalkulasjonspriser Og Skattekostnad I Samfunnsøkonomiske Analyser. R-14/99. Available online: https://www.regjeringen.no/globalassets/upload/fin/vedlegg/okstyring/rundskriv/arlige/1999/r_14_1999.pdf (accessed on 16 January 2024).
46. Kingdon, J. *Agendas, Alternatives and Public Policy*, 2nd ed.; Pearson: New York, NY, USA, 2010.
47. Asphjell, A.; Asphjell, Ø.; Kvisle, H.H. *2013 Elbil på Norsk*; Transnova: Hong Kong, 2013; ISBN 978-82-7704-142-1. (In Norwegian)
48. *Ikke Unntak, Faktaboks I Artikkel: Staten Stopper Forskning*; News article, Avgift 160000; VG: Oslo, Norway, 1985.

49. Figenbaum, E. *Elbiler–Teknikk, Forskning og utvikling, Infrastruktur*; Marked. Rapport nr. 271 (01) 1993; Teknologisk Institutt: Oslo, Norway, 1993.
50. Figenbaum, E. *Testprogram for Elbiler–Elbiler i Norge–Brukererfaringer*; Report; Teknologisk Institutt: Oslo, Norway, 1994.
51. Figenbaum, E. Norwegian Experience. In *Electric & Hybrid Vehicle Technology 95*; UK & International Press: London, UK, 1995.
52. Figenbaum, E. *Testprogram for Elbiler–Hovedrapport*; Report; Teknologisk Institutt: Oslo, Norway, 1995.
53. Figenbaum, E. *Miljøvennlige Biler i Flåter–Kartlegging av bilflåter*; Report; Teknologisk Institutt: Oslo, Norway, 1996.
54. Figenbaum, E. *Miljøvennlige Biler i Flåter–Hovedrapport*; Report; Teknologisk Institutt: Oslo, Norway, 1997.
55. Figenbaum, E. *PIVCO Markedsintroduksjon–Ytelsestester*; Report; Teknologisk Institutt: Oslo, Norway, 1997.
56. *Berge Lover Nedsatt El-Bilavgift*; News article; NTB Tekst: Oslo, Norway, 1989.
57. Ministry of Finance. National Budget Documents 1990. St.prp nr. 1 Statsbudsjettet 1990 (1989–1990), Document 1: Statsbudsjettet Medregnet Folketrygden. Document 2: Skatter Og Avgifter Til Statskassen. 1989. Available online: https://stortinget.no/no/Saker-og-publikasjoner/Stortingsforhandlinger/Lesevisning/?p=1989-90&paid=1&wid=a&psid=DIVL801&pgid=a_0271 (accessed on 16 January 2024).
58. Parliament. 1990 National Budget Negotiation. Decision in the Parliament to Temporary Exempt BEVs from the Vehicle Registration Taxa s proposed by the Government in: St.prp nr. 1 1989-90, Dokument 1: Statsbudsjettet Medregnet Folketrygden, Dokument 2: Skatter Og Avgifter Til Statskassen. 1989. Available online: https://stortinget.no/no/Saker-og-publikasjoner/Stortingsforhandlinger/Lesevisning/?p=1989-90&paid=1&wid=a&psid=DIVL801&pgid=a_0047 (accessed on 16 January 2024).
59. Ministry of Finance. National Budget Vehicle Taxation Reform Document. St prp nr 1 Tillegg nr 3. 1995. Available online: https://www.stortinget.no/no/Saker-og-publikasjoner/Stortingsforhandlinger/Lesevisning/?p=1995-96&paid=1&wid=c&psid=DIVL2868 (accessed on 16 January 2024).
60. Parliament. 1995 National Budget Vehicle Reform Negotiation Meeting Minutes. Stortingsbehandling Av Avgiftsvedtak for 1996. Stortinget 8 Desember 1995. Available online: https://www.stortinget.no/no/Saker-og-publikasjoner/Stortingsforhandlinger/Lesevisning/?p=1995-96&paid=7&wid=a&psid=DIVL14&pgid=b_0517&s=True (accessed on 16 January 2024).
61. Ministry of Finance. National Budget Documents. 1995. Available online: https://www.stortinget.no/no/Saker-og-publikasjoner/Saker/Sak/?p=7815 (accessed on 24 October 2023).
62. Ministry of Finance. National Budget Document 2023. 2022. Available online: https://www.regjeringen.no/no/statsbudsjett/2023/id2927365/ (accessed on 16 January 2024).
63. Parliament. *Budsjett-Innst. S. NR. 1 (2003–2004) Innstilling Til Stortinget FRA Finanskomiteen st.meld. NR. 1 (2003–2004), st.prp. NR. 1 (2003–2004) OgSt.prp. nr. 1 Tillegg nr. 1-11 (2003–2004)*; National Budget 2004 Negotiation in the Parliament, Meeting Minutes; Stortinget: Oslo, Norway, 2003.
64. Parliament. *Nyere, Sikrere Og Mer Miljøvennlig Bilpark*; Parliament Vehicle Taxation Policy Settlement, Bilavgiftsforlik; Agreement between the Majority of Parties in the Parliament: Oslo, Norway, 2015.
65. Parliament. Minutes of the National Budget for 2017 Debate in the Parliament. Meeting 5. Des. 2016, 10.00. 1. Finansministerens Redegjørelse Om Regjeringens Forslag Til Statsbudsjett Og Om Nasjonalbudsjettet for 2017 899, 2. Nasjonalbudsjettet 2017 Og Forslag Til Statsbudsjett for 2017. 2016. Available online: https://www.stortinget.no/globalassets/pdf/referater/stortinget/2016-2017/refs-201617-12-05.pdf (accessed on 16 January 2024).
66. Ministry of Finance. National Budget Documents 2018. 2017. Available online: https://www.regjeringen.no/no/statsbudsjett/2018/id2570841/ (accessed on 16 January 2024).
67. Ministry of Finance. National Budget Documents 2022. 2021. Available online: https://www.regjeringen.no/no/statsbudsjett/2022/id2871447/ (accessed on 16 January 2024).
68. Simmones, H.; Thronsen, M. *Et Norsk Eventyr: Norsk Elbilforening i 25 år*; Dinamo Forlag: Bærum, Norway, 2020.
69. *Gratis Oslo Parkering for Elbiler*; News article; Aftenposten: Oslo, Norway, 1997.
70. *Norske Elbiler Til California*; News article; Aftenposten: Oslo, Norway, 1995.
71. Parliament. Minutes of Meeting. Decision in the Parliament to Change the Law to Enable the Toll Road Exemption. Stortinget Innst. S. nr. 74. (1996–1997). Innstilling Fra Samferdselskomiteen Om Forslag Fra Stortingsrepresentant Lars Sponheim Om Regelendringer Som Gjør Det Mulig for Lokale Styresmakter å Frita Elektriske Biler for Bompengeavgift. 1996. Available online: https://www.stortinget.no/no/Saker-og-publikasjoner/Publikasjoner/Innstillinger/Stortinget/1996-1997/inns-199697-074/?lvl=0 (accessed on 16 January 2024).
72. Figenbaum, E. TH!NK, a Unique Citycar. In *Electric & Hybrid Vehicle Technology*; UK & International Press: Surrey, UK, 1998; 98p.
73. Figenbaum, E. *PIVCO Markedsintroduksjon-Sluttrapport*; Report; Teknologisk Institutt: Oslo, Norway, 1998.
74. Hoogma, R.; Kemp, R.; Schot, J.; Truffer, B. *Experimenting for Sustainable Transport*; The Approach of Strategic Niche Management; Chapter 4: Experiments in Electrifying Mobility, The PIVCO Experience: Ecological Product Differentiation; Routledge: Hoboken, NJ, USA, 2002; ISBN 978-0-415-27116-5.
75. *Samferdselsministeren Lokker Me Gratis Parkering: En Trøm Av El-Biler Til Oslo*; News article; Aftenposten: Oslo, Norway, 1998.
76. *Gratis P-Plass for El-Biler*; News article; Aftenpostn: Oslo, Norway, 1999.
77. *Gratis Parkering for El-Drevne Biler*; News article; Aftenposten Aften: Oslo, Norway, 1998.
78. Lovdata. Law Change Implementation. Forskrift om Offentlig Parkeringsregulering og Parkeringsgebyr (Parkeringsforskriften). FOR-1993-10-01-921. 1993, Revision 19 Jan. 1999, Forskrift 139. Available online: https://lovdata.no/dokument/SFO/forskrift/1993-10-01-921/KAPITTEL_2#KAPITTEL_2 (accessed on 16 January 2024).

79. Lovdata. Law Change Implementation. Forskrift Om Endring Av Forskrift Fastsatt Av Finans-Og Tolldepartementet Til Utfylling Og Gjennomføring Mv. Av Skatteloven Av 26. Mars 1999 nr. 14. Kunngjort 03.02.2000. Ikrafttredelse Fra Og Med Inntektsåret. 2000. Available online: https://lovdata.no/dokument/LTI/forskrift/2000-01-10-19 (accessed on 16 January 2024).
80. *Nye Tiltak for økt Bruk Av Elbiler*; Government press release 12.11.1999; Ministry of Trade and Industry: Oslo, Norway, 1999.
81. Lovdata. Law Change Implementation. Forskrift Om Endring i Forskrift Til Utfylling Og Gjennomføring Mv. Av Skatteloven Av 26. Mars 1999 nr. 14. Kunngjort 23.12.2004. Ikrafttredelse: Inntektsåret 2004, Inntektsåret 2005. 2004. Available online: https://lovdata.no/dokument/LTI/forskrift/2004-12-17-1709 (accessed on 16 January 2024).
82. Lovdata. Law Change Implementation. Forskrift Om Endring i Forskrift Til Utfylling Og Gjennomføring Mv. Av Skatteloven Av 26. Mars 1999 nr. 14. Kunngjort 16.12.2008. Ikrafttredelse 12.12.2008 Med Virkning Fra Og Med Inntektsåret 2009. 2008. Available online: https://lovdata.no/dokument/LTI/forskrift/2008-12-12-1329 (accessed on 16 January 2024).
83. Lovdata. Law Change Implementation. Forskrift Om Endring i Forskrift Til Utfylling Og Gjennomføring Av Skatteloven Av 26. Mars 1999 nr. 14. 2017. Available online: https://lovdata.no/dokument/LTI/forskrift/2017-12-14-2101 (accessed on 16 January 2024).
84. Parliament. Innst. 3 S (2021–2022). Innstilling Fra Finanskomiteen Om Skatter, Avgifter Og Toll 2022. Includes the Parliament Decision on Changes to Company Car Tax Incentive. 2022. Available online: https://www.stortinget.no/globalassets/pdf/innstillinger/stortinget/2021-2022/inns-202122-003s.pdf (accessed on 7 December 2023).
85. Lovdata. Law Change Implementation. Forskrift Om Endring i Forskrift Til Utfylling Og Gjennomføring Mv. Av Skatteloven Av 26. Mars 1999 nr. 14. Kunngjort 11 April 2022. Ikrafttredelse 7 April 2022 Med Virkning Fra Inntektsåret 2004. 2022. Available online: https://lovdata.no/dokument/LTI/forskrift/2022-04-07-564 (accessed on 16 January 2024).
86. NOU 1990:11. Generell Merverdiavgift På Omsetning Av Tjenester (nb.no). Available online: https://www.regjeringen.no/no/dokumenter/nou_1990-11/id115324/ (accessed on 16 January 2024).
87. *Board Meeting Minutes of the EV Association*; Norstart—Norsk Elbilforening Møtereferat; EVA: Oslo, Norway, 2000.
88. Parliament. Minutes of Meeting in Parliament. 2001 National Budget Agreement. Budsjettavtale for Statsbudsjettet for 2001 Mellom Arbeiderpartiet, Kristelig Folkeparti, Senterpartiet Og Venstre Inngått 18 November 2000. Publisert Av Finansdepartementet. 2000. Published by Ministry of Finance 30.05.2009. Available online: https://www.regjeringen.no/no/dokument/dep/fin/statsbudsjettet/budsjett-2001/budsjettavtale-for-statsbudsjettet-for-2/id411925/ (accessed on 16 January 2024).
89. Parliament. Minutes of Meeting in Parliament. Zero VAT-Rate Decision in the Parliament. Innstilling Fra Finanskomiteen Om Lov Om Endringer i Lov 19 June 1969 nr. 66 om Merverdiavgift (Merverdiavgiftsloven) m.v. (Merverdiavgiftsreformen 2001). Innst. O. nr. 24—2000–2001. 2000. Available online: https://www.stortinget.no/globalassets/pdf/innstillinger/odelstinget/2000-2001/inno-200001-024.pdf (accessed on 16 January 2024).
90. Ministry of Finance. National Budget Documents 2001. 2000. Available online: https://www.regjeringen.no/no/statsbudsjett/2001/id489537/ (accessed on 16 January 2024).
91. Revised National Budget 2015. Ministry of Finance. 2015. Available online: https://www.regjeringen.no/no/statsbudsjett/2015/rnb/id2409228/ (accessed on 16 January 2024).
92. ESA. Efta Surveillance Authority Decision of 21 April 2015 on the State Aid Measures in Favour of Electric Vehicles (Norway). EEA EFTA Surveillance Authority Decision. Case No: 76399, Document No: 746191, Decision No: 150/15/COL. 2015. Available online: https://www.eftasurv.int/state-aid/state-aid-register/state-aid-measures-favour-electric-vehicles (accessed on 16 January 2024).
93. ESA. EFTA Surveilance Authority Decision. Tax Reductions on Zero Emission Vehicles. EEA EFTA Surveillance Authority, 19 December 2017. Case No: 81341, Document No: 877360, Decision No: 228/17/COL. 2017. Available online: https://www.eftasurv.int/state-aid/state-aid-register/tax-reductions-zero-emission-vehicles (accessed on 16 January 2024).
94. ESA. EFTA Surveilance Authority Decision. Prolongation of the Zero VAT Rating for Zero-Emission Vehicles 2021–2022. EEA EFTA Surveillance Authority, 16 December 2020. Case No: 85854, Document No: 1158139, Decision No: 148/20/COL. 2020. Available online: https://www.eftasurv.int/cms/sites/default/files/documents/gopro/College%20Decision%20148%2020%20COL%20-%20State%20aid%20-%20Norway%20-%20Prolongation%20of%20zero%20VAT%20rating%20for%20zero-emission%20vehicles%202021-.pdf (accessed on 16 January 2024).
95. RNB. Revised National Budget Document 2022. Ministry of Finance. 2022. Available online: https://www.regjeringen.no/no/statsbudsjett/2022/rnb/id2909908/ (accessed on 16 January 2024).
96. Revised National Budget 2022 Agreement in Parliament. Parliament. 2022. Available online: https://www.stortinget.no/no/Saker-og-publikasjoner/Saker/Sak/?p=89362 (accessed on 16 January 2024).
97. Low Emission Committee. Et Klimavennlig Norge. Official Norwegian Report (NOU). 2006. Available online: https://www.regjeringen.no/contentassets/56ae831eec35484881c6b237c2e817ac/no/pdfs/nou200620060018000dddpdfs.pdf (accessed on 16 January 2024).
98. *Minibussene Ut Av Kollektivfeltene*; News article; Dagbladet: Oslo, Norway, 2001.
99. *NITO Og NIF Aksjonerer for Think*; News article; Teknisk Ukeblad: Oslo, Norway, 2002.
100. *Transcript of Radio News Broadcast on Channel P4*; News article; P4: Lillehammer, Norway, 2002.
101. *Elbil i Kollektivfeltene*; News article; Teknisk Ukeblad: Oslo, Norway, 2003.
102. *Står i Kø for å Slippe Køkjøring: Tidoblet Salg Av Elbiler*; News article; Aftenposten: Oslo, Norway, 2003.
103. Bech, C. Why Think Nordic believes that BEVs are part of the future? Think Nordic 2004. In Proceedings of the 2004 European Ele-Drive Transportation Conference & Exhibition on Urban Sustainable Mobility, Estoril, Portugal, 17–20 March 2004.

104. *Minibusser Ut Av Kollektivfelt Fra Nyttår*; News article; VG: Oslo, Norway, 2008.
105. *El-Biler i Kollektivfeltene i Oslo Og Akershus*; Conference Speech, Tale/Innlegg; Ministry of Transport: Oslo, Norway, 2003.
106. Ministry of Transport. *Nullutslippsbiler: Kan Kjøre i Kollektivfelt Fra 1. Juni*; Press release; Samferdselsdepartementet: Oslo, Norway, 2005.
107. *Vil Drøfte Kollektivfeltets Fremtid*; News article; cFædrelandsvennen: Oslo, Norway, 2014.
108. *Vegvesenet Vil Kreve Medpassasjer*; Budstikka: Oslo, Norway, 2015.
109. *Hver Tredje Elbil Er Borte*; Budstikka: Oslo, Norway, 2015.
110. Ministry of Finance. Bilavgifter. Rapport fra en arbeidsgruppe. Avgitt til Finansdepartementet 30. april 2003. Oslo, Norway. Available online: https://www.regjeringen.no/globalassets/upload/kilde/fin/rap/2003/0003/ddd/pdfv/177320-rapport_bil.pdf (accessed on 16 January 2024).
111. Selvig, E.; Stølan, A.; Flugsrud, K. *Helse- og Miljønytte av Innfasing av 0-Utslippskjøretøy i Norge*; Civitas: Oslo, Norway, 2003.
112. *Elbileiernes Reisevaner*; Rapport 2006-040; ECON Analyse: Oslo, Norway, 2006.
113. Electrification Resource Group. Handlingsplan for Elektrifisering Av Veitransport. Rapport Fra Ressursgruppe Nedsatt Av Samferdselsdepartementet. 2009. Available online: https://www.regjeringen.no/no/dokumenter/handlingsplan-for-elektrifisering-av-vei/id560916/ (accessed on 16 January 2024).
114. *Nye Og Ambisiøse Mål Om Reduksjon Av Klimagassutslepp Få Nye Bilar*; Press release; Ministry of Transport: Oslo, Norway, 2007.
115. Ministry of the Environment. Climate Policy Bill. Norsk Klimapolitikk. St.Meld. nr. 34. 2007. Available online: https://www.regjeringen.no/no/dokumenter/Stmeld-nr-34-2006-2007-/id473411/ (accessed on 17 October 2023).
116. Parliament. Climate Policy Settlement. Avtale Om Klimameldingen. 17 January 2008. Arbeiderpartiet, Sosialistisk Venstreparti, Senterpartiet, Høyre, Kristelig Folkeparti Og Venstre er Enige Om Nedenstående Merknader Til St.Meld. nr. 34 (2006–2007) Norsk Klimapolitikk. Utover Det Som Framgår Av Avtalen Slutter Partene Seg Til Klimameldinga. 2008. Available online: https://www.regjeringen.no/contentassets/fbe5a5829a5d468fab6e4eec0a39512d/avtale_klimameldingen_2008_01_17.pdf (accessed on 17 October 2023).
117. European Union. Regulation (EC) No 443/2009 of The European Parliament and of the Council of 23 April 2009 Setting Emission Performance Standards for New Passenger Cars as Part of the Community's Integrated Approach to Reduce CO2 Emissions from Light-Duty Vehicles. Available online: https://www.eumonitor.eu/9353000/1/j9vvik7m1c3gyxp/vitgbgiqypyz (accessed on 16 January 2024).
118. Ministry of Government Administration and Reform 2008. Nye Reisesatser for Ansatte i Staten. Fornyings-Og Administrasjonsdepartementet. Pressemelding | Dato: 16 April 2008. Available online: https://www.regjeringen.no/no/dokumentarkiv/stoltenberg-ii/fad_2006-2009/nyheter-og-pressemeldinger/pressemeldinger/2008/nye-reisesatser-for-ansatte-i-staten/id508085/ (accessed on 16 January 2024).
119. Ministry of Transport. *Gratis Med Elbiler På Riksvegferjer*; Press release, Pressemelding Nr: 95/08; Samferdselsdepartementet: Oslo, Norway, 2008.
120. *Gratisferje-Idé Kom Frå Pendlar*; News article; Firda: Byrknesøy, Norway, 2008.
121. Ministry of Finance. Nye Tiltak for å Bekjempe økt Arbeidsledighet. Ministry of Finance Press Release 11/2009, 26/02/09. 2009. Available online: https://www.regjeringen.no/globalassets/upload/fin/statsbudsjettet/tiltak09/pressehefte_fin.pdf (accessed on 16 January 2024).
122. *Om Endringer i Statsbudsjettet 2009 Med Tiltak for Arbeid. St. prp. Nr. 37 (2008–2009)*; Ministry of Finance: Oslo, Norway, 2009.
123. Bilnorge. News Article. Zero Lader. 3 March 2009. Available online: https://www.bilnorge.no/artikkel.php?aid=34729 (accessed on 16 January 2024).
124. TU. News Article. Statoil Selger Strøm. 17 November 2011. Available online: https://www.tu.no/artikler/statoil-selger-strom/239104 (accessed on 16 January 2024).
125. Hagman, R.; Selvig, E. Environmentally Friendly Vehicles. In *Experiences and Definitions*; TemaNord 2007:531; Nordic Council of Ministers: Copenhagen, Denmark, 2007.
126. Ministry of Finance. Bilavgifter. Rapport fra en Interdepartemental Arbeidsgruppe. In *Hvordan kan en på best Mulig måte Prise de Samfunnsøkonomiske Kostnadene som Veitrafikken Forårsaker*; Avgitt til Finansdepartementet 20; Ministry of Finance: Oslo, Norway, 2007.
127. Green Highway Elbil- Och Laddhybridguide. April 2009. Green Highway Interreg. Project Report. Available online: https://temfunderingar.wordpress.com/2009/04/30/elbil-och-laddhybridguide-fran-april-2009/ (accessed on 16 January 2024).
128. *SPØRREUNDERSØKELSE OM BRUK AV OG HOLDNINGER TIL ELBILER I NORSKE STORBYER*; NOTAT; Asplan Viak: Bærum, Norway, 2009.
129. Mathisen, T.A.; Solvoll, G.; Smith, K.H. Bruk av elbiler. In *Forventninger og Tilfredshet*; SIB rapport nr. 6/2010; Handelshøyskolen i Bodø, Senter for Innovasjon og Bedriftsøkonomi: Bodø, Norway, 2010.
130. Jørgensen, F.; Mathisen, T.A.; Solvoll, G. *Elbil eller Konvensjonell bil? Økonomiske Analyser*; SIB rapport nr. 2/2010; Handelshøyskolen i Bodø, Senter for Innovasjon og Bedriftsøkonomi: Bodø, Norway, 2010.
131. *Et nytt Transportparadigme i Emning*; Econ Pöyry Report R-2010-095; Econ Pöyry /Ministry of Transport: Oslo, Norway, 2010.

132. Klimakur2010. Klimakur2020. Tiltak Og Virkemidler for å Nå Norske Klimamål Mot 2020. Miljødirektoratet TA 2590/2010. Statens Vegvesen, Avinor, Jernbaneverket, Sjøfartsdirektoratet, Kystverket, Klima-Og Forurensningsdirektoratet. 17 March 2010. Available online: https://www.miljodirektoratet.no/ansvarsomrader/klima/klimatiltak/klimakur/klimakur-2020/ (accessed on 16 January 2024).
133. Figenbaum, E.; Eskeland, G.S.; Leonardsen, J.A.; Hagman, R. *85 g CO_2 per km i 2020—Er Det Mulig?* TØI-rapport 1264/2013; Transportøkonomisk institutt: Oslo, Norway, 2013. Available online: https://www.toi.no/publications/85-g-co2-km-in-2020-is-that-achievable-article31927-29.html (accessed on 16 January 2024).
134. Ministry of the Environment. Meld. St. 21 (2011–2012). Norsk Klimapolitikk. Det Kongelige Miljøverndepartement. 25 April 2012. Available online: https://www.regjeringen.no/no/dokumenter/meld-st-21-2011-2012/id679374/ (accessed on 16 January 2024).
135. European Union. Regulation (EU) no 333/2014 of the European Parliament and of the Council of 11 March 2014 Amending Regulation (EC) No 443/2009 to Define The Modalities for Reaching the 2020 Target to Reduce CO_2 Emissions from New Passenger Cars. Available online: https://eur-lex.europa.eu/legal-content/EN/TXT/?uri=uriserv:OJ.L_.2014.103.01.0015.01.ENG (accessed on 16 January 2024).
136. Parliament. Climate Policy Settlement. Stortinget. Innst. 390 S (2011–2012). In *Innstilling Til Stortinget Fra Energi-Og Miljøkomiteen Meld. St. 21 (2011–2012)*; Parliament: Oslo, Norway, 2012. Available online: https://www.stortinget.no/no/Saker-og-publikasjoner/Publikasjoner/Innstillinger/Stortinget/2011-2012/inns-201112-390/ (accessed on 16 January 2024).
137. Ministry of Finance. *News Note on Government Web Page: Feil Om El-Biler*; Nyhet: Oslo, Norway, 2011.
138. Political Platform for a Government Formed by the Conservative Party and the Progress Party. Sundvolden, 7 October 2013. Government Declaration for 2013–2016. 2013. Available online: https://www.regjeringen.no/contentassets/a93b067d9b604c5a8 2bd3b5590096f74/politisk_platform_eng.pdf (accessed on 16 January 2024).
139. Parliament Petition to the Government. Innst. 2 S (2013–2014). In *Innstilling Fra Finanskomiteen Om Nasjonalbudsjettet for 2014 Og Forslaget Til Statsbudsjett for 2014*; Parliament: Oslo, Norway, 2013. Available online: https://www.stortinget.no/globalassets/pdf/innstillinger/stortinget/2013-2014/inns-201314-002.pdf (accessed on 16 January 2024).
140. Parliament Petition to the Government. Innst. 4 L (2013–2014). Innstilling Fra Finanskomiteen Om Skatter, Avgifter Og Toll 2014—Lovsaker. 2013. Available online: https://www.stortinget.no/no/Saker-og-publikasjoner/Publikasjoner/Innstillinger/Stortinget/2013-2014/inns-201314-004/?lvl=0 (accessed on 16 January 2024).
141. Thema. Report. Utvikling Og Nedtrapping Av Ladbare Bilers Virkemidler. ZERO November 2013. Available online: https://zero.no/wp-content/uploads/2016/05/utvikling-og-nedtrapping-av-ladbare-bilers-virkemidler.pdf (accessed on 16 January 2024).
142. Assum, T.; Kolbenstvedt, M.; Figenbaum, E. *The Future of Electromobility in Norway—Some Stakeholder Perceptives*; TØI rapport 1385/2014; Institute of Transport Economics: Oslo, Norway, 2014. Available online: https://www.toi.no/publications/the-future-of-electromobility-in-norway-some-stakeholder-perspectives-article33109-29.html (accessed on 16 January 2024).
143. Ministry of Finance. Notification (Letter) from the Royal Ministry of Finance to ESA Subject: Notification of VAT Zero Rate for Electric Vehicles. 4 November 2014. Available online: https://www.regjeringen.no/contentassets/0b202c86bcd4423ca6044b81f3 590f81/notificationvat_el.pdf (accessed on 16 January 2024).
144. The Norwegian Tax Administration. Revidert Nasjonalbudsjett 2015—Endringer i Merverdiavgiftsloven Merverdiavgiftsfritaket for Elbiler. Avgitt 23 June 2015. Available online: https://www.skatteetaten.no/rettskilder/type/skattedirektoratets-meldinger/revidert-nasjonalbudsjett-2015--endringer-i-merverdiavgiftsloven-merverdiavgiftsfritaket-for-elbiler/ (accessed on 16 January 2024).
145. Green Tax Committee. NOU 2015: Sett Pris På Miljøet. 2015. Available online: https://www.regjeringen.no/no/dokumenter/nou-2015-15/id2465882/ (accessed on 16 January 2024).
146. *Nyere, Sikrere Og Mer Miljøvennlig Bilpark*; Vehicle Tax Policy Settlement, Bilavgiftsforlik; Parliament: Oslo, Norway, 2015.
147. Rasmussen, I. *Virkninger av endringer i insentiver for kjøp og bruk av ladbare biler*; Rapport 2011/30; Vista Analyse: Oslo, Norway, 2011.
148. Figenbaum, E.; Kolbenstvedt, M. *Electromobility in Norway-Experiences and Opportunities with Electric Vehicles*; TØI rapport 1281/2013; Institute of Transport Economics: Oslo, Norway, 2013. Available online: https://www.toi.no/publications/electromobility-in-norway-experiences-and-opportunities-with-electric-vehicles-article32104-29.html (accessed on 16 January 2024).
149. Laurikko, J.; Granström, R.; Haakana, A. Realistic estimates of EV range based on extensive laboratory and field tests in Nordic climate conditions. *World Electr. Veh. J.* **2013**, *6*, 192–203. [CrossRef]
150. Klöckner, C.A.; Nayum, A.; Mehmetoglu, M. Positive and negative spillover effects from electric car purchase to car use. *Transp. Res. Part* **2013**, *21*, 32–38. [CrossRef]
151. Ryghaug, M.; Toftaker, M. A Transformative Practice? Meaning, Competence, and Material Aspects of Driving Electric Cars in Norway. *Nat. Cult.* **2014**, *9*, 146–163. [CrossRef]
152. Figenbaum, E.; Assum, T.; Kolbenstvedt, M. Electromobility in Norway-experiences and opportunities. *Res. Transp. Econ.* **2015**, *50*, 29–38. [CrossRef]
153. Klimakur2030. 2020. Tiltak Og Virkemidler Mot 2030. M-1625/2020. Utarbeidet Av: Miljødirektoratet, Enova, Statens Vegvesen, Kystverket, Landbruksdirektoratet, NVE. Utgitt Av: Miljødirektoratet. 2020. Available online: https://www.miljodirektoratet.no/klimakur (accessed on 16 January 2024).
154. Bruvoll, A.; Babri, S.; Bråthen, S.; Lilliestråle, A.; Støle, Ø.H. Toll Road Committee. 2020. På Veg Mot Et Bedre Bomsystem. Utfordringer Og Muligheter i Det Grønne Skiftet. Rapport Fra Utvalg. Overlevert Samferdselsministeren. 14 September 2020. Available online: https://www.regjeringen.no/contentassets/cbeb78d09109475aba2bfa9df488971a/1409_utvalgsrapport-framtidige-inntekter-i-bomringene.pdf (accessed on 16 January 2024).

155. *Subject: Notification of Tax Measures for Electric Vehicles*; Letter to EFTA Surveillance Authority 2017; Ministry of Finance: Oslo, Norway, 2017.
156. Transport Authorities. National Transport Plan Proposal. Nasjonal Transportplan 2018–2029. Grunnlagsdokument. 2016. Available online: https://www.regjeringen.no/no/dokumenter/grunnlagsdokument-nasjonal-transportplan-2018-2029/id2477391/ (accessed on 16 January 2024).
157. Government. UNFCCC Paris Agreement. Norway's First NDC 2016. Available online: https://unfccc.int/NDCREG?gclid=CjwKCAjw1t2pBhAFEiwA_-A-NJSS5k7rJrSzRZnwq8kC8LeoYWwi1_rknEn9w0eBs2Du2f0iT4vwZxoCS3sQAvD_BwE (accessed on 16 January 2024).
158. Norwegian Environment Agency. Report. Kunnskapsgrunnlag for Lavutslippsutvikling. Miljødirektoratet Rapport M229–2014. 2014. Available online: https://www.miljodirektoratet.no/globalassets/publikasjoner/M229/M229.pdf (accessed on 16 January 2024).
159. Ministry of Transport. *Meld. St. 33 (2016–2017), Nasjonal Transportplan 2018–2029*; Samferdselsdepartementet: Oslo, Norway, 2017. Available online: https://www.vegvesen.no/fag/fokusomrader/nasjonal-transportplan/tidligere-nasjonale-transportplaner/nasjonal-transportplan-2018-2029/ (accessed on 16 January 2024).
160. Parliament. Innst. 460 S (2016–2017). 2017. Available online: https://www.stortinget.no/no/Saker-og-publikasjoner/Publikasjoner/Innstillinger/Stortinget/2016-2017/inns-201617-460s/?all=true (accessed on 16 January 2024).
161. *Innspill Til Helhetlig Gjennomgang Av Kjøretøy-Og Drivstoffavgiftene*; Sendt til Samferdselsdepartementet 4 August 2014; EVA: Oslo, Norway, 2014.
162. Lovdata. Law Change Implementation. Lov Om Eierseksjoner (Eierseksjonsloven). Dato LOV-2017-06-16-65. Ministry of Local Government and Regional Development. 2017. Available online: https://lovdata.no/dokument/NL/lov/2017-06-16-65#KAPITTEL_5 (accessed on 16 January 2024).
163. Lovdata. Law Change Implementation. Lov Om Endringar i Burettslagslova mv. (Rett Til å Setje Opp Ladepunkt for Elbil, Samanslåing av Eigarseksjonssameige) Kunngjort 4 December 2020. Ikrafttredelse: Kongen Fastset. Available online: https://lovdata.no/dokument/LTI/lov/2020-12-04-137 (accessed on 16 January 2024).
164. Ministry of Transport. Prop. 87 S (2017–2018) Nokre Saker Om Luftfart, Veg, Særskilde Transporttiltak, Kyst Og Post Og Telekommunikasjonar. Tilråding Frå Samferdselsdepartementet 15. Mai 2018, Godkjend i Statsråd Same Dagen. (Regjeringa Solberg). 2018. Available online: https://www.regjeringen.no/contentassets/c6a173c941ca442eb3ffcd4340c9762a/nn-no/pdfs/prp201720180087000dddpdfs.pdf (accessed on 16 January 2024).
165. NPRA. Statens Vegvesen. Riksregulativ for Ferjetakster Gjeldende Fra 1 March 2018. Statens Vegvesen. 2018. Available online: https://www.fosennamsos.no/getfile.php/138587-1519810692/FosenNamsos%20Sj%C3%B8/Bilder/Artikkelbilder/Takster/Takstbestemmelser%20010318%20ferje.pdf (accessed on 13 April 2021).
166. Elbil.no. News Note. Her Får Du Gratis Elbil-Parkering—Og Her Må Du Betale. Elbil.no Artikkel. 6 January 2022. Available online: https://elbil.no/norge-rundt-fa-oversikt-over-gratis-parkering/ (accessed on 16 January 2024).
167. Lovdata. Forskrift Om Vilkårsparkering for Allmennheten Og Håndheving Av Private Parkeringsreguleringer (Parkeringsforskriften). FOR-2016-03-18-260. 2016. Available online: https://lovdata.no/dokument/SF/forskrift/2016-03-18-260 (accessed on 16 January 2024).
168. *News Notice in Aftenposten 30 November 2008 Stating that the City Council Would Finance 400 Chargers in Oslo up to 2011*; Aftenposten: Oslo, Norway, 2008.
169. Transnova. Forslag Til: Nasjonal Strategi Og Finansieringsplan for Infrastruktur for Elbiler. Transnova. 2014. Available online: https://www.regjeringen.no/globalassets/upload/sd/vedlegg/transnovas_ladestrategi.pdf (accessed on 16 January 2024).
170. *Strategi for Ladestasjoner Og Infrastruktur for Elbil 2015–2016*; Enova SF: Trondheim, Norway, 2015.
171. Civitas Stavn. Helhetlig Utbyggingsplan for Infrastruktur Til Ladbare Biler i Fylkene Akershus, Hedmark, Oppland and Østfold. Civitas Stavn Report; Oslo, Norway. 15 May 2012. Available online: https://www.yumpu.com/no/document/view/19971846/last-ned-og-les-rapporten-her-pdf-civitas (accessed on 16 January 2024).
172. *Strategi for Ladeinfrastruktur 2014 2020 Oppfølging Av Moss Kommunes Energi Og Klimaplan*; Moss Kommune: Moss, Norway, 2014.
173. *Ladepunktstrategi for Skedsmo kommune 2015–2020*; Skedsmo, Norway. 2014. Available online: https://klimaostfold.no/wp-content/uploads/2017/01/Skedsmo-kommune-Ladepunktstrategi.pdf (accessed on 16 January 2024).
174. Ministry of Transport. Handlingsplan for Infrastruktur for Alternative Drivstoff i Transport. 2019. Available online: https://www.regjeringen.no/no/dokumenter/handlingsplan-for-infrastruktur-for-alternative-drivstoff-i-transport/id2662448/ (accessed on 16 January 2024).
175. European Union. Directive 2014/94/EU of the European Parliament and of the Council of 22 October 2014 on the Deployment of Alternative Fuels Infrastructure). Available online: https://eur-lex.europa.eu/legal-content/en/TXT/?uri=CELEX:32014L0094 (accessed on 16 January 2024).
176. Jeløya. Government Declaration. Politisk Plattform for En Regjering Utgått Fra Høyre, Fremskrittspartiet Og Venstre. Jeløya. 14 January 2018. Available online: https://www.venstre.no/assets/v-h-frp-politisk-plattform-2018.pdf (accessed on 19 October 2023).
177. *Notification of Zero Rate VAT for Electric Vehicles*; Ministry of Finance: Oslo, Norway, 2020.
178. Ministry of Climate and Environment. Meld. St. 13 (2020–2021) Melding Til Stortinget. Klimaplan for 2021–2030. 2021. Available online: https://www.regjeringen.no/no/dokumenter/meld.-st.-13-20202021/id2827405/ (accessed on 16 January 2024).

179. Parliament. Minutes of Parliament Debate over Bill to the Parliament on the Right to Charge for Flat Owners. Endringar i Burettslagslova mv. (Rett Til å Setje Opp Ladepunkt for Elbil, Samanslåing Av Eigarseksjonssameige) Prop. 144 L (2019–2020), Innst. 78 L (2020–2021), Lovvedtak 21 (2020–2021). 2020. Available online: https://www.stortinget.no/no/Saker-og-publikasjoner/Saker/Sak/?p=81159 (accessed on 16 January 2024).
180. Government. UNFCCC Paris Agreement. Norway's Second Update to the NDC. 2022. Available online: https://unfccc.int/sites/default/files/NDC/2022-11/NDC%20Norway_second%20update.pdf (accessed on 16 January 2024).
181. Parliament. Resolution (Vedtak) 792, during the Discussion of Meld. St. 13 (2020–2021) Klimaplan for 2021–2030. 2021. Available online: https://www.stortinget.no/no/Saker-og-publikasjoner/Vedtak/Vedtak/Sak/?p=82805 (accessed on 16 January 2024).
182. Figenbaum, E. *Electromobility Status in Norway: Mastering Long Distances—The Last Hurdle to Mass Adoption*; TØI 1627/2018; Transportøkonomisk institutt: Oslo, Norway, 2018. Available online: https://www.toi.no/publications/electromobility-status-in-norway-mastering-long-distances-the-last-hurdle-to-mass-adoption-article34903-29.html (accessed on 16 January 2024).
183. *Videokonferansehøring: Meld. St. 13 (2020-2021) Klimaplan for 2021–2030*; Parliament: Oslo, Norway, 2021.
184. *Resolution (Vedtak) 1232 during the Discussion of Meld. St. 20 (2020-2021) Nasjonal Transportplan 2022–2033*; Parliament: Oslo, Norway, 2021.
185. SVV. Kunnskapsgrunnlag Om Hurtigladeinfrastruktur for Veitransport. Statens Vegvesen Og Miljødirektoratet 2022. Rapport M-2232 Miljødirektoratet. 1 March 2022. Available online: https://www.miljodirektoratet.no/publikasjoner/2022/mars/kunnskapsgrunnlag-om-hurtigladeinfrastruktur-for-veitransport/ (accessed on 16 January 2024).
186. SVV. Invitasjon Til å Bidra Med Innspill Til Regjeringens Ladestrategi. Meeting the 21 March 2022. Available online: https://www.regjeringen.no/no/aktuelt/invitasjon-til-a-bidra-med-innspill-til-regjeringens-ladestrategi/id2902693/ (accessed on 16 January 2024).
187. Parliament. *Agreement in the Parliament on the Implementation of VAT on BEVs Purchase Price Exceeding 500,000 NOK*; Revidert nasjonalbudsjett 2022: Enighet mellom Ap, Sp og SV; Arbeiderpartiet: Oslo, Norway, 2022. Available online: https://www.arbeiderpartiet.no/aktuelt/revidert-nasjonalbudsjett-2022-enighet-mellom-ap-sp-og-sv/ (accessed on 16 January 2024).
188. Lovdata. Lov Om Endinger i Merverdiavgiftsloven. Kunngjort 20 December 2022. Ikrafttredelse 1 January 2023. 2022. Available online: https://lovdata.no/dokument/LTI/lov/2022-12-20-108 (accessed on 16 January 2024).
189. Parliament. *Agreement on the National Budget for 2023*; Parliament: Oslo, Norway, 2022. Available online: https://www.stortinget.no/no/Saker-og-publikasjoner/Statsbudsjettet/statsbudsjettet-2023/ (accessed on 16 January 2024).
190. Takstretningslinjer for Bompengeprosjekt På Offentlig Veg. Retningslinjer. Fastsatt i Vegdirektoratet Juli 2023. Available online: https://www.autopass.no/siteassets/takstretningslinjer/lars/takstretningslinjer-for-bompengeprosjekt-pa-offentlig-veg---juli-2023.pdf (accessed on 16 January 2024).
191. *Kunnskapsgrunnlag om Hurtigladeinfrastruktur for Veitransport*; Miljødirektoratet og Statens Vegvesen: Oslo, Norway, 2022. Available online: https://www.regjeringen.no/no/aktuelt/kunnskapsgrunnlaget-om-utbygging-av-ladeinfrastruktur-er-klart/id2902606/ (accessed on 16 January 2024).
192. Fridstrøm, L.; Alfsen, K. Norway's Path to Sustainable Transportation. TØI report 1321/2014. 2014. Available online: https://www.toi.no/publications/norway-s-path-to-sustainable-transport-article32521-29.html (accessed on 16 January 2024).
193. Jong, W.; van der Linde, V. Clean Diesel and Dirty Scandal: The Echo of Volkswagen's Dieselgate in an Intra-Industry Setting. *Public Relat. Rev.* **2022**, *48*, 102146. [CrossRef]
194. Samset, K.; Bukkestein, I. Hvordan Stoppe Dårlig Begrunnete Prosjekter På Et Tidlig Tidspunkt. Concept-Programmet Norges Teknisk-Naturvitenskapelige Universitet 7491 NTNU—Trondheim. 2020. Available online: https://www.ntnu.no/documents/1261860271/1261996393/Hvordan+stoppe+d%C3%A5rlig+begrunnete+prosjekter+WEB+(1).pdf/6d566e4d-8310-e104-2dec-b748b82c6fc8?t=1606199737906 (accessed on 20 December 2023).
195. Ministry of Finance. National Budget Documents 2024. 2023. Available online: https://www.regjeringen.no/no/statsbudsjett/2024/id2994174/ (accessed on 16 January 2024).
196. Agreement on National Budget for 2024. Budsjettforlik Mellom Ap/Sp Og SV 2024. Versjon, 3 December 2023., kl. 08:30. Avtaletekst. Available online: https://www.sv.no/wp-content/uploads/2023/12/031223-budsjettforlik-verbaler-beriktiget-041223.pdf (accessed on 19 December 2023).
197. Figenbaum, E. Chapter on Norway in the IEA HEV TCP Annual Report 2023. The Electric Drive Ramps Up. Available online: https://ieahev.org/wp-content/uploads/2023/06/Digital_HEVTCP_Annual_Report_2023_v0.3-3-1.pdf (accessed on 19 December 2023).

World Electric Vehicle Journal

MDPI

Communication

Subcooled Liquid Hydrogen Technology for Heavy-Duty Trucks

Enrico Pizzutilo [1], Thomas Acher [2,*], Benjamin Reuter [1,*], Christian Will [1] and Simon Schäfer [2]

[1] Daimler Truck AG, Fasanenweg 10, 70771 Leinfelden-Echterdingen, Germany; enrico.pizzutilo@daimlertruck.com (E.P.)
[2] Linde GmbH, Linde Hydrogen FuelTech, Dr.-Carl-von-Linde-Straße 6-14, 82049 Pullach, Germany
* Correspondence: thomas.acher@linde.com (T.A.); benjamin.reuter@daimlertruck.com (B.R.)

Abstract: Subcooled liquid hydrogen (sLH2) is an onboard storage, as well as a hydrogen refueling technology that is currently being developed by Daimler Truck and Linde to boost the mileage of heavy-duty trucks, while also improving performance and reducing the complexity of hydrogen refueling stations. In this article, the key technical aspects, advantages, challenges and future developments of sLH2 at vehicle and infrastructure levels will be explored and highlighted.

Keywords: hydrogen mobility; FCEV; heavy-duty trucks; subcooled liquid hydrogen; sLH2; liquid hydrogen; LH2; hydrogen refueling station

Citation: Pizzutilo, E.; Acher, T.; Reuter, B.; Will, C.; Schäfer, S. Subcooled Liquid Hydrogen Technology for Heavy-Duty Trucks. *World Electr. Veh. J.* **2024**, *15*, 22. https://doi.org/10.3390/wevj15010022

Academic Editor: Joeri Van Mierlo

Received: 17 November 2023
Revised: 14 December 2023
Accepted: 20 December 2023
Published: 8 January 2024

1. Introduction

On the way toward carbon-neutral road transport mobility, heavy-duty trucks (HDTs) are one of the most challenging applications to decarbonize [1]. In this context, truck original equipment manufacturers (OEMs) are exploring a dual technology-open strategy, with both battery electric vehicles (BEVs) and fuel cell electric vehicles (FCEVs) being developed and adopted as complementary solutions [2,3].

BEVs are considered the best choice for short distances, with plannable routes and a lighter load. On the other hand, FCEVs are the preferred technology for cases of high mileage and energy consumption, such as long-haul and on-demand applications [4]. Furthermore, FCEVs are projected to be an attractive option when flexibility is also required and where there are local grid constraints [3].

In an FCEV, one of the main components is the onboard hydrogen storage system. Despite having a high gravimetric energy density, hydrogen has a very low volumetric density when stored at an ambient temperature and pressure. Therefore, in order to reach the mileage targets (as in Figure 1), hydrogen needs to be either stored at a higher pressure or lower temperature. To this end, several potential candidates for onboard hydrogen storages can be considered [5–7], namely the following:

(1) Compressed hydrogen gas (CHG) at room temperature and high pressures;
(2) Cryo-compressed hydrogen (CcH2) at low temperatures and high pressures;
(3) Liquid hydrogen (LH2) at very low temperatures (<20 K) and low pressures (<10 bar).

Each of these storage technologies has a different storage pressure, as well as density (Figure 2).

While CHG hydrogen can only reach storage densities of up to ~40 kg/m^3 (at 700 bar and 15 °C), subcooled liquid hydrogen (sLH2) can reach up to ~62 kg/m^3 (at ~16 bar and −245 °C). By combining higher pressures (e.g., 350 bar) and low-to-cryogenic temperature (e.g., −250/−200 °C), it is possible to reach even higher energy densities (e.g., ~72 kg/m^3). However, the storage technologies of cryo-compressed hydrogen (CcH2) are more complex and currently have a lower technology readiness level (TRL) compared to the previously mentioned two technologies, as they need to handle both very low temperatures as well as high pressure in both the tank system and the refueling line (pipes, connectors, etc.). Hence, in this paper, we will focus on CHG and LH2.

Figure 1. Technology fit for commercial vehicles based on different use requirements: battery electric trucks, eActros (**left**) and eActros LongHaul (**middle**), and the GenH2 truck using liquid hydrogen in a fuel cell (**right**) (source: Daimler Truck Capital Market Day—November, 2021).

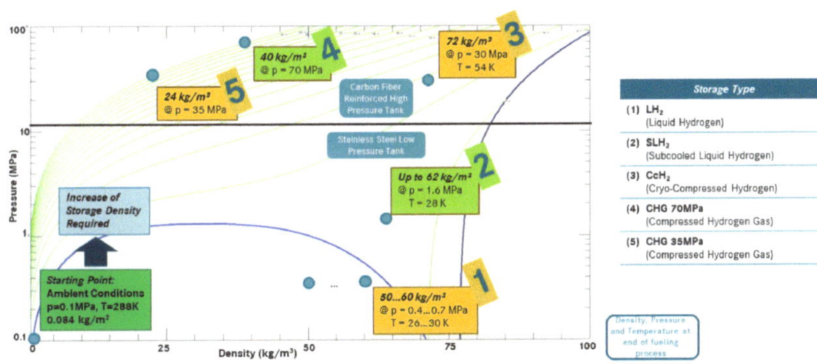

Figure 2. Density vs. pressure diagram for different on-board storage technologies.

Today, compressed hydrogen at 70 MPa (CHG70) and 35 MPa (CHG35) is used in light-duty vehicles [8] and for buses, respectively. With respect to the more challenging HDT use cases, however, OEMs are currently pursuing different concepts [6]. Considering that such a technology choice has a large impact on the whole hydrogen value chain, it is of utmost importance that OEMs and infrastructure players collaborate and work closely together.

In this context, Daimler Truck and Linde are jointly developing a new storage and refueling solution, namely "subcooled liquid hydrogen" (sLH2). Thanks to an improved tank and interface design encompassing an increased pressure (up to ~20–25 bar), sLH2 enhances refueling performances while reducing the complexity of protocols and hardware at the hydrogen refueling stations (HRS) [9]. Some of the key parameters/advantages at the vehicle and HRS levels will be detailed in the following subsections.

2. Vehicle Advantages

The transition of HDT toward zero-emission vehicles implies a profound transformation of vehicle architecture. In FCEVs, one substantial challenge is the integration of large tank systems to achieve range and payload target. Considering sLH2 and CHG70 as the reference technologies for heavy-duty long-haul trucks, the architecture of the respective tank systems, as well as their integration in vehicles, will differ substantially, resulting in different vehicle characteristics (Figure 3).

Key Assessment Criteria		
Capacity on Truck	> 80 kg (4x2)	~60 kg (4x2)
Operating pressure	4-8bar	20-700bar
Target pressure	16bar	700bar (p_target depend on p-initial and Tamb)
Range	> 1000 km	~600-700 km
Hold Time (w/o Boil off)	SOC 100% ≈ 10 h [1] **SOC 80% ≈ 130 h [1]** SOC 50 % ≈ 200 h [1]	n.a.
Handling / Refueling	**Multiple tank filling possible, no return gas, no communication**	Multiple tank filling possible, no return gas, **communication required**
Tank Type	Stainless steel without additional reinforcement, vacuum insulated	Typ IV (liner + carbon/glass fibers)

[1] Assuming 5W heat transfer and depending on operating pressure

Figure 3. Comparison of the main vehicle characteristic of a 4 × 2 HDT equipped with sLH2 and CHG70 tank systems (**left** and **right** respectively); SOC refers to the vehicle state of charge (see details below).

sLH2 has an approximately 50 % higher density (up to 62 kg/m^3 at p = 16 bar and T = 28 K) compared to CHG70 (40 kg/m^3 at p = 700 bar and T = T_{amb}). At the same time, an insulated stainless-steel, low-pressure tank is sufficient to store sLH2 compared to Type IV high-pressure tanks reinforced with carbon fibers, typically used in a CHG70 configuration [10].

This results in lighter (approximately 20–30% less weight per stored kg of hydrogen) and cheaper (approximately 40–50% lower costs per stored kg of hydrogen) tanks with lower volumes, higher stored mass of hydrogen and mileage (sLH2 showcases approximately 50% range increase, from ~700 km of CHG70 to more than 1000 km of sLH2, depending on the consumption profile) [10].

Overall, we conclude that the sLH2 technology has clear advantages in terms of ranges, vehicle investment costs and payloads compared to the more common CHG technology. Furthermore, despite being a novel technology, the necessary know-how to develop sLH2 tanks is quite similar to the wide-spread liquid natural gas (LNG) tank, resulting in multiple potential suppliers and/or manufacturers that can scale-up and industrially produce such tank systems.

Despite such clear advantages, one challenge with sLH2 on the vehicle side is the boil-off onboard. However, internal simulations and tests indicate that boil-off kicks in after approximately 10 h if the state of charge (SOC) is 100% and only after more than 160 h when the tank is half empty (~50% SOC). However, considering that HDTs are normally driven on a daily basis, these values suggest that boil-off would be a rather rare event during normal operation.

3. Refueling Protocol and HRS Advantages

The sLH2 refueling process is based on improved LH2 refueling, without back-gas or limitation toward multiple tanks or back-to-back refueling. To achieve this, Linde developed a novel sLH2 refueling station including an sLH2 pump with a flow rate exceeding 400 kg/h, with a target pressure of 16 bar during refueling [10]. Fueling times of less than 10 min for typical HDTs can be realized with this configuration.

Thanks to the higher density of LH2 and the lower required pressure during refueling, the hydrogen delivery, as well as the storage and compression at the station, is not only easier compared to gaseous compressed hydrogen, but also noticeably more compact, as can be seen from the example in Figure 4.

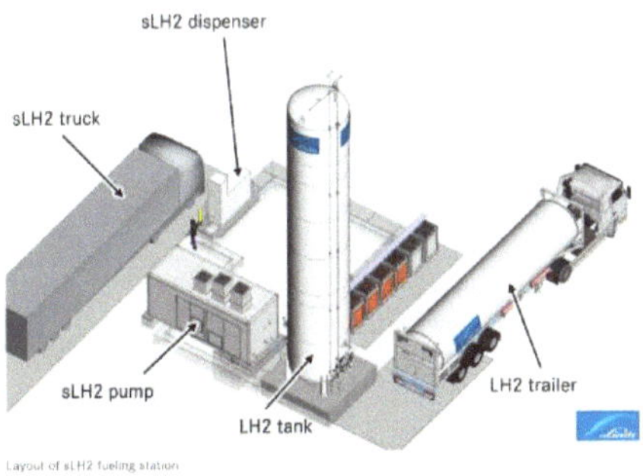

HRS characteristics	sLH2 specification
Flow rate	400 – 500 kg/h
Fueling time	10 – 15 min
HRS energy demand	0.05 kWh/kg H_2 dispensed
Station footprint	50 m² w/o dispenser
HRS – vehicle interaction	no data transmission no back-gas
Storage tank size	4000 kg LH2 – 10 h fueling can be adapted to customer needs
Fueling modes	parallel and back-to-back

Figure 4. Exemplary layout of a small sLH2 HRS with one dispenser.

At the HRS level, beside the smaller footprint, the advantages of sLH2 and liquid delivery are outlined qualitatively in Figure 5.

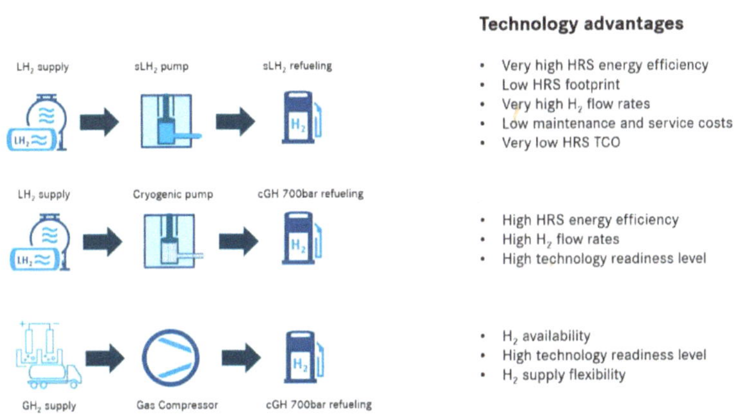

Figure 5. Technology advantages of different fueling technologies (sLH2 on top).

The improved refueling performance with sLH2 fueling leads to a very low TCO for the HRS, as well as a high HRS energy efficiency (0.05 kWh/kg H2), and footprint/complexity reductions are quite remarkable compared to CHG70 [10].

In this respect, also considering the advantages at the vehicle level, sLH2 is a highly attractive technology for customers in the trucking sector and beyond. However, there are still a few steps remaining before sLH2 becomes widely accepted within the industry. Besides the market availability and low cost of liquid hydrogen (a discussion that is out of scope for the current paper), one of the remaining hurdles is the standardization process that will be discussed in the next section.

4. Standardization

Linde and Daimler Truck are not proprietary of the technology and are promoting the advantages of using sLH2 in HDT in order to expand the technology adoption by other OEMs, as well as more infrastructure providers. In order to achieve that, a white paper process was initiated in 2021 [11,12]. The resulting specifications for the fueling and hardware interface, after the conclusion of the activities within the Clean Energy Partnership (CEP) in 2022, are now under standardization at the ISO level.

The CEP sLH2 white paper activities saw the participation of multiple stakeholders from the trucking and infrastructure sectors, and resulted in two papers being developed:

(1) LH2 fueling from the station into the truck is well known from former projects, but has some disadvantages, e.g., gas return from the tank to the fueling station, and fueling

stops only based on signals from the truck. Therefore, the first white paper focuses on sLH2 (subcooled liquified hydrogen) fueling to avoid gas return from the vehicle tank, and defines fueling stops without data communication required. sLH2 fueling is a process in which the liquefied hydrogen is subcooled and can be used in this state to fill the vehicle tank.

The fueling procedure is subdivided into three steps:

- Pre-fueling (incl. purging and leakage testing, pressure system determination, etc.);
- Main fueling (with two fueling steps, one with a reduced flow rate for the cooldown of piping and storage system and a second with a target fueling rate of 400 kg/h);
- Post-fueling (after the p_{target} is reached, further purging and leakage testing needs to be conducted before the nozzle is disconnected).

The flow, pressure and temperature profiles during a typical refueling event are shown in Figure 6.

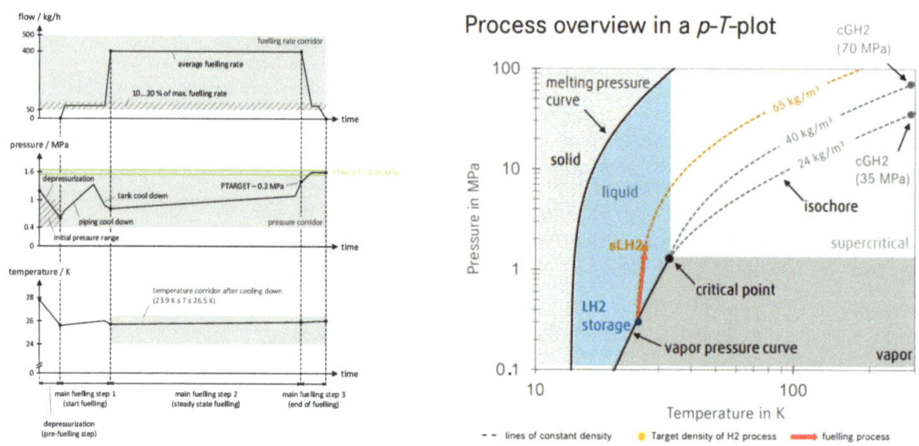

(**a**) (**b**)

Figure 6. (**a**) Exemplarily flow, pressure and temperature profiles during sLH2 fueling; (**b**) p–T plot of hydrogen sLH2.

(2) Furthermore, having the vehicle storage system, connected to the propulsion unit, on one hand and the fueling unit on the other, a component joining both units for hydrogen transfer is required. Therefore, the goal of the second white paper is the development of a subcooled liquid hydrogen fueling interface applied in trucks, of which the main hardware components are shown in Figure 7. This coupling component shall be easily reproducible in a series of production process.

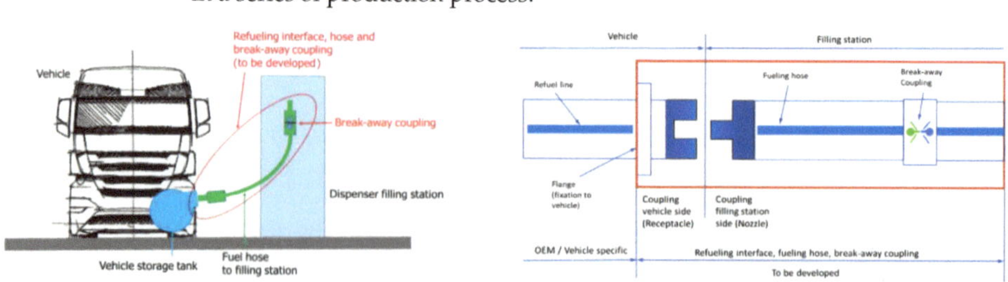

(**a**) (**b**)

Figure 7. (**a**) Overview of the fueling interface-components; (**b**) Overview of system boundaries and interface with focus of standardization effort highlighted in red box.

Within the documents, a complete set of information on controlling, testing dimensioning, geometry, design and further requirements (e.g., environmental, electrical, operational) is provided. Since early 2023, the sLH2 protocol and interface are being discussed within the ISO activities (TC 197—Hydrogen Technologies), with the target of achieving a global standard. Within these activities, revision of the following documents has been proposed and is expected to be completed by 2026:

- ISO 13984: liquid H2—land vehicle fueling protocol [13];
- ISO 13985: liquid H2—land vehicle fuel tanks [14];
- ISO 19886: liquid H2—land vehicle fueling connectors [15].

5. Conclusions

In the present paper, the advantages of sLH2 technologies for vehicles as well as refueling stations are shown. Overall, sLH2 features a significant commercial advantage for HDTs and HRSs, while also reducing the space requirements, thanks to the higher energy density of liquid hydrogen and reduced amount of equipment. At the same time, the refueling protocol, that is currently undergoing a standardization process, solves some of the critical challenges for fueling vehicles with liquid hydrogen. Considering also the initial positive testing results, we are confident that sLH2 will be a standard solution in the future portfolio of heavy-duty road transport and non-road transport applications.

Author Contributions: Conceptualization, E.P., B.R. and T.A.; writing—original draft preparation, E.P., B.R. and T.A.; writing—review and editing, C.W. and S.S. All authors have read and agreed to the published version of the manuscript.

Funding: The project was entirely funded by Linde and Daimler Truck.

Data Availability Statement: Data is contained within the paper.

Conflicts of Interest: All authors are/were employed by either Linde or Daimler Truck. These companies have a commercial interest related to this technology but both companies agreed to publish information on this technology to effect an adoption beyond these companies.

References

1. DeStatis, Road Transport Emission. Available online: https://www.destatis.de/Europa/EN/Topic/Environment-energy/CarbonDioxideRoadTransport.html#:~:text=Passenger%20cars%20and%20motorcycles%20accounted,trucks%20for%20a%20further%2013%25 (accessed on 23 October 2023).
2. Ledna, C.; Muratori, M.; Yip, A.; Jadun, P.; Hoehne, C. Decarbonizing Medium- & Heavy-Duty On-Road Vehicles: Zero-Emission Vehicles Cost Analysis, NREL, 2022. Available online: https://www.nrel.gov/docs/fy22osti/82081.pdf (accessed on 24 October 2023).
3. Hydrogen Council. Roadmap towards Zero Emissions: BEVs and FCEVs, 2021. Available online: https://hydrogencouncil.com/wp-content/uploads/2021/10/Transport-Study-Full-Report-Hydrogen-Council-1.pdf (accessed on 24 October 2023).
4. Daimler Truck #HydrogenRecordRun: Mercedes-Benz GenH2 Truck Cracks 1000 Kilometer Mark with One Fill of Liquid Hydrogen. Available online: https://media.daimlertruck.com/go/HydrogenRecordRun (accessed on 13 November 2023).
5. Rivard, E.; Trudeau, M.; Zaghib, K. Hydrogen Storage for Mobility: A Review. *Materials* **2019**, *12*, 1973. [CrossRef] [PubMed]
6. H2Mobility, Overview Hydrogen Refuelling for Heavy Duty Vehicles, 2021. Available online: https://h2-mobility.de/wp-content/uploads/sites/2/2021/08/H2-MOBILITY_Overview-Hydrogen-Refuelling-For-Heavy-Duty-Vehicles_2021-08-10.pdf (accessed on 23 October 2023).
7. DOE Hydrogen and Fuel Cell Technology Office. Hydrogen Storage. Available online: https://www.energy.gov/eere/fuelcells/hydrogen-storage (accessed on 23 October 2023).
8. SAE. Fueling Protocols for Light Duty Gaseous Hydrogen Surface Vehicles J2601_202005, 2020. Available online: https://www.sae.org/standards/content/j2601_202005/ (accessed on 23 October 2023).
9. Maus, S. Technology Pitch: Subcooled Liquid Hydrogen (sLH2), NOW & CEP Heavy Duty Event, 21 April 2021. Available online: https://www.now-gmbh.de/wp-content/uploads/2021/05/Heavy-Duty-Event-Subcooled-Liquid-Hydrogen-sLH2-Schaefer-Linde-Maus-Daimler.pdf (accessed on 24 October 2023).
10. Pizzutilo, E.; Acher, T. Subcooled Liquid Hydrogen (sLH2). In Proceedings of the 8th International Workshop on Hydrogen Infrastructure, Brussels, Belgium, 13 September 2022.

11. CEP. White Paper sLH2 Fueling Specification—Release 1.8, 2021. Available online: https://cleanenergypartnership.de/wp-content/uploads/2022/03/CEP_20210914_Whitepaper_sLH2-Fueling-Specification_Release_1.8_clean-min.pdf (accessed on 3 January 2024).
12. CEP. White Paper sLH2 Interface Specification—Release 1.12, 2021. Available online: https://cleanenergypartnership.de/wp-content/uploads/2022/03/CEP_WhitePaper_sLH2-Interface-Specification_Release1.12-min.pdf (accessed on 3 January 2024).
13. *ISO 13984*; International Organization for Standardization: Liquid Hydrogen—Land Vehicle Fuelling System Interface. International Organization for Standardization, ISO Central Secretariat: Geneva, Switzerland, 1999. Available online: https://www.iso.org/standard/23570.html (accessed on 14 November 2023).
14. *ISO 13985*; International Organization for Standardization: Liquid Hydrogen—Land Vehicle Fuel Tanks. International Organization for Standardization, ISO Central Secretariat: Geneva, Switzerland, 2006. Available online: https://www.iso.org/standard/39892.html (accessed on 13 November 2023).
15. *ISO 19886*; International Organization for Standardization: Liquid Hydrogen—Land Vehicle Fuelling Connector. International Organization for Standardization, ISO Central Secretariat: Geneva, Switzerland, to be released. Available online: https://www.iso.org/standards.html (accessed on 10 November 2023).

World Electric Vehicle Journal

MDPI

Article

Speed Change Pattern Optimization for Improving the Electricity Consumption of an Electric Bus and Its Verification Using an Actual Vehicle

Yiyuan Fang [1,*], Wei-hsiang Yang [1], Yushi Kamiya [1], Takehito Imai [2], Shigeru Ueki [2] and Masayuki Kobayashi [2]

1 Faculty of Science and Engineering, Waseda University, Tokyo 169-8555, Japan
2 Organization for the Promotion of Low Emission Vehicles, Tokyo 160-0004, Japan
* Correspondence: fyy@aoni.waseda.jp

Abstract: In this study, we focused on the eco-driving of electric vehicles (EVs). The target vehicle is an electric bus developed by our research team. Using the parameters of the bus and speed pattern optimization algorithm, we derived the EV's eco-driving speed pattern. Compared to the eco-driving of internal combustion engine vehicles (ICVs), we found several different characteristics. We verified these characteristics with actual vehicle driving test data of the target bus, and the results confirmed its rationality. The EV's eco-driving method can improve electricity consumption by about 10–20% under the same average speed.

Keywords: energy consumption; efficiency; EV (electric vehicle); simulation; optimization

Citation: Fang, Y.; Yang, W.-h.; Kamiya, Y.; Imai, T.; Ueki, S.; Kobayashi, M. Speed Change Pattern Optimization for Improving the Electricity Consumption of an Electric Bus and Its Verification Using an Actual Vehicle. *World Electr. Veh. J.* **2024**, *15*, 16. https://doi.org/10.3390/wevj15010016

Academic Editors: Joeri Van Mierlo and Genevieve Cullen

Received: 28 November 2023
Revised: 22 December 2023
Accepted: 2 January 2024
Published: 4 January 2024

1. Introduction

The energy efficiency of the transportation sector has become a key factor to reduce greenhouse gas emissions and fuel consumption in response to the negative impacts of global warming [1–3]. As a method of energy conservation and environmental sustainability, eco-driving has attracted considerable research interest over the past two decades [4–6]. Eco-driving is an emerging research field, and its definition is not yet strictly defined. However, it generally refers to the practice of driving vehicles in a way that improves fuel economy [7–9].

Many studies have shown that eco-driving is a low-cost, high-efficiency method of energy conservation and emission reduction [1,10,11]. Eco-driving has been widely discussed and applied worldwide due to the aforementioned advantages. German scholars were the first to focus on this field in 2001. As of 2020, scholars from the United States and China have contributed the most publications in this field (total papers—percentage: 178—23% (US), 117—15% (China)) [4,12,13]. Numerous studies from around the world have shown the enormous potential of eco-driving in energy conservation, emission reduction, and other aspects [14–16]. Eco-driving has also been summarized into some specific and easy-to-implement principles that are promoted worldwide. In European countries including England, Germany, Italy, and Finland, eco-driving methods such as the golden rules of eco-driving have been regarded as part of the driving license examination [4,17]. In Japan, the 10 recommendations for eco-driving promoted by government departments such as the ministry of the environment are well-known to the public [18].

Many popular eco-driving principles, including gentle acceleration and quick shifting up, are usually based on ICVs [19,20]. With the popularization of EVs, research on EV eco-driving becomes more and more important. Many researchers study eco-driving as an optimization problem. For example, a study conducted by Mensing et al. shows that using optimization techniques at a fixed distance and time to adjust the driver's operations significantly improves the energy efficiency of the ICV [21]. This fixed distance and time method is convenient to clarify the energy consumption improvement effect of eco-driving under the same driving conditions, so we also adopted it in our research. However, the power system characteristics of EVs and ICVs are different, and the applicability of EVs needs further verification. And a study conducted by Sundström et al. introduces a generic dynamic programming function for Matlab [22], which can be used in vehicle power consumption optimization problems. Referring to this research, we built a speed change pattern optimization simulator by combining our developed accuracy-proven vehicle simulator with an optimization algorithm and used it to develop EV eco-driving. In addition, an eco-driving optimization study often focuses on algorithms and lacks the verification of actual vehicle experiments [23,24]. In this regard, after deriving the optimal EV eco-driving, we verified its characteristics using the driving test data of a small electric bus that was developed by our research team.

The purpose of this study is to explore eco-driving strategies that are applicable to EVs. Currently, many eco-driving views for cars are based on ICVs. Are they still applicable to EVs, which have undergone significant changes in their powertrain systems and are rapidly becoming popular [25]? We want to find out what kind of driving strategies will improve the electricity consumption of EVs. For this purpose, we selected a self-developed electric bus as the object vehicle, constructed a simulator that can accurately calculate the power consumption of the vehicle during operation, and combined it with an optimization method to derive the EV eco-driving speed change pattern, which is the speed change pattern that results in the best electricity consumption under the set conditions. After investigating it, we obtained some eco-driving strategies that are applicable to EVs and discovered the differences between them and the eco-driving strategies that are applicable to ICVs. Then, we verified the correctness of these eco-driving strategies through the experimental driving data of the object vehicle and investigated the reasons why these eco-driving strategies can improve the electricity consumption.

2. Target Vehicle and Simulator

2.1. Target Vehicle and Simulation Conditions

In this study, the target vehicle is a small electric bus, the Waseda Electric Bus-3Advanced (WEB-3A). This vehicle was created by converting a small diesel bus using Hino Motors into a remodeled small electric bus with standard specifications. Table 1 summarizes the basic specifications.

Table 1. Basic specifications of WEB-3A.

	Base Diesel Bus	WEB-3A
Manufacturer/Type	Hino/Poncho (BDG-HX6JLAE)	
Capacity	31 persons	
Curb/Gross weight [kg]	5710/7415	5990/7695
Engine or Motor	132 kW Engine	145 kW/400 Nm (PMSM)
Transmission	5 speed AT	Fixed
Battery [kWh]/[V]	None	40/331 (TOSHIBA "SCiB™")
Exterior photograph	None	

Since we focused on the aforementioned street bus in this study, we optimized the speed change pattern, in which "driving distance" and "average speed" are fixed from start to stop, while considering the distance between bus stops and the schedule [26,27]. Our purpose was to cover a total distance of 400 m in three intervals (acceleration, coasting, and deceleration) at an average speed of 30 km/h. In addition, we also focused on the double travel distance when stops were skipped (800 m total with an average speed of 30 km/h). In this study, we assumed that there would be no impact from traffic lights or congestion.

2.2. Vehicle Driving Energy Calculation Simulator and the Speed Change Pattern Optimization Method

Figure 1 shows the schematic of the backward simulator used to calculate the driving energy of WEB-3A. The power consumed by the battery is obtained by inputting the vehicle's speed. The vehicle simulator was constructed using methods that are commonly used in electric vehicle simulations. It can simulate the power consumption of a vehicle during operation by using information on the vehicle's speed and road gradient. The vehicle simulation considers the driving resistance of the vehicle (acceleration resistance, air resistance, rolling resistance, and slope resistance), the transmission efficiency and the motor/inverter efficiency during driving and regeneration (transmission efficiency is a constant value, while motor/inverter efficiency comes from the efficiency map), and the power consumption of the vehicle's auxiliary equipment.

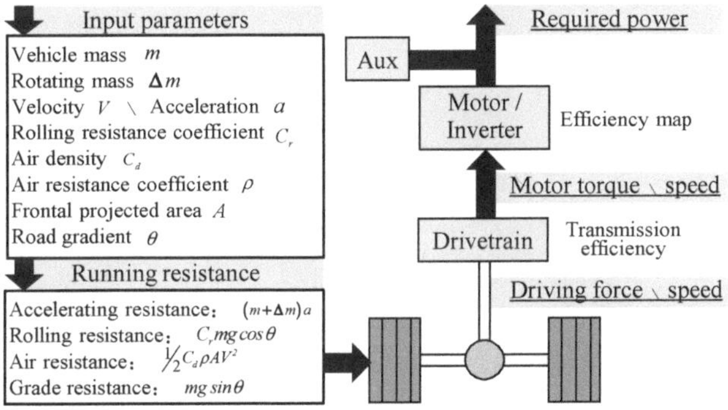

Figure 1. Image of vehicle running energy calculation simulator.

The vehicle simulator constructed using the above method can achieve high-precision calculation of instantaneous electricity consumption and comprehensive electricity consumption for the entire journey. Figure 2 shows the comparison between the actual measured motor power and the simulated calculated motor power of the object vehicle on a certain route (which is described in detail in Section 4). It can be seen that the simulation results are highly consistent with the measured values, and the comprehensive electricity consumption error of the simulation is within 5%.

We optimized the drive of 400 m (or 800 m) with an average speed of 30 km/h, as mentioned above. First, we define a cost function to search for the speed change pattern that consumes least energy, as shown in Equation (1).

$$C = \int_{t_{start}}^{t_{end}} P(j)\mathrm{d}t \tag{1}$$

Here, C [kWh] is the consumed energy, t [s] is time, P [kW] is consumed power, and j [m/s^3] is the jerk (control variable).

Figure 3 shows a schematic of the optimization method used in this study (dynamic programming) (x [m] represents position, v [m/s] represents speed, and a [m/s^2] represents acceleration). The following are the constraints and convergence conditions: (a) maximum jerk of ±1 m/s^3, (b) maximum acceleration (deceleration) of ±0.2 G, (c) starting (stopping) speed of 0 km/h, and (d) maximum speed of 60 km/h.

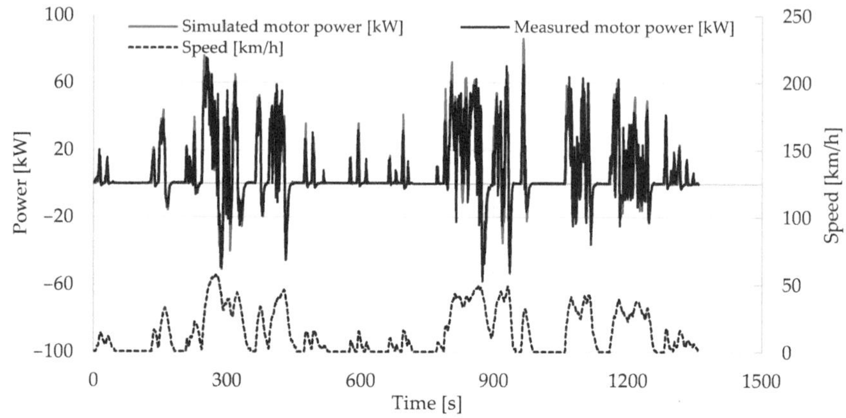

Figure 2. Comparison of simulated and measured motor power consumption.

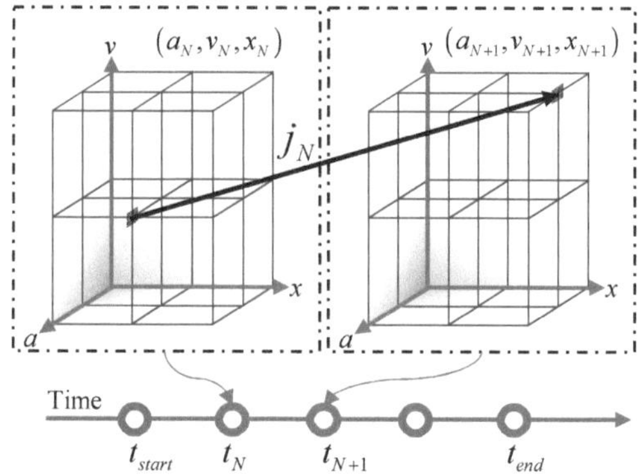

Figure 3. Image of dynamic programming.

The speed change pattern is optimized by incorporating the proposed optimization method into the vehicle's driving energy calculation simulator. Calculations are performed in the following order: (a) determine the relationship among acceleration, speed, position, and time as state variables and jerk as the control variable, (b) input the state variables of each tiny time period into the vehicle simulator to calculate the battery electricity consumption, and (c) search for the combination that minimizes the cost function.

3. Investigation and Trial Calculation of the Electricity Consumption Optimization Speed Change Pattern

3.1. Derivation of the Electricity Consumption Optimization Speed Change Pattern

In this section, we summarize the optimization of the speed change pattern for various conditions.

The vehicle loss conditions are listed in Table 2. In addition, we investigate the use of "coasting", which has gained attention recently for improving the electricity consumption of electric vehicles. In most cases, coasting is not advantageous in terms of fuel efficiency or safety in internal combustion vehicles; thus, it is not employed in regular driving. However, it is widely employed in trains as an eco-driving method. In some cases, coasting has been implemented in electric vehicles. For example, some EVs using a one-pedal accelerator in the neutral range of pedal opening, to account for the driver's unintentional fine operation, set a dead zone to keep the output of the motor at 0 Nm, so that the vehicle maintains coasting [28], while others maintain coasting by releasing the accelerator pedal [29]. Coasting is possible by reducing the motor torque to 0 Nm while the inverter is operating [30] or disconnecting the inverter from the motor [31]. In the current study, we employed the latter "inverter off coasting control (with coasting control)". Finally, as the second analytical condition, we employed "without coasting control".

Table 2. List of various data used for vehicle loss calculation.

Transmission Efficiency	Auxiliary Equipment Consumption Power	Motor/Inverter Efficiency (Using Efficiency Map Data)
98%	1.5 kW	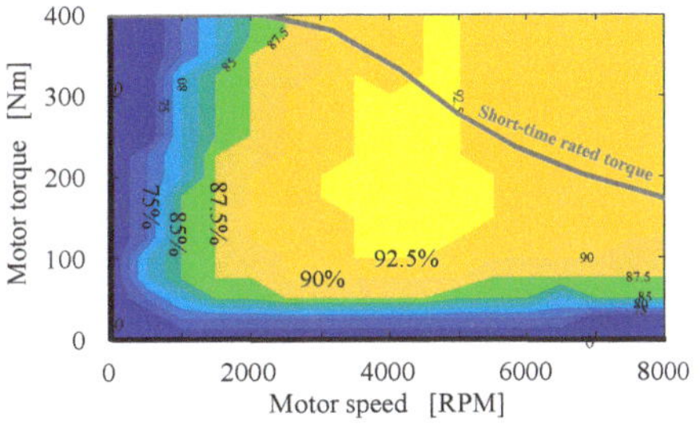

Figure 4 illustrates the simulator's speed change pattern optimization result. The following section summarizes the details of "with coasting control (Co)" and "without coasting control (W/O Co)".

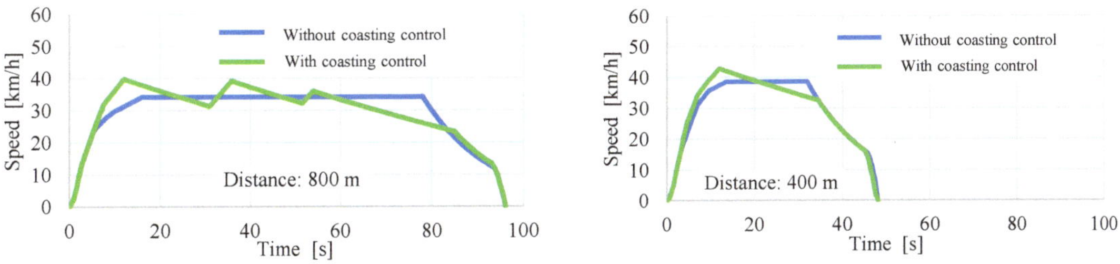

(**a**) (**b**)

Figure 4. Optimized speed change patterns in different settings: (**a**) distance: 800 m, time: 96 s, average speed: 30 km/h; (**b**) distance: 400 m, time: 48 s, average speed: 30 km/h.

3.2. Discussion on the Details of the Derived Electricity Consumption Optimization Speed Change Pattern

This section examines the results of the "with coasting control (inverter OFF coasting control)" and "without coasting control" settings, which are derived in the previous section. For detailed discussions, driving is divided into three parts: acceleration, cruising, and deceleration. Due to space constraints, we only present the discussion on the 800 m drive.

First, we consider the acceleration interval. Figure 5 illustrates the details of the acceleration interval in optimized speed change patterns. Both types of controls "should accelerate strongly" compared to the typical internal combustion engine vehicle's eco-driving acceleration pattern [19,20]. In particular, the vehicle starts near the maximum allowable acceleration (0.2 G) based on the optimization calculation, then eases slightly, but remains close to full acceleration. This strong acceleration can reduce the cruising speed under the situation of fixed driving distance and time, thereby reducing the energy that is required for acceleration and the air resistance loss of the entire driving trip. When performing similar acceleration, for an internal combustion engine vehicle, the engine must be revved high while the gear remains low, leading to poor fuel efficiency. However, the motor is resistant to load changes while maintaining good efficiency across a wide range of operating points. Therefore, strong acceleration is not a major issue in terms of electricity consumption. We can see this from the motor operating points of Figure 5, which demonstrate that good efficiency is maintained. For a diesel bus, if the bus "accelerates slowly" while leaving a bus stop, it may disrupt traffic flow and potentially cause accidents. Thus, there is a safety concern. However, with an electric bus, while passenger comfort is important, relatively strong acceleration to merge safely into the traffic does not cause a major issue in terms of electricity consumption.

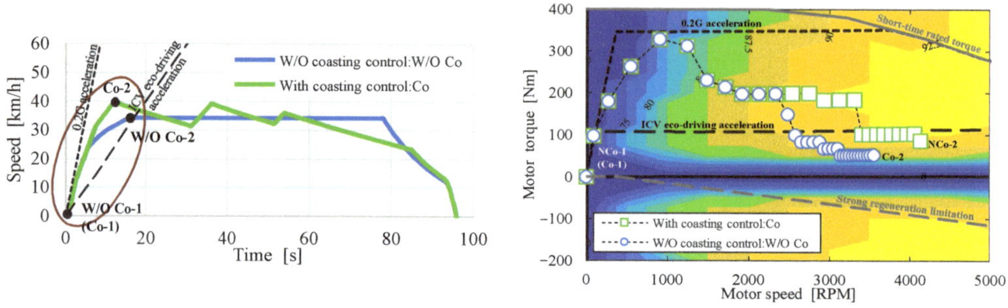

(a) (b)

Figure 5. Details of acceleration interval in optimized speed patterns (distance: 800 m): (**a**) speed—time profile; (**b**) motor torque—speed profile.

Next, we consider the cruising interval. Figure 6 shows the details of the cruising interval in optimized speed change patterns. "With coasting control" is "repetition of acceleration and coasting", while "without coasting control" is "constant speed driving", which is also recommended for heavy internal combustion engine vehicles as well. From the motor operating points of Figure 6, in some cases, a repetition of acceleration and coasting may be preferable to a constant speed of driving in the cruising interval (depending on the loss when the motor operating point is at 0 Nm). This conclusion is similar to the "coasting-powering operation" being recommended for trains.

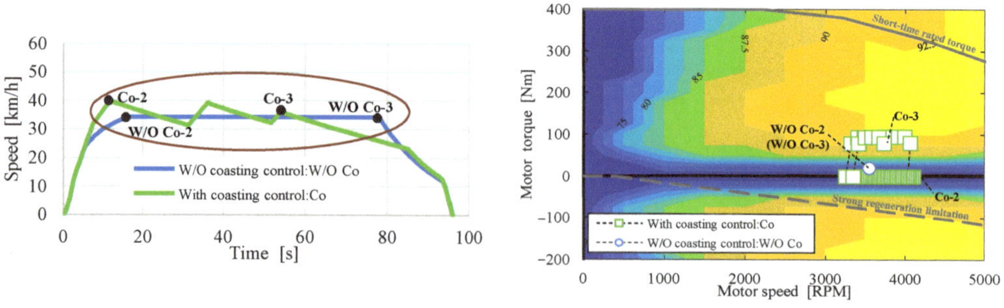

(a) (b)

Figure 6. Details of cruising interval in optimized speed patterns (distance: 800 m): (**a**) speed—time profile; (**b**) motor torque—speed profile.

Finally, considering the deceleration interval, Figure 7 shows the details of the deceleration interval in optimized speed change patterns. Both types of coasting controls were described as "deceleration while maintaining the maximum regeneration". To maximize regenerative energy recovery, this is a speed change along the vehicle-set regenerative braking line (the break line in the motor's operating points of Figure 7). Energy dissipation due to mechanical braking in the same interval can be prevented, thereby contributing substantially to improved efficiency. Note that when using "with coasting control", coasting deceleration has advantages over regenerative deceleration in energy saving and is therefore preferred. Afterwards, it is switched to regenerative deceleration for a stronger deceleration. After nearly reaching the minimum regenerative speed, it decelerates or stops using mechanical braking. This operation is comparable to that of a diesel bus.

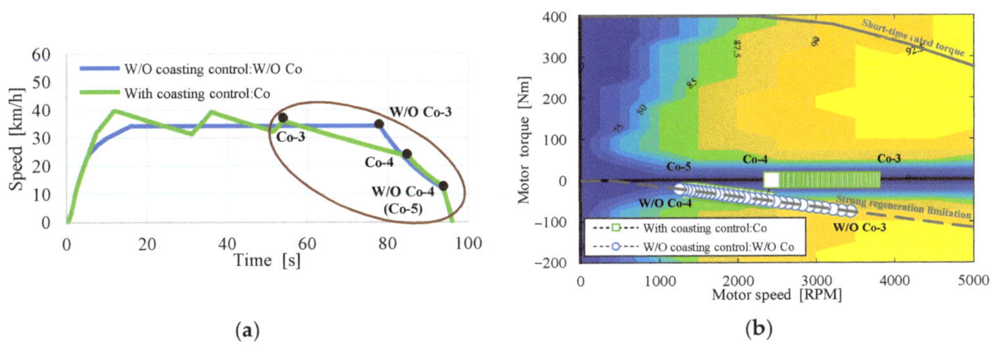

Figure 7. Details of deceleration interval in optimized speed patterns (distance: 800 m): (**a**) speed—time profile; (**b**) motor torque—speed profile.

3.3. Calculation of the Improvements in Electricity Consumption with the Derived Electricity Consumption Optimization Speed Change Pattern

In this section we compare the electricity consumption when the target vehicle, WEB-3A, is driven with the various electricity consumption optimization speed change patterns. Figure 8 summarizes the speed change patterns. We specifically used the electricity consumption during (a) the cruising zero style (constant acceleration interval and constant deceleration interval without cruising) as the reference and compared this value to the (b) ICV eco-driving speed change pattern for diesel buses and the optimization speed change pattern when the two types of coasting control mentioned above were used ((c) without coasting control and (d) with coasting control). The (b) ICV eco-driving is based on relevant reference studies [19,20]. The three internal combustion engine vehicle's eco-driving principles were considered as follows: (i) limiting acceleration: ICV eco-driving uses a smaller acceleration of approximately 0.06 G to limit the acceleration based on gentle acceleration and a quick shift up; (ii) constant speed cruising: ICV eco-driving uses cruise control to reduce unnecessary acceleration and deceleration and to maintain a constant speed while cruising; (iii) engine braking: ICV eco-driving simulates the engine braking of diesel buses by using a smaller deceleration when slowing down.

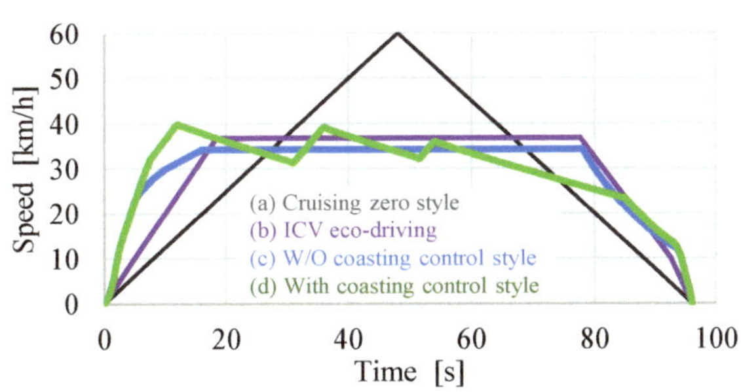

Figure 8. Various speed change patterns in different settings.

Table 3 compares the electricity consumption derived from the vehicle driving energy calculation simulator. We can quantitatively see that driving with the electricity consumption optimization speed change pattern derived in this study improves electricity consumption.

Table 3. Electricity consumption comparison of various speed change patterns in different settings.

	Electricity Consumption [kWh/km]	Electricity Consumption [%]
(a) Cruising zero style	0.408	(Benchmark)
(b) ICV eco-driving	0.382	−6.2%
(c) W/O coasting control style	0.370	−10.0%
(d) With coasting control style	0.318	−24.2%

At this point, the three strategies of EV eco-driving can be confirmed again as follows: acceleration, regenerative braking, and coasting. Firstly, acceleration: At the same average speed, a faster acceleration can reduce the maximum speed/cruise speed of a trip, thereby reducing the energy required for acceleration and the air resistance loss of the entire driving trip. Secondly, regenerative braking: Using regenerative braking as much as possible can greatly improve the energy efficiency of the deceleration interval (without coasting control), convert kinetic energy into electrical energy, and reduce the energy loss of mechanical braking. Thirdly, coasting: The energy efficiency of coasting is very high. Therefore, using coasting to drive when allowed can effectively improve the energy efficiency of the vehicle, for example, cruising by repetition of acceleration and coasting or decelerating by coasting.

4. Verification of Derived Speed Change Pattern Optimization Based on the Public Road Driving Test Data

In this chapter, we verify the validity of the speed change pattern optimization derived in the previous chapter based on the public road driving test data. The optimization resulted in the following order (without coasting control): "acceleration interval with acceleration strongly", "cruising interval with constant speed", and "deceleration interval with maintaining the maximum regeneration and mechanical braking". We compared the optimization result to the measured value for each interval.

4.1. Public Road Driving Test

Our research group conducted a 12-month driving test in Tonomachi, Kawasaki City, Japan, using the electric bus WEB-3A (December 2015 to November 2016). This test was conducted four times daily covering a distance of ~5.5 km one way. The vehicle route is shown in Figure 9, and an illustration of the changes in vehicle speed and elevation along the route is shown in Figure 10. The route includes a bridge and the slope changes around it; however, the remainder of the route is flat. In the following test, we extracted various data from the verification test for analysis. We excluded areas with a change in slope. There was no change in the number of passengers, because it was a trial operation.

The driving test was conducted in Kawasaki City, which is in the Tokyo metropolitan area. This area is highly developed, with a high road density and traffic congestion. There are many occassions for acceleration and deceleration when driving a car and few situations for long-term cruising. Therefore, strategies related to acceleration and deceleration are more applicable, while strategies related to cruising are less applicable. If the traffic is smooth and there are more situations for free cruising in a city or road scene, the applicability of the above results may change. To maintain consistency with the optimal settings and to avoid a decrease in generality caused by road slope characteristics, we chose this relatively flat urban road to verify the optimization results. The maximum speed allowed on this route is 60 km/h, but due to the influence of traffic signals and traffic congestion, there are more instances of acceleration and deceleration when starting and stopping, and about one-third of the time is spent in a stationary state, resulting in a slow average speed of only about 15 km/h.

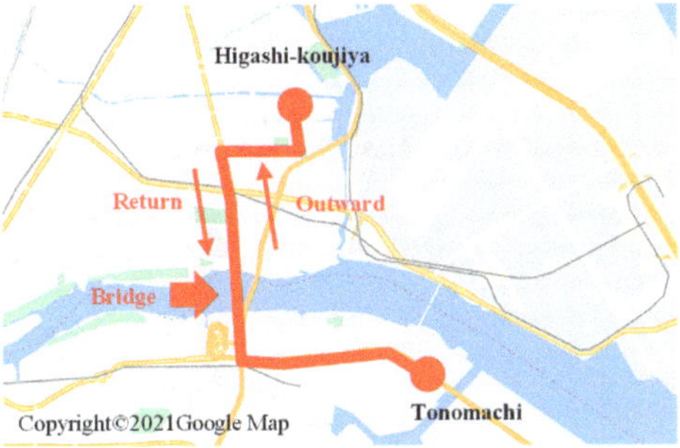

Figure 9. Route profile of Tonomachi/Higashi-koujiya shuttle route.

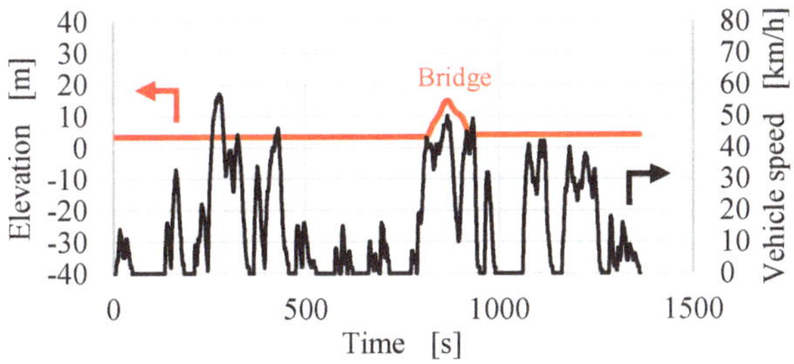

Figure 10. Running profile (from Tonomachi to Higashi-koujiya, 13 September 2016-2nd).

4.2. Verification of Derived Speed Change Pattern Optimization

4.2.1. Comparison of the Optimization Result and Measured Value in the Acceleration Interval

The optimization result was "acceleration interval with acceleration strongly". Figures 11 and 12 show the comparison with the measured value for the speed change pattern and motor operating point, respectively. The four types of values shown with a dotted line are the measured results (e.g., 0712_Trip55 is the 55th trip data from 12 July), the two types of optimization results are shown with a solid line (e.g., W/O Co means the optimization without coasting control), and the ICV eco-driving acceleration pattern is shown with break line. The most similar to the optimization results and ICV eco-driving acceleration pattern were extracted from the test data.

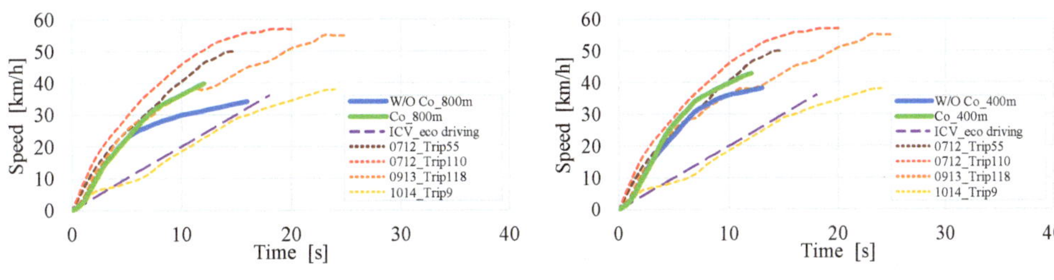

Figure 11. Speed—time profile at acceleration interval: (**a**) 800 m; (**b**) 400 m.

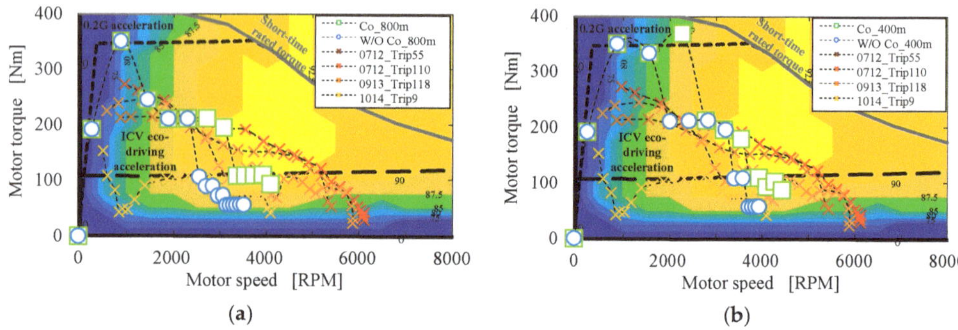

Figure 12. Motor torque—speed profile at acceleration interval: (**a**) 800 m; (**b**) 400 m.

Figures 13 and 14 compare the average motor efficiency (motor output/motor input) and the average vehicle efficiency (powertrain output/battery output). The figure shows 16 types of acceleration data, obtained on the same test day (12 July), as well as four different types of measured values to increase generality. The average efficiency was calculated from start to 30 km/h.

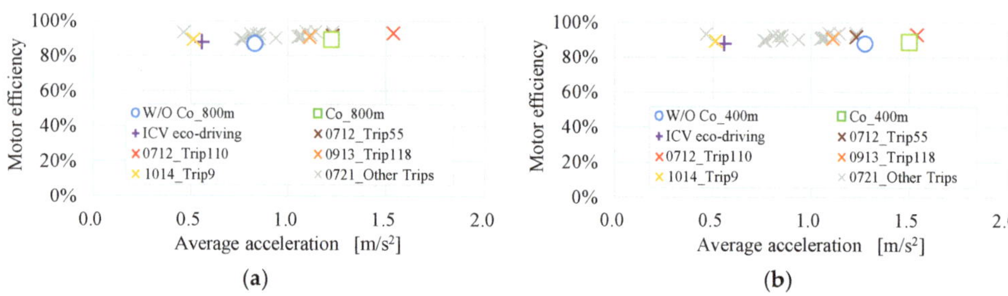

Figure 13. Relationship between motor efficiency and average acceleration: (**a**) 800 m; (**b**) 400 m.

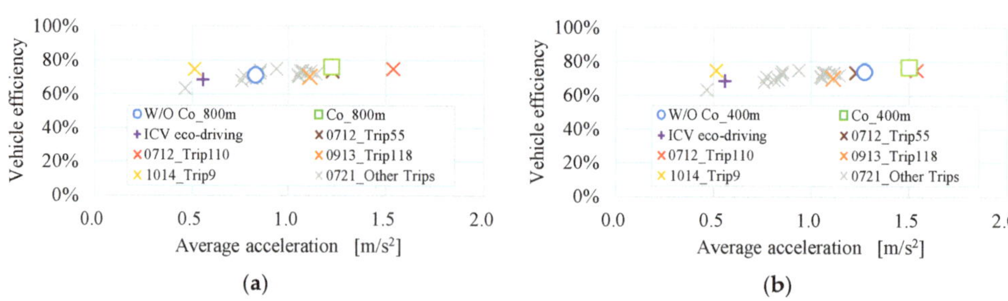

Figure 14. Relationship between vehicle efficiency and average acceleration: (**a**) 800 m; (**b**) 400 m.

The optimization result and the measured result were consistent. Specifically, efficiency remained rather constant regardless of acceleration, indicating that it is quite different from the property of internal combustion engine vehicles [19,20]. These results verify the previous optimization result: even if the electric vehicle performs strong acceleration, there will be no deterioration in efficiency.

4.2.2. Comparison of the Optimization Result and Measured Values in the Cruising Interval

The WEB-3A adopts the "without coasting control" setting, so the optimization result for this type of control was "cruising interval with constant speed". Figure 15 shows the comparison of electricity consumption and motor operating point with the measured and optimized values. The figures illustrate 14 types of data obtained on the same test day (October 14), when the speed change was within ±2 km/h, and the acceleration was within ±1 km/h/s. In Figure 15a, the solid line represents the theoretical electricity consumption of a vehicle driven at a constant speed. The optimization result without coasting control is consistent with both the theoretical consumption and measured consumption. Furthermore, the conclusion of the previous section, "acceleration interval with acceleration strongly", has the effect of bringing the vehicle speed in the subsequent cruising interval closer to the theoretical minimum electricity consumption (about 30 km/h); thus, it was a valid optimization result.

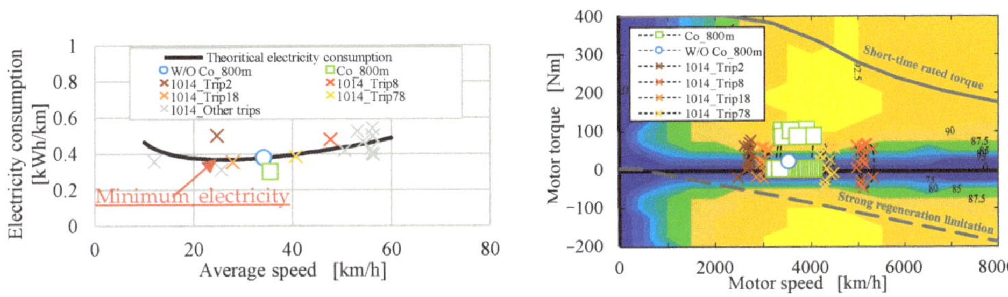

(a) (b)

Figure 15. Various comparisons of cruising interval: (**a**) relationship between electricity consumption and average speed; (**b**) motor torque—speed profile.

4.2.3. Comparison of the Optimization Result and Measured Value in the Deceleration Interval

The optimization result was "deceleration while maintaining the maximum regeneration". Here, we continue the comparison of "deceleration with maximum regenerative drive". Figures 16 and 17 show the comparison of the speed change pattern and motor operating point with the measured value, respectively. Figures 18 and 19 are comparisons of energy regeneration efficiency, with the former representing the average deceleration dependency and the latter representing the deceleration speed band notation. These are equivalent to the regenerative system efficiency (to the motor power generation unit) [32], which is derived by dividing the regenerative energy that was actually generated by the theoretically generatable regenerative energy. In order to broaden the scope, we collected 39 different types of deceleration data (other trips) in addition to the four measured values. Furthermore, for comparison, we included six different types of measured energy regeneration efficiency when using both regenerative and mechanical brakes. Overall, the optimization result and measured value were consistent, demonstrating the efficacy of "deceleration while maintaining the maximum regeneration" in electric buses. Additionally, the measured data showed that the energy regeneration efficiency (74–96% with a mean of 85%) improved significantly compared to using both regenerative and mechanical brakes (33–49% with a mean of 41%).

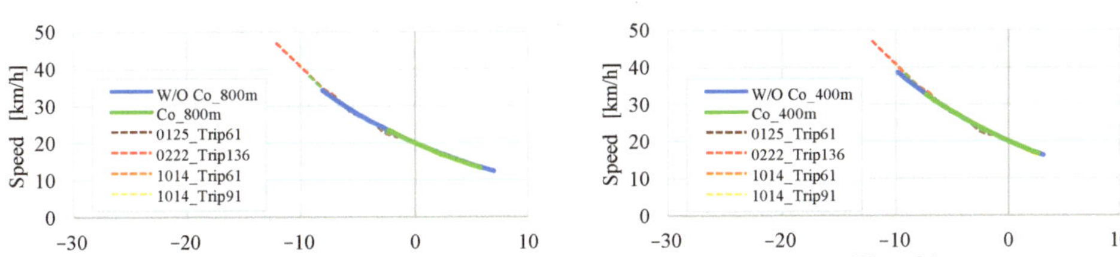

(a) (b)

Figure 16. Speed—time profile at deceleration interval: (**a**) 800 m; (**b**) 400 m.

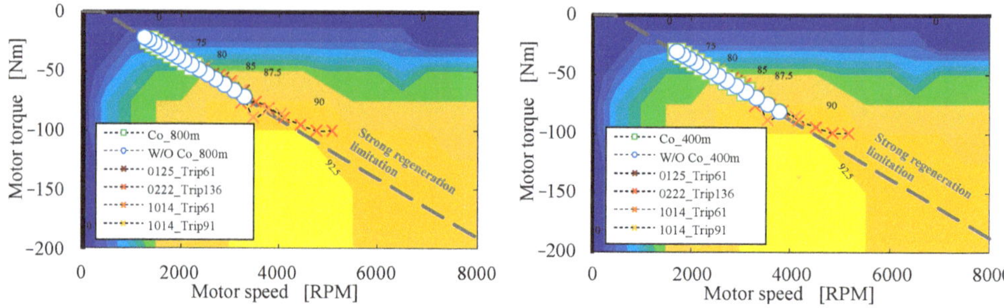

(a) (b)

Figure 17. Motor torque—speed profile at deceleration interval: (**a**) 800 m; (**b**) 400 m.

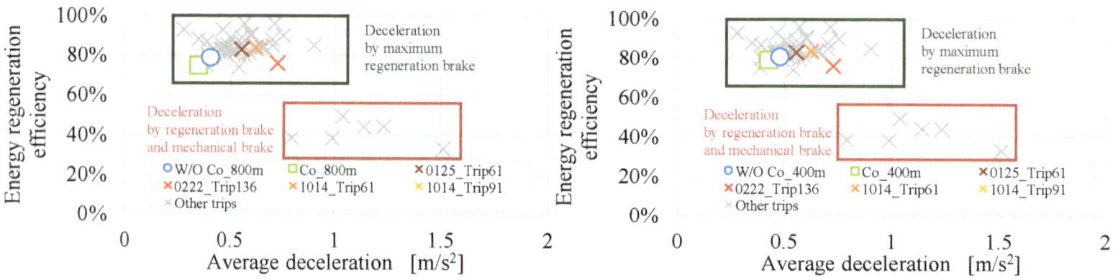

(a) (b)

Figure 18. Relationship between energy regeneration efficiency (up to the motor generator) and average deceleration: (**a**) 800 m; (**b**) 400 m.

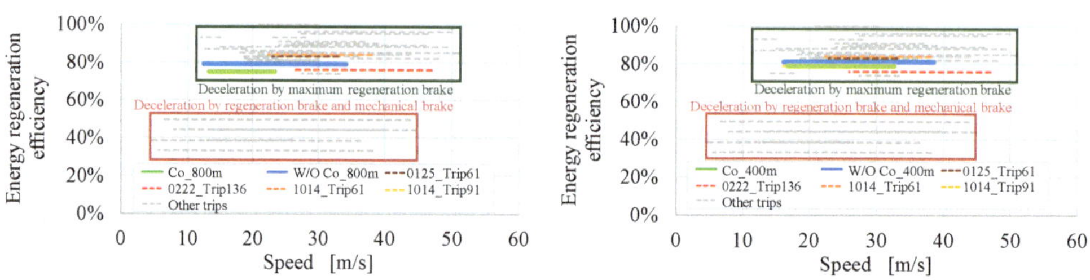

(a) (b)

Figure 19. Relationship between energy regeneration efficiency (up to the motor generator) and speed zone: (**a**) 800 m; (**b**) 400 m.

5. Conclusions

We report an electric vehicle driving energy calculation simulator with a speed change optimization function that is proposed in this study. We were able to derive a speed change pattern that optimizes electricity consumption while performing various types of coasting controls using the designed simulator.

Based on the optimization calculation with the simulator, the optimal speed change pattern (EV eco-driving) was derived for electric buses "without coasting control" and "with coasting control" (assume "inverter off coasting control"). When the target vehicle is driven in the EV eco-driving speed change pattern, according to our trial calculation, this method can improve the electricity consumption by about 10–20% under the same average speed.

To confirm the validity of the optimization results of the speed change pattern derived, mentioned above, we used the object vehicle's road driving test data. The optimization result is in the following order (without coasting control): "acceleration interval with acceleration strongly", "cruising interval with constant speed driving", and "deceleration while maintaining the maximum regeneration". We verified these results by comparing them to actual measured data, which are the speed change in each interval, and found that they were consistent.

Specifically, we examined the details of the "acceleration interval with acceleration strongly", which was significantly different from that of internal combustion engine vehicles, and confirmed with our measured data that the previous optimization result is valid: even if an electric bus performs strong acceleration, there will be no deterioration in efficiency. Internal combustion engines have large variations in fuel consumption during acceleration, but the properties of an electric bus, whose efficiency does not depend on the pattern of acceleration change, contributes to eliminating variations in electricity consumption during acceleration.

Finally, we summarized the three eco-driving strategies that are applicable to EVs and mentioned above and anticipated their expected application scenarios in the real world: no need to limit acceleration, use regenerative braking, and use coasting. They are, respectively, suitable for city roads with frequent starts and stops and intercity roads (or highways) that are mainly for cruising.

No need to limit acceleration: EVs and ICVs have significant differences in their powertrain systems, so eco-driving methods based on ICVs may not be applicable to EVs. Limiting acceleration based on gentle acceleration and quick shifting up may improve the efficiency of the internal combustion engine but has no effect on the efficiency of the motor/inverter. At the same average speed, a faster acceleration can reduce the maximum speed/cruise speed of a trip, thereby reducing the energy that is required for acceleration and the air resistance loss of the entire driving trip. Therefore, from the perspective of eco-driving, there is no need to consider acceleration limits when driving EVs.

Regenerative braking: Using regenerative braking as much as possible can greatly improve the energy efficiency of the deceleration interval, convert kinetic energy into electrical energy, and reduce the energy loss of mechanical braking. Actively using regenerative braking can convert most of the deceleration kinetic energy into electrical energy for future driving. The mean energy regeneration efficiency is 85% when only using regenerative braking for deceleration, while the mean energy regeneration efficiency is 41% when using both regenerative braking and mechanical brakes. If regenerative braking is not used at all, all of this energy will be converted into the thermal losses of the mechanical brakes. When the two strategies mentioned above are applied to city road driving with frequent starts and stops, the effect is particularly significant, with an expected improvement of about 10% in electricity consumption.

Coasting: Coasting has already been widely used as a basic eco-driving method in railway transportation. The energy efficiency of coasting is very high. Therefore, using coasting to drive when allowed can effectively improve the energy efficiency of the vehicle, for example, by cruising by repetition of acceleration and coasting. Additionally, from the perspective of eco-driving, when road traffic conditions permit, coasting should be the first choice for deceleration, followed by regenerative braking. This method is particularly effective when driving on city-to-city roads or highways with fewer vehicles, with an expected improvement of about 10% in electricity consumption.

Author Contributions: Conceptualization, all authors; methodology, Y.F., W.-h.Y. and Y.K.; software, Y.F. and W.-h.Y.; validation, Y.F. and Y.K.; formal analysis, Y.F., W.-h.Y. and Y.K.; investigation, Y.F., W.-h.Y. and Y.K.; resources, Y.K.; data curation, Y.F. and Y.K.; writing—original draft preparation, Y.F.; writing—review and editing, Y.K., T.I., S.U. and M.K.; visualization, Y.F. and W.-h.Y.; supervision, Y.K.; project administration, Y.K.; funding acquisition, Y.K., T.I., S.U. and M.K. All authors have read and agreed to the published version of the manuscript.

Funding: This research received no external funding.

Data Availability Statement: The data presented in this study are available on request from the corresponding author.

Conflicts of Interest: The authors declare no conflicts of interest.

References

1. Ho, S.H.; Wong, Y.D.; Chang, V.W.C. What Can Eco-Driving Do for Sustainable Road Transport? Perspectives from a City (Singapore) Eco-Driving Programme. *Sustain. Cities Soc.* **2015**, *14*, 82–88. [CrossRef]
2. Huang, Y.; Ng, E.C.Y.; Zhou, J.L.; Surawski, N.C.; Chan, E.F.C.; Hong, G. Eco-driving technology for sustainable road transport: A review, Renew. *Sustain. Energy Rev.* **2018**, *93*, 596–609. [CrossRef]
3. Xu, Y.; Li, H.; Liu, H.; Rodgers, M.O.; Guensler, R.L. Eco-Driving for Transit: An Effective Strategy to Conserve Fuel and Emissions. *Appl. Energy* **2017**, *194*, 784–797. [CrossRef]
4. Chen, Z.; Xiong, S.; Chen, Q.; Zhang, Y.; Yu, J.; Jiang, J.; Wu, C. Eco-Driving: A Scientometric and Bibliometric Analysis. *IEEE Trans. Intell. Transp. Syst.* **2022**, *23*, 22716–22736. [CrossRef]
5. He, M.; Lin, T.; Wu, X.; Luo, J.; Peng, Y. A Systematic Literature Review of Reverse Logistics of End-of-Life Vehicles: Bibliometric Analysis and Research Trend. *Energies* **2020**, *13*, 5586. [CrossRef]
6. Barkenbus, J.N. Eco-Driving: An Overlooked Climate Change Initiative. *Energy Policy* **2010**, *38*, 762–769. [CrossRef]
7. Wang, Z.; Wu, G.; Barth, M.J. A Review on Cooperative Adaptive Cruise Control (CACC) Systems: Architectures, Controls, and Applications. In Proceedings of the 2018 21st International Conference on Intelligent Transportation Systems (ITSC), Maui, HI, USA, 4–7 November 2018; pp. 2884–2891. [CrossRef]
8. Caban, J.; Vrábel, J.; Šarkan, B.; Ignaciuk, P. About Eco-Driving, Genesis, Challenges and Benefits, Application Possibilities. *Transp. Res. Procedia* **2019**, *40*, 1281–1288. [CrossRef]
9. Caban, J. A Study of Eco-Driving Possibilities in Passenger Cars Used in Urban Traffic. *Arch. Automot. Eng.* **2021**, *91*, 37–48. [CrossRef]
10. Zhou, M.; Jin, H.; Wang, W. A Review of Vehicle Fuel Consumption Models to Evaluate Eco-Driving and Eco-Routing. *Transp. Res. Part D Transp. Environ.* **2016**, *49*, 203–218. [CrossRef]
11. Alam, M.S.; McNabola, A. A Critical Review and Assessment of Eco-Driving Policy & Technology: Benefits & Limitations. *Transp. Policy* **2014**, *35*, 42–49. [CrossRef]
12. Mongeon, P.; Paul-Hus, A. The Journal Coverage of Web of Science and Scopus: A Comparative Analysis. *Scientometrics* **2016**, *106*, 213–228. [CrossRef]
13. Gandia, R.M.; Antonialli, F.; Cavazza, B.H.; Neto, A.M.; de Lima, D.A.; Sugano, J.Y.; Nicolai, I.; Zambalde, A.L. Autonomous Vehicles: Scientometric and Bibliometric Review. *Transp. Rev.* **2019**, *39*, 9–28. [CrossRef]
14. Mata-Carballeira, O.; del Campo, I.; Asua, E. An Eco-Driving Approach for Ride Comfort Improvement. *IET Intell. Transp. Syst.* **2022**, *16*, 186–205. [CrossRef]
15. Čulík, K.; Štefancová, V.; Hrudkay, K.; Morgoš, J. Interior Heating and Its Influence on Electric Bus Consumption. *Energies* **2021**, *14*, 8346. [CrossRef]
16. Linkov, V.; Perůtka, J.; Zaoral, A.; Tučka, P.; Ťápal, A.; Zůvala, R.; Řezáč, P. The Speed Behavior of Czech Professional Drivers According to Ordinary vs. Variable Speed Limit Signs: An on-Road and Driving Simulation-Based Comparison. *Perner's Contacts* **2019**, *14*, 35–43.
17. The Golden Rules of Ecodriving. Available online: https://www.ecodrive.org/en/what_is_ecodriving/the_golden_rules_of_ecodriving/ (accessed on 20 December 2023).
18. 10 Recommendations for Eco-Driving. Available online: https://www.mlit.go.jp/report/press/sogo10_hh_000171.html (accessed on 22 December 2023).
19. Allen, R.W.; Rosenthal, T.J.; Park, G. An Overview of Eco-Driving Theory, Capability Evaluation, and Training Applications. *Sensors* **2021**, *21*, 6547.
20. Fleming, J.; Yan, X.; Lot, R. Incorporating Driver Preferences into Eco-Driving Assistance Systems Using Optimal Control. *IEEE Trans. Intell. Transp. Syst.* **2021**, *22*, 2913–2922. [CrossRef]
21. Mensing, F.; Trigui, R.; Bideaux, E. Vehicle trajectory optimization for application in ECO-driving. In Proceedings of the 2011 IEEE Vehicle Power and Propulsion Conference, Chicago, IL, USA, 6–9 September 2011; pp. 9–14. [CrossRef]
22. Sundström, O.; Guzzella, L. A generic dynamic programming Matlab function. In Proceedings of the 2009 IEEE Control Applications, (CCA) & Intelligent Control, (ISIC), St. Petersburg, Russia, 8–10 July 2009; pp. 1625–1630. [CrossRef]
23. Koch, A.; Bürchner, T.; Herrmann, T.; Lienkamp, M. Eco-driving for different electric powertrain topologies considering motor efficiency. *World Electr. Veh. J.* **2021**, *12*, 6. [CrossRef]
24. Maamria, D.; Gillet, K.; Colin, G.; Chamaillard, Y.; Nouillant, C. On the use of Dynamic Programming in eco-driving cycle computation for electric vehicles. In Proceedings of the 2016 IEEE Conference on Control Applications (CCA), Buenos Aires, Argentina, 19–22 September 2016; pp. 1288–1293. [CrossRef]
25. International Energy Agency. Global EV Outlook 2023. Available online: https://www.iea.org/data-and-statistics/data-product/global-ev-outlook-2023 (accessed on 20 December 2023).

26. Tokyo Metropolitan Bureau of Transportation. 2019 Transport Statistics Annual Report. Available online: https://www.kotsu.metro.tokyo.jp/about/information (accessed on 16 April 2022).
27. Koyano, S.; Okamura, H.; Miyagi, M.; Kokuryo, K. *Reduction Effect of CO_2 by Idling Stop in the Case of Route Buses*; Tokyo Metropolitan Research Institute for Environmental Protection: Tokyo, Japan, 2009; pp. 76–85.
28. Naoki, M.; Ikuma, S.; Tatsuya, S.; Keigo, A.; Yohei, N. e-Pedal system which provides simple driving by capacious throttle pedal controllability with electric brake control. In Proceedings of the JSAE Congress, Yokohama, Japan, 24–26 May 2017; pp. 217–222.
29. Volkswagen: Brake or Coast? The ID.4's Intelligent Energy Recuperation Concept. Available online: https://motor-fan.jp/article/photo/100003361 (accessed on 21 January 2021).
30. Kawai, H.; Sunohara, T.; Tasaka, Y.; Fukasawa, S. Permanent-Magnet Synchronous Motor Propulsion System for Tokyo Metro Ginza Line Trains. *Toshiba Rev.* **2008**, *63*, 45–49.
31. Hisanori, Y.; Sho, K.; Yoshinori, Y.; Kota, T. Development and Future Prospects of PMSM Drive Control Technologies for Railway Rolling Stock. *Mitubishi Denki Giho* **2016**, *90*, 513–516.
32. Fang, Y.; Xu, T.; Yang, W.; Ihara, Y.; Kamiya, Y. Detailed Analysis of Regenerative Energy When the Electric Bus Driving on Expressways. In Proceedings of the FISITA 2021 World Automotive Congress, Virtual, 14–16 September 2021; pp. 1–10.

World Electric Vehicle Journal

MDPI

Article

A Novel Approach for a Predictive Online ECMS Applied in Electrified Vehicles Using Real Driving Data

Felix Deufel *, Malte Freund and Frank Gauterin

Institute of Vehicle System Technology, Karlsruhe Institute of Technology, 76131 Karlsruhe, Germany; frank.gauterin@kit.edu (F.G.)
* Correspondence: felix.deufel@kit.edu

Abstract: To increase the efficiency of electrified vehicles, many energy management strategies (driving strategies) have been proposed. These include both offline optimization techniques to identify a system's theoretical optimum and online optimization techniques created for onboard use in the vehicle. In the field of online optimization, predictive approaches can achieve additional savings. However, predictions are challenging, and robust usability in all driving situations of the vehicle is not guaranteed. In this study, a new approach for a predictive energy management strategy is presented. It is demonstrated how this so-called predictive Online Equivalent Consumption Minimization Strategy (ECMS) can achieve additional fuel savings compared to a non-predictive Online ECMS by predicting recuperation events using map data. As long as the route is known, map data are available, and the current position of the global navigation satellite system (GNSS) is given, the predictive Online ECMS can be applied. If these requirements are not met, the non-predictive basic implementation can still be used to ensure robust functionality. The methodology is investigated using a backward simulation model of a D-segment vehicle powered by a 48 V hybrid electric system in a P2 topology. A dataset including real driving cycles including map data from Open Street Map (OSM) is used. However, the investigations are limited to the consideration of traffic signal (TS) positions on the upcoming route. Simulation results focus on the interaction between the energy management strategy (EMS) and usable battery energy. More than 1 % average saving potentials compared to a non-predictive implementation are shown. The highest saving potentials are found with a usable battery energy of 100 Wh.

Keywords: electrified powertrains; 48 V system; equivalent consumption minimization strategy (ECMS); model predictive control (MPC); li-ion battery; global navigation satellite system (GNSS); real driving cycles

Citation: Deufel, F.; Freund, M.; Gauterin, F. A Novel Approach for a Predictive Online ECMS Applied in Electrified Vehicles Using Real Driving Data. *World Electr. Veh. J.* **2023**, *14*, 353. https://doi.org/10.3390/wevj14120353

Academic Editors: Joeri Van Mierlo and Genevieve Cullen

Received: 29 September 2023
Revised: 5 December 2023
Accepted: 14 December 2023
Published: 18 December 2023

1. Introduction

Due to emission regulations and an increase in environmental consciousness in general, a broad variety of alternative drive systems have been developed. These include 48 V hybrid electric vehicles (HEVs), which have the benefit of decreasing CO_2 emissions at moderate system expense, especially for inner-city driving. A 48 V system is described by component dimensioning, topology, and an energy management strategy (EMS) [1]. The EMS has to guarantee a robust operation in various driving situations. An overview of the most common methods provided within [1–8] shows that EMS development has been extensively researched over last years. In this paper, a novel approach for a predictive Online ECMS is presented using real driving cycles. It is demonstrated how, in the case of a known journey and the availability of map data, a predictive Online ECMS is established by using the present Global Navigation Satellite System (GNSS) position. This is the case for instance, when manually entering a route into a navigation system or returning to a previously traveled path which is identified by intelligent algorithms. It is shown how a predictive Online ECMS can achieve additional fuel savings compared to a non-predictive

Online ECMS by predicting recuperation events due to traffic signals (TS). The simulation results focus on the interaction between the EMS and battery.

2. Related Work

The EMS can be subdivided in multiple ways. They can be categorized into rule-based, optimization-based, and learning-based techniques. Mixed forms also exist. Furthermore, offline and online methods can be distinguished. Offline strategies are defined by the need for prior knowledge of the whole driving profile. With this, a specific hybrid design is described, for instance, in terms of possible fuel consumption savings for a certain cycle. The global optimum is determined for benchmark analysis. Online strategies require only limited prior knowledge of the upcoming driving path. The ECMS that is investigated in this work can be assigned to optimization-based EMS concepts. Depending on the particular implementation, an ECMS is associated with either offline or online techniques. Using an Offline ECMS, the global optimum for time-invariant systems is found due to the equivalence to Pontryagin's Minimum Principle (PMP) [9,10], where a so-called equivalence factor λ is found iteratively to solve the optimization problem [11,12]. This Offline ECMS is frequently utilized to find the global optimum in offline applications, especially due to the low computing effort [13]. The 2D-ECMS has been created to investigate topologies with two traction motors [14]. For an Online ECMS, which was first published by Paganelli et al. [15], the idea of a Stae Of Charge (SOC)-dependent control of the equivalence factor λ was effectively implemented [16–26]. In addition, there are predictive Online ECMS techniques where predictions are essential for the ECMS's fundamental operation [18,27]. However, it has also been demonstrated that the introduction of predictive information can enhance non-predictive Online ECMS implementation. The following publications should be mentioned:

- In [28], step functions are used for adjusting λ by taking into account the future energy demand. A 10% improvement compared to a non-predictive Online ECMS solution was reached.
- In [29], optimal recuperation is realized by predictive charging and discharging of the battery. A 6% improvement compared to a non-predictive Online ECMS solution is achieved.
- In [30], velocity prediction using a Convolutional Neural Network (CNN) for optimal λ determination is realized. A 0.2% to 0.5% improvement compared to the non-predictive Online ECMS is presented.
- In [31], velocity prediction is used to determine SOC nodes. A 9.7% improvement compared to the non-predictive Online ECMS solution is given.
- In [32], λ adaptation is realized considering future energy demand with a dynamic prediction horizon. An improvement between 0.3% and 4% compared to the non-predictive Online ECMS is achieved.
- In [33], velocity prediction at intersections considering traffic signal (TS) state and traffic flow leads to an improvement of 0–2% compared to the non-predictive Online ECMS.

The presented prediction approaches show a wide range of possibilities for the development of predictive driving strategies. When comparing the results, it should be noted that different boundary conditions, vehicle models, and types of ECMS were used in each publication. The approach of [28], for example, has only been tested on previously known very hilly routes and is therefore only useful in very specific scenarios. Therefore, a comparison and evaluation of the results is of limited value. However, for the authors of the paper, these investigations form the basis for developing their own predictive approach. In [34], an approach utilizing the recuperation potential has already been published. It was shown that noticeable CO_2 reduction potentials occur, in particular, with limited battery capacity. Detailed investigations regarding predicting torque for predictive EMS are presented in [35,36]. However, it was shown that the prediction of the future torque is very difficult and often only possible with a certain degree of uncertainty. Therefore, this paper presents a novel approach for a predictive Online ECMS considering recuperation potentials using map data without the need for torque predictions. Additionally, a comparison to a non-predictive Online ECMS is provided.

3. Modeling

In vehicle simulation, forward and backward simulation can be distinguished. Forward simulation models are based on the physical causality of the system by comparing the target velocity with the actual vehicle velocity using a driver model. A velocity can then be calculated for each time step based on the acceleration brought on by the control input of the driver model. In contrast, a backward simulation model presupposes that the vehicle adheres to a predetermined profile of acceleration and velocity. Therefore, no driver model is necessary [4]. The verified backward calculation model of a 48 V HEV (P2 topology, see Figure 1) with an Offline ECMS and an iteratively calculated λ from the work of [13] is applied.

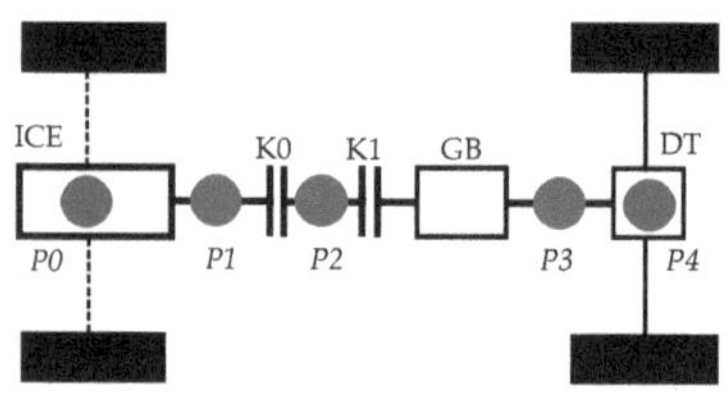

Figure 1. Topologies of HEVs in parallel configuration from [34].

In this model, torque is calculated from the longitudinal dynamics of the vehicle. Hereby, the wheel radius as well as the transmission ratios of the vehicle are taken into account. The correlations from longitudinal dynamics are shown below. The different parameters and their corresponding units are listed in Table 1.

$$F_{Wheel} = F_{air} + F_{roll} + F_{acc} + F_{slp} \quad (1)$$

$$F_{air} = c_w \cdot A \cdot \frac{\rho}{2} \cdot v^2 \quad (2)$$

$$F_{roll} = m \cdot g \cdot cos\alpha \cdot f_R \quad (3)$$

$$F_{acc} = m \cdot a \quad (4)$$

$$F_{slp} = m \cdot g \cdot sin\alpha \quad (5)$$

Table 1. Parameters and units of the driving resistances.

Drag Coefficient	c_w	0.3
Projected Frontal Area	A	2.5 m^2
Air Density	ρ	1.2 kg/m^3
Vehicle Mass	m	1600 kg
Gravitational Acceleration	g	9.81 m/s^2
Rolling Resistance Coefficient	f_R	0.012

An internal combustion engine (ICE), electric motor (EM), and gearbox (GB) are modeled using stationary maps. The 48 V battery is represented by a simple inner resistance model. Equations (6) and (7) are used to compute the battery voltage under load U_{bat} and the corresponding battery current I_{bat}. Therefore, the battery power P_{em}, the battery losses $P_{em,loss}$, and the power from auxiliary consumers P_{aux} are considered. Moreover, the open-circuit voltage U_{OCV} and the inner resistance R_i are required. In addition, as a measure of energy deviation from the starting conditions, an energy deviation dE from reference SOC is calculated (Equation (8)). It is used as a criterion for a neutral energy balance [34]

$$I_{bat} = \frac{P_{em} + P_{em,loss} + P_{aux}}{U_{bat}} \tag{6}$$

$$U_{bat} = U_{OCV}(SOC) - R_i(SOC) \cdot I_{bat} \tag{7}$$

$$dE = \int U_{bat} I_{bat} dt \tag{8}$$

The battery is of a nickel–mangan–cobalt/graphite cell type. R_i and U_{OCV} are calculated using SOC-specific component data. However, for simplification a large 48 V battery (>10 kWh) with constant SOC characteristics (SOC = 70%) is used within the investigations. Other effects, such as degradation of the battery and its impact on CO_2 emissions, are neglected [13]. The recuperated energy is determined using a simplified logic considering the limits of the electrical components and the application of the mechanical brake.

The studies are based on real driving cycles including four different drivers. These real driving cycles were already used and presented in [35,36]. Hereby, relevant map data from Open Street Map (OSM) was matched with the original GNSS tracks according to Figures 2 and 3. For detailed information on the preprocessing of the driving data, please refer to [36].

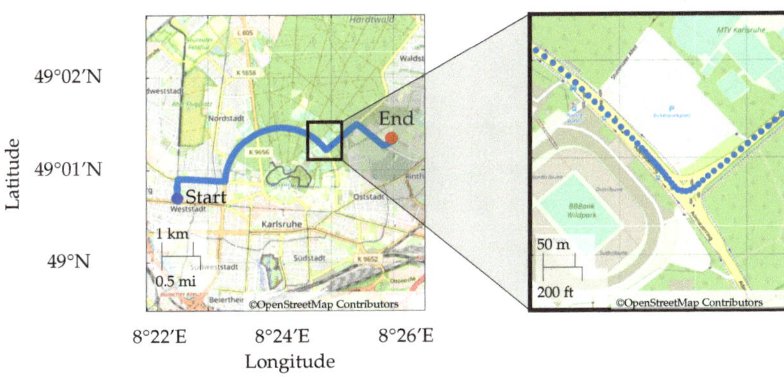

Figure 2. Visualization of available driving data. From [36].

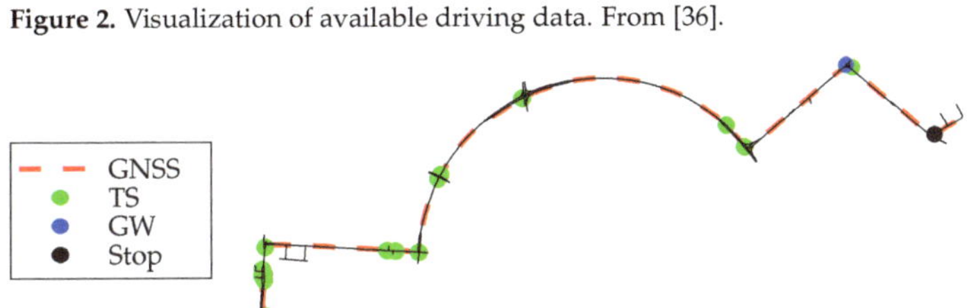

Figure 3. Visualisation of identified Open Street Map (OSM) data including GNSS track, traffic signal (TS), give way (GW), and stop. From [36].

This publication is limited to the cycles of Driver 1, which covers 63 cycles of city driving, country road driving, and highway driving of a total duration of 30 h and almost 3000 km (for more information, see [35]). For the design of a non-predictive Online ECMS, three representative cycles are selected for each road type. These nine cycles should represent real operation as good as possible. The most important characteristics are shown below (Table 2).

In Figure 4, the traffic signal (TS) density is shown for the 63 cycles. Cycles marked in dark gray are selected for the exemplary application of the newly developed predictive Online ECMS approach. The chosen cycles are characterized by at least one TS per km.

Table 2. Real driving cycles to parametrize non-predictive Online ECMS.

Road Type	Avg. vel. in km/h	Max. vel. in km/h	Dist. in km	Dur. in h	Stand-still in %
City	28	69	6	0.2	26
	19	60	4	0.2	39
	25	62	11	0.4	21
Country Road	73	110	39	0.5	1
	57	90	17	0.3	6
	67	118	41	0.6	5
Highway	108	168	164	1.5	2
	116	189	162	1.4	1
	101	176	74	0.7	6

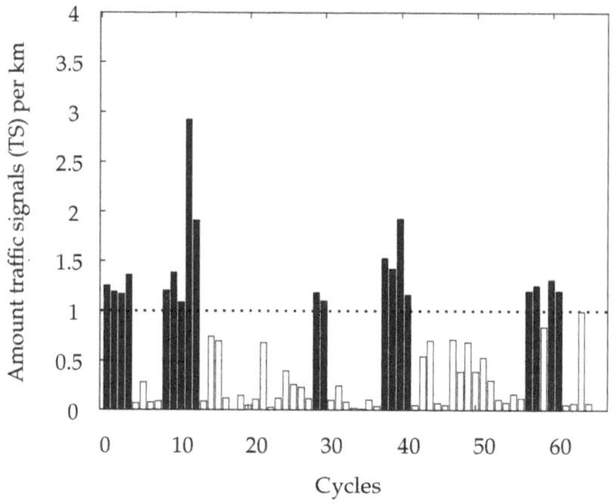

Figure 4. Overview of the 63 driving cycles. As the investigations are limited to the consideration of TS positions, TS per km are shown for each cycle. Cycles marked in dark gray are selected for the exemplary application of the newly developed predictive Online ECMS approach. These cycles are characterized by at least one TS per km.

4. Methodology

In the concept of an ECMS, an equivalent fuel consumption is calculated taking into account the fuel's lower heating value, Q_{lhv} and an equivalence factor λ to convert battery power into fuel power. Using the equivalent fuel consumption a cost function J is stated, where the optimization problem P is written as follows [13]:

$$P : \min_{u} \int J(u,x)dt \tag{9}$$

$$J(u,x) = \dot{m}_{fuel} + \lambda \frac{P_{bat}}{Q_{lhv}} \tag{10}$$

The local constrains are given as follows:

$$SOC_{min} \leq SOC(t) \leq SOC_{max} \tag{11}$$

$$P_{bat,min} \leq P_{bat}(t) \leq P_{bat,max} \tag{12}$$

$$T_{ICE,min} \leq T_{ICE}(t) \leq T_{ICE,max} \tag{13}$$

$$T_{EM,min} \le T_{EM}(t) \le T_{EM,max} \quad (14)$$

$$n_{ICE,min} \le n_{ICE}(t) \le n_{ICE,max} \quad (15)$$

$$n_{EM,min} \le n_{EM}(t) \le n_{EM,max} \quad (16)$$

The presented inequalities represent the SOC-limits and maximum battery power. Also, the limitations of both torque and speed from ICE and EM are considered. At each time step, the ideal torque split (control variable in the optimization problem) is determined by minimizing P. As a state variable the SOC is used. In an Offline ECMS, a constant λ is found iteratively for time-invariant systems. For an online-capable implementation of the ECMS, the idea of a SOC-dependent control of the equivalence factor λ was implemented in several studies [16–26]. In this work, an average equivalence factor $\lambda_{Offline,avg}$ is used for the Online ECMS. However, this does not guarantee charge-sustaining (CS) behavior in online operations: depending on the cycle, the SOC trajectories result in an excessive charging or discharging of the battery. Therefore, a penalty term is added. According to $dSOC$, (difference between the real SOC and the reference SOC), the value of the energy (λ) is either raised or lowered. As concluded in [25,37], the trigonometric penalty function is better than a proportional penalty function: it allows tiny deviations from the reference SOC but strongly penalizes significant deviations. Therefore, the penalty term consists of the penalty factor kp_{SOC} multiplied by the cubic derivation of SOC $dSOC^3$ (see Equation (17)). In terms of CS operation, the deviation of the battery's energy content at the end of the cycle is limited to a specific value. These presumptions are used to establish the proper kp_{SOC} for the non-predictive Online ECMS.

$$\lambda(t) = \lambda_{Offline,avg} - kp_{SOC} \cdot dSOC(t)^3 \quad (17)$$

In this paper, a novel approach for a predictive Online ECMS that considers map data to achieve saving potentials compared to the non-predictive Online ECMS implementation is presented. The investigations are limited to the consideration of TS positions on the upcoming route. The appearance of a TS within the upcoming horizon (represented by $flag_{TS}$) has a direct impact on λ using an additional parameter kp_{TS}:

$$\lambda(t) = \lambda_{Offline,avg} - kp_{SOC} \cdot dSOC(t)^3 - kp_{TS} \cdot flag_{TS} \quad (18)$$

A summary of the applied methodology is given by Figure 5.

Figure 5. Applied methodology for predictive Online ECMS.

5. Results

In the investigations fuel consumption is minimized, whereby there is a proportional relationship between fuel consumption and CO_2 emissions. The CO_2 values presented in this work are calculated with the relation 1 l/100 km = 23.2 gCO_2/km. First, the Offline ECMS is used to iteratively determine the optimum $\lambda_{Offline}$ value for each of the nine cycles selected (Table 2). An overview is given in Table 3. $\lambda_{Offline}$ ranges from 2.55 to 2.88.

The lowest λ values occur during city driving. The highest lambda values, on the other hand, occur on highways.

Table 3. Results from Offline ECMS for real driving cycles from Table 2.

Road Type	$\lambda_{Offline}$	CO_2 (g/km)
City	2.61 2.55 2.61	134.16 139.87 144.42
Country Road	2.74 2.70 2.64	124.69 122.31 172.21
Highway	2.81 2.75 2.88	148.40 168.92 169.08

The non-predictive Online ECMS is parametrized according to [34]. Cycles with high TS density are typically city driving cycles. Therfore, a $\lambda_{Offline,avg,city}$ of 2.60 is chosen (Table 4). Parameter studies, which will not be discussed in detail, result in a kp_{SOC} of 3.57 to achieve charge-sustaining (CS) operation. For further information the reader is referred to [34]. A battery-specific parametrization of the non-predictive Online ECMS is waived in this publication.

Table 4. Final parametrization for non-predictive Online ECMS.

$\lambda_{Offline,avg,city}$	2.60
kp_{SOC}	3.57

In the next step, a predictive Online ECMS is to be parametrized to show additional saving potentials for routes with a high density of TS (Figure 4). In contrast to [34], the route itself and the Global Navigation Satellite System (GNSS) position are assumed to be known for this predictive Online ECMS. It is also expected that the appropriate map data are available. Both parameters kp_{TS} and horizon length $t_{horizon}$ have to be specified. The investigations will be carried out for different battery sizes. Parameter ranges to identify the best parametrization of kp_{TS} and $t_{horizon}$ are given in Table 5.

Table 5. Ranges to identify optimal parameters kp_{TS} and horizon length $t_{horizon}$ of the predictive Online ECMS for a usable battery energy of 25 Wh, 50 Wh, 75 Wh, 100 Wh, 200 Wh, 300 Wh, 400 Wh, 500 Wh and 1000 Wh.

	Min	Max
kp_{TS}	0	5
$t_{horizon}$ in s	5 s	100 s

In Figure 6, CO_2 reduction potentials in % over kp_{TS} for different $t_{horizon}$ in the case of a usable battery energy of 25 Wh are given for the selected cycles with high TS density from Figure 4. For formatting reasons, the plots are restricted to 16 out of 19 cycles. For each cycle, there exists an individual kp_{TS} with a corresponding $t_{horizon}$ which leads to the best results. It can also be seen that above a certain value of kp_{TS}, there is no further influence on CO_2. To ensure robust applicability, a parametrization for the overall largest CO_2 savings potential for each battery size can be determined based on these investigations.

A behavior similar to that shown in Figure 6 is seen for a usable battery energy of 100 Wh in Figure 7.

In Figure 8, the velocity, the presence of a TS, and the SOC trajectories are presented over time for an exemplary cycle with 100 Wh usable battery energy. This includes both the non-predictive Online ECMS and the predictive Online ECMS for the chosen $\lambda_{Offline,avg,city}$ = 2.60 using the ideal kp_{TS} and $t_{horizon}$ setting. Additionally, the SOC trajectory is given for a non-predictive Online ECMS with a λ of 2.70.

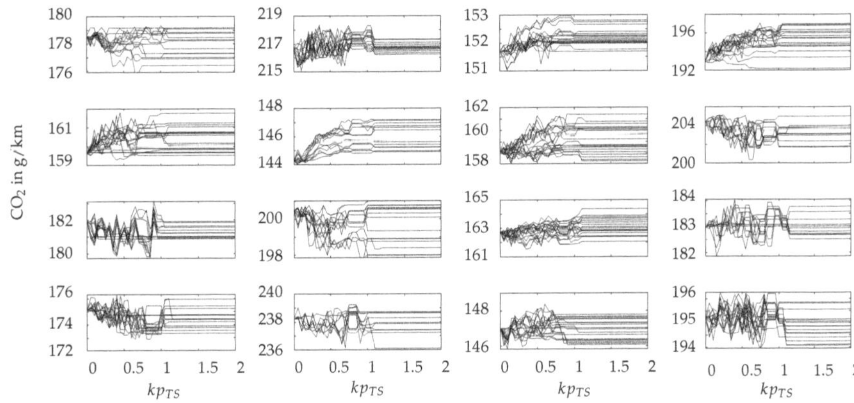

Figure 6. Usable battery energy 25 Wh: CO_2 over kp_{TS} of several $t_{horizon}$ for predictive Online ECMS (16/19 cycles). Each graph represents a specific $t_{horizon}$.

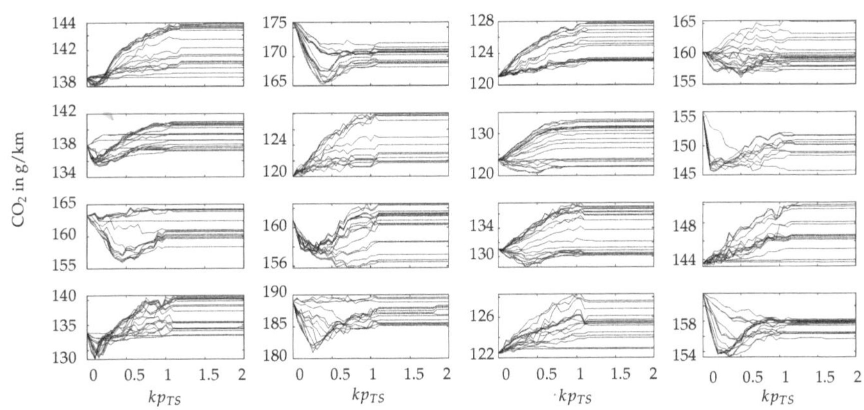

Figure 7. Usable battery energy 100 Wh: CO_2 over kp_{TS} of several $t_{horizon}$ for predictive Online ECMS (16/19 cycles). Each graph represents a specific $t_{horizon}$.

In a first step, the analysis focuses on both the non-predictive and the predictive Online ECMS for $\lambda_{Offline,avg,city}$ = 2.60. At t = 340 s as well as at t = 370 s and t = 500 s there is a correlation between the traffic light position and the speed. Speed is reduced in all three cases and leads to corresponding recuperation phases. At time points t = 370 s and t = 500 s, significantly higher recuperable energies are observed in the SOC trajectory for the predictive Online ECMS. At t = 340 s, on the other hand, there is no increase in recuperable energy. In contrast to those three points mentioned above, however, it is also possible that no significant reduction in speed and therefore no recuperation phase occurs despite the presence of a traffic light. This is shown, for example, at t = 200 s and t = 280 s. This can happen, for example, when the traffic light is green. While the recuperable energy remains unchanged at t = 280 s, CO_2 emissions can increase locally at t = 200 s compared to the non-predictive ECMS: the battery state of charge is kept longer with the predictive Online ECMS, which is associated with additional ICE operation. It is concluded that the predicitve Online ECMS can have both positive and negative effects on the optimality of the EMS. At some timesteps a local improvement is achieved when applying the predictive Online ECMS by better taking into account recuperation potentials. At other timesteps, a local increase in CO_2 emissions is achieved due to additional operation of the ICE. There are also situations where there is no impact on the optimality of the EMS. Ultimately, the decisive factor is which effects predominate. Overall, a well-parametrized predictive Online ECMS leads to a reduction in fuel consumption compared to the non-predictive Online ECMS.

In a second step, the SOC trajectory for λ of 2.70 should also be considered. A closer look at the non-predictive SOC trajectorys of $\lambda_{Offline,avg,city}$ = 2.60 and λ = 2.70 reveals that CO_2 reduction potentials by applying a predictive Online ECMS are highly dependent on the chosen non-predictive basic implementation: At t = 380 s, the additional energy hub for the use of recuperated energy is much higher for λ = 2.70 than for $\lambda_{Offline,avg,city}$ = 2.60. Anyway, both non-predictive Online ECMS implementations reach the upper SOC limit multiple times and therefore both λ seem to be too high for the shown driving cycle. It is concluded that significant saving potentials can already be achieved by an adequate choice of non-predictive Online ECMS. At the same time, however, the additional savings from the proposed predictive Online ECMS using recuperation potentials compared to a non-predictive implementation are reduced.

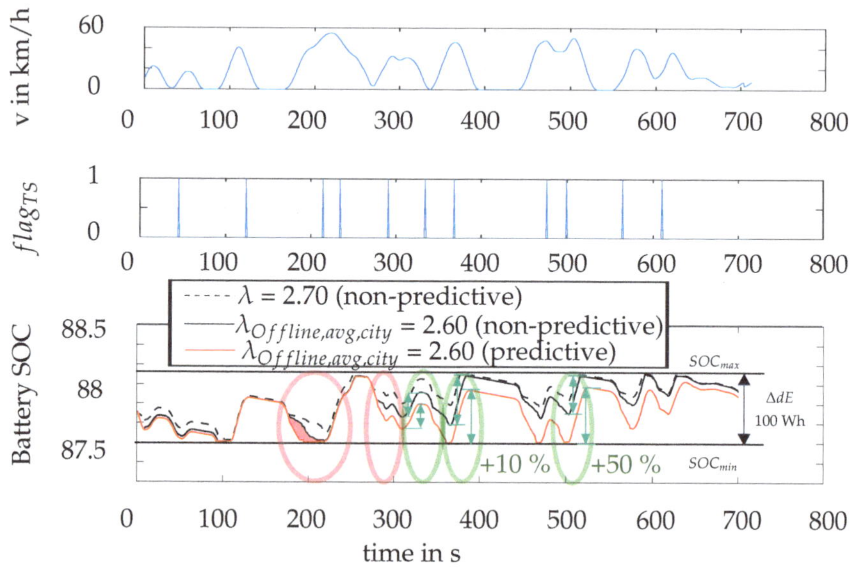

Figure 8. Usable battery energy 100 Wh: velocity (upper graph), $flag_{TS}$ (middle graph), and battery SOC over time (lower graph). Both for non-predictive Online ECMS (black) and predictive Online ECMS (orange) with $t_{horizon}$: 65 s, kp_{TS} = 0.2. In addition, the course for the non-predictive Online ECMS with λ = 2.70 is plotted (black dashed).

As already stated, Figure 8 reveals that λ reductions also occur when the battery is already discharged before a recuperation phase is initiated (t = 200 s). Therefore, a dependence of kp_{TS} on SOC is introduced in a follow-up work. Thus, when SOC is around the lower limit, no reduction in the value of the electric energy (λ) is allowed. Apart from this measure, a dependence of kp_{TS} on the occurrence of the recuperation potential in the prediction horizon could also be added. If the TS is quite close, the influence should be large. If the TS is in the later part of the studied horizon, the influence is reduced.

In contrast to Figures 6 and 7, there are no additional saving potentials for a predictive implementation when a large battery (usable battery energy 1000 Wh) is used, see Figure 9. Here, a significant deterioration is observed for all kp_{TS}. This is in line with the findings already made in the context of [34] that considering recuperation potentials in a predictive

Online ECMS does not lead to any noticeable saving potential for large batteries compared to a well-parametrized non-predictive Online ECMS.

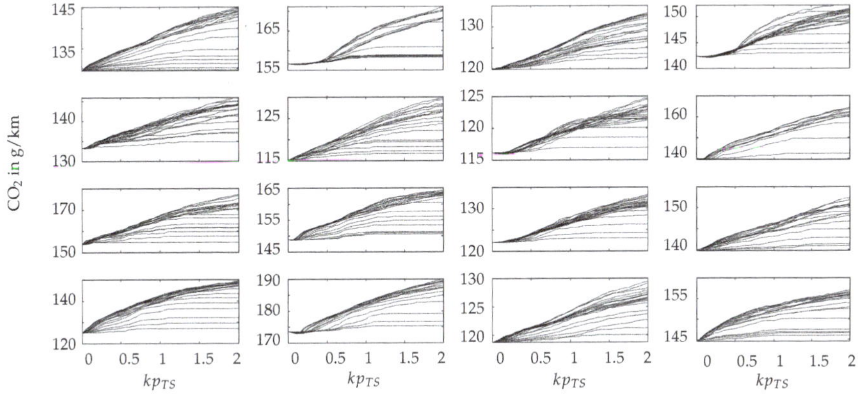

Figure 9. Usable battery energy 1000 Wh: CO_2 over kp_{TS} of several $t_{horizon}$ for predictive Online ECMS (16/19 cycles). Each graph represents a specific $t_{horizon}$.

As shown in Figure 10, an overall improvement is observed when applying a predictive Online ECMS. However, the saving potentials depend strongly on the usable energy content of the battery. The highest saving potentials exist with a usable battery energy of 100 Wh. With lower battery capacities, the saving potentials using a predictive implementation become less. For a usable battery energy larger than 100 Wh, no more significant saving potentials are found. The corresponding parameters for each battery size are listed in Table 6.

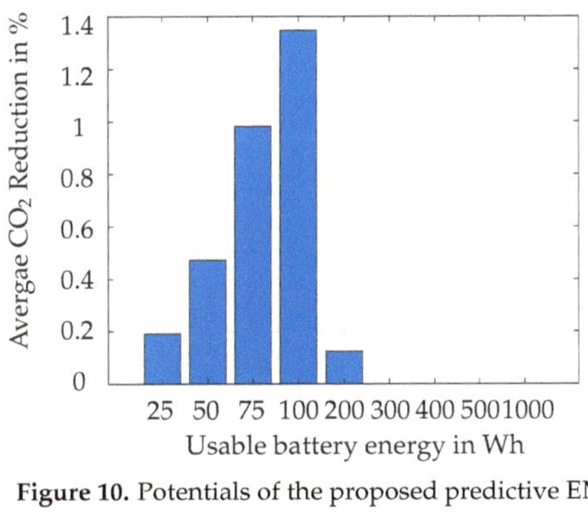

Figure 10. Potentials of the proposed predictive EMS for different usable battery energies: Average CO_2 reduction in % when applying the predictive Online ECMS compared to the non-predictive Online ECMS.

Table 6. Optimal overall parameters of the predictive Online ECMS for different battery sizes including average CO_2 reduction potentials compared to the non-predictive Online ECMS (Figure 10).

Usable Energy in Wh	kp_{TS}	Horizon in s	Reduction CO_2 %
25	0.35	30	0.19
50	0.15	40	0.47
75	0.35	45	0.98
100	0.20	50	1.35
200	0.05	25	0.12
300	No additional CO_2 reduction potentials by applying the proposed predictive Online ECMS considering TS		
400			
500			
1000			

6. Conclusions

In this study, a new approach for a predictive energy management strategy (EMS) was presented, which complements the existing field and provides starting points for future studies. It was demonstrated how a predictive Online Equivalent Consumption Minimization Strategy (ECMS) can achieve additional fuel savings compared to a non-predictive Online ECMS by predicting recuperation events using map data. Within the investigations, TS from the upcoming road profile are considered in the predictive Online ECMS, whereby more than 1 % average saving potentials compared to a non-predictive implementation were shown. The highest saving potentials are found with a usable battery energy of 100 Wh. With lower usable battery energy, the saving potentials decrease using the proposed predictive implementation. For batteries larger than 100 Wh, no more significant saving potentials are found. Furthermore, a big dependence of the added value by implementing a predictive Online ECMS from the basic non-predictive Online ECMS is revealed. In a follow-up work, a dependency of kp_{TS} on SOC could be introduced. Thus, if the battery state of charge is already at SOC_{min}, no additional reduction in the value of the electrical energy (λ) is allowed. Furthermore, a dependence of kp_{TS} on the occurrence of the recuperation potential in the predicted horizon can be implemented. If the recuperation occurs early in the time horizon, a large influence is aimed at; if it occurs late in the horizon, a small influence should be realized. Apart from that, the predictive Online ECMS could be enhanced by using additional map data, telemetry data or information from Radar, Lidar, and camera. Also, Car-to-Car (C2C) and Car-to-X (C2X) communication could be used to consider the status of the traffic signal.

To apply the predictive Online ECMS the route must be given and the current position of the global navigation satellite system (GNSS) must be known. Subsequent studies could investigate, how intelligent methods can be used to better estimate the current position of the vehicle or to predict the route. In order to validate the proposed predictive Online ECMS, an implementation in the real vehicle is required. For such an implementation in a real vehicle, the upcoming velocity has to be approximated to transfer map information from the distance domain to the time domain. Alternatively, a specific future distance could be used instead of the specified time horizon in the proposed predicitve Online ECMS.

Author Contributions: Conceptualization, F.D.; methodology, F.D.; software, F.D. and M.F.; validation, F.D., formal analysis, F.D.; investigation, F.D.; resources, F.D.; data curation, F.D.; writing—original draft preparation, F.D.; writing—review and editing, F.D.; visualization, F.D.; supervision, F.G. All authors have read and agreed to the published version of the manuscript.

Funding: This research was funded by Mercedes-Benz AG.

Institutional Review Board Statement: Not applicable.

Informed Consent Statement: Not applicable.

Data Availability Statement: Data are contained within the article.

Acknowledgments: The authors would like to thank Mercedes-Benz AG for funding this project.

Conflicts of Interest: The authors declare that this study received funding from Mercedes-Benz AG. The funder was not involved in the study design, collection, analysis, interpretation of data, the writing of this article or the decision to submit it for publication.

References

1. Silvas, E.; Hofman, T.; Murgovski, N.; Etman, P.; Steinbuch, M. Review of Optimization Strategies for System-Level Design in Hybrid Electric Vehicles. *IEEE Trans. Veh. Technol.* **2016**, *66*, 57–70. [CrossRef]
2. Tran, D.D.; Vafaeipour, M.; El Baghdadi, M.; Barrero, R.; van Mierlo, J.; Hegazy, O. Thorough state-of-the-art analysis of electric and hybrid vehicle powertrains: Topologies and integrated energy management strategies. *Renew. Sustain. Energy Rev.* **2020**, *119*, 109596. [CrossRef]
3. Serrao, L. A Comparative Analysis Of Energy Management Strategies For Hybrid Electric Vehicles. Ph.D. Thesis, Ohio State University, Columbus, OH, USA, 2009.
4. Onori, S.; Serrao, L.; Rizzoni, G. *Hybrid Electric Vehicles*; Springer: London, UK, 2016.
5. Rizzoni, G.; Onori, S. Energy Management of Hybrid Electric Vehicles: 15 years of development at the Ohio State University. *Oil Gas Sci. Technol.-Rev. D'IFP Energ. Nouv.* **2015**, *70*, 41–54. [CrossRef]
6. Salmasi, F.R. Control Strategies for Hybrid Electric Vehicles: Evolution, Classification, Comparison, and Future Trends. *IEEE Trans. Veh. Technol.* **2007**, *56*, 2393–2404. [CrossRef]
7. Xu, N.; Kong, Y.; Chu, L.; Ju, H.; Yang, Z.; Xu, Z.; Xu, Z. Towards a Smarter Energy Management System for Hybrid Vehicles: A Comprehensive Review of Control Strategies. *Appl. Sci.* **2019**, *9*, 2026. [CrossRef]
8. Jiang, Q.; Ossart, F.; Marchand, C. Comparative Study of Real-Time HEV Energy Management Strategies. *IEEE Trans. Veh. Technol.* **2017**, *66*, 10875–10888. [CrossRef]
9. Kim, N.; Cha, S.; Peng, H. Optimal Control of Hybrid Electric Vehicles Based on Pontryagin's Minimum Principle. *IEEE Trans. Control Syst. Technol.* **2011**, *19*, 1279–1287.
10. Kim, N.; Rousseau, A. Sufficient conditions of optimal control based on Pontryagin's minimum principle for use in hybrid electric vehicles. *Proc. Inst. Mech. Eng. Part D J. Automob. Eng.* **2012**, *226*, 1160–1170. [CrossRef]
11. Serrao, L.; Onori, S.; Rizzoni, G. ECMS as a realization of Pontryagin's minimum principle for HEV control. In Proceedings of the 2009 American Control Conference, St. Louis, MO, USA, 10–12 June 2009; IEEE: Piscataway Township, NJ, USA, 2009; pp. 3964–3969.
12. Zheng, C.H. Numerical Comparison Of Ecms And Pmp-Based Optimal Control Strategy In Hybrid Vehicles. *Int. J. Automot. Technol.* **2014**, *15*, 1189–1196. [CrossRef]
13. Foerster, D.; Decker, L.; Doppelbauer, M.; Gauterin, F. Analysis of CO_2 reduction potentials and component load collectives of 48 V-hybrids under real-driving conditions. *Automot. Eng. Technol.* **2021**, *6*, 45–62. [CrossRef]
14. Mayer, A. Two-Dimensional ECMS for System Analysis of Hybrid Concepts featuring Two Electric Traction Motors. In Proceedings of the 2019 International Symposium on Systems Engineering (ISSE), Edinburgh, UK, 1–3 October 2019.
15. Paganelli, G.; Delprat, S.; Guerra, T.M.; Rimaux, J.; Santin, J.J. Equivalent consumption minimization strategy for parallel hybrid powertrains. In Proceedings of the IEEE 55th Vehicular Technology Conference, VTC Spring 2002 (Cat. No.02CH37367), Birmingham, AL, USA, 6–9 May 2002; IEEE: Piscataway Township, NJ, USA, 2002; pp. 2076–2081.
16. Onori, S.; Serrao, L. On Adaptive-ECMS strategies for hybrid electric vehicles. In Procedings of the Les Rencontres Scientifiques d'IFP Energies Nouvelles—International Scientific Conference on Hybrid and Electric Vehicles—RHEVE 2011, Columbus, OH, USA, 6–7 December 2011.
17. Onori, S.; Serrao, L.; Rizzoni, G. *Adaptive Equivalent Consumption Minimization Strategy for Hybrid Electric Vehicles*; Ohio State University: Columbus, OH, USA, 2010.
18. Musardo, C.; Rizzoni, G.; Staccia, B. A-ECMS: An Adaptive Algorithm for Hybrid Electric Vehicle Energy Management. In Proceedings of the 44th IEEE Conference on Decision and Control, Seville, Spain, 12–15 December 2005; IEEE Operations Center: Piscataway, NJ, USA, 2005.
19. Kessels, J.; Koot, M.; van den Bosch, P.; Kok, D.B. Online Energy Management for Hybrid Electric Vehicles. *IEEE Trans. Veh. Technol.* **2008**, *57*, 3428–3440. [CrossRef]
20. Liu, T.; Zou, Y.; Liu, D.x.; Sun, F.c. Real-time control for a parallel hybrid electric vehicle based on Pontryagin's Minimum Principle. In Proceedings of the 2014 IEEE Conference and Expo Transportation Electrification Asia-Pacific (ITEC Asia-Pacific), Beijing, China, 31 August–3 September 2014; IEEE: Piscataway Township, NJ, USA, 2014; pp. 1–5.
21. Ouddah, O.; Adouane, L.; Abdrakhmanov, R. From Offline to Adaptive Online Energy Management Strategy of Hybrid Vehicle Using Pontryagin's Minimum Principle. *Int. J. Automot. Technol. Vol.* **2017**, *19*, 571–584. [CrossRef]
22. Sivertsson, M. Adaptive Control Using Map-Based ECMS for a PHEV. *IFAC Proc. Vol.* **2012**, *45*, 357–362. [CrossRef]
23. Zhang, F.; Xu, K.; Li, L.; Langari, R. Comparative Study of Equivalent Factor Adjustment Algorithm for Equivalent Consumption Minimization Strategy for HEVs. In Proceedings of the 2018 IEEE Vehicle Power and Propulsion Conference (VPPC), Chicago, IL, USA, 27–30 August 2018; IEEE: Piscataway Township, NJ, USA, 2018; pp. 1–7.

24. Fu, Z.; Liu, X. Equivalent Consumption Minimization Strategy Based on a Variable Equivalent Factor. In Proceedings of the IEEE 2017 Chinese Automation Congress (CAC), Jinan, China, 20–22 October 2017. [CrossRef]
25. Enang, W.; Bannister, C. Robust proportional ECMS control of a parallel hybrid electric vehicle. *Proc. Inst. Mech. Eng. Part D J. Automob. Eng.* **2017**, *231*, 99–119. [CrossRef]
26. Chasse, A.; Sciarretta, A.; Chauvin, J. *Online Optimal Control of a Parallel Hybrid with Costate Adaptation Rule*; Institut Français du Pétrole: Rueil Malmaison, France, 2010.
27. Gao, A.; Deng, X.; Zhang, M.; Fu, Z. Design and Validation of Real-Time Optimal Control with ECMS to Minimize Energy Consumption for Parallel Hybrid Electric Vehicles. *Math. Probl. Eng.* **2017**, *2017*, 3095347. [CrossRef]
28. Han, J.; Kum, D.; Park, Y. Synthesis of Predictive Equivalent Consumption Minimization Strategy for Hybrid Electric Vehicles Based on Closed-Form Solution of Optimal Equivalence Factor. *IEEE Trans. Veh. Technol.* **2017**, *66*, 5604–5616. [CrossRef]
29. Kural, E.; Güvenc, B.A. Predictive-Equivalent Consumption Minimization Strategy for Energy Management of A Parallel Hybrid Vehicle for Optimal Recuperation. *J. Polytech.* **2015**, *18*, 113–124.
30. Zhang, F.; Xi, J.; Langari, R. Real-Time Energy Management Strategy Based on Velocity Forecasts Using V2V and V2I Communications. *IEEE Trans. Intell. Transp. Syst.* **2017**, *18*, 416–430. [CrossRef]
31. Chen, D.; Kim, Y.; Stefanopoulou, A.G. Predictive Equivalent Consumption Minimization Strategy With Segmented Traffic Information. *IEEE Trans. Veh. Technol.* **2020**, *69*, 14377–14390. [CrossRef]
32. Kazemi, H.; Fallah, Y.P.; Nix, A.; Wayne, S. Predictive AECMS by Utilization of Intelligent Transportation Systems for Hybrid Electric Vehicle Powertrain Control. *IEEE Trans. Intell. Veh.* **2017**, *2*, 75–84. [CrossRef]
33. Bouwman, K.R.; Pham, T.H.; Wilkins, S.; Hofman, T. Predictive Energy Management Strategy Including Traffic Flow Data for Hybrid Electric Vehicles. *IFAC-PapersOnLine* **2017**, *50*, 10046–10051. [CrossRef]
34. Deufel, F.; Gießler, M.; Gauterin, F. Optimal Control of Electrified Powertrains in Offline and Online Application Concerning Dimensioning of Li-Ion Batteries. *Vehicles* **2022**, *4*, 464–481. [CrossRef]
35. Deufel, F.; Gießler, M.; Gauterin, F. A Generic Prediction Approach for Optimal Control of Electrified Vehicles Using Artificial Intelligence. *Vehicles* **2022**, *4*, 182–198. [CrossRef]
36. Deufel, F.; Jhaveri, P.; Harter, M.; Gießler, M.; Gauterin, F. Velocity Prediction Based on Map Data for Optimal Control of Electrified Vehicles Using Recurrent Neural Networks (LSTM). *Vehicles* **2022**, *4*, 808–824. [CrossRef]
37. Görke, D. *Untersuchungen zur Kraftstoffoptimalen Betriebsweise von Parallelhybridfahrzeugen und Darauf Basierende Auslegung Regelbasierter Betriebsstrategien*; Springer: Wiesbaden, Germany, 2016. [CrossRef]

World Electric Vehicle Journal

MDPI

Article

Investigating Investment Plans for Expanding Battery and Electric Vehicle Production in Europe

Hong Yang [1,*] and Lewis Fulton [2]

1 Institute of Transportation Studies, University of California, Davis, CA 95616, USA
2 Energy Futures Program, Institute of Transportation Studies, University of California, Davis, CA 95616, USA; lmfulton@ucdavis.edu
* Correspondence: hhyang@ucdavis.edu

Abstract: There has been significant EV sales growth in Europe, benefiting from its policies for promoting electric vehicles (EVs) and investments in manufacturing. This study investigates the investment announcements for EV and battery production announced by manufacturers and compares them to four scenarios with different EV penetration levels in Europe. This study projects the required capacities and estimates the investment needs to meet different EV sale targets in each scenario. The investigations show that, for Europe to achieve 60% new EV sales by 2030 and to be on track for 100% by 2035, its 4.8 million planned production capacity of EVs would fall short of the needed 9.2 million in 2030. The gap could close to 2.0 million when tentative announcements are counted. The results for batteries indicate that tentative plans are adequate and firm plans can satisfy most scenarios by 2030. More investments into EV production, along with policy support and incentives, are needed for more rapid scenarios.

Keywords: investment; electric vehicle; vehicle assembly; original equipment manufacturer; Europe; battery production

Citation: Yang, H.; Fulton, L. Investigating Investment Plans for Expanding Battery and Electric Vehicle Production in Europe. *World Electr. Veh. J.* **2023**, *14*, 347. https://doi.org/10.3390/wevj14120347

Academic Editors: Joeri Van Mierlo and Genevieve Cullen

Received: 13 November 2023
Revised: 9 December 2023
Accepted: 12 December 2023
Published: 14 December 2023

1. Introduction

From 1990 to 2020, greenhouse gas (GHG) emissions have decreased by 32% in the European Union (EU) [1]. The reduction in GHG emissions is likely to continue, especially with strong regulatory support. For example, the legislative proposal "Fit for 55" aims to reduce the EU's GHG emissions by 55 percent by 2030. The REPowerEU Plan presented by the European Commission aims to promote energy saving, clean energy production, and the diversification of energy suppliers. All these will help bring the 2030 EV sales target and climate ambition within reach. Looking ahead to address the ambitious 2030 and 2050 target, measures for serious emission reduction are still needed, especially in the transportation sector, which emits the most GHG emission [2].

Transportation electrification has been considered a promising pathway to decarbonization in the road transportation sector in the long term. Worldwide, the sales of electric vehicles (EVs) in 2021 hit a new record of 6.6 million; in Europe, EV sales increased by two-thirds year-on-year to 2.3 million. Germany remained the largest EV market in Europe in terms of the number of EVs sold; Norway had the highest market share of new EV sales in Europe, followed by Iceland and Sweden [3]. The growth in EV sales has grown significantly, thanks in part to strong policy support [4], and Europe is determined to reduce GHG emissions and to retake the lead in EV transition. More recently, the EU agreed on legislation that could ban new internal combustion engine vehicle sales beginning in 2035 [5].

Despite regulatory support, most EU member states provide financial support via different institutions and programs to support a strong EV uptake [6]. The European Investment Bank (EIB) has provided funding for strengthening the electric charging network in

Germany [7] and Italy [8]. Lienert and Bellon argued that global automakers are spending nearly USD 515 billion in investment into the EV and related battery industry through 2030 [9], and European automakers are leading EV and battery investment in total dollar amounts. While the importance of decarbonizing the road transportation sector is well known, as is the need for significant capital investments in the EV and battery industries, there is a lack of detailed estimates of the overall investment needs for expanding the production of EVs and batteries to achieve different transitional targets in the EV transition. This research effort to track announced investments and compare them to requirements fills a void in this area.

In particular, this study collects and examines publicly available original equipment manufacturers (OEMs) and battery suppliers' announcements for expanding the production of EVs and batteries in Europe between 2020 and 2022 and out to 2030. This paper is interested in two particular questions, as follows:

- What is the planned production capacity of EVs and batteries in Europe towards 2030, given investment announcements made by OEMs and battery suppliers?
- How fast will upfront capital investment into EVs and batteries need to scale up to meet different EV penetration levels and transitional targets, and are the announcements and plans adequate for this picture?

For the purpose of clarification, EV in this paper refers to battery electric vehicles (BEV) and plug-in hybrid electric vehicles (PHEV) in the light-duty vehicle (LDV) sector. Due to investment announcement data availability, battery production in this paper only covers the downstream battery production, including battery cell production and battery pack assembly. Processes like acquiring and processing minerals for batteries are not included. Lastly, Europe in this paper includes the 27 EU members, the European Free Trade Association (Iceland, Norway, Liechtenstein, and Switzerland), and the United Kingdom.

2. Background

To achieve a successful and rapid diffusion of EVs in the future, countries and regions have been offering various incentives to attract capital investment into not only EV assembly and battery production but also other automotive parts and battery recycling. There has been an increasing emphasis on "localizing" the EV supply chain in major EV supply markets, including Europe [10]. From the government's perspective, localization helps locate the automotive industry's value-creation efforts, including creating more high-skilled jobs and capturing innovation spillovers [11].

From the view of OEMs, having assembly plants close to critical parts supply (such as the supply of EV batteries, which account for between 30% and 40% of the total cost of an EV [12]) can help reduce their supply chain risks. Yet localizing the whole EV supply chain and relying less on EV imports is not easy. OEMs need to align their investments in EV production with battery production and, on the battery side, must secure enough raw materials to produce the batteries. Nevertheless, the demand for EVs needs to be spurred in parallel so that OEMs have the confidence to continue investing.

As the EV transition continues to accelerate, the question that needs to be asked has shifted from "Will the automotive future be electric" to "How large will the EV market be", and "Can the EV market scale up fast enough?" Previously, Slowik and Lutsey evaluated plants that are manufacturing EVs currently and some major carmakers' newly announced commitments in the US but did not further investigate how much total investment might be needed in the future, given different sales targets or other market growth assumptions [13]. Since 2021, there has been some grey literature containing EV-related investment data, including CIC energiGUNE [14], Environmental Resources Management (ERM) [15], and Atlas Public Policy [16]. In the past several months, as Reuters documented, more companies have invested in EV battery factories in Europe [17]. S&P Global also observed an increase in private equity investment in the European EV industry [18]. Despite weaker sales growth due to high interest rates and slow economic growth, analysts at Automotive News claim that EV investment will stay strong in Europe [19].

To the authors' knowledge, no academic research in current literature has provided a detailed picture of how much investment for EV and battery production has been committed in Europe towards 2035 and if the planned production capacities behind these investments align with Europe's fast EV growth plans. Therefore, this is the first known research paper that fills the aforementioned research gap by investigating the planned production capacity of EV assembly and battery production in Europe to 2030 and estimating how fast the investment for EV and battery production needs to scale up to meet various EV penetration levels and transitional targets. Moreover, by comparing the announced investment to the investment need under each scenario, this paper also contributes by providing critical benchmarks and timely insights on whether the current industry ambition and plans are adequate for each different picture in the future.

3. Methodology

In the Cobb–Douglas production function, total production is a function of labor and capital input constrained by the total factor productivity [20]. In Equation (1), Y is total production, L is labor input, K is capital input, A is total factor productivity, α and β are the output elasticities of capital and labor, respectively. This study assumes L^{β} is sufficient and constant and investigates how EV assembly and final battery production capacity will be affected by changes in K in Europe. Therefore, this study focuses on investigating the relationship between capital investment and EV assembly and final battery production capacity in Europe.

$$\mathrm{Y} \overset{\mathrm{def}}{=} \mathrm{A} \times L^{\beta} \times K^{\alpha} \tag{1}$$

As such, this study starts with collecting publicly available OEM investment announcements that mention both the amount of investment and the production capacity and end up with 57 OEM investment announcements of expanding the production of EVs and batteries in Europe made by OEMs and suppliers from 2020 to 2022 towards 2030 which are documented in [21]. As shown in Figure 1, this study uses these announced investment plans to calculate the average investment per unit of production capacity of EVs and batteries in Europe through 2030, which will be explained in Section 3.1.

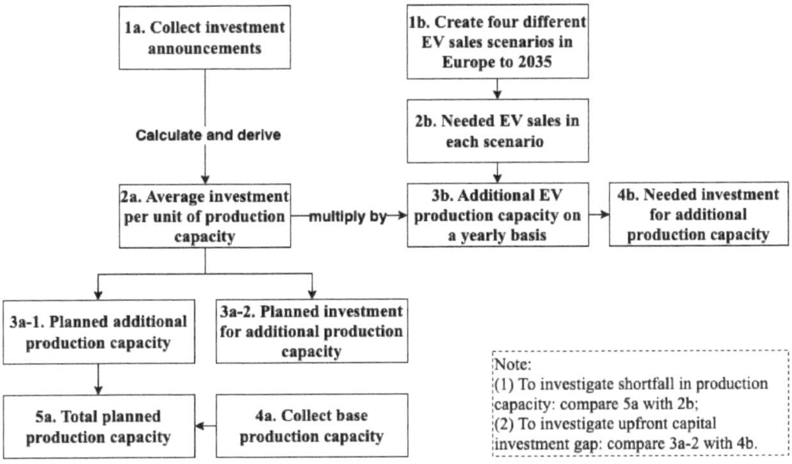

Figure 1. Flowchart of research framework.

Based on the average investment per unit of production capacity, this study then estimates either the investment or the production capacity in announcements where such information is not clarified. For both EVs and batteries, this study adds their respective planned production capacity to their base production capacity; the base production capacity of EVs was estimated based on historical EV sales data in Europe. Finally, this study compares the planned capacities to the required EV sales needs to meet different EV penetration levels and transitional targets. How different EV sales scenarios are developed will be explained in Section 3.2.

3.1. Planned EV and Battery Production Capacity

The total planned production capacity is derived as the sum of the base production capacity and the planned production capacity. This paper chooses the highest annual EV sales number between 2010 and 2021 annual EV sales in Europe and uses this number as the base production capacity of EVs in Europe. According to the International Energy Agency (IEA), 2284 thousand units of EVs were sold in Europe in 2021, which is assumed as the base production capacity of EVs in Europe for this study. The base production capacity of EVs is not further categorized by OEM to avoid missing any production capacity provided through non-major OEMs.

In this study, investment plans are divided into either firm or tentative announcements. Firm announcements are usually very specific and disclose information, including the use of the investment, the anticipated start time of the actual production, and the plant's rated capacity and/or the investment size. On the other hand, tentative investment plans may not disclose the use or the amount of the investment or when production can start. For example, some tentative investment plans are in the "advanced discussion" or "proposal submission" stage as of the time of writing. In summary, there are 41 firm announcements (11 for EV and 30 for battery) and 16 tentative announcements (7 for EV and 9 for battery). All currencies in this study are in the 2022 US dollar.

3.1.1. Planned Battery Production Capacity

The process of estimating how many EV batteries can be produced in the future based on relevant investment announcements can be complicated due to factors like the future EV driving range, size of EV, and EV sales composition (BEV versus PHEV). In this study, it is assumed that impacts from all possible factors are eventually reflected in one value—the average battery capacity per vehicle in Europe, regarding the new EV sales.

According to the most recent research, EV battery size varies between 52 kWh/vehicle and 85 kWh/vehicle in Europe [22]. For example, the Tesla Model Y was the best-selling EV in Europe in 2022 [23], and the new entry-level Tesla Model Y has a 55 kWh/vehicle battery [24]. Furthermore, there is a trend that sports utility vehicles (SUVs) and large BEV models are dominating current EV options [25]. Therefore, looking out to 2030 and 2035, this study selects the upper quantile between 52 kWh/vehicle and 85 kWh/vehicle (which is 74 kWh/vehicle) and assumes the battery capacity per vehicle Is 74 kWh/vehicle on average through 2035.

Announced investment for battery production in Europe ranges from USD 0.48 billion to USD 8.80 billion. Equation (2) explains how the investment per unit of production capacity for batteries on average is calculated. Thereinto, n is the number of announcements that mention both the investment and the capacity, which are documented in [21], I_i is the announced amount of investment, and C_i is the rated production capacity announced by OEMs and suppliers. This study uses the interquartile range (IQR) to detect outliers. If any outlier exists, it will fall out of the IQR in a boxplot. If identified, outliers are excluded to calculate $IC_{battery}$.

$$IC_{battery} \stackrel{\text{def}}{=} \frac{\sum_{i=1}^{n} I_i}{\sum_{i=1}^{n} C_i} \tag{2}$$

In this study, $IC_{battery}$ is USD 91/kWh/year. As indicated in an earlier section, this value only covers the final battery production process. This paper then uses $IC_{battery}$ to calculate the planned capacity of a battery plant (i.e., the number of EVs that a facility can build to support) and the committed amount of investment behind the announcement in the cases where they are not disclosed in other announcements, following Equations (3) and (4).

Thereinto, CC_i is the calculated battery production capacity, and CI_i is the calculated investment for a certain battery production announced by OEMs and suppliers.

$$CC_i = \frac{I_i}{IC_{battery}} \quad (3)$$

$$CI_i = C_i \times IC_{battery} \quad (4)$$

3.1.2. Planned EV Production Capacity

Likewise, this paper uses the announcements that mention the investment and the capacity to calculate the average investment per unit of EV production capacity in Europe. and these announcements are documented in [21]. The amount of announcements is labeled as m in Equation (5); $Invest_j$ is the announced amount of the investment; and $Capacity_j$ is the rated production capacity of EVs announced by OEMs.

$$IC_{assembly} \stackrel{\text{def}}{=} \frac{\sum_{j=1}^{m} Invest_j}{\sum_{j=1}^{m} Capacity_j} \quad (5)$$

It is found that $IC_{assembly}$ is about USD 5699/vehicle/year in Europe. Following the same logic as described in Equations (3) and (4), this is further used to calculate the planned EV production capacity or the investment in cases where they are not disclosed in some announcements.

3.2. Required Annual New EV Sales

The new EV sales towards 2035 are projected based on multiple sources. Bloomberg New Energy Finance (BNEF) projects annual sales of new EVs in Europe to reach around 4.3 million in 2025, which is about 28% of all passenger vehicle sales [26]. It is also assumed in BNEF's accelerated scenario that all new sales of passenger vehicles will be 100% electric by 2035, which aligns with the aforementioned legislation that can potentially ban all sales of new internal combustion engine vehicles.

In IEA's Announced Policy Scenario, the annual new EV sales in Europe are projected to reach 7.6 million by 2030, representing 52% of EV sales share [27]. This study estimates that 100% of the annual new EV sales share can be translated into 15.4 million and 14.6 million annual new EV sales in volume based on BNEF and IEA's projections, respectively. Therefore, this paper chooses the mean between these two projections, 15 million, as the volume that corresponds to 100% EV sales share for this study.

In this study, four EV sales scenarios (Table 1) are created based on two BNEF scenarios and four scenarios used in Mock and Díaz's study. Annual new EV sales share is specified every 5 years. The annual new EV sales are derived by multiplying 15 million new sales of light-duty vehicles (LDVs) with the new EV sales share.

3.3. Investment Needs for EV and Battery Production

In each scenario, the estimated annual investment needs for EV production are derived as the production of $IC_{Assembly}$ and the annual additional EV production capacity needs. The estimated annual investment needs for battery production are derived as the production of $IC_{Battery}$ and the average battery capacity per vehicle, and multiplied by the annual additional capacity, which is the same as the annual additional capacity of EV production because this study assumes one battery per EV.

3.4. Sensitivity Analysis

The planned production capacities can vary, depending on OEM and supplier commitments. Therefore, a tentative scenario is created so that firm and tentative OEM investment announcements are both taken into account for calculating the future planned production capacities. The tentatively planned production capacities are also compared with the required production capacities.

Table 1. EV sales scenarios in Europe.

Scenario	Definition	Annual New EV Sales Share
Fastest Scenario	This scenario assumes that no new internal combustion engine vehicles will be sold by 2030. The annual new EV sales will stay constant from 2030 to 2035.	29% in 2025 100% in 2030 100% in 2035
High Ambition Scenario	This scenario assumes more supportive policies will be in place to push for much faster EV adoption, no new internal combustion engine vehicle will be sold after 2035, and 70% lower worldwide harmonized light vehicle test procedure (WLTP [1]) CO_2 emissions by 2030.	22% in 2025 61% in 2030 100% in 2035
Accelerated Ambition Scenario	In this scenario, EV adoption is assumed to slow down in the early 2030s due to the saturation of some segments.	20% in 2025 50% in 2030 85% in 2035
Moderate Ambition Scenario	In this scenario, it is assumed that the fleet average WLTP CO_2 emission target in 2025 is improved to a 20% reduction.	20% in 2025 42% in 2030 64% in 2035

[1] WLTP is a laboratory test that is used to measure fuel consumption and CO_2 emissions from passenger cars and vans, and their pollutant emissions.

As mentioned in Section 3.1.1, estimating how many EV batteries can be produced in the future can be affected by various factors. To complement the baseline scenario, a lower case is created where this study assumes the battery capacity per vehicle in Europe is 62.5 kWh/vehicle on average. This value is derived based on the assumption of new EV sales and the assumption of the average battery capacity per PHEV in Europe. This study assumes that the new BEV sales to the new PHEV sales are about 4 to 1 based on current market trends and expert suggestions; this study also assumes that the average battery capacity per PHEV in Europe is about 16.5 kWh/vehicle, according to Mock and Díaz's study. Additionally, a higher case is created to reflect a potential future market where more BEVs than PHEVs are sold, or the EV driving range is higher. In this case, after consulting with industry experts, the average battery capacity per vehicle is adjusted to 92.5 kWh/vehicle, representing a 25% increase over 74 kWh in the base case. The investment needs for battery production will thus vary accordingly, and this study calculates the investment needs for all four scenarios in the lower and higher case.

4. Results

4.1. Compare the Annual Planned Production Capacities to the Annual New EV Sales

As depicted in Figure 2, the new EV sales are likely to reach 9.2 million in Europe by 2030 in the High Ambition scenario, representing about 60% of EV sales share. In the Accelerated Ambition scenario and Moderate Ambition scenario, the new EV sales could reach 7.5 million and 6.3 million by 2030, respectively.

As mentioned in Section 3.1, the base production capacity of batteries in Europe in this study is about 2.3 million. According to [28], battery plants take about 5 years to ramp up to their full capacity. Therefore, this paper assumes that the battery plant's capacity factor is 20% in the first year and increases linearly until it reaches 100% in the fifth year.

If only firm investment announcements are considered, the planned battery production capacity in Europe will be able to support 12.7 million EVs by 2030, with an average of 74 kWh/vehicle battery capacity per vehicle; this is sufficient for the High Ambition scenario (Figure 3). The planned battery production capacity from 2030 to 2035 is currently flat because the investment plans announced so far only cover through 2030. More investment plans for battery production are anticipated. Notably, though the planned production capacity of batteries may not be enough for the fastest scenario by 2030, it seems to be fully sufficient for all scenarios through the late 2020s.

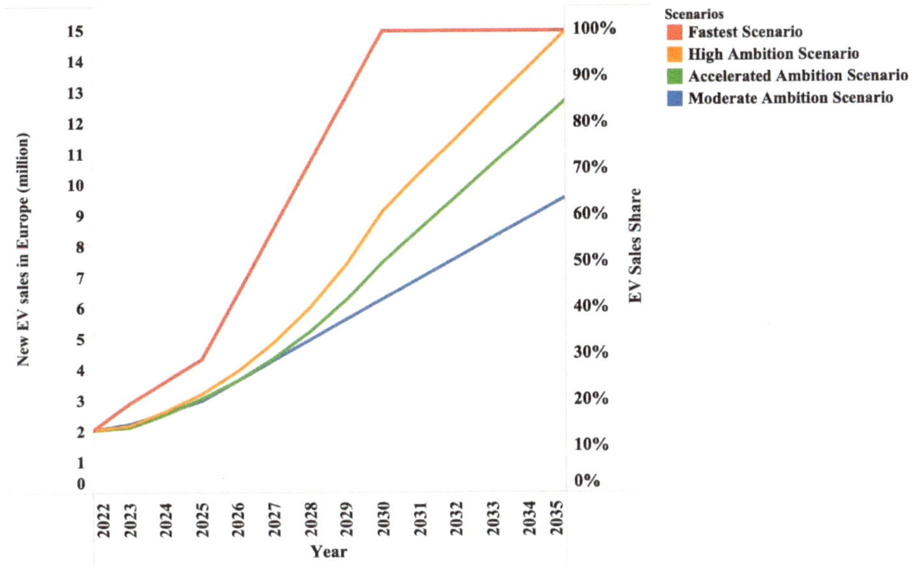

Figure 2. Annual new EV sales in Europe (million), 2022–2035.

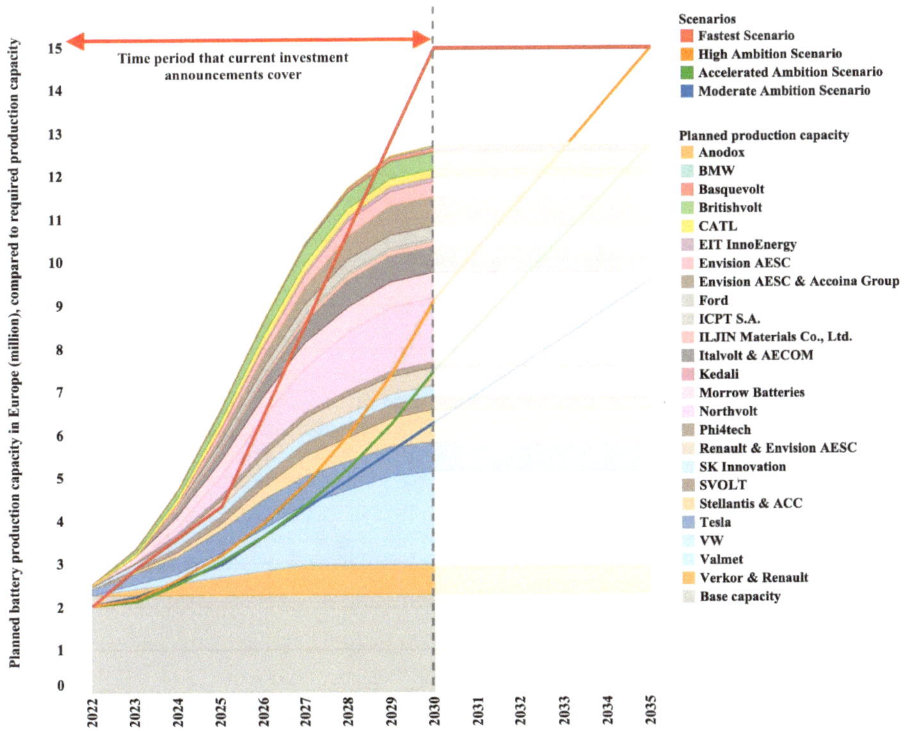

Figure 3. Planned production capacity of batteries in Europe (million) compared to the required production capacity, 2022–2035.

The planned production capacity of EVs will be 4.8 million EVs in Europe by 2030, which is about 1.5 million short of the Moderate Ambition in Europe by 2030. As indicated in Figure 4, the production capacity of EVs in Europe may be lagging for most of the scenarios in this paper before 2030.

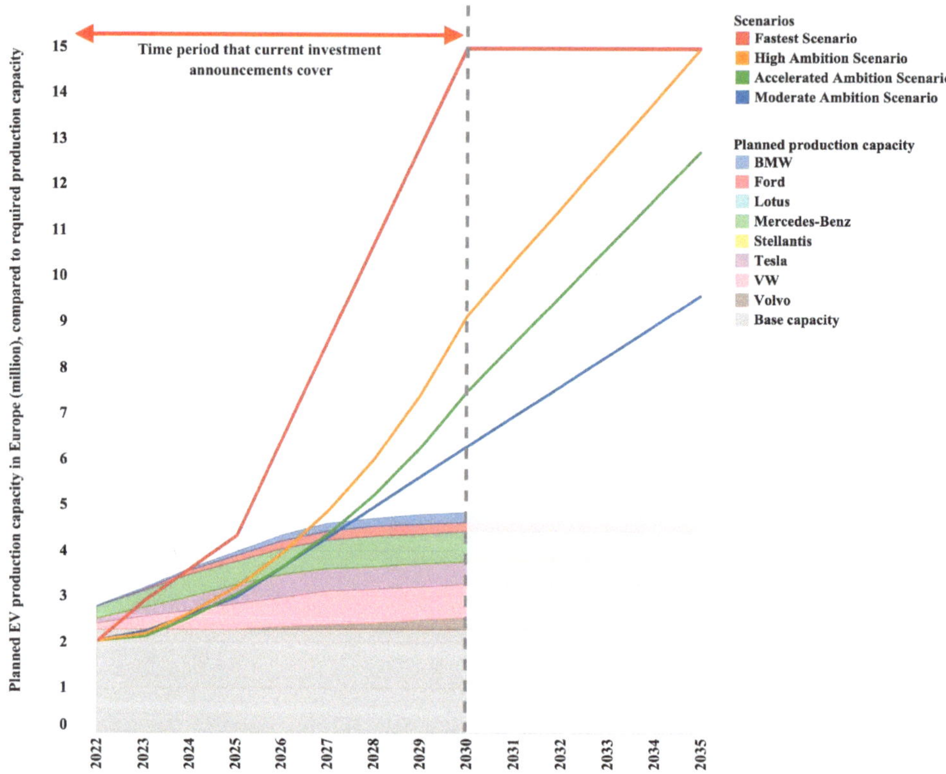

Figure 4. Planned production capacity of EVs in Europe (million) compared to the required production capacity, 2022–2035.

4.2. Compare the Investment Needs to the Announced Investment for Expanding EV and Battery Production

To achieve an ambitious EV sales scenario, more investment will be needed soon (Appendix A). In the Fastest scenario, investment needs can reach USD 14.4 and USD 12.1 billion in 2026 to support additional production capacity of EVs and batteries, respectively; such annual additional investments may be needed at least through 2030 to sustain this scenario. The High Ambition scenario may face its first uptake in investment needs in 2031, with investment for additional EV and battery production reaching about USD 11.6 and USD 9.8 billion, respectively.

As shown in Table 2, about USD 69 billion for battery production has been announced in Europe between 2020 and 2022 towards 2030 based on firm announcements. This can be translated into many batteries that can support about 12.7 million EVs with a battery capacity of 74 kWh/vehicle, enabling Europe to meet the High Ambition scenario. If the fastest scenario is pursued, about USD 18 billion in investment would be needed before 2030 or earlier. For EV production, USD 15 billion has been announced through firm announcements in Europe towards 2030. With tentative investments included, the investment amounts to about USD 28 billion and the planned production capacity of EVs is sufficient for the Moderate Ambition scenario by 2030.

Table 2. Compare the announced investments to the investment needs for expanding the production capacity of EVs and batteries in Europe by 2030 and 2035 (billion USD).

EV Sales Scenario	Cumulative Investment Announced	Cumulative Investment Needs	
		By 2030	By 2035
Battery production			
Fastest	USD 69 (USD 110 [1])	USD 87	USD 87
High Ambition		USD 39	USD 83
Accelerated Ambition		USD 37	USD 72
Moderate Ambition		USD 29	USD 51
EV production			
Fastest	USD 15 (USD 28 [2])	USD 74	USD 74
High Ambition		USD 33	USD 70
Accelerated Ambition		USD 31	USD 61
Moderate Ambition		USD 24	USD 43

[1] USD 110 billion is the total announced investment through 2030 for battery production based on firm and tentative announcements. [2] USD 28 billion is the total announced investment through 2030 for EV production based on firm and tentative announcements.

4.3. Sensitivity Analysis Results

When tentative investment announcements are also taken into consideration, the planned production capacity of batteries in Europe by 2030 can increase to nearly 19.7 million EVs being supported, exceeding the need in the fastest scenario by 4.2 million (Figure A2). For EV production, the planned production capacity of EVs can increase to about 7.2 million, which is sufficient for the Moderate Ambition scenario and is only 0.3 million short of the Accelerated Ambition scenario by 2030 (Figure A3).

As suggested in Table 3, with a lower battery capacity per vehicle on average, the planned battery production capacity can increase to 14.6 million by 2030 based on firm announcements, which can put Europe very close to meeting the fastest scenario. As indicated in the results from the higher case, the planned production capacity of batteries is nearly 16.2 million by 2030 based on both firm and tentative announcements; and when only firm announcements are considered, Europe can meet the High Ambition scenario.

Table 3. Planned battery production capacity in Europe to 2030 in the Lower and Higher case.

Capacity Needed in 2030			Planned Production Capacity in 2030		
			Average battery capacity per vehicle (kWh/vehicle)		
High Ambition scenario	Fastest scenario		Lower case: 62.5	Base case: 74	Higher case: 92.5
9.2 million	15.0 million	Firm	14.6 million	12.5 million	10.6 million
		Firm and tentative	22.9 million	19.2 million	16.2 million

This paper also examines the investment needs in the lower and the higher case (Table 4). In the lower case, the firm announced investment for battery production (i.e., USD 69 billion) almost reaches the investment needs in the fastest scenario (i.e., USD 74 billion). In the higher case, the firm announced investment would put Europe on track for the High Ambition scenario; with tentative investments included, the fastest scenario could be met by 2030.

Table 4. Investment needs for the production capacity of batteries in Europe in the Lower and Higher case (billion USD).

Scenario	Investment Announced So Far to 2030	Cumulative Investment Needs			
		Lower Case: 62.5 kWh/vehicle		Higher Case: 92.5 kWh/vehicle	
		By 2030	By 2035	By 2030	By 2035
Fastest	USD 69 (USD 110)	USD 74	USD 74	USD 109	USD 109
High Ambition		USD 33	USD 70	USD 49	USD 103
Accelerated Ambition		USD 31	USD 61	USD 46	USD 90
Moderate Ambition		USD 24	USD 43	USD 36	USD 64

5. Discussion

By tracking the announced investment plans for expanding the production capacity of EVs and batteries in Europe, translating them into future production capacities, and comparing these to different EV transitional targets towards 2030 and 2035, this paper provides a better understanding of whether these announced investment plans appear to be adequate to meet the EV sale targets and emission reduction goals in Europe. This paper also gives a sense of how fast future investments will need to scale up to be sufficient. There are several implications of this study, as follows.

In general, this study suggests that if Europe is committed to meeting the High Ambition scenario (on track for 100% EV sales share by 2035), the planned production capacity of batteries seems to be adequate, but the production capacity of EVs will require a lot more investment that has yet been announced. This provides practical insights for automakers and automotive suppliers and helps improve the EV supply chain "visibility" (knowledge of sourcing options and potential supply scale). The results of this study may help automakers and suppliers better understand where the EV assembly and battery production is going within Europe, what is happening elsewhere, and whether trouble of shortage might be brewing in the supply chain [29].

In terms of the source of investment, while existing OEMs and battery companies will likely continue expanding their investments directly, this study suggests that some support for financing could be provided by the governments through different fiscal incentives and subsidies, including vehicle purchase incentives which can help spur EV demand growth and supply-side investments. In addition to direct incentives, regulatory policies that can help de-risk the market and help boost OEMs' confidence that EV demand will be spurred to achieve these targets, may be needed as well to keep incentivizing production investments.

For investment in battery production, this paper suggests that OEMs and battery suppliers seem on track to achieve the required targets in the 2030 time frame, even for the most ambitious scenario. This means that some battery investments announced so far may come from OEM and battery suppliers' anticipation of future vehicle investments, investments for a future market of PHEVs, potential export of EVs, or battery swapping.

However, such a ramp-up in battery manufacturing needs to be treated carefully. If the ramp-up in battery production capacity happens before significant EV demand is in place, battery plants may run the risk of low operational efficiency. Operational efficiency is one of the major concerns for gigafactories. For example, if a 50 GWh plant only achieves 66% of its planned annual output, it can lose about USD 500 million in value annually; the loss can also be translated to a modeled profit of 6% to 8% [30]. To mitigate risks of low operation efficiency of battery plants, it may be reasonable to consider export as an important destination for the over-production of batteries.

Though more investment is desired in general, the results in this paper also point at one potential risk: the gap between investments in different parts of the EV supply chain. The results in this paper suggest that in Europe, more investment plans and production capacities have been announced for batteries than for EV production so far. One explanation may be that OEMs and battery suppliers are concerned about the supply chain shortage, especially the scarcity of certain raw materials for producing batteries. European governments and automakers have invested heavily to push the expansion of battery production in Europe in recent years, to manage such supply chain risk and to avoid relying on battery imports from Asia [13]. Therefore, OEMs prefer securing enough battery capacity (or securing a strategic partnership with a battery supplier who can secure enough raw materials for producing batteries) before investing in any EV production facility close by. The increasing demand for batteries may seem more certain for the industry at this moment.

To bridge the production capacity gap, as well as the investment gap, between EV production and battery production, this study suggests that an additional investment of USD 5 billion to USD 18 billion may be needed to increase the production capacity of EVs in Europe to reach the High Ambition scenario by 2030. In the meantime, it may be beneficial to consider importing EVs from other countries and regions into Europe as well. The European Automobile Manufacturers' Association, the American Automotive Policy Council, the Truck and Engine Manufacturers Association, and the Alliance for Automotive Innovation have made a joint statement in support of the US-EU Transatlantic Trade and Technology Council to revive the coordination on issues arising from the nexus of trade with the supply chain. In the tentative scenario, Europe would be about 0.3 million short of the Accelerated Ambition scenario (50% EV sales share by 2030 and 85% by 2035). Such volume is within the range of EV imports seen in 2020 as 30% of 3.0 million EVs were imported into Europe [31].

It has to be stressed that the results presented in this research rely on various assumptions that reflect the current announced investment plans and understanding of the key trends in the EV market. For example, this study assumes that the average battery capacity per vehicle in Europe towards 2035 is about 74 kWh/vehicle based on the current range of EV battery size in Europe and the trend of preferring larger BEV models and SUVs. This study also assumes that all firms announced investment plans will come through on time, while OEMs and suppliers can change their plans anytime due to reasons like changes in the overall company strategy and geopolitical tension. Hence, the analysis should be updated as more investment plans are announced and updated.

In addition to the exploration of possible variations presented in the sensitivity analyses, the authors also recognize other factors that may lead to potential variations in the estimated needs for battery and EV production presented in this paper. Firstly, this study could overestimate the investment needs for additional battery and EV production, if economies of scale and learning curve are considered. Further, if the EV assembly factories find ways to retrofit their existing factories instead of constructing new assembly facilities in the near future, then the estimated investment needs for EV production in this paper could be overestimated. However, if the investment cost for construction materials increases, the estimated investment needs presented in this paper can be underestimated. Last but not least, as the EV market is driven by ambitious policy targets and incentives, changes in EV sale targets and policy incentives can increase the volatility of the EV market, which further increases the variations in the estimates in this paper.

Additionally, this study has not covered the full EV supply chain. There are other important auto parts and materials, and all industries and stakeholders should work collaboratively to ensure enough materials and production capacity are available at all stages. In addition to issues around the production and supply, OEMs are also watching the charging infrastructure roll-out and are waiting for the EV sales to ramp up as recharging networks develop. However, building up the charging infrastructure requires significant amounts of planning and capital investment upfront as well. This research effort will be extended to estimate the investment needs in charging infrastructure and other possible areas to better understand the investment needs for an EV-dominated vehicle future by 2035.

6. Conclusions

The analysis presented in this paper provides several contributions. Firstly, it provides a timely picture of the future production capacity of EVs and batteries in Europe, based on publicly available investment announcements made by OEMs and suppliers from 2020 to 2022. The results in this paper indicate that the planned production capacity of batteries will be sufficient for the High Ambition scenario towards 2030, putting Europe on the trajectory of achieving 100% new EV sales share by 2035. However, the planned production capacity of EVs currently falls short of meeting the High Ambition scenario, or even the Moderate Ambition scenario (42% EV sales share towards 2030) in the same timeframe.

Secondly, the results of the sensitivity analyses in this paper show that, with tentative investment plans considered, the planned production capacity of batteries may exceed the Fastest scenario by 2030. However, the planned production capacity of EVs may still risk facing a 0.3 million shortage for the Accelerated Ambition scenario and a 2.0 million shortage for the High Ambition scenario. This study acknowledges different factors and uncertainties (e.g., changes in policies and incentives, technology improvement, and economies of scale) in the future EV market, leading to potential over- or underestimations in this study. It is anticipated that more investment in EV production will still come in the future. Yet with time for hitting 2030 targets getting short, it may be challenging to achieve the more ambitious scenarios without importing vehicles from other world regions.

Therefore, this study suggests that policies that help spur faster investment for EV production are needed in Europe; at the same time, consideration of potential trade dynamics with other major EV supply markets, such as importing EVs due to a shortage in domestic capacity, may be needed as well. There is also a completely different risk: markets will not develop quickly, and investment proves to be greater than market needs. In this case, opportunities to sell excess production (such as batteries) into other markets may be welcome.

Overall, this paper is the first known effort that shows how fast upfront capital investment for EV and battery production will need to scale up in Europe under four different EV sales scenarios. The interpretation of capital investment signals future expected capacity and gives an idea of the final cost of EVs. These signals can inform the industry of the current landscape of investment opportunities and help governments formulate regulatory and fiscal incentives that would accelerate the pace of achieving the EV sales targets by 2030 in Europe.

It is important to keep track of such investment announcements and update the results in this paper when more announcements become available in the next few years. While this research provides insights into the planned capacity and investment needs for future battery and EV production in Europe, it has not included the investment needs for other parts of the EV supply chain, which can be investigated in the future. This paper does not discuss the source of upfront investment, though it may not need to be a new investment but redirections. For future research, it may be interesting to consider trade dynamics between Europe and other major EV markets or extend such research to the other transportation sectors, such as the medium- and heavy-duty vehicle sectors.

Author Contributions: Conceptualization: H.Y. and L.F.; Methodology: H.Y. and L.F.; Software: H.Y.; Data curation: H.Y.; Validation: H.Y.; Formal analysis: H.Y.; Writing—original draft preparation: H.Y.; Writing—review and editing: H.Y. and L.F.; Visualization: H.Y.; Supervision: L.F.; Project administration: L.F.; Funding acquisition: L.F. All authors have read and agreed to the published version of the manuscript.

Funding: This research was funded by Sustainable Transportation Energy Pathways (STEPS) and Energy Futures program funding, within the Institute of Transportation Studies at UC Davis, and sponsors of these programs.

Data Availability Statement: Data are contained within the article and can be found in Yang, H. (2023). OEM investment announcements for future electric vehicles and related battery production in Europe from 2020 to 2022 towards 2030. Mendeley Data, 1. https://doi.org/10.17632/fwd9hz2cy2.1, accessed on 12 December 2023.

Acknowledgments: We would like to thank Aditya Ramji, Ian Sutherland, Thiemo Schalk, Timothy Wallington, Chris Birdsall, and Aaron P.R. Eberle for reviewing the manuscript. We also sincerely appreciate all valuable comments and suggestions from the editor and reviewers.

Conflicts of Interest: The authors declare no conflict of interest.

Appendix A

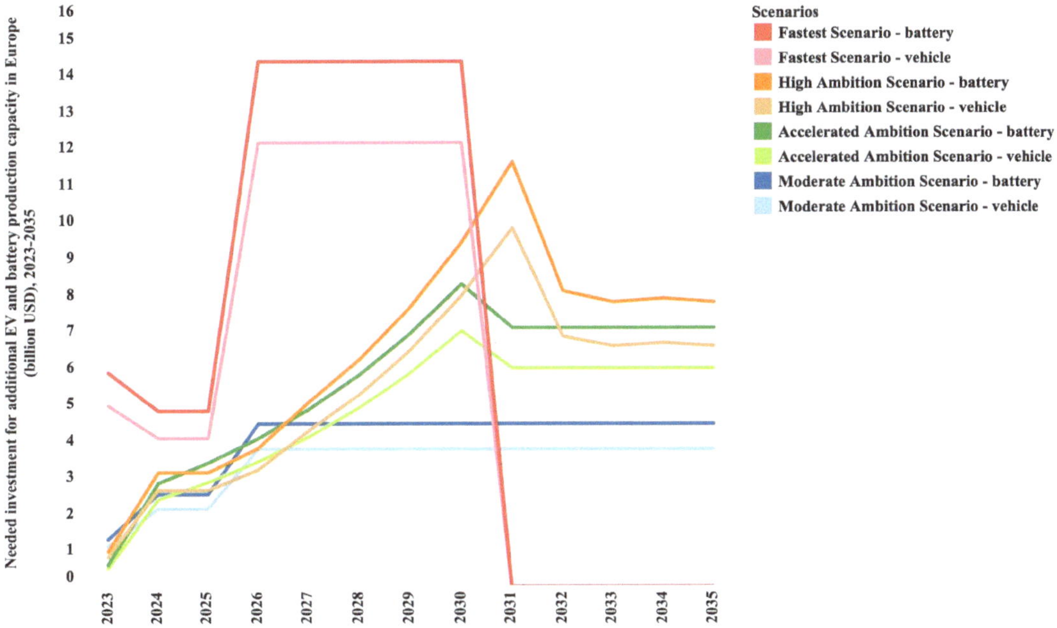

Figure A1. Annual investment needs for EV and battery production capacity in Europe from 2022 to 2035 (billion USD).

Appendix B

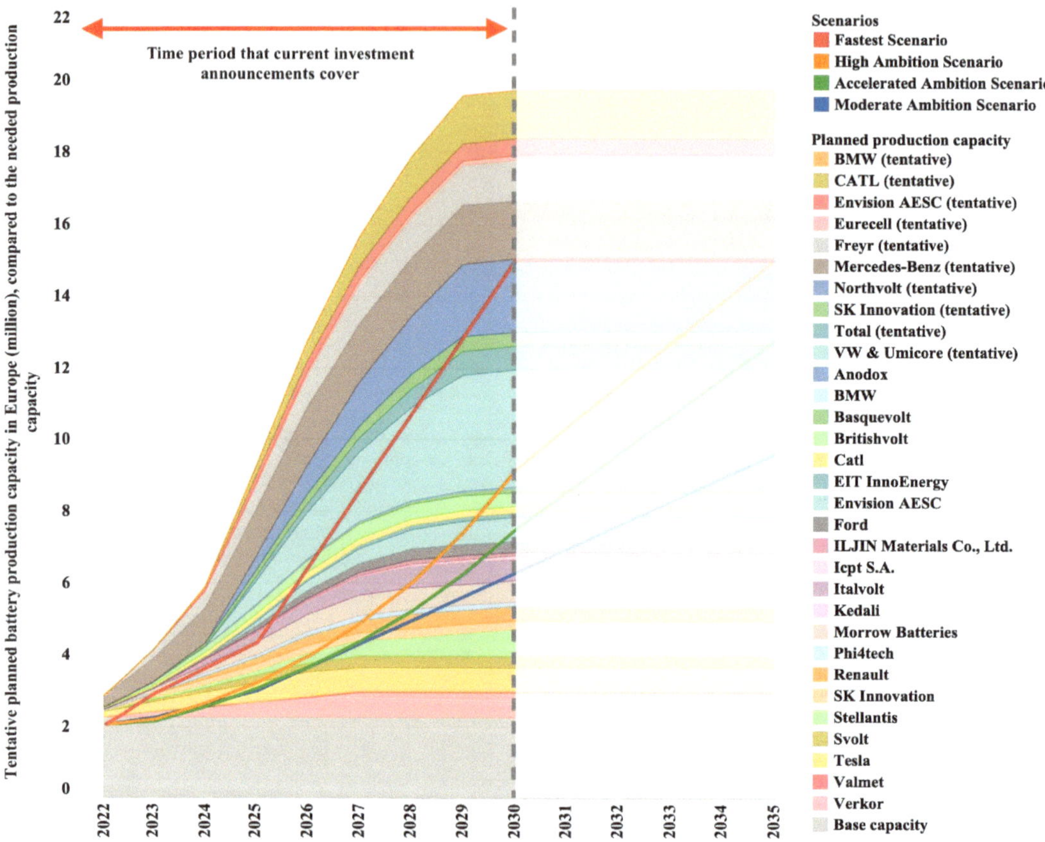

Figure A2. Tentative planned battery production capacity in Europe compared to the required capacity (million), 2022–2035.

Appendix C

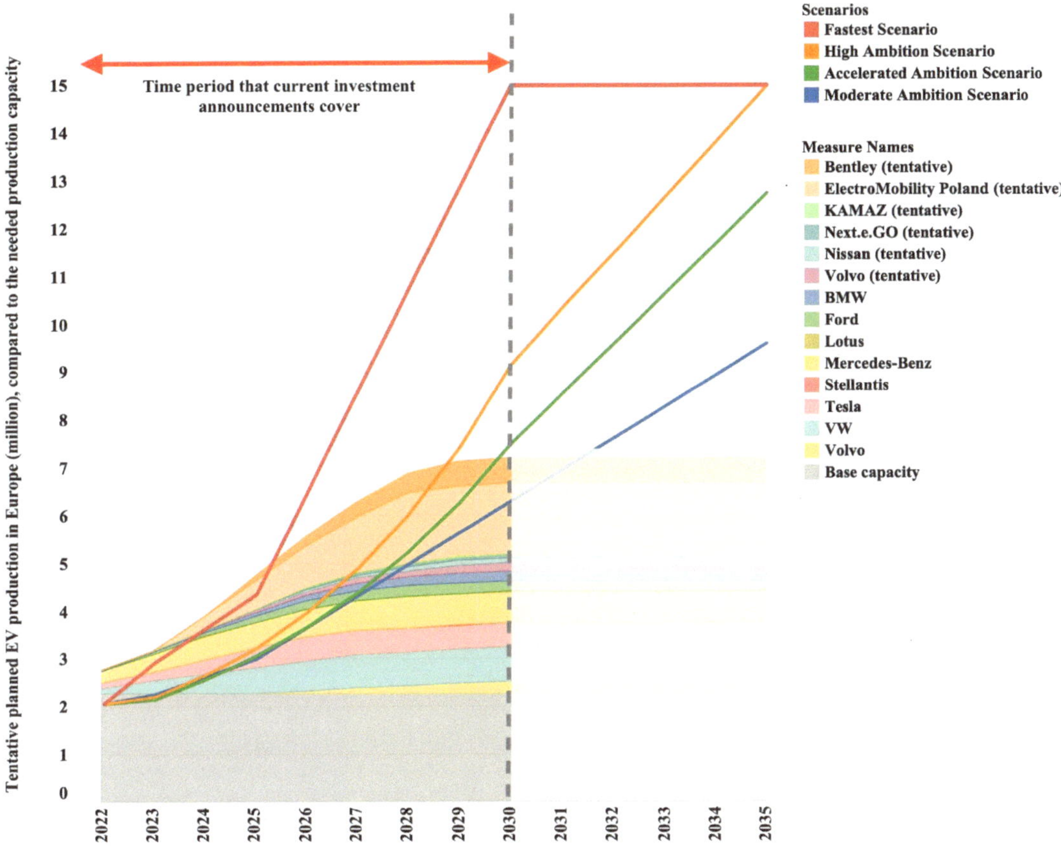

Figure A3. Tentative planned EV production capacity in Europe compared to the required capacity (million), 2022–2035.

References

1. European Environment Agency. Total Greenhouse Gas Emission Trends and Projections in Europe (8th EAP). Available online: https://www.eea.europa.eu/ims/total-greenhouse-gas-emission-trends (accessed on 26 October 2022).
2. d'Aprile, P.; Engel, H.; Helmcke, S.; Hieronimus, S.; Nauclér, T.; Pinner, D.; van Gendt, G.; Walter, D.; Witteveen, M. Europe's Path to Decarbonization. 2022. Available online: https://www.mckinsey.com/capabilities/sustainability/our-insights/how-the-european-union-could-achieve-net-zero-emissions-at-net-zero-cost (accessed on 8 December 2023).
3. International Energy Agency. Global EV Outlook 2022. 2022. Available online: https://iea.blob.core.windows.net/assets/ad8fb04c-4f75-42fc-973a-6e54c8a4449a/GlobalElectricVehicleOutlook2022.pdf (accessed on 8 December 2023).
4. Bernard, M.R.; Hall, D.; Lutsey, N. Update on Electric Vehicle Uptake in European Cities. 2021. Available online: https://theicct.org/sites/default/files/publications/ev-uptake-eu-cities-oct21.pdf (accessed on 8 December 2023).
5. Abnett, K. *EU Approves Effective Ban on New Fossil Fuel Cars from 2035*; Reuters: London, UK, 2022. Available online: https://www.reuters.com/markets/europe/eu-approves-effective-ban-new-fossil-fuel-cars-2035-2022-10-27/ (accessed on 28 October 2022).
6. European Automobile Manufacturers Association. Electric Vehicles: Tax Benefits & Purchase Incentives. Available online: https://www.acea.auto/files/Electric_vehicles-Tax_benefits_purchase_incentives_European_Union_2021.pdf (accessed on 24 November 2021).

7. European Investment Bank. Germany: EIB Provides the Mobility House with €15 Million for Smart Charging Technology. Available online: https://electricenergyonline.com/article/energy/category/ev-storage/143/850052/germany-eib-provides-the-mobility-house-with-15-million-for-smart-charging-technology-.html (accessed on 19 August 2020).
8. European Investment Bank. Italy: €26 Million to Duferco to Strengthen the National Electric Charging Network. Available online: https://www.eib.org/en/press/all/2022-120-energy-eur26-million-to-duferco-to-strengthen-the-national-electric-charging-network (accessed on 9 March 2022).
9. Lienert, P.; Bellon, T. Global Carmakers Now Target $515 Billion for EVs, Batteries. Available online: https://www.reuters.com/business/autos-transportation/exclusive-global-carmakers-now-target-515-billion-evs-batteries-2021-11-10/ (accessed on 10 November 2021).
10. Gibbs, N. Ukraine War Redrawing Europe's Auto Supply Chain Map. Available online: https://www.autonews.com/suppliers/ukraine-war-redrawing-europes-auto-supply-chain-map (accessed on 23 June 2022).
11. Eddy, J.; Pfeiffer, A.; van de Staaij, J. *Recharging Economies: The EV-Battery Manufacturing Outlook for Europe*; McKinsey: Chicago, IL, USA, 2019. Available online: https://www.mckinsey.com/~/media/McKinsey/Industries/Oil%20and%20Gas/Our%20Insights/Recharging%20economies%20The%20EV%20battery%20manufacturing%20outlook%20for%20Europe/Recharging-economies-The-EV-battery-manufacturing-outlook-for-Europe-vF.pdf (accessed on 8 December 2023).
12. Institute for Energy Research. Electric Vehicle Battery Costs Soar. Available online: https://www.instituteforenergyresearch.org/renewable/electric-vehicle-battery-costs-soar/ (accessed on 25 April 2022).
13. Bui, A.; Slowik, P.; Lutsey, N. Power Play: Evaluating the U.S. Position in the Global Electric Vehicle Transition. 2021. Available online: https://theicct.org/sites/default/files/publications/us-position-global-ev-jun2021.pdf (accessed on 8 December 2023).
14. CIC energiGUNE. Europe's Large Gigafactory Industry Begins to Take Shape. Available online: https://cicenergigune.com/en/blog/europe-large-gigafactory-industry-take-shape (accessed on 15 February 2022).
15. ERM. Electric Vehicle Market Update. 2022. Available online: https://blogs.edf.org/climate411/wp-content/blogs.dir/7/files/2022/04/electric_vehicle_market_report_v6_april2022.pdf (accessed on 8 December 2023).
16. Taylor, T.; Gabriel, N. The EV Transition: Key Market and Supply Chain Enablers. 2022. Available online: https://atlaspolicy.com/wp-content/uploads/2022/12/2022-EV-Transition-Key-Market-and-Supply-Chain-Enablers.pdf (accessed on 8 December 2023).
17. Reuters. Companies Invest in EV Battery Factories in Europe. Available online: https://www.reuters.com/business/autos-transportation/companies-invest-ev-battery-factories-europe-2023-05-18/ (accessed on 19 July 2023).
18. Vidal, K.; Sabater, A. Private Equity Investment in European EV Industry up in Q1. Available online: https://www.spglobal.com/marketintelligence/en/news-insights/latest-news-headlines/private-equity-investment-in-european-ev-industry-up-in-q1-75925619 (accessed on 25 May 2023).
19. Gibbs, N. EV Investment Will Stay Strong in Europe Despite Weaker Sales Growth, Ratings Agency Says. Available online: https://europe.autonews.com/automakers/europe-remains-bullish-ev-investment-despite-weaker-sales (accessed on 22 November 2023).
20. Cobb, C.W.; Douglas, P.H. A Theory of Production. *Am. Econ. Rev.* **1928**, *18*, 139–165. Available online: https://www.jstor.org/stable/1811556 (accessed on 8 December 2023).
21. Yang, H. OEM Investment Announcements for Future Electric Vehicles and Related Battery Production in Europe from 2020 to 2022 towards 2030. Mendeley Data, V1. 2023. Available online: https://doi.org/10.17632/fwd9hz2cy2.1 (accessed on 12 December 2023).
22. Mock, P.; Díaz, S. Pathways to Decarbonization: The European Passenger Car Market in the Years 2021–2035. 2021. Available online: https://theicct.org/wp-content/uploads/2021/06/decarbonize-EU-PVs-may2021.pdf (accessed on 8 December 2023).
23. Basterra, J. Europe's Best-Selling Electric Cars in 2022. Available online: https://www.electromaps.com/en/blog/europes-best-selling-electric-vehicles-2022 (accessed on 10 March 2023).
24. Kane, M. BYD-Powered Tesla Model Y Receives EU Approval. Available online: https://insideevs.com/news/604025/byd-tesla-modely-eu-approval/ (accessed on 12 August 2022).
25. International Energy Agency. Global EV Outlook 2023: Catching up with Climate Ambitions. 2023. Available online: https://iea.blob.core.windows.net/assets/dacf14d2-eabc-498a-8263-9f97fd5dc327/GEVO2023.pdf (accessed on 8 December 2023).
26. BloombergNEF & Transport & Environment. Hitting the EV Inflection Point. 2021. Available online: https://www.transportenvironment.org/wp-content/uploads/2021/08/2021_05_05_Electric_vehicle_price_parity_and_adoption_in_Europe_Final.pdf (accessed on 8 December 2023).
27. International Energy Agency. Global EV Data Explorer. Available online: https://www.iea.org/data-and-statistics/data-tools/global-ev-data-explorer (accessed on 22 September 2022).
28. International Energy Agency. Global EV Outlook 2021. 2021. Available online: https://iea.blob.core.windows.net/assets/ed5f4484-f556-4110-8c5c-4ede8bcba637/GlobalEVOutlook2021.pdf (accessed on 8 December 2023).
29. Irwin, J. The Auto Industry Is Craving Greater Supply Chain Visibility. Available online: https://www.autonews.com/suppliers/auto-industry-craving-greater-supply-chain-visibility (accessed on 23 June 2022).

30. Hensley, R.; Laczkowski, K.; Möller, T.; Schwedhelm, D. Can the Automotive Industry Scale Fast Enough? Available online: https://www.mckinsey.com/industries/automotive-and-assembly/our-insights/can-the-automotive-industry-scale-fast-enough (accessed on 12 May 2022).
31. Eurostat. Trade in Electric and Hybrid Electric Cars on the Rise. Available online: https://ec.europa.eu/eurostat/web/products-eurostat-news/-/ddn-20210524-1 (accessed on 24 May 2021).

World Electric Vehicle Journal

MDPI

Article

Incremental Profitability Evaluation of Vehicle-to-Grid-Enabled Automated Frequency Restoration Reserve Services for Semi-Public Charging Infrastructure: A Case Study in Belgium

Andrei Goncearuc [1,2,*], Nikolaos Sapountzoglou [1], Cedric De Cauwer [1], Thierry Coosemans [1], Maarten Messagie [1] and Thomas Crispeels [2]

[1] ETEC Department & MOBI Reserach Group, Vrije Universiteit Brussels (VUB), Pleinlaan 2, 1050 Brussels, Belgium; nikolaos.sapountzoglou@vub.be (N.S.); cedric.de.cauwer@vub.be (C.D.C.); thierry.coosemans@vub.be (T.C.); maarten.messagie@vub.be (M.M.)

[2] BUTO Department, Vrije Universiteit Brussel, Pleinlaan 2, 1050 Brussels, Belgium; thomas.crispeels@vub.be

* Correspondence: goncearuc.andrei@vub.be

Abstract: The current paper defines a framework for the introduction of automated frequency restoration reserve services, enabled by vehicle-to-grid technology, into the business model of an entity owning and operating a network of semi-public Electric Vehicle Supply Equipment. It assesses the profitability of this introduction by performing a case study based on the real-life electric vehicle charging data from the EVSE network located in a hospital parking lot. From the results of the study, it is clearly visible that the introduction of vehicle-to-grid-enabled automated frequency restoration reserve services has a significant positive incremental profitability; however, this is heavily dependent on the plug-in ratio of the charging network, determined by electric vehicle users' behavior.

Keywords: vehicle-to-grid; business model; infrastructure; electric vehicle supply equipment; market development

Citation: Goncearuc, A.; Sapountzoglou, N.; De Cauwer, C.; Coosemans, T.; Messagie, M.; Crispeels, T. Incremental Profitability Evaluation of Vehicle-to-Grid-Enabled Automated Frequency Restoration Reserve Services for Semi-Public Charging Infrastructure: A Case Study in Belgium. *World Electr. Veh. J.* **2023**, *14*, 339. https://doi.org/10.3390/wevj14120339

Academic Editor: Grzegorz Sierpiński

Received: 20 October 2023
Revised: 1 December 2023
Accepted: 4 December 2023
Published: 6 December 2023

1. Introduction

1.1. Context

Recent years have shown a significant increase in the popularity of electric vehicles (EVs), which, in combination with the renewable energy supply, is generally considered a positive trend, leading to reduced pollution and a cleaner environment (e.g., reduced oil consumption and CO_2 emissions) [1,2]. At the same time, the growing number of EVs on the roads brings certain challenges. One of these challenges is the increasing pressure on electricity grids [3]. However, EVs can also provide a solution to this issue by means of vehicle-to-grid (V2G) technology [4], allowing for bidirectional energy transfer between the EV battery and the electricity grid, and thus providing the opportunity not only to consume and store energy in EV batteries but also to inject it back into the grid. Moreover, this creates additional opportunities both for grid operators, who would potentially benefit from a solution to grid-balancing issues, and for the participants of the EV charging business ecosystem, which could potentially benefit from the additional revenue streams.

Based on a number of previous studies defining the EV charging business ecosystem [5], the business models of its participants [5,6], the introduction of V2G technology into these business models [5] and the initial opportunities of V2G service organizers in grid-balancing markets [7], the current study makes a step further into the investigation of the V2G potential in grid-balancing services. Namely, this study assesses the incremental profitability of the introduction of V2G-enabled automated Frequency Restoration Reserve (aFRR) services, into the business model of an entity owning, managing, and maintaining a semi-public EV charging infrastructure.

1.2. Literature Overview

The vehicle-to-grid (V2G) concept was primarily introduced by the research of Kempton et al. [8], outlining the technical and financial opportunities enabled by the bidirectional energy flow to and from the EV battery. One of these opportunities, further elaborated by a number of follow-up studies [9–14], is the potential application of V2G technology to energy-grid-balancing services. The participants of the EV charging business ecosystem, managing and maintaining the network of Electric Vehicle Supply Equipment (ESVE) would, in this case, take over the role of the grid-balancing service providers (BSP), aggregating a number of V2G EVSE and providing power and energy to the central grid in case of necessity [9,10] in exchange for remuneration based on their bidding strategies in a grid-balancing auction [11]. It is important to mention that the application of V2G technology is not limited by the central grid-balancing services and includes numerous applications, including peak shaving, local load-balancing and others [12,13]. However, the V2G-enabled grid-balancing services are particularly valuable in the light of the potential future necessity of electric grid reinforcements [14].

The initial business model of the participants of the EV charging business ecosystem, managing and maintaining the network of EVSE, is mainly based on the provision of EV charging services as the core value proposition at present, covering the needs of the EV users as the main customer segment and receiving EV charging fees as the main revenue stream [5,6,15].

However, the V2G-enabled transformation of this business model introduces an additional value proposition: grid-balancing services. The new value proposition targets a new customer segment, namely transmission system operators (TSO) (entities responsible for managing and maintaining a high-voltage electricity grid). At the same time, the currently existing main customer segment—the EV users—takes the role of the key partner, providing the EV batteries for the V2G-enabled grid-balancing services [5,7,16].

According to Elia [17], the Belgian TSO, there are three types of grid-balancing services designed to avoid frequency deviations from a predefined constant level (e.g., 50 Hz in Belgium):

- Frequency Containment Reserve (FCR): primary reserve, which is automatically fully activated within a timeframe of 30 s in case of a significant frequency deviation and stabilizes the frequency fluctuations [18].
- Automated Frequency Restoration Reserve (aFRR): secondary reserve, which is automatically fully activated within a timeframe between 30 s and 7.5 min, in order to restore the frequency at the predefined level [19].
- Manual Frequency Restoration Reserve (mFRR): tertiary reserve, which is manually activated on demand within 15 min, in order to restore frequency at the predefined level in case of major imbalances [20].

According to the recent study, performed by Elia [21], EVs can be mainly used to provide FCR and aFRR services, as the provision of these services requires a relatively fast automatic activation and can be performed with limited energy resources. Moreover, according to [5], the inclusion of grid balancing services in the list of their value propositions can become a significant additional revenue stream for the participants in the EV charging business ecosystem.

Automated Frequency Restoration Reserve (aFRR)

From the revenue point of view, the aFRR service is particularly interesting for entities willing to engage themselves in the energy balancing market, since it opens two additional revenue streams: balancing power capacity and balancing energy remunerations [19].

In practice, the rules and procedures related to the provision of aFRR services differ from one TSO to another. However, even though the current study focuses on the Belgian TSO Elia, the procedural differences are not critical, and the results could be extrapolated to other geographical regions with minor adjustments.

It is also important to mention that the aFRR market was initially designed for large electricity-generating entities (e.g., gas and hydroelectric power plants), and still has substantial regulative barriers for small and medium enterprises (SMEs), CPOs, and other smaller prosumers willing to participate in the provision of the service [22]. The main barriers are:

- Minimum amount of 1 MW of power for capacity bid and 1 MWh for energy bid [19].
- Pay-as-bid auction principle, where the TSO pays exactly the amount indicated in the elected bid. The problem with this principle is that smaller entities rarely have sufficient resources for efficient continuous market analytics and are simply not able to indicate an up-to-date adequate price [23].
- Expensive specialized metering equipment, which must be installed at every delivery point aiming to provide aFRR services [24].

However, recent years show a visible decentralization trend in the grid balancing market, indicating that these regulatory barriers can be diminished in the near future. For instance, the provision of FCR services does not require the installation of additional specialized metering equipment and requires only a standard digital meter [25]. Moreover, the FCR power capacity auctions are transferred to the pay-as-cleared principle, where all the elected bids from different BSPs receive equivalent remuneration based on the highest price from the elected bids [23]. These changes in the regulatory framework of the FCR services can be seen as the first step towards the decentralization of the whole grid-balancing market, including aFRR services.

1.3. Contribution

Since the V2G technology has not reached its maturity phase and the opportunities provided by the technology are not yet widely applied, the existing literature still lacks studies related to the profitability assessment of V2G-enabled aFRR. Therefore, the aim of the current study is to address this gap by defining the framework for the introduction of the aFRR services into the business model of an entity owning and operating an Electric Vehicle Supply Equipment (EVSE) network and assess the incremental profitability of this introduction based on a case-study of semi-public EV charging infrastructure.

2. Methodology

2.1. Model

The revenue streams of an entity owning, operating, and managing a network of EVSE is mainly represented by the fees received from the provision of EV charging services, which form the core value proposition of its initial business model. The cost structure, however, comprises numerous elements, including the cost of the supplied energy, depreciation of EVSE, human resources (HR) remunerations, and others [6,15].

As mentioned before, the introduction of V2G-enabled grid-balancing services is able to diversify the list of value propositions, entering a new market with a new customer segment and creating additional revenue streams. The focus of the current study lies in the assessment of the incremental profitability of the provision of V2G-enabled aFRR, which is the difference between the additional revenues and expenses caused by the introduction of the service. The factors influencing the incremental profitability of the provision of V2G-enabled aFRR are described in the current section in Equations (1)–(5).

The revenue generated by the provision of V2G-enabled aFRR (R_{aFRR}) consists of two components, and can be defined as follows (Equation (1)):

$$R_{aFRR} = CR_{aFRR} + ER_{aFRR}, \tag{1}$$

- CR_{aFRR}: power capacity remuneration;
- ER_{aFRR}: energy remuneration.

The provision and remuneration of the aFRR service are based on the auction principle. After concluding the contract with a TSO, a BSP is able to make power capacity bids on a day-ahead auction. Moreover, there are two types of power capacity auctions—"all-CCTU" and "per-CCTU". The abbreviation CCTU means the Capacity Contracting Time Unit: the 4 h block when the power capacity bid made by the BSP can be activated by the TSO. Thus, in the first auction type, the bids are made for the whole 24 h, while in the second, bids are made for the 4 h blocks beginning from midnight (00:00 to 04:00; 04:00 to 08:00; 08:00 to 12:00, etc.). The BSP has to choose the suitable auction and CCTU(s) (in case of "per CCTU" auction) and make a power capacity bid, indicating the amount of power it is able to provide on the next day and the price of the desired service in EUR per MW of indicated power per hour (EUR/MW/h). The maximum amount of power the BSP is able to bid is defined beforehand by means of a prequalification test performed by the TSO. The bids are elected by the TSO based on the forecast-balancing power necessary for the next day and the "cheapest available" principle. If the bid made by the BSP is elected, the BSP receives the remuneration for the reserved amount of power (per MW) for the reserved time period (per hour) [19].

It is also important to mention that participation in the provision of aFRR services involves a certain risk of penalties in case of non-compliance with the contractual obligations of the BSP. The penalties can occur either due to the failure of spontaneous availability and/or activation tests performed by the TSO, or due to actual failure to provide the service during the activation. However, the maximum penalty should not exceed the remuneration of the respective month. Additionally, there is also a risk-mitigation opportunity, a so-called Transfer of Obligations (TO) procedure, allowing for the transfer of the power capacity obligations made by one BSP to another at the last hour before the due time, in case of any unexpected problems [19]. However, this procedure is based on agreements between the BSPs and can be costly for the demanding side.

Thus, for the V2G-enabled aFRR, the remuneration for the reserved power capacity (CR_{aFRR}) mechanism can be formulated as follows (Equation (2)):

$$CR_{aFRR} = aFRR_{Capacity\ Bid} \times \sum\nolimits_{y=1}^{Z} Ny \times Ky \times T_{reserved} \times \left(P_{plug\text{-}in} - P_{failure} \times F_{failure}\right) - P_{TO} \times C_{TO} \quad (2)$$

- $aFRR_{Capacity\ Bid}$: aFRR capacity bid (in €/MW/h) for the considered time period ($T_{reserved}$);
- y: type of EVSE (from 1 to Z) (e.g., uni/bi-directional; AC/DC; EVSE power level);
- Ny: number of EVSE types y participating in the provision of aFRR services;
- Ky: power level of EVSE type y;
- $T_{reserved}$: reservation time period of the available BSP power capacity;
- $P_{plug\text{-}in}$: probability that the EVSE type y is going to be plugged into an EV during the reservation time period ($T_{reserved}$);
- $P_{failure}$: risk factor, indicating the probability that the BSP will fail and be penalized;
- $F_{failure}$: the multiplication factor forming aFRR penalties, which is the factor to be multiplied with the price of the missing MW of power the BSP was not able to deliver;
- P_{TO}: risk factor, indicating the probability of the necessity of opting for the transfer of obligations (TO) service;
- C_{TO}: cost of TO service.

During the CCTU for which the balancing power capacity was reserved, the TSO can actually activate the bid, and its activation initiates the second type of aFRR remuneration—balancing energy remuneration (ER_{aFRR}). In order to provide (for aFRR+) (or decrease for aFRR−) the necessary power capacity, the BSP has to inject (or consume, in case of aFRR−) energy into the grid during the whole activation period, while the TSO will pay for this balancing energy. The balancing energy remuneration is also based on the auction principle, but is intra-day in this case. The BSP, whose power capacity bid was elected on the prior day, makes another intra-day energy bid, indicating the amount of energy (in MWh) and the price. In this case, the BSP receives the remuneration (cost reduction, for aFRR−) only in case of activation, based on the actual amount of MWhs injected (consumed, for aFRR−) into the TSO grid [19]. Thus, the V2G-enabled aFRR energy remuneration can be formulated as follows (Equation (3)):

$$ER_{aFRR} = aFRR_{Energy\ Bid} \times \sum_{y=1}^{Z} Ny * PLy * T_{activated} \quad (3)$$

- $aFRR_{Energy\ Bid}$: aFRR energy bid (in €/MWh) for the considered time period ($T_{activated}$);
- $T_{activated}$: activation time period of the available BSP power capacity.

The influence of the introduction of V2G-enabled aFRR services on the cost structure mainly involves the increase in infrastructure depreciation costs related to the difference between unidirectional and V2G EVSE prices, along with the necessary precise metering equipment to be installed at every delivery point (EVSE or EVSE hub). Thus, the additional costs related to the provision of V2G-enabled aFRR services can be defined as follows (Equation (4)):

$$C_{aFRR} = \sum_{y=1}^{Z} \frac{\Delta Py}{Ly} + \sum_{m=1}^{N} \frac{Pm}{Lm} \quad (4)$$

- ΔPy: difference in price between uni- and bidirectional EVSE with comparable power level;
- Ly: useful lifetime of EVSE type y;
- m: number of aFRR delivery points (from 1 to N) in the EVSE network;
- Pm: price of specialized aFRR metering equipment;
- Lm: useful lifetime of specialized aFRR metering equipment.

Defining the incremental profits for the provision of V2G-enabled aFRR service (IP_{aFRR}) as the difference between the additional revenues (R_{aFRR}) and costs (C_{aFRR}) results in the following formula (Equation (5)):

$$IP_{aFRR} = aFRR_{Capacity\ Bid} \times \sum_{y=1}^{Z} Ny \times Ky \times T_{reserved} \times \left(P_{plug\text{-}in} - P_{failure} \times F_{failure}\right) - P_{TO} \times C_{TO} + aFRR_{Energy\ Bid} \times \sum_{y=1}^{Z} Ny \times Ky \times T_{activated} - \sum_{y=1}^{Z} \frac{\Delta Py}{Ly} + \sum_{m=1}^{N} \frac{Pm}{Lm} \quad (5)$$

2.2. *V2G-Enabled aFRR Use-Case*

2.2.1. General Provisions

In order to assess the incremental profitability of the V2G-enabled aFRR services, the current research applies the defined model, generating a case-study based on real-life data and a set of grounded assumptions.

In general, the process of the provision of V2G-enabled aFRR services can be compared with the use of stationary batteries for similar purposes. The EV battery increases (for aFRR+) or decreases (for aFRR−) the power level of the TSO grid in case of need, while the TSO pays for the reserved capacity and the activated energy.

However, the reserved capacity bids for aFRR+ are, on average, higher than the aFRR− bids, while the V2G technology allows not only for energy to be consumed at a lower price (for aFRR−) but also to be injected and sold through energy bids by aFRR+ [26]. Moreover, according to the internal EV charging data, in most cases, the EVs plug in at >50% state of charge (SOC), while participation in aFRR− requires buffer space in the EV battery. Finally, due to this need for additional buffer battery space, the EV is not able to charge during the CCTU outside the activation periods, solely relying on aFRR− activation periods to charge. At the same time, the expected parking time is typically longer than the time needed to charge, creating the opportunity to compensate for the depleted energy in aFRR+.

Considering all the above-mentioned issues, the case-study generated by the current research is focused on the provision of V2G-enabled aFRR+ services.

The provision of aFRR+ can be performed in two ways, depending on the power baseline set by the BSP before the activation. Either, during the activation, the BSP stops consuming energy from the grid, reducing its own power and increasing the power in the TSO grid compared to the declared baseline (while consuming), or the BSP injects energy into the TSO grid, increasing the power in grid compared to the idle-state baseline, as shown in Figure 1.

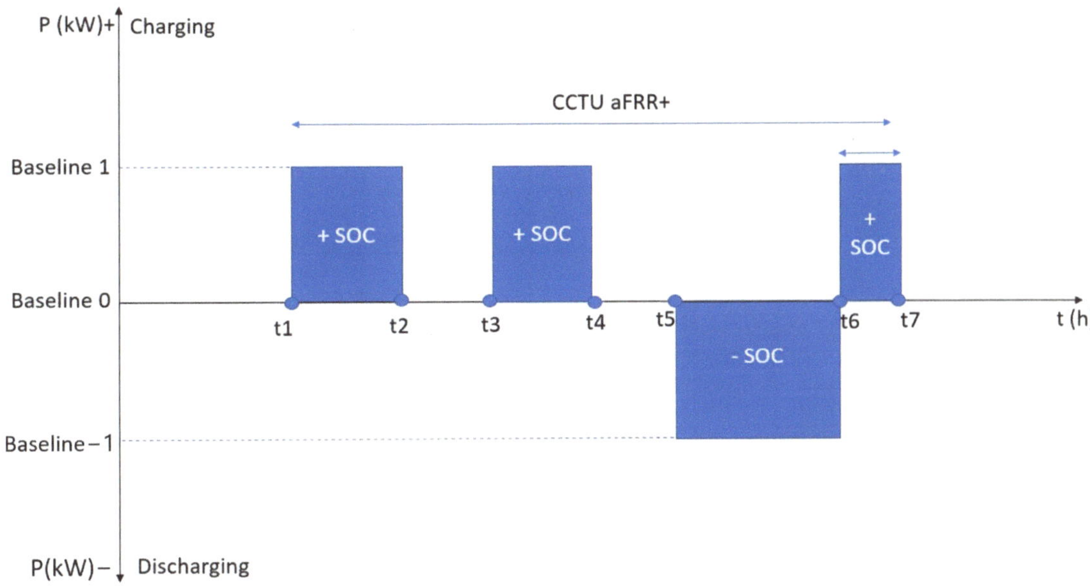

Figure 1. Example of the V2G-enabled aFRR+ provision process.

Figure 1 shows an example of the V2G-enabled aFRR+ provision process with time in hours on the *x*-axis and power in kW on the *y*-axis. The reserved CCTU begins at time t1 with the declared power baseline 1. At this point, the reservation period begins, but aFRR+ is not activated, so EVs connected to the EVSE network and engaged in the provision of the service are consuming energy and increasing their SOC. At timepoint t2, the TSO activates aFRR+ and the BSP stops consuming, dropping the power baseline to 0. The activation ends at timepoint t3, and connected EVs can continue to charge until timepoint t4, when they reach 100% SOC and remain plugged-in, but idle. At timepoint t5, the TSO initiates another activation, but this time the EVs are not able to stop charging, as they are idle and the power baseline is at level 0. Thus, the EVs begin to discharge, injecting energy into the grid. At timepoint t6, the activation ends, and EVs can begin to recharge the discharged energy, and at timepoint t7, the reserved CCTU ends. It is important to notice that, in case aFRR+ power capacity is provided by the reduction in or stopping of consumption, the BSP does not receive the energy remuneration, as no energy was actually injected into the grid.

Regarding the resulting SOC after the end of aFRR+ CCTU, due to the opportunity for service provision via stopping or reducing consumption, in the worst case, the additional ΔSOC would be equal to 0%, meaning that the EV would remain at the same state of charge as before CCTU. Therefore, a time buffer should be created after the CCTU to bring the EV to the SOC desired by the EV user.

However, on average, the probability of the occurrence of ΔSOC = 0% is less likely. By analyzing the open access data retrieved from Elia [26], the average aFRR+ activation time per CCTU (4 h) is 103 min (ex., injecting ~17.2 kWh of energy to the grid through 10 kW V2G charger), while, according to the internal EV charging data, the average time to reach 100% SOC is around 51 min (the vast majority of EVs plug in with 60–80% SOC). By subtracting 103 min from 4 h, it becomes clear that a time buffer of 137 min of non-active time within a CCTU is already present, making it easy to cover the time needed to reach 100% SOC.

2.2.2. Coping with Uncertainties for aFRR Capacity Remuneration

In addition the SOC, there is another important factor that plays a role. Unlike stationary batteries, the EV batteries move together with the vehicles, while the successful provision of V2G-enabled aFRR services requires every participating EVSE to be connected to an EV during the elected CCTU. Moreover, as the capacity bids are made on the day-ahead auction, the plug-in probabilities ($P_{plug\text{-}in}$) of the EVSE network for the elected CCTU should be known at least one day beforehand. This creates an area of uncertainty, consisting of the probability of using the costly TO risk-mitigation technique (P_{TO}) and the probability of failing to deliver the service and receiving the penalty ($P_{failure}$). Therefore, an accurate forecasting technique is of major importance for the successful implementation of the service.

The current research applies an EV-charging data-driven forecasting method, limiting the risk of failure. By making use of the historic EV charging data retrieved from the EVSE, which is meant to be engaged in the provision of V2G-enabled aFRR, the study defines a set of plug-in probabilities ($P_{plug\text{-}in}$) for every minute of the day. This allows for the CCTU(s) with the highest $P_{plug\text{-}in}$ to be elected, limiting the risk of failure.

After defining the CCTU(s) with the highest $P_{plug\text{-}in}$, the risk could be further mitigated by the TO option. This could be achieved by comparing how accurately the $P_{plug\text{-}in}$ values retrieved from the EVSE, which is meant to be engaged in the provision of V2G-enabled aFRR, one hour before and at the beginning of the elected CCTU(s) that correspond with each other (% of correspondence), and double checked by means of statistical analysis methods (e.g., *t*-test; ANOVA) (BSP can opt for a TO at the final hour before the CCTU). The high retrieved value indicates the high accuracy of the forecast and allows for the result of $(1 - P_{plug\text{-}in})$ to be used as the P_{TO} value.

Finally, the probability of failure ($P_{failure}$), despite all the risk-mitigation techniques, can be retrieved by calculating the joint forecasting accuracy of every CCTU timestep, adjusted for $P_{plug\text{-}in}$ at the beginning of CCTU.

2.2.3. Values of the Model Parameters

After outlining the general provisions of the case study and describing the methods used to cope with uncertainties, it is relevant to define the values for a number of parameters that actually participate in the calculations.

As shown in Table 1, the values of the parameters are divided into three subgroups. The first subgroup represents the values retrieved from external data sources. It is important to note the importance of ΔPy variable, as, according to [7], the profitability of the whole business model is very sensitive to the price of V2G EVSE. The ΔPy value presented in Table 1 is retrieved from the difference in the privately retrieved price quotes for a 10 kW DC bidirectional charger and a unidirectional AC charger of a similar power level.

Table 1. Values of the model parameters.

	Parameter	Symbol	Value	Units
External data source	EVSE type	y	DC V2G	/
	EVSE power level [27]	Ky	0.01	MW
	Difference between uni- and bidirectional EVSE price [27–30]	ΔPy	3000	€
	aFRR capacity bid [26]	$aFRR_{Capacity\ Bid}$	65.07	€/MW/h
	aFRR energy bid [31]	$aFRR_{Energy\ Bid}$	282.60	€/MWh
	CCTU time [19]	$T_{reserved}$	4	H
	Average activation time per CCTU [32]	$T_{activated}$	103	minutes
	EVSE useful lifetime [15,33]	Ly	10	Years
	Metering equipment cost [34]	Pm	2000	€
	Metering equipment useful lifetime [34]	Lm	10	Years
	Failure factor [19]	$F_{failure}$	1.3	/

Table 1. *Cont.*

	Parameter	Symbol	Value	Units
EV charging data	Plug-in probability during CCTU	$P_{plug\text{-}in}$	[0.136; 0.99]	/
	Probability of failure	$P_{failure}$	[0.009; 0.32]	/
	Probability of TO	P_{TO}	[0.01; 0.864]	/
Assumptions	Cost of TO	C_{TO}	1.2×Capacity remuneration	€
	EVSE network size	Ny	250	Units

The second represents the ranges of probabilities retrieved by means of calculations from the available charging dataset, which are discussed in more detail in the results of the study (Section 3). The third subgroup is the values that are part of the assumptions list designed explicitly for the current case study.

2.2.4. Design and Assumptions of the Case Study

The current research assesses the annual incremental profitability of the V2G-enabled aFRR+ services by means of a case study of semi-public EVSE infrastructure located in a hospital parking lot (UZ Bussel). The dataset includes 9344 charging sessions from 12 EV chargers over a two-year period (January 2020–January 2022), filtered to include the workdays only, assuming the highest probability of EVs remaining plugged in for a longer period of time during working hours. Moreover, the current case study generates results for participation in only one CCTU per day, namely CCTU 4 (12:00–16:00), which is the one with the highest plug-in probabilities and lowest risk of failure.

Following the application of the model defined in Section 2.1, the case study adopts the following assumptions:

(a) The costs of TO are defined by the bilateral contracts between the BSPs and are therefore not disclosed. The current study assumes this cost to be 120% of the capacity remuneration, as it is slightly lower than the one that is applicable for penalties.
(b) The average EV battery capacity of the EVs charging at the respective EVSE is 50 kWh.
(c) The provided case study does not include any bidding strategies, assuming all the power capacity bids are to be elected based on the average market price.

2.2.5. Scenarios

As is clear from the previous sections, the successful implementation of the V2G-enabled aFRR services is heavily dependent on the EV users' charging behavior, determining the $P_{plug\text{-}in}$ at a certain point in time. Therefore, the current study provides three different modeling scenarios, considering different types of behavior and interactions with EV users, which affect the $P_{plug\text{-}in}$ and its derivatives (P_{TO}; $P_{failure}$):

- Scenario 1: Natural behavior. The EV user agrees to the fact that his/her EV is going to be used for V2G-enabled aFRR services (or is unaware of this fact), but does not change his/her charging behavior and acts naturally. This scenario is based purely on the historical real-life data of EV charging patterns determining $P_{plug\text{-}in}$, P_{TO}, and $P_{failure}$. The EV user is not bound by any obligations and is able to unplug the EV at any time. At the same time, the EV user receives no shared revenues from the provision of V2G-enabled aFRR services.
- Scenario 2: Binding contract. The EV users receive binding day-ahead contracts, offering 20% of the aFRR+ capacity revenues for the permission to use their EV batteries for grid-balancing purposes. In this case, the EV would be plugged in and blocked for a period of 6 h, beginning 1 h before the elected CCTU (allowing fpr the user to opt for the TO option in case of emergency) and ending 1 h after the CCTU (ensuring that 100% SOC is reached for the EV after the provision of the service). In

case of a violation of contract terms (e.g., not plugging in or unplugging before the contractually defined moments), the EV user pays a penalty equivalent to the penalty the BSP would receive for missing the MW (securing the BSP from losses in case of contract violations). This allows for a the situation where $P_{TO} = P_{failure} = 0$. This can be seen as another risk-mitigation method, cutting out the additional expenses related to uncertainties by sharing 20% of capacity revenues with the EV users.

- Scenario 3: Non-binding contract. The EV users receive non-binding day-ahead contracts, offering 20% of aFRR+ capacity revenues for the permission to use their batteries for grid-balancing purposes. This contract type is a non-binding commercial offering that does not involve any penalties in case the EV user is not plugged-in during the defined period of time. Thus, in the worst case, the violation of the contract terms by the EV user would mean that no remuneration is received. In this scenario, 20% of the contracted users are assumed to violate the non-binding contract on average, creating losses related to TO and penalties for the BSP. This scenario can be seen as another risk-mitigation method, although less efficient than the one described in Scenario 2 in absolute terms for the BSP, but it is also less binding, and thus more attractive for EV users. In this case, the P_{TO} and $P_{failure}$ are limited to 20% of their initial value.

3. Results

Before proceeding to the actual results of the study, determining the incremental profitability of the V2G-enabled aFRR+ services for an entity owning, managing, and maintaining EVSE infrastructure, it is important to discuss the results of the $P_{plug\text{-}in}$ calculations and its derivatives, which play a crucial role in the successful implementation of the service. By making use of the method described in Section 2.2.2 and a real-life dataset retrieved from the EVSE network located in a hospital parking lot, the current study has defined the $P_{plug\text{-}in}$ distribution, as presented in Figure 2:

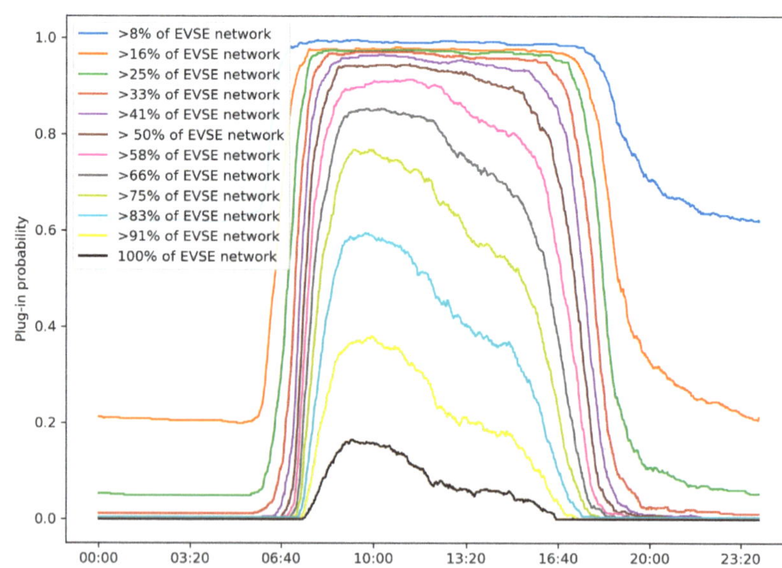

Figure 2. Plug-in probabilities of EVSE network during the working day.

Figure 2 shows the $P_{plug\text{-}in}$ (y-axis) of the EVSE network during the time of the day (x-axis). Every curve on the graph represents the probability that at least a certain percentage of the EVSE network (indicated in the legend) is connected to an EV (and can potentially be used for grid-balancing) at a certain point in time. With regard to the location and the nature of the given EVSE network, it is clear the $P_{plug\text{-}in}$ drastically increases around 07:00 and decreases around 17:00, indicating the average working hours of the hospital. This observatio directly points to the fact that the CCTU for the provision of grid-balancing services should be elected within this timeframe. Considering these conditions, there are two options regarding the CCTU choice: CCTU 3 (08:00–12:00) and CCTU 4 (12:00–16:00). However, there is also another point of attention, namely, P_{TO}. As mentioned before, the BSP can opt for TO at final hour before the elected CCTU, while the $P_{plug\text{-}in}$ values at 07:00 and 08:00 have significant differences, making the TO forecast inaccurate. At the same time, the $P_{plug\text{-}in}$ values at 11:00 and 12:00 match each other very well. Therefore, the optimal risk-limiting choice is to opt for CCTU 4 (12:00–16:00) in this case. Another important observation is that the higher the considered percentage of the EVSE network, the lower the chance of having this percentage simultaneously plugged into the EVs. However, the $P_{plug\text{-}in}$ density of up to 50% of EVSE network engagement remains quite high.

Thus, the first, and main, risk-mitigation method is an analysis of the historical plug-in data, as the ability to provide V2G-enabled aFRR services is the combination of the availability of V2G EVSE and the plugged-in EV. Therefore, the increase in the EVSE network engagement without the respective increase in EVSE occupation rate (increasing the potential plug-in probability) would only lead to losses. Also, as mentioned before, there is a TO option, serving as an official risk-mitigation method, limiting the potential losses related to penalties. Finally, the inclusion of the EV users in the contractual obligations, as described in scenarios provided in Section 2.2.5, serves as an additional, final risk-mitigation technique.

The incremental profitability of a V2G-enabled aFRR+ service for every EVSE network engagement level and every scenario defined in Section 2.2.5 is provided in Figure 3:

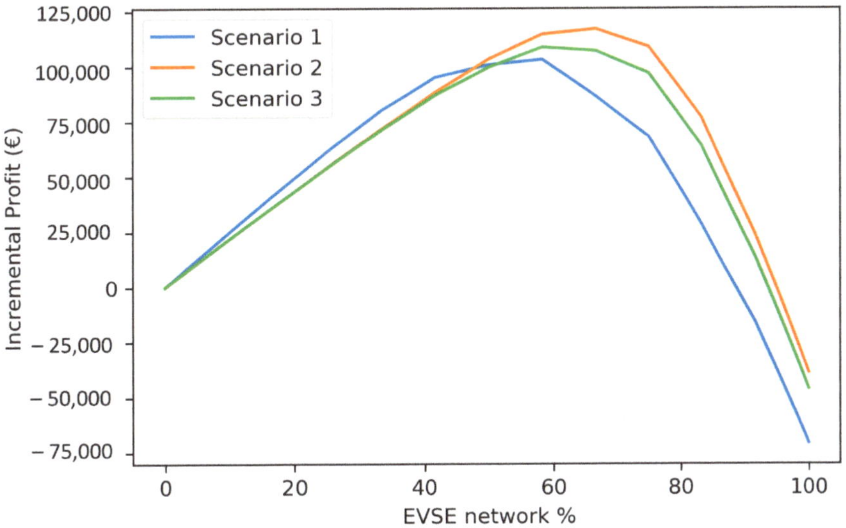

Figure 3. V2G-enabled aFRR incremental profit in the function of EVSE network % engaged in the provision of the service.

It is noticeable from Figure 3 that all the modeled scenarios show a positive incremental profit growth until the engagement of 60% (70% for Scenario 3) is reached by the EVSE network in aFRR+ services. These results are particularly interesting in light of the previously conducted research on the profitability of the provision of EV charging services only [7], showing negative profitability results (namely, −76,738 EUR) for this EVSE network size (250 EVSE units) caused by the high fixed costs and high electricity prices for smaller consumers. At the same time, it is clearly visible from Figure 3 that the incremental profits from the provision of V2G-enabled aFRR+ services are able to cover these losses, allowing for the break-even point to be reached on this relatively small network size.

Furthermore, after reaching the peak, the incremental profits begin to fall, eventually becoming negative at above 90% of network engagement. This behavior is explained by the lowering $P_{plug\text{-}in}$ that goes along with the increasing network engagement (clearly visible on Figure 2), and resulting increase in P_{TO} and $P_{failure}$. Moreover, even though the potential penalty is capped by the power capacity revenues, the negative incremental profit is caused by the additional expenses related to the provision of the service.

It is also noticeable that, at lower EVSE network engagement values (up to 50%), Scenario 1 (blue curve) is more profitable than the other scenario. This can be explained by the lower P_{TO} and $P_{failure}$, which lead smaller expenses compared to the EV users' remuneration. However, after 60% of EVSE network engagement, Scenario 1 shows a strong negative trend, reaching negative values faster than other scenarios. The reason for this is that the BSP in Scenario 1 does not mitigate the P_{TO} and $P_{failure}$ by means of contracts with EV users, and bares more risks when the plug-in probability of the chosen percentage of the EVSE network begins to fall.

4. Conclusions

The current research has defined the framework for the introduction of the V2G-enabled aFRR services into the business model of an entity owning and operating an EVSE network, and used the defined framework for an assessment of its profitability based on a case study of EVSE infrastructure located in a hospital parking lot.

From the performed analysis, based on real-life data and a set of modeling assumptions, it becomes clear that the introduction of V2G-enabled aFRR services into the business model of an entity owning, managing, and operating a network of semi-public EVSE can have a significant positive incremental profitability.

However, it is important to bear in mind that the provision of aFRR services is heavily related to the plug-in probability of the EVSE network, influencing the potential network engagement in the service and the probability of costly risk-mitigation techniques and penalties. As is visible from the results of the case study, the profits increase up to 60–70% of the EVSE network engagement in aFRR service, with a relatively high simultaneous plug-in probability. Up to this level, the increasing additional revenues are able to cover the expenses. At higher levels of network engagement, the simultaneous plug-in probability of the network is significantly lower, resulting in a higher probability of TO, penalties, and diminishing profitability.

By comparing different scenarios from the case study, it becomes clear that above 50% EVSE network engagement it becomes more profitable to conclude contracts and share profits with the EV users. Even non-binding contracts (assumed to be violated in 20% of cases) partially mitigate the penalty and TO risks born by the BSP and increase profitability at the higher levels of EVSE network engagement.

Finally, it should be pointed out that even though the defined framework is applied to the semi-public EVSE network in the current research, its application (with minor adjustments) can be extrapolated to public and private EVSE infrastructures as well. Moreover, the framework can be applied to the unidirectional smart charging infrastructure; however, this would remove the opportunity to make incremental balancing energy bids and limit the direction of power-balancing. Therefore, the results of these potential use cases could be significantly different, and are interesting topics for future research.

Author Contributions: Conceptualization, A.G.; methodology, A.G.; software, A.G.; validation, C.D.C., N.S. and T.C. (Thomas Crispeels); formal analysis, A.G.; investigation, A.G.; writing—original draft preparation, A.G.; writing—review and editing, C.D.C., N.S. and T.C. (Thomas Crispeels); visualization, A.G.; supervision, C.D.C., N.S. and T.C. (Thomas Crispeels); project administration, C.D.C.; funding acquisition, T.C. (Thierry Coosemans) and M.M. All authors have read and agreed to the published version of the manuscript.

Funding: This research was funded by H2020 Research and Innovation Programme, grant number 875131.

Data Availability Statement: Data are contained within the article.

Conflicts of Interest: The authors declare no conflict of interest. The funders had no role in the design of the study; in the collection, analyses, or interpretation of data; in the writing of the manuscript; or in the decision to publish the results.

References

1. International Energy Agency. Electric Vehicles. Available online: https://www.iea.org/reports/electric-vehicles (accessed on 11 April 2023).
2. Koroma, M.S.; Costa, D.; Philippot, M.; Cardellini, G.; Hosen, M.S.; Coosemans, T.; Messagie, M. Life cycle assessment of battery electric vehicles: Implications of future electricity mix and different battery end-of-life management. *Sci. Total Environ.* **2022**, *831*, 154859. [CrossRef] [PubMed]
3. Abousleiman, R.; Scholer, R. Smart Charging: System Design and Implementation for Interaction between Plug-in Electric Vehicles and the Power Grid. *IEEE Trans. Transp. Electrif.* **2015**, *1*, 18–25. [CrossRef]
4. Sovacool, B.K.; Axsen, J.; Kempton, W. The Future Promise of Vehicle-to-Grid (V2G) Integration: A Sociotechnical Review and Research Agenda. *Annu. Rev. Environ. Resour.* **2017**, *42*, 377–406. [CrossRef]
5. Goncearuc, A.; Sapountzoglou, N.; De Cauwer, C.; Coosemans, T.; Messagie, M.; Crispeels, T. An integrative approach for business modelling: Application to the EV charging market. *J. Bus. Res.* **2022**, *143*, 184–200. [CrossRef]
6. Goncearuc, A.; Sapountzoglou, N.; De Cauwer, C.; Coosemans, T.; Messagie, M.; Crispeels, T. Business model quantification framework for the core participants of the ev charging market. *World Electr. Veh. J.* **2021**, *12*, 229. [CrossRef]
7. Goncearuc, A.; Sapountzoglou, N.; De Cauwer, C.; Coosemans, T.; Messagie, M.; Crispeels, T. Profitability Evaluation of Vehicle-to-Grid-Enabled Frequency Containment Reserve Services into the Business Models of the Core Participants of Electric Vehicle Charging Business Ecosystem. *World Electr. Veh. J.* **2023**, *14*, 18. [CrossRef]
8. Kempton, W.; Tomic, J.; Letendre, S.; Brooks, A.; Lipman, T. *Vehicle-to-Grid Power: Battery, Hybrid, and Fuel Cell Vehicles as Resources for Distributed Electric Power in California*; UC Davis: Davis, CA, USA, 2001.
9. Venegas, F.G.; Petit, M.; Perez, Y. Active integration of electric vehicles into distribution grids: Barriers and frameworks for flexibility services. *Renew. Sustain. Energy Rev.* **2021**, *145*, 111060. [CrossRef]
10. Ali, H.; Hussain, S.; Khan, H.A.; Arshad, N.; Khan, I.A. Economic and Environmental Impact of Vehicle-to-Grid (V2G) Integration in an Intermittent Utility Grid. In Proceedings of the 2020 2nd International Conference on Smart Power and Internet Energy Systems, SPIES 2020, Bangkok, Thailand, 15–18 September 2020. [CrossRef]
11. Sortomme, E.; El-Sharkawi, M.A. Optimal combined bidding of vehicle-to-grid ancillary services. *IEEE Trans. Smart Grid* **2012**, *3*, 70–79. [CrossRef]
12. Steward, D. *Critical Elements of Vehicle-to-Grid (V2G) Economics*; No. DSEV1030; National Renewable Energy Laboratory, U.S. Department of Energy: Denver, CO, USA, 2017.
13. Cundeva, S.; Dimovski, A. Vehicle-to-grid system used to regulate the frequency of a microgrid. In Proceedings of the 17th IEEE International Conference on Smart Technologies, EUROCON 2017—Conference Proceedings, Ohrid, North Macedonia, 6–8 July 2017. [CrossRef]
14. Knezović, K.; Marinelli, M.; Codani, P.; Perez, Y. Distribution grid services and flexibility provision by electric vehicles: A review of options. In Proceedings of the Universities Power Engineering Conference, Stoke on Trent, UK, 1–4 September 2015. [CrossRef]
15. Madina, C.; Zamora, I.; Zabala, E. Methodology for assessing electric vehicle charging infrastructure business models. *Energy Policy* **2016**, *89*, 284–293. [CrossRef]
16. Borne, O. Vehicle-to-Grid and Flexibility for Electricity Systems: From Technical Solutions to Design of Business Models. Available online: https://theses.hal.science/tel-02101210 (accessed on 1 December 2023).
17. Elia. Keeping the Balance. Available online: https://www.elia.be/en/electricity-market-and-system/system-services/keeping-the-balance (accessed on 11 April 2023).
18. Elia. Terms and Conditions for Balancing Service Providers for Frequency Containment Reserve (FCR). Available online: https://www.elia.be/-/media/project/elia/elia-site/public-consultations/2020/20200317-tc-bsp-fcrfinal-consulten.pdf (accessed on 15 August 2022).
19. Elia. Terms and Conditions for Balancing Service Providers for Automatic Frequency Restoration Reserve (aFRR). April 2020. Available online: https://www.elia.be/-/media/project/elia/elia-site/electricity-market-and-system---document-library/balancing---balancing-services-and-bsp/2020/20200303_tc-bsp-afrr_en.pdf (accessed on 11 April 2023).
20. Elia. Terms and Conditions for Balancing Service Providers for Manual Frequency Restoration Reserve (mFRR). April 2019. Available online: https://www.elia.be/-/media/project/elia/elia-site/public-consultations/20191004_terms_and_conditions_mfrr/tc-bsp-mfrrfeb2020.pdf (accessed on 11 April 2023).
21. Elia. Using Electric Vehicles to Balance the Network. Available online: https://innovation.eliagroup.eu/projects/v2g/ (accessed on 23 August 2022).

22. Lauinger, D.; Vuille, F.; Kuhn, D. A review of the state of research on vehicle-to-grid (V2G): Progress and barriers to deployment. In Proceedings of the European Battery, Hybrid and Fuel Cell Electric Vehicle Congress, Geneva, Switzerland, 14–16 March 2017.
23. Elia. Remuneration of mFRR and aFRR Capacity: Pay-as-Bid & Pay-as-Cleared. 2020. Available online: https://www.elia.be/en/public-consultation/20200901_public-consultation-on-the-study-on-pay-as-bid-vs-pay-as-cleared (accessed on 11 April 2023).
24. Elia. Technical Documentation Concerning the Provision of Ancillary Services. 2021. Available online: https://www.elia.be/-/media/project/elia/elia-site/electricity-market-and-system/system-services/how-to-become-provider-documents-technical/20221109_afrr-ems-requirements.pdf (accessed on 11 April 2023).
25. Elia. Market Facilitation. Available online: https://innovation.eliagroup.eu/innovation-pillars/market-facilitation/ (accessed on 18 August 2022).
26. Elia. Individual aFRR Capacity Bids. 2023. Available online: https://opendata.elia.be/explore/dataset/ods125/information/ (accessed on 14 April 2023).
27. Izvia. Bornes de Recharge V2G Entreprise. Available online: https://www.izivia.com/bornes-electriques-lieu-de-travail/bornes-de-recharge-V2G-entreprise (accessed on 21 August 2022).
28. Sovacool, B.K.; Kester, J.; Noel, L.; de Rubens, G.Z. Actors, business models, and innovation activity systems for vehicle-to-grid (V2G) technology: A comprehensive review. *Renew. Sustain. Energy Rev.* **2020**, *131*, 109963. [CrossRef]
29. Indra. Vehicle-to-Grid (V2G) Chargers. Available online: https://www.indra.co.uk/v2g (accessed on 11 August 2022).
30. Alfen. EV Charging Stations. Available online: https://alfen.com/en-be/ev-charging-stations (accessed on 23 August 2022).
31. Elia. Individual Incremental Balancing Energy Bids (Historical Data). 2023. Available online: https://opendata.elia.be/explore/dataset/ods068/information/ (accessed on 14 April 2023).
32. Elia. Activated Balancing Energy Volumes per Minute (Historical Data). 2023. Available online: https://opendata.elia.be/explore/dataset/ods061/information/ (accessed on 1 December 2023).
33. Nelder, C.; Rogers, E. Reducing EV Charging Infrastructure Costs. 2020. Available online: https://www.researchgate.net/profile/Chris-Nelder/publication/338660434_Reducing_EV_Charging_Infrastructure_Costs/links/5e2237d6299bf1e1fab9ed9c/Reducing-EV-Charging-Infrastructure-Costs.pdf (accessed on 23 August 2022).
34. Elia. General Technical Requirements for Private Measurement. 2021. Available online: https://www.elia.be/-/media/project/elia/elia-site/electricity-market-and-system/system-services/how-to-become-provider-documents-technical/technical-requirements-for-private-measurement-devices-final-v3_28102021.pdf (accessed on 1 December 2023).

World Electric Vehicle Journal

Article

A System for the Efficient Charging of EV Fleets

Tobias Fleck [1,†], Sascha Gohlke [2,†] and Zoltan Nochta [2,*,†]

[1] SAP Deutschland, 69190 Walldorf, Germany; tobias.fleck@sap.com
[2] Data-Centric Software Systems (DSS) Research Group, Institute of Applied Research (IAF), Karlsruhe University of Applied Sciences, 76133 Karlsruhe, Germany; sascha.gohlke@h-ka.de
* Correspondence: zoltan.nochta@h-ka.de
† These authors contributed equally to this work.

Abstract: Smart charging is a means of monitoring and actively controlling EV chargers to optimize the distribution and consumption of energy with a focus on peak-load avoidance. This paper describes the most important requirements that have influenced the design and implementation of the "Smart Charging System" (SCS). It presents the architecture and main functional building blocks of the SCS, which have been realized in an iterative development process as an extension component of the already existing open-source solution "Open e-Mobility". We also provide details on the functionality of the core smart charging algorithm within SCS and show how various data sources can be utilized by the system to increase the safety and efficiency of EV charging processes. Furthermore, we describe our iterative approach to developing the system, introduce the real-world testbed at SAP Labs France in Mougins/France, and share evaluation results and experiences gathered over a three-year period.

Keywords: fleet; smart charging; infrastructure; ICT; load management

Citation: Fleck, T.; Gohlke, S.; Nochta, Z. A System for the Efficient Charging of EV Fleets. *World Electr. Veh. J.* **2023**, *14*, 335. https://doi.org/10.3390/wevj14120335

Academic Editors: Joeri Van Mierlo and Genevieve Cullen

Received: 18 October 2023
Revised: 17 November 2023
Accepted: 21 November 2023
Published: 2 December 2023

1. Introduction

In the past decade, the global market share of electric vehicles (EV) has been growing rapidly. A significant proportion of EVs of all types, including cars, delivery vans, trucks, buses, etc., belong to corporate fleets. For example, in Germany 58% of all electric cars sold in 2021 were registered to companies [1]. Companies are increasingly using their EV fleets for business-related and sometimes even mission-critical purposes as EVs prove to be more and more reliable. To ensure the high operational readiness of EVs and reduce dependency on publicly accessible charging stations, many companies build and operate their own EV charging infrastructure (CI) on their premises. Those facilities are also often used by employees to charge privately owned EVs at work. Establishing and operating a CI poses a number of economic challenges to a company, including high capital and operating costs (TCO), volatile and less predictable utilization (during and outside business hours), complex tax regulations, etc. [2–4]. In addition, businesses must take several technical boundaries at typical parking areas into consideration, such as missing or insufficient cabling, grid power limitations, bad network connectivity, etc. A properly designed software system can help enterprises master many of the operational challenges during the entire life cycle of charging stations and other related assets. A crucial task thereby is to optimize the distribution of available and in many cases limited amount of power among multiple, often heterogeneous EVs and chargers in a safe and cost-efficient manner. Smart charging algorithms can also help increase driver satisfaction by maximizing the average state of charge (SoC) across multiple simultaneously charged EVs at a given location [5]. In addition, ref. [3] shows that an intelligent charging strategy can almost double the utilization of the infrastructure and the available power compared to an uncontrolled baseline charging strategy. Further related work in the research area of smart charging is summarized in various studies. For example, ref. [6] reviewed seven case studies related to smart charging. In the context of commercial EV fleets, ref. [7] identified challenges in return-to-base scenarios. A comprehensive overview of smart charging applications together with an overview of publicly known pilot projects is provided in [8]. Case studies often include real-world testing in technically limited environments with a very small number of EVs. For example, in [9], a possible design of charging infrastructure for company locations is presented while considering charging preferences and trip data of a bakery in Germany. A three-day experiment in a so-called "mobility house" (containing student housing, a grocery store, and a parking garage) showed that rule-based peak shaving and load demand forecasting can reduce load demand peaks by 25.4–38.5% while ensuring a minimum SoC of 50% [10]. The applicability of smart charging approaches that were designed specifically for charge-at-work scenarios, such as [11,12], and various related scheduling strategies, e.g., refs. [13–17] are usually evaluated in simulations rather than operational environments. The same holds for [4], which proposed a charging simulation model to support the design of a corporate charging infrastructure based on employees' driving data. Further challenges in the context of scheduling charging processes and related requirements for a software system are presented by [2]. Other operational challenges that require the usage of additional hardware, e.g., to recognize vehicles that are not connected to the system (e.g., unplugged EVs, conventional vehicles that block EV parking lots, etc.), are not in the scope of our work. An approach to deal with such problems can be the integration of sensors, such as LiDAR systems, and related data platforms that help connected automated vehicles detect and handle certain situations, including searching suitable parking lots [18,19].

In this paper, we present the Smart Charging System (SCS), which is a software system mainly designed to serve companies that operate EV fleets and have one or more parking areas at their sites equipped with charging stations. In contrast to the above-mentioned research, our smart charging approach and system implementation has been deployed and actively used to power a large number of EVs in resource-constrained environments. Research results are often validated with the help of simulations, because research institutions usually do not operate EV fleets and EV fleet operators do not provide researchers with access to their business facilities and data. Nevertheless, simulation is a useful instrument, for example, to initially test new features or the applicability of system improvements before putting those into operation. The same holds for pilot projects, which usually only run for a limited time with a small number of EVs in a lab environment (e.g., known vehicle properties, selected hardware, controllable user behaviour). The findings presented in this work are based on the iterative implementation process and evaluation in a real-world testbed. We discuss requirements and approaches to fulfil them with the help of our SCS, such as dealing with missing data, connectivity issues, etc., in the real-world, which are usually not covered in smaller test scenarios or in theoretical research backed by simulations. Technically, the SCS is an optional extension component of the open-source system "Open e-Mobility" (OE) [20]. OE is currently used to manage thousands of charging stations at different locations of various companies. It can be deployed and operated as an on-premises system or as a containerized cloud solution and communicate with several other systems via the provided interfaces. The deployment and activation of the presented SCS within an OE instance helps manage energy distribution, infrastructure protection, and other related other requirements in an automated manner. Without activating the SCS in an OE system, related tasks must be carried out by the operator, who manually enters power limits for each charging point (CP), for example. The remainder of the paper is organized as follows: In Section 2, we describe our iterative approach to develop the software system, summarize the identified main functional and non-functional requirements, and introduce the real-world testbed at SAP Labs France in Mougins (France) that was used for technical evaluation. In Section 3, we present the architecture and main functional building blocks of the SCS, explain the functionality of the core EV charging algorithm, and show its integration with various data sources that are supported in the current implementation. We also discuss particular advantages of the incrementally added features and data sources that helped increase the efficiency of the charging infrastructure. Finally, we draw conclusions and outline directions for our future work in Section 4.

2. Materials and Methods

2.1. Iterative Software Development Approach

The SCS, along with other software components of OE, was realized and tested in multiple iterations in a period of three years. The development process included multiple phases and related test cycles. In each development phase, a new, encapsulated and independent component was added to the already existing SCS so that respective improvements were feasible and measurable. When designing and implementing the SCS, we focused in particular on the following capabilities as main requirements for the system:

- **Infrastructure protection:** During the simultaneous charging of EVs, huge demand peaks can occur, damaging the infrastructure or even leading to outages. The SCS must deal with several related thresholds at the same time, such as the mains connection power of the site, limitations of the local electrical infrastructure according to fuse hierarchies, capacity of individual power lines and transformers, etc. In addition, it is important to communicate with the local energy management system (EMS), if deployed on site, to quickly react on fluctuations of the available power caused by electricity consuming devices (e.g., machinery, HVAC devices) or by energy producing assets (e.g., PV, CHP).
- **Management of heterogeneous charging equipment:** A company's CI can contain AC and DC chargers of various vendors, types and versions. Considering only "abstracted" equipment in the software system can lead to severe problems, because "real" devices behave differently with respect to their, e.g., charging curve characteristics, in/output ratios, means of data provisioning, interfaces, configurable parameters, etc. The larger the CI, the greater the cumulative effect of these factors can be.
- **Support of EV-specific charging:** During a charging session, the EV's battery management system may autonomously increase (or lower) the demanded power. As a reaction, the SCS may limit the maximum available power or provide the EV with additional power, e.g., by rescheduling other EVs' charging sessions. Accordingly, the SCS requires up-to-date information about connected vehicles, including the maximum allowed current/power, number of phases used, etc.
- **Context-aware prioritization:** In the business context, a prioritization of charging sessions is often needed: A salesperson, who wants to visit a customer and needs a "full" battery within two hours, has higher priority than another employee, who leaves the office at the evening. To determine prioritization, data items from different sources are required, e.g., planned arrival time, estimated departure time, capacity of EV batteries, current SoC, etc.
- **Interoperability and scalability:** The SCS must seamlessly interact with other system components over available interfaces and network protocols. It should also be able to serve CIs of different size and allow adding (removing) locations to the overall setup.
- **Flexibility:** CI sites have different properties and characteristics, for example, with regard to the number and type of served EVs, usual charging times, local infrastructure limitations, etc. Consequently, the structure and operational complexity of the SCS also varies between deployment sites. In order to address this, the SCS needs to be built to be modular and thus adaptable to the given infrastructure, EV fleet, user needs, and prioritization requirements. In general, the SCS must be able to work in different complexity levels and enable adding/removing components independently from each other.
- **Exception handling:** In case of errors, e.g., due to malfunctioning charging stations or EVs, a proper exception handling in near real time is needed. Thereby, vendor- and device-specific error messages must be captured and properly interpreted. It must also be ensured that failing or bad network connectivity (HTTP, WebSocket, TCP/IP)

does not jeopardize running charging sessions and missing data are handled when planning new sessions. If there is an outage in the local electrical system, a safe restart of charging procedures is required.

2.2. Method of System Evaluation

The evaluation of the system in an operative environment took place at the premises of SAP Labs France in Mougins, France. The charging infrastructure on site has been initially set up and maintained by local team members. Experiments with the SCS began on 1 April 2020, which was a good time to start, because the local CI was less stressed as usual (due to COVID-19) and users were therefore more tolerant of potential technical problems. As time went on, the number of both charging stations and charging operations increased steadily, so that the scalability of the system could be tested as well. The testbed currently comprises 38 charging points (31 AC and 7 DC) at 22 charging stations of different vendors, including Schneider Electric, Legrand, ABB, Delta, IES, Webasto, Ebee, Mennekes, Keba, StarCharge, Wall Box Chargers, and Joinon. In the evaluation period, the system served in total over 650 employees to charge 291 company cars of various vendors including Tesla, Jaguar, Kia, Renault, Volkswagen, Audi, Mercedes, Hyundai, BMW, Fiat, Volvo, and Nissan. In total, more than 25,000 EV charging sessions were executed successfully, consuming approximately 700 MWh energy with a combined session time of almost 3400 days. The system protected the power-constrained infrastructure well since it never experienced critical overload situations throughout the entire test period. In addition to the real-world tests, a tool [21] that simulates the behaviour of multiple OCPP-compliant charging points was also used to frequently test SCS. It especially helped avoid technical and safety-related problems that would have been occurred due to errors (bugs) in the software implementation.

3. Results and Discussion

3.1. System Design

The high-level architecture of the SCS contains four main functional components as shown in Figure 1. The main task of the component *Smart Charging Core* is to calculate and dynamically adjust the distribution of available power among the active charging sessions in the given CI (see details in Section 3.2). The *Data Manager* stores permanent data, such as the system configuration and master data about the capabilities of the installed charging stations. It also maintains temporary information needed to carry out calculations, for example. As part of OE, the SCS interacts with other components of the entire charging-point management system, termed as "Internal Components" in Figure 1. For instance, the SCS logs relevant technical events using the *Logging* interface of OE. The SCS also provides information about the status of active charging sessions for EV drivers via the *Mobile App* as well as for the CI's technical operator via *Browser/Portal*. The SCS can communicate and exchange data with further "External Components", including *EMS*, enterprise resource planning (*ERP*) or EV vendors' *Vehicle Backend*, if they are available and made accessible within the CI owner's IT environment. These external systems are mainly used by the SCS as data sources to support ongoing calculations of charging plans. The required connections to these systems and charging stations on site, including protocol- and API-specific messaging, are handled by the *Communication Manager*. The component *Integration Layer* is mainly responsible for collecting the required data from the different connected sources in a synchronous or asynchronous way, and also for the preparation of the gathered data for further processing by the core component (see details in Section 3.3).

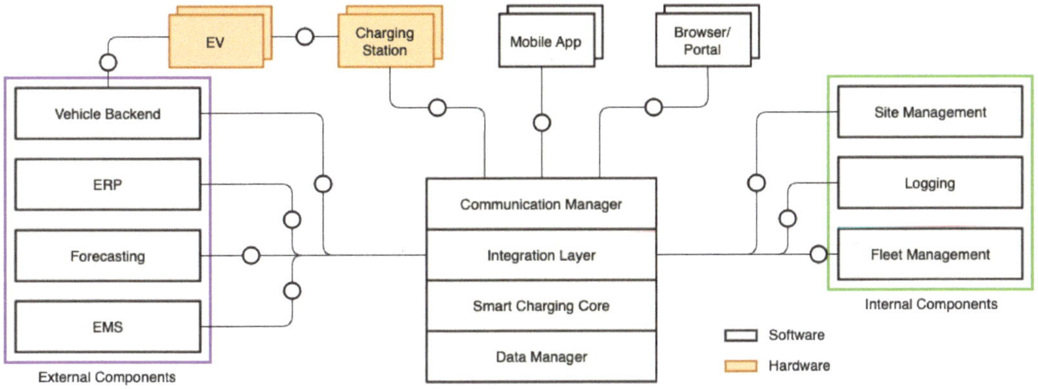

Figure 1. High-level architecture diagram of the Smart Charging System. The four main components in the middle can fetch data from various external data sources and also interact with other parts of Open e-Mobility.

The deployment of the four main components, i.e., without the above-mentioned additional data sources, is sufficient to operate the SCS with basic functionality. In this case, the *Smart Charging Core* can work with predefined configuration values, such as fixed safety limits for power consumption, and it does not take into account dynamic information, such as instantaneous solar power generation. The above-mentioned additional internal and external components can be added (activated) optionally and independently from each other helping to adapt the SCS to specific requirements in the given scenario.

3.2. Smart Charging Core

The current version of the SCS implements a scheduling procedure illustrated by the flowchart in Figure 2. The initial concept is presented in [13,22], and the corresponding implementation is available online on GitHub [23]. The main goal of the overall process (see also the pseudo-code in Algorithm 1) is to share the basically limited charging power at a given location among the connected EVs in a fair manner. The SCS triggers the calculation when a new charging session starts to meet the additional demand, or when an ongoing session ends to redistribute the released capacity. The scheduling procedure can also be executed periodically (e.g., every 15 min) to adjust the power consumption of ongoing sessions, as well as on demand, when significant changes in the amount of available energy are detected (e.g., through additional solar production). The output of the calculation is a charging plan that determines which of the connected EVs should receive power in the next k time slots without violating the local site's safety limits (see also Algorithm 2 for details). During charging plan creation, EVs can be prioritized (using Algorithm 3), which is especially useful when the total (aggregated) power demand would exceed the maximum available power in one or more time slots. As shown in Figure 2 several optional data sources (represented as rectangles) can be utilized during the calculation in order to adapt the system's behaviour. It is important to note that the scheduling process is able to work without receiving data from those sources, e.g., due to a technical problem. For example, in case the actual battery type and capacity of a plugged-in EV is unknown because the Fleet Management component is temporarily unavailable, the SCS can run using preset values.

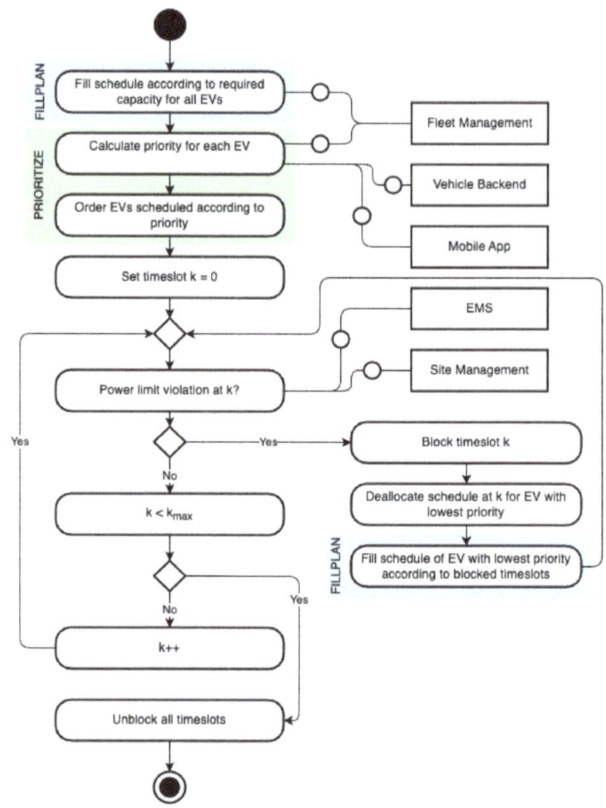

Figure 2. Flowchart diagram of the scheduling procedure based on [13] with the main calculation steps and optionally involved data sources (Fleet Management, Vehicle Backend, Mobile App, EMS, Site Management).

Algorithm 1 Scheduling procedure

```
procedure SCHEDULE(evList, tsList)
    for i ← 1 to evList.length do
        FILLPLAN(ev[i], tsList)                          ▷ Algorithm 2 called
    end for
    PRIORITIZE(evList)                                   ▷ Algorithm 3 called
    for k ← 1 to tsList.length do
        sumI_ts ← 0
        for i ← 1 to evList.length do
            sumI_ts ← sumI_ts + evList[i].tsList[k].I
        end for
        index ← 1
        while sumI_ts ≥ fuse limit do                    ▷ Check if total current exceeds limit
            tsList[k] ← blocked                          ▷ Block time slot for rescheduling
            sumI_ts ← sumI_ts − evList[index].tsList[k].I
            FILLPLAN(evList[index], tsList)              ▷ Reschedule EV with lowest priority
            index + +
        end while
    end for
end procedure
```

Algorithm 2 Procedure to fill EV charge plans

1: **procedure** FILLPLAN($ev, tsList$)
2:     **for** $k \leftarrow 1$ to $tsList.length$ **do**
3:         **if** $tsList[k]$ **not** $blocked$ & $ev.cap_{init} + ev.cap_{cha}(k) \leq ev.cap_{max}$ **then**
4:             $ev.tsList[k].I \leftarrow min(ev.I_{max}, cp.I_{max})$ ▷ Assign lower value of CP/EV max current
5:         **else**
6:             $ev.tsList[k].I \leftarrow 0$
7:         **end if**
8:     **end for**
9: **end procedure**

Algorithm 3 Prioritization procedure

procedure PRIORITIZE($evList$)
    for $i \leftarrow 1$ to $evList.length$ **do**
        $mCap_{minSoC} \leftarrow evList[i].cap_{des} - (evList[i].cap_{init} + evList[i].cap_{cha})$
        $\Delta t \leftarrow evList[i].t_{dep} - evList[i].t_{now}$
        if $mCap_{minSoC} \geq 0$ **then** ▷ Assign higher priority if EV is below minimum SoC
            $evList[i].priority \leftarrow mCap_{minSoC} / ((\Delta t * evList[i].I_{max}) + 1e - 8)$
        else
            $mCap_{maxSoC} \leftarrow evList[i].cap_{max} - (evList[i].cap_{init} + evList[i].cap_{cha})$
            $evList[i].priority \leftarrow mCap_{maxSoC} / ((\Delta t * evList[i].I_{max}) + 1e - 8) - 1000$
        end if
    end for
    sort $(evList, priority)$ ▷ Sort according priority
    return
end procedure

The scheduling algorithm (see Algorithm 1) initially creates a "greedy" charging plan for each ev in $evList$ for n time slots of duration d represented in $tsList$. In a practical setup, for example, with $n = 96$ and $d = 0.25$ hours, a charging plan for the next 24 h can be created.

By executing Algorithm 2 for each EV (see Lines 2–4 in Algorithm 1), the maximum possible charging current will be assigned to each EV, according to the limitations of the given EV ($ev.I_{max}$) and the charging point ($cp.I_{max}$).

This is repeated for the next time slots until the sum of the EV's initial charge capacity $ev.cap_{init}$ (measured in Ah) and charged capacity $ev.cap_{cha}$ reaches/exceeds the battery's maximum capacity $ev.cap_{max}$. Note that $ev.cap_{cha}$ is calculated based on the charging current I assigned to the EV and the total duration of passed k time slots.

To face potential conflicts that could occur if the total scheduled charging power within one or more time slots exceeds power limitations of the charging infrastructure, some EVs' initially created charging plans must be adjusted, i.e., delayed. For that purpose, EVs are ranked by executing Algorithm 3 (see Line 5 in Algorithm 1). In order to determine the critical time slots, $sumI_{ts}$, the sum of charging currents assigned to all EVs in $evList$ in each time slot, is calculated. A particular time slot will be *blocked* (see Line 13 in Algorithm 1) if the resulting value is not below the relevant technical limitation of the charging site's electrical system (called *fuse limit*). The $sumI_{ts}$ is reduced by the previously given charging current I of the lowest ranked EV (see Line 14 in Algorithm 1), whose charging plan will be refilled. Afterwards, the EV with the lowest priority is rescheduled by applying Algorithm 2 (see line 15 in Algorithm 1). Reducing the charging current to zero in all blocked time slots (see Line 6 in Algorithm 2) leads to a delayed/prolonged charging of the particular EV, because the intended cap_{max} value cannot be reached otherwise. This shifting procedure is repeated for the next ranked EVs until the violation of the fuse limit within the time slot is solved. Note that an adjustment of the charging current in the last unblocked time slot to match fuse limits more exactly is implemented but not included in the pseudo-code due to readability and space reasons.

The aforementioned prioritization of EVs for being potentially rescheduled is performed in Algorithm 3. To rank EVs in $evList$, the missing capacity to reach the minimum SoC $mCap_{minSoC}$ (measured in Ah) is calculated for each EV. This is the difference between the EV's desired charge capacity cap_{des} (calculated from the desired SoC, in %, as entered by the user) and the sum of its initial capacity cap_{init} on arrival and the capacity charged cap_{cha} since then (see Line 3 in Algorithm 3). The urgency of charging depends on the available time Δt between departure time t_{dep} (e.g., entered by the EV driver) and current time t_{now}. If the minimum SoC is not yet reached, the priority is calculated based on $mCap_{minSoC}$, the urgency Δt, and the maximum charging current I_{max} of the EV (see Line 6 in Algorithm 3). The applied formula basically ensures that EVs/drivers with higher energy demand and less time for charging will receive a higher priority in average and thus will not be taken as first candidates for being "shifted". The other EVs that already reached their minimum expected SoC will be ranked based on the charge capacity that is missing to reach the maximum capacity of the vehicle's battery $mCap_{maxSoC}$. The chosen formula (see Line 9 in Algorithm 3) gives in average a higher rank for those EVs with higher energy demand and less available time to fully charge their batteries. Accordingly, first candidates for rescheduling will be those EVs that almost reached their batteries' maximum capacity and still have time to wait.

3.3. Integration Layer

To leverage the capabilities of the generic *Smart Charging Core* component and to configure the implemented algorithms properly, information from several heterogeneous data sources with regard to, e.g., available APIs, security settings, data formats, etc., must be gathered. In case these sources are not deployed in the given environment and/or (temporarily) unavailable, the algorithms must be provided with preset values to ensure operational safety at any point in time. Similarly, a calculated charging plan must be transmitted to all connected charging stations and the respective EVs, so that they can interpret received messages (commands) and set configuration parameters or return data as requested. These data-oriented tasks are carried out by the *Integration Layer*. This component basically allows the adaptation of the *Smart Charging Core* to the given context and operational environment. Note that the SCS currently supports Open Charge Point Protocol (OCPP) version 1.6 [24]. Accordingly, the *Integration Layer* creates and maintains a charging profile for each connected, OCPP-compatible charging station within the CI. A fundamental task thereby is to handle misbehaving or incompatible charging stations. For that purpose, the *Integration Layer* monitors and reflects the current status of the CI, and it reacts on events that occurred. It can also collect data on ongoing charging sessions in near real time and help redistribute the available power according to the actual power consumption of ongoing sessions. Below, the tasks and functionality of the *Integration Layer* are explained in more detail.

- **Error and Exception Handling:** The SCS presupposes a proper implementation of OCPP by the charging stations and the support of OCPP charging profiles. However, OCPP implementations vary by charging station manufacturer and model. Compatibility problems often appear in specific setups and cases, which were not known upfront. The *Integration Layer* offers different mechanisms to master such situations. When collecting data to properly configure the core scheduling algorithm, the capabilities of connected stations are checked. It is especially examined whether each station is able to work with the generated OCPP profiles. If not, the given charging station will be excluded from the optimization, because it cannot be limited. In order to not endanger the electrical infrastructure, the SCS will automatically adjust infrastructure limits for the next optimization cycles by subtracting the maximum power that the incompatible station can draw. The adjustment of these limits only happens if the affected charging station is charging. Otherwise, the full capacity can be considered by the optimization. A similar mechanism is applied if a charging station is rejecting or not answering to charging profile requests, e.g., due to network issues. In this case, the faulty stations are collected and handled as incompatible charging stations in a separate optimization cycle. At the same time, a notification framework informs the administrators about the stations, which are not working correctly in order to take action if the issue persists.

- **AC/DC Handling:** The *Integration Layer* supports both AC and DC charging sessions according to their specifics. AC sessions can use one, two, or all three phases depending on the given charging station and connected EV. When triggering the *Smart Charging Core*, this information must be taken into account to determine the demanded charging current per phase. DC stations usually use all three phases, which makes phase assignments redundant. For DC chargers, however, the efficiency values need to be taken into account because the conversion from AC to DC is carried out by the charging station and not by the EV (as in case of AC chargers).
- **Vendor-specific handling:** Charging station vendors tend to vary in how they handle OCPP charging profiles, for example, by using their preferred measurement units (kW or Ampere). Therefore, the *Integration Layer* provides a framework and mechanism to adapt a generic charging profile template to vendor specific requirements.
- **EV-specific handling:** With the help of the *Fleet Management* component, the SCS is able to retrieve data about converters and batteries of almost every EV-model on the market, by using the Electric Vehicle Database [25] and other similar repositories. To keep the vehicle data up to date, synchronization jobs with the respective data sources are implemented. The data can be used to instantiate vehicles of a certain type in *Fleet Management*. By assigning these vehicles to users, the system can determine which EV model is charging at which station, without the need to establish a communication channel to the EV itself. The extracted information (converter data, battery size) is used to send power limits and battery capacities to the *Smart Charging Core* instead of waiting for actual monitored consumption data and adopting power limits later. In addition, the system provides implementations of service interfaces offered by OEMs, such as Mercedes, and also by third-party service platforms, like Tronity, to receive live information about the current state of charge during AC sessions. The stored data of the EVs are extended with this information and can be provided for different purposes such as priority handling.
- **Real-time behaviour adaptation:** The process of EV charging (both AC and DC) can be influenced by many factors. The charging curve, i.e., the power drawn over time, depends not only on the type, age, and condition of the hardware on the vehicle side but also on external parameters, such as temperature. In some cases, significant deviations from the expected model-specific behaviour can be observed when charging a particular EV. A negative implication can be that EVs consume less power than expected and thus allocated to the session when it started. DC chargers especially manage power usage actively by monitoring the connected battery's charging status. An efficient charging system must react to such varying (in general, unpredictable) power curves. The *Integration Layer* captures the momentary power drawn in the charging sessions and supports the SCS in calculating the real charge demand of a particular vehicle. The implemented mechanism puts a buffer on top of the observed power output of the charging station and uses the increased value as a new power limit for the session, whereby the new limit remains below the connector's maximum limit. By supervising whether the EV uses the buffer, the algorithm can determine if the car would be able to draw more power and provide it to the session if available. This way, it is also ensured that incorrect or missing vehicle data does not lead to the allocation of later unused power capacity.
- **Dynamic power limits:** In most cases, charging stations are operated in combination with other energy consuming or producing devices. The amount of demanded and produced power within an electrical system can heavily vary depending on season,

time of the day, temperature, weather, etc. Setting a fixed, safety-oriented power limit for the CI could make it basically independent from the fluctuations but lead to lower throughput and efficiency. For this reason, the SCS can be integrated with external EMS that monitors and controls the overall electrical infrastructure on site. This integration should be as flexible as possible, to support as many EMS providers as possible. Thus, the SCS provides an API endpoint to push energy data, but also integrates with external APIs to pull/request data from. By taking into account EMS data, it is possible to dynamically adapt the available power for the CI according to the current solar production, building consumption, etc.

- **Priority Handling:** To support the prioritization of EVs (as shown in Algorithm 3) and related charging sessions, the SCS collects as much information as possible. For instance, the *Mobile App* provides a dialogue for the driver to enter their planned departure time, required state of charge, and also the current state of charge (if the data are not provided by another integrated source or service). After the data are collected, they are processed and passed to the *Smart Charging Core*, which uses the received parameters to determine which EVs are prioritized and can thus charge faster. This ensures a fair sharing of power among trustworthy EV drivers and helps minimize inactivity times.

3.4. Implementation and Deployment in a Real-World Testbed

The SCS was implemented using TypeScript as programming language and the NodeJS runtime, which are utilised by other OE components as well. The integration with components of the existing OE was carried out in accordance with the programming guidelines of the overall project. To integrate optional, encapsulated components, OE uses a feature-oriented approach and a *Component Manager*. It enables the selection and individual activation, deactivation and configuration of certain functional artefacts in a flexible manner [26]. An example for such an optional feature in the context of OE is the roaming functionality, which can be switched on or off according to end users' requirements. The flexible combination of technically encapsulated system components is fundamentally limited, mainly due to semantic (sometimes mutual) dependencies between the components. These dependencies arise when components require the functionality of other components to work properly.

For example, the *Smart Charging Core* component relies on the component *Site Management*, which is used in OE to maintain the configuration of sites (company locations), site areas (parking facilities at a given site), and related charging station assignments. At the same time, *Site Management* does not depend on an activated *Smart Charging Core* component. Therefore, when manually enabling or disabling components in the admin user interfaces (UIs), the *Component Manager* verifies whether the known dependencies are met. If a violation was found, the *Component Manager* prevents the change and evaluates necessary actions to resolve the violation, providing the system administrator with the required information. If all (known) dependencies are satisfied, the configurations are stored in the database. When a component is disabled, the corresponding database entry is deleted. Storing component configurations in the database provides flexibility during runtime. The activation status of a component can be determined by a database request, and according to the concepts in [27,28], feature flag variables in the code can be used to determine whether the component's implementation should be executed or not. Validated changes take effect immediately, but the user needs to log out and log in again to update the user context and properly display configuration menus. Furthermore, the implementation offers multiple variants for each component. For example, the smart charging algorithm used in *Smart Charging Core* can be replaced with an alternative algorithm while keeping other components unchanged. This flexibility is achieved through the use of a factory pattern [29], which allows for different implementations of the same component. The communication and data exchange with external data sources adheres to commonly accepted practices like REST [30] using standardized web communication protocols, such as HTTPS and WebSocket Secure (WSS), and the related commands. The system further ensures the safety of the electrical infrastructure by employing fallback mechanisms. In the event of API call failures, it uses default values regarding battery sizes, state of charge, departure time, etc. Any issues that may arise are handled by a sophisticated error handling system, which makes sure to not exceed infrastructure limits. Additionally, every step performed in the SCS can be logged by calling the logging component within OE, enabling advanced troubleshooting in case of any unintended behaviour.

Before deploying the SCS in the testbed, as described in Section 2.2, it was possible to exceed the preset maximum power limit, for example, if a large number of EVs had to be charged simultaneously. With the roll-out of the first version of the SCS core system with its main components (i.e., without using any other external data sources), it was ensured for the first time that the maximum power limit of the local infrastructure could no longer be exceeded. This "safety-first" strategy did not take into account the actual power limits of the vehicles' converters. Instead, the algorithm assigned to each charging session the maximum charging current, which was derived from the chargers' maximum output power, e.g., 22 kW in case of most AC chargers. The actual assignment of the determined power to a particular charger takes place in updating the charger's OCPP Charging Profile using the message *SetChargingProfile.req*. As a result, the fixed maximum power limit of the CI was reached quickly, so only a few chargers could operate in parallel while the other charging stations received no power. The disadvantage of this approach is also illustrated in the upper part of Figure 3. In the example, a Tesla Model 3 charges constantly at 11 kW, although the connector has a maximum power of 22 kW. Without adjusting the limit to the actually demanded power, the SCS statically allocates 22 kW for the entire session duration. The unused yet blocked 11 kW are "wasted", i.e., they cannot not be given to other stations at the same time. For instance, in a CI segment created for testing with a power limit of 110 kW, only five EVs could be charged simultaneously.

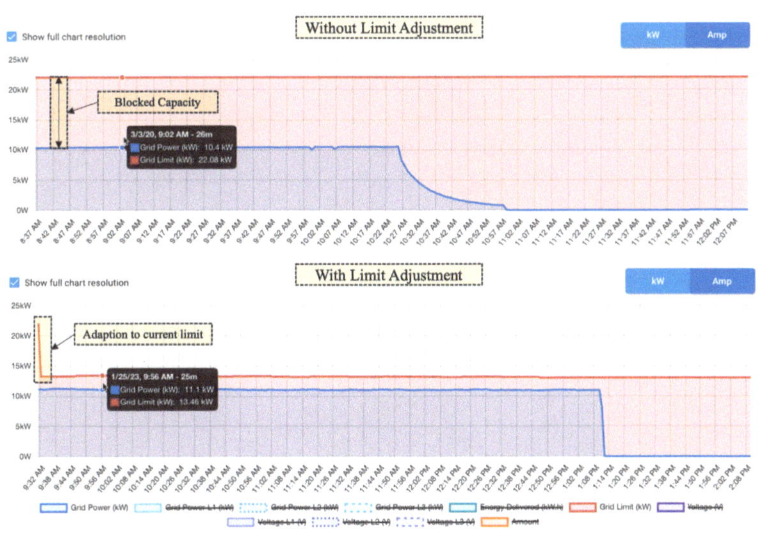

Figure 3. Power limit adaption to current consumption (screenshots from "Open e-Mobility").

Such inefficiencies motivated the incorporation of additional information into the charging power calculation. The required data sources were added step by step by continuously extending and enhancing the *Integration Layer* and related other components. When retrieving the connected EV's actual demanded power at the beginning of a charging session (using the OCPP message *MeterValues.req*), the allocated power limit can be adjusted (lowered) by updating the OCPP profile limit of the station. This adaptive adjustment of the power limit for a session is shown in the lower part of Figure 3. As a result, the charging algorithm can redistribute the otherwise unused power among other charging stations. For technical and safety reasons, the actual limits per charger were calculated by adding a safety buffer to the observed power consumption. In the example shown in Figure 3, the buffer is set to approx. 20%. Accordingly, the limit for the charging session of the example Tesla Model 3 is set to 13.5 kW. Using this enhancement, the number of parallel powered sessions increased significantly, since eight (instead of only five) EVs with a power consumption of 11 kW each could be charged.

However, at the beginning of each session, the maximum connector power remains allocated and thus blocked for other sessions at least until the next execution of the scheduling algorithm. The applied safety buffer per station (approx. 2.5 kW in case of the exemplary Tesla) will not be usable by any other session at all.

To address this issue, the *Fleet Management* component was introduced. It provides model-specific master data about EVs and enables the assignment of particular EVs to drivers. When starting a charging session at a charging station, the driver is authenticated and thus a linkage to the data about the respective EV is established. By retrieving the electrical properties of the vehnicle from the database, the maximum charging power of the EVs can be used in the optimization from the beginning. It was now possible to assign 11 kW as the definite limit to the exemplary Tesla Model 3 without allocating any additional safety buffer. Figure 4 shows a comparison of charging the same EV with the above discussed limit adoption (in the upper part) and with the initially set model-specific power limit (in the lower part). In essence, it became possible to utilize the freed power at other stations in parallel. On the above mentioned 110 kW infrastructure segment 10 (average) EVs could be charged at the same time without safety risks.

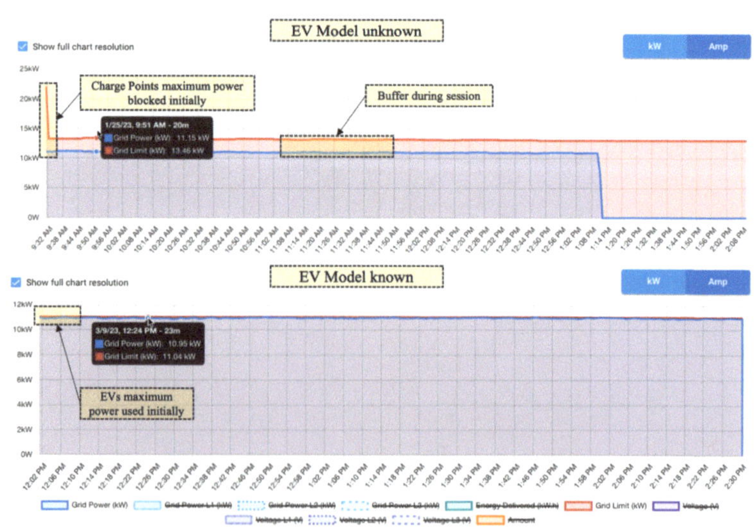

Figure 4. Utilization of vehicle data in charging limit calculations (screenshots from "Open e-Mobility").

At that stage, the SCS was only able to efficiently distribute power within the CI according to a fixed maximum power that was set as a strict upper limit. The limit was determined, as a proportion of the maximum power consumed by the entire facility (mainly office buildings). Thereby, neither the actual power consumption nor the power provided by the building's PV system were considered as input parameters. Since many energy-consuming devices are not always in operation and/or do not constantly draw a high amount of power, ignoring their actual energy consumption leads to a rather low power limit assigned to the CI. Similarly, considering the actual on-site power generation can help safely increase the CI's maximum power consumption limit.

The integration of the SCS with the locally installed *EMS* solved this issue. The EMS vendor provides a REST API to query collected data about all connected and monitored devices, including solar panels and the stationary battery installed in the building. The continuous retrieval of the actual power consumption and production on site allowed the dynamic updating of the CI's maximum power limit. Using this integration feature, it was possible to allocate 50 kW additional power in average to the charging infrastructure. On the above-mentioned 110 kW infrastructure it was now possible to charge up to 15 EVs at the same time on average. Figure 5 shows the power distribution of the testing facility while taking into account building consumption, solar power production, and charging station consumption.

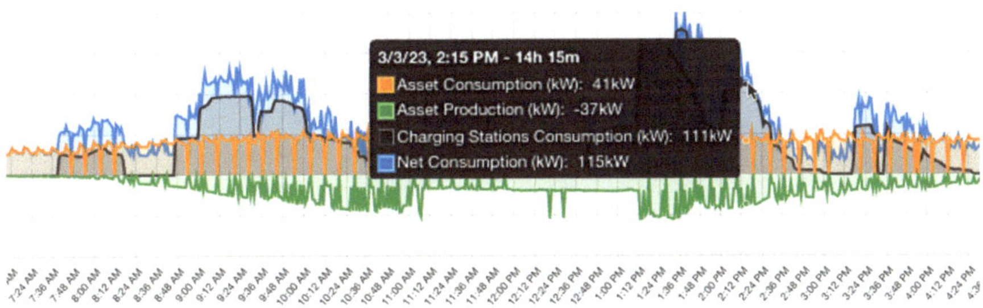

Figure 5. EV charging as part of the electrical infrastructure (screenshot from "Open e-Mobility").

By combining all of the above system components and associated "live" data, the SCS was able to efficiently distribute power while treating each EV charging session equally. This approach is beneficial in some use cases, for example, when a logistics company's delivery vehicles must be recharged during the night. However, in other settings, some vehicles must be served faster and/or charge a higher amount of energy than others to fulfil business-related requirements. Some EVs can have a longer stay at the charging facility and thus more time to charge than others. The vehicles' total charging demand may vary depending on the planned driving distances or specific routes the users need to drive till the next charge can occur. To meet these requirements and preferences, the incorporation of further data items, such as the given EV's current and target SoC, as well as its planned departure time, is required. These parameters can be provided, for example, by the user manually, via the *Mobile App* (see in Figure 1). If user authentication is performed without using the app, e.g., by presenting a personalized RFID card at the charging station, default values for the above-mentioned parameters are taken. By passing the collected data to the *Smart Charging Core*, the scheduling of sessions can be carried out according to users' actual needs, and energy can also be provided/distributed in a more efficient way. Figure 6 shows how the prioritization effects the start of powering a charging session in the system according to the users' known planned departure time.

In the depicted example, two EVs, EV_1 and EV_2, arrive at 2:00 p.m. and start charging at the same time. EV_1 can stay till 6:00 p.m., while EV_2 must leave earlier, at 4:00 p.m. Due to the limited available power of approximately 11 kW (see the red line), only one EV can be charged at its maximum current. If EV_1 would be charged before EV_2, EV_2 would not have enough time to charge until it must leave, and EV_1 will be inactive after it was fully charged. If the system can take the planned departure times into account, it schedules EV_2 first, allowing to charge to full capacity before it has to leave at 4:00 p.m. EV_1 can start afterwards and have another two hours to charge until 6:00 p.m. Viewing it from the involved drivers' perspective, in this particular example, the EV prioritization helps double the efficiency of the power-constrained infrastructure.

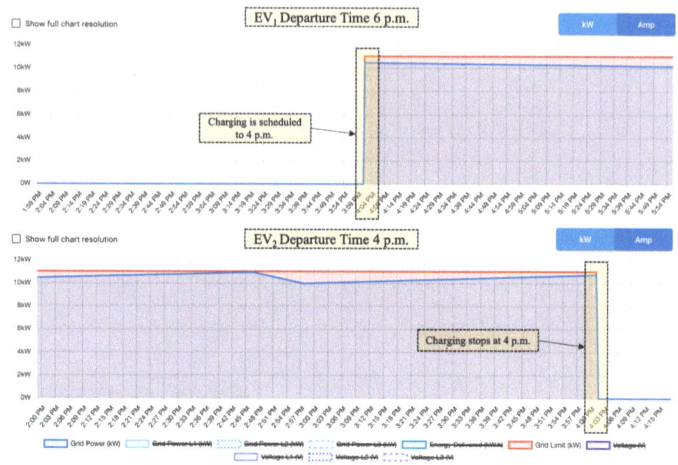

Figure 6. Effects of prioritization on two concurrent sessions (screenshot from "Open e-Mobility").

In addition to the support of rather passive AC chargers, the SCS is also able to deal with DC chargers that actively control charging processes while connected to an EV's battery. Figure 7 shows an example of how the SCS combines different data to dynamically adjust the power limit (in red) during an ongoing DC charging session. The information about the plugged vehicle's battery (in the example, a Jaguar I-PACE EV400 with battery capacity of 90 kWh, and initial SoC of 30% which corresponds to 27 kWh) is used to initially set the maximum allowed power, which is 104 kW in this case. The DC charger in the example, which is capable to deliver up to 150 kW, is regulated accordingly. The remaining 46 kW can be distributed among other chargers (as long as the resulting total power does not violate other thresholds). Over time, the car's battery management system automatically reduces the power drawn, in order to protect the battery. As a consequence, the power limit in the example session is readjusted (lowered) three times, making an increasing amount of power available for other (newly started or ongoing) charging sessions. The battery's increasing SoC is further used to recalculate prioritization decisions (the effects of those are not depicted here).

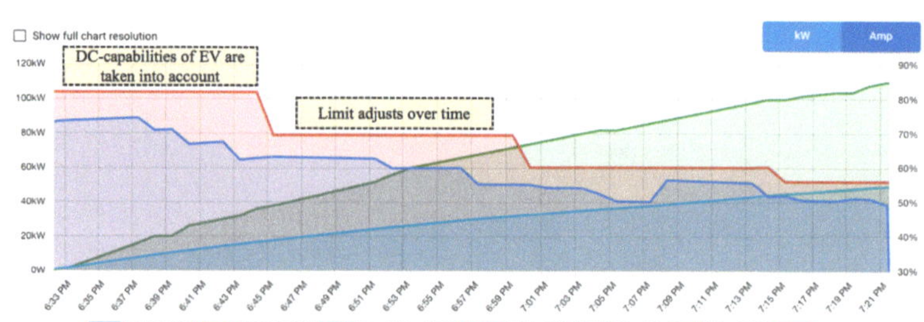

Figure 7. Example charging process on a Delta Ultra-fast Charger (screenshot from "Open e-Mobility").

As illustrated above, the SCS, in combination with the external components and data, can almost triple the efficiency of the power-limited charging infrastructure. To achieve similar results with a non-controlled hardware solution, the infrastructure limit would have to be tripled. For the above test infrastructure, this could require an increase in transformer power by 200 kW, which would lead to very high costs.

4. Conclusions and Future Work

In this paper, we presented an extension to the open-source charging-point management system "Open e-Mobility" to enable the intelligent control and power scheduling of EV charging at enterprise sites. The extension called "Smart Charging System" has already been successfully deployed and used in various charging infrastructures. Thanks to the modular system architecture and the realized multiple interfaces to external data sources, various factors and data can be incorporated in the calculation of charging plans for both AC and DC charging stations. We validated the positive impact of this flexible design in a real-world environment at SAP Labs France in Mougins, France. As illustrated by examples, the usage of various data sources and specific information led to better power utilization and helped increase the overall effectiveness of the charging infrastructure.

Since the SCS has been continuously extended and enhanced in an agile and iterative development approach, the positive effects generated by newly added features, functions, and data sources were measurable. However, many of those realized improvements, such as the increased average SoC at the end of charging sessions, cannot be clearly attributed to a single dedicated SCS feature nor to the usage of a given dataset. Rather, at any point of time it was possible to observe the combined effects of all deployed SCS functions, data sources, etc., and compare the performance of the system with the previous state, i.e., without the respective new features, data, etc. Therefore, at this point we cannot give a reliable recommendation regarding which of the features or data sources a CI operator should incorporate first and/or in which order to maximize the benefits. Nevertheless, the interested community (researchers, developers, operators) can immediately benefit from the work and our reported experience: The source code related to smart charging functionality has been made available under open-source licence, similar to the other parts of Open e-Mobility, and we have proven the long-term practicality of the implemented ideas and approaches. Since the described software system can basically only help better control and optimize EV charging processes, it can especially be useful in resource-constrained environments, in which overload situations could occur. In environments without such limitations, there is only less or even no need to establish and run a sophisticated software system.

The SCS will be enhanced by several features and functions in the near future. Currently, for example, it is not possible to create charging plans that can take advantage of variable or time-dependent electricity tariffs. The mainly economic impact of such tariffs on the calculation of charging plans has been studied theoretically in numerous publications but has hardly been implemented in practice. Another aspect concerns the realization and integration of predictive algorithms to deal with short-term uncertainties in a company's charging infrastructure [31]. The current scheduler implementation assigns power limits to ongoing EV charging sessions based on actual information, i.e., previously set data, without taking potential future data and related uncertainties into account. Regarding the communication with charging stations, it is also planned to support OCPP version 2.0 and profit from related enhancements.

Author Contributions: Conceptualization, T.F. and Z.N.; methodology, Z.N.; software, T.F.; validation, T.F. and S.G.; formal analysis, Z.N.; investigation, S.G.; data curation, T.F. and S.G.; writing—original draft preparation, T.F., S.G. and Z.N.; writing—review and editing, T.F., S.G. and Z.N.; visualization, T.F.; supervision, Z.N.; project administration, Z.N.; funding acquisition, Z.N. All authors have read and agreed to the published version of the manuscript.

Funding: As part of the research project *Green4EVer* [32], this work has been funded by the German Federal Ministry for Economic Affairs and Climate Action (BMWK, Ref. 01ME20009).

Data Availability Statement: The data related to the present research are owned by SAP and cannot be shared at the point of publication.

Acknowledgments: We especially thank to our colleagues in the SAP Labs France team led by Hanno Klausmeier for their continuous support and Oliver Frendo for his expertise provided.

Conflicts of Interest: Author Tobias Fleck was employed by the company SAP Deutschland. The remaining authors declare that the research was conducted in the absence of any commercial or financial relationships that could be construed as a potential conflict of interest.

Abbreviations

The following abbreviations are used in this manuscript:

API	Application Programming Interface
CHP	Combined Heat and Power
CI	Charging Infrastructure
CP	Charging Point
EMS	Energy Management System
ERP	Enterprise Resource Planning
EV	Electric Vehicle
OCPP	Open Charge Point Protocol
REST	Representational State Transfer
RFID	Radio Frequency Identification
SCS	Smart Charging System
SoC	State of Charge
TCO	Total Cost of Ownership

References

1. Kraftfahrt-Bundesamt. Statistik—Neuzulassungen Alternative Antriebe. Available online: https://www.kba.de/DE/Statistik/Produktkatalog/produkte/Fahrzeuge/fz28/fz28_gentab.html?nn=354746 (accessed on 28 September 2023).
2. Bodenschatz, N.; Eider, M.; Berl, A. Challenges and Requirements of Electric Vehicle Fleet Charging at Company Parking Sites. In Proceedings of the 2021 11th International Conference on Advanced Computer Information Technologies (ACIT), Deggendorf, Germany, 15–17 September 2021; pp. 623–628. [CrossRef]
3. Frendo, O.; Karnouskos, S.; Gaertner, N.; Kipouridis, O.; Rehman, K.; Verzano, N. Charging Strategies and Implications for Corporate Electric Vehicle Fleets. In Proceedings of the 2018 IEEE 16th International Conference on Industrial Informatics (INDIN), Porto, Portugal, 18–20 July 2018; IEEE: Piscataway, NJ, USA, 2018; pp. 466–471. [CrossRef]
4. Voss, M.F.; Haveman, S.P.; Bonnema, G.M. In-Company Smart Charging: Development of a Simulation Model to Facilitate a Smart EV Charging System. *Energies* **2021**, *14*, 6723. [CrossRef]
5. Frendo, O.; Graf, J.; Gaertner, N.; Stuckenschmidt, H. Data-Driven Smart Charging for Heterogeneous Electric Vehicle Fleets. *Energy AI* **2020**, *1*, 100007. [CrossRef]
6. Bera, T.K.; Bohre, A.K.; Ahmed, I.; Bhattacharya, A.; Yadav, A. Smart Charging for Electric Vehicles (EVs): A Short Review. In Proceedings of the 2022 IEEE Global Conference on Computing, Power and Communication Technologies (GlobConPT), New Delhi, India, 23–25 September 2022; pp. 1–6. [CrossRef]
7. Al-Hanahi, B.; Ahmad, I.; Habibi, D.; Masoum, M.A.S. Charging Infrastructure for Commercial Electric Vehicles: Challenges and Future Works. *IEEE Access* **2021**, *9*, 121476–121492. [CrossRef]
8. Deb, S.; Pihlatie, M.; Al-Saadi, M. Smart Charging: A Comprehensive Review. *IEEE Access* **2022**, *10*, 134690–134703. [CrossRef]
9. Waclaw, A.; Aloise, T.; Lienkamp, M. Charging Infrastructure Design for Commercial Company Sites with Battery Electric Vehicles: A Case Study of a Bavarian Bakery. In Proceedings of the 2020 Fifteenth International Conference on Ecological Vehicles and Renewable Energies (EVER), Monte-Carlo, Monaco, 10–12 September 2020; pp. 1–10. [CrossRef]
10. Wallberg, A.; Flygare, C.; Waters, R.; Castellucci, V. Peak Shaving for Electric Vehicle Charging Infrastructure—A Case Study in a Parking Garage in Uppsala, Sweden. *World Electr. Veh. J.* **2022**, *13*, 152. [CrossRef]
11. Ma, T.; Mohammed, O.A. Optimal Charging of Plug-in Electric Vehicles for a Car-Park Infrastructure. *IEEE Trans. Ind. Appl.* **2014**, *50*, 2323–2330. [CrossRef]
12. Zhang, G.; Tan, S.T.; Wang, G.G. Real-Time Smart Charging of Electric Vehicles for Demand Charge Reduction at Non-Residential Sites. *IEEE Trans. Smart Grid* **2018**, *9*, 4027–4037. [CrossRef]
13. Frendo, O.; Gaertner, N.; Stuckenschmidt, H. Real-Time Smart Charging Based on Precomputed Schedules. *IEEE Trans. Smart Grid* **2019**, *10*, 6921–6932. [CrossRef]
14. Jiang, W.; Zhen, Y. A Real-Time EV Charging Scheduling for Parking Lots with PV System and Energy Store System. *IEEE Access* **2019**, *7*, 86184–86193. [CrossRef]
15. Deb, S.; Goswami, A.K.; Harsh, P.; Sahoo, J.P.; Chetri, R.L.; Roy, R.; Shekhawat, A.S. Charging Coordination of Plug-In Electric Vehicle for Congestion Management in Distribution System Integrated with Renewable Energy Sources. *IEEE Trans. Ind. Appl.* **2020**, *56*, 5452–5462. [CrossRef]
16. Rottondi, C.; Neglia, G.; Verticale, G. Complexity Analysis of Optimal Recharge Scheduling for Electric Vehicles. *IEEE Trans. Veh. Technol.* **2016**, *65*, 4106–4117. [CrossRef]

17. Sepetanc, K.; Pandzic, H. A Cluster-Based Model for Charging a Single-Depot Fleet of Electric Vehicles. *IEEE Trans. Smart Grid* **2021**, *12*, 3339–3352. [CrossRef]
18. Meng, Z.; Xia, X.; Xu, R.; Liu, W.; Ma, J. HYDRO-3D: Hybrid Object Detection and Tracking for Cooperative Perception Using 3D LiDAR. *IEEE Trans. Intell. Veh.* **2023**, *8*, 4069–4080. [CrossRef]
19. Xia, X.; Meng, Z.; Han, X.; Li, H.; Tsukiji, T.; Xu, R.; Zheng, Z.; Ma, J. An Automated Driving Systems Data Acquisition and Analytics Platform. *Transp. Res. Part C Emerg. Technol.* **2023**, *151*, 104120. [CrossRef]
20. SAP Labs France. Open E-Mobility. Available online: https://github.com/sap-labs-france (accessed on 28 September 2023).
21. SAP. E-Mobility Charging Stations Simulator. Available online: https://github.com/SAP/e-mobility-charging-stations-simulator (accessed on 28 September 2023).
22. Frendo, O.; Gaertner, N.; Stuckenschmidt, H. Open Source Algorithm for Smart Charging of Electric Vehicle Fleets. *IEEE Trans. Ind. Inform.* **2021**, *17*, 6014–6022. [CrossRef]
23. SAP. Emobility-Smart-Charging. Available online: https://github.com/SAP/emobility-smart-charging (accessed on 16 November 2023).
24. Open-Charge-Alliance. OCPP 1.6. Available online: https://www.openchargealliance.org/protocols/ocpp-16/ (accessed on 28 September 2023).
25. EV Database. Electric Vehicle Database. Available online: https://ev-database.org/ (accessed on 28 September 2023).
26. Apel, S.; Kästner, C. An Overview of Feature-Oriented Software Development. *J. Object Technol.* **2009**, *8*, 49. [CrossRef]
27. Fowler, M. FeatureFlag. Available online: https://martinfowler.com/bliki/FeatureFlag.html (accessed on 16 November 2023).
28. Hodgson, P. Feature Toggles (Aka Feature Flags). Available online: https://martinfowler.com/articles/feature-toggles.html (accessed on 16 November 2023).
29. Gamma, E. (Ed.) *Design Patterns: Elements of Reusable Object-Oriented Software*, 39th ed.; Addison-Wesley Professional Computing Series; Addison-Wesley: Boston, MA, USA, 2011.
30. Fielding, R.T. Architectural Styles and the Design of Network-based Software Architectures. Ph.D. Thesis, University of California, Irvine, CA, USA, 2000.
31. Gohlke, S.; Nochta, Z. Towards a Short-Term Forecasting Framework to Efficiently Charge Company EV Fleets. In Proceedings of the 7th E-Mobility Power System Integration Symposium (EMOB2023), Copenhagen, Denmark, 25 September 2023; pp. 191–198. [CrossRef]
32. Green4EVer. Project Website. Available online: https://www.h-ka.de/en/iaf/green4ever (accessed on 28 September 2023).

World Electric Vehicle Journal

MDPI

Article

Energy Cost Analysis and Operational Range Prediction Based on Medium- and Heavy-Duty Electric Vehicle Real-World Deployments across the United States

Yin Qiu *, Cristina Dobbelaere and Shuhan Song

CALSTART, Pasadena, CA 91106, USA; cdobbelaere@calstart.org (C.D.); ssong@calstart.org (S.S.)
* Correspondence: yqiu@calstart.org

Abstract: While the market for medium- and heavy-duty battery-electric vehicles (MHD EVs) is still nascent, a growing number of these vehicles are being deployed across the U.S. This study used over 2.3 million miles of operational data from multiple types of MHD EVs across various regions and operating conditions to address knowledge gaps in total cost of ownership and operational range. First, real-world energy cost savings were determined: MHD fleets should experience energy cost savings each year from 2021 to 2035, regardless of vehicle platform, with the greatest savings seen in transit buses (up to USD 4459 annually) and HD trucks (up to USD 3284 annually). Second, to help fleets across various geographies throughout the U.S. assess the suitability of EVs for their year-round operating needs, operational range was modeled using the XGBoost algorithm (R^2: 70%) given 22 input features relevant to vehicle efficiency. Finally, this paper recommends (1) that MHD fleets apply energy-saving practices to minimize the impacts of cold temperatures and high congestion levels on vehicle efficiency and range, and (2) that local hauling fleets select trucks with a nominal range nearly double the expected maximum daily range to account for range losses under local, urban driving conditions.

Keywords: BEV (battery electric vehicle); heavy-duty; medium-duty; cost; range; energy efficiency; machine learning

Citation: Qiu, Y.; Dobbelaere, C.; Song, S. Energy Cost Analysis and Operational Range Prediction Based on Medium- and Heavy-Duty Electric Vehicle Real-World Deployments across the United States. *World Electr. Veh. J.* **2023**, *14*, 330. https://doi.org/10.3390/wevj14120330

Academic Editors: Joeri Van Mierlo and Genevieve Cullen

Received: 24 October 2023
Revised: 22 November 2023
Accepted: 25 November 2023
Published: 30 November 2023

1. Introduction

Electrifying the transportation sector has become one of many global strategies to combat climate change and improve air quality, along with the adoption of other zero-emission technologies. Medium- and heavy-duty (MHD) electric vehicles (EVs) have the advantage of being more energy efficient than diesel vehicles, in addition to producing zero tailpipe greenhouse gas emissions. In an experimental driving cycle evaluation study, three HD EV platforms, namely a step van, a yard tractor, and a Class 8 truck, consumed 3–6 times less energy than diesel counterparts [1]. MHD EVs are now capable of meeting certain commercial duty cycles and replacing internal combustion engine vehicles, given current technologies. An assessment using MHD vehicle trip data indicates that Class 2b–7 EVs can support 62–76% of commercial vehicle travel demand in California [2]. In recent years, the number of MHD EV options available on the market has significantly increased, up 36% globally since 2021 [3]. Despite rapid improvements in MHD EV energy efficiency and model availability, the adoption of these vehicles has occurred more slowly due to barriers like high up-front costs, range and charging limitations [4–6], and public skepticism that MHD EVs can meet fleet duty cycle requirements [2,7]. This paper seeks to advise fleets on two major barriers to EV adoption: total cost of ownership and range.

Compared to diesel vehicles, EVs offer reduced energy costs that significantly benefit their total cost of ownership. A preliminary model-based comparison [8] showed that MHD EVs were 2–4 times more energy efficient than diesel vehicles, while a 2018 California Air Resources Board (CARB) meta-analysis using data from real deployments found that battery-electric trucks and buses were 3–6 times as efficient as diesel counterparts, with a vehicle's precise estimated energy efficiency ratio (EER) depending on its vehicle platform and duty cycle, with greater efficiency at lower average speeds [9]. Given that electricity is consistently cheaper than diesel per unit of energy [10] and that heavier vehicles tend to consume more energy per mile than light vehicles [11], fleets switching from diesel to electric MHD vehicles should experience energy cost savings, which helps reduce total cost of ownership. In addition, past research has shown that electric truck ownership becomes more economical as load capacity increases, with energy savings as a function of weight [12]. This study not only supported these previous model- and data-based findings, but also estimated the energy cost savings associated with improved efficiency.

To address users' uncertainty about real-world EV performance, predictive models have been widely used to project EV energy consumption, efficiency, and range and to understand their determinants and trade-offs (Table 1). A recent study on 40-ft and 60-ft battery-electric buses found that bus speed significantly affects average energy consumption per mile [13]. Previous light-duty EV research has successfully adopted simulation-based models, machine learning models (e.g., regression, PCA, and tree-based models), and neural networks to identify features that most strongly impact vehicle efficiency to guide fleets' actions. Energy efficiency and range were found to be strongly correlated with a vehicle's battery capacity [14,15], speed profile [15–18], weight [15], acceleration [15], and road profile [17]. While light-duty EV energy efficiency has been widely studied using real-world big data–driven methodologies, there remains a knowledge gap in predicting the energy efficiency and range of MHD EVs. The methodologies used to study light-duty EVs can be applied to MHD EVs to better understand the key determinants of vehicle efficiency and make predictions on efficiency and range under real-world physical conditions. Findings from such analyses can help ease fleet uncertainty regarding EV performance before procurement and can improve MHD EV efficiency in operation given fleet-specific duty cycles and vehicle model selections.

Table 1. Methods and significant features from previous research modeling energy efficiency of light-duty EVs.

Literature	Model	Features That Significantly Impacted Light-Duty EV Energy Efficiency
Qi et al., 2017 [16]	PCA, decision tree, ANN	Negative kinetic energy, positive kinetic energy, speed, traffic
Fetene et al., 2017 [14]	Regression	Speed, acceleration, trip distance, season, rush hour, battery level when trip starts, temperature, precipitation, wind speed, visibility
Modi et al., 2019 [19]	CNN	Significant features not specified, but the selected model used the following features: speed, road elevation, tractive effort
Weiss et al., 2020 [20]	Regression	Vehicle weight
Xu et al., 2020 [17]	Simulation-based inference model	Speed, road type
Ahmed et al., 2022 [15]	Regression	Speed, acceleration, vehicle weight

Research regarding MHD EVs' performance in real-world deployment settings has been scarce [21], and industry stakeholders struggle with a lack of information and data to understand MHD EVs' actual duty cycle suitability, total cost of ownership, and performance in the face of variables like climate, terrain, and driving speed [7]. The Medium- and Heavy-Duty Electric Vehicle Data Collection project, funded by the U.S. Department of Energy (DOE), collected data from 144 MHD vehicles across six vehicle platforms and nine U.S. states and made it publicly available for researchers. Using this diversified and robust real-world vehicle performance dataset, this paper aims to fill the knowledge gap surrounding the in-use energy efficiency of MHD EVs, refining the methodology and expanding upon a conference paper submitted and presented at the 36th Electric Vehicle Symposium & Exposition (EVS 36) [22]. This study (1) compared the energy costs of MHD EVs and their conventional diesel internal combustion engine (ICE) counterparts, (2) generated a machine learning model to predict energy efficiency and highlight significantly impactful features, and (3) applied the model to predict operational range for transit buses and HD trucks in both local and regional duty cycles in four U.S. cities.

2. Materials and Methods

2.1. Materials and Data

Onboard data loggers, either from third party suppliers or pre-installed by vehicle manufacturers, were used to collect data directly from vehicles' Controller Area Network. Data was aggregated by day or by trip, depending on each data logger's frequency of reporting. Data validation and cleaning were conducted to prepare the data for analysis: measurement errors and outliers were eliminated, metric units were standardized, and missing values were imputed. For example, when a vehicle's energy consumption data was not usable due to data quality concerns (i.e., Fleet10), it was calculated using the vehicle's battery capacity and state of charge (SOC) used. The resulting vehicle performance dataset, which covered a total of 144 vehicles from six different vehicle platforms operated by 28 fleets across 16 U.S. cities, contained 37,352 vehicle-days and 2.3 million miles traveled. Table 2 and Figure 1 summarize the makeup, status, and geographic distribution of the on-road vehicle dataset.

Table 2. Summary of vehicles included in this study.

Vehicle Platform	Gross Vehicle Weight Rating (lbs.)	Number of Vehicles in Analysis	Number of Vehicle-Days in Analysis
Transit Bus	>33,000	90	28,093
Type C School Bus	>33,000	17	1809
Class 8 Day Cab Tractor	>33,000	14	1269
Class 7 Box Truck	26,001–33,000	7	1144
Class 6 Box Truck	19,501–26,000	6	2025
Class 4 Step Van	14,001–16,000	10	3012
Total		144	37,352

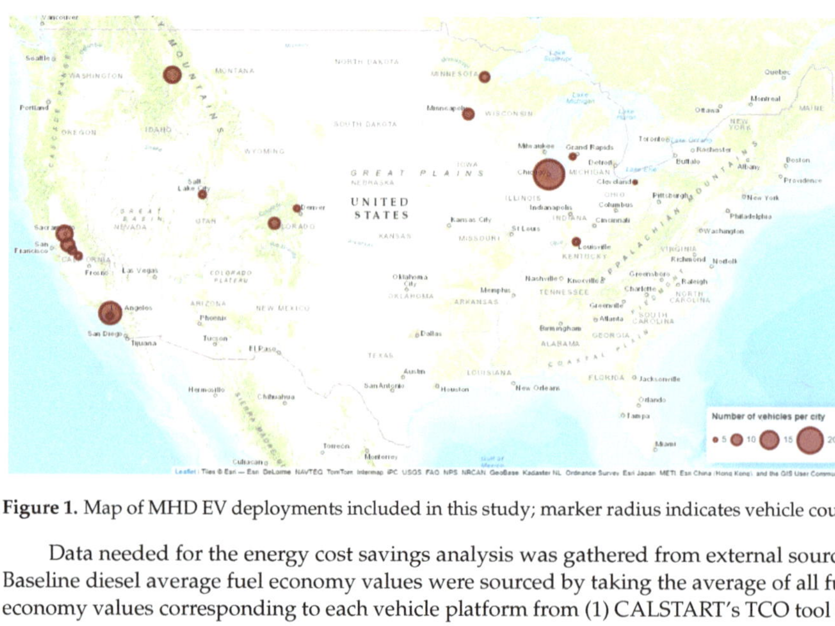

Figure 1. Map of MHD EV deployments included in this study; marker radius indicates vehicle count.

Data needed for the energy cost savings analysis was gathered from external sources. Baseline diesel average fuel economy values were sourced by taking the average of all fuel economy values corresponding to each vehicle platform from (1) CALSTART's TCO tool [8] and (2) the U.S. DOE Alternative Fuels Data Center's average fuel economy dataset [23], where available. The price of diesel (USD/gallon) was gathered from the U.S. Energy Information Administration's (EIA) diesel price forecast dataset [24]. The price of electricity (USD/kilowatt-hour (kWh)) was gathered from (1) the EIA's electricity price forecast dataset [24] and (2) levelized costs of delivered electricity USD 0.17–0.38 per kWh estimated by the National Renewal Energy Laboratory (NREL) given a set of 20 scenarios, ranging from kilowatt- to megawatt-scale charging and accounting for variations in location type, utilization rate, cost of electric vehicle supply equipment (EVSE) installation and upgrades, and various utility rates [25].

Some data parameters corresponding to input features for the vehicle energy efficiency model in Section 3.2 were not directly collected by onboard data loggers; in these cases, data were downloaded from external sources (Table 3).

Table 3. Features as inputs to the energy efficiency predictive model.

Feature Groups		Features	Sources
Duty Cycle		Average Driving Speed, Total Distance, Total Run Time, Driving Time, Idling Time Percentage	MHD EV Data Collection (CALSTART, 2023)
Vehicle Configuration		Manufacturer, Model Name, Model Year, Weight Class, Vehicle Platform, Body Style, Rated Energy, Nominal Range, Estimated Payload	MHD EV Data Collection (CALSTART, 2023); ZETI Database (CALSTART, 2023) [26]
Use Case		Vocation, Sector	MHD EV Data Collection (CALSTART, 2023)
Geography		Region, State	MHD EV Data Collection (CALSTART, 2023)
City Profile	Climate	Average Ambient Temperature, Average Precipitation	NOAA daily average temperatures [27]; NLDAS-2 hourly dataset [28]; ERA-5-Land hourly dataset [29]
	Road	Average Road Grade	R package {slopes} [30] applied on OpenStreetMap network [31]
	Congestion	Annual Hours of Delay (general, highway)	Urban Mobility Report Congestion Data (Texas A&M Transportation Institute, 2021) [32]

For each vehicle in the dataset, a climate profile consisting of temperature and precipitation data was gathered. When not collected by onboard data loggers, daily average ambient temperatures were downloaded from the National Oceanic and Atmospheric Administration (NOAA) [27]. Trip-level temperatures were downloaded from the National Aeronautics and Space Administration's (NASA) NLDAS-2 dataset [28] at the midpoint location and time of the trip. Hourly precipitation was downloaded per city for 2018–2022 from the ERA-5-Land hourly dataset [29] and summed by day or trip, depending on the granularity of the corresponding vehicle's data.

When downloading annual congestion data, 2019 data were used to avoid the exogenous impact of the COVID-19 pandemic [32]. The metric of annual hours of delay for general roads was used for buses and local hauling trucks, while annual hours of delay for highways was used for regional hauling trucks. For cities not covered by the congestion dataset, metrics were collected for each city's nearest neighbor by physical distance.

City road slope was computed using road network data from Open Street Map [31], 1 arc-second Digital Elevation Model from the U.S. Geological Survey (USGS) TNM database [33], and the R package {slopes} [30]. Road segments were filtered to only include primary, secondary, tertiary, trunk, residential, and link roads for all above road types excluding residential. Road grade for each road segment was computed, and an aggregated mean over road grades of all road segments was used in modeling for each city.

Since actual payload data were not available, maximum payload per vehicle model was obtained from CALSTART's Zero-Emission Technology Inventory (ZETI) database [26], which contains vehicle specification data for 843 models of MHD trucks and buses [34].

When payload was measured in units other than weight (e.g., passengers or volume), these units were converted to weight using assumptions indicated in the Urban Bus Toolkit [35]. For example, the number of passengers that could be carried in a bus was assumed to be 1.75 times the number of bus seats to represent both seated and standing passengers. Payloads of buses were calculated by first converting seat capacities to passenger capacities and then multiplying passenger capacities by the 178-lb average adult weight.

2.2. Methods

2.2.1. Energy Efficiency Comparison and Energy Cost Savings Analyses

Figure 2 below shows the procedure used for the energy efficiency comparison analysis and energy cost savings analysis. In this study, energy cost was defined as the cost of fuel in U.S. dollars (USD) needed to drive a vehicle one mile. Maintenance costs were not included due to a lack of sufficient historical maintenance data to accurately assess an EVs' longer-term maintenance needs.

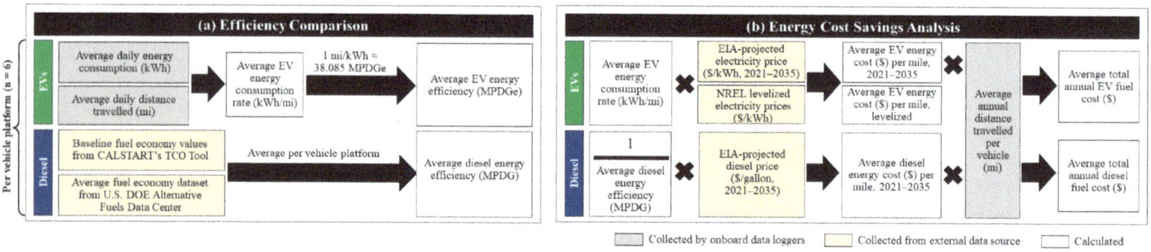

Figure 2. Procedures of (**a**) efficiency comparison analysis and (**b**) energy cost savings analysis.

First, a comparison of energy efficiency between each EV platform and its diesel counterpart was conducted (Figure 2a). Real-world energy consumption rate (kWh/mi) and energy efficiency in miles per diesel gallon equivalent (MPDGe) were determined for each EV platform from average daily energy consumed and average daily distance traveled, using baseline diesel comparison average fuel economy from CALSTART's TCO tool [8] and the U.S. DOE Alternative Fuels Data Center's average fuel economy dataset [23].

For each vehicle platform, average energy cost savings per mile were (1) projected from 2021–2035 using EIA price projections [24] and (2) calculated using the average levelized electricity costs estimated by NREL [25] with 2022 diesel price projections [24] (Figure 2b).

Together, these complementary sources of electricity prices presented a more nuanced understanding of EVs' energy costs: while the EIA source provided price projections on a per-year basis over a broad time period, NREL's estimates, despite their lack of temporal granularity, accounted for the real-world variability of charging costs associated with 20 diverse charging infrastructure scenarios.

2.2.2. Vehicle Efficiency Prediction: Model Selection, Feature Engineering and Model Training

Knowing the mechanisms that affect vehicle efficiency can inform fleets' operations by predicting efficiency performance and ultimately range. When selecting from a wide array of machine learning algorithms, we considered the tradeoff between interpretability and performance. On one end of the spectrum, linear models are the most interpretable but are generally weak in predictive performance, especially when dealing with high-dimensional data and non-linear relationships. On the other end, neural networks can achieve higher predictive performance at the expense of high computation costs and low interpretability, as they are essentially "black box" models. Tree-based algorithms stood out to best fit our use case, as they offer a balance between interpretability and predictive performance and can be trained and tuned reasonably quickly.

For this study, three tree-based algorithms (i.e., XGBoost, Gradient Boosted Trees or Gradient Boosting, and Random Forest) were selected to train the efficiency prediction model. These algorithms adopted a range of ensemble methods, such as bagging and boosting, to help overcome model overfitting, which is commonly seen in decision trees. Additionally, two linear models that use L1 and L2 regularization techniques, also known as Lasso and Ridge Regression, were adopted as baseline models in this study, given their ability to perform automatic feature selection in high dimensional datasets.

Before training the machine learning models, exploratory data analysis and feature engineering were conducted to select and transform 22 features as inputs for the models (Table 3). Figure 3 illustrates the feature engineering procedure. Since vehicle types and regions were imbalanced in the data, we applied stratified sampling when splitting train and test data to ensure the test score properly reflected predictive performance of all categories of interest. K-Nearest-Neighbor (KNN) imputation was used to fill in missing numerical features with the mean of five nearest neighbors, followed by rescaling to meet linear model requirements. Although tree-based models generally perform well with imbalanced data, SMOGN resampling [36] was applied on the training data for all models to further improve model performance on underrepresented areas of datapoints. Then, one-hot encoding and ordinal encoding were applied, resulting in 75 features in total. Finally, quadratic terms of ambient temperature and driving speed were added to the linear models to better fit their non-linear relationships with the target variable (i.e., energy consumption rate), but it was unnecessary to add these terms for tree-based models.

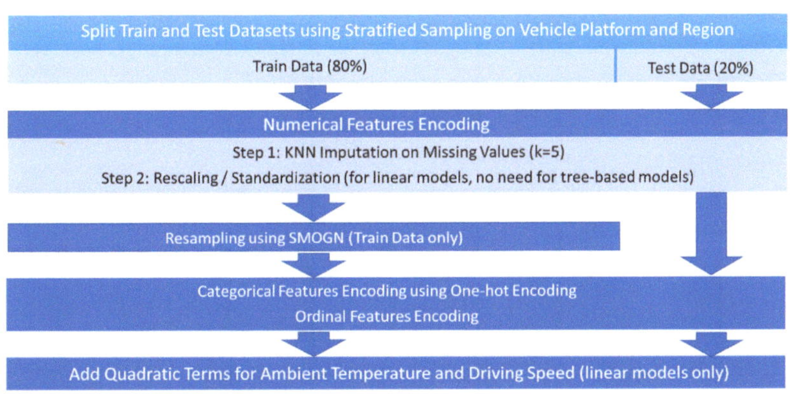

Figure 3. Feature engineering procedure on train and test datasets.

In model training, this study applied five commonly used machine learning algorithms for comparison to predict vehicle efficiency, calculated as total energy consumption divided by driving distance and measured by energy consumption rate (kWh/mi). Using Scikit-Learn [37] and other Python packages, the study was able to tune the hyperparameters with the random search method and perform k-fold cross-validation to avoid overfitting on the training set. Mean Absolute Error (MAE) was the key evaluation metric used in training since MAE assigns equal weights to all errors, which is less sensitive to the impact of outliers.

2.2.3. Operational Range Prediction: One Year of Duty Cycle Simulation and Range Forecast

It is critical for fleets to assess how MHD EVs will accommodate their operations and duty cycle needs when planning procurement. Predicting operational range values in real-world operating conditions under vehicle type–specific duty cycles can help fleets gauge the maximum range a vehicle might achieve versus manufacturer specification. The efficiency model developed in Section 3.2 was used to address this issue by predicting and visualizing the operational range of MHD EVs based on hypothetical operating conditions, manufacturer-rated battery capacities, and an assumed 90% SOC battery buffer (Equations (1) and (2)). We chose three different vehicle types (i.e., transit bus, local HD truck, and regional HD truck) in four different cities (i.e., Los Angeles, Louisville, Missoula, and Chicago) to assess the impact of real-world operating conditions and duty cycles on MHD EV ranges. The 2022 BYD K9M was selected as the vehicle model for transit buses, while the 2021 Freightliner eCascadia was chosen for local and regional HD trucks. Vehicles were assumed to be brand new and operating at full State of Health. City profile data were gathered using the same methodology as described in Section 2.1.

$$\text{Operational Range (mi)} = \text{Usable Battery Capacity (kWh)}/\text{Vehicle Efficiency (kWh/mi)} \quad (1)$$

$$\text{Usable Battery Capacity (kWh)} = \text{Nominal Battery Capacity (kWh)} \times \text{Battery State of Health (\%)} \times \text{Battery State of Charge Buffer (\%)} \quad (2)$$

One year of operating duty cycle data was simulated in R. Using our real-world data as a benchmark, we summarized monthly and weekly averages of daily total distance, total run time, and driving time for each of the three simulated vehicle types (i.e., transit bus, local HD truck, regional HD truck). For each pair of month and day of week, 200 data points were simulated using the averages and standard deviations of residuals, assuming a normal distribution. The simulated data pool was then cleaned by removing outliers and negative data points. For each day in 365 days, one data point was randomly sampled from the simulated data pool based on day of week and month. Forecasting with the R package {forecast} was used if data were missing or underrepresented in a certain time in the 356 days. Daily average driving speed and idling time percentage were calculated from the simulated features. All duty cycle features were engineered and validated to have ranges and distributions similar to the real-world data.

3. Results and Discussion

3.1. Energy Efficiency Advantages Indicate Energy Cost Savings

3.1.1. Energy Efficiency Comparison Analysis

The distribution of the real-world energy consumption rate for each of the six vehicle platforms is shown in Figure 4. When comparing the real-world energy efficiency of EVs and the fuel economy of baseline vehicles, MHD EVs performed an average of 3.4–5.8 times as well as their conventional counterparts, mirroring CARB's estimated EER results [9] (Table 4).

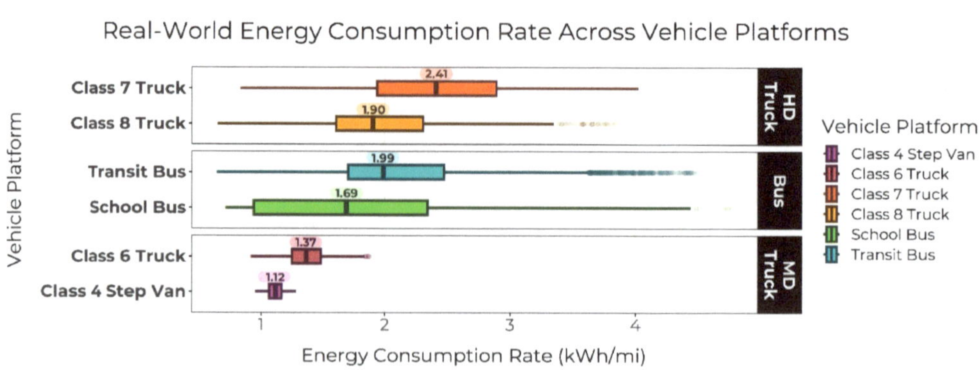

Figure 4. Boxplot illustrating daily energy consumption rate found across each of the six vehicle platforms, with annotations marking median energy consumption rate. Higher energy consumption rate indicates worse efficiency performance and reduced MPDGe.

Table 4. Average and 95% confidence interval of energy efficiency by vehicle type and platform.

Vehicle Type	Vehicle Platform	Average EV Energy Efficiency (MPDGe)	Average Baseline Fuel Economy (MPDG)	Energy Efficiency Ratio (EER)
Medium-Duty Truck	Class 4 Step Van	34.18 (±0.22)	9.04	3.8
	Class 6 Truck	28.09 (±0.18)	8.21	3.4
Heavy-Duty Truck	Class 7 Truck	16.89 (±0.35)	4.40	3.8
	Class 8 Truck	20.58 (±0.40)	3.56	5.8
Bus	Type C School Bus	27.16 (±0.73)	7.06	3.8
	35–40-ft Transit Bus	19.07 (±0.08)	3.83	5.0

HD trucks and transit buses had the highest estimated EERs, while MD trucks and school buses—the most efficient vehicle platforms for both fuel types—had lower EERs. Vehicle platforms maintained similar efficiency rankings relative to each other regardless of fuel type, aside from Class 8 trucks, which were the least efficient diesel vehicles but third least efficient EVs, behind Class 7 trucks and transit buses. Although it is expected that Class 8 trucks may experience worse real-world efficiency than Class 7 trucks, which have lower maximum payloads than Class 8 trucks, external factors such as climate, percent idling time, and driver behavior may have impacted these two truck platforms' relative real-world performance.

3.1.2. Energy Cost Savings Comparison Analysis

EIA 2022 price projections indicated that MD trucks, HD trucks, school buses, and transit buses had estimated average cost savings of USD 0.195, USD 0.493, USD 0.201, and USD 0.529 per mile, respectively; by 2035, these per-mile projected cost savings are projected to increase by 14.2% on average, to USD 0.224, USD 0.552, USD 0.238, and USD 0.589 per mile, respectively.

In a 2024 cross-section of these results (Figure 5), energy cost savings were smaller when using electricity prices based on NREL's breakeven costs relative to the EIA's national average electricity price projections. However, for both estimates, the average cost per mile was consistently lower for EVs than for baseline vehicles. Thus, even when accounting for the installation and maintenance of EVSE infrastructure, fueling MHD EVs is still less expensive per mile on average than fueling their diesel counterparts.

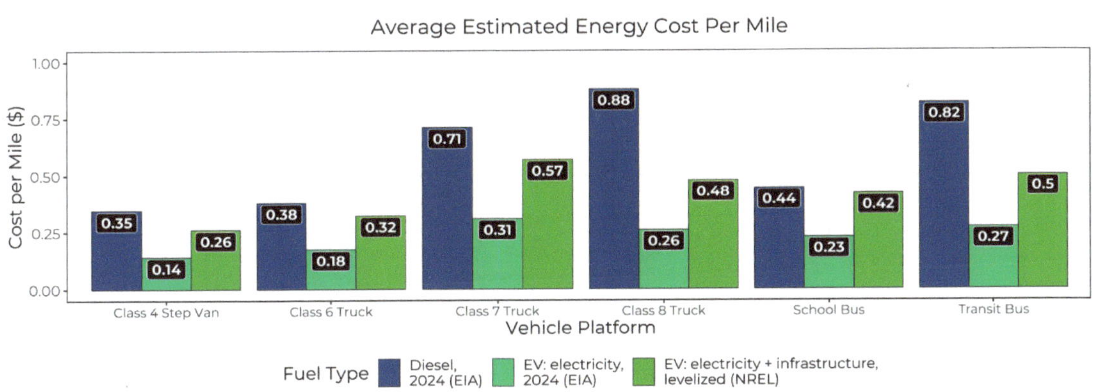

Figure 5. Average estimated energy cost per mile for baseline diesel vehicles in 2024 and real-world EVs, calculated using either EIA's projected 2024 U.S. average electricity price or NREL's levelized electricity prices.

Finally, for each vehicle platform in the real-world dataset, estimated total annual fuel cost savings were determined using EIA-projected average cost per mile and average annual distance traveled per vehicle in each vehicle platform (Figure 2b). Because of the combination of their high per-mile fuel cost savings and high annual distance traveled, transit buses and HD trucks had high estimated annual fuel cost savings (Figure 6). Transit buses, which had the highest per-vehicle average annual mileage (7570 miles per year), experienced the greatest fuel cost savings, followed by Class 8 and Class 7 trucks, which had local/regional duty cycles and traveled an average of 4937 and 4779 miles per year, respectively.

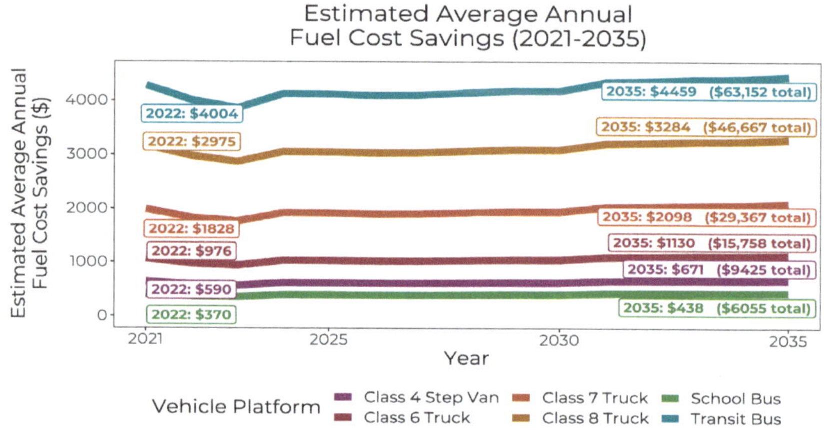

Figure 6. Estimated average annual fuel cost savings by vehicle platform from 2021 to 2035. Annotations indicate 2022 and 2035 cost savings estimates, as well as cumulative total estimated cost savings by 2035.

These results support previous DOE findings that a vehicle's duty cycle strongly impacts total cost of ownership [7]: although electric school buses had 43% better energy efficiency performance (MPDGe) than electric transit buses, their lower annual average distance (1837 miles) resulted in 90% lower cumulative total fuel cost savings. Thus, switching from diesel to electric is much more cost-effective for higher-mileage than lower-mileage vehicle platforms.

3.2. *Vehicle Efficiency Predictions Based on Known Real-World Factors*

Many factors affect actual EV efficiency, including ambient temperature, driving speed, topography, and manufacturing configurations. However, studies determining these variables' relative impacts are lacking. This paper incorporated real-world data from these factors and developed machine learning models on in-use performance data to estimate energy consumption rate (kWh/mi).

3.2.1. Model Performance Evaluation

Each of the five machine learning models was evaluated using the following metrics: R^2, Mean Absolute Error (MAE), Mean Squared Error (MSE) and Root Mean Squared Error (RMSE) (Table 5). Among the five models, tree-based models (XGBoost, Random Forest, and Gradient Boosted Trees) had better performance than linear models (Lasso and Ridge Regression). While the three tree-based models produced R^2 values of 69–70%, XGBoost had the highest R^2 (70%) and was selected as the best model to predict operational range in Section 3.3. The XGBoost model can explain 70% of the variations in the target variable (energy efficiency), which is good performance considering the large scale and diversified sources of real-world data.

Table 5. Model performance evaluation metrics.

Regression Models	R^2	Mean Absolute Error (MAE)	Mean Squared Error (MSE)	Root Mean Squared Error (RMSE)
Lasso (L1 Regularization)	0.550	0.351	0.236	0.486
Ridge (L2 Regularization)	0.567	0.339	0.227	0.476
Gradient Boosted Trees (GBR)	0.694	0.252	0.161	0.401
Random Forest (RFR)	0.699	0.255	0.158	0.397
XGBoost (XGB)	0.702	0.253	0.156	0.396

3.2.2. Model Result Analysis

A preliminary analysis indicated that MHD EVs were most efficient when operated at daily average speeds between 20 and 40 mph compared to lower speeds. At speeds below 20 mph, a higher percentage of idling time versus driving time was observed, which likely contributed to worse efficiency. This analysis also indicated that MHD EVs driving more than 100 miles per day achieved a higher average efficiency than those traveling less. Again, a higher percentage of idling time was observed in shorter trips, resulting in worse efficiency. The ideal operating environment included minimal traffic, mild to warm ambient temperatures (50–80 °F) [38], and relatively flat terrain. Finally, decreases in vehicle size and weight significantly increased vehicle efficiency.

While these results were not unexpected, further analysis was conducted to reveal the most important factors in the XGBoost model. The SHAP (Shapley Addictive exPlanations) value [39] was examined to determine the predictive impact of each feature on vehicle efficiency (Figure 7). Clear horizontal separation (red dots on one side and blue on the other) shows the direction and magnitude of the impact each feature has on the output. For example, high driving speed values had a negative effect on the output (kWh/mi) and thus are associated with improved efficiency. Among the top 10 features, all features except model year showed clear efficiency trends, with consistent impacts on the magnitude and direction of change in efficiency. Specifically, higher average driving speed, average ambient temperature, and total distance were associated with improved energy efficiency of MHD EVs. In contrast, lower congestion hour delay, rated energy (i.e., battery capacity), idling time percentage, payload, and total run time were associated with reduced efficiency. Model year was one of the important features, but it is unclear whether older or newer models were more efficient in general.

All tree-based models achieved similar R^2 scores. Each model's feature importance ranking was slightly different, but all three models included average driving speed, average ambient temperature, total distance, and congestion in their respective top features (Table 6). While the algorithm identified the original equipment manufacturer (OEM) Proterra as a significant feature, this is likely a result of the selection bias in the data sample from MHD EV early deployments, where there is a disproportionately high number of Proterra buses—about 45% of vehicle-days and 37.5% of vehicle count. Therefore, the significance of this feature might not be generalizable to the overall U.S. MHD EV population as the diversity of OEMs in real-world deployments increases.

Table 6. SHAP identified top features impacting the prediction on vehicle efficiency.

Top Features	XGBoost	Random Forest	Gradient Boosted Trees
Average driving speed	#1	#2	#3
Average ambient temperature	#2	#3	#1
Manufacturer Proterra	#3	#1	#6
Total distance	#4	#5	#5
Congestion hour delay	#5	#6	#2

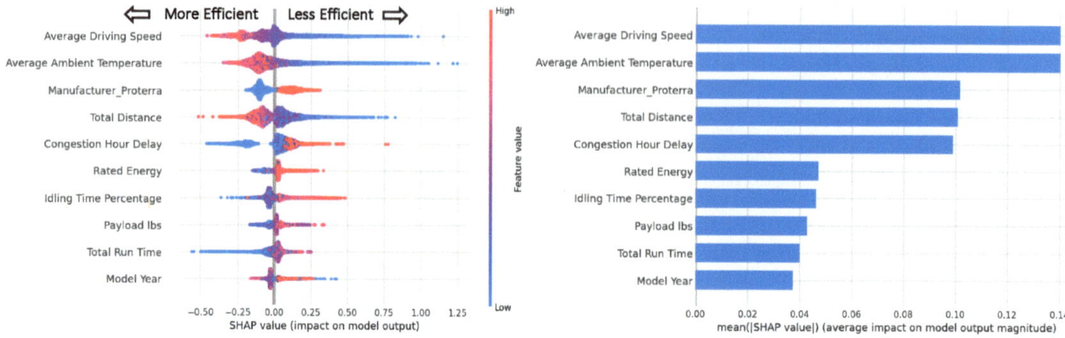

Figure 7. The XGBoost model's top 10 features ordered by feature importance (left: bee swarm plot to show the direction and magnitude of the impact each feature has on vehicle efficiency; right: bar plot to show the mean absolute impact of each feature on vehicle efficiency). In the bee swarm plot, positive SHAP values indicate datapoints with feature values (red: high feature value, blue: low feature value) that are associated with more energy use or lower efficiency. In contrast, negative SHAP values signify datapoints with feature values that are associated with less energy or higher efficiency.

Average driving speed was consistently among the top important features across all models, meaning it had a critical effect on efficiency. Energy efficiency of transit buses became less optimized and substantially more variable when average driving speed was less than 10 mph (Figure 8). HD trucks were more likely to have energy efficiency as high as 4 kWh/mi when average driving speed was less than 15 mph. However, for both vehicle types, when average speed reached 20–40 mph, the efficiency converged to a narrow range of values and stabilized around 1.5–2 kWh/mi.

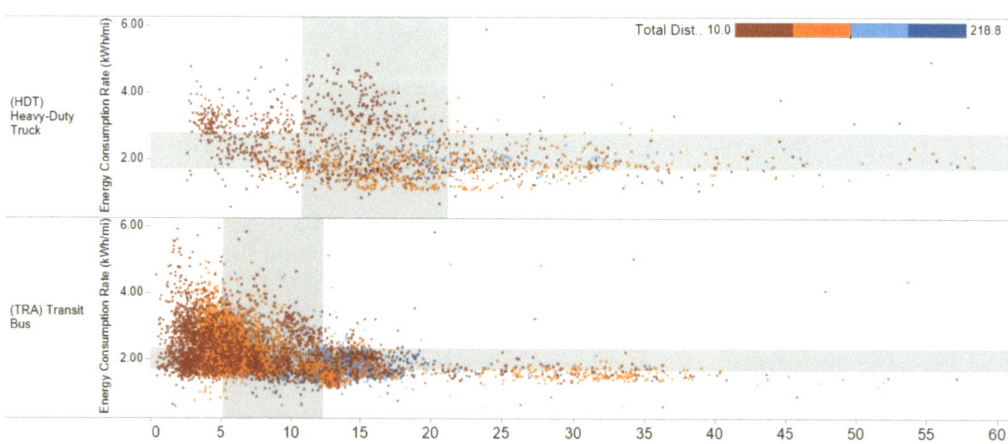

Figure 8. Scatter plot of vehicle efficiency and daily average driving speed for HD trucks (**top**) and transit buses (**bottom**).

The average driving speed feature was aggregated by day, which must be understood within the context of fleet operations. Throughout a real-world operational day, vehicles drive at a range of speeds and alternate among driving, idling, and off statuses. Vehicles may idle in traffic, run on the highway, or stop-and-go on local city roads. Lower daily average speed may indicate a larger share of driving in urban congested areas with frequent or longer stops and shorter total distance traveled. These driving conditions are commonly observed in urban delivery trucks, city bus circulators, and school buses. A daily average speed of 20–40 mph may imply a duty cycle with fewer stops and less traffic or loading time, and MHD trucks operating at these average speeds were observed to achieve higher energy efficiency. Future studies on MHD EVs may tailor efforts to further understand mechanisms behind their energy efficiencies at different speeds.

3.3. Operational Range Predictions

A summary of simulated year-long duty cycles for transit buses and local and regional HD trucks are presented in Table 7. In the vehicles' simulated duty cycles, transit buses traveled the farthest with the longest run time and driving time but had the lowest daily average driving speed due to frequent stops or residential speed limits. Local HD trucks traveled the shortest distance with the shortest driving time and highest idling time percentage. Regional HD trucks traveled long distances with the highest speed and lowest idling time percentage. In the simulated data, the maximum distance traveled in a day was 177 miles for a regional HD truck and 103 miles for a local HD truck. Regional HD trucks spent a greater fraction of time driving, indicating that they tend to travel on highways and have fewer stops.

Table 7. Averages and 95% confidence intervals of simulated duty cycle features.

Vehicle Type	Total Distance (miles)	Driving Time (hours)	Total Run Time (hours)	Average Driving Speed (mph)	Idling Time Percentage (%)
Transit bus	83.5 (±3.8)	5.6 (±0.2)	8.4 (±0.4)	15.6 (±0.7)	25.2 (±2.6)
Local HD truck	45.3 (±1.4)	2.8 (±0.1)	4.1 (±0.2)	18.0 (±0.9)	28.5 (±2.0)
Regional HD truck	71.3 (±4.0)	3.2 (±0.2)	4.3 (±0.2)	22.7 (±1.3)	23.3 (±1.5)

For transit buses, operational range was modeled across four U.S. cities with different climates, congestion levels, and hilliness (Table 8, Figure 9). For each city, congestion and hilliness remained constant throughout the year, while climate variables changed seasonally. Average ambient temperature was the feature with the strongest impact on operational range. The modeled transit bus in Los Angeles, with the warmest winters, showed the most consistent operational range throughout the year, despite a high congestion hour delay that was 30 times that of Louisville. The operational range of the transit bus in Missoula dropped significantly in cold winter months, during which average ambient temperature fell as low as 6 °F. In the summer, when ambient temperature was no longer the limiting factor, transit buses in Missoula had a longer average operating range than in the other regions, likely thanks to Missoula's light traffic. In Chicago, a city with low average ambient temperatures and high congestion levels, transit buses were predicted to have low operating range throughout the year compared to transit buses in other cities.

Table 8. Profiles of four U.S. cities.

City	Average Ambient Temperature (°F)	Precipitation (Inches)	Congestion Hour Delay (h)	Average Road Grade (%)
Los Angeles, CA	65.7 (±1.0; 46–86)	0.002 (±0.0004)	952,183,000	2.1
Louisville, KY	59.6 (±1.7; 22–86)	0.006 (±0.0005)	30,610,000	1.7
Missoula, MT	41.8 (±1.6; 6–74)	0.003 (±0.0002)	2,263,000	1.4
Chicago, IL	53.2 (±2.0; 10–85)	0.005 (±0.0005)	331,657,000	0.5

The comparison between the local HD truck and the regional HD truck highlighted the impact of duty cycle on operational range when climate, congestion, and road slope are held constant (Table 7, Figure 10). Throughout a year, local HD trucks consistently had a lower operational range than regional HD trucks, due to lower daily average driving speed, shorter total distance traveled, and a higher percentage of idling time. This could be a result of local HD trucks operating in urban areas and thus spending more time idling or in traffic. From the model estimates, a local-haul HD truck fleet may need to deploy trucks with a nominal range nearly double the expected daily range to meet duty cycles in colder months. While the same truck model had a longer range as a regional HD truck overall, there were still days when the regional truck's predicted operational range dropped to about 65% of its nominal range. In summary, fleets need to select proper MHD EV models to be prepared for these rare occasions when transitioning to a fully electric fleet.

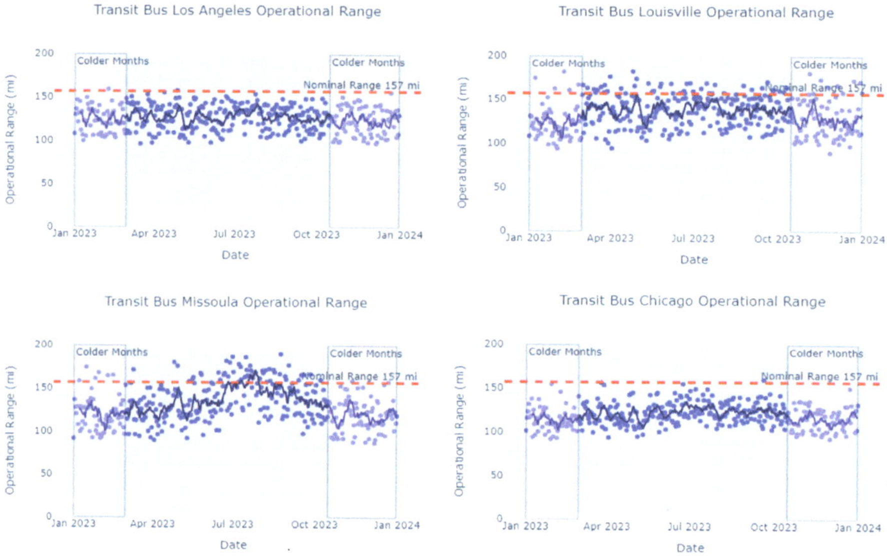

Figure 9. Transit bus year-round operational range predictions (blue points) for the 2022 BYD K9M vehicle model in four U.S. cities (**top left**: Los Angeles, CA; **top right**: Louisville, KY; **bottom left**: Missoula, MT; **bottom right**: Chicago, IL). A trend line (dark blue line) showing a seven-day moving average of predicted range is added to each scatter plot to illustrate the corresponding city's seasonal pattern and the impact of temperature on operational range. A reference line (dashed red line) is added to compare predicted operational range with the transit bus's nominal range.

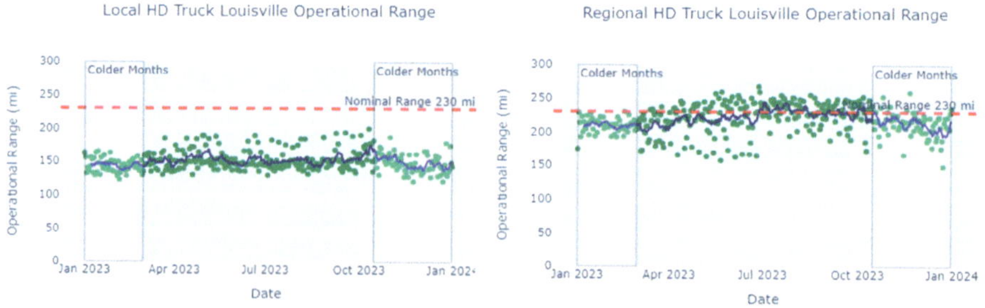

Figure 10. HD truck year-round operational range predictions (green points) for the 2021 Freightliner eCascadia vehicle model in Louisville, KY (**left**: local duty cycle; **right**: regional duty cycle). A trend line (dark blue line) of a seven-day moving average of predicted range is added to each scatter plot to illustrate the corresponding city's seasonal pattern and the impact of temperature on operational range. A reference line (dashed red line) is added to compare nominal range to the predicted operational range.

4. Conclusions

As EV adoption grows, the value of a publicly accessible operational dataset from early MHD EV deployments will only increase. This study made use of such a dataset to (1) provide a high-level understanding of energy cost savings across various types of MHD EVs and (2) execute a novel approach employing the predictive power of machine learning to model MHD EVs' energy efficiency. The outcome of this analysis could help fleets across various geographies throughout the U.S. assess the suitability of EVs for their operational needs.

4.1. Energy Efficiency Comparison and Energy Cost Savings Analyses

MHD EVs were found to perform an average of 3–6 times as efficiently as their diesel ICE counterparts, demonstrating that theoretical efficiency advantages associated with EVs hold true in practice. By using EVs instead of diesel vehicles, fleets should experience significant energy cost savings from 2021 to 2035, regardless of vehicle platform, with the greatest savings expected for fleets with transit buses (up to USD 4459 per bus annually) and HD trucks (up to USD 3284 per truck annually), especially those with high-mileage duty cycles. Even when accounting for the additional costs associated with installing and maintaining EVSE infrastructure, fueling MHD EVs was still projected to be less expensive per mile on average than fueling diesel MHD vehicles.

4.2. Vehicle Efficiency Prediction and Year-round Operational Range Forecast

This study found that a vehicle's operational range could be substantially lower than its nominal range under driving conditions with low temperatures, high congestion, and local duty cycles, and thereby highlighted the importance of estimate operational range when choosing a MHD EV. Using the efficiency model presented in Section 3.2, fleets can forecast a vehicle's year-round operational range to evaluate whether it meets their operating needs. Based on these results, there are two notable considerations that fleets should anticipate before purchasing and operating MHD EVs.

1. Because temperature and congestion can significantly impact EVs' efficiency and range, fleets should select vehicle models that can satisfy most of their range needs throughout an entire year, while extending operational range in colder months and congested areas by applying energy-saving practices. For example, fleets should plan to pre-heat vehicle cabin and keep vehicle doors closed as much as possible, charge midday on extremely cold days, and optimize routes and schedules to avoid heavy traffic where possible.
2. Due to variations in duty cycle characteristics, local-haul operations (less than 100 miles daily) can have 25% lower operational range than regional-haul operations (100–300 miles daily), despite using the same vehicle model in the same example city. Furthermore, local HD truck fleets may need to deploy trucks with a nominal range nearly double their expected maximum daily range to meet route needs under more extreme driving conditions, such as colder temperatures, and local duty cycle requirements, such as the high idling time percentage and traffic levels found in urban delivery duty cycles. Alternatively, fleets can consider downsizing HD trucks to MD trucks or vans if they have sufficient payload.

4.3. Limitations and Future Work

While this study addressed several critical issues for fleets, it also had limitations. The energy cost savings analyses were based on average efficiency values, average miles driven per vehicle platform, and average price estimates, and EIA fuel prices did not account for EVSE installation or maintenance costs. As a result, an individual vehicle may experience a different real-world efficiency and different cost savings from those estimated in this study. Additionally, electricity demand charges and vehicle efficiency improvement rates can be incorporated into future scenario analyses.

When modeling energy efficiency, predictions for trucks were limited to local and regional haul (less than 300 miles per day) and were not generalized to long-haul duty cycles. Compared to route-based energy consumption modeling, our model required less granular inputs, both in terms of time (i.e., duty cycle at vehicle-day level) and geography (i.e., city served as the geographic area of operation for all climate inputs). The energy efficiency model is therefore best used to quickly estimate a vehicle's efficiency in a given city or to compare a vehicle's performance across cities or duty cycles. However, the model can still be improved with additional computational resources and data. Incorporating a higher number of features and more detailed features would enable better predictions. For example, using actual cargo weight data rather than a maximum payload constant for each vehicle model would improve the payload feature's explanatory power, especially for trucks. Similarly, incorporating a targeted route as an input would provide details about actual road grade and traffic level that are not decipherable from city-level approximations (i.e., average road slope and congestion level).

Future work can use the output of the efficiency model to understand energy costs for fleets given their selected vehicle model, use case, and city profile. Finally, we plan to build a user-friendly, web-based tool that employs the model to help fleets predict operational capabilities of MHD EVs operating in their regions, thereby boosting fleets' confidence in the EV transition. This tool will be a resource for accelerated MHD EV deployment; by addressing EV performance knowledge gaps in an intuitive, accessible manner, it will enable a better understanding of real-world MHD EV efficiency and range among fleet managers, policymakers, and the public.

Author Contributions: Conceptualization, Y.Q.; methodology, Y.Q., C.D. and S.S.; validation, Y.Q., C.D. and S.S.; formal analysis, Y.Q., C.D. and S.S.; investigation, Y.Q., C.D. and S.S.; resources, Y.Q., C.D. and S.S.; data curation, C.D. and S.S.; writing—original draft preparation, Y.Q., C.D. and S.S.; writing—review and editing, Y.Q., C.D. and S.S.; visualization, Y.Q. and C.D.; supervision, Y.Q.; project administration, Y.Q. All authors have read and agreed to the published version of the manuscript.

Funding: This research was funded by the Office of Energy Efficiency and Renewable Energy ("EERE"), an office within the United States Department of Energy, with the grant award No. DE-EE0008891.

Data Availability Statement: Publicly available datasets were analyzed in this study. Data owned by the U.S. Department of Energy from the Medium- and Heavy-Duty EV Deployment: Data Collection project can be found on the LiveWire Data Platform site: https://livewire.energy.gov/project/calstart (accessed on 1 October 2023).

Acknowledgments: The authors appreciated the support and guidance received from several CALSTART colleagues, including Chase LeCroy, Kevin Leong, Jasna Tomic, Mark Hill, Lily Paul, and Lauren Thie.

Conflicts of Interest: The authors declare no conflict of interest.

References

1. Sato, S.; Jiang, Y.J.; Russell, R.L.; Miller, J.W.; Karavalakis, G.; Durbin, T.D.; Johnson, K.C. Experimental driving performance evaluation of battery-powered medium and heavy duty all-electric vehicles. *Int. J. Electr. Power Energy Syst.* **2022**, *141*, 108100. [CrossRef]
2. Forrest, K.; Mac Kinnon, M.; Tarroja, B.; Samuelsen, S. Estimating the technical feasibility of fuel cell and battery electric vehicles for the medium and heavy duty sectors in California. *Appl. Energy* **2020**, *276*, 115439. [CrossRef]
3. CALSTART. Zero-Emission Truck and Bus Market Update. October 2022. Available online: https://globaldrivetozero.org/zeti-data-explorer/ (accessed on 20 October 2022).
4. Prasad, S.L.; Gudipalli, A. Range-Anxiety Reduction Strategies for Extended-Range Electric Vehicle. *Int. Trans. Electr. Energy Syst.* **2023**, *2023*, 7246414. [CrossRef]
5. Chakraborty, P.; Parker, R.; Hoque, T.; Cruz, J.; Du, L.; Wang, S.; Bhunia, S. Addressing the range anxiety of battery electric vehicles with charging en route. *Sci. Rep.* **2022**, *12*, 5588. [CrossRef] [PubMed]
6. Giuliano, G.; Dessouky, M.; Dexter, S.; Fang, J.; Hu, S.; Miller, M. Heavy-duty trucks: The challenge of getting to zero. *Transp. Res. Part D Transp. Environ.* **2021**, *93*, 102742. [CrossRef]

7. Smith, D.; Ozpineci, B.; Graves, R.L.; Jones, P.T.; Lustbader, J.; Kelly, K.; Walkowicz, K.; Birky, A.; Payne, G.; Sigler, C.; et al. *Medium- and Heavy-Duty Vehicle Electrification: An Assessment of Technology and Knowledge Gaps (No. ORNL/SPR-2020/7)*; Oak Ridge National Laboratory (ORNL): Oak Ridge, TN, USA; Available online: https://doi.org/10.2172/1615213 (accessed on 29 April 2022).
8. CALSTART. California HVIP Total Cost of Ownership Estimator. Available online: https://californiahvip.org/tco/ (accessed on 24 October 2022).
9. CARB. Battery Electric Truck and Bus Energy Efficiency Compared to Conventional Diesel Vehicles, May 2018. Available online: https://ww2.arb.ca.gov/sites/default/files/2018-11/180124hdbevefficiency.pdf (accessed on 20 March 2023).
10. U.S. DOE Alternative Fuels Data Center. Average Retail Fuel Prices in the United States, October 2022. Available online: https://afdc.energy.gov/fuels/prices.html (accessed on 24 October 2022).
11. CALSTART. MHD EV Data Visualization Dashboard Version 1.5. Available online: https://calstart.org/projects/medium-heavy-duty-ev-deployment-data/ (accessed on 11 October 2023).
12. Nykvist, B.; Olsson, O. The feasibility of heavy battery electric trucks. *Joule* **2021**, *5*, 901–913. [CrossRef]
13. Perugu, H.; Collier, S.; Tan, Y.; Yoon, S.; Herner, J. Characterization of battery electric transit bus energy consumption by temporal and speed variation. *Energy* **2023**, *263*, 125914. [CrossRef]
14. Fetene, G.M.; Kaplan, S.; Mabit, S.L.; Jensen, A.F.; Prato, C.G. Harnessing big data for estimating the energy consumption and driving range of electric vehicles. *Transp. Res. Part D Transp. Environ.* **2017**, *54*, 1–11. [CrossRef]
15. Ahmed, M.; Mao, Z.; Zheng, Y.; Chen, T.; Chen, Z. Electric Vehicle Range Estimation Using Regression Techniques. *World Electr. Veh. J.* **2022**, *13*, 105. [CrossRef]
16. Qi, X.; Wu, G.; Boriboonsomsin, K.; Barth, M.J. Data-driven decomposition analysis and estimation of link-level electric vehicle energy consumption under real-world traffic conditions. In Proceedings of the 14th World Conference of Transport-Research-Society (WCTRS), Shanghai, China, 10–15 July 2016; pp. 36–52.
17. Xu, X.; Aziz, H.A.; Liu, H.; Rodgers, M.O.; Guensler, R. A scalable energy modeling framework for electric vehicles in regional transportation networks. *Appl. Energy* **2020**, *269*, 115095. [CrossRef]
18. Ullah, I.; Liu, K.; Yamamoto, T.; Zahid, M.; Jamal, A. Electric vehicle energy consumption prediction using stacked generalization: An ensemble learning approach. *Int. J. Green Energy* **2021**, *18*, 896–909. [CrossRef]
19. Modi, S.; Bhattacharya, J.; Basak, P. Estimation of energy consumption of electric vehicles using Deep Convolutional Neural Network to reduce driver's range anxiety. *ISA Trans.* **2019**, *98*, 454–470. [CrossRef] [PubMed]
20. Weiss, M.; Cloos, K.C.; Helmers, E. Energy efficiency trade-offs in small to large electric vehicles. *Environ. Sci. Eur.* **2020**, *32*, 46. [CrossRef]
21. Yuksel, T.; Michalek, J.J. Effects of Regional Temperature on Electric Vehicle Efficiency, Range, and Emissions in the United States. *Environ. Sci. Technol.* **2015**, *49*, 3974–3980. [CrossRef] [PubMed]
22. Qiu, Y.; Dobbelaere, C.; Song, S. Real-world Energy Efficiency Analysis and Implications: Medium- and Heavy-Duty EV Deployments Across the U.S. In Proceedings of the 36th International Electric Vehicle Symposium and Exhibition, Sacramento, CA, USA, 11–14 June 2023; Available online: http://evs36.com/wp-content/uploads/finalpapers/FinalPaper_Qiu_Yin_Dobbelaere_Cristina.pdf (accessed on 1 October 2023).
23. U.S. DOE Alternative Fuels Data Center. Average Fuel Economy by Major Vehicle Category, February 2020. Available online: https://afdc.energy.gov/data/10310 (accessed on 22 February 2023).
24. U.S. EIA. Short-Term Energy Outlook, November 2021. Available online: https://www.eia.gov/analysis/projection-data.php#annualproj (accessed on 20 January 2023).
25. National Renewable Energy Laboratory. Estimating the Breakeven Cost of Delivered Electricity to Charge Class 8 Electric Tractors. 2022. Available online: https://www.nrel.gov/docs/fy23osti/82092.pdf (accessed on 9 March 2023).
26. CALSTART. Drive to Zero's Zero-Emission Technology Inventory (ZETI) Tool Version 8.2. 2023. Available online: https://globaldrivetozero.org/tools/zero-emission-technology-inventory/ (accessed on 11 October 2023).
27. NOAA. Global Historical Climatology Network Daily (GHCNd). Available online: https://www.ncei.noaa.gov/products/land-based-station/global-historical-climatology-network-daily (accessed on 26 October 2022).
28. NASA. NLDAS-2: North American Land Data Assimilation System Forcing Fields. Available online: https://developers.google.com/earth-engine/datasets/catalog/NASA_NLDAS_FORA0125_H002 (accessed on 26 October 2022).
29. Muñoz Sabater, J. ERA5-Land Monthly Averaged Data from 1981 to Present. 2023. Available online: https://developers.google.com/earth-engine/datasets/catalog/ECMWF_ERA5_LAND_HOURLY (accessed on 6 February 2023).
30. Lovelace, R.; Felix, R.; Talbot, J. Slopes package v1.0.0. Available online: https://ropensci.github.io/slopes/index.html (accessed on 13 January 2023).
31. OpenStreetMap Data Extracts. Available online: https://download.geofabrik.de/index.html (accessed on 13 January 2023).
32. Texas A&M Transportation Institute. Urban Mobility Report. 2021. Available online: https://mobility.tamu.edu/umr/congestion-data (accessed on 10 January 2023).
33. USGS, TNM Download v2.0. Available online: https://apps.nationalmap.gov/downloader/ (accessed on 13 January 2023).
34. CALSTART. Drive to Zero's Zero-Emission Technology Inventory Data Explorer Version 3.3. 2023. Available online: https://globaldrivetozero.org/zeti-data-explorer/ (accessed on 11 October 2023).

35. Urban Bus Toolkit. Percent Seated Capacity. Available online: https://ppiaf.org/sites/ppiaf.org/files/documents/toolkits/UrbanBusToolkit/assets/1/1c/1c27.html (accessed on 28 January 2023).
36. Branco, P.; Torgo, L.; Ribeiro, R. SMOGN: A Pre-Processing Approach for Imbalanced Regression. *Proc. Mach. Learn. Res.* **2017**, *74*, 36–50. Available online: http://proceedings.mlr.press/v74/branco17a/branco17a.pdf (accessed on 1 February 2023).
37. Pedregosa, F.; Varoquaux, G.; Gramfort, A.; Michel, V.; Thirion, B.; Grisel, O.; Blondel, M.; Prettenhofer, P.; Weiss, R.; Dubourg, V.; et al. Scikit-learn: Machine Learning in Python. *J. Mach. Learn. Res.* **2011**, *12*, 2825–2830. Available online: https://scikit-learn.org/stable/index.html (accessed on 23 March 2023).
38. Qiu, Y.; Leong, K.; Ichien, D.; Dobbelaere, C.; LeCroy, C. A Cross-Country Analysis of Medium-Duty and Heavy-Duty Electric Vehicle Deployments. In Proceedings of the 35th International Electric Vehicle Symposium and Exhibition, Oslo, Norway, 11–15 June 2022; Available online: https://calstart.org/wp-content/uploads/2023/11/EVS35Paper_DOE_v2.pdf (accessed on 29 November 2023).
39. Lundberg, S.M.; Erion, G.; Chen, H.; DeGrave, A.; Prutkin, J.M.; Nair, B.; Katz, R.; Himmelfarb, J.; Bansal, N.; Lee, S.-I. From local explanations to global understanding with explainable AI for trees. *Nat. Mach. Intell.* **2020**, *2*, 56–67. [CrossRef]

World Electric Vehicle Journal

MDPI

Article

Impact of V2G Flexibility on Congestion Management in the German Transmission Grid

Manuel Ruppert *, Nora Baumgartner, Alexandra Märtz and Tim Signer

Institute of Industrial Production, Karlsruhe Institute of Technology (KIT), Hertzstrasse 16, 76133 Karlsruhe, Germany
* Correspondence: manuel.ruppert@kit.edu

Abstract: In this study, we investigate the effect of vehicle-to-grid (V2G) flexibility potential on solving transmission grid congestion in Germany using congestion management measures. We extend existing work on effects of V2G on transmission grid congestion by determining the flexibility provided for improving grid operation based on mobility behavior and findings on V2G user requirements from real-world electric vehicle users. Furthermore, the impact on transmission grid operation is analyzed using an optimal congestion management model with high temporal and spatial resolution. Using a scenario for the year 2030 with ambitious targets for European renewable generation development and electrification of private vehicles, our findings show that by enabling the available fleet of V2G vehicles to participate in congestion management, cost and amount can be reduced by up to 11%. However, the required capacity is shown to be lower than installed capacities in ambitious future scenarios, implying that a limited number of vehicles close to congestion centers will be utilized for transmission grid operation. Our results further suggest that high numbers of vehicles with low availability of V2G for grid operation purposes can lead to an increase in congestion management measures, while V2G proves beneficial for congestion management emissions and cost in all scenarios.

Keywords: electric vehicle; energy storage; optimization; smart charging; V2G (vehicle to grid)

Citation: Ruppert, M.; Baumgartner, N.; Märtz, A.; Signer, T. Impact of V2G Flexibility on Congestion Management in the German Transmission Grid. *World Electr. Veh. J.* **2023**, *14*, 328. https://doi.org/10.3390/wevj14120328

Academic Editors: Joeri Van Mierlo and Genevieve Cullen

Received: 6 October 2023
Revised: 16 November 2023
Accepted: 25 November 2023
Published: 29 November 2023

1. Introduction

The ongoing transformation towards a more sustainable energy system is driven by concerns about the impact of traditional energy sources on the environment and climate change. To mitigate these concerns, alternative sources of energy and ways to improve energy efficiency are necessary. One of the most promising options is electrification across different sectors in combination with increased electricity generation from renewable energy sources (RESs), which are becoming increasingly cost-competitive. In the European electricity system, the share of renewable energy sources has been rising steadily in the last few years, and several countries have set ambitious targets to increase this share further. In addition to the increasing the RES share in electricity systems, the electrification of different sectors, such as transportation, heating and cooling, is gaining momentum. In the private transportation sector, electric vehicles (EVs) are increasingly popular due to governmental subsidies, reduced carbon emissions, declining cost and increasing range of the vehicles' battery package. The diffusion rate of EVs is expected to continue, with many governments setting targets for EV adoption. For example, the newly elected government has formulated a new medium-term target of 15 million EVs in Germany by the year 2030 [1]. However, the anticipated increasing electrification of privately owned vehicles presents new challenges for electricity grids, as uncontrolled charging of EVs can lead to synchronous charging behavior, resulting in significant electricity demand peaks and additional stress on the grid [2]. Vehicle-to-grid (V2G) technology has been proposed as a possible solution to this challenge. V2G technology provides a decentralized source of flexibility that can mitigate the increase in existing peaks in today's electricity patterns and accordingly improve RES integration potential when used effectively in grid congestion management. In addition, V2G technology can partially substitute the role of redispatch power provision of thermal power plants and other storage technologies [3] and lead to more economical and ecological congestion management. In previous studies on V2G technology, many aspects of the impact on the energy system have been investigated. While the majority of these works focused on electricity market integration and balancing markets, the impact on ultra-high voltage and high voltage grid expansion requirements has been studied in [4]. Refs. [5,6] have investigated the potential of EVs for congestion management in Germany and Austria, each using an aggregated model of the transmission grid. Furthermore, EV flexibility has been previously investigated as an element of aggregated decentralized flexibility potential for transmission grid operation [7].

In this study, we investigate transmission grid effects as well as detailed EV flexibility potential by combining approaches of EV flexibility modeling and an optimal congestion management model. We use the model cascade to investigate the potential impact of V2G technology on the German transmission grid for a future energy scenario. We take into account results from a study on user requirements for V2G performed in the project and present results for the impact of V2G technology on congestion management using a case study of the European electricity system in 2030.

2. Modeling of V2G in Transmission Grids

To assess the potential impact of large-scale use of V2G on the transmission grid, a model cascade was developed by combining project findings on user requirements and diffusion modeling of projected EV uptake. This approach allows for the estimation of time-dependent V2G flexibility potential. To quantify the potential benefits of V2G technology, the flexibility shifting potential of charging operations was analyzed in the simulation of the transmission grid. In this last step, all previously generated input data are utilized in an optimal power flow formulation using a minimal congestion management formulation. The results allow for the identification of the most congested grid areas, where V2G could be most effective in alleviating grid stress.

2.1. Model Overview

To investigate transmission grid congestion management actions, first, a detailed modeling of generation, consumption and flexibility behavior is required on a nodal level. Using these data, the grid operation and utilization of connected flexibility options are often analyzed using optimal power flow formulations. In this work we present an approach that includes V2G flexibility in an optimal congestion management formulation based on the basic optimal power flow. To generate the necessary data for EV charging behavior and flexibility potential for V2G operation, three submodels are used. The resulting modeling framework and the information flow is illustrated in Figure 1, which provides a schematic overview of the model cascade. The framework begins with a detailed analysis of V2G user requirements, which takes the individual preferences of EV owners into account. This information is then combined with diffusion modeling of EV uptake, predicting future EV adoption. The resulting time-dependent V2G flexibility potential is then utilized to perform the simulation of the German transmission grid and evaluate the benefits of reducing congestion management measures in different scenarios.

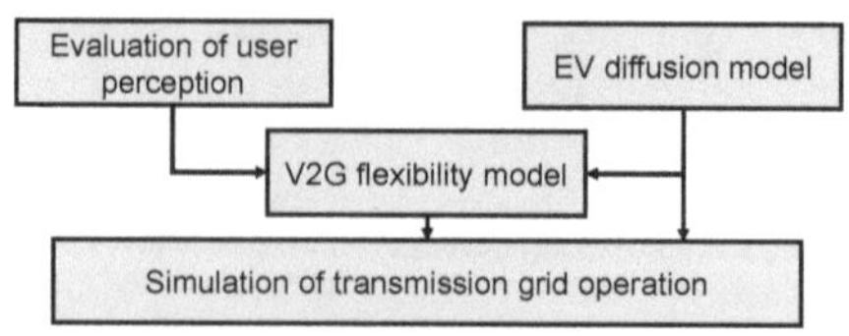

Figure 1. Schematic representation of the different model parts and input/output flow.

2.2. User Requirements

The bidirectional charging process directly involves the EV user, which makes the EV user one of the primary actors within a V2G system [8]. To enable a successful implementation of V2G technology, it is therefore important to actively engage the user while dismantling perceived barriers, such as a perceived loss of control over the charging process [9–11] or concerns of a shortened battery life due to V2G [10]. One way to foster user acceptance is to account for charging requirements, which can be defined by the EV user. The minimum range is such a requirement. We define it as the minimum necessary range that EVs must always be able to cover in unpredictable cases, for example, an emergency case [12]. It is also an essential parameter from an aggregator's point of view, as it defines the flexibility potential made available by the EV user.

In this study, we account for user requirements by integrating the results of the minimum range from a representative survey (n = 1196) conducted in January 2021 to investigate users' willingness to pay (WTP) and minimum range requirements in the context of a V2G charging tariff. Specifically, by building a mediation model, the study evaluates the importance of three charging strategies on users' WTP and minimum range requirements. The study reveals EV owners' preference for a climate-neutral charging strategy, leading to a higher readiness to accept lower minimum ranges and lower monetary savings [12]. As previous studies highlight the importance of EV experience to create informed decisions about issues in the realm of V2G [13,14], we addressed our survey to three stakeholder groups with different levels of EV experience (see [12]) and asked respondents to provide their minimum range (SoC_{min}) requirements in an open-ended question. The question referred to a BMW i3 with a range of 270 km.

The results in Table 1 show that EV users indicated the lowest SoC_{min} values, which is equivalent to approximately 40% of the battery capacity of a BMW i3. Previous field studies with EV participants found similar values [15]. In this study, we report the average minimum range (SoC_{min} = 40%) for the EV owner group (N_{high} = 264), as this group is the most experienced with EVs and therefore provides the most realistic values (see [12]).

Table 1. EV owners' minimum range requirements.

Sample	(in km)							
	M	SD	SE	Min	Max	$q_{0.25}$	$q_{0.5}$	$q_{0.75}$
N = 1196	119.01	98.37	2.84	0	500	50	100	150
N_{low} = 691	119.75	97.91	3.73	0	500	50	100	150
N_{med} = 241	126.05	104.78	6.75	15	500	50	100	150
N_{high} = 264	110.64	93.16	5.73	1	500	50	100	120

2.3. EV Diffusion

The technology ramp-up of electric vehicles in Germany was assessed using the Bass diffusion modeling approach, similar to [16]. The Bass diffusion model is a commonly used approach for assessing the adoption of new technologies [17]. The model is based on the assumption that the spread of new technologies often follows an S-curve pattern. The interplay between present and potential adopters, called innovators (q) and imitators (p), is central to the Bass diffusion model. The market potential is denoted by M, and t represents the index for the specific year being considered. The model forecasts fleet sizes for every year since the start year t_0, where the difference between the current year t and the start year t_0 is $t - t_0 = 0$. A formal description of the Bass diffusion model can be found in Equation (1), whereby $N(t)$ represents the number of cumulative adoptions up to a given time t.

$$N(t) = m\,\frac{1 - e^{-(p+q)(1-t_0)}}{1 + \frac{p}{q}e^{-(p+q)(1-t_0)}} \tag{1}$$

The innovation coefficient q and imitation coefficient p of the model are estimated by fitting the S-curve derived from the Bass diffusion model to historic annual EV stock data [18] in Germany and planned registration targets for the year 2030. The delta between the S-curve and the input data is minimized by using a non-linear regression method. More precisely, a Levenberg–Marquardt numerical optimization algorithm was employed in the OriginPro Solver to estimate the parameters of the Bass EV diffusion model.

2.4. EV Flexibility

The V2G flexibility model was designed to generate representative, synthetic charging and flexibility profiles and thus estimate the V2G flexibility potential of EVs in Germany [19]. An overview of the methodological approach is illustrated in Figure 2. In the first step, parking and mobility profiles were created based on data from the German Mobility Panel [20]. The underlying dataset contains plausible data from 1850 households with a total of 3074 persons and 70,252 trips. Subsequently, the charging behavior of the EV was simulated by the additional user-specific input data on EV and information on the charging infrastructure. Battery capacity, energy consumption as well as the availability of charging points per location and associated charging power (selectable charging power of 3.7 kW, 11 kW, 22 kW and 55 kW) per charging point were set as parameters at the beginning of the simulation. The input parameters were assumed to be identical for all EVs and the time resolution is 10 min.

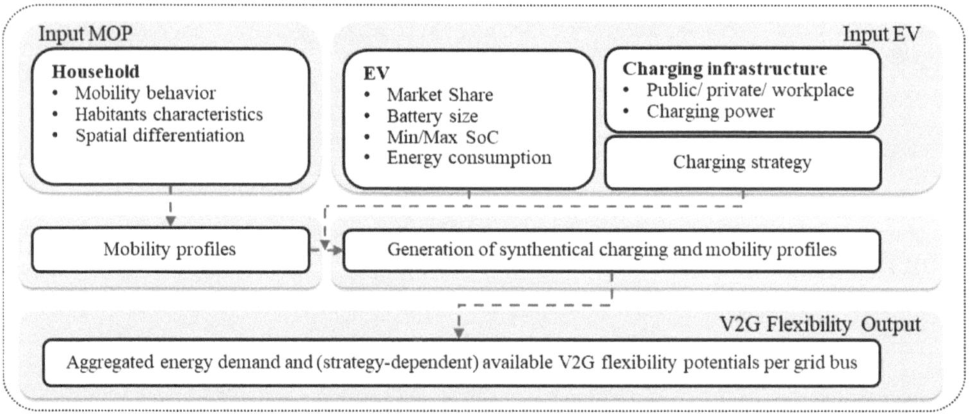

Figure 2. V2G flexibility model.

In addition, different charging strategies were implemented and shown in Figure 3. One charging strategy is the so-called as soon as possible (ASAP) strategy. Here, the vehicle is charged immediately with the maximum SoC-dependent charging power available at the charging location up to the maximum SoC level or until the departure time for the next trip. Another strategy is based on the assumption that EVs will start charging as late as possible during parking periods with charging opportunities while simultaneously considering user restrictions. We accounted for user restrictions by including the minimum range from Section 2.1, which is henceforth called the security range. The safety range (the range to which charging should take place as soon as possible after arrival at a charging station) and the target range (the range to which charging should take place as soon as possible before a journey until the start of the journey) were taken into account. After arrival at a charging station, charging takes place on the one hand as early as possible to a safety range and on the other hand as late as possible to a target range (target range $\geq$ safety range), which is to be reached at the time of departure.

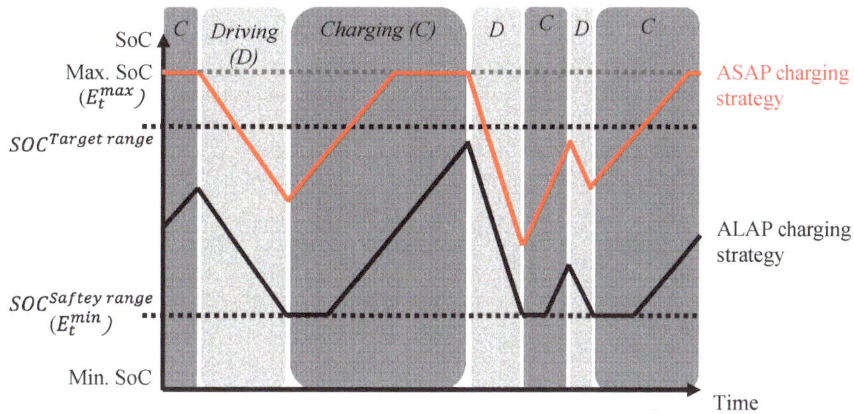

Figure 3. Schematic representation of the V2G flexibility potential and upper and lower bounds.

The amounts of energy required for the journeys are determined based on the distances driven and the energy consumption. This results in the necessary energy demand for the charging processes. The maximum amount of energy that can be charged is then determined for each time step. This depends on the parking time, the charging status of the vehicle battery and the available charging infrastructure at the respective locations of the vehicles [19].

Synthetic charging and mobility profiles are derived based on the mobility profiles and by simulating the charging behavior. These representative charging profiles can then be evaluated and interpreted regarding energy demand and V2G flexibility of the charging process. To integrate flexibility in the grid model, user requirements and the corresponding EV market ramp-up are considered in the flexibility model in the user-specific EV input data scope. Based on the user requirements and the EV market penetration, the model can be used to estimate the flexibility potential. The V2G flexibility potential can be estimated considering the implemented charging strategies. The ASAP charging strategy sets the upper limit for the allowed SOC. The second charging strategy sets the lower SoC limit. The area between the charging states of the two extreme SoC levels represents the permissible range for the SOC and, combined with the available charging power, describes the flexibility potential.

2.5. Transmission Grid

Using a multi-objective optimization approach, we have developed a framework to investigate the optimal congestion management in the interconnected European transmission grid. The approach allows us to examine the role of EVs that need to be considered in the liberalized power market, such as congestion cost, additional carbon emissions, as well as deviations from market-based dispatch results, based on a formulation developed in [21,22]. The model is applied to the central European electricity market, with the grid simulation focusing on congestion management measures in Germany. We utilized highly spatially resolved time series of renewable generation and demand using data and methodology described in [23].

To model the interaction between the electricity market and congestion management, we used a two-step approach. In the first step, we determined the optimal dispatch of electricity generation in the interconnected market using linear programming. This widely used and described economic dispatch approach [24] considers various parameters such as fuel prices, generator capacities and transmission constraints to identify the most cost-efficient solution for meeting electricity demand. The results of this step provide the minimal-cost, copperplate-based dispatch solution for the electricity market on a national level, with the objective function shown in Equation (2). For every timestep t, each of the system's elements are assigned a variable cost term C that is multiplied by the amount of generation p, with the set of thermal and hydraulic generation G, renewable generation source RES, decentral flexibility elements F and electricity demand D. In the case of the last-mentioned, cost occurs when load shedding LS is required. A more detailed description of the formulation can be found in [22].

$$\begin{gathered} min \sum_{g \in G, t \in T} C_g \times p_{g,t} + \sum_{res \in RES, t \in T} C_{RES} \times p_{res,t} + \sum_{f \in F, t \in T} C_F \times p_{f,t} \\ + \sum_{d \in D, t \in T} C_{LS} \times p_{d,t} \\ \forall g \in G,\ res \in RES,\ f \in F,\ d \in D,\ t \in T \end{gathered} \tag{2}$$

The linear formulation of a storage system can be modeled using the generalized formulation shown in Equation (3). The available energy $e_{s,t}$ of storage s in time step t is determined by the available energy in the previous time step $t-1$, charged power $p_{g,t}^{in}$ and discharged $p_{g,t}^{out}$ with their respective efficiency η and external energy inflows $\zeta_{s,t}^{in}$ and outflows $\zeta_{s,t}^{out}$.

$$e_{s,t} = e_{s,t-1} + p_{g,t}^{in} \times \eta_{g,in} - p_{g,t}^{out} / \eta_{g,out} + \zeta_{s,t}^{in} - \zeta_{s,t}^{out}\ \forall s \in S,\ t \in T \tag{3}$$

When applying Equation (3) to V2G charging, the available energy is provided by the car battery, and efficiency is determined by losses within the vehicle and in auxiliary equipment such as the wallbox, while the mobility demand results in an irregular outflow. For single vehicles, charging and discharging power is zero during driving or when they are not plugged into a charger. Using the fleet flexibility potential aggregation of the V2G flexibility model shown in Figure 3, this can be expressed by Equations (4)–(7), where the bounds of the EV fleet storage state e_t and charging and discharging capacity p_t are determined by the time-variant upper and lower bounds depending on the composition of plugged-in and unavailable EVs. The external energy outflow $\zeta_{,t}^{out}$ is defined as the energy used at the time of plug in E_t^{mob} for mobility requirements since the previous plug-in. Using a fleet-wide aggregation of V2G flexibility can lead to the violation of individual storage state constraints but also implicates a large advantage in computational complexity compared to a discrete modeling approach.

$$E_t^{min} \le e_t \le E_t^{max}\ \forall\ t \in T \tag{4}$$

$$0 \le p_t^{in} \le P_t^{in,max}\ \forall\, t \in T \tag{5}$$

$$0 \le p_t^{out} \le P_t^{out,max}\ \forall\, t \in T \tag{6}$$

$$\zeta_t^{out} = E_t^{mob}\ \forall\, t \in T \tag{7}$$

In the second step, we determine the required dispatch adjustments using a linearized optimal power flow formulation. This step accounts for the V2G flexibility potential developed by implementing available capacities and bounds from the V2G flexibility model previously described. In the linearized optimal power flow formulation, the nonlinear branch flow equations are simplified by an approximation which assumes a lossless system with constant voltage levels [25]. The resulting linearized power flow equation is shown in Equation (8), with the active power flow $P_{i,j}$ between nodes i and j dependent on the respective bus voltage angles φ.

$$P_{i,j} = b_{i,k}(\varphi_k - \varphi_i) \tag{8}$$

To determine the optimal congestion management measures, we formulated the objective function as a minimization of the total amount of congestion measure volume in an analogous manner to [22], as shown in Equation (9). While all variables have an additional bus index in the following due to the additional spatial dimension, the index is omitted for the sake of simplicity. Each congestion management action consists of the deviation from the market result, denoted by Δp. As both positive and negative measures are contributing in a uniform manner to the objective, the objective function consists of the absolute value of change. This is not required for load adjustment, where reduction in load always results in a positive contribution, which is additionally penalized by the load shedding penalty factor $C_{LS,grid}$. The corresponding bounds are shown in Equations (10)–(13). The available potential for reduction or increase in generation for conventional, renewable and flexibility generation is subject to the technical minimum and maximum limits P_{min}. and P_{max} as well as the market dispatch P_t which results from Equation (2). It should be noted that for volatile RES generation from solar and wind, no generation increase potential remains, as $P_{res,max}$ equals $P_{res,t}$.

$$min \sum_{g\in G,t\in T} \left|\Delta p_{g,t}\right| + \sum_{res\in RES,t\in T} \left|\Delta p_{res,t}\right| + \sum_{f\in F,t\in T} \left|\Delta p_{f,t}\right| + \sum_{d\in D,t\in T} C_{LS,grid} \times \Delta p_{d,t} \quad \forall g \in G,\ res \in RES,\ f \in F,\ d \in D,\ t \in T \tag{9}$$

$$P_{g,t} - P_{g,min} \le \Delta p_{g,t} \le P_{g,max} - P_{g,t} \ \forall\ t \in T,\ g \in G \tag{10}$$

$$P_{res,t} - P_{res,min} \le \Delta p_{res,t} \le P_{res,max} - P_{res,t} \ \forall\ t \in T,\ res \in RES \tag{11}$$

$$P_{f,t} - P_{f,min} \le \Delta p_{f,t} \le P_{f,max} - P_{f,t} \ \forall\ t \in T,\ f \in F \tag{12}$$

$$0 \le \Delta p_{d,t} \le P_{d,t} \ \forall\ t \in T,\ d \in D \tag{13}$$

To include V2G flexibility, Equations (3)–(7) can be applied in an analogous manner by adding a spatial component on a nodal basis, with the nodal EV density being determined by the regionalization developed in [23]. The calculation is performed for 8760 timesteps with consecutive weekly optimization horizons, ensuring an optimization time-horizon long enough to allow the activation of available flexibility from individual mobility demand patterns. Overall, this two-step approach provides a comprehensive framework for modeling the interaction between the electricity market and transmission grid operation.

3. Case Study

Using the methodology presented previously, a case study was conducted to evaluate the possibility of deploying V2G to solve grid congestion. The study was carried out for the German high-voltage transmission grid in the year 2030 using a scenario developed in the project ENSURE [26]. The scenario "Storyline B" assumes an ambitious path towards decarbonization of the electricity sector, with high growth for RES generation until 2030. The generation capacities are shown in Figure 4. While both onshore and offshore wind generation, as well as solar generation, increase compared to today's state, power generation from hard coal and lignite has been phased out by 2030 under the assumptions. The corresponding transmission grid scenario for the future date and the corresponding spatial distribution of generation and demand is shown in Figure 5. A very detailed description of the scenario and its regionalization, as well as the extension of the future energy scenarios to 2050, can be found in [26].

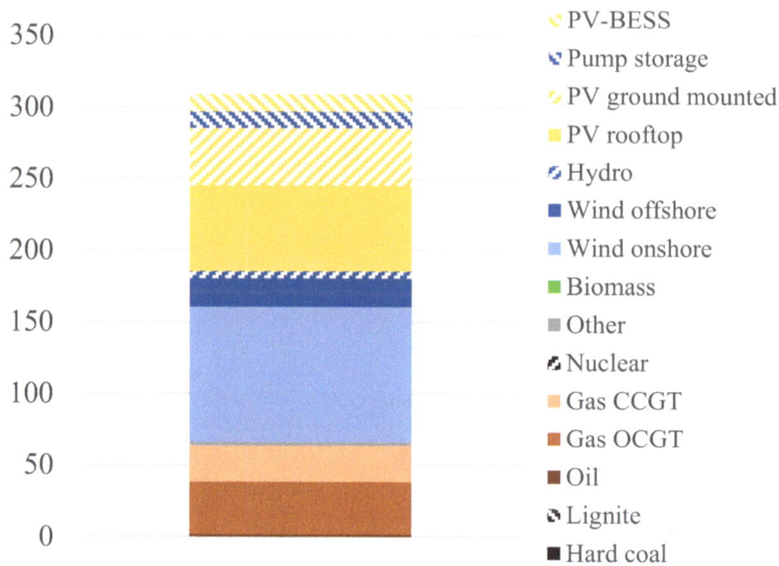

Figure 4. Generation capacities in Germany from central and decentralized sources in the case study for the year 2030 [20].

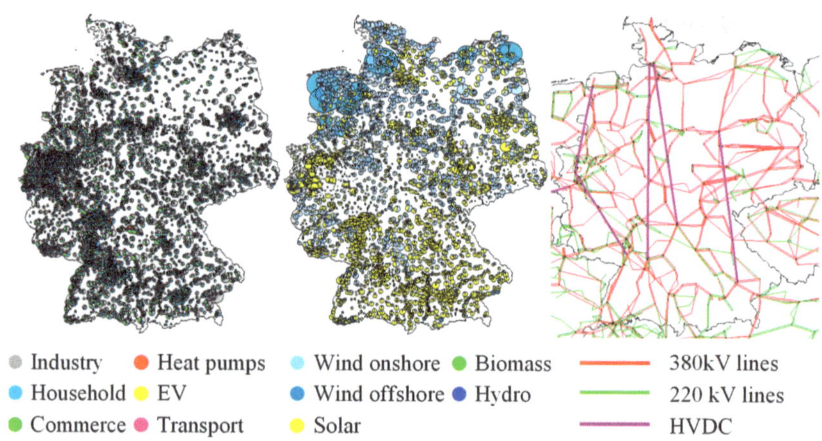

Figure 5. Demand allocation (**left**), RES allocation (**center**) on high voltage level and transmission grid model (**right**) of Germany.

To determine the scenario-dependent V2G flexibility potential, the input parameters shown in Table 2 are defined. Here, the minimum range from Section 2.1 is taken into account. At the same time, two different market shares are included in the analyses, which result from the results of the Bass diffusion model. Altogether, we investigate the impact of V2G in four scenarios, with three alternative parameter sets from the Base scenario: The scenario Work extends bidirectional charging availability from purely home charging to workplace charging, which significantly reduces peak charging demand in case of immediate charging (ASAP) as can be seen in Figure 5. Furthermore, available flexibilities during working hours are increased for market and grid utilization.

Table 2. Scenario-related input data.

Scenario	BEV Count	Charging Location	Charging Power	Battery Capacity	Consumption	Efficiency	Availability Market	Availability Grid Operation	Safety Range	Target Range
Base	15 million	Home charging	11 kW	50 kWh	15 kWh/100 km	90%	100%	100%	40%	56.2%
Work		Home- and workplace charging								
Reduced	10 million	Home charging								
Grid	15 million	Home charging					20%			

The available flexibility potential for the scenarios Base and Work is shown in Figure 6. In the home-charging scenario, available charging power decreases, with only half of the total capacity during midday on business days. On the EV fleet-averaged level, available SoC upper bound levels remain consistently very high, as most of the charging unavailabilities are not connected to driving but parking at locations without charging equipment, which can be seen in the visualization of mobility behavior in Figure 6. Consequently, the relative change in total available charging power at midday is more significant than the change in the upper and lower SoC bounds for the scenario Work as shown in Figure 7.

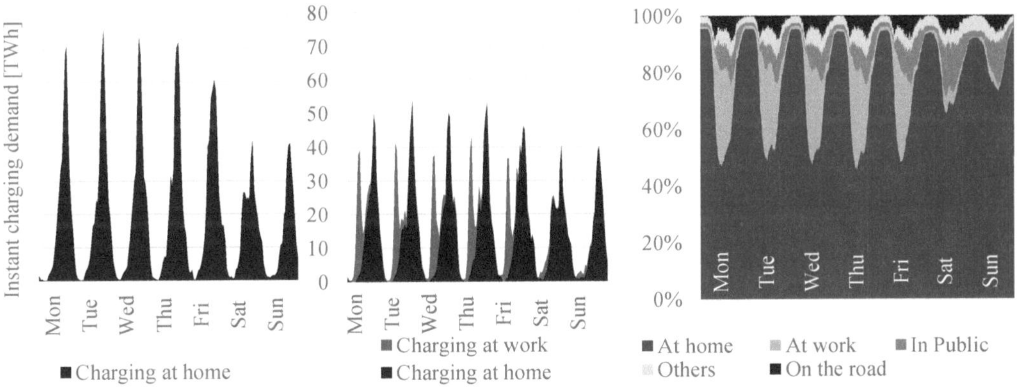

Figure 6. V2G charging demand in ASAP mode for scenario Base averaged over the vehicle fleet, one week (**left**) and scenario Work (**middle**) and distribution of EV location over one week (**right**).

To estimate the innovation and imitation coefficient, a non-linear regression method was applied to both historical EV fleet data of the German Ministry of Transport [18] and future EV fleet size targets of the German government [1]. Using these inputs, two EV ramp-up scenarios were developed. The first scenario Base aligns with the current government's objective of reaching 15 million EVs by 2030, while the second reduced transition speed scenario Reduced was devised with the aim of achieving a number of 10 million EVs by 2030. Both variants assume that, eventually, all conventional vehicles will be replaced by EVs. EVs are expected to be the primary choice for meeting vehicle emission reduction targets, supported by increasing investments in charging infrastructure and major vehicle manufacturers' upcoming lineups of EVs. Additionally, the German vehicle fleet size is assumed to remain constant. However, trends such as autonomous driving and car sharing could lead to smaller vehicle fleets in the long term. As quantifying such effects is challenging and rapid changes in the individual mobility sector until 2030 seem unlikely, in the investigated scenarios, the fleet size is assumed to remain constant. Figure 8 displays the forecasted yearly EV fleet sizes for both scenarios.

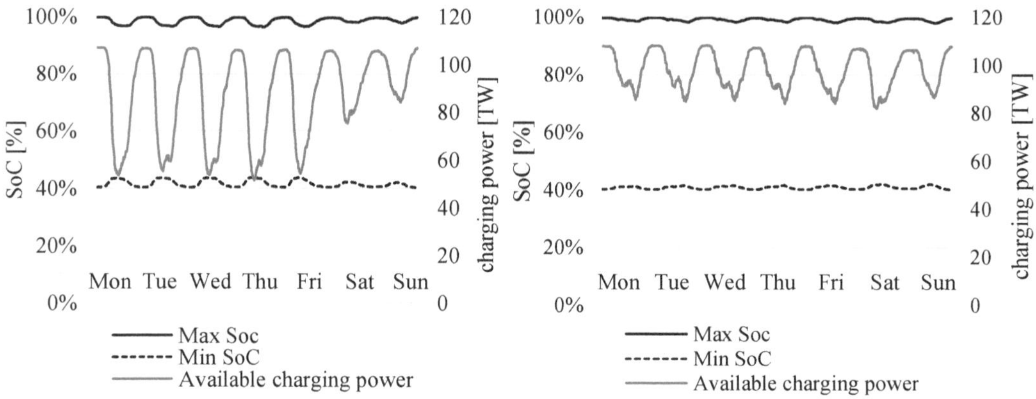

Figure 7. (**left**): V2G flexibility potential for Base (fleet-averaged, one week; charging availability at home), (**right**): V2G flexibility potential (fleet-averaged, one week; charging availability at home and at work).

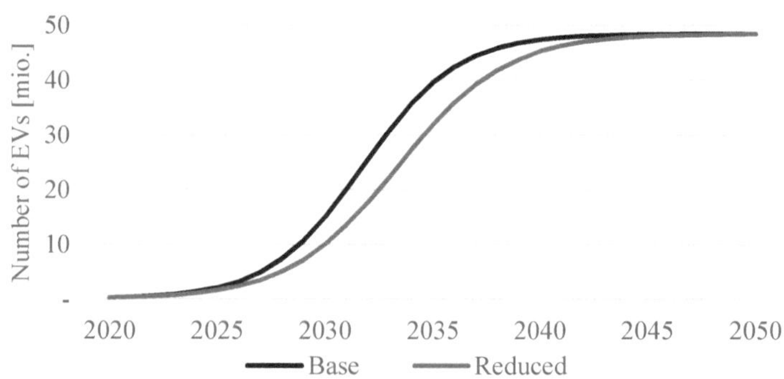

Figure 8. Development of EV adoption in Germany for scenarios Base and Reduced.

Though the current diffusion of EVs is still in its early stages, the model predicts that the adoption rate will speed up, especially after the year 2025. The model results also suggest that by the end of the 2030s, market saturation can be anticipated, leading to a reduction in the number of new EVs entering the market. Based on the two scenarios considered, nearly the entire German car fleet of over 48 million vehicles will be replaced by EVs between 2042 and 2045. Considering the predicted annual vehicle registrations of up to 5.5 million in the base scenario and up to 4.8 million annual EV registrations in the reduced scenario, the scenarios can be considered as an optimistic upper bound when compared to yearly historical passenger car registrations in Germany, which averaged at about 3.5 million annual vehicle registrations [27]. In the fourth scenario Grid, the participation rate of V2G vehicles in the electricity market is reduced to 20% by limiting the available charging and discharging power uniformly. Charging and discharging power for transmission grid operation remains the same, thus assuming an option for the transmission grid operator to utilize available flexibility when it is needed due to transmission grid congestion, even if the user does not participate in the electricity market.

On the transmission grid level, a dataset for Germany, including overhead lines and cables above 200 kV, AC and HVDC lines connected to busbars and the present state of the grid with projected expansions until 2030 is used. The grid dataset is connected to the regionalized data on the high voltage level via transformers from extra high voltage levels to high voltage levels between 60 and 150 kV using the methodology described in [23]. The data include the present state of the transmission grid as well as projected expansion measures in terms of deconstruction, replacement and construction of substations, busbars, lines and transformers until the year 2030, as detailed in the German network development plan. Technical data were derived from publicly available sources or approximated based on comparable equipment.

4. Results

The underlying assumptions in the energy scenario assumed in this case study lead to increased utilization of the German transmission grid, as the phase-out of coal generation and general reduction in available thermal generation capacities go hand-in-hand with increased renewable generation, especially wind generation in Northern Germany. Subsequently, the increased interconnection capacities with neighboring countries are used extensively, as spatial differences in renewable generation favor higher exchange volumes. The resulting required congestion management measures without V2G flexibility for grid operation are shown in Table 3. As adjustment of exchange flows is penalized, the main elements of congestion management in the scenario are positive thermal redispatch and curtailment of RES generation. This is due to wind onshore and offshore generation in Northern Germany being the main reason for the observed congestion. The left part of Figure 9 shows the spatial distribution of lines with active bounds in the optimization result, where congestion management measures have remediated line overloadings in the congestion-free solution. Here, the structural overloading of transmission lines in the north–south direction is observable. Negative thermal redispatch is the inferior solution when minimizing the volume of adjustments, as RES generation at the source of the congestion is more efficient in most hours. This result might differ when congestion alleviation costs are included in the objective function, as RES generation does not have variable costs, while the reduction in thermal generation units is economically beneficial. Maximum positive dispatch adjustment ranges from 3392 MW in the scenario Work to 5745 MW in the scenario Reduced, while minimum negative assignments range from −2797 MW in the same scenario to −4700 MW in scenario Work. The hourly ordered distribution of dispatch adjustments can be found in the right part of Figure 9. The maximum simultaneous demand for congestion management is limited compared to the total available capacity from the entire EV fleet. A primary reason for this is that due to the wide distribution over the entire grid area, only limited capacities at suitable nodes are available.

Table 3. Congestion management measures without V2G flexibility.

(TWh)	Positive Thermal Redispatch	Negative Thermal Redispatch	Positive Hydro Redispatch	Negative Hydro Redispatch	RES Curtailment	Exchange Adjustment
Congestion management	16.10	−0.20	0.35	−0.38	−16.02	0.21

The impact of including V2G as an additional source of flexibility in the model can be found in Figure 10. As expected, the volume of congestion management measures decreases for all scenarios. While the Reduced scenario results in the most considerable reduction, this scenario also reduces the EV electricity consumption and thus might lower congestion before flexibility usage. Both Work and Base scenarios lead to a comparable volume decrease. Both perform better than the Grid scenario with a lower participation factor when determining the national dispatch. This leads to the assumption that market-oriented dispatch of V2G is generally beneficial for reducing grid congestion, and additional measures are required when the initial V2G dispatch is lowered. The effect on CO_2 emissions and costs differs for the Work scenario on the one hand and the Base and Grid scenario on the other hand. While relative cost and CO_2 emission changes correlate very well for each scenario, both increase for the Work scenario, while they decrease otherwise. This can be explained by the higher correlation between conventional electricity demand and the availability of charging at work, which is not beneficial for transmission grid operation in this scenario.

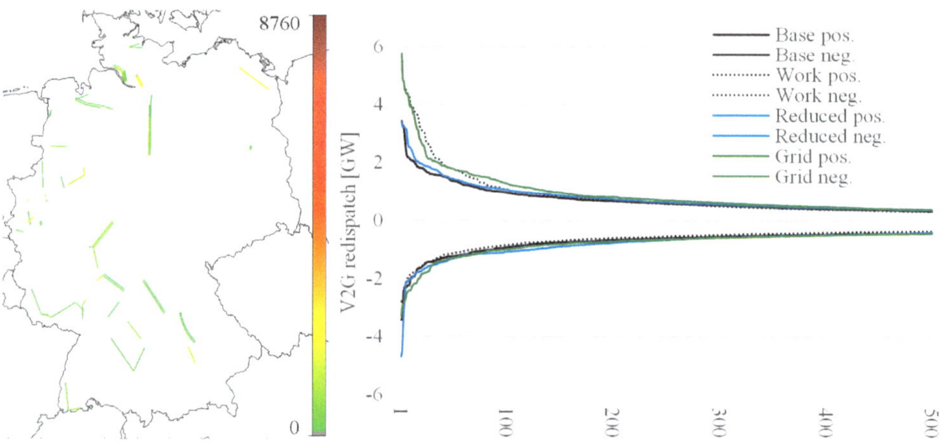

Figure 9. Congested transmission grid lines without V2G flexibility (**left**) and ordered hourly positive and negative V2G congestion management utilization for each scenario (**right**).

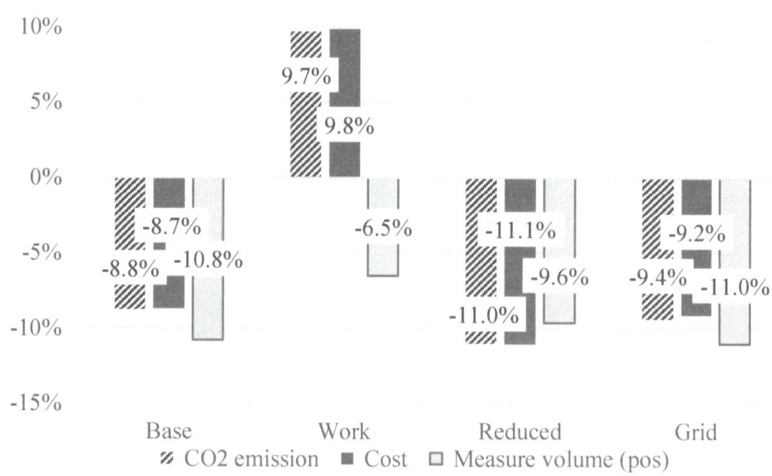

Figure 10. Results of V2G flexibility scenarios in comparison to reference case.

The model results for the sensitivity of the V2G share, as shown in Figure 11, allow a more detailed interpretation of how transmission grid congestion varies depending on the V2G adaptation rate. In the Grid scenario, the V2G share on the market side was reduced to 20%. Here, however, the V2G share of the total EV fleet has been varied synchronously. The 0% and 100% cases are represented by the Reference and Base scenarios, respectively. It is observed that a proportional reduction in V2G share leads to an increased requirement for congestion management measures. This can be explained by the fact that market actions are based on national capacities, while grid measures require spatial alignment to address overloads. However, this effect is not observed in terms of emissions and costs, where increasing V2G shares lead to reduced outcomes for both. As expected, the greatest reduction in congestion management measures is observed in the Base scenario.

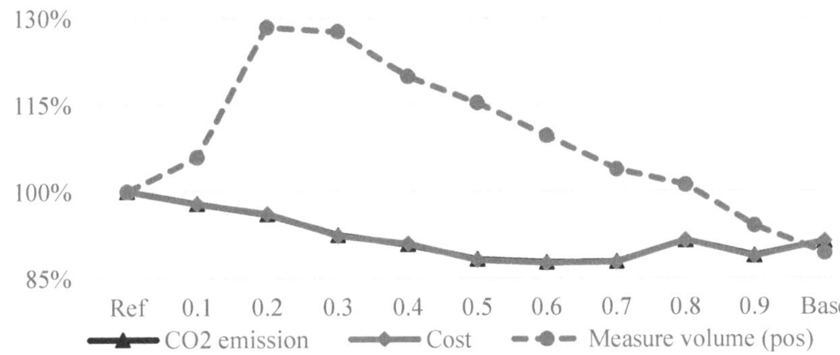

Figure 11. Sensitivity of the congestion management results to the V2G share of total EVs.

5. Conclusions

In this paper, we presented a model framework to investigate the impact of V2G flexibility on congestion management in the German transmission grid. Our approach extends the existing literature by including user requirements for V2G as well as a highly spatially disaggregated flexibility modeling allowing a combination with a detailed future transmission grid model. We showed the effect of V2G for an ambitious scenario in the year 2030. The model cascade includes an analysis of V2G user requirements and diffusion modeling of projected EV uptake. These data have been used in the V2G flexibility modeling approach to derive time-dependent V2G flexibility potential representing the input data and boundaries for the transmission grid optimization model. The simulation of the German transmission grid was conducted to identify the congested grid areas where V2G could be most effective in alleviating grid stress. The results show that including V2G in congestion management can reduce the required number of redispatch measures by more than 10%. This is a conservative estimation compared to the results for eight million EVs in [5] but can be explained by the more detailed spatial modeling in this approach, leading to fewer EVs being able to effectively contribute to eliminating transmission grid congestions. The results are also in the same range compared to the study on redispatch in Austria [6]. However, we cannot observe a strong negative effect of EV flexibility participating in the electricity market previously for the German case. The presented approach also accounts for the minimum range requirements of EV owners and assesses the adoption of EVs in Germany using the Bass diffusion modeling approach. The study shows that congestion management measures such as positive thermal redispatch and curtailment of RES generation are necessary to ensure the grid's stability. However, the introduction of V2G as an additional source of flexibility can significantly reduce the volume of congestion management measures. The results suggest that market-oriented dispatch of V2G is generally beneficial for reducing grid congestion. Nonetheless, additional measures may be required when the initial V2G dispatch is lowered. The impact of V2G on CO_2 emissions and costs varies depending on the scenario, with the Work scenario showing an increase in both, while the Base and Grid scenarios show a decrease. In future work, further decentralized flexibilities and interconnections between the European countries and their EV transition plans can be included to investigate the role of V2G for transmission grid operation. Especially, interdependence with other battery storage applications could be helpful, if a high technical and spatial level of modeling detail can be sustained. While the role of individual EV users has been included in this work, the high importance of spatial alignment of flexibility requirements and V2G usage points to a need for research on flexibility potential on an individual level, determining the type and location of future V2G potential needed, as not every EV can contribute to congestion relief equally.

Author Contributions: Conceptualization, M.R.; methodology, M.R., N.B., A.M. and T.S.; software, M.R. and A.M.; validation, M.R.; formal analysis, All; investigation, All; resources, M.R.; data curation, A.M. and M.R.; writing—original draft preparation, All; writing—review and editing, M.R.; visualization, All; supervision, M.R.; project administration, M.R.; funding acquisition, M.R. All authors have read and agreed to the published version of the manuscript.

Funding: The authors gratefully acknowledge funding by the German Federal Ministry of Education and Research (BMBF) within the Kopernikus project ENSURE "New Energy Grid Structures for the German Energiewende" under the funding number 03SFK1F0-2 and funding by the German Federal Ministry of Economic Affairs and Climate Action (BMWK) within the project Bidirectional Charging Management (BCM) under the funding number 01MV18004H.

Data Availability Statement: The data related to the presented research are owned by KIT and cannot be shared at the point of publication.

Conflicts of Interest: The authors declare no conflict of interest.

References

1. SPD. Bündnis 90/Die Grünen und FDP, Koalitionsvertrag: Mehr Fortschritt Wagen. Bündnis für Freiheit, Gerechtigkeit und Nachhaltigkeit. 2021. p. 178. Available online: https://www.bundesregierung.de/resource/blob/974430/1990812/1f422c60505b6a88f8f3b3b5b8720bd4/2021-12-10-koav2021-data.pdf?download=1 (accessed on 24 November 2023).
2. Blumberg, G.; Broll, R.; Weber, C. The impact of electric vehicles on the future European electricity system–A scenario analysis. *Energy Policy* **2022**, *161*, 112751. [CrossRef]
3. Kern, T.; Kigle, S. Modeling and evaluating bidirectionally chargeable electric vehicles in the future European energy system. *Energy Rep.* **2022**, *8*, 694–708. [CrossRef]
4. Slednev, V.; Jochem, P.; Fichtner, W. Impacts of electric vehicles on the European high and extra high voltage power grid. *J. Ind. Ecol.* **2021**, *26*, 824–837. [CrossRef]
5. Staudt, P.; Schmidt, M.; Gärttner, J.; Weinhardt, C. A decentralized approach towards resolving transmission grid congestion in Germany using vehicle-to-grid technology. *Appl. Energy* **2018**, *230*, 1435–1446. [CrossRef]
6. Loschan, C.; Schwabeneder, D.; Lettner, G.; Auer, H. Flexibility potential of aggregated electric vehicle fleets to reduce transmission congestions and redispatch needs: A case study in Austria. *Int. J. Electr. Power Energy Syst.* **2023**, *146*, 108802. [CrossRef]
7. Hoffrichter, A.; Offergeld, T.; Blank, A. Simulation of Transmission Grid Operation Incorporating Flexibility at Distribution Level. In Proceedings of the 2019 16th International Conference on the European Energy Market (EEM), Ljubljana, Slovenia, 18–20 September 2019.
8. Noel, L.; de Rubens, G.Z.; Kester, J.; Sovacool, B.K. *Vehicle-To-Grid*; Springer: Cham, Switzerland, 2019.
9. Delmonte, E.; Kinnear, N.; Jenkins, B.; Skippon, S. What do consumers think of smart charging? Perceptions among actual and potential plug-in electric vehicle adopters in the United Kingdom. *Energy Res. Soc. Sci.* **2020**, *60*, 101318. [CrossRef]
10. Krueger, H.; Cruden, A. Integration of electric vehicle user charging preferences into Vehicle-to-Grid aggregator controls. *Energy Rep.* **2020**, *6*, 86–95. [CrossRef]
11. Yilmaz, S.; Cuony, P.; Chanez, C. Prioritize your heat pump or electric vehicle? Analyzing design preferences for Direct Load Control programmes in Swiss households. *Energy Res. Soc. Sci.* **2021**, *82*, 102319. [CrossRef]
12. Baumgartner, N.; Kellerer, F.; Ruppert, M.; Hirsch, S.; Mang, S.; Fichtner, W. Does experience matter? Assessing user motivations to accept a vehicle-to-grid charging tariff. *Transp. Res. Part D Transp. Environ.* **2022**, *113*, 103528. [CrossRef]
13. Noel, L.; Carrone, A.P.; Jensen, A.F.; de Rubens, G.Z.; Kester, J.; Sovacool, B.K. Willingness to pay for electric vehicles and vehicle-to-grid applications: A Nordic choice experiment. *Energy Econ.* **2018**, *78*, 525–534. [CrossRef]
14. Chen, C.-F.; de Rubens, G.Z.; Noel, L.; Kester, J.; Sovacool, B.K. Assessing the socio-demographic, technical, economic and behavioral factors of Nordic electric vehicle adoption and the influence of vehicle-to-grid preferences. *Renew. Sustain. Energy Rev.* **2020**, *121*, 109692. [CrossRef]
15. Ensslen, A.; Ringler, P.; Dörr, L.; Jochem, P.; Zimmermann, F.; Fichtner, W. Incentivizing smart charging: Modeling charging tariffs for electric vehicles in German and French electricity markets. *Energy Res. Soc. Sci.* **2018**, *42*, 112–126. [CrossRef]
16. Ensslen, A.; Will, C.; Jochem, P. Simulating Electric Vehicle Diffusion and Charging Activities in France and Germany. *World Electr. Veh. J.* **2019**, *10*, 73. [CrossRef]
17. Jochem, P.; Vilchez, J.J.G.; Ensslen, A.; Schäuble, J.; Fichtner, W. Methods for forecasting the market penetration of electric drivetrains in the passenger car market. *Transp. Rev.* **2018**, *38*, 322–348. [CrossRef]
18. Kraftfahrt-Bundesamt. Number of Electric Vehicles in Germany from 2012 to 2022. Available online: https://www.kba.de/DE/Statistik/Produktkatalog/produkte/Fahrzeuge/fz13_b_uebersicht.html?nn=3514348 (accessed on 24 November 2023).

19. Ried, S.; Dengiz, T.; Soldner, S.; Jochem, P. Aggregating load shifting potentials of electric vehicles for energy system models. In Proceedings of the 2020 17th International Conference on the European Energy Market (EEM), Stockholm, Sweden, 16–18 September 2020.
20. Ecke, L.; Chlond, B.; Magdolen, M.; Vortisch, P. Deutsches Mobilitätspanel (MOP)—Wissenschaftliche Begleitung Und Auswertungen Bericht 2019/2020: Alltagsmobilität Und Fahrleistung, Institute for Transport Studies (KIT). 2020. Available online: https://daten.clearingstelle-verkehr.de/192/233/Bericht_MOP_20_21.pdf (accessed on 24 November 2023).
21. Ruppert, M.; Slednev, V.; Ardone, A.; Fichtner, W. Dynamic Optimal Power Flow with Storage Restrictions Using Augmented Lagrangian Algorithm. In Proceedings of the Power Systems Computation Conference, Dublin, Ireland, 11–15 June 2018.
22. Ruppert, M.; Slednev, V.; Finck, R.; Ardone, A.W. Fichtner. Utilizing distributed flexibilities in the european transmission grid. In *Advances in Energy System Optimization*; Springer International Publishing: New York, NY, USA, 2020.
23. Slednev, V.; Bertsch, V.; Ruppert, M.; Fichtner, W. Highly resolved optimal renewable allocation planning in power systems under consideration of dynamic grid topology. *Comput. Oper. Res.* **2018**, *96*, 281–293. [CrossRef]
24. Happ, H.H. Optimal Power Dispatch—A Comprehensive Survey. *IEEE Trans. Power Appar. Syst.* **1977**, *96*, 841–854. [CrossRef]
25. Frank, S.; Rebennack, S. An introduction to optimal power flow: Theory, formulation, and examples. *IE Trans.* **2016**, *48*, 1172–1197. [CrossRef]
26. Perau, C.; Slednev, V.; Ruppert, M.W.; Fichtner, W. Regionalization of four storylines for the decarbonization of the European power system including flexibilities. In Proceedings of the IEWT Conference, Vienna, Austria, 8–10 September 2021.
27. VDA, Number of Private Vehicle Registrations in Germany from 1955 to 2022. 2023. Available online: https://www.vda.de/en/news/facts-and-figures/annual-figures/new-registrations (accessed on 24 November 2023).

World Electric Vehicle Journal

MDPI

Article

Acceptance of E-Motorcycles: A Longitudinal Survey at Loewensteiner Platte, South Germany

Daniela Ludin [1,*], Norbert Schreier [2], Erika Mueller [1], Wanja Wellbrock [1], Patrick Melber [1] and Jonas Sollner [1]

[1] Faculty of Economics, Heilbronn University, Max-Planck-Str. 39, 74081 Heilbronn, Germany; erika.mueller@hs-heilbronn.de (E.M.); wanja.wellbrock@hs-heilbronn.de (W.W.); patrick.melber@outlook.com (P.M.); jonas.sollner@t-online.de (J.S.)

[2] Faculty of Mobility and Technology, Esslingen University, Kanalstraße 33, 73728 Esslingen, Germany; norbert.schreier@hs-esslingen.de

* Correspondence: daniela.ludin@hs-heilbronn.de

Abstract: The acceptance of e-motorcycles among German motorcyclists is the focus of this quantitative longitudinal study. By comparing survey results from 2017 and 2022, questions about changes in perception of e-motorcycles over time as well as possible stimulating factors are analyzed. The research design is built upon literature research, a secondary literature analysis, and a survey of motorcyclists. Statistical procedures are used for data analysis and interpretation. The literature analysis enables the present study to be integrated into the current state of research. The findings show that the willingness to consider an e-motorcycle as the next purchase was low in 2017 and dropped from 20% to 5% in 2022, which contrasts with the rising sales figures of e-motorcycles in the German market. Based on these findings, conclusions are drawn about the market potential of e-motorcycles in Germany and an overview of the general assessments and concerns of motorcyclists is provided.

Keywords: sustainable mobility; electro-mobility; e-motorcycle technology; motorcyclists; social acceptance; behavioral economics

Citation: Ludin, D.; Schreier, N.; Mueller, E.; Wellbrock, W.; Melber, P.; Sollner, J. Acceptance of E-Motorcycles: A Longitudinal Survey at Loewensteiner Platte, South Germany. *World Electr. Veh. J.* **2023**, *14*, 326. https://doi.org/10.3390/wevj14120326

Academic Editors: Joeri Van Mierlo and Genevieve Cullen

Received: 27 September 2023
Revised: 22 November 2023
Accepted: 24 November 2023
Published: 28 November 2023

1. Introduction

One of the biggest problems associated with the operation of conventional vehicles is the pollution emitted by combustion engines. Globally, internal combustion engines in vehicles are now responsible for a large part of air pollution [1]. Electrically powered vehicles are seen as a means of reducing the consumption of oil and gasoline and lowering the emission of pollutants from individual traffic. Electro-mobility is playing an increasingly important role in Germany. Car manufacturers such as Volkswagen, with its id-models, are establishing a market segment. However, the focus is mainly on e-cars. While motorcycles have been appearing in the e-mobility segment for some time, the public has hardly noticed them. Currently, there is no state support for e-motorcycles in the form of purchase premiums or environmental bonuses. In this respect, e-motorcycles can represent an alternative to the e-car that has received little attention in Europe to date. Pollutant emissions can be reduced by replacing conventional motorcycles with e-motorcycles. This can be an important factor, especially in cities with high levels of air pollution. Looking at the environmental impact of e-motorcycles compared to conventional motorcycles, it should be noted that although the use of e-motorcycles reduces pollutant emissions, the production of e-motorcycles consumes significantly more energy than the production of conventional motorcycles [2]. While research on acceptance of e-motorcycles is growing, it is rarely examined in the German context, e.g., [2–6]. Several studies have investigated the specific technological, environmental, political, and economic factors of e-motorcycles worldwide [3]. In these studies, it became obvious that the assessments and concerns of motorcyclists play an important role in the acceptance of e-motorcycles. This paper first seeks to gain empirical insights into the acceptance level of e-motorcycles, obtained by questioning German motorcyclists at a popular motorcycling spot, and then to detect possible changes over time by executing the same survey at two different points in time. The study aims to fill a research gap by answering the following two research-guiding questions:

RQ 1: Are there differences in e-motorcycle acceptance when considering gender and age?

RQ 2: Has e-motorcycle acceptance changed between 2017 and 2022?

This study is divided into theoretical, methodological, and empirical parts. In the theoretical part, the relevant scientific fields are presented and an overview of the current state of research is provided. The methodological part explains the procedures used to answer the research questions. Finally, the empirical part presents and evaluates the results of the survey on acceptance of e-motorcycles.

2. Literature Review

2.1. Motorcycle Technology

According to German Federal Motor Vehicle Office (Kraftfahrzeugbundesamt) [7] in accordance with Directive 2002/24/EG [8] and EU Regulation No. 168/2013 [9], motorcycles are classified as two- or three-wheelers with an internal combustion engine of more than 50 ccm or a speed of more than 45 km/h. Looking at the technology of e-motorcycles, the electric motor and high-voltage-battery are the main parts that determine the vehicle concept, and therefore the whole driving experience. While the weight of the battery is considerable, the electric engine delivers maximum torque instantly, whereas the torque curve of a combustion engine only reaches its maximum at a specific rotation speed [10]. There is no remarkable engine sound, no need for a gearbox, and the overall concept is simpler. Due to less wear on parts, the effort required for maintenance and repair is lower than for conventional motorcycles [11]. Looking at the ecological sustainability of e-motorcycles, there is substantial potential to reduce in-use exhaust emissions worldwide [1]. E-motorcycles beat conventional motorcycles in terms of greenhouse gas emissions as well as energy consumption in all categories: tank-to-wheel, wheel-to-wheel, and over the full lifecycle [2].

2.2. Sustainable Mobility

In addition to the avoidance of pollutants, the operation of electric motorcycles offers other advantages that sustainably reduce the burden on the environment. For example, electric drives are significantly more efficient than conventional gasoline-powered engines, as the energy in electric drives is converted directly into motive power while internal combustion engines convert part of the energy into heat. In addition, electric motorcycles are significantly quieter than conventional motorcycles, actively reducing the noise pollution caused by traffic [12].

Despite the environmental benefits that would result from a switch to electrically powered motorcycles, there has not yet been any great success on the market for e-motorcycles. This is due not least to their higher acquisition costs and existing technical problems. Their lower range compared to conventional motorcycles, together with limited charging infrastructure, has prevented the sustainable success of e-motorcycles to date [1].

In light of these issues, technological, economic, and social developments are important for the sustainable market development of e-motorcycles. A growing market, especially in Asia, is leading to falling prices due to economies of scale and increasing competitive pressure. Because of falling prices, the market attractiveness of e-motorcycles is increasing. Market developments in East Asia show that falling electricity prices together with rising gasoline prices and an increased demand for mobility can accelerate the market for electrically powered vehicles. Thus far, however, developments have not resulted in an increasing market share for e-motorcycles in either Europe or Asia. The main problem is that e-motorcycles do not offer their buyers any direct benefits for their higher price compared to conventional motorcycles. In contrast, e-bikes offer their customers an additive benefit through an additional e-drive compared to normal bicycles [2].

2.3. Social Acceptance and Behavioral Economics

Social acceptance is a key element in many debates surrounding the sustainability transition [13,14]. Fournis and Fortin described a socio-technical paradox known from the context of wind energy, where everybody supports wind energy while nobody likes it in the nearby vicinity: "The social dimension has become a factor of equal importance to technology in the wind farms implementation" [13]. This paradox can probably be applied to sustainable mobility and the acceptance of e-motorcycles. In addition, various studies have concluded that social acceptance is vital in terms of technology and innovation transfer [15–18].

According to Upham et al. social acceptance can be understood as "A favorable or positive response (including attitude, intention, behavior and—where appropriate—use) relating to a proposed or in situ technology or socio-technical system, by members of a given social unit (country or region, community or town and household, organization)" [17] and p. 102 in [18]. Social-acceptance can be divided into three dimensions, as proposed by Wüstenhagen et al. [14] and p. 3 in [18]:

- Socio-political acceptance points to the overall "Societal climate towards a technology or innovation within a society" (p. 3 in [18]). In the case of this study, this dimension would refer to how approaches to achieve sustainable mobility, in particular the diffusion of e-motorcycles, are positively or negatively perceived by the public and opinion leaders [14,18].
- Community or local acceptance focuses on "Attitudes and behaviors exhibited by those indirectly affected" (p. 3 in [18]). Relating to e-motorcycles, this could describe charging infrastructure located near residents.
- Market acceptance can be understood as "The process of market adoption of an innovation" (p. 2685 in [14]). Stakeholders such as consumers and investors are relevant in this dimension. Market acceptance is measurable, for example, in the market share of motorcycles and related purchasing behavior (p. 3 in [18]).

Whether a general transition towards sustainable mobility, including an increase in electric mobility, will be successful is dependent on acceptance levels in the above-mentioned three dimensions, among other things. For this study, the sociopolitical and market acceptance dimensions are the focus.

Looking at the situation in Germany from a behavioral economics perspective, Augenstein [19] stated there seems to be a dichotomy of opinion. On the one hand, people have a generally positive attitude towards the topic and diffusion process of electro-mobility. On the other hand, there is a blockade against recognizing electro-mobility as a holistic substitute for existing drive systems [19].

However, this study focuses specifically on the e-motorcycle vehicle segment. In Southeast Asia, where transportation is dominated by motorcycles, the economic aspect of the customer plays an essential role, as, apart from the European view, the motorcycle represents an essential medium for locomotion [1]. Because an e-motorcycle has lower consumption and emission values than a conventional motorcycle, one would think that in these countries there is a fundamental acceptance and will to opt for an e-motorcycle. However, a societal dichotomy is evident here as well. A study by Guerra [20] which looked at the acceptance of e-motorcycles in Indonesia showed important technical prerequisites that must be in place to promote social acceptance. The time required for recharging, recharging infrastructure, and higher purchase costs in relation to conventional motorcycles do not yet overlap with the ideas of potential customers. From Guerra's empirical analysis, it is possible to depict individuals who possess the following characteristics and demonstrate acceptance regarding e-motorcycles. This applies to younger people who are critically concerned with environmental influences and lead a healthy lifestyle [20].

Despite the ecological advantages, market success is not seen apparent due to additional costs and technological immaturity. With a share of nearly 2% of the German motorcycle sales in 2021, e-motorcycles are a small yet fast-growing segment, with an increase of 149% in the first half of 2022 [21]. Their relatively low range in combination with limited charging infrastructure is seen as a main disadvantage for e-motorcycles [1]. Looking at Germany, the generally positive attitude towards e-mobility contrasts with a low willingness to change mobility patterns and technologies [18,19]. Focusing on the motorcycle segment in Germany, purely economic criteria cannot explain customer behavior.

Several German studies focus on the social acceptance of electric vehicles on the demand side, especially by private users [18,22,23]; however, until now research has largely neglected the role of motorcyclists in the transition to e-mobility [2,5].

3. Methodology

The research design was built around literature research, a secondary literature analysis, and a survey of motorcyclists. Statistical procedures were used for data analysis and interpretation. A literature analysis enabled the integration of the present study into the current state of research [24,25].

A longitudinal quantitative survey based on a non-probabilistic convenience sample was used to build the empirical core approach of this study [26–28]. By using standardized closed questions, comparable data should be generated to answer the research questions. The questions included in the questionnaire can be divided into two basic types. On the one hand, there were identification questions, the aim of which was to identify the respondent, for example, by gender and age. Participants were asked about their gender (male/female) and their age. Regarding the latter, they were asked to assign themselves to one of the following age groups: younger than 25, 25–29 years, 40–59 years, or older than 60 years. No identification questions asked about personal information which could endanger the anonymity of the respondent. On the other hand, the selection type questions provided alternative options to answer a question. Regarding these questions, the respondent decided on a combination of a yes–no type, where only one yes or no answer can be selected, for example, Q2 "Would you consider an e-motorbike for your next motorbike purchase"? and Q5 "Would you buy an e-motorbike if a state bonus of approximately EUR 4000 were introduced, as in Italy/Austria"? A variation of the selection type question was used for Q1, in which respondents could choose one of several options about whether or not they already have experience with e-motorbikes. In addition, scale was used for several questions, allowing respondents to develop a certain tendency with respect to a statement concerning the maintenance costs, noise level, and reduced environmental impact of e-bikes (Q3). This scale was used to learn about possible concerns the participants might have related to e-motorbikes regarding state-of-the-art of the technology, charging times, and charging infrastructure (Q4). Participants could choose on the following scale: strongly agree, somewhat agree, somewhat disagree, strongly disagree, and no response. As a third type, multiple-choice questions were used in the context of the selection type of question, in which more than two answer categories were selectable [29–31]. This question type applies to Q6, in which participants were asked whether they already own an e-bike, e-car, or e-scooter. The content of the questions was derived from the literature review mentioned above, e.g., other studies about acceptance aspects [18–20].

With the help of a scientific questionnaire, we analyzed the acceptance of e-motorcycles. The Löwensteiner Platte in the town of Löwenstein in Baden-Württemberg was selected as the location for the survey. This choice was based on the high visitor frequency of motorcyclists who rest at this location. In addition, the surveys were conducted at the end of the regular motorcycle season, on 28 October 2017 and again on 22 October 2022. Both times, 41 people were surveyed. This included 33 male and 8 female motorcyclists. The survey contained seven questions, two of which were used to identify the respective age classification and gender of the person. The remaining questions were intended to elicit clarifying results on acceptability in the areas of vehicle technology, sustainability, and behavioral economics. The set of questions was mostly identical for the two survey dates in order to allow a comparison of changes in acceptance of e-motorcycles. Two questions were added to the survey in 2022 to investigate willingness to buy an e-motorcycle if it were supported with a state premium and whether the participants already owned another type of e-vehicle.

4. Findings

4.1. Differences in E-Motorcycle Acceptance Considering Gender and Age

The mean value of the communicated answers was classified into four asymmetric categories when evaluating the questionnaires. Thus, the significance of the results in columns one and four of Table 1 of a smaller interval should be strengthened.

Table 1. Result classification of the mean values (own table).

No Acceptance	Rather No Acceptance	Rather High Acceptance	High Acceptance
$X \leq 1.5$	$1.5 < X \leq 2.5$	$2.5 < X \leq 3.5$	$3.5 < X \leq 4$

The results of the survey yielded a cumulative mean value of 2.1 in 2017 and 1.7 in 2022, which means that there is a rather negative consensus regarding the acceptance of e-motorcycles within the scope of this study (Table 2).

Table 2. Result classification of the mean values for 2017 and 2022 (own table).

Mean Value	2017	2022
Frequency 1 ($\leq$ 1.5)	4	14
Frequency 2 ($1.5 < x \leq 2.5$)	29	25
Frenquency 3 ($2.5 < x \leq 3.5$)	8	2
Frequency 4 ($x > 3.5$)	0	0
	41	41

In order to answer the first research question ("Are there differences in e-motorcycle acceptance by considering gender and age"?), the means difference test between gender and the total sum of the mean value was used. In this way, a low link can be found with a value of −0.27 for 2017 and 0.17 in 2022. Thus, the results for women were never in the range of positive acceptance. Men, on the other hand, tended to rate acceptance positively, with a percentage of 24%. It was striking that no results could be assigned to a high acceptance of a value greater than 3.5 in either year. When considering the means difference test between the criterion of age and the mean value, a higher degree of link of 0.40 can be determined for 2017, whereas in 2022 the link dropped to −0.15. Only male persons with an age of over 60 years were classified as having a rather high level of acceptance, which was indicated by a value greater than 3.0.

4.2. Differences in E-Motorcycle Acceptance between 2017 and 2022

To answer the second research question ("Is there a difference in e-motorcycle acceptance between 2017 and 2022"?), the participants' answers that showed tendencies towards acceptance or non-acceptance were analyzed.

The results reveal a rather critical view of e-motorcycles. Even though the prominence and visibility of e-motorcycles among participants increased by 25% in 2022, the willingness to buy such a vehicle decreased by 27% compared to 2017 (Table 3).

The willingness to consider an e-motorcycle as the next purchase was low in 2017, and dropped even further from 20% (eight survey participants) to 5% (two survey participants) in 2022. This is interesting, as knowledge about e-motorcycles grew significantly over this period. The participants seemed to be less tentative or show greater acceptance with regard to other kinds of electrically powered vehicles, as 15% of the participants already owned an e-bike, e-scooter, or e-car.

Table 3. Result classification of the mean values for 2017 and 2022 by survey question (own table).

Year		Age	Q1	Q2	Q3.1	Q3.2	Q3.3	Q4.1	Q4.2	Q4.3	Q5	Q6	Average MV (Without Q5+6)
2017	Mean value	2.05	1.39	1.59	2.61	1.80	2.63	2.19	2.08	1.73			2.09
2022	Mean value	2.88	1.73	1.15	1.87	1.39	2.25	1.81	1.79	1.48	1.15	0.20	1.67
Delta 2022–2017		0.83	0.34	−0.44	−0.74	−0.41	−0.38	−0.38	−0.28	−0.26			−0.42
Change in % compard to 2017		40%	25%	−27%	−28%	−23%	−15%	−17%	−14%	−15%			−20%

In 2022, 78% of the participants considered the technology to be in the development stage. The lower costs of ownership in comparison with a conventional motorcycle were not seen as a countable advantage by 74% of respondents. The greater environmental friendliness and the lower noise emission were not relevant factors for 58% of the motorcyclists. On the contrary, the missing acoustic profile of a combustion engine seemed to be a negative factor for the acceptance of e-motorcycles for 88% of respondents. Charging duration and low range, together with limited charging infrastructure, were seen as weaknesses for e-motorcycles. Adding to the picture, only 5% of those questioned would buy an e-motorcycle in the case of a state purchase bonus of approximately EUR 4000, such as is available in Italy or Austria.

In conclusion, the acceptance scores consistently decreased slightly between 14% and 28% (Q3.1 to Q4.2) (Table 3). In addition, the general acceptance indicator (average of all MV) decreased by 20% when comparing the results of the 2017 and 2022 surveys. These are remarkable results, as the sales figures of e-motorcycles in the German market rose significantly in the first half of 2022 [1].

When looking at the results of the two additional questions used in the 2022 survey, social acceptance remains low even in the scenario where incentives are in place. Only two of the 41 survey participants (5%) would be willing to buy an e-motorcycle if they were to receive a state subsidies of EUR 4000 such as the one provided in Austria and Italy. Thus, when considering the second research question, the social acceptance of e-motorcycles among participants remained low and even dropped from 2017 to 2022. Presumably this is because the diverging driving experience of e-motorcycles, continuing technical immaturity, and high acquisition costs have not been resolved from the viewpoint of the survey participants.

5. Discussion

Unlike in other countries, e.g., Asian countries, two-wheeled vehicles play a minor role in everyday traffic in Germany. In many Asian countries, motorcycles and scooters form an important means of transport, as they are cheaper, space saving, and often replace cars [2,4]. In Germany, on the other hand, the car is the main means of transportation for a large part of the population with regard to commuting. In 2020, 68% of Germans regularly used a car to reach their workplace. Motorcycles fell into the other 1% of vehicles [32]. Even though urban areas in Germany are electrified, two-wheeled vehicles are increasingly used for fulfilling micro-mobility needs or in the context of sharing offers in more sustainable way, e.g., e-scooters or e-bike-sharing, while riding a motorcycle is predominantly a leisure activity [1,2]. The power of the engine and sound of the motor are part of the experience for many motorcyclists, and might not be same when riding an e-motorcycle. For many users riding such a powerful and loud vehicle is part of their lifestyle. The "move into the wild" syndrome, as described for car drivers by Viola [33], could be applied to motorcyclists as well. The results of this survey support the above-mentioned statements, as the participants viewed the missing or varying features of e-motorcycles concerning the motor sound and noise as negative attributes. As motorcycling in Germany is mainly a leisure activity, the importance of positive emotions such as happiness, flow, or self-image congruency as part of the driving experience might be higher than in other countries in which motorcycles are mainly used for transport and commuting. In addition to terrain characteristics, vehicle dynamics are important features, while range and accompanying infrastructure might contribute to the driving experience as well [34,35]. Thus, the driving experience of e-motorcycles might be different from riding a combustion engine motorcycle, potentially leading to lower acceptance of e-motorcycles among the participants.

Market acceptance, i.e., the adoption process of market innovations, seems to be slow for e-motorcycles when looking at the "traditional" consumer segments of motorcyclists.

There is only a limited supply of e-motorcycles from the leading motorcycling companies; in particular, the German manufacturer and market leader BMW does not offer e-motorcycles. Therefore, there is no electric equivalent model for the best-selling conventional motorcycle in Germany, the BMW GS 1250. At the same time, there is strong growth in sales figures driven by new market competitors. It would be interesting to find out whether new target groups are responsible for the growing sales figures of e-motorcycles and how this increase can be explained.

6. Conclusions

Our analysis of motorcycle technology, ecological sustainability, and behavioral economics reveals insights into the structure of the e-motorcycle segment. The relevant aspects were addressed through two surveys during 2017 and 2022, showing that motorcyclists' acceptance of e-motorcycles was relatively low and declined over the study period. Our results further suggest that technical immaturity cannot be overcome through incentives.

Considering limitations, the surveys only asked about the lower maintenance costs of e-motorcycles as a possible important factor for acceptance. The potentially higher purchase price of e-motorcycles could be another impeding factor, along with technical immaturity; however, this factor was not considered in the survey. In addition, it has to be mentioned that this study only involved motorcyclists in South Germany, and used a non-probabilistic convenience sample. Apart from these limitations, the sample size is very small. Due to these framework conditions, the power of generalization of these findings is limited. Further research steps should consist of sending the questionnaire to other European regions.

In terms of the social implications and the dimension of socio-political acceptance of e-motorcycles, the results of this study suggest that even though there is a political desire for an increase in electrically powered vehicles and traffic, technical immaturity leads to them finding little acceptance in the market among classic motorcyclists.

Regarding practical and managerial implications, the results of this study could help e-motorcycle manufacturers to enhance their marketing and product strategies. Trying to sell e-motorcycles to classic motorcycle customers using established marketing concepts and selling propositions does not seem to be very promising, while investing in research and development to create the best e-motorcycle to meet classic customer requirements might not succeed either. Instead, the challenge may involve creating new marketing strategies against the background of e-motorcycle technology, sustainable mobility, social acceptance, and customer behavior. One promising way to increase acceptance of e-motorcycles could be to create new types of two-wheelers which may not be direct competitors to classic motorcycles. This could be achieved by using additional engineering degrees of freedom available thanks to not having to integrate a combustion engine. In addition, new ways of promoting e-motorcycles should be developed and tested, for instance, not rationally as better means of ecological and sustainable transport, but emotionally charged, for example, as a joyful and exciting adventure for the young and brave.

From the managerial perspective, it is important to examine whether the traditional motorcycle brands have the potential to be expanded to e-mobility or whether new brands have to be created. To reach new customer types, e-motorcycle manufacturers need better overviews as well as insights from possible focus groups. The findings of this study could help decision-makers to better understand the mindset of motorcyclists. It could be a promising approach to use the results of this study to identify and analyze these customer segments. Theoretical implications could be developed in further research by following up on the questions around whether traditional motorcyclists are reluctant to buy e-motorcycles and what the profiles of the new e-motorcycle buyers are from marketing and behavioral economics perspectives.

Author Contributions: Conceptualization, D.L. and N.S.; methodology, E.M.; Writing—original draft, P.M. and J.S.; Writing—review & editing, W.W. All authors have read and agreed to the published version of the manuscript.

Funding: This research received no external funding.

Data Availability Statement: Data is contained within the article.

Conflicts of Interest: The authors declare no conflict of interest.

References

1. Amjad, S.; Rudramoorthy, R.; Sadagopan, P.; Neelakrishnan, S. Implementation and evaluation of change-over speed in plug-in hybrid electric two wheeler. *Energy* **2016**, *109*, 858–865. [CrossRef]
2. Weiss, M.; Dekker, P.; Moro, A.; Scholz, H.; Patel, M.K. On the electrification of road transportation—A review of the environmental, economic, and social performance of electric two-wheelers. *Transp. Res. Part D Transp. Environ.* **2015**, *41*, 348–366. [CrossRef] [PubMed]
3. Liu, Y.; Lai, I.K.W. The Effects of Environmental Policy and the Perception of Electric Motorcycles on the Acceptance of Electric Motorcycles: An Empirical Study in Macau. *SAGE Open* **2020**, *10*, 2158244019899091. [CrossRef]
4. Chang, C.-W.; Chang, S.-H.; Chiu, H.-J.; Liu, Y.-C. Understanding consumers' intention to switch to electric motorcycles: A transaction cost economics perspective. *Australas. J. Environ. Manag.* **2022**, *29*, 7–23. [CrossRef]
5. Eccarius, T.; Lu, C.-C. Powered two-wheelers for sustainable mobility: A review of consumer adoption of electric motorcycles. *Int. J. Sustain. Transp.* **2020**, *14*, 215–231. [CrossRef]
6. Peters, A.; Dütschke, E. How do consumers perceive electric vehicles? A comparison of German consumer groups. *J. Environ. Policy Plan.* **2014**, *16*, 359–377. [CrossRef]
7. Kraftfahrzeugbundesamt. *Methodische Erläuterungen zu Statistiken über Fahrzeugzulassungen (FZ)*; Stand: Flensburg, Germany, 2022.
8. Regulation (EU), No.168/2013 of the European Parliament and of the Council of 15 January 2013 on the Approval and Market Surveillance of Two- or Three-Wheel Vehicles and Quadricycles. Available online: https://eur-lex.europa.eu/LexUriServ/LexUriServ.do?uri=OJ:L:2013:060:0052:0128:en:PDF (accessed on 15 October 2022).
9. Directive 2002/24/EC of the European Parliament and of the Council of 18 March 2002 Relating to the Type-Approval of Two or Three-Wheel Motor Vehicles and Repealing Council Directive 92/61/EEC. Available online: https://eur-lex.europa.eu/LexUriServ/LexUriServ.do?uri=CONSLEG:2002L0024:20081211:EN:PDF (accessed on 15 October 2022).
10. Baumann, M.; Buchholz, M.; Dietmayer, K. A Two-wheel Driven Power Train for Improved Safety and Efficiency in Electric Motorbikes. *World Electr. Veh. J.* **2016**, *8*, 102–111. [CrossRef]
11. Requia, W.J.; Adams, M.D.; Arain, A.; Koutrakis, P.; Ferguson, M. Carbon dioxide emissions of plug-in hybrid electric vehicles: A life-cycle analysis in eight Canadian cities. *Renew. Sustain. Energy Rev.* **2017**, *78*, 1390–1396. [CrossRef]
12. Zero Motorcycles. Available online: https://www.zeromotorcycles.com/technology (accessed on 15 October 2022).
13. Fournis, Y.; Fortin, M.-J. From social 'acceptance' to social 'acceptability' of wind energy projects: Towards a territorial perspective. *J. Environ. Plan. Manag.* **2016**, *60*, 1133406. [CrossRef]
14. Wüstenhagen, R.; Wolsink, M.; Bürer, M.J. Social acceptance of renewable energy innovation: An introduction to the concept. *Energy Policy* **2007**, *35*, 2683–2691. [CrossRef]
15. Mallett, A. Social acceptance of renewable energy innovations: The role of technology cooperation in urban Mexico. *Energy Policy* **2007**, *35*, 2790–2798. [CrossRef]
16. Cohen, J.J.; Reichl, J.; Schmidthaler, M. Re-focussing research efforts on the public acceptance of energy infrastructure: A critical review. *Energy* **2014**, *76*, 4–9. [CrossRef]
17. Upham, P.; Oltra, C.; Boso, À. Towards a cross-paradigmatic framework of the social acceptance of energy systems. *Energy Res. Soc. Sci.* **2015**, *8*, 100–112. [CrossRef]
18. Burghard, U.; Scherrer, A.; Dütschke, E.; Globisch, J. *Social Acceptance of Electric Mobility in Germany*; Working Paper Sustainability and Innovation; No. S 12/2020; Fraunhofer-Institut für System- und Innovationsforschung ISI: Karlsruhe, Germany, 2020. Available online: https://www.isi.fraunhofer.de/content/dam/isi/dokumente/sustainability-innovation/2020/WP-12-2020_Social%2520acceptance%2520electric%2520mobility%2520Germany_ubu_als_de_glj.pdf (accessed on 15 October 2022).
19. Augenstein, K. Analysing the potential for sustainable e-mobility—The case of Germany. *Environ. Innov. Soc. Transit.* **2015**, *14*, 101–115. [CrossRef]

20. Guerra, E. Electric vehicles, air pollution, and the motorcycle city: A stated preference survey of consumers' willingness to adopt electric motorcycles in Solo, Indonesia. *Transp. Res. Part D Transp. Environ.* **2019**, *68*, 52–64. [CrossRef]
21. European Association of Motorcycle Manufacturers (ACEM). Available online: https://www.acem.eu/images/publiq/2022/ACEM_statistical_release_-_Jan_-_June_2022.pdf (accessed on 15 October 2022).
22. Franke, T.; Günther, M.; Trantow, M.; Krems, J.F. Does this range suit me? Range satisfaction of battery electric vehicle users. *Appl. Ergon.* **2017**, *65*, 191–199. [CrossRef]
23. Halbey, J.; Philipsen, R.; Schmidt, T.; Ziefle, M. Range Makes All the Difference? Weighing up Range, Charging Time and Fast-Charging Network Density as Key Drivers for the Acceptance of Battery Electric Vehicles. In *Advances in Human Aspects of Transportation, Proceedings of the AHFE 2017 International Conference on Human Factors in Transportation, The Westin Bonaventure Hotel, Los Angeles, CA, USA, 17–21 July 2017*; Springer: Cham, Switzerland, 2017; pp. 939–950.
24. Snyder, H. Literature review as a research methodology: An overview and guidelines. *J. Bus. Res.* **2019**, *104*, 333–339. [CrossRef]
25. Linnenluecke, M.K.; Marrone, M.; Singh, A.K. Conducting systematic literature reviews and bibliometric analyses. *Aust. J. Manag.* **2020**, *45*, 175–194. [CrossRef]
26. Nardi, P.M. *Doing Survey Research: A Guide to Quantitative Methods*; Routledge: New York, NY, USA, 2018.
27. Creswell, J.W.; Creswell, J.D. *Research Design: Qualitative, Quantitative, and Mixed Methods Approaches*; SAGE Publications: Thousand Oaks, CA, USA, 2022.
28. Maruyama, G.; Ryan, C.S. *Research Methods in Social Relations*; Wiley-Blackwell: Chichester, UK, 2014.
29. Roopa, S.; Rani, M.S. Questionnaire designing for a survey. *J. Indian Orthod. Soc.* **2012**, *46*, 273–277. [CrossRef]
30. Atteslander, P. *Methoden der Empirischen Sozialforschung*; Erich Schmidt: Berlin, Germany, 2010.
31. Kuß, A.; Wildner, R.; Kreis, H. *Marktforschung: Grundlagen der Datenerhebung und Datenanalyse*; Springer Gabler: Wiesbaden, Germany, 2014.
32. Destatis, 68% der Erwerbstätigen Fuhren 2020 mit dem Auto zur Arbeit, Pressemitteilung Nr. N 054 vom 15. September 2021. Available online: https://www.destatis.de/DE/Presse/Pressemitteilungen/2021/09/PD21_N054_13.html (accessed on 15 October 2022).
33. Viola, F. Electric Vehicles and Psychology. *Sustainability* **2021**, *13*, 719. [CrossRef]
34. Kruger, S. Soul searching on the wings of my wheels: Motorcyclists' happiness. *J. Psychol. Afr.* **2018**, *28*, 218–223. [CrossRef]
35. Will, S.; Metz, B.; Hammer, T.; Pleß, R.; Mörbe, M.; Henzler, M.; Harnischmacher, F. Relation between riding pleasure and vehicle dynamics—Results from a motorcycle field test. *Appl. Ergon.* **2021**, *90*, e103231. [CrossRef] [PubMed]

World Electric Vehicle Journal

MDPI

Article

Modular and Scalable Powertrain for Multipurpose Light Electric Vehicles

Mehrnaz Farzam Far [1,*], Damijan Miljavec [2], Roman Manko [2], Jenni Pippuri-Mäkeläinen [1], Mikaela Ranta [1], Janne Keränen [1], Jutta Kinder [3] and Mario Vukotić [2]

1 VTT Technical Research Centre of Finland, P.O. Box 1000, FI-02044 Espoo, Finland; jenni.pippuri-makelainen@vtt.fi (J.P.-M.); mikaela.ranta@vtt.fi (M.R.)

2 Faculty of Electrical Engineering, University of Ljubljana, Trzaska 25, 1000 Ljubljana, Slovenia; damijan.miljavec@fe.uni-lj.si (D.M.); roman.manko@fe.uni-lj.si (R.M.); mario.vukotic@fe.uni-lj.si (M.V.)

3 THIEN eDrives, GmbH, Millennium Park 11, 6890 Lustenau, Austria; jutta.kinder@thien-edrives.com

* Correspondence: mehrnaz.farzamfar@vtt.fi

Abstract: Light electric vehicles are best suited for city and suburban settings, where top speed and long-distance travel are not the primary concerns. The literature concerning light electric vehicle powertrain design often overlooks the influence of the associated driving missions. Typically, the powertrain is initially parameterized, established, and then evaluated with an ex-post-performance assessment using driving cycles. Nevertheless, to optimize the size and performance of a vehicle according to its intended mission, it is essential to consider the driving cycles right from the outset, in the powertrain design. This paper presents the design of an electric powertrain for multipurpose light electric vehicles, focusing on the motor, battery, and charging requirements. The powertrain design optimization is realized from the first stages by considering the vehicle's driving missions and operational patterns for multipurpose usage (transporting people or goods) in European urban environments. The proposed powertrain is modular and scalable in terms of the energy capacity of the battery as well as in the electric motor shaft power and torque. Having such a possibility gives one the flexibility to use the powertrain in different combinations for different vehicle categories, from L7 quadricycles to light M1 vehicles.

Keywords: charging; driving cycles; electric powertrain design; induction motor; light electric vehicle

Citation: Farzam Far, M.; Miljavec, D.; Manko, R.; Pippuri-Mäkeläinen, J.; Ranta, M.; Keränen, J.; Kinder, J.; Vukotić, M. Modular and Scalable Powertrain for Multipurpose Light Electric Vehicles. *World Electr. Veh. J.* **2023**, *14*, 309. https://doi.org/10.3390/wevj14110309

Academic Editors: Joeri Van Mierlo and Genevieve Cullen

Received: 9 October 2023
Revised: 1 November 2023
Accepted: 3 November 2023
Published: 11 November 2023

1. Introduction

Transportation is one of the fastest-growing sources of greenhouse gas emissions, accounting for 78% of the rise in emissions from 1990 to 2019 [1]. In 2020, transportation emissions declined by 14% solely due to the COVID-19 pandemic but witnessed a swift 12% increase in 2021 [2] following the relaxation of lockdown measures. Meanwhile, the European Union and governments worldwide have implemented various regulations and measures aimed at reducing transportation emissions on a global scale [3–7]. For instance, on 19 April 2023, the European Union and the Council modified Regulation (EU) 2019/631 to Regulation (EU) 2023/851 to enhance the CO_2 emission performance criteria for new passenger cars and new light commercial vehicles, aligning them with the heightened climate goals of the European Union. Notably, this amendment bolsters the emission targets and establishes a goal of 100% emission reduction for both cars and vans starting in 2035 [8].

Replacing internal combustion engine (ICE) vehicles with electric vehicles (EV) is a step in the right direction towards reducing emissions and supporting climate targets; however, this action alone is not enough to solve the entire problem. Additional efforts such as integrating clean energy sources, optimizing powertrains based on vehicles' missions, promoting shared vehicle usage, battery reusing, and interoperable charging technology [9] are required.

Light electric vehicles (LEVs) are particularly suitable for urban and suburban environments, where high speed and long range are not the main priorities. LEVs have lower energy consumption and higher vehicle weight-to-payload ratios, and they require fewer resources for production compared to other EVs [10]. This makes them more efficient and affordable, especially for large-scale production and shared mobility scenarios.

Table 1 provides an overview of the specifications of some L7e-C category EVs available on the market. All the vehicles listed in this table meet the European regulations outlined in (EU) No 168/2013 [11], which specify that L7e-C vehicles must have a maximum continuous rated power of no more than 15 kW and a maximum speed of no more than 90 km/h.

Table 1. Specifications of some L7e-C category EVs available on the market.

Name	Power (kW)	Max Torque (Nm)	Max Speed (km/h)	Voltage (V)	Battery Type, Capacity (kWh)	Range (km)
Honda Micro Commuter	15	N/A	80	N/A	Li-ion, N/A	96
Mahindra e-Supro van	15	90	60	72	Li-ion, 15	115
Mahindra Treo	7.5	42	55	48	Li-ion, 7.3	141
Microlino	11	89	90	N/A	Li-NMC, 14	230
Renault Twizy	13	57	80	58	Li-ion, 6.1	100
Regis Epic0 Compact	15	N/A	75	144	Li-NMC, 15.2	140
PILOTCAR, P-1000	10	102	55	48	Li-ion, 26	220
Tazzari Zero	15	150	90	80	Li-ion, 14.2	150

The design of a powertrain starts by defining various quantities such as power, torque, speed, voltage, battery capacity, and range of the vehicle. The power and torque requirements of an electric traction motor depend on the vehicle's desired performance characteristics, size, and weight. An accurate definition of the performance characteristics and, therefore, a reduction in the vehicle's energy consumption require knowledge of the vehicle's actual driving cycles and mission. As an example, Lindh et al. [12] used an actual driving cycle of a bus route in Lappeenranta to investigate the speed and torque requirements for a heavy-duty vehicle. Based on these requirements, the authors proposed a permanent magnet traction motor design suitable for hybrid buses. However, the literature on LEV powertrain design rarely considers the associated driving missions. Instead, the powertrain is first parametrized and settled, and then, an ex-post-performance assessment is conducted using various driving cycles [13–16].

Different battery technologies for EVs are reviewed and compared in [17]. Lithium-ion (Li-ion) batteries are the most frequently utilized battery type in EVs due to their significantly higher energy density (Wh/kg) compared to other alternatives. The two most used Li-ion chemistries are NMC (nickel–manganese–cobalt) and LFP (lithium–iron–phosphate). NMC cells provide a higher energy density and have a better charging performance at low temperatures, whereas LFP cells are more affordable and have a longer cycle life. The high energy density of NMC is vital for providing sufficient ranges for everyday electric driving, especially for LEVs, in which the mass and physical dimensions of the battery have strict limitations. On the other hand, the price range of LFP would better fit with LEVs. However, if an LEV is to be charged at a reasonable speed in cold-climate environments, expensive and heavy thermal management systems would be required in the case of LFP batteries. Hence, optimal cell chemistry is a compromise of price, energy density, and cold-climate performance. The battery capacity is determined based on various factors, such as the vehicle's energy consumption, the expected driving conditions, the desired driving range, the weight of the vehicle, and the efficiency of the electrical motor. Battery cost is a significant factor contributing to the price of EVs. The cost breakdown of batteries can be classified into four primary segments: materials, labor, assembly, and overhead. Materials constitute the largest portion, representing approximately 60% of the total battery cost. Anticipated reductions in battery cell and pack costs are expected to occur gradually due to advancements in battery material chemistry, improvements in battery cell design, and a decrease in assembly expenses resulting from increased production volumes and learning. For instance, it is predicted that the global price per kWh of a battery pack will decrease by approximately 64% in 2050 compared to 2019 [9]. Nevertheless, the current high cost of batteries makes small LEVs a more cost-effective choice compared to larger EVs, given their smaller battery size.

The voltage levels of LEVs available on the market typically range from 48 to 80 V. The specific voltage level of an LEV is determined during the design stage, considering several factors such as the power and torque requirements of the motor, the desired range of the vehicle, and the weight and cost constraints of the battery pack. Using higher voltages in electric vehicles can be more cost-effective for energy distribution as lower currents require smaller cable cross-sections and connectors. In addition, higher voltage levels offer greater powertrain scalability towards higher vehicle categories with superior performance characteristics, such as increased torque, power, and vehicle speed.

In this paper, we present the design of a modular and scalable electric powertrain for L7e-C category EVs with multipurpose usages (transporting people and goods), with a focus on the motor, battery, and charging requirements. The modularity and scalability of the proposed electric powertrain allow one to modify each component according to the vehicle's mission, without affecting the overall powertrain concept. This approach facilitates the adaptation of the powertrain for use in higher classes of vehicles, such as M1. To ensure that the powertrain components are rightly sized according to the vehicle's mission, the associated driving missions and operational patterns are considered in the design from the very first stages.

This research has been presented at the EVS36 Symposium in Sacramento, USA, in June 2023.

2. Materials and Methods

2.1. Vehicle Requirements and Specifications

The first step in powertrain design involves setting the base specifications and requirements of the vehicle. As mentioned previously, this paper focuses on L7e light four-wheel EVs. To comply with European regulations, the vehicle properties and specifications were defined based on regulation (EU) No 168/2013 [11], as follows: a mass in running order of up to 600 kg (without the battery and payload), a top speed of 90 km/h, and a maximum continuous power of 15 kW. The total gross weight of the vehicle is assumed to be 1200 kg (150 kg of which is the battery and 450 kg of which is the payload). The 1-D vehicle model was described using basic parameters with the following values: wheel's radius $R_w = 0.31$ m, vehicle's front area $A_v = 2.17$ m^2, wind drag coefficient $C_x = 0.3$, tires' pressure $p_t = 3$ bar, mass density of air $\rho = 1.2$ kg/m^3, and rolling resistance coefficient C_r, which is a function of vehicle speed and whose value varies between 0.0083 and 0.011.

Other basic properties required for the powertrain design include the vehicle's driving range and its torque and power profiles. To ensure that the powertrain components are appropriately sized according to the vehicle's mission, the associated driving missions and operational patterns should be considered from the very first stages of design. For this purpose, we used a driving cycle generated based on typical trips in a European city, Helsinki, to study the vehicle's driving range, torque–speed profile, and power requirements. This driving cycle was dynamically created using an in-house simulation tool [18] taking into account the speed limits, traffic lights, road curvature, historical traffic data, and predefined limits for acceleration and deceleration. The specific driving cycle was chosen from a large set of routes and their corresponding driving cycles to represent a typical trip in the Helsinki region, with a mix of urban and suburban driving conditions. It is worth noting that the presented driving cycle also included terrain shape in the form of road elevation changes during the velocity profile. The speed and elevation profiles of this driving cycle are illustrated in Figure 1.

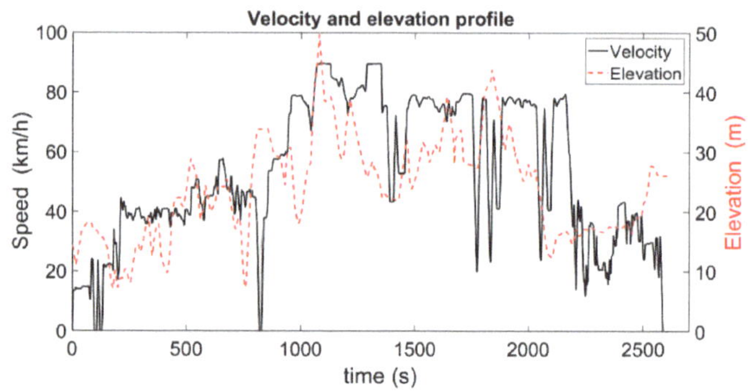

Figure 1. Simulated speed and elevation profiles of the Helsinki region with a mix of urban and suburban driving conditions.

According to the highest allowable vehicle speed (90 km/h), a wheel radius of 0.31 m (R15), and the maximum electric motor rotational speed (10,000 rpm), we defined a gear ratio of 12:1 with a mean efficiency of 95%. Applying the identified Helsinki driving cycle and incorporating the upper boundary conditions, we calculated the vehicle's needed torque and shaft power requirements. Figures 2 and 3 present the power and torque profiles of an electrical machine for the Helsinki driving cycle, without and with considering a 450 kg payload in the vehicle's weight, respectively.

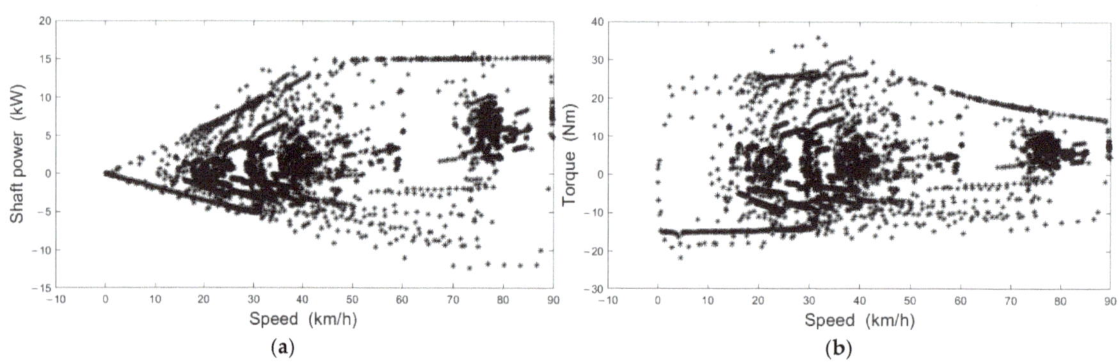

Figure 2. (**a**) Power–speed and (**b**) torque–speed profiles of an electrical machine for the Helsinki driving cycle with a total vehicle weight of 750 kg.

Further, the energy consumption at each driving point and the cumulative needed energy within one cycle were calculated. The overall energy consumed in the case of an unloaded vehicle in one driving cycle (38.2 km), including the road profile in the form of elevation (Figure 1), is 2.73 kWh. In the case of a fully loaded vehicle (with a payload of 450 kg), the consumed energy is 3.56 kWh. To achieve a range of approximately 100 km and enable shorter charging times even with standard three-phase home charging facilities, the battery must provide enough energy to allow the vehicle to complete at least three Helsinki driving cycles, covering a total distance of 114.6 km. For such a distance, the electric energy consumed by the battery ranges between 8.16 kWh and 10.68 kWh. If we assume a 30% battery safety margin, the battery should have an energy capacity in the range of 15 kWh.

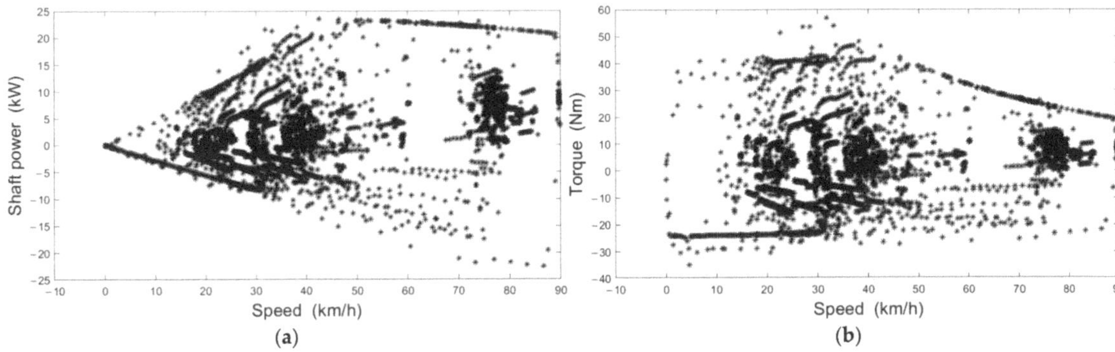

Figure 3. (**a**) Power–speed and (**b**) torque–speed profiles of an electrical machine for the Helsinki driving cycle with a total vehicle weight of 1200 kg (including a 450 kg payload).

Based on the calculated torque and power profiles (Figures 2 and 3), the following electro-mechanical requirements were established for the design of the electrical machine:

- The electrical machine should be capable of delivering a shaft power of 15 kW in the constant power operational region, with this region extending up to a speed of 90 km/h corresponding to a motor speed of 9240 rpm (Figure 2a).
- The machine should have a nominal torque of approximately 40 Nm (Figure 2b) in the case of an unloaded vehicle. Meanwhile, the maximum power and maximum torque should reach up to 23 kW and 60 Nm, respectively, due to the vehicle's full payload (Figure 3).
- The electric machine's torque–speed curve should have a corner point between 35 and 50 km/h, corresponding to a motor speed of around 3600–5100 rpm, to ensure that the torque demands remain within the region of highest efficiency for the electric machine.
- The variation of the torque between 40 Nm and 60 Nm should depend on the vehicle's payload and the driving cycle's starting condition.

2.2. Traction Motor Design

Induction motors (IMs) and permanent magnet synchronous motors (PMSMs) are the most used electric traction motors in EVs. IMs are robust, reliable, and cost-effective, and their torque characteristics are proportional to the current. PMSMs, on the other hand, have higher efficiency, power density, and torque density than IMs. PMSMs perform better at high speeds and can produce high torque at low speeds, but they are more complex and expensive to produce due to the use of critical raw materials.

The type of traction motor chosen for an EV depends on the specific requirements. PMSMs are often used for high-performance EVs, while IMs are preferred for low-cost EVs. In this work, we chose an IM for the traction motor due to its robustness and to avoid the use of rare earth materials.

The electromagnetic field of an induction machine is governed by

$$\nabla \times \left(\frac{1}{\mu} \nabla \times \boldsymbol{A} \right) = \boldsymbol{J}, \tag{1}$$

where μ is the permeability; $\boldsymbol{A}$ is the magnetic vector potential; and $\boldsymbol{J}$ is the current density. In a 2D analysis, $\boldsymbol{A}$ and $\boldsymbol{J}$ have components only in the axial direction. The right-hand side of (1) equals zero in the air gap and in the laminated iron of an electrical machine as the current density is zero or close to zero in those areas. Considering eddy currents induced in the rotor's winding, one can present the field equation of the rotor bars as

$$\sigma \frac{\mathrm{d}\boldsymbol{A}}{\mathrm{d}t} + \nabla \times \left(\frac{1}{\mu_0} \nabla \times \boldsymbol{A} \right) = \frac{\sigma \Delta v_{\mathrm{m}}}{l} \boldsymbol{e}_z, \tag{2}$$

σ being the rotor bar's electrical conductivity, Δv_{m} being the voltage drop between the ends of the mth bar, l being the total length of the bar, and $\boldsymbol{e}_z$ being the unit vector in the axial direction.

The traction motor was simulated using the 2D transient magnetic finite element method in the Altair Flux software. The end-region elements (end-winding and end-ring resistances and leakage inductance) were analytically calculated and added as lumped elements to the external electrical circuit. The boundary conditions and requirements for designing the motor were selected for the gear ratio 12:1 as follows:

- A continuous power of 15 kW through the whole speed range (Figure 2a) and a peak power of 23 kW (Figure 3a).
- A nominal torque of 40 Nm (Figure 2b) and a maximum torque of 60 Nm (Figure 3b)
- A maximum torque of 70 Nm at zero speed to overcome the curb (calculated based on the wheel's size, the vehicle's mass, and a curb height of 15 cm).
- A nominal phase RMS voltage in the range of 125 V to 175 V, allowing lower electric currents and smaller electric power wires cross-sections.
- The maximum rotational speed of the motor should be 10,000 rpm, due to the 90 km/h maximum vehicle speed limitation, wheel size, and gearbox ratio).
- A maximum motor efficiency > 92%.

To minimize the motor's size and maximize its overload operational capabilities, we selected a liquid cooling system instead of an air-cooling system. The liquid cooling system offers a higher heat removal capability, especially when the vehicle is heavily loaded at lower rotational speeds.

Considering these conditions and requirements, we simulated and studied various IM designs. A comprehensive analysis was conducted to determine the optimal number of stator slots and rotor bars to minimize the torque pulsations at the nominal operating point [19,20]. The analysis showed that the lowest torque pulsation occurred with 36 stator slots, 50 rotor bars, and 2 pole pairs. However, using 50 rotor bars would result in a thin rotor tooth. To ensure stable manufacturing and adequate rotor strength, the design with the second-lowest torque pulsations, namely, 30 rotor bars, was chosen. This decision aligns with our cost-effectiveness objectives, as the combination of 36 stator slots and 30 rotor bars can be produced using conventional methods and readily available electrical laminations. In addition, the shape of the rotor bars was chosen to enhance the critical torque value and reduce factors such as eddy current losses, magnetizing currents, and manufacturing complexity. The lamination design and 3D form of the proposed traction motor with liquid cooling are presented in Figure 4a and Figure 4b, respectively. The winding layout, shown in Figure 4c, has a double-layer fractional slot winding with a coil span of eight slots and a winding factor of 0.945.

To reach a cost-effective scalability, the cross-sections of the stator and rotor lamination were fixed, and the axial length of the motor (with proper adjustment of the number of turns in the stator winding) was varied according to the voltage and power requirements. Different axial lengths were investigated to determine the optimized length for the motor. Table 2 presents the main parameters of three of these designs. The results of these three designs, along with the analysis of the crossroad dynamics in the form of acceleration rates, will be presented in Section 3.

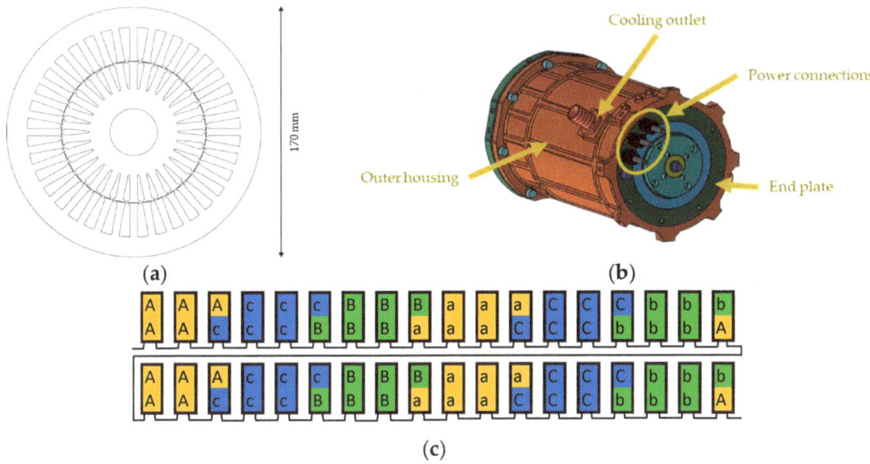

Figure 4. (**a**) 2D cross-section, (**b**) schematic 3D view, and (**c**) winding layout (with three phases A, B, and C, small letter denoting the opposite direction of a coil side) of the proposed traction motor.

Table 2. Design specifications of the proposed traction motors.

Motor Parameter	Motor 1	Motor 2	Motor 3
Continuous power (kW)		15	
Core material		Cogent M270-35A	
Material of stator winding and rotor bars		Copper	
Number of pole pairs		2	
Number of stator slots/rotor bars		36/30	
Stator core outer diameter (mm)		170	
Stator core inner diameter (mm)		96.8	
Air gap width (mm)		0.4	
Shaft diameter (mm)		32	
Thickness of lamination (mm)		0.35	
Rotor end ring segment resistance (Ω, 100 °C)		3.04×10^{-6}	
Rotor end ring segment inductance (H)		4.82×10^{-9}	
Core length stator/rotor (mm)	200	150	100
Number of turns per coil	5	6	8
Stator winding one-phase resistance (Ω, 100 °C)	0.054	0.0672	0.1011
Stator end winding inductance (H)	4.14×10^{-5}	5.98×10^{-5}	1.07×10^{-4}
Peak power (kW)	47.5	40.6	29.9
Base speed (rpm)	4000	4380	4680
Peak torque (Nm)	127.0	97.3	66.9
Maximum efficiency (%)	94.5	93.9	92.5
Total weight (kg)	46.1	38.3	30.8

2.3. Battery and Frequency Inverter

Based on the above-defined nominal phase RMS voltage level, the maximum allowable battery voltage could range from 340 V to 475 V. Another voltage limit was set by the off-the-shelf frequency inverter needed to drive the designed traction motor. We decided to use a SEVCON Gen4Size8 frequency inverter [21] with a maximum voltage limit of 400 V. To achieve the maximum voltage level of 400 V, we used 96 Li-ion NMC battery cells from KOKAM (model: SLPB100216216H [22]) connected in series, each having a capacity of 40 Ah. When using cells with an average voltage of 3.7 V, the battery stack delivers a maximum energy of 14 kWh. The proposed battery assembly provided sufficient energy to fulfill the defined range of the vehicle.

The battery designed for the vehicle had integrated BMS, contactors, fuses, DC/DC converter (400 V/12 V, 1.2 kW), and air-cooling fans in the battery box (Figure 5). This battery design featured an energy storage system with an energy capacity of 14 kWh, fulfilling all the requirements regarding voltage, vehicle range, overload capability, fast charging, and safety concerns (battery's structural integrity). Moreover, the battery was built modularly, so its energy capacity could be scaled up to 25 kWh, for example, for the M1 vehicle type just by substituting the battery cells with the ones that have a higher 60 Ah capacity, such as the KOKAM Model: KCL216060EN1 [23].

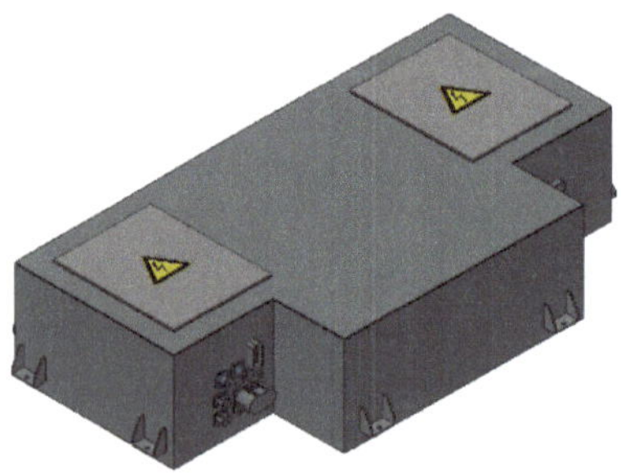

Figure 5. Modular battery for different vehicle classes (L7 to M1).

2.4. Charging Requirements

A detailed analysis of the vehicle's utilization was conducted to determine the required power of the charging system. In this work, vehicle utilization is referred to as duty cycles and encompasses the trips that a vehicle is expected to make during the day. The duty cycles are presented as schedule-format descriptions of the driven route's origins and destinations, non-driving related activities (such as cargo loading and unloading), and breaks from all activities when the vehicle is not occupied. The cycle profiles can be generated via mesoscopic simulations such as activity-based transport modeling (ABTM) [24]. We implemented the duty cycles into an in-house microscopic simulation tool [18] to study the charging requirements of the vehicle while considering the driving cycles, vehicle dynamics, and charging module. In the simulation, it was assumed that a fleet of about 28,000 vehicles with a utility rate of 50% had been distributed around the capital region of Helsinki (i.e., Helsinki, Espoo, Vantaa, and Kauniainen), providing a shared fleet for 24/7 transportation (e.g., people during the day and goods at night).

Figure 6a shows the simulated daily distance driven by each vehicle in the fleet. As shown in this Figure, the average distance exceeds 200 km, and some vehicles travel over 300 km per day. Figure 6b displays the maximum available charging time between each trip, assuming that charging is feasible at all locations where the vehicles park.

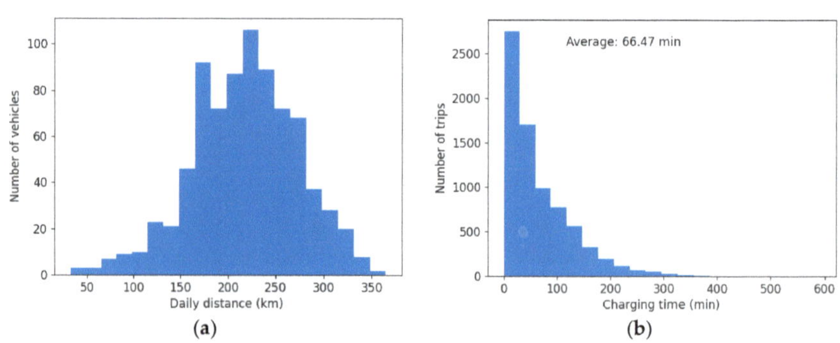

Figure 6. (**a**) Distribution of the simulated daily distance of each vehicle and (**b**) maximum available charging time between the trips.

To gain an understanding of the vehicle's operation in a more realistic situation where charging is not always possible, the parking locations of the vehicles from the initial simulations were analyzed and the 100 most-often-used locations were set to be equipped with a charger, while the other parking locations were left without charging facilities. Out of the 100 charging locations, 20% were equipped with fast 7 kW chargers and the rest with slower 3.7 kW chargers. The results of these two charging scenarios are compared in Section 3.

3. Results and Discussion

Figure 7 presents the efficiency maps of the three motor designs, which were calculated using the 2D finite element method. The efficiency maps show the maximum achievable power values without any electric current limitations. As can be seen from the efficiency maps, Motor 1, with a corner point at 4000 rpm, has the largest efficiency area in comparison to the other motors.

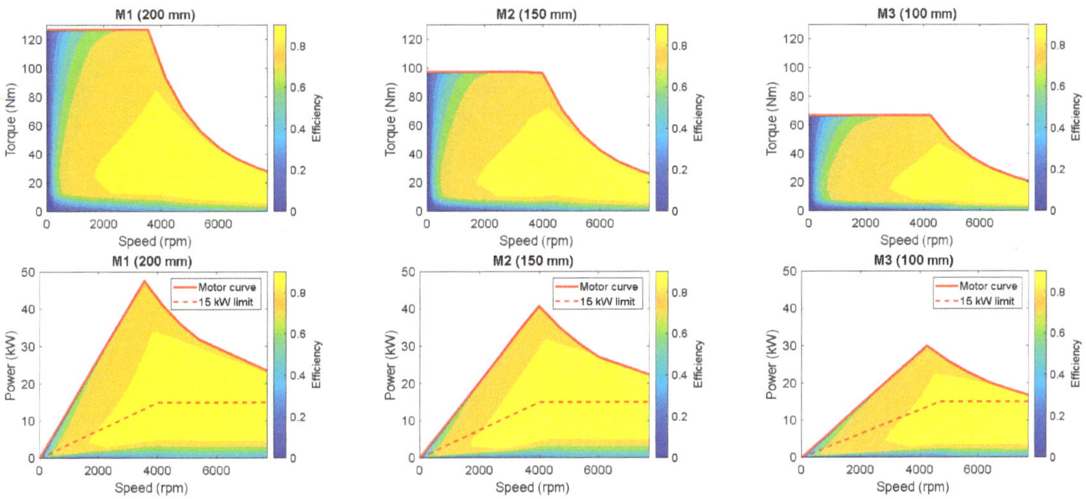

Figure 7. Efficiency maps of motor variants with different core lengths: (**a**) Motor 1 (200 mm); (**b**) Motor 2 (150 mm); and (**c**) Motor 3 (100 mm). The dashed lines represent the power–speed profiles respecting the 15-kW continuous power limit.

The vehicle's dynamics were simulated with these three motor designs. In the simulation, the resistive forces acting on the vehicle (e.g., slope, friction, air drag) and the total inertia, scaled to the electrical motor, were considered. For the gearbox and differential, a constant efficiency level of 95% was assumed. The comparison of the motors was carried out on roads with slopes ranging from 0% to 26% (maximum road slope), in terms of the maximum achievable speed and acceleration times from 0 to 40 km/h. The acceleration test results are presented in Table 3. According to these results, Motor 1 achieves the highest speeds on all the slopes and outperforms the other two motors in the acceleration tests.

Table 3. Maximum achievable speed on different slopes for a fully loaded vehicle (1200 kg).

Motor Variant	Top Speed 0% Slope	Top Speed 26% Slope	Acceleration 0% Slope	Acceleration 26% Slope
Motor 1	90 km/h	41.7 km/h	4.00 s	14.12 s
Motor 2	90 km/h	39.6 km/h	5.14 s	N/A
Motor 3	90 km/h	38.9 km/h	7.57 s	N/A

The torque's range was established by analyzing the driving cycles, as described in Section 2.1. Nevertheless, to validate the torque requirements and to make any necessary adjustments, the real-world urban driving cycle acceleration time was considered as well. This will further permit the adjustment of the torque characteristics of the traction motor and gearbox ratio, if needed, to fulfill the high dynamics of an EV in urban driving conditions.

The acceleration time in the urban driving condition was measured by conducting an acceleration test on a vehicle with an ICE that had similar size and weight characteristics to the vehicle under investigation. The acceleration test resulted in a time of approximately 4 s to reach from 0 km/h to 40 km/h. The results of this test were then utilized to define the acceleration torque requirements by inputting the data into a MATLAB R2020b software program that was designed based on the vehicle's equation of motion.

The developed MATLAB R2020b software allows for the simulation of changes in the dynamic performance of the EV powertrain, considering factors such as the torque–speed characteristic of the traction motor, the gearbox ratio, vehicle mass (1200 kg), vehicle shape (drag coefficient), wheel diameter, slope coefficient, and tire pressure. The output of the software is the calculated speed of the vehicle over time, from which the acceleration can be determined.

Figure 8 presents the power–speed and torque–speed curves, along with the transient acceleration behavior, of all three motors on the 0% slope, assuming a fully loaded vehicle with a total mass of 1200 kg. According to the results, the vehicle with Motor 1 fulfills the acceleration requirements from 0 to 40 km/h in 4 s and, therefore, is chosen as the traction motor of the proposed vehicle.

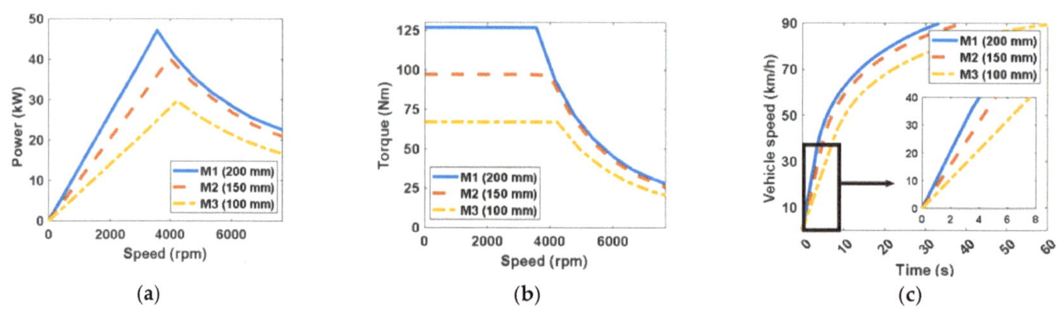

Figure 8. Influence of (**a**) power–speed and (**b**) torque–speed curves on (**c**) the vehicle's acceleration.

Table 4 summarizes the results of the following two charging scenarios: one with charging available at all parking locations and the other with charging available at only 100 locations. In the first scenario, all the scheduled trips were fulfilled without the battery state of charge (SOC) dropping below 35%. In the second scenario, some vehicles could not complete their trips due to the high daily distances. Nevertheless, on average, 92% of the daily target distance was covered, and the issues were mostly related to the very last trip of the day. It is worth noting that the simulations did not consider certain driver behaviors, such as selecting a vehicle with a higher SOC or choosing a parking spot with a charger instead of one without. Accounting for these factors would likely increase the likelihood of completing all the trips. Therefore, 7 kW was chosen as the maximum charging power of the vehicle. With such power, the battery could be charged from zero to 100% SOC in slightly more than 2 h.

In this paper, we highlight the importance of tailoring the powertrain of the vehicle according to its anticipated driving mission, resulting in optimized energy efficiency. While standard driving cycles can serve this purpose, they typically do not include complex and variable driving conditions such as traffic congestion, elevation changes, and weather conditions. By using real-world driving cycles, we can incorporate these variables when designing the powertrain. In cases where real-world driving cycles are not accessible, accurately simulated driving cycles, such as the one used in this paper, offer a suitable alternative. To illustrate the performance and advantages of our approach, we will prototype an LEV with the proposed powertrain and test its efficiency under various driving cycles, including the standard, real-world, and simulated scenarios.

Table 4. Summary of simulated duty cycle with two charging scenarios.

Charging Scenario	Completed Daily Distance, (Mean (Min–Max))	Trips/day, (Mean (Min–Max))	Completed Distance vs. Target
Slow charging (3.7 kW) available at all parking locations	217.6 km (32.1–364.7) km	10.5 (1–20)	100%
Slow charging (3.7 kW) at 80 locations and fast charging (7 kW) at 20 locations	196.6 km (32.1–326.6) km	9.8 (1–17)	92.4%

While our research has primarily addressed the technical and operational aspects of the powertrain, it is important to acknowledge that cost considerations play a significant role in the real-world adoption of electric vehicles. Future studies should delve deeper into the economic aspects of electric vehicle development, exploring cost-effective strategies for component design, manufacturing, and charging infrastructure. This will help bridge the gap between the technical advancements presented in this paper and the practical affordability of electric vehicles, ultimately contributing to their wider adoption and sustainability.

4. Conclusions

This paper presents the design of an electric powertrain for multipurpose light electric vehicles, focusing on the motor, battery, and charging requirements. The first step of the design involved studying a driving cycle from a typical European city such as Helsinki, to derive the vehicle's driving range, torque–speed profile, and power requirements. This ensures that the proposed design is optimized according to the vehicle's driving missions and operational patterns. We explored several cross-section designs for IMs, each with varying pole pair and slot numbers, and ultimately chose the design that aligned with the driving cycle requirements. In the next step, to achieve economical scalability, we maintained consistent cross-sections of the motor while adjusting its axial length. We explored various axial lengths and selected the most optimal motor length based on the torque and power profile and the vehicle's dynamics. The most efficient motor was found to meet the vehicle dynamics' criterion of accelerating from 0 to 40 km/h in 4 s. The battery capacity was estimated based on the driving range. The maximum charging power of the vehicle was set to 7 kW after a detailed analysis of the vehicle's expected utilization throughout the day.

Author Contributions: Conceptualization, all authors; methodology, all authors; software, D.M., R.M, M.R., and M.V.; validation, D.M., R.M., M.R., and M.V.; formal analysis, D.M., R.M., M.R., and M.V.; investigation, D.M., R.M., M.R., and M.V.; resources, D.M. and J.P.-M.; data curation, D.M. and M.R.; writing—original draft preparation, M.F.F.; writing—review and editing, all authors; visualization, all authors; supervision, D.M. and J.P.-M.; project administration, J.P.-M.; funding acquisition, J.P.-M., D.M., J.K. (Janne Keränen), and M.F.F. All authors have read and agreed to the published version of the manuscript.

Funding: This work was fully supported by the EU REFLECTIVE project, funded by the European Union's Horizon 2020 research and innovation program under grant agreement No. 101006747.

Data Availability Statement: Data are available on request due to restrictions.

Acknowledgments: The authors wish to thank Petr Hajduk for generating the driving cycles used in this work.

Conflicts of Interest: All the authors declare that the research was conducted in the absence of any commercial or financial relationships that could be construed as a potential conflict of interest.

References

1. Friedrich, M.; Ge, J.; Vigna, L. 4 Charts Explain Greenhouse Gas Emissions by Countries and Sectors. Available online: https://www.wri.org/insights/4-charts-explain-greenhouse-gas-emissions-countries-and-sectors# (accessed on 9 October 2023).
2. Carbon Dioxide Emissions from the Transportation Sector Worldwide from 1970 to 2022. Available online: https://www.statista.com/statistics/1291615/carbon-dioxide-emissions-transport-sector-worldwide/ (accessed on 1 November 2023).
3. Regulation (EU) 2019/631 of the European Parliament and of the Council of 17 April 2019 Setting CO_2 Emission Performance Standards for New Passenger Cars and for New Light Commercial Vehicles, and Repealing Regulations (EC) No. 443/2009 and (EU) No 510/2011. Available online: https://eur-lex.europa.eu/legal-content/EN/TXT/?uri=CELEX%3A02019R0631-20210301 (accessed on 1 November 2023).
4. Greene, D.L.; Park, S.; Liu, C. Public Policy and the Transition to Electric Drive Vehicles in the U.S.: The Role of the Zero Emission Vehicles Mandates. *Energy Strategy* **2014**, *5*, 66–77. [CrossRef]
5. Li, C.; Negnevitsky, M.; Wang, X.; Yue, W.L.; Zou, X. Multi-criteria Analysis of Policies for Implementing Clean Energy Vehicles in China. *Energy Policy* **2019**, *129*, 826–840. [CrossRef]
6. Åhman, M. Government Policy and the Development of Electric Vehicles in Japan. *Energy Policy* **2006**, *34*, 433–443. [CrossRef]
7. Allwood, J.M.; Dunant, C.F.; Lupton, R.C.; Cleaver, C.J.; Serrenho, A.C.H.; Azevedo, J.M.C.; Horton, P.M.; Clare, A.; Low, H.; Horrocks, I.; et al. *Absolute Zero: Delivering the UK's Climate Change Commitment with Incremental Changes to Today's Technologies*, 1st ed.; University of Cambridge: Cambridge, UK, 2019.
8. CO_2 Emission Performance Standards for Cars and Vans. Available online: https://climate.ec.europa.eu/eu-action/transport/road-transport-reducing-co2-emissions-vehicles/co2-emission-performance-standards-cars-and-vans_en (accessed on 1 November 2023).
9. Zhou, W.; Cleaver, C.J.; Dunant, C.F.; Allwood, J.M.; Lin, J. Cost, Range Anxiety and Future Electricity Supply: A Review of How Today's Technology Trends May Influence the Future Uptake of BEVs. *Renew. Sustain. Energy Rev.* **2023**, *173*, 113074. [CrossRef]
10. Ehrenberger, S.; Dasgupta, I.; Brost, M.; Gebhardt, L.; Seiffert, R. Potentials of Light Electric Vehicles for Climate Protection by Substituting Passenger Car Trips. *World Electr. Veh. J.* **2022**, *13*, 183. [CrossRef]
11. Regulation (EU) No 168/2013 of the European Parliament and of the Council of 15 January 2013 on the Approval and Market Surveillance of Two- or Three-Wheel Vehicles and Quadricycles (Text with EEA Relevance). Available online: http://data.europa.eu/eli/reg/2013/168/2020-11-14 (accessed on 9 October 2023).
12. Lindh, P.; Gerami Tehrani, M.; Lindh, T.; Montonen, J.H.; Pyrhönen, J.; Sopanen, J.T.; Niemelä, M.; Alexandrova, Y.; Immonen, P.; Aarniovuori, L.; et al. Multidisciplinary Design of a Permanent-Magnet Traction Motor for a Hybrid Bus Taking the Load Cycle into Account. *IEEE Trans. Ind. Elect.* **2016**, *63*, 3397–3408. [CrossRef]
13. Ruuskanen, V.; Nerg, J.; Parviainen, A.; Rilla, M.; Pyrhönen, J. Design and Drive-cycle Based Analysis of Direct-driven Permanent Magnet Synchronous Machine for a Small Urban Use Electric Vehicle. In Proceedings of the 16th European Conference on Power Electronics and Applications, Lappeenranta, Finland, 26–28 August 2014.
14. Ruuskanen, V.; Nerg, J.; Pyrhönen, J.; Ruotsalainen, S.; Kennel, R. Drive Cycle Analysis of a Permanent-Magnet Traction Motor Based on Magnetostatic Finite-Element Analysis. *IEEE Trans. Veh. Technol.* **2015**, *64*, 1249–1254. [CrossRef]
15. Łebkowski, A. Light Electric Vehicle Powertrain Analysis. *Sci. J. Silesian Univ. Technol. Ser. Transp.* **2017**, *94*, 123–137. [CrossRef]
16. Chirca, M.; Dranca, M.A.; Breban, S.; Oprea, C.A. PMSM Evaluation for Electric Drive Train for L6e Light Electric Vehicles. In Proceedings of the International Conference and Exposition on Electrical and Power Engineering (EPE), Iasi, Romania, 22–23 October 2020.
17. Houache, M.; Yim, C.-H.; Karkar, Z.; Abu-Lebdeh, Y. On the Current and Future Outlook of Battery Chemistries for Electric Vehicles—Mini Review. *Batteries* **2022**, *8*, 70. [CrossRef]
18. Anttila, J.; Todorov, Y.; Ranta, M.; Pihlatie, M. System-Level Validation of an Electric Bus Fleet Simulator. In Proceedings of the IEEE Vehicle Power and Propulsion Conference (VPPC), Hanoi, Vietnam, 14–17 October 2019.
19. Joksimović, G.; Levi, E.; Kajević, A.; Mezzarobba, M.; Tessarolo, A. Optimal Selection of Rotor Bar Number for Minimizing Torque and Current Pulsations Due to Rotor Slot Harmonics in Three-Phase Cage Induction Motors. *IEEE Access* **2020**, *8*, 228572–228585. [CrossRef]
20. Tapani, J.; Hrabovcova, V.; Pyrhonen, J. *Design of Rotating Electrical Machines*, 2nd ed.; John Wiley & Sons: West Sussex, UK, 2013.
21. AC Motor Controller Gen4 Size 8. Available online: https://cdn.borgwarner.com/docs/default-source/default-document-library/gen4-size-8.pdf?sfvrsn=4b2b5a3d_7 (accessed on 9 October 2023).
22. Superior Lithium Ion Battery. Available online: https://www.master-instruments.com.au/files/battery_information/kokam_cell_brochure2021.pdf (accessed on 9 October 2023).
23. Kokam Li-ion Cell, Kokam. 2021. Available online: https://kokam.com/en/product/cell/lithium-ion-battery (accessed on 9 October 2023).
24. Needell, Z.A.; Trancik, J.E. Efficiently Simulating Personal Vehicle Energy Consumption in Mesoscopic Transport Models. *Transp. Res. Rec.* **2018**, *2672*, 163–173. [CrossRef]

World Electric Vehicle Journal

MDPI

Article

Utility Factor Curves for Plug-in Hybrid Electric Vehicles: Beyond the Standard Assumptions

Karim Hamza * and Kenneth P. Laberteaux

Toyota Motor North America, Ann Arbor, MI 48105, USA; ken.laberteaux@toyota.com
* Correspondence: khamza@umich.edu

Abstract: The utility factor (UF) of a plug-in hybrid electric vehicle (PHEV) refers to the ratio of miles traveled in electric mode to the total miles traveled. Standard UF curves provide a prediction of the expected achievable UF by a PHEV given its all-electric range (AER), but such predictions entail assumptions about both the driving patterns (distance traveled and energy intensity) and charging behavior. Studies have attempted to compare the real-world UF achieved by PHEVs to their standard values, but deviations can stem from deviations in assumptions about: (i) achievable electric range, (ii) travel distance and (iii) charging frequency. In this paper, we derive analytical models for modified utility factor curves as a function of both AER and charging behavior. We show that average charging frequency is insufficient to exactly predict UF but can still estimate bounds. Our generalized model can also provide insights into the efficacy of PHEVs in reducing carbon emissions.

Keywords: PHEV (plug-in hybrid electric vehicle); utility factors; charging; travel distance; regulations codes; standards (RCS)

Citation: Hamza, K.; Laberteaux, K.P. Utility Factor Curves for Plug-in Hybrid Electric Vehicles: Beyond the Standard Assumptions. *World Electr. Veh. J.* **2023**, *14*, 301. https://doi.org/10.3390/wevj14110301

Academic Editors: Joeri Van Mierlo and Genevieve Cullen

Received: 3 October 2023
Revised: 25 October 2023
Accepted: 27 October 2023
Published: 31 October 2023

1. Introduction

Plug-in hybrid electric vehicles (PHEVs) have powertrains that combine the traits of battery electric vehicles (BEVs) and hybrid electric vehicles (HEVs). Driving energy in HEVs, much like conventional vehicles, comes from fuel that powers an internal combustion engine (ICE) [1], but HEVs also have electric motor(s) and batteries that can assist in powering the vehicle (or completely power for short periods) and/or recapture energy during deceleration. The distinguishing feature of PHEVs [2] compared to HEVs is having larger capacity batteries that allow the vehicle to travel appreciable distances, known as the all-electric range (AER), without turning on its ICE, plus the capability to charge the battery from grid electricity like BEVs.

PHEVs have two main modes of driving: charge depletion (CD), in which electric energy from the battery is the main source of power, and charge sustaining (CS), in which fuel for the ICE is the main source of power. The utility factor (UF) of a PHEV generally refers to the ratio of miles traveled in CD mode to its total miles traveled [3]. It is important to acknowledge that different configurations and designs of PHEVs exist [4], and that under certain conditions and/or some PHEV powertrain designs, CD mode can involve small amounts of fuel consumption [5]. However, the current work in this paper focuses on usage cases where fuel consumption in CD mode is mostly negligible. As such, within the context of this paper, CD mode can be interchangeably referred to as "EV mode" and CS mode as "HEV mode".

Different standards exist [3,6,7] that aim at predicting the expected UF of a PHEV as a function of its AER, often presented as "UF curves", with the most prominent of which being SAE J2841 [3]. Accurate prediction of the UF for a PHEV carries significant importance because the UF is often used as a simplified metric for estimating tailpipe carbon emissions and thus has implications for present-day and future regulatory policies. However, standard UF curves entail underlying assumptions that may not necessarily be met by real-world PHEVs. Several modeling studies [8–10] have attempted to assess the effect of different drive patterns than those in SAE J2841, while other studies [11–13] have compared standard UF values with those that could be inferred from real-world PHEV data. As discussed in [12,13] and illustrated in Figure 1, three main categories of reasons why real-world UF may not align with standard UF curves include mismatches in assumptions about: (i) attained AER, (ii) actual travel distance and (iii) charging frequency.

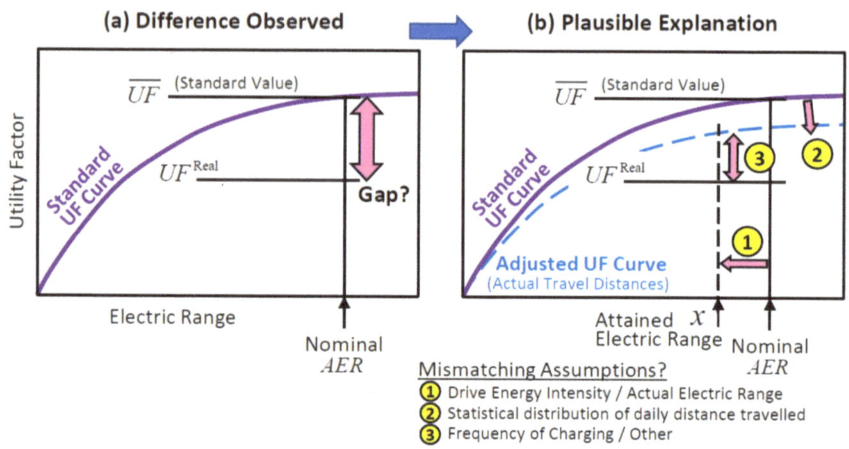

Figure 1. Illustration of three categories of assumptions mismatch between standard UF curves and real-world.

While examining the illustration in Figure 1, a number of considerations ought to be kept in mind:

- The gap between standard and real UF values (Figure 1a) as well as the categories of reasons 1–3 (Figure 1b) were drawn in the "negative" direction (i.e., real UF being less than the standard UF). However, this is primarily for illustrative purposes. In reality, it is plausible for any of the three categories of reasons or the overall gap to be in either the positive (i.e., better UF than the standard rating) or negative directions.
- Each of the three main categories of reasons may include several sub-reasons; for example, category #1 (real-world attained AER) could be affected by the acceleration rate and speed driving style of vehicle owners, ambient temperature (which in turn affects both the efficiency of the electric powertrain as well as the heating/cooling power consumption for climate control of the passenger cabin), weight of passengers and cargo, gradient of the terrain (uphill/downhill), or towing load.
- It is also important to note that those three categories of reasons, while understood to be the main contributors to the UF gap, are not the only contributing reasons, nor is it necessarily true that they are linearly independent. For example, some PHEV designs may utilize electric power to warm up the battery during a cold climate, while others might utilize an alternative approach such as briefly turning on the engine, which in turn might affect the observable miles traveled in CD or EV mode.
- The (real-world) attained AER is not necessarily a static number like the nominal AER that is published by regulatory agencies such as the US EPA [14]. In fact, the attained AER can change from day to day depending on the vehicle usage conditions, and such daily variations in the attained AER can have interactions with the other two categories of reasons (charging frequency and distance traveled). Nonetheless, to avoid over-complicating the problem, secondary interactions between the reasons and "all other/unknown" reasons are often lumped with one of the three main categories of reasons.

While the SAE J2841 standard [3] has laid out the methodology for deriving UF curves from any dataset that provides a statistical distribution of vehicle miles traveled per day, the "standard UF values" quoted in the literature (such as [8,9,12,13]) often refer to the UF curves shown in [3], which were derived from the vehicle daily miles traveled profile in the 2001 National Household Travel Survey (NHTS-2001) [15]. Several studies [6,8–10] have shown alternative UF curves for different datasets of real-world vehicle miles traveled, and in particular, Paffumi et al. [10] considered models for additional daytime charging, thereby relaxing the typical assumption of "exactly once-only charging event (to full) before every drive day" in [3].

To the best of the authors' knowledge, the current work in this paper is the first of its kind at attempting to lay out mathematical formulations for relaxing the typical once-only per drive day assumption to cases where the charging frequency is less than once per drive day. Similar to previous work [10], when considering charging frequencies different than exactly once per drive day, other assumptions come into play that result in a "family of plausible UF curves", for which we also propose mathematical models that estimate the upper and lower bounds. Aside from cases of charging frequencies less than once per drive day, we also propose mathematical models for plausible bounds on UF curves when the charging frequency is more than once per drive day. The outcome from the extended families of UF curves can then serve as a means of gauging the carbon emissions reduction benefit via PHEVs under a broader set of usage conditions.

This paper started with a brief overview of relevant literature as well as a framing of the scope and motivation for the current work. The rest of the manuscript is organized as follows: Section 2 provides details of the mathematical models, along with examples of corresponding UF curves that illustrate how the curves change with various changes in assumptions. Section 3 provides a more in-depth discussion of the findings, along with some estimates of carbon emission reduction benefits. The manuscript then concludes with some highlights of key findings and expected future work directions.

2. Mathematical Model

2.1. Notations and Assumptions

Similar to the SAE J2841 standard [3], the primary input to the mathematical model is a dataset for daily miles traveled by a sample of vehicles. Vehicles in the dataset don't necessarily need to be PHEVs or any particular type of powertrain, and as such, the methodology in [3] does not require any real-world charging information to be available in the dataset. In [3], UF for a hypothetical PHEV is calculated with the assumption that it would drive the same miles traveled profile, with a charging frequency of exactly one full charge before every drive-day (which we will denote as $\lambda = 1$) and no charging during daytime (which we will denote as $\mu = 0$). Section 2.2 in this manuscript will derive mathematical models for estimating the UF in cases when the charging is less than once before every drive day and no daytime charging ($0 < \lambda < 1$, $\mu = 0$), while Section 2.3 will derive mathematical models for estimating the UF in cases when the charging is once before every drive day, *plus* some daytime charging ($\lambda = 1$, $\mu > 0$). To be considered more closely representative of the real-world, the horizontal axis for such UF curves ought to be the real-world attained AER, which will be denoted by the symbol x. In such a setup, it is also implicitly assumed that "all other/unknown" reasons for the UF gap are lumped with the Figure 1b category of reasons #1 (attained AER).

To better streamline the UF calculations in the mathematical model, for every vehicle sample i (i = 1 to N), the statistical profile for daily miles traveled is discretized and represented as a matrix of the number of driving days per year D_{ij}, with the first index (i) referring to days of travel by a certain vehicle, while the second index j refers to a range of miles traveled per day, depending on a discretization parameter δ. For example, if $\delta = 0.5$ miles, then $j = 10$ refers to the range between 4.5 and 5.0 miles per day of travel distance. In this paper, we utilize the 2010–2012 California Household Travel Survey (CHTS) dataset [16], from which the matrix D_{ij} has been extracted. To make it easier to replicate the work in the current paper, a copy of the matrix D_{ij} has been placed in shared/public-accessible cloud storage [17]. Assuming the discretization parameter (δ), which was chosen at 0.5 miles, provides sufficient resolution, the annual miles traveled (l_{ij}) by a vehicle sample i on days with daily miles between $(j-1)\delta$ and $j\delta$ is estimated (using upper bounds for daily miles interval) via Equation (1) as follows:

$$l_{ij} = D_{ij} \cdot \delta \cdot j \tag{1}$$

It follows that the total annual miles traveled by a vehicle (L_i) and the probability density mass for a fraction of the vehicle's miles traveled (p_{ij}) on days with daily miles between $(j-1)\delta$ and $j\delta$ can be calculated as:

$$L_i = \sum_{j=1}^{J} l_{ij} \tag{2}$$

$$p_{ij} = \frac{l_{ij}}{L_i} \tag{3}$$

where J is the maximum number of discretized bins of miles per day of travel. In the current work, with $\delta = 0.5$ miles, J was set at a value of 2000, which means that days with more than 1000 miles of travel (very rare occurrences in the CHTS dataset) were treated as if they were days with 1000 miles of travel.

A reference $\overline{UF}_i$ value, conforming to the charging behavior assumptions of [3] (i.e., $\lambda = 1$, $\mu = 0$), can then be calculated for each vehicle sample given the real-world attained AER (x) as:

$$\overline{UF}_i(x) = \sum_{j=1}^{J} p_{ij} \left(\frac{\min(x, \delta \cdot j)}{\delta \cdot j} \right) \tag{4}$$

In order to go from calculating a UF value for each vehicle in the dataset to a population-wide estimate, SAE J2841 defines two different UF metrics [3]: (i) multi-day individual utility factor (MDIUF) and (ii) fleet utility factor (FUF). Simply put, MDIUF is a "simple average" of the UF values calculated for each vehicle in the dataset, while FUF effectively weighs each vehicle in the dataset by its total miles traveled. The purpose and usage of the two metrics are different. While MDIUF represents the expected value of UF for a randomly drawn sample vehicle, FUF, on the other hand, is an estimate of the total fraction of miles traveled by all vehicles that can be electrified. Furthermore, some datasets (CHTS included) can have representative weights for each of the vehicle samples that are often based on household demographics [18], so weighing each sample by its representative weight (w_i) effectively scales it up to the population that the dataset is intended to represent (all of California in the case of the CHTS dataset). The calculation of MDIUF and FUF can thus be performed as follows:

$$\overline{MDIUF}(x) = \frac{1}{W} \sum_{i=1}^{N} w_i \cdot \overline{UF}_i(x) \tag{5}$$

$$\overline{FUF}(x) = \frac{1}{Y} \sum_{i=1}^{N} L_i w_i \cdot \overline{UF}_i(x) \tag{6}$$

where W and Y are respectively the sum of vehicle sample weights and total weighed miles traveled, calculated as:

$$W = \sum_{i=1}^{N} w_i \tag{7}$$

$$Y = \sum_{i=1}^{N} L_i w_i \tag{8}$$

For datasets that don't have vehicle sample weights, the typical assumption is to treat all vehicles equally (i.e., by setting all $w_i = 1$ in Equations (5)–(7)). For illustration purposes, the reference MDIUF and FUF curves (per Equations (5) and (6)) for the CHTS dataset are plotted in Figure 2 for the range of real-world attained AER (x) between 0 and 100 miles. Also shown in Figure 2 are the reference values from SAE J2841 [3] that are based on NHTS-2001. One notable observation in Figure 2 is that the reference UF curves (both MDIUF and FUF) via CHTS seem to have larger UF numbers than the reference UF curves via NHTS-2001 at any given value of x. This implies that the recorded vehicle travel in CHTS generally had fewer miles per drive day than in NHTS-2001. One plausible explanation for this could be the timing of data collection, where CHTS (data collected between 2010 and 2012) could have been affected by the 2008–2009 recession period in the US. Another plausible explanation could be due to the method of data collection, where NHTS-2001 utilized self-reported trip length data, while CHTS data utilized in this paper came from on-board device (OBD) logging of the sample vehicles. However, regardless of what dataset is used, the mathematical modeling approach in this paper could still be applied.

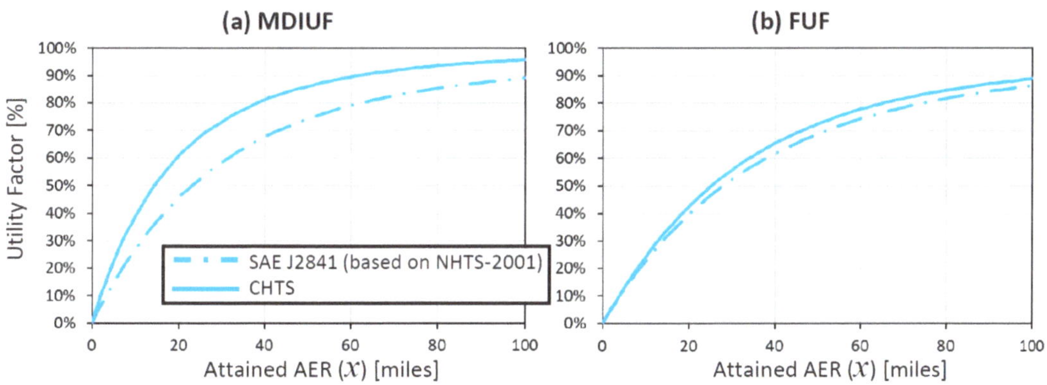

Figure 2. Reference UF curves via CHTS and SAE J2841 (based on NHTS-2001).

2.2. *Charging Frequency Less Than Once per Drive Day*

2.2.1. Overview

As a note, when considering mathematical modeling for charging frequency less than once per drive day, we are specifically referring to the case where the overnight charging before every drive day is less than once per day (i.e., $0 < \lambda < 1$) and there is no daytime charging. To reduce clutter, we no longer mention daytime charging, but it should be understood that it is assumed that ($\mu = 0$) within Section 2.2. It also ought to be noted that while λ is assumed to be strictly greater than zero and strictly less than 1, this is only to exclude the trivial case of $\lambda = 0$, for which UF is zero, and the case of $\lambda = 1$, for which UF can be calculated via Equations (5) and (6). Before building a mathematical model that addresses the fully general case of ($0 < \lambda < 1$), we first consider a few special cases.

2.2.2. Special Case: Binary Charging Behavior

This special case considers a mathematical model where the charging frequency for each individual vehicle in the data set (λ_i) can only take a value of either 0 or 1. In other words, while the overall average for the population of vehicles (λ) is between 0 and 1, this average is only attained via one set of vehicles (A) always charging (i.e., $\lambda_i = 1, i \in A$), while some vehicles are never charging (i.e., $\lambda_i = 0$, $i \notin A$). This can be mathematically expressed as:

$$UF_i^{Bin}(x) = \lambda_i \cdot \overline{UF}_i(x) \tag{9}$$

$$\lambda = \frac{1}{W}\sum_{i\in A} w_i \cdot \lambda_i \tag{10}$$

where UF_i^{Bin} is the UF value for vehicle i with binary charging behavior, while $\overline{UF}_i$ is the reference UF value, computed via (Equation (4)). When considering going from UF values for each vehicle (per Equation (9)) to MDIUF and FUF estimates, it can make a big difference to the UF, *which vehicles* are in the set (A). As such, it is possible to estimate an expected value with the assumption of independent likelihood (i.e., any vehicle in the dataset is equally likely to be in the set A), as well as upper and lower bounds.

Expected Value with Independent Likelihood: when every vehicle in the dataset is equally likely to be in set (A), it can be shown that the UF for each vehicle i reduces to its reference UF value multiplied by the population average charging frequency (λ), which in turn allows for computing MDIUF and FUF as:

$$MDIUF^{Bin,Ind}(x) = \lambda \cdot \overline{MDIUF}(x) \tag{11}$$

$$FUF^{Bin,Ind}(x) = \lambda \cdot \overline{FUF}(x) \tag{12}$$

Upper and Lower Bounds: to compute upper and lower bounds for MDIUF and FUF, we setup optimization problems for maximizing/minimizing UF values with "which vehicles are in the set A" (i.e., which vehicle samples have corresponding λ_i = 1) as the decision variables. The optimization problem has the form:

$$\text{For MDIUF}: \text{ Minimize/Maximize } f_{MDIUF}^{Bin}(x) = \frac{1}{W}\sum_{i=1}^{N} \lambda_i \cdot w_i \cdot \overline{UF}_i(x) \tag{13}$$

$$\text{For FUF}: \text{ Minimize/Maximize } f_{FUF}^{Bin}(x) = \frac{1}{Y}\sum_{i=1}^{N} \lambda_i \cdot w_i L_i \cdot \overline{UF}_i(x) \tag{14}$$

$$\text{Subject To}: \frac{1}{W}\sum_{i=1}^{N} w_i \cdot \lambda_i = \lambda \tag{15}$$

$$\lambda_i \in \{0,1\} \tag{16}$$

As a note, this optimization setup has the form of a linear program [19] (objective and constraints are linear in the decision variables λ_i), with all other quantities in Equations (13)–(15) being constant or possible to pre-compute before running the optimization problem to determine the upper/lower bounds for MDIUF or FUF. However, in the case of binary charging behavior, it is an integer-linear program (per the constraint in Equation (16)). While a generic integer linear program can be challenging to solve, when the number of vehicle samples in the dataset is sufficiently large, the equality constraint in Equation (15) can be satisfied within reasonable tolerance while relaxing the optimization problem to only solve its linear program version. The results of this model (independent, upper and lower bounds) for λ = 0.5 are shown in Figure 3a.

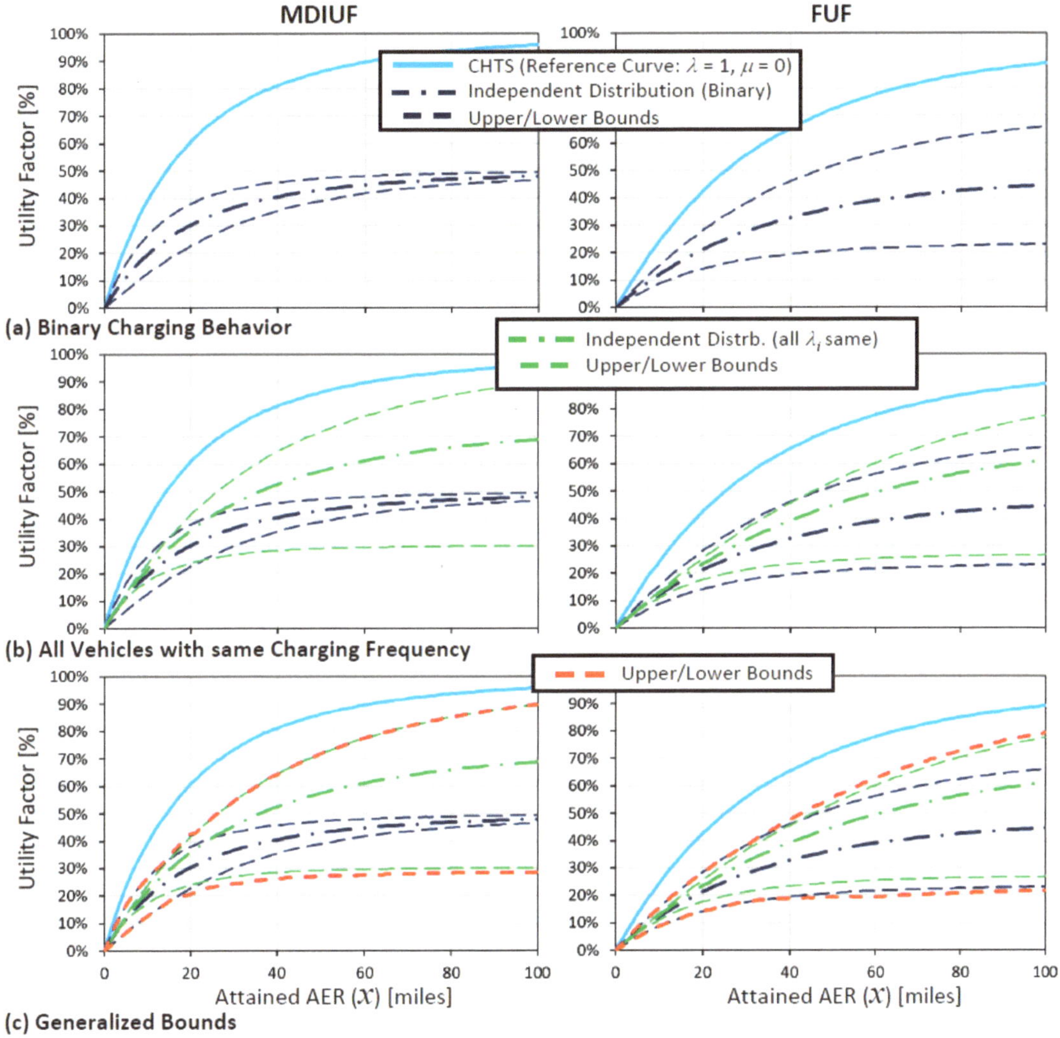

Figure 3. Reference UF curves and various models for $\lambda = 0.5$.

2.2.3. Special Case: All Vehicles with the Same Charging Frequency

This special case considers a mathematical model where the charging frequency for each individual vehicle in the data set (λ_i) is exactly equal to the charging frequency of the population average (λ). In other words, this is a case where all vehicles in a population are behaving exactly the same in terms of frequency of charging. For this special case, going from vehicle UF to population MDIUF and FUF is fairly straight-forward (similar to Equations (5) and (6)) as:

$$MDIUF^{Same}(x) = \frac{1}{W}\sum_{i=1}^{N} w_i \cdot UF_i^{Same}(x, \lambda_i) \quad (17)$$

$$FUF^{Same}(x) = \frac{1}{Y}\sum_{i=1}^{N} L_i w_i \cdot UF_i^{Same}(x, \lambda_i) \quad (18)$$

However, the calculation of UF for each vehicle sample as a function of its charging frequency (λ_i = λ) requires some further modeling assumptions. For this, we consider the calculation of an expected UF value based on an independent probability distribution for which days have a charging event before them, as well as upper and lower bounds.

Expected Value with Independent Probability: when considering a set of drive days by a vehicle sample i such that any drive day has an equal chance as any other drive day of having an overnight charging, the chance of a charging event becomes a Bernoulli experiment [20] with a chance of success equal to the overall average (which is λ_i = λ). Under such conditions, the UF of a vehicle sample can be expressed as:

$$UF_i^{Same,Ind}(x,\lambda_i) = \lambda_i \cdot \overline{UF}_i(x) + \lambda_i(1-\lambda_i) \cdot \overline{TDUF}_i(x) + \ldots \quad (19)$$

where the expression λ_i $(1 - \lambda_i)$ that is multiplied by the second term in Equation (19) represents the probability of the current driving day not having a charging event after the previous driving day had a charging event. We don't consider the rest of the terms after the second term in Equation (19) as they would be multiplied by λ_i $(1 - \lambda_i)^2$ (or even smaller numbers), so we consider them negligible compared to the first two terms. The equivalent UF for two days in a row after one overnight charging event is calculated as:

$$\overline{TDUF}_i(x) = \sum_{j=1}^{J} \left(\frac{p_{ij}}{\delta \cdot j} \sum_{k=1}^{J} d_{ik} \cdot \min(\delta \cdot j, \max(0, x - \delta \cdot k)) \right) \quad (20)$$

where d_{ij} is the probability density mass for vehicle i having a day with miles traveled within a certain range of miles per day:

$$d_{ij} = \frac{D_{ij}}{D_i} \quad (21)$$

$$D_i = \sum_{j=1}^{J} D_{ij} \quad (22)$$

As a sanity check, one could confirm that the expression in (Equation (19)) converges to the same expression in (Equation (9)) as λ_i approaches a value of either 0 or 1.

Lower Bound: when considering a set of drive days by a vehicle sample, the temporal distribution of the charging events relative to driving days can have a significant effect on the attained UF. Generally speaking, when the charging events are mostly uniformly spaced, this results in better UF than non-uniformly spaced. For example, if λ_i = 0.5 with uniform spacing, this means that a charging event happens exactly one per two drive days, which can maximize utilization of each charging event. The opposite, least favorable temporal distribution is when/if all the charging events happen on back-to-back drive days while leaving a long gap of days without any charging events. Furthermore, if the stacked-up charging events are occurring before drive days that have the least contribution to attained electric miles, it would represent the lower bound for UF. To compute this lower bound, we set up an integer linear programming optimization problem similar to the setup in Section 2.2.2, but with the decision variables (v_j) controlling the temporal distribution of charging events.

$$\text{Minimize } f^{Same,LB}(x,\lambda_i) = \sum_{j=1}^{J} v_{ij} \cdot p_{ij} \left(\frac{\min(x, \delta \cdot j)}{\delta \cdot j} \right) \quad (23)$$

$$\text{Subject To}: \frac{1}{D_i} \sum_{j=1}^{J} v_{ij} \cdot d_{ij} = \lambda_i \quad (24)$$

$$v_{ij} \in \{0,1\} \quad (25)$$

Upper Bound for $\lambda_i = 0.5$: in this estimate of an upper bound for UF, it is assumed that the temporal distribution of charging events is exactly evenly spaced at one charging event per two drive days, while being statistically independent from the number of miles traveled on any given day. The UF can then be calculated in a similar manner as Equation (19), but with the drive days either having a charging event or a charging event one day earlier (and no three or more drive days without a charging event):

$$UF_i^{Same,UB,Half}(x) = 0.5 \cdot \overline{UF}_i(x) + 0.5 \cdot \overline{TDUF}_i(x) \tag{26}$$

Upper Bound: For cases where $\lambda_i < 0.5$, we consider an upper bound for the UF via linear interpolation between zero and the upper bound value obtained from Equation (26). This corresponds to a temporal distribution of charging events where a portion of the time horizon has evenly spaced charging events at a frequency of one per two drive days, while the rest of the time horizon has no charging events. Likewise, for cases when $\lambda_i > 0.5$, we consider an upper bound for the UF via linear interpolation between the value obtained from Equation (26) and the case when $\lambda_i = 1$. This corresponds to a temporal distribution of charging events where a portion of the time horizon has evenly spaced charging events at a frequency of one per two drive days, while the rest of the time horizon has one charging event for every drive day. Combining the cases, the formula for the upper bound is summarized as:

$$UF_i^{Same,UB}(x, \lambda_i) = \begin{cases} 2\lambda_i \cdot UF_i^{Same,UB,Half}(x) & \lambda_i \le 0.5 \\ (2\lambda_i - 1) \cdot \overline{UF}_i(x) + (2 - 2\lambda_i) \cdot UF_i^{Same,UB,Half}(x) & \lambda_i > 0.5 \end{cases} \tag{27}$$

The results of this model (independent, upper and lower bounds) for $\lambda = 0.5$ are shown in Figure 3b.

2.2.4. Generalized Upper and Lower Bounds

Given some value (λ) for the overall average frequency of charging, Section 2.2.2 highlighted plausible variations in UF by considering sensitivity to *which vehicles* within the population, while Section 2.2.3 highlighted plausible variations in UF by considering sensitivity to the temporal distribution of charging events relative to daily miles traveled by each vehicle. We now set up a set of more generic optimization problems, whose optimal solution provides the overall upper/lower bounds for MDIUF and FUF.

$$\text{For upper bound MDIUF}: \ \text{Maximize } f_{MDIUF}^{Gen,UB}(x) = \frac{1}{W}\sum_{i=1}^{N} w_i \cdot UF_i^{Same,UB}(x, \lambda_i) \tag{28}$$

$$\text{For lower bound MDIUF}: \ \text{Minimize } f_{MDIUF}^{Gen,LB}(x) = \frac{1}{W}\sum_{i=1}^{N} w_i \cdot UF_i^{Same,LB}(x, \lambda_i) \tag{29}$$

$$\text{For upper bound FUF}: \ \text{Maximize } f_{FUF}^{Gen,UB}(x) = \frac{1}{Y}\sum_{i=1}^{N} w_i L_i \cdot UF_i^{Same,UB}(x, \lambda_i) \tag{30}$$

$$\text{For lower bound FUF}: \ \text{Minimize } f_{FUF}^{Gen,LB}(x) = \frac{1}{Y}\sum_{i=1}^{N} w_i L_i \cdot UF_i^{Same,LB}(x, \lambda_i) \tag{31}$$

$$\text{Subject To}: \ \frac{1}{W}\sum_{i=1}^{N} w_i \cdot \lambda_i = \lambda \tag{32}$$

$$0 \le \lambda_i \le 1 \tag{33}$$

To calculate the upper or lower bounds for MDIUF or FUF, one needs to utilize the appropriate objective (among Equations (29)–(31)) along with the constraints in Equations (32) and (33). Though the constraints are linear and the decision variables are continuous (which generally makes the optimization problem easier to solve than integer programming), the objective functions, which include terms defined in Section 2.2.3, can have nonlinear terms in (λ_i), such as the terms defined via Equation (19) or even sub-optimization problems, such as when estimating lower bounds for an individual vehicle sample via Equations (23)–(25). As such, the optimization approach we tested for solving Equations (29)–(33) is the nonlinear programming technique known as successive linear programming [21]. The calculated upper and lower bound UF curves for λ = 0.5 are shown in Figure 3c.

A brief overview of the successive linear programming (SLP) optimization approach [21] is that it is a technique that attempts to find the optimum value within a solution domain of continuous variables (all λ_i in this case) for a nonlinear objective and/or constraints via solving a series of linear programming optimization problems. In each iteration of SLP, a linear approximation is constructed for the nonlinear objective and/or constraints (via function value and gradient at the "current solution" point). The linear approximation is then solved via linear programming techniques, with the solution of the linear programming becoming the new "current solution" point, and the process is repeated until convergence, which is typically when iterations of SLP can no longer find a better solution satisfying the problem constraints. In implementation for the current problem (solving for all λ_i), the constraints (Equations (32) and (33)) are actually linear, which means that the successive iterations of SLP always retain a feasible solution. Since the results of SLP, much like gradient-following optimization techniques, can be dependent on the "starting point", a multi-start point strategy is employed, with two special cases (from Sections 2.2.2 and 2.2.3) included among the starting points. This ensures that the solution returned by SLP (as shown in Figure 3c) is always "outside the envelope" of either of the two special cases (shown in Figure 3a,b).

2.3. Charging Frequency: More Than Once per Drive Day

There seems to be a perception that additional charging events beyond once per drive day tend to have diminishing returns. Part of this perception may come from some daytime charging events being limited by available time (e.g., charging a PHEV while the owner is shopping) and therefore not completely topping off the battery/restore full electric driving range. The other issue about daytime charging events is the timing relative to miles traveled on a given day. Even a top-off charging event may not add many electric miles if it happens too early or too late during a drive day. Furthermore, if a daytime charging event occurs on a day where the miles traveled are less than the electric range of the PHEV, there would be no "additional" electric miles on that day. As such, one may find corner cases where a lower bound on UF for (λ = 1, μ > 0) isn't noticeably better than the reference case of (λ = 1, μ = 0). Thus, in this section, we focus more on plausible estimates of UF than minimum/maximum upper and lower bounds.

We develop mathematical models for the estimation of the UF factor for vehicles that charge to full before every drive day and gain a second full top-off charging event on *some* of the drive days, in other words, the case of (λ = 1, 0 < $\mu \leq$ 1). We assume the daytime charging events have an independent probability of occurring (i.e., any drive day is equally likely to have a second charging event), which allows us to formulate a generic estimate of the UF of a sample vehicle as follows:

$$UF_i^{Daytime,Ind}(x,\mu_i) = \sum_{j=1}^{J} p_{ij}\left(\left(\frac{\min(x,\delta j)}{\delta j}\right) + \mu_i\left(\frac{\Phi(\delta j)}{\delta j}\right)\right) \tag{34}$$

where the function $\Phi(\delta j)$ in the second term of Equation (34) represents the expected value for additional electric miles on a day with travel distance equal to (δj) due to the occurrence of a daytime charging event. The additional electric miles are a natural function of both the total miles traveled during that day as well as when the daytime charging event occurs and can be expressed as:

$$\Phi(\delta j) = \begin{cases} 0 & \delta j \leq x \\ \int_{y=0}^{\delta j} \phi(y) \cdot g(y) dy & \delta j > x \end{cases} \tag{35}$$

$$\phi(y) = \begin{cases} \min(y, \delta j - x) & y \leq x \\ x & x < y \leq \delta j - x \\ \delta j - y & y > \delta j - x \end{cases} \tag{36}$$

where the function $g(y)$ is the probability density function for occurrence of the daytime charging event after how many miles (y) have been traveled. In demonstration of this model (as shown in Figure 4), we utilize the exponential distribution [20], which is a typical assumption for random arrivals in queueing systems. The distribution is also shifted such that the daytime charging event wouldn't occur until half the electric range has been traveled, and the exponential parameter is set to have the average occurrence of the daytime charging event between half the electric range and the full electric range. We also note that Equation (34) can be used to estimate UF if all vehicles in the dataset had the same frequency for daytime charging (which we consider a plausible scenario), or μ_i can have binary (0 or 1) values and an integer linear optimization framework (similar to Section 2.2.2) can be applied to estimate upper and lower bounds, as shown in Figure 4.

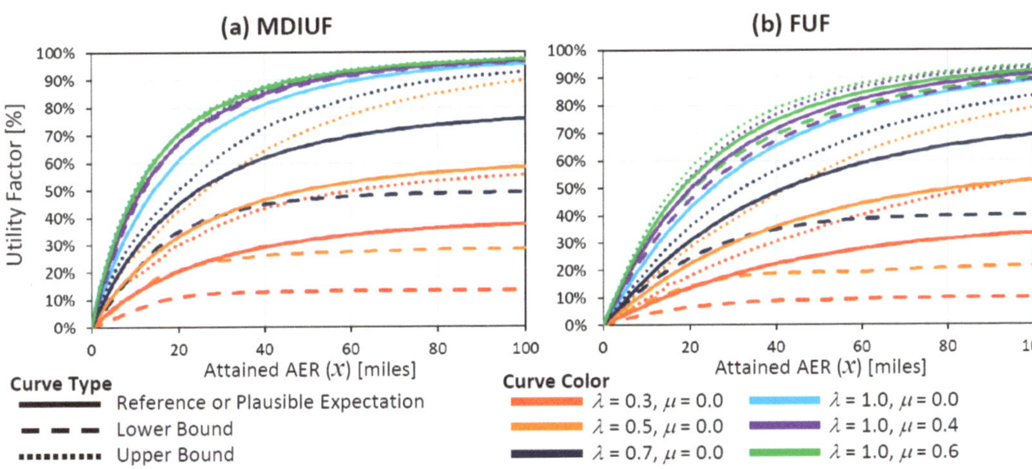

Figure 4. Plausible and Upper/Lower UF curves for select modelled charging frequency.

2.4. Summary of Modelled Cases

A summary list of the modelled cases for charging behavior in this work is provided in Table 1. To further enhance the modeling perspective, it is noted that the actual timing of a charging event (during nighttime or daytime) isn't what drives the mathematical model. Rather, a charging event (fully topping the electric range in the battery) between two drive days is the mathematical equivalent of an "overnight" charging event, while an "additional" charging event within a one-day window after a previous charging event is the mathematical equivalent of a "daytime" charging event. In which case, as long as there is no significant temporal overlap between the charging events, one may use the sum of λ and μ as a proxy estimator for the fully generalized case (as is performed in the discussion in Section 3). However, detailed derivation for the fully generalized case is beyond the scope of current work.

Table 1. Summary of Charging Behavior Modelled Cases.

Case	Discussed	Description
$\lambda_i \in \{0, 1\}$, $\mu = 0$	Section 2.2.2	No daytime charging. Overnight charging behavior is binary; some vehicles always charge, others never charge.
All $\lambda_i = \lambda, 0 \leq \lambda \leq 1$, $\mu = 0$	Section 2.2.3	No daytime charging. The overnight charging frequency is the same for all vehicles.
$0 \leq \lambda_i \leq 1$, $\mu = 0$	Section 2.2.4	No daytime charging. Generalized case for overnight charging, where some vehicles always charge, some never charge, others somewhere in-between.
All $\lambda_i = 1, 0 \leq \mu_i \leq 1$	Section 2.3	Vehicles always charge overnight. Some vehicles also gain one additional charging event during the day.
$0 \leq \lambda_i \leq 1, 0 \leq \mu_i$	Section 3, future work	Fully generalized case where overnight charging frequency for each individual vehicle can be anywhere between always and never, while at the same time, each individual vehicle may have additional one or more charging events during the day

3. Discussion

British statistician George Box once famously wrote *"All models are wrong, some are useful"*. When one reflects on when the concept of UF and UF curves were originally being developed (with standard assumptions), one of the main appeal points was that UF curves were relatively "easy" to use and understand. However, when standard assumptions are called into question in order to create more realistic real-world behavior models, it becomes apparent that real-world UF observations can be affected by many complex details. Perhaps the primary contribution in this work is not so much the calculation of UF curves for various charging frequencies as it is bringing into focus how wide the difference between upper and lower bounds can be in some cases, as observed in Figure 4. It also ought to not be forgotten that the modeling work in Sections 2.2 and 2.3 is still involving several simplifying assumptions, such as independent probabilities or consideration of daytime charging only after overnight charging before a drive day has been fulfilled, when in reality, some vehicle owners may not necessarily charge before every drive day, but they can still occasionally do daytime charging. It may be possible to further verify the degree of validity of such assumptions by comparing them with real-world PHEVs data, though this would require some much more detailed vehicle data than is typically available via public travel surveys. Such validation work is beyond the scope of the current paper.

Another issue about the concept of UF that is often ignored in favor of it being easy to use is that UF can be taken as a proxy estimate of only the tail-pipe emissions from a PHEV. In some instances, deviations in the real-world from ideal UF curves are being used as a reason to suggest lowering the regulatory carbon emissions reduction benefit of PHEVs [22,23]. However, this implicitly ignores the fact that electric miles also have equivalent carbon emissions depending on the fuel mix for electricity generation. What could be a more relevant metric for gauging the benefit of PHEVs is considering how much well-to-wheels [24] reduction of carbon emissions they can bring compared to an equivalent conventional vehicle. For a present-day scenario demonstrating such a concept, we use the EPA label values [14] for the 2022 RAV4 and RAV4 Prime (respectively at 29 MPG, 38 MPG in CS mode and 0.36 kWh/mile in CD mode), along with the 2021 average carbon intensity for the US electric grid at ~450 g-CO_2/kWh (calculated via fuel mix information from the US EIA [25] and GREET model [24]) and 10,778 g-CO_2/gal for E10 gasoline. We also consider scenarios for the year 2050 with projected carbon intensity for electricity at 180 g-CO_2/kWh, as well as 50% reduced carbon intensity for gasoline. Using plausible FUF values (from Figure 4b) at different charging frequencies, carbon emissions offset results for the considered scenarios are shown in Figure 5.

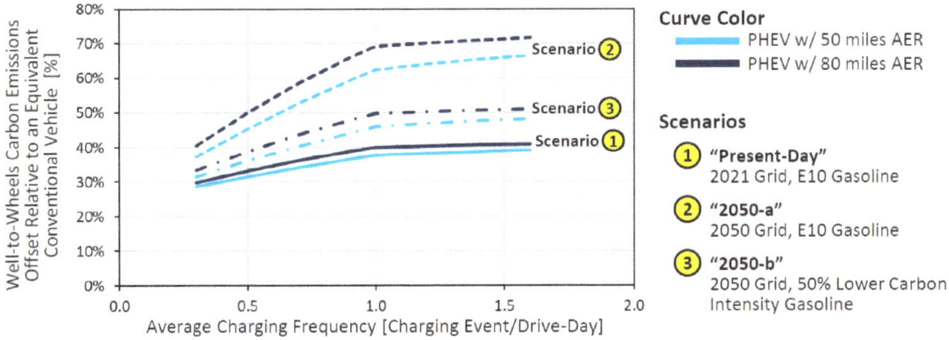

Figure 5. Select carbon emissions offset scenarios relative to an equivalent conventional vehicle.

Considering the bigger picture of well-to-wheels analysis (Figure 5) can dramatically change the sensitivity of carbon emission reductions (relative to an equivalent conventional vehicle) to the charging frequency of PHEVs. In fact, at the present-day average US electric grid, the carbon emissions in CD mode (electric) miles are only about 25% less than the carbon emissions per mile traveled in CS mode on the E10 gasoline blend. While a 50-mile AER PHEV (scenario 1 in Figure 5) offsets about 37% of the carbon emissions of an equivalent conventional vehicle when charged exactly once per drive day, the same PHEV can still offset about 31% of the carbon emissions when its frequency of charging is 0.5 (once per two drive days on average). The difference between a 50- and an 80-mile AER PHEV also appears minimal in (scenario 1 Figure 5). With lower carbon intensity in the electric grid and no change in the gasoline blend (scenario 2 in Figure 5), the difference between 50- and 80-mile AER, as well as the sensitivity to charging frequency, becomes more pronounced. However, it also stands to reason that future liquid fuels would have an increased fraction of biofuels and/or carbon-capture/synthetic fuels that are lower in carbon intensity. In such cases (scenario 3 in Figure 5), it can still be an attractive option across a broad range of usage conditions.

4. Conclusions

This paper considered an extension of the standard assumptions for generating UF curves, with the aim of encompassing a broader range of realistic charging behavior by PHEV owners. Mathematical models were proposed that address a specific number of cases, as well as an optimization framework that can be utilized to estimate upper and lower bounds. Though UF is a fairly simple concept, it has the drawback of not being indicative of the bigger picture (such as well-to-wheels or full lifecycle) of carbon emissions. When considering well-to-wheels, it can be shown that the carbon emissions reduction benefits of a PHEV relative to an equivalent conventional vehicle are not very sensitive to the frequency of charging or longer than 50 miles AER unless the difference in carbon intensity between CD mode and CS mode is high (which doesn't occur except with very low carbon electricity and high fossil content liquid fuels). Future extensions of this work may include comparison with detailed real-world PHEV data in order to gauge the realism of the various assumptions in the proposed mathematical models, as well as further extension of the mathematical model to consider the interaction/supplementation effects of daytime charging when overnight charging is less frequent than every night.

Author Contributions: Conceptualization, K.H. and K.P.L.; methodology, K.H. and K.P.L.; software, K.H.; validation, K.H. and K.P.L.; formal analysis, K.H. and K.P.L.; investigation, K.H. and K.P.L.; resources, K.P.L.; data curation, K.H.; writing—original draft preparation, K.H.; writing—review and editing, K.P.L.; visualization, K.H.; supervision, K.P.L.; project administration, K.P.L.; funding acquisition, K.P.L. All authors have read and agreed to the published version of the manuscript.

Funding: This research received no external funding.

Data Availability Statement: All data utilized in this manuscript comes from publicly accessible sources that have been cited among the references.

Conflicts of Interest: The authors declare no conflict of interest.

References

1. US Department of Energy. How Hybrids Work. Available online: https://www.fueleconomy.gov/feg/hybridtech.shtml (accessed on 16 November 2022).
2. US Environmental Protection Agency. Electric & Plug-In Hybrid Electric Vehicles. Available online: https://www.epa.gov/greenvehicles/explaining-electric-plug-hybrid-electric-vehicles (accessed on 16 November 2022).
3. Society of Automotive Engineers (SAE). *J2841_201009: Utility Factor Definitions for Plug-In Hybrid Electric Vehicles Using Travel Survey Data*; SAE International: Warrendale, PA, USA, 2010.
4. US Department of Energy. Plug-In Hybrids. Available online: https://www.fueleconomy.gov/feg/phevtech.shtml (accessed on 16 November 2022).
5. US Environmental Protection Agency. Explaining Electric & Plug-In Hybrid Electric Vehicles. Available online: https://19january2017snapshot.epa.gov/greenvehicles/explaining-electric-plug-hybrid-electric-vehicles_.html (accessed on 1 March 2023).
6. United Nations Economic Commission for Europe. United Nations Global Technical Regulation on Worldwide Harmonized Light Vehicles Test Procedures (WLTP). Available online: https://unece.org/fileadmin/DAM/trans/main/wp29/wp29wgs/wp29gen/wp29registry/ECE-TRANS-180a15am4e.pdf (accessed on 17 March 2023).
7. Liu, X.; Zhao, F.; Hao, H.; Chen, K.; Liu, Z.; Babiker, H.; Amer, A. From NEDC to WLTP: Effect on the Energy Consumption, NEV Credits, and Subsidies Policies of PHEV in the Chinese Market. *Sustainability* **2020**, *12*, 5747. [CrossRef]
8. Bradley, T.; Quinn, C. Analysis of plug-in hybrid electric vehicle utility factors. *J. Power Sources* **2010**, *195*, 5399–5408. [CrossRef]
9. Wu, X.; Aviquzzaman, M.; Lin, Z. Analysis of plug-in hybrid electric vehicles' utility factors using GPS-based longitudinal travel data. *Transp. Res. Part C* **2015**, *57*, 1–12. [CrossRef]
10. Paffumi, E.; De Gennaro, M.; Martini, G. Alternative utility factor versus the SAE J2841 standard method for PHEV and BEV applications. *Transp. Policy* **2018**, *68*, 80–97. [CrossRef]
11. Smart, J.; Bradley, T.; Salisbury, S. Actual Versus Estimated Utility Factor of a Large Set of Privately Owned Chevrolet Volts. *Int. J. Altern. Powertrains* **2014**, *3*, 30–35. [CrossRef]
12. Raghavan, S.; Tal, G. Plug-in hybrid electric vehicle observed utility factor: Why the observed electrification performance differ from expectations. *Int. J. Sustain. Transp.* **2020**, *16*, 105–136. [CrossRef]
13. Hamza, K.; Laberteaux, K.; Chu, K.-C. On inferred real-world fuel consumption of past decade plug-in hybrid electric vehicles in the US. *Environ. Res. Lett.* **2022**, *17*, 104053. [CrossRef]
14. US Environmental Protection Agency. Fuel Economy Guide. Available online: https://www.fueleconomy.gov/feg/printGuides.shtml (accessed on 18 March 2023).
15. US Department of Transportation. National Household Travel Survey. Available online: https://nhts.ornl.gov/download.shtm (accessed on 18 March 2023).
16. National Renewable Energy Laboratory. 2010–2012 California Household Travel Survey. Available online: https://www.nrel.gov/transportation/secure-transportation-data/tsdc-california-travel-survey.html (accessed on 18 March 2023).
17. Utility Factors Extended (Data Available on Google Drive Cloud Storage). Available online: https://drive.google.com/drive/folders/1UL3FTffKbgbgi_EMVu0bWt_CQTkS2_hr?usp=sharing (accessed on 18 March 2023).
18. NuStats LLC. 2010–2012 California Household Travel Survey Final Report. Available online: https://www.nrel.gov/transportation/secure-transportation-data/assets/pdfs/calif_household_travel_survey.pdf (accessed on 19 March 2023).
19. Hu, T.; Kahng, A. *Linear and Integer Programming Made Easy*; Springer: Berlin/Heidelberg, Germany, 2016.
20. Collani, E.; Draeger, K. *Binomial Distribution Handbook for Scientists and Engineers*; Springer: Berlin/Heidelberg, Germany, 2001.
21. Bazaraa, M.; Sherali, H.; Shetty, C. *Nonlinear Programming: Theory and Algorithms*; Wiley: New York, NY, USA, 2006.
22. Plötz, P.; Moll, C.; Bieker, G.; Mock, P. From lab-to-road: Real-world fuel consumption and CO_2 emissions of plug-in hybrid electric vehicles. *Environ. Res. Lett.* **2021**, *16*, 054078. [CrossRef]
23. US Environmental Protection Agency. Multi-Pollutant Emissions Standards for Model Years 2027 and Later Light-Duty and Medium-Duty Vehicles. 2023. Available online: https://www.regulations.gov/search?documentTypes=Proposed%20Rule&filter=EPA-HQ-OAR-2022-0829 (accessed on 23 October 2023).
24. Argonne National Laboratory. GREET Model. Available online: https://greet.es.anl.gov/ (accessed on 20 March 2023).
25. US Energy Information Administration. Available online: https://www.eia.gov/electricity/annual/ (accessed on 3 October 2023).

World Electric Vehicle Journal

MDPI

Article

Zero-Emission Truck Powertrains for Regional and Long-Haul Missions

Mikko Pihlatie *, Mikaela Ranta, Pekka Rahkola and Rafael Åman

VTT Technical Research Centre of Finland, P.O. Box 1000, 02044 VTT Espoo, Finland; mikaela.ranta@vtt.fi (M.R.); pekka.rahkola@vtt.fi (P.R.); rafael.aman@vtt.fi (R.Å.)

* Correspondence: mikko.pihlatie@vtt.fi; Tel.: +358-400-430395

Abstract: Zero-emission trucks for regional and long-haul missions are an option for fossil-free freight. The viability of such powertrains and system solutions was studied conceptually in project ESCALATE for trucks with GVW of 40 tonnes and beyond through various battery electric and fuel cell prime mover combinations. The study covers battery and fuel cell power sources with different degrees of battery electric as well as H_2 and fuel cell operation. As a design basis, two different missions with a single-charge/H_2 refill were analysed. The first mission was the VECTO long-haul profile repeated up to 750 km, whereas the second was a real 520 km on-road mission in Finland. Based on the simulated energy consumption on the driving cycle, on-board energy demand was estimated, and the initial single-charge and H_2 refill operational scenarios were produced with five different power source topologies and on-board storage capacities. The traction motors of the tractor were dimensioned so that a secondary mission of GVW up to 76 tonnes on a shorter route or a longer route with more frequent battery recharge and/or H_2 refill can be operated. Based on the powertrain and vehicle model, various infrastructure options for charging and H_2 refuelling strategies as well as various operative scenarios with indicative total cost of ownership (TCO) were analysed.

Keywords: electric vehicle; battery; fuel cell; charging

Citation: Pihlatie, M.; Ranta, M.; Rahkola, P.; Åman, R. Zero-Emission Truck Powertrains for Regional and Long-Haul Missions. *World Electr. Veh. J.* **2023**, *14*, 253. https://doi.org/10.3390/wevj14090253

Academic Editors: Joeri Van Mierlo and Genevieve Cullen

Received: 14 August 2023
Revised: 25 August 2023
Accepted: 31 August 2023
Published: 8 September 2023

1. Introduction

The European Green Deal outlines the climate change mitigation targets as follows: "all 27 EU Member States committed to turning the EU into the first climate neutral continent by 2050". They pledged to reduce emissions by at least 55% by 2030, compared to 1990 levels, and to reach a fossil-free society by 2050 [1]. The transportation sector is responsible for roughly one-quarter of the total greenhouse emissions in the EU with road vehicles contributing to over 60% of the emissions. Lorries, buses, and coaches are responsible for about a quarter of carbon dioxide (CO_2) emissions from road transport in the EU and for some 6% of total EU emissions. Despite some improvements in fuel consumption efficiency in recent years, these emissions are still rising, mainly due to increasing road freight traffic. This requires tremendous efforts in the coming years to introduce zero-emission powertrains and energy infrastructure into regional and long-haul trucking operations. Vehicle emission regulations and other policy measures will pave the way for the transition towards zero-emission transport curbing the total EU emissions of CO_2; although, for heavy-duty vehicles, the current regulation is still quite mild: −15% from 2025 onwards and −30% from 2030 onwards, compared with the 2019/2020 level [2].

Even though electrification is only one tool among others to reduce vehicle emissions, it appears to be the most efficient and feasible technology. This has been studied by now in several research papers where many earlier ones found the status of technology not sufficient for competitive electric truck operations, for example, insufficient energy density and battery capacity on-board and slow charging capability [3]. Later studies have concluded that the business case for electric trucks is emerging but the techno-economic competitiveness is depending a lot on the commodity being transported and the related use case and system level, with medium-duty trucks being able to operate battery electric for a large share of road freight haulage, whereas for heavy-duty and longer missions, the fraction is smaller [4]. Continuous technology development, especially in batteries and charging equipment, has been the main driver towards competitiveness of heavy-duty trucks. The combination of shorter charging times and, therefore, higher vehicle availability through fast charging can improve competitiveness of battery electric trucks in an increasing share of transport missions [5]. In the developing market, there is a dilemma between uptake of electric heavy-duty vehicles (e-HDVs) and the building up of the charging infrastructure. To resolve the chicken-and-egg problem of e-mobility, the three largest truck manufacturers joined in the project for establishing charging infrastructure for electric long-distance freight transport [6]. The uptake of e-HDVs has been most feasible in conventional city deliveries that typically, like e-Buses, drive pre-defined work cycles, their energy consumption can be estimated quite accurately, and they can be charged overnight in their dedicated parking depots, while their daily driving range varies only a little. However, in long-haul operations and on shorter missions, if there are more variables in the logistics assignment, there should be flexibility in the system to use opportunity charging (fast charging, high-power charging, HPC) in addition to depot charging. Driving in multiple shifts creates the demand for HPC instead of slow depot charging. Ad hoc assignments (e.g., courier service, construction transports, maintenance and utility vehicles), logistics in greater metropolitan areas, long-haul transportation, etc., can be based on the depot but additionally would require HPC. Roughly put, the heavier the (articulated) vehicle is, the more the demand for HPC during the work shift will be. According to recent analysis on long-haul operations in Europe, about 40,000 overnight and 9000 fast megawatt chargers will be needed to support battery electric long-haul freight [7].

In Nordic countries, the long-haul trucks are typically in a heavier weight class that is up to GVW 76 tonnes. These national regulations make Nordics a small and specific market area so that the newest innovations would rather need to be implemented by national or joint activities. Combining the harsh weather conditions with the higher vehicle masses makes the Nordics an ideal location for piloting new innovations. In 2021–2022 piloting case examples in the electrification of heavy transports, it was reported in Sweden that High-Capacity Transport (HCT)-articulated vehicles of 64 and 80 GVW tonnes can be electrified in certain routes and drive cycles [8,9]. Also, it was reported that the transportation safety norms for dangerous goods (ADR) can be fulfilled using e-HDVs [10].

The strict definition of zero-emission road transport requires the prime mover to be electricity (battery electric vehicle, BEV) or hydrogen (fuel cell vehicle, FCV), or a hybridisation of the two, such as a battery electric truck with a H_2 fuel cell range extender. While battery electric trucks are, in terms of technology and market readiness, several years ahead of fuel-cell-powered trucks, research on the pros and cons of each and optimal ways to combine them from powertrain to systems merit a proper analysis. Published research suggests that for many freight operations, battery electric trucks offer the lowest total cost of ownership, but fuel cell trucks can still be a viable additional option in some use cases [11]. When assessing the system-level viability for zero-emission trucking including vehicle use cases and missions, energy consumption and sensitivity to, e.g., payload, GVW, driving cycle, and conditions, and the viability of the different powertrain and infrastructure configurations, modelling, and simulation provide an invaluable tool to support decision making. As shown by previous research, an unfavourable combination of conditions can reduce the available operative range of battery electric trucks by 41–47% [12]. The mitigation measures to reduce the risk for trucking operators from such a range reduction should contain elements from the powertrain, vehicular design, and infrastructure implementation. Relevant topics include sufficient margin in designed battery or H_2 tank capacity and powertrain efficiency, vehicular aspects such as minimising driving resistances and auxiliary consumption, as well as location, availability, and capacity of charging and H_2 refilling infrastructures.

The present paper deals with conceptualising and designing modular zero-emission powertrains suitable for regional and long-haul missions using trucks with GVW of 40 tonnes and beyond. In Europe, this relates to VECTO vehicle groups 4–12, and in the US, it relates to Class 8 trucks. The conceptual design object is a 6 × 4 tractor (VECTO 12, overview [13]) prototype with a modular zero-emission powertrain capable of multiple zero-emission missions in regional and long-haul operations. The powertrain has an intermediate-sized traction battery for electric operation, fuel cell system capable of providing average power for selected missions, and a strategy for battery-fuel cell hybrid operations. The modular powertrain, therefore, enables three energy and operation strategies to be analysed in one demonstration setting: BEV operation, FCV operation, and fuel cell range-extended BEV (FCRE) operation.

In line with the European 2050 goals, the present paper has been produced through the project ESCALATE, which aims to demonstrate high-efficiency zero-emission HDV powertrains (up to 10% increase) for long-haul applications that will provide a range of 800 km without refuelling/recharging and cover at least 500 km average daily operation (6+ months) in real conditions. ESCALATE is built on the novel concepts around three main innovation areas, which are (i) standardised, well-designed, cost-effective modular and scalable multi-powertrain components; (ii) fast fuelling and grid-friendly charging solutions; and (iii) Digital Twin (DT) and AI-based management tools considering capacity, availability, speed, and nature of the charging infrastructures as well as the fleet structures. Throughout the project lifetime, five pilots, five DTs, and five case studies on TCO (with the target of 10% reduction), together with their environmental performance via LCA, will be performed.

2. Materials and Methods

Two design basis driving cycles for the GVW 40-tonne prototype tractor demonstrator were used. Vehicle configuration was as is seen in Figure 1 (left). The first driving cycle is a single-charge and refill mission of 750 km based on the long-haul mission profile of the Vehicle Energy Consumption calculation TOol VECTO of the European Commission [14], and the second one is a real roundtrip mission in Finland of 520 km exposed to various Nordic road conditions. The real route runs from the port of Helsinki in the south of Finland up to Jyväskylä in central parts of Finland, along the TEN-T core corridor. Modular powertrain and vehicle model was constructed to support the conceptual design, and driving cycles for both the VECTO long-haul and the real mission were constructed, utilising open road network and speed limit data. Charging and H_2 refuelling sites were planned to support the missions. Energy consumption on the said driving cycles and loading were estimated through simulation, and this information was used for preliminary dimensioning of the powertrain. The electric drives of the tractor were dimensioned so that a secondary mission of GVW up to 76 tonnes configuration, Figure 1 (right), on a shorter route can be operated. The basic parameters of the powertrain and the vehicle combinations used in the simulation are given in Table 1.

Figure 1. Tractor and semitrailer (prototype) with the nominal GVW 40 tonnes configuration (**left**) and a HCT configuration of tractor, semitrailer, dolly, and another semitrailer with GVW of 76 tonnes (**right**).

In the piloting phase, the electric truck (in BEV and/or FCEV configuration) will operate on a flexible time schedule. The vehicle will be depot-charged in Jyväskylä. It is possible to drive directly to Vuosaari port in Helsinki without need for opportunity charging on the road. The driver's resting hours will be well enough to make each leg without additional breaks due to possible charging events. The time schedule allows the driver to unload the cargo plus having the lawful break in port before heading to second leg. Meanwhile, the truck can be opportunity-charged (high-power charging, HPC).

Table 1. Main vehicle and powertrain parameters of the simulated vehicle combinations (GVW 40 tonnes tractor and semitrailer, GVW 76 tonnes tractor, semitrailer, dolly, and semitrailer).

Parameter	
Total mass	40 tonnes/76 tonnes
Empty mass	16 tonnes/26 tonnes
Tractor axle configuration	6 × 4
Motor nominal power	430 kW
Motor nominal speed	1200 rpm
Number of gears	5
Battery efficiency	97%
Inverter efficiency	98%
Driveline efficiency	93%
Aerodynamic factor C_dA	7.96 m^2/12.0 m^2
Rolling resistance factor	0.0065
Maximum speed	80 km/h

While the work cycle in the planned piloting phase offers high flexibility, it is crucial to design and validate the configuration also in work cycles of heavier gross vehicle weights. Typically, the long-haul trucks drive in three-shift work only by changing the driver by the road. The EC regulations for driver's rest times require one 45 min break after each 4.5 h period of driving [15]. The 45 min break can be split into 15 + 30 min, of which the 30 min need to take place after each 4.5 h of driving. Thus, it is important that the vehicle supports HPC in a way that enough energy can be charged for at least 2–2.5 h of driving. The validation of the functionality of the charging and terminal operations during the fast charging will be covered in other phases of the project.

The energy consumption of electric trucks was evaluated by means of simulations. For this purpose, the VTT's in-house simulation platform 'Smart eFleet', originally developed for urban buses [16] and validated based on measurements in [17], was utilised. The simulation platform models the longitudinal dynamics of a vehicle travelling on a specific route. The route is in the simulations divided into short segments, of which each includes data on the topology, traffic lights, road curvature, speed limit, and length obtained from open data sources. In the simulation, a speed reference is formed for each vehicle based on the characteristics of the route, i.e., the speed limit and the road curvature. In addition to this, a traffic component can be included to model the impact of congestion. The speed of the vehicle is controlled by a PI controller. As the power flow of the simulation model is forward-facing, the powertrain design parameters automatically set limits on the acceleration, and the simulation model is well suited for cases where no speed measurement data are yet available.

Two different powertrain options for the zero-emission truck were modelled, a pure battery electric powertrain and a battery electric with a fuel cell acting as a range extender. The electric motor is modelled as an efficiency look-up table dependent on the rotational speed and the torque. The efficiencies of the gearbox, the battery, and the inverter are assumed to be constant. The power rating of the electric drive was dimensioned to enable operation with GVW of 76 tonnes and to meet the power requirement of 5 kW/GVW-tonne. A simple efficiency curve was implemented for the fuel cell, and the power of the fuel cell system will be scaled based on the degree of FC hybridisation. Estimated mass of the power source components will be taken into account as well. For battery use, a simple limitation of available output power on battery state of charge was implemented.

To ensure the traction performance of the vehicle combination, the mechanical driveline includes a 5-speed gearbox. The gear change logic uses fixed traction motor speeds for up and down shifting, keeping the traction motor speed in a range with sufficient power output capability and the highest possible efficiency. The traction power is delivered to the road using tandem-driven bogie axles. For the operation with 40 tonnes GVW, the tandem-driven axles would not be needed, but this selection is made to enable also the operation with 76 tonnes GVW. Losses in the mechanical driveline are taken into account using efficiency factors for the gearbox and driving axles. The road load model includes the gravity force due to slope and driving resistance forces for tire rolling resistance and aerodynamic drag.

3. Results and Discussion

The results from the simulations are shown in Figure 2 for the VECTO long-haul mission profile and in Figure 3 for the actual long-haul mission with 40 tonnes configuration. In the synthesised results, the VECTO profile (Figure 2) is repeated until the design basis of a 750 km mission is reached. The average energy consumption for the VECTO long-haul driving cycle was 1.83 kWh/km, resulting in a total energy of 1373 kWh drawn from the battery in pure battery electric mode. Six different power source combinations to fulfil the 750 km mission are shown in Table 2, and the corresponding simulation results for each variation are shown in Figure 3. The fuel cell was selected individually for each powertrain combination. The maximum power and the average power on the VECTO profile are included in Table 2. The H_2 tank indicates the minimum amount of hydrogen required for the VECTO long-haul mission, whereas the fuel cell is selected to be able to provide enough power also for a truck with a maximum weight of 76 tonnes. Therefore, the fuel cell does not operate at maximum power in the first three options, and the fuel cell efficiency is relatively good. The efficiency of the fuel cell system, including all auxiliary devices and cooling of the fuel cell, is shown in Figure 4.

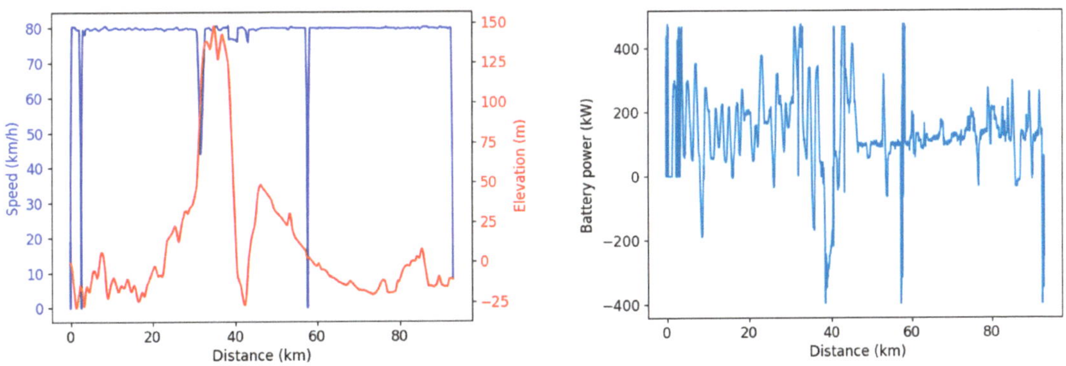

Figure 2. VECTO long-haul driving cycle with elevation and driving speed (**left**) and battery power (**right**) with 40 tonnes configuration.

Conceptual powertrain design configurations for the power source capacities are given in Table 2 for the VECTO long-haul profile. The design basis analysis assumes that the entire mission is carried out without intermediate or opportunity charging or H_2 refilling, in other words, energy storages are full at the start of the mission and will be depleted at the end.

For the second use case, the driving cycle consists of a roundtrip, as shown in Figure 5 (520 km). The route was simulated with the same powertrain configurations as in Table 2, and the resulting energy consumption for the nominal GVW 40 tonnes configuration in pure battery electric mode was 1.82 kWh/km. The consumption in the direction Vuosaari—Jyväskylä was 1.85 kWh/km, and in the opposite direction, it was 1.78 kWh/km. The battery state of charge and fuel tank level are illustrated in Figure 6. An additional vehicle configuration was analysed based on a GVW of 76 tonnes, as shown in the right side of Figure 1, and using the same powertrain as previously described. The energy consumptions stated above can be considered to be slightly on the conservative side to provide sufficient margins at the preliminary design phase.

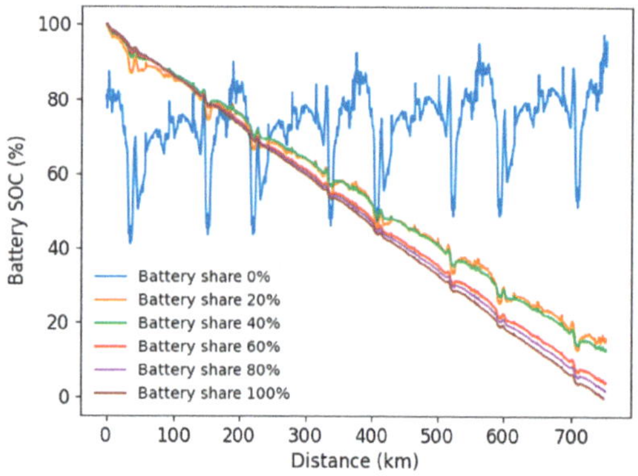

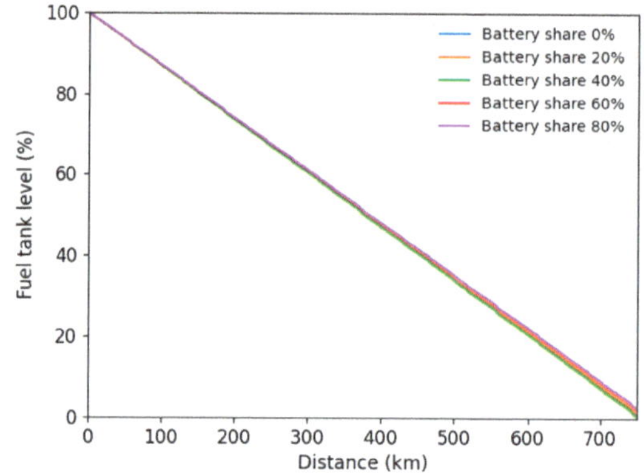

Figure 3. Battery state of charge (SOC, **left**) and hydrogen tank level (**right**) during the VECTO 750 km long-haul mission for the different powertrain variants with 40 tonnes configuration.

Table 2. Power source combinations for 750 km VECTO long-haul mission with 40 tonnes configuration.

Share of Battery Electric Operation	Battery Size	H_2 Tank	Fuel Cell Maximum Power	Average Fuel Cell Power	Weight (Fuel Cell, Hydrogen Storage, Battery)
0%	50 kWh	85 kg	230 kW	133 kW	2500 kg
20%	275 kWh	71 kg	180 kW	110 kW	3210 kg
40%	549 kWh	58 kg	110 kW	86 kW	4120 kg
60%	824 kWh	42 kg	57 kW	57 kW	5030 kg
80%	1098 kWh	21 kg	29 kW	29 kW	5950 kg
100%	1373 kWh	0 kg	-	-	6860 kg

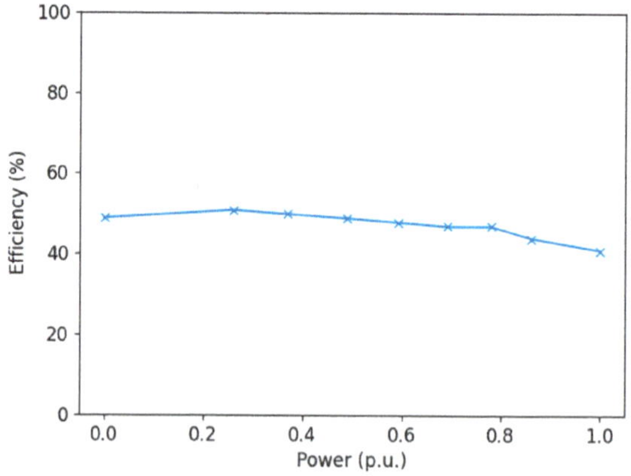

Figure 4. Efficiency of the fuel cell system. The fuel cell output power is expressed relative to the maximum power.

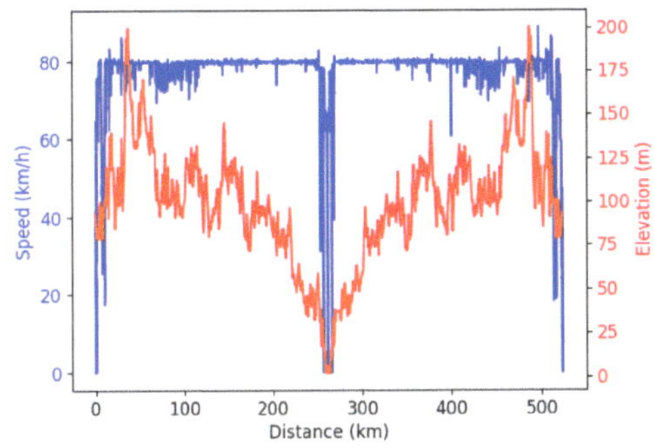

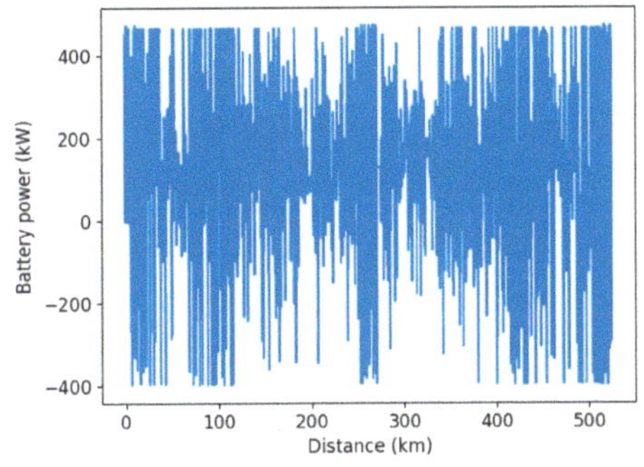

Figure 5. Speed and elevation of the Jyväskylä—Vuosaari roundtrip (**left**) and corresponding battery power in pure BEV mode (**right**) with 40 tonnes configuration.

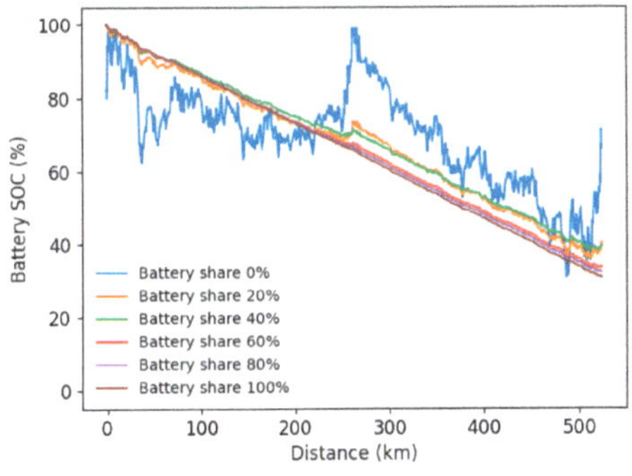

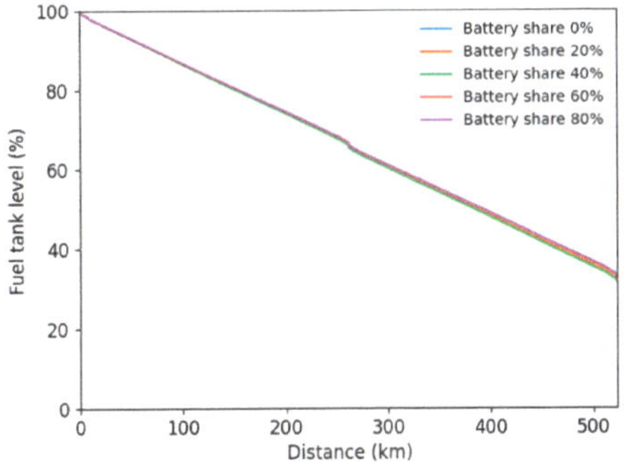

Figure 6. Simulated battery state of charge (SOC, **left**) and hydrogen fuel tank level (**right**) on the Jyväskylä—Vuosaari route (roundtrip) with 40 tonnes configuration.

The results with the 76 tonnes configuration are illustrated in Figure 7. The battery was charged in Vuosaari for roughly 45 min with a charging power of maximum 1 MW, and the fuel cell power was raised to the maximum level for all powertrain configurations. In addition, a larger H_2 tank was used for the first two powertrain configurations operating with 0% and 20% battery shares, respectively. H_2 refilling in Vuosaari could be an option to minimise the H_2 tank. The resulting total energy consumption levels on the Jyväskylä—Vuosaari route are shown in Table 3. It is to be noted that these consumption numbers are not fully optimised. The energy management strategy could be tuned based on the mission to prioritise battery electric energy and minimise the use of hydrogen, especially in the case that it is more expensive than charged electricity. The fuel cell efficiency varied in the range 48–53% in the simulations, while the battery efficiency was 97%. In other words, 1 kg of hydrogen corresponds to 16–17 kWh of usable energy. The energy management strategy should be selected based on the available recharging infrastructure. The strategy chosen here allows the vehicle to operate on long distances on the expense of high energy consumption when relying heavily on the fuel cell.

Further simulations were performed on the Vuosaari—Jyväskylä route. Operation in pure battery electric mode is possible with intermittent charging halfway. The simulated results, when charging at a power of 1 MW is available in Jyväskylä, are shown in Figure 8. The charging break is assumed to be roughly 45 min with a couple of minutes reserved for connecting and disconnecting. The smallest batteries are obviously not enough for this case, while a battery of minimum 549 kWh is sufficient.

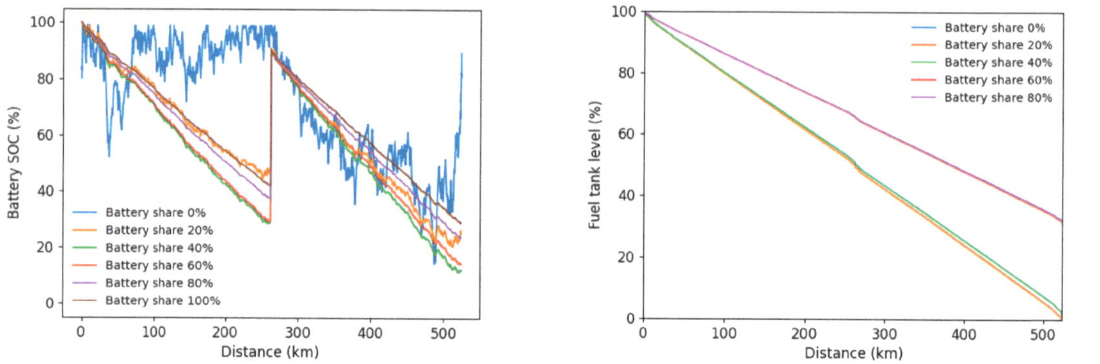

Figure 7. Battery state of charge (**left**) and hydrogen tank level (**right**) when operating on the Jyväskylä—Vuosaari route with 76 tonnes configuration. The battery is charged in Vuosaari for about 45 min with a maximum power of 1 MW.

Table 3. Energy consumption for the 520 km on-road roundtrip long-haul mission for 40 tonnes and 76 tonnes configurations. An intermediate recharging/refuelling at the turning point the mission assumed for the 76 tonnes configuration.

Share of Battery Electric Operation	GVW 40 Tonnes Nominal Case				GVW 76 Tonnes HCT Case			
	Jyväskylä—Vuosaari		Vuosaari—Jyväskylä		Jyväskylä—Vuosaari		Vuosaari—Jyväskylä	
	Battery Electric Energy (kWh/km)	Hydrogen Energy (kWh/km)	Battery Electric Energy (kWh/km)	Hydrogen Energy (kWh/km)	Battery Electric Energy (kWh/km)	Hydrogen Energy (kWh/km)	Battery Electric Energy (kWh/km)	Hydrogen Energy (kWh/km)
0%	−0.04	3.72	0.05	3.70	−0.04	7.59	0.02	7.52
20%	0.30	3.10	0.35	3.09	0.47	5.94	0.67	5.89
40%	0.62	2.52	0.67	2.51	1.42	3.63	1.63	3.60
60%	1.02	1.84	1.08	1.83	2.20	1.88	2.38	1.86
80%	1.39	0.94	1.46	0.93	2.60	0.96	2.78	0.95
100%	1.78	0.00	1.85	0.00	2.99	0.00	3.21	0.00

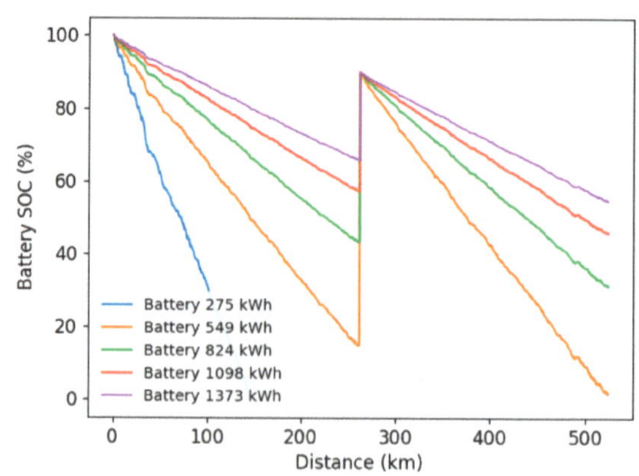

Figure 8. Operation in pure battery electric mode with charging in Vuosaari on the Jyväskylä—Vuosaari roundtrip with 40 tonnes configuration.

The current study is partly based on conservative vehicle parameter values to ensure that the vehicle will meet the required performance criteria. In future studies concentrating on detailed system design and operational optimisation, the impact of the design parameters will be studied and the potential to minimise energy consumption will be evaluated. Various energy management strategies will be evaluated and the battery models will be enhanced to properly take into account the power variation during high-power charging. Secondly, in the case of a hybrid battery-fuel cell power source, control strategies will be investigated to optimise various parameters, such as total energy consumption, operational reliability to complete given missions, or the total cost of ownership. Furthermore, the impact of environmental conditions and driving behaviours, such as statistical variation in temperature, wind conditions, and driving resistances, will be analysed in more detail once reliable information on these based on statistics, operator data, or real piloting operations is available.

The final and optimal choice for the power source and prime mover split depends on additional factors such as infrastructure availability, electricity and hydrogen prices, required payload capacity, and system level availability and productivity. These data will become available when the final design of the demonstrator vehicle as the fuel cell range-extended electric truck is manufactured and put in trial operation. The final prototype design is expected to have battery and fuel cell capacities in mid-area, between the 40% and 60% rows of Tables 2 and 3. The testing and data from the piloting operations will include operation in both purely electric and fuel cell modes as well as their various combinations.

As part of this subsequent analysis, an additional element of the research approach will be to assess and compare the system-level techno-economics of the powertrain and system configurations in the said use cases and missions. The analysis is upcoming based on the results of the vehicle and mission simulations and related technical and operational data. The methodology is based on earlier total cost of ownership (TCO) analysis on electric city buses [18] and the related literature.

4. Conclusions

Approach and methodology for the conceptual design of zero-emission truck powertrains intended for regional and long-haul missions are presented. Various scenarios with the developed vehicle and powertrain models were analysed, taking into account infrastructure and charging/refuelling all along the missions, as well as their impact on the operative planning.

The approach starts from the design basis of an uninterrupted 750 km mission in the VECTO long-haul profile, and secondly, a 520 km mission on a real route in southern Finland. The powertrain designed is capable for vehicle combinations flexibly from a GVW of 40 tonnes all the way up to 76 tonnes. The energy infrastructure analysed included overnight depot charging to start the driving missions with a 100% charged battery and an intermediate fast charging halfway through the roundtrip. Energy use for two truck configurations was estimated through simulation for both use cases.

Six different conceptual powertrain designs with varying degrees of charged electric fuel cell operation were presented. For fully electric operation, the estimated battery capacity required 1373 kWh of traction battery capacity, whereas the other extreme of power source design with H_2 as the prime mover gives a hydrogen storage capacity of 83 kg. The four intermediate powertrain options combine battery and H_2 tank capacities in various ways. In terms of total energy consumption (tank to wheel), the smallest overall mission energy consumption is with fully electric operation—this depends on the relative efficiencies of battery electric and fuel cell electric powertrains.

The conceptual pre-design analysis shows that operation of the GVW 40 tonnes truck is viable in the nominal design basis driving cycles in a battery electric mode with a fuel cell range extender. Additionally, a number of modular powertrain concepts were proposed to meet the design criteria. The work also provides requirements for the energy infrastructure to support operations, pointing to megawatt-level opportunity charging being mandatory for operating the heaviest payload case in purely electric mode.

This research has been presented at the EVS36 Symposium in Sacramento, USA, in June 2023.

Author Contributions: Conceptualisation, M.P. and R.Å.; methodology, M.P., M.R. and P.R.; software, M.R.; validation, M.R. and P.R.; formal analysis, M.P. and M.R.; investigation, M.R. and M.P.; resources, M.P. and R.Å.; data curation, M.R.; writing—original draft preparation, M.R. and M.P.; writing—review and editing, All; visualisation, All; supervision, M.P.; project administration, M.P.; funding acquisition, M.P. All authors have read and agreed to the published version of the manuscript.

Funding: The research was carried out in the ESCALATE project, which has received funding from the European Union's Horizon Europe research and innovation programme under the Grant Agreement No. 101096598.

Data Availability Statement: The data related to the present research are owned by VTT and other relevant ESCALATE partners and cannot be shared at the point of publication.

Acknowledgments: The work was supported by project ESCALATE and consortium partners of the project, especially Sisuauto, Rauanheimo, Transport Jylhä, Kempower, and Valmet Automotive. This support is gratefully acknowledged.

Conflicts of Interest: The authors declare no conflict of interest.

References

1. Delivering the European Green Deal. Available online: https://ec.europa.eu/info/strategy/priorities-2019-2024/european-green-deal/delivering-european-green-deal_en (accessed on 27 October 2022).
2. Reducing CO_2 Emissions from Heavy-Duty Vehicles. Available online: https://climate.ec.europa.eu/eu-action/transport-emissions/road-transport-reducing-co2-emissions-vehicles/reducing-co2-emissions-heavy-duty-vehicles_en (accessed on 27 October 2022).
3. Çabukoglu, E.; Georges, G.; Küng, L.; Pareschi, G.; Boulouchos, K. Battery electric propulsion: An option for heavy-duty vehicles? Results from a Swiss case-study. *Transp. Res. Part C Emerg. Technol.* **2018**, *88*, 107–123. [CrossRef]
4. Liimatainen, H.; van Vliet, O.; Aplyn, D. The potential of electric trucks—An international commodity-level analysis. *Appl. Energy* **2019**, *236*, 804–814. [CrossRef]
5. Nykvist, B.; Olsson, O. The feasibility of heavy battery electric trucks. *Joule* **2021**, *5*, 901–913. [CrossRef]
6. The Future is Electric—Turning a European Infrastructure for Long-Distance Freight Transport into Reality. Traton Group, 2022. Available online: https://traton.com/en/newsroom/current-topics/the-future-is-electric.html (accessed on 26 January 2023).
7. Shoman, W.; Yeh, S.; Sprei, F.; Plötz, P.; Speth, D. Battery electric long-haul trucks in Europe: Public charging, energy, and power requirements. *Transp. Res. Part D Transp. Environ.* **2023**, *121*, 103825. [CrossRef]
8. Scania Builds Extremely Heavy and Extra-Long Electrified Truck for Jula Logistics. Scania CV AB, 2021. Available online: https://www.scania.com/group/en/home/newsroom/press-releases/press-release-detail-page.html/4101348-scania-builds-extremely-heavy-and-extra-long-electrified-truck-for-jula-logistics (accessed on 26 January 2023).
9. The World's First Electric Timber Truck Has Been Delivered in Sweden, and It Can Haul 80 tons. Electrek, 2022. Available online: https://electrek.co/2022/07/07/electric-timber-truck-sweden/ (accessed on 26 January 2023).
10. Wibax' Electric Truck Is Fully Operational. Wibax AB, 2022. Available online: https://www.wibax.com/en/wibax-eldrivna-lastbil-ar-i-full-drift/ (accessed on 26 January 2023).
11. Rout, C.; Li, H.; Dupont, V.; Wadud, Z. A comparative total cost of ownership analysis of heavy duty on-road and off-road vehicles powered by hydrogen, electricity, and diesel. *Heliyon* **2022**, *8*, e12417. [CrossRef]
12. Alonso-Villar, A.; Davíðsdóttir, B.; Stefánsson, H.; Ásgeirsson, E.I.; Kristjánsson, R. Electrification potential for heavy-duty vehicles in harsh climate conditions: A case study based technical feasibility assessment. *J. Clean. Prod.* **2023**, *417*, 137997. [CrossRef]
13. VECTO—Overview. Available online: https://climate.ec.europa.eu/system/files/2018-12/201811_overview_en.pdf (accessed on 30 March 2023).
14. VECTO—Tool. Available online: https://climate.ec.europa.eu/eu-action/transport/road-transport-reducing-co2-emissions-vehicles/vehicle-energy-consumption-calculation-tool-vecto_en (accessed on 30 March 2023).
15. Driving Time and Rest Periods. Available online: https://transport.ec.europa.eu/transport-modes/road/social-provisions/driving-time-and-rest-periods_en (accessed on 30 March 2023).
16. Ranta, M.; Karvonen, V.; Potter, J.J.; Pasonen, R.; Pursiheimo, E.; Halmeaho, T.; Ponomarev, P.; Pihlatie, M. Method Including Power Grid Model and Route Simulation to Aid Planning and Operation of an Electric Bus Fleet. In Proceedings of the 2016 IEEE Vehicle Power and Propulsion Conference (VPPC), Hangzhou, China, 17–20 October 2016; pp. 1–5. [CrossRef]

17. Anttila, J.; Todorov, Y.; Ranta, M.; Pihlatie, M. System-Level Validation of an Electric Bus Fleet Simulator. In Proceedings of the 2019 IEEE Vehicle Power and Propulsion Conference (VPPC), Hanoi, Vietnam, 14–17 October 2019; pp. 1–5. [CrossRef]
18. Pihlatie, M.; Kukkonen, S.; Halmeaho, T.; Karvonen, V.; Nylund, N.-O. Fully Electric City Buses—The Viable Option. In Proceedings of the 2014 IEEE International Electric Vehicle Conference (IEVC), Florence, Italy, 17–19 December 2014. [CrossRef]

MDPI AG
Grosspeteranlage 5
4052 Basel
Switzerland
Tel.: +41 61 683 77 34

World Electric Vehicle Journal Editorial Office
E-mail: wevj@mdpi.com
www.mdpi.com/journal/wevj

MDPI

www.ingramcontent.com/pod-product-compliance
Lightning Source LLC
Chambersburg PA
CBHW042014260926
51850CB00020B/3401
9783725839391